Engineering
Fluid
Mechanics

John A. Roberson

WASHINGTON STATE UNIVERSITY, PULLMAN

Clayton T. Crowe

WASHINGTON STATE UNIVERSITY, PULLMAN

HOUGHTON MIFFLIN COMPANY   BOSTON   TORONTO

DALLAS   GENEVA, ILLINOIS   PALO ALTO   PRINCETON, NEW JERSEY

Fifth Edition

ENGINEERING FLUID MECHANICS

Sponsoring Editor: Rodger H. Klas
Development Editor: June Goldstein
Project Editor: Susan Lee-Belhocine
Production/Design Coordinator: Martha Drury
Senior Manufacturing Coordinator: Marie Barnes
Marketing Manager: Michael Ginley

Cover photograph courtesy of Amoco Chemical Company.

A sailboat is a combination of beauty and fluid mechanics. The billowing sails yield the force to propel the vessel, and the hull is streamlined to minimize drag.

Printed in the U.S.A.

Library of Congress Catalog Card Number: 92-72399

ISBN: 0-395-63789-9

23456789-DH-96 95 94 93

To
Amy and Linda

Contents

Preface

Philosophy

We think that most students who take their first course in fluid mechanics have better intuition and confidence about solid mechanics than they do about fluid mechanics. Therefore, it is necessary to be especially deliberate in helping them to get the "feel" for flow patterns, pressure variation, and shear stress in fluid flow. We have tried to present these concepts in a clear and elementary way so that students will develop confidence in the subject and will be able to proceed to more complex topics with relative ease. We do this in our text by

- Generously using photographs and line drawings in descriptions of fluid flow phenomena, as well as in the statements of problems.
- Including almost all steps in the derivations of equations, along with complete word descriptions to assist the student in following the derivation. We have tried to be especially thorough in derivations involving the basic control volume equation, continuity equation, energy equation, and momentum equation.
- Including a wide range of problems. Some problems focus on conceptual ideas with minimal analysis; these help to "steer" the student in the right direction with regard to basic concepts and principles.
- Including many example problems.

Organization

The first half of the text is devoted primarily to basic principles, the second half to applications of these principles to engineering problems. However, even in the sections on basic principles, practical applications are introduced through examples and problems. In this way additional motivation is provided to the student.

Special Features

In this fifth edition we have made the following changes:

- Added new material on
  −critical flow in channels of non-rectangular cross section
  −hydraulic jump in channels of non-rectangular cross section
  −open-channel transitions
  −spillways
  −flow in streams at flood stage
  −flow resistance in rock-bedded streams.

- Expanded material on
  –flow in culverts and sewers
  –flow in pipe networks
  –water surface profiles
  –flow over broad-crested weirs.

- Added 88 new problems, and revised 177 problems.

Use of the Book and Course Planning

It is our experience that students find this text very readable; therefore, the instructor can be very creative in lectures. The instructor does not have to spend a lot of time interpreting the text for the students, so lecture sessions can be used to *complement* the text. In this regard we strongly suggest that students see some of the films listed at the end of Chapters 2 and 5. They are very helpful in conveying the "feel" of fluid flow to the student.

We also suggest that a basic course in fluid mechanics include the material in Chapters 1 through 11 and Chapter 13. Chapters 12, 14, and 15 can be optional, depending upon the students' majors and instructor's interests. Chapter 16 could be included in a second course or taken as a self-study "minicourse" for gifted students. If the text is used in a two-quarter sequence, all of the material could be included.

Acknowledgments

Special recognition is given to the late Charles L. Barker, who introduced John Roberson to the field of fluid mechanics and motivated him to write a text on the subject, and to Hunter Rouse, who inspired him to further studies in the field.

We wish to thank our many colleagues and students—both at Washington State University and at other institutions—who made valuable suggestions for improving the text. We wish to express our appreciation to the reviewers of this fifth edition: M. M. Aral, Georgia Institute of Technology; Xiaoyan Huang, Washington State University; Gerard P. Lennon, Lehigh University; Robert L. Meserve, Northeastern University; Paul R. Munger, University of Missouri, Rolla; and A. Jacob Odgaard, University of Iowa. Their in-depth reviews and suggestions were most helpful.

Finally, we owe a debt of gratitude to Cheri Yost, who did most of the typing and manuscript preparation. We are also grateful to our families for their encouragement and understanding during the writing and editing of the text.

John A. Roberson
Clayton T. Crowe

Engineering
Fluid
Mechanics

Introduction

The results of fluid flow in nature are manifested in many ways. These dunes are a result of air flow (wind) passing over the sands of the Sahara Desert. The dune sands move by rolling, bouncing and by suspension in the very turbulent air next to the sand surface. (Courtesy H. F. Garner)

Physical Characteristics of Fluids

Fluid mechanics is the science that deals with the action of forces on fluids. In contrast to a solid, a fluid is a substance the particles of which easily move and change their relative position. More specifically, a fluid is defined as a substance that will continuously deform—that is, flow under the action of a shear stress—no matter how small that shear stress may be. A solid, on the other hand, can resist a shear stress, assuming, of course, that the shear stress does not exceed the elastic limit of the material. The rate of deformation of the fluid is related to the applied shear stress by *viscosity*, a property of the fluid. Thus very viscous fluids, such as honey and cold oils, flow very slowly for a given shear stress. An example of this effect is observed when one pours a cup of oil or honey on an inclined surface. It takes a considerable length of time for these very viscous fluids to flow down the incline, whereas a cup of water would flow down and off very rapidly.

Distinction Between Solids, Liquids, and Gases

A fluid can be either a gas or a liquid. The molecular structure of liquids is such that the spacing between molecules is essentially constant (the spacing changes only slightly with temperature and pressure), so that a given mass of liquid occupies a definite volume of space. Therefore, when one pours a liquid into a container, it assumes the shape of the container for the volume it occupies. The molecules of solids also have definite spacing. However, the solid's molecules are arranged in a specific lattice formation and their movement is restricted, whereas liquid molecules can move with respect to each other when a shearing force is applied. The spacing of the molecules of gases is much wider than that of either solids or liquids, and it is also variable. Thus a gas completely fills the container in which it is placed. The gas molecules travel in straight lines through space until they either bounce off the walls of the container or are deflected by interaction with other gas molecules.

Fluid as a Continuum

In considering the action of forces on fluids, one can either account for the behavior of each and every molecule of fluid in a given field of flow or simplify the problem by considering the average effects of the molecules in a given volume. In most problems in fluid dynamics the latter approach is possible, which means that the fluid can be regarded as a *continuum*—that is, a hypothetically continuous substance.

The justification for treating a fluid as a continuum depends on the physical dimensions of the body immersed in the fluid and on the number of molecules in a given volume. Let us say that we are studying the flow of air past a sphere with a diameter of 1 cm. A continuum is said to prevail if the number of molecules in a volume much smaller than the sphere's is sufficiently great so that the average effects (pressure, density, and so on) within the volume either are constant or change smoothly with time. The number of molecules in a cubic meter of air at room temperature and sea-level pressure is about 10^{25}. Thus the number of molecules in a volume of 10^{-19} m^3 (about the size of a dust particle, which is very much smaller than the sphere) would be 10^6. This number of molecules is so large that the average effects within the microvolume are indeed virtually constant. On the other hand, if the 1-cm sphere were at an altitude of 305 km, there would be only one chance in 10^8 of finding a molecule in the microvolume, and the concept of an average condition would be meaningless. In this case, the continuum assumption would not be valid. It may thus be concluded that the assumption of a continuum is valid for fluid flow except in the rarest conditions, such as those encountered in outer space.

1.2   Flow Classification

The subject of fluid mechanics can be subdivided into two broad categories: hydrodynamics and gas dynamics. *Hydrodynamics* deals primarily with the flow of fluids for which there is virtually no density change, such as the flow of liquid or the flow of gas at low speeds. Hydraulics, for example—the study of liquid flows in pipes or open channels—falls within this category. The study of fluid forces on bodies immersed in flowing liquids or in low-speed gas flows can also be classified as hydrodynamics.

Gas dynamics, on the other hand, deals with fluids that undergo significant density changes. High-speed flows of gas through a nozzle or over a body, the flow of chemically reacting gases, and the movement of a body through the low-density air of the upper atmosphere fall within the general category of gas dynamics.

An area of fluid mechanics not classified as either hydrodynamics or gas dynamics is *aerodynamics,* which deals with the flow of air past aircraft or rockets, whether low-speed incompressible flow or high-speed compressible flow.

1.3 Historical Note

The science of fluid mechanics began with the need to control water for irrigation purposes in ancient Egypt, Mesopotamia, and India. Although these civilizations understood the nature of channel flow, there is no evidence that any quantitative relationships had been developed to guide them in their work. It was not until 250 B.C. that Archimedes discovered and recorded the principles of hydrostatics and flotation. Although the empirical understanding of hydrodynamics continued to improve with the development of fluid machinery, better sailing vessels, and more intricate canal systems, the fundamental principles of classical hydrodynamics were not set forth until the seventeenth and eighteenth centuries. Newton (4), Daniel Bernoulli (2), and Euler (9) made the greatest contributions to establishing these principles.

Classical hydrodynamics, though a fascinating subject that appealed to mathematicians, was not applicable to many practical problems because the theory was based on inviscid fluids. The practicing engineers at that time needed design procedures that involved the flow of viscous fluids. Consequently, they developed empirical equations that were usable but narrow in scope. Thus, on the one hand, the mathematicians and physicists developed theories that in many cases could not be used by the engineers, and on the other, the engineers used empirical equations that could not be used outside the limited range of application from which they were derived.

Near the beginning of the twentieth century, however, it was necessary to merge the general approach of physicists and mathematicians with the experimental approach of engineers to bring about significant advances in the understanding of flow processes. Osborne Reynolds' (7) paper in 1883 on turbulence and later papers on the basic equations of motion contributed immeasurably to the development of fluid mechanics. After the turn of the century, Ludwig Prandtl (6) proposed the concept of the boundary layer. This concept not only paved the way to sophisticated analyses needed in the development of the airplane, but also resolved many of the paradoxes involved with the flow of a low-viscosity fluid.

Gas dynamics is a relatively new field, in that the earliest works on it did not appear until the nineteenth century. Riemann (8) published his paper on compression (shock) waves in 1876, and 20 years later Mach (3) observed such waves on supersonic projectiles. Once again it was Prandtl who organized the systematic study of gas dynamics around the turn of the century (5). Interest in gas dynamics increased tremendously after World War I, which led to the development of supersonic wind tunnels before World War II and supersonic flight soon after. The advent of space flight led to still another area of gas dynamics, called rarefied flow, in which the density of the air is so low that the motion and impact of individual molecules must be considered. This is in contrast to the treatment of fluid as a continuum, which is what we do in most of our earthbound problems.

FIGURE 1.1

Leonhard Euler (1707–1783) (left), Professor of Physics and Mathematics. Euler had a greater interest in mathematics than Bernoulli and, in fact, Euler formalized the equation we now call Bernoulli's. Euler developed the basic equations of fluid motion.

Daniel Bernoulli (1700–1782) (right), Professor of Mathematics. Bernoulli published works on the statics and dynamics of fluids and made the first observations and notes relating to the equation that bears his name. (Culver Pictures)

1.4 Significance of Fluid Mechanics

The significance of fluid mechanics becomes apparent when we consider the vital role it plays in our everyday lives. When we turn on our kitchen faucets, we activate flow in a complex hydraulic network of pipes, valves, and pumps. When we flick on a light switch, we are drawing energy either from a hydroelectric source that operates by the flow of water through turbines or from a thermal power source derived from the flow of steam past turbine blades. When we drive our cars, pneumatic tires provide suspension, hydraulic shock absorbers reduce road shocks, gasoline is pumped through tubes and later atomized, and air resistance creates a drag on the auto as a whole; and when we stop, we are confident in the operation of the hydraulic brakes. Very complex fluid processes are also involved in the manufacture of the paper on which this book is printed. And our very lives depend on a very important fluid mechanic process—the flow of blood through our veins and arteries.

Some of the most significant environmental problems facing society today involve fluid mechanics. For example, coastal cities often discharge their wastewater (usually treated) into the sea, near the sea bed, far enough from shore so that the wastes become sufficiently diluted with the ambient sea water to render the resulting mixture harmless. The process involves mixing the wastewater with the ambient liquid, a complex turbulence phenomenon. The degree of mixing is a function of the characteristics of the wastewater and the ambient liquid (such as density) as well as the discharge velocity of the wastewater. Also involved in this process are the velocity and pattern of coastal currents. In addition to the fluid mechanics of such a problem, the contaminants in the mixture may change both chemically and biologically in the process. Thus sophisticated models linking the basic flow model with other aspects of the problem are required to design a satisfactory waste disposal system. Such models are generally developed and used by multidisciplinary teams that may include engineers, mathematicians, chemists, and bioscientists. There is an increasing need for engineers who have the ability and mathematical skills to assist in the generation of, and to use, sophisticated computational models of this type. Other problems, similar in nature, that involve fluid mechanics include air pollution and underground hazardous waste problems. By mastering the subject matter in this text, you will acquire the foundation necessary for study of the more involved processes noted here.

1.5   Trends in Fluid Mechanics

Modern developments in fluid mechanics, as in all fields, involve the use of high-speed computers in the solution of problems. Remarkable progress is being made in this area, and the use of the computer in fluid dynamic design is increasing. In the design of aircraft, computers are used to predict the flow over engine nacelles and appendages in order to select configurations that minimize aerodynamic drag. The NASA publication on wind tunnels (1) explains the role of computers in aircraft design. Computational solutions for wind forces on buildings and structures are used to complement measurements on wind tunnel models to insure the safety and structural integrity of the full-scale structures.

The ever-increasing speed and memory capacity of modern computers are leading to even more exciting applications of computers in fluid mechanics. Computer solutions for the motion of terrestrial winds and weather fronts are leading to more accurate forecasting of local weather conditions. The coupling of fluid mechanics with heat transfer and chemical kinetics in computational solutions will lead to improved designs for industrial power and propulsion systems. As space stations and space travel become more feasible, computers will play a vital role in the design of flow systems in microgravity environments that are difficult to examine through terrestrial experimentation. The application of computers to the analysis of flows in biological systems is only beginning, but it will continue to grow as the mechanics of flows in these systems becomes better understood.

The science of fluid mechanics is developing at a rapid rate. Armed with more detailed measurements and numerical models, fluid mechanicians have developed higher levels of understanding that have led to sophisticated designs and applications of fluid systems. Still, there are many areas in which only rudimentary information and physical models are available. Turbulence is a prime example. Even though we presently have high-speed computers at our disposal, the solutions are only as valid as the equations we use to describe the basic flow phenomena. And there is currently no general analytic model that completely describes the nature of turbulence. We have good data on turbulence in straight pipes, so reliable empirical formulas have been developed to describe the turbulence in such a simple case. But turbulence in high-shear flows, buoyant flows, and compressible flows is still the subject of extensive study. Analyses of the flow of multiphase mixtures such as solids in a liquid (slurries) and bubbles in a liquid still rely heavily on empiricism. In oil recovery operations, the engineer is confronted with the problem of the flow of immiscible liquids, such as oil in water, which is not well understood. These are areas which represent exciting challenges to current and future practitioners of fluid mechanics.

References

1. Baals, D. B., and W. R. Corliss, *Wind Tunnels of NASA,* Sup. of Doc., U.S. Govt. Printing Office, Wash. D.C., 1981.

2. Bernoulli, Daniel, *Hydrodynamics,* and Bernoulli, Johann, *Hydraulics.* Trans. Thomas Carmody and Helmut Kobus. Dover Publications, New York, 1968.

3. Mach, E., and P. Salcher. "Photographische Fixierung der durch Projektile in der Luft eingeleiteten Vorgänge." *S.B. Akad. Wiss. Wien.,* 95 (1887), 764–780.

4. Newton, Isaac S. *Philosophie Naturalis Principia Mathematica.* Trans. Florian Cajori. University of California Press, Berkeley, 1934.

5. Oswatitsch, K. *Gas Dynamics.* Trans. G. Kuerti. Academic Press, New York, 1956.

6. Prandtl, L. "Über Flussigkeitsbewegung bei sehr kleiner Reibung." In *Verhandlungen des III. Internationalen Mathematiker Kongresses.* Leipzig, 1905.

7. Reynolds, O. "An Experimental Investigation of the Circumstances Which Determine Whether the Motion of Water Shall Be Direct or Sinuous, and of the Law of Resistance in Parallel Channels." *Phil. Trans. Roy. Soc. London,* 174 (1883).

8. Riemann, B. "Über die Fortpflanzung ebener Luftwellen von endlicher Schwingungsweite." In *Gesammelte Werke.* Leipzig, 1876.

9. Rouse, H., and S. Ince. *History of Hydraulics.* Iowa Institute of Hydraulic Research, State University of Iowa, 1957.

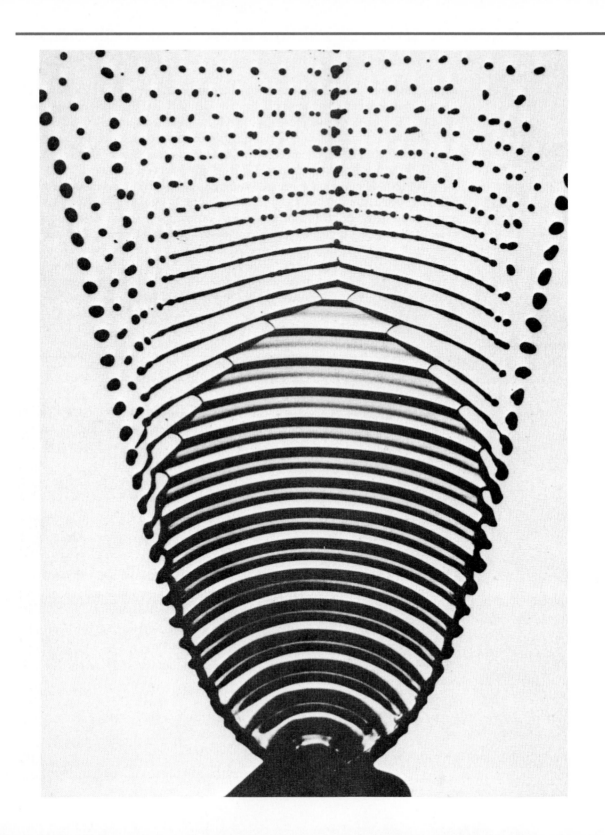

Fluid Properties

A sheet of liquid is discharged from the nozzle at the bottom of the photo. The waves in the liquid sheet are produced by vibrating the nozzle. The sheet eventually breaks up into droplets under the action of surface tension. (Courtesy Milton Van Dyke, Stanford University)

2.1 Basic Units

Every fluid has certain characteristics by which its physical condition may be described. We call such characteristics *properties* of the fluid. These properties are expressed in terms of a limited number of basic dimensions (length, mass or force, time, and temperature), which in turn are quantified by basic units. The traditional system of units in the United States has been the foot–pound–second system. However, because all the engineering societies are urging the adoption of the SI (Système International) system, this text uses both SI and traditional units. Approximately half of the problems and the majority of the examples are given in the SI system; the remainder are given in the foot–pound–second system.

SI System of Units

The basic unit of temperature in the SI system is the Kelvin (K). The Kelvin scale is 0 K at absolute zero and 273.16 K at the freezing point of water. The Celsius scale (°C) is 0°C at the freezing point of water. Therefore the conversion formula is

$$K = 273 + {}^\circ C \tag{2.1}$$

The basic units of mass, length, and time in the SI system are the kilogram (kg), meter (m), and second (s). The corresponding unit of force is derived from Newton's second law: the force required to accelerate a kilogram at one meter per second per second is defined as the *newton* (N). The acceleration due to gravity at the earth's surface is 9.81 m/s². Thus the weight of a kilogram at the earth's surface is

$$W = Mg$$

$$= (1)(9.81)\text{kg} \cdot \text{m/s}^2 = 9.81 \text{ N} \tag{2.2}$$

From Eq. (2.2) we can also determine the units for kilograms in terms of the meter–newton–second units. From Eq. (2.2),

$$M = \frac{W}{g} \frac{N}{m/s^2}$$

Thus the mass M of a body in kilograms is given by its weight in newtons at the earth's surface divided by 9.81 m/s² (the acceleration due to gravity at the earth's surface). As shown, the units of mass may also be expressed in equivalent meter–newton–second units as $N \cdot s^2/m$.

The unit of work and energy in the SI system is the *joule* (J), which is a newton meter (N · m). The unit of power is the *watt* (W), which is a joule per second.

The prefixes used in the SI system to indicate multiplication of units by powers of 10 are

$$G \text{ (giga)} = 10^9 \qquad c \text{ (centi)} = 10^{-2}$$
$$M \text{ (mega)} = 10^6 \qquad m \text{ (milli)} = 10^{-3}$$
$$k \text{ (kilo)} = 10^3 \qquad \mu \text{ (micro)} = 10^{-6}$$

For example, km stands for kilometer, or 1000 meters, and mm signifies millimeter, or 0.001 meter.

Traditional Units

The system of units that preceded SI units in several countries is the so-called English system. The fundamental units of length and mass are the foot (ft), equal to 30.48 cm, and the slug, equal to 14.59 kg. The time unit of second is common to both systems. The force required to accelerate a mass of one slug at one foot per second per second is one pound force (lbf). The mass unit more common to mechanical engineers in the traditional system is the pound mass (lbm). The conversion factor for changing pounds mass to slugs is g_c, which is equal to 32.2 lbm/slug.

EXAMPLE 2.1 What is the weight of a pound mass on the earth's surface, where the acceleration due to gravity is 32.2 ft/s², and on the moon's surface, where the acceleration is 5.31 ft/s²?

Solution By Newton's second law,

$$W = Mg$$

where the unit for W is lbf; for M, slugs; and for g, ft/s². The mass of one pound mass in slugs is by definition

$$M = \frac{1 \text{ lbm}}{g_c} = \frac{1 \text{ lbm}}{32.1 \text{ lbm/slug}} = \frac{1 \text{ slugs}}{32.2}$$

Therefore, the weight on the earth's surface is

$$W = \frac{1 \text{ slug}}{32.2} \times 32.2 \frac{\text{ft}}{\text{s}^2} = 1 \text{ lbf} \qquad \blacktriangleleft$$

and on the moon's surface is

$$W = \frac{1 \text{ slug}}{32.2} \times 5.31 \frac{\text{ft}}{\text{s}^2} = 0.165 \text{ lbf} \qquad \blacktriangleleft$$

Thus the weight of one pound mass on the earth's surface is one pound force. However a pound force does *not* equal a pound mass because they have different units.

The traditional unit for temperature is the degree Fahrenheit (°F), which is $\frac{5}{9}$ of the Celsius degree. The corresponding absolute temperature scale is in degrees Rankine (°R). The Fahrenheit temperature at the freezing point of water is 32°F, and the formula for conversion between °F and °R is

$$°R = 460 + °F \qquad (2.3)$$

2.2   System; Extensive and Intensive Properties

From Chapter 4 through the remainder of the text, wide use is made of the control-volume approach. In its application we use the concept of a system of particles and the intensive and extensive properties related to this concept. A *system*, in fluid mechanics and thermodynamics, is defined as a given quantity of matter. To illustrate, if at an instant a quantity of matter were designated as a system and dyed red to distinguish it from the remaining matter, this red matter would always constitute the system, even though it moved through space and might change in shape and volume. Because a system always consists of the same matter, the mass of a given system is constant. Properties related to the total mass of the system are called *extensive properties* and are usually represented by upper-case letters—for example, mass M and weight W. Properties that are independent of the amount of fluid are called *intensive properties* and are often designated by lowercase letters, such as pressure p (force per unit area) and *mass density* ρ (mass per unit volume). The distinction between extensive and intensive properties is very important in the derivation and application of the control-volume approach introduced in Chapter 4.

2.3 Properties Involving the Mass or Weight of the Fluid

Specific Weight, γ

The gravitational force per unit volume of fluid, or simply the weight per unit volume, is defined as *specific weight*. It is given the symbol γ (gamma). Water at 20°C has a specific weight of 9.79 kN/m³. In contrast, the specific weight of air at the same temperature and at standard atmospheric pressure is 11.8 N/m³. Specific weights of common liquids are given in Table A.4 in the Appendix.

Mass Density, ρ

The mass per unit volume is *mass density;* hence it has units of kilograms per cubic meter, or N · s²/m⁴. Because specific weight is weight per unit volume, mass density is given by the specific weight at the earth's surface divided by g. The mass density of water at 4°C is $9810/g = 1000$ kg/m³. For air at 20°C and at standard pressure, the mass density is 1.20 kg/m³. Mass density, often simply called *density,* is given the Greek symbol ρ (rho). The densities of common fluids are given in Tables A.2 to A.5.

Variation in Density

The density of some fluids is more easily changed than that of others. For example, air can easily be compressed with a consequent density change, whereas a very large pressure is needed to effect a relatively small density change in water. For most applications, water can be considered incompressible and, in turn, can be assumed to have constant density. Air, on the other hand, is a relatively compressible fluid with variable density. However, there are even some air-flow problems for which the air density changes only slightly; one case is wind forces acting on buildings.

Incompressibility does not always imply constant density. For example, a mixture of salt in water changes the density of the water without changing its volume. Therefore there are some flows, such as in estuaries, in which density variations may occur within the flow field even though the fluid is essentially incompressible. Such a fluid is termed *nonhomogeneous.* This text emphasizes the flow of *homogeneous* fluids, so the term *incompressible* used throughout the text implies constant density.

Specific Gravity, S

The ratio of the specific weight of a given liquid to the specific weight of water at a standard reference temperature is defined as *specific gravity*, S. The standard reference temperature for water is often taken as 4°C, where the specific weight of water at atmospheric pressure is 9810 N/m³. With this reference, the specific gravity of mercury at 20°C is

$$S_{Hg} = \frac{133 \text{ kN/m}^3}{9.81 \text{ kN/m}^3} = 13.6$$

Because specific gravity is a ratio of specific weights, it is dimensionless and, of course, independent of the system of units used.

Equation of State and Density of Gases

The fundamental equation of state for an ideal gas is

$$p \forall = n R_u T$$

where p is the absolute pressure*, $\forall$ is the volume, n is the number of moles, R_u is the universal gas constant (the same for all gases), and T is absolute temperature. The equation of state can be rewritten as

$$p = \frac{n \mathcal{M}}{\forall} \frac{R_u}{\mathcal{M}} T$$

where $\mathcal{M}$ is the molecular weight of the gas. The product of the number of moles and the molecular weight is the mass of gas. Thus $n\mathcal{M}/\forall$ is the mass per unit volume, or density. The quotient $R_u/\mathcal{M}$ is the gas constant, R. Thus the equation of state can be expressed as

$$p = \rho R T \tag{2.4}$$

Actually no gas is ideal. However, a gas far removed from the liquid phase, which is generally what we encounter in gas-flow problems, behaves like an ideal gas. Values of R for a number of gases are given in Table A.2. To determine the mass density of a gas, we simply solve Eq. (2.4) for ρ:

$$\rho = \frac{p}{RT} \tag{2.5}$$

*We discuss pressure in detail in Chapter 3.

EXAMPLE 2.2   Air at standard sea-level pressure ($p = 101$ kN/m²) has a temperature of 4°C. What is the density of the air?

Solution   We apply Eq. (2.5) to solve for ρ:

$$\rho = \frac{p}{RT}$$

$$= \frac{101 \times 10^3 \text{ N/m}^2}{287 \text{ J/kg K} \times (273 + 4)\text{K}} = 1.27 \text{ kg/m}^3 \qquad \blacktriangleleft$$

2.4   Properties Involving the Flow of Heat

Specific Heat, c

The property that describes the capacity of a substance to store thermal energy is called *specific heat*. By definition, it is the amount of heat that must be transferred to a unit mass of substance to raise its temperature by one degree. The specific heat of gases depends on the process accompanying the change in temperature. If the specific volume v of the gas ($v = 1/\rho$) remains constant while the temperature changes, then the specific heat is identified as c_v. However, if the pressure is held constant during the change in state, then the specific heat is identified as c_p. The ratio c_p/c_v is given the symbol k. Values for c_p and k for various gases are given in Table A.2.

Specific Internal Energy, u

The energy that a substance possesses because of the state of the molecular activity in the substance is termed *internal energy*. Internal energy is usually expressed as a specific quantity—that is, internal energy per unit mass. In the SI system, the *specific internal energy, u*, is given in joules per kilogram. The internal energy is generally a function of temperature and pressure. However, for an ideal gas, it is a function of temperature alone.

Specific Enthalpy, h

The combination $u + p/\rho$ is encountered frequently in the equations for thermodynamics and compressible flow; it has been given the name *specific enthalpy*. For an ideal gas, u and p/ρ are functions of temperature alone. Consequently their sum, specific enthalpy, is also a function solely of temperature.

2.5 | **Viscosity**

The most important distinction between a solid such as steel and a viscous fluid such as water or air is that shear stress in a solid material is proportional to shear strain, and the material ceases to deform when equilibrium is reached, whereas the shear stress in a viscous fluid is proportional to the *time rate* of strain. The proportionality factor for the solid is the shear modulus. The proportionality factor for the viscous fluid is the *dynamic,* or *absolute, viscosity.* For example, the shear stress of a fluid near a wall is given by

$$\tau = \mu \frac{dV}{dy} \tag{2.6}$$

where τ (tau) is the shear stress, μ (mu) is the dynamic viscosity, and dV/dy is the time rate of strain, which is also the velocity gradient normal to the wall. Thus the definition of the viscosity, μ, is the ratio of the shear stress to the velocity gradient, $\mu = \tau/(dV/dy)$.

Consider the flow shown in Fig. 2.1. This velocity distribution is typical of that for laminar (nonturbulent) flow next to a solid boundary. Several observations relating to this figure will help you appreciate the interaction between viscosity and velocity distribution. First, the velocity gradient at the boundary is finite. The curve of velocity variation cannot be tangent to the boundary because this would imply an infinite velocity gradient and, in turn, an infinite shear stress, which is impossible. Second, a velocity gradient that becomes less steep (dV/dy becomes smaller) with distance from the boundary has a maximum shear stress at the boundary, and the shear stress decreases with distance from the boundary. Also note that *the velocity of the fluid is zero at the stationary boundary. This is characteristic of all flows dealt with in this basic text. That is, at the boundary surface the fluid has the velocity of the boundary—no slip occurs.*

Many of the equations of fluid mechanics include the combination μ/ρ in them. Because it occurs so frequently, this combination has been given the spe-

FIGURE 2.1 _____

Velocity distribution next to a boundary.

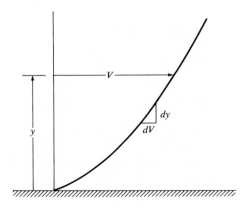

FIGURE 2.2

*Conveyor-belt
transportation system.*

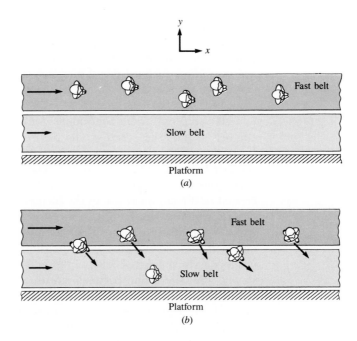

cial name *kinematic viscosity* (so called because the force dimension cancels out in the combination μ/ρ). The symbol used to identify kinematic viscosity is ν (nu).

Whenever shear stress is applied to fluids, motion occurs. This is the basic difference between fluids and solids. Solids can resist shear stress in a static condition, but fluids deform continuously under the action of a shear stress. Another important characteristic of fluids is that the viscous resistance is independent of the normal force (pressure) acting within the fluid. In contrast, for two solids sliding relative to each other, the shearing resistance is totally dependent on the normal force between the two.

The manner in which viscous forces are produced can be seen in the conveyor-belt analogy. Consider a type of transit system in which people are carried from one part of a city to another on conveyor belts (Fig. 2.2*a*). People ride the fast-moving belt from left to right—an equilibrium condition exists. Next visualize the action when the people step off the fast belt onto a slower-moving belt (Fig. 2.2*b*). Before they step off the fast-moving belt, each possesses a certain amount of momentum in the x direction. But as soon as they acquire the speed of the slower belt, their momentum in the x direction is reduced by a significant amount. It is known from basic mechanics that a change in momentum of a body results from an external force acting on that body. In our example, it is the slower belt that exerts a force in the negative x direction as each person steps on the belt. Conversely, as each person steps on the slower belt, a force is exerted on the belt in the positive x direction. Now, if the people step from the faster belt to the slower belt at a rather steady rate, then a rather continuous force is exerted on the slower belt. In effect, by the action of the people moving in the negative y direction, they produce a force on the slow belt in the positive x direction. One may think of this as a shear force in the x direction. In a similar manner, it can be visualized that if people stepped from the slow-moving belt to the faster one, a "shear force" in the negative x direction would be imposed on the faster belt.

If the people were continuously going both ways (back and forth) from one belt to the other, there would be, in effect, a continuous augmenting force (force in the direction of motion) on the slow belt and a like retarding force on the fast belt. Furthermore, as the relative speed between the belts changes (analogous to a change in velocity gradient), the shear force is increased or decreased in direct proportion to the increase or decrease in relative velocity. Thus if both belts were made to have the same speed, the shear force would be zero.

In fluid flow we can think of streams of fluid traveling in a given general direction, such as in a pipe, with the fluid nearer the pipe center traveling faster (analogous to the faster belt) while the fluid nearer the wall is traveling more slowly. The interaction between streams, in the case of gas flow, occurs when the molecules of gas travel back and forth between adjacent streams, thus creating a shear stress in the fluid. Because the rate of activity (back-and-forth motion) of the gas molecules increases with an increase in temperature, it follows that the viscosity of a gas should increase with the temperature of the gas. Such is indeed the case, as can be seen in Fig. 2.3.

For liquids, the shear stress is involved with the cohesive forces between molecules. These forces decrease with temperature, which results in a decrease in viscosity with an increase in temperature (see Fig. 2.3). The variation of viscosity (dynamic and kinematic) for other fluids is given in Figs. A.2 and A.3 in the Appendix.

Units of Viscosity

From Eq. (2.6) it can be seen that the units of μ are $N \cdot s/m^2$.

$$\mu = \frac{\tau}{dV/dy} = \frac{N/m^2}{(m/s)/m} = N \cdot s/m^2$$

The units of kinematic viscosity ν are m^2/s.

$$\nu = \frac{\mu}{\rho} = \frac{N \cdot s/m^2}{N \cdot s^2/m^4} = m^2/s$$

EXAMPLE 2.3   A board 1 m by 1 m that weighs 25 N slides down an inclined ramp (slope = 20°) with a velocity of 2.0 cm/s. The board is separated from the ramp by a thin film of oil with a viscosity of 0.05 N $\cdot$ s/m². Neglecting edge effects, calculate the spacing between the board and the ramp.

FIGURE 2.3

*Kinematic viscosity for air
and crude oil.*

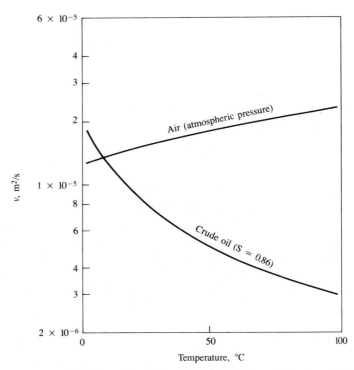

Solution   The board and ramp (left) and a free body of the board (right) are
shown below. For a constant sliding velocity, the resisting shear force is equal to
the component of weight parallel to the inclined ramp. Therefore,

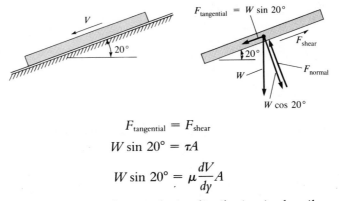

$$F_{\text{tangential}} = F_{\text{shear}}$$

$$W \sin 20° = \tau A$$

$$W \sin 20° = \mu \frac{dV}{dy} A$$

In this case we can assume a linear velocity distribution in the oil, so dV/dy can
be expressed as $\Delta V/\Delta y$, where ΔV is the velocity of the board and Δy is the spacing between the board and the ramp. We then have

$$W \sin 20° = \mu \frac{\Delta V}{\Delta y} A$$

or
$$\Delta y = \frac{\mu \Delta V A}{W \sin 20°}$$

$$= \frac{0.05 \ \text{N} \cdot \text{s/m}^2 \times 0.020 \ \text{m/s} \times 1 \ \text{m}^2}{25 \ \text{N} \times \sin 20°}$$

$$= 0.000117 \ \text{m}$$

$$= 0.117 \ \text{mm} \qquad \blacktriangleleft$$

Newtonian versus Non-Newtonian Fluids

Fluids for which the shear stress is directly proportional to the rate of strain are called *Newtonian fluids.* Because shear stress is directly proportional to the shear strain, dV/dy, a plot relating these variables (see Fig. 2.4) results in a straight line passing through the origin. The slope of this line is the value of the dynamic viscosity. For some fluids the shear stress may not be directly proportional to the rate of strain; these are called non-Newtonian fluids. One class of non-Newtonian fluids, pseudo plastics, has the interesting property that the ratio of shear stress to shear strain decreases as the shear strain increases (see Fig. 2.4). Some polymer solutions are pseudo plastics. Another type of non-Newtonian fluid, called a *Bingham plastic,* acts like a solid for small values of shear stress and then behaves as a fluid at higher shear stress. The shear stress versus shear strain rate for a Bingham plastic is also shown in Figure 2.4. This book will be limited to theory and applications involving Newtonian fluids. For more information on the theory of flow of non-Newtonian fluids, please see references (1), (2), and (4).

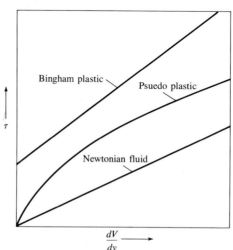

FIGURE 2.4

Shear-stress relations for different types of fluids.

2.6   Elasticity

When pressure is applied to a fluid, it contracts; when the pressure is released, it expands. The elasticity of a fluid is related to the amount of deformation (expansion or contraction) for a given pressure change. The elasticity is often called the compressibility of the fluid. Quantitatively, the degree of elasticity is given by E_v, the definition of which is

$$dp = -E_v \frac{d\mathbb{V}}{\mathbb{V}} \tag{2.7}$$

or
$$E_v = -\frac{dp}{d\mathbb{V}/\mathbb{V}} \tag{2.8}$$

where E_v is the bulk modulus of elasticity, dp is the incremental pressure change, $d\mathbb{V}$ is the incremental volume change, and $\mathbb{V}$ is the volume of fluid. Because $d\mathbb{V}/\mathbb{V}$ is negative for a positive dp, a negative sign is used in the definition to yield a positive E_v. An alternative form of Eq. (2.8) is

$$E_v = \frac{dp}{d\rho/\rho} \tag{2.9}$$

By comparing Eqs. (2.8) and (2.9), one can see that $d\rho/\rho = -d\mathbb{V}/\mathbb{V}$. We can verify this equality by considering a given mass M of fluid, where

$$M = \rho\mathbb{V} \tag{2.10}$$

If we differentiate both sides of Eq. (2.10), we have

$$dM = \rho\,d\mathbb{V} + \mathbb{V}\,d\rho \tag{2.11}$$

But $dM = 0$ because the mass is constant. Hence we find that

$$\mathbb{V}\,d\rho = -\rho\,d\mathbb{V} \qquad \text{or} \qquad \frac{d\rho}{\rho} = -\frac{d\mathbb{V}}{\mathbb{V}}$$

The bulk modulus of elasticity of water is approximately 2.2 GN/m², which corresponds to 0.05% change in volume for a change of 1 MN/m² in pressure. Obviously, the term *incompressible* is justifiably applied to water.

The elasticity of an ideal gas is proportional to the pressure. For an isothermal (constant temperature) process,

$$\frac{dp}{d\rho} = RT$$

so

$$E_v = \rho\frac{dp}{d\rho} = \rho RT = p$$

For an adiabatic process, $E_v = kp$, where k is the ratio of specific heats, c_p/c_v.

2.7 | Surface Tension

According to the theory of molecular attraction, molecules of liquid considerably below the surface act on each other by forces that are equal in all directions. However, molecules near the surface have a greater attraction for each other than they do for molecules below the surface. This produces a surface on the liquid that acts like a stretched membrane. Because of this membrane effect, each portion of the liquid surface exerts "tension" on adjacent portions of the surface or on objects that are in contact with the liquid surface. This tension acts in the plane of the surface, and its magnitude per unit length is defined as *surface tension,* σ (sigma). Surface tension for a water–air surface is 0.073 N/m at room temperature. The effect of surface tension is illustrated for the case of capillary action in a small tube (Fig. 2.5). Here the end of a small-diameter tube is put into a reservoir of water, and the characteristic curved water surface occurs within the tube. The relatively great attraction of the water molecules for the glass causes the water surface to curve upward in the region of the glass wall. Then the surface-tension force acts around the circumference of the tube, in the direction indicated. It may be assumed that θ (theta) is equal to 0° for water against glass. This produces a net upward force on the water that causes the water in the tube to rise above the water surface in the reservoir. An example gives a quantitative illustration of the principle.

FIGURE 2.5 _____

Capillary action in a small tube.

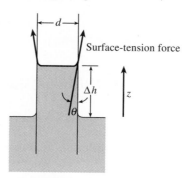

Surface-tension force

EXAMPLE 2.4   To what height above the reservoir level will water (at 20°C) rise in a glass tube, such as that shown in Fig. 2.5, if the inside diameter of the tube is 1.6 mm?

Solution   By taking the summation of forces in the vertical direction on the water in the tube that has risen above the reservoir level, we have

$$F_{\sigma,z} - W = 0$$

$$\sigma \pi d \cos \theta - \gamma(\Delta h)(\pi d^2/4) = 0$$

However, θ for water against glass is so small it can be assumed to be $0°$; therefore $\cos \theta \approx 1$. Then

$$\sigma \pi d - \gamma(\Delta h)\left(\frac{\pi d^2}{4}\right) = 0$$

or

$$\Delta h = \frac{4\sigma}{\gamma d} = \frac{4 \times 0.073 \text{ N/m}}{9790 \text{ N/m}^3 \times 1.6 \times 10^{-3} \text{ m}} = 18.6 \text{ mm} \qquad \blacktriangleleft$$

Other manifestations of surface tension include the excess pressure (over and above atmospheric pressure) created inside droplets and bubbles, the transformation of a liquid jet into droplets, and the binding together of wetted granular material, such as fine, sandy soil.

Surface-tension forces for some applications are shown in Fig. 2.6.

FIGURE 2.6

Surface-tension forces for several different cases.

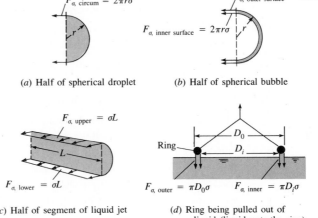

(a) Half of spherical droplet   (b) Half of spherical bubble

(c) Half of segment of liquid jet   (d) Ring being pulled out of a liquid (liquid wets the ring)

2.8 Vapor Pressure

The pressure at which a liquid will boil is called its *vapor pressure*. This pressure is a function of temperature (vapor pressure increases with temperature). In this context we usually think about the temperature at which boiling occurs. For example, water boils at 212°F at sea-level atmospheric pressure (14.7 psia). However, in terms of vapor pressure, we can say that by increasing the temperature of water at sea level to 212°F, we increase the vapor pressure to the point at which it is equal to the atmospheric pressure (14.7 psia), so that boiling occurs. When we think of incipient boiling in terms of vapor pressure, it is easy to visualize that boiling can also occur in water at temperatures much below 212°F if the pressure in the water is reduced to its vapor pressure. For example, the vapor pressure of water at 50°F (10°C) is 0.122 psia (approximately 1% of standard atmospheric pressure). Therefore, if the pressure within water at that temperature is reduced to that value, the water boils.* Such boiling often occurs in flowing liquids, such as on the suction side of a pump. When such boiling does occur in flowing liquids, vapor bubbles start growing in local regions of very low pressure and then collapse in regions of higher pressure downstream. This phenomenon, which is called *cavitation,* is discussed in Chapter 5.

Table A.5 in the Appendix gives values of vapor pressure for water.

Problems

2.1   Meteorologists often refer to air masses in forecasting the weather. Estimate the mass of 1 mi^3 of air in slugs and kilograms. Make your own reasonable assumptions with respect to the conditions of the atmosphere.

2.2   Determine the density of air, helium, and carbon dioxide at an absolute pressure of 140 kN/m^2 (20 psia) and a temperature of 30°C (86°F).

2.3   Calculate the density and specific weight of carbon dioxide at 400 kN/m^2 absolute and 60°C.

2.4   Calculate the density and specific weight of helium at 300 kN/m^2 absolute and 60°C.

2.5   Natural gas is stored in a spherical tank at a temperature of 10°C. At a given initial time, the pressure in the tank is 100 kPa gage, and the atmospheric pressure is 100 kPa absolute. Some time later, after considerably more gas is pumped into the tank, the pressure in the tank is 300 kPa gage, and the temperature is still 10°C. What will be the ratio of the mass of air in the tank when $p = 300$ kPa gage to that when the pressure was 100 kPa gage?

2.6   At a temperature of 40°C and an absolute pressure of 103 kN/m^2, what is the ratio of the density of water to the density of air, ρ_w/ρ_a?

2.7   What is the weight of a 4-ft^3 tank of oxygen if the oxygen is pressurized to 300 psia, the tank itself weighs 100 lbf, and the temperature is 50°F?

2.8   What are the specific weight and density of air at an absolute pressure of 345 kPa and a temperature of 38°C?

2.9   What are the density, specific weight, dynamic viscosity, and kinematic viscosity of carbon dioxide at a pressure of 500 kN/m^2 absolute and a temperature of 20°C?

*Actually, boiling can occur at this vapor pressure only if there is a gas–liquid surface present to allow the process to start. Boiling at the bottom of a pot of water is usually initiated in crevices in the material of the pot, in which minute bubbles of air are entrapped even when the pot is filled with water.

2.10 Find the kinematic and dynamic viscosities of air and water at a temperature of 10°C (50°F) and an absolute pressure of 103 kPa (15 psia).

2.11 Classify the following according to whether they are extensive or intensive properties: specific weight γ, density ρ, mass M, surface tension σ, vapor pressure p_v, weight W, velocity V, and acceleration a.

2.12 For an ideal gas, it can be shown that the difference between the specific heat at constant pressure and that at constant volume is the gas constant R; that is, $c_p - c_v = R$. Derive expressions for c_p and c_v in terms of k and R.

2.13 What is the change in the viscosity and density of water between 10°C and 70°C? What is the change in the viscosity and density of air between 10°C and 70°C? Assume standard atmospheric pressure ($p = 101 \text{ kN/m}^2$ absolute).

2.14 What is the change in the kinematic viscosity of air between 10°C and 50°C? Assume standard atmospheric pressure.

2.15 Find the dynamic and kinematic viscosity of kerosene, SAE 10W motor oil, and water at a temperature of 38°C (100°F).

2.16 What is the ratio of the dynamic viscosity of air to that of water at standard pressure and a temperature of 20°C? What is the ratio of the kinematic viscosity of air to that of water for the same conditions?

2.17 Two plates are separated by a 1/4-in. space. The lower plate is stationary; the upper plate moves at a velocity of 10 ft/s. Oil (SAE 10W-30, 100°F), which fills the space between the plates, has the same velocity as the plates at the surface of contact. The variation in velocity of the oil is linear. What is the shear stress in the oil?

2.18 Given $u = 10 \, y^{1/6}$, where u is the velocity of water (20°C) in meters per second and y is the distance from the boundary as shown below. Determine the shear stress in the water at $y = 2$ mm.

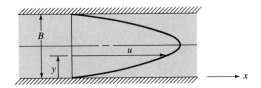

PROBLEMS 2.18, 2.19, 2.20, 2.21

2.19 The velocity distribution for the flow of crude oil at 150°F ($\mu = 7 \times 10^{-5}$ lbf-s/ft²) between two walls is given by $u = 100 \, y(0.1 - y)$ ft/s, where y is measured in feet and the space between the walls is 0.1 ft. Plot the velocity distribution and determine the shear stress at the walls.

2.20 A liquid flows between parallel boundaries as shown above. The velocity distribution near the lower wall is given in the following table.

y in mm	V in m/s
1.0	1.0
2.0	1.99
3.0	2.98

    a. If the viscosity of the liquid is 10^{-2} N · s/m², what is the maximum shear stress in the liquid?

    b. Where will the minimum shear stress occur?

2.21   Suppose that glycerin is flowing ($T = 20°C$) and that the pressure gradient dp/dx is -1.6 kN/m³. What are the velocity and shear stress at a distance of 12 mm from the wall if the space B between the walls is 5.0 cm? What are the shear stress and velocity at the wall? The velocity distribution for viscous flow between stationary plates is

$$u = \frac{-1}{2\mu}\frac{dp}{dx}(By - y^2)$$

2.22   A laminar flow occurs between two horizontal parallel plates under a pressure gradient dp/ds (p decreases in the positive s direction). The upper plate moves left (negative) at velocity u_t. The expression for local velocity u is given as

$$u = -\frac{1}{2}\frac{y}{\mu}\frac{dp}{ds}(Hy - y^2) + u_t\frac{y}{H}$$

Is the magnitude of the shear stress greater at the moving plate ($y = H$) or at the stationary plate ($y = 0$)?

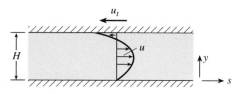

PROBLEMS 2.22

2.23   For the conditions of Problem 2.22, derive an expression for the y position of zero shear stress.

2.24   For the conditions of Problem 2.22, derive an expression for the plate speed u_t required to make the shear stress zero at $y = 0$.

2.25   Consider the ratio μ_{100}/μ_{50}, where μ is the viscosity of oxygen and the subscripts 100 and 50 are the temperatures of the oxygen in degrees Fahrenheit. Does this ratio have a value a) less than 1, b) equal to 1, or c) greater than 1?

2.26   A solid circular cylinder of diameter d and length ℓ slides inside a vertical smooth pipe that has an inside diameter D. The small space between the cylinder and the pipe is lubricated with an oil film that has a viscosity μ. Derive a formula for the rate of descent of the cylinder in the vertical pipe. Assume that the cylinder has a weight W and is concentric with the pipe as it falls. Use the formula to find the rate of descent of a cylinder 100 mm in diameter that slides inside a 100.5-mm pipe. The cylinder is 200 mm long and weighs 20 N. The lubricant is SAE 20W oil at 10°C.

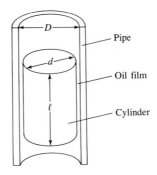

PROBLEMS 2.26, 2.27

2.27 Consider the pipe, cylinder, and oil described in Prob. 2.26. Suppose that the cylinder has a downward velocity of 0.5 m/s and is observed to be decelerating at a rate of 14 m/s². What is its weight?

2.28 The device shown consists of a disk that is rotated by a shaft. The disk is positioned very close to a solid boundary. Between the disk and the boundary is viscous oil.

    a. If the disk is rotated at a rate of 2 rad/s, what will be the ratio of the shear stress in the oil at $r = 2$ cm to the shear stress at $r = 3$ cm?

    b. If the rate of rotation is 2 rad/s, what is the speed of the oil in contact with the disk at $r = 3$ cm?

    c. If the oil viscosity is 0.01 N · s/m² and the spacing y is 2 mm, what is the shear stress for the conditions noted in part (b)?

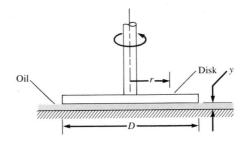

PROBLEMS 2.28, 2.29

2.29 What torque is required to rotate the disk of Prob. 2.28 at a rate of 2 rad/s, with $D = 8$ cm and with the same viscosity and spacing as in part (c)?

2.30 Some instruments having angular motion are damped by means of a disk connected to their shaft. The disk, in turn, is immersed in a container of oil, as shown. Derive a formula for the damping torque as a function of the disk diameter D, spacing S, rate of rotation ω, and oil viscosity μ.

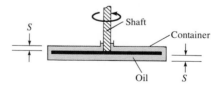

PROBLEM 2.30

2.31 A pressure of 3×10^6 N/m^2 is applied to a mass of water that initially filled a 1000-cm volume. Estimate its volume after the pressure is applied.

2.32 What pressure increase must be applied to water to reduce its volume by 1%?

2.33 Which of the following is the formula for the gage pressure within a very small spherical droplet of water? a) $p = \sigma/d$, b) $p = 4\sigma/d$, c) $p = 8\sigma/d$.

2.34 A spherical soap bubble has an inside radius R, a film thickness t, and a surface tension σ. Derive a formula for the pressure within the bubble relative to the outside atmospheric pressure. What is the pressure difference for a bubble with a 3-mm radius? Assume σ is the same as for pure water.

2.35 A water bug is suspended on the surface of a pond by surface tension. The bug has six legs, and each leg is in contact with the water over a length of 1 cm. What is the maximum mass (in g) of the bug if it is to avoid sinking?

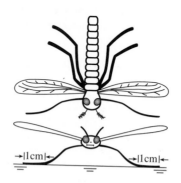

PROBLEM 2.35

2.36 A water column in a glass tube is used to measure the pressure in a pipe. The tube is 1/4 in. (6.35 mm) in diameter. How much of the water column is due to surface-tension effects? What would be the surface-tension effects if the tube were 1/8 in. (3.2 mm) or 1/32 in. (0.8 mm) in diameter?

PROBLEM 2.36

2.37   Calculate the maximum capillary rise of water (at 10°C) between two vertical glass plates spaced 0.50 mm apart.

2.38   What is the pressure within a 1-mm spherical droplet of water relative to the atmospheric pressure outside?

2.39   Consider a soap bubble 2 mm in diameter and a droplet of water, also 2 mm in diameter, that are falling in air. If the value of the surface tension for the film of the soap bubble is assumed to be the same as that for water, which has the greatest pressure inside it? a) the bubble, b) the droplet, c) neither, the pressure is the same for both.

2.40   The vapor pressure of water at 100°C is 101 kN/m², because water boils under these conditions. The vapor pressure of water decreases approximately linearly with decreasing temperature at a rate of 3.1 kN/m²°C. Calculate the boiling temperature of water at an altitude of 3000 m, where the atmospheric pressure is 69 kN/m² absolute.

References

1. Harris, J. *Rheology and Non-Newtonian Flow,* Longman, New York, 1977.

2. Schowalter, W. R. *Mechanics of Non-Newtonian Fluids,* Pergamon Press, New York, 1978.

Films

3. Lumley, John L. *Deformation of Continuous Media.* National Committee for Fluid Mechanics Films, distributed by Encyclopaedia Britannica Educational Corporation.

4. Markovitz, Hershel. *Rheological Behavior of Fluids.* National Committee for Fluid Mechanics Films, distributed by Encyclopaedia Britannica Educational Corporation.

5. Trefethen, Lloyd. *Surface Tension of Fluids.* National Committee for Fluid Mechanics Films, distributed by Encyclopaedia Britannica Educational Corporation.

Fluid Statics

Arch dams, such as this one (the 140-m-high Gordon Arch Dam in Tasmania), are the thinnest of all types of dams. The hydrostatic forces that act on the dam are transferred to the rock walls of the canyon. If a narrow canyon with solid rock walls exists at the dam site, an arch dam is almost always more economical to construct than other types. (Courtesy of the Hydroelectric Commission of Australia)

I n general, fluids exert both normal and shearing forces on surfaces that are in contact with them. However, only fluids with velocity gradients produce shearing forces. For fluids at rest, only normal forces exist. These normal forces in fluids are called *pressure forces.*

3.1   Pressure

Definition of Pressure

At every point in a static fluid a certain pressure intensity exists. Specifically, this pressure intensity, usually simply called pressure, is defined as follows:

$$p = \lim_{\Delta A \to 0} \frac{\Delta F}{\Delta A} = \frac{dF}{dA} \tag{3.1}$$

where F is the normal force acting over the area A. Pressure intensity is a scalar quantity; that is, it has magnitude only and acts equally in all directions. This is easily demonstrated by considering the wedge-shaped element of fluid in equilibrium in Fig. 3.1. The forces that act on the element are the surface forces and the weight force.

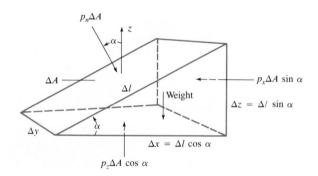

FIGURE 3.1

Pressure forces on a fluid element in equilibrium.

By writing the equation of equilibrium for the x direction, we obtain

$$(p_n \Delta y \Delta \ell) \sin \alpha - p_x (\Delta y \Delta \ell \sin \alpha) = 0$$

or
$$p_n = p_x \qquad \qquad (3.2)$$

For the z direction, we obtain

$$-(p_n \Delta y \Delta \ell) \cos \alpha + p_z (\Delta y \Delta \ell \cos \alpha) - \tfrac{1}{2} \gamma \Delta \ell \cos \alpha \Delta \ell \sin \alpha \Delta y = 0$$

Now, when we divide this equation by the product $\Delta \ell \Delta y \cos \alpha$ and shrink the element to a point ($\Delta \ell \to 0$), the last term disappears. Thus we have

$$p_n = p_z \qquad \qquad (3.3)$$

Combining Eqs. (3.2) and (3.3), we finally arrive at the result

$$p_n = p_x = p_z \qquad \qquad (3.4)$$

Since the angle α (alpha) is arbitrary and p_n is independent of α, we conclude that the pressure at a point in a static fluid acts with the same magnitude in all directions:

$$p_n = p_x = p_y = p_z$$

Pressure Transmission

In a closed system a pressure change produced at one point in the system will be transmitted throughout the entire system. The principle is known as Pascal's law after Blaise Pascal, the French scientist who first stated it in 1653. This phenomenon of pressure transmission, along with the ease with which fluids can be moved, has led to the widespread development of hydraulic controls for operating equipment such as aircraft-control surfaces, heavy earthmoving equipment, and hydraulic presses. Figure 3.2 is an illustration of the application of this principle in the form of a hydraulic lift used in service stations. Here air pressure

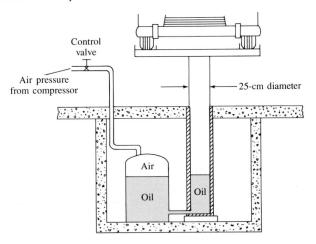

FIGURE 3.2

Hydraulic lift.

from a compressor establishes the pressure in the oil system, which in turn acts against the piston in the lift. It can be seen that if a pressure of 600 kN/m², for example, acts on the 25-cm piston, then a force equal to pA, or 29.45 kN, will be exerted on the piston. To handle larger or smaller loads it is necessary only to increase or decrease the pressure.

EXAMPLE 3.1   A hydraulic jack has the dimensions shown in the accompanying figure. If one exerts a force F of 100 N on the handle of the jack, what load, F_2, can the jack support?

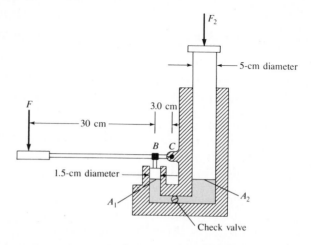

Solution   The force F_1 exerted on the small piston is obtained by taking moments about C. Therefore,

$$(0.33 \text{ m}) \times (100 \text{ N}) - (0.03 \text{ m})F_1 = 0$$

$$F_1 = \frac{0.33 \text{ m} \times 100 \text{ N}}{0.03 \text{ m}} = 1100 \text{ N}$$

Because the small piston is in equilibrium, this force is equal to the pressure force on the piston, or

$$p_1 A_1 = 1100 \text{ N}$$

Hence
$$p_1 = \frac{1100}{A_1} = \frac{1100}{\pi d^2/4} = 6.22 \times 10^6 \text{ N/m}^2$$

Now we know the pressure in the liquid. Therefore, we can solve for the force on the large piston. Since $p_1 = p_2$,

$$F_2 = p_1 A_2$$

where A_2 is the area of the large piston. Finally,

$$F_2 = 6.22 \times 10^6 \, \frac{\text{N}}{\text{m}^2} \times \frac{\pi}{4} \times (0.05 \text{ m})^2 = 12.22 \text{ kN} \qquad \blacktriangleleft$$

Absolute Pressure, Gage Pressure, and Vacuum

In a region such as outer space, which is virtually void of gases, the pressure is essentially zero. Such a condition can be approached very nearly in the laboratory when a vacuum pump is used to evacuate a bottle. The pressure in a vacuum is called *absolute zero,* and all pressures referenced with respect to this zero pressure are termed *absolute pressures.* Therefore, atmospheric pressure at sea level on a particular day might be given as 101 kN/m², which is equivalent to 760 mm of deflection on a mercury barometer.

Many pressure-measuring devices measure not absolute pressure but only differences in pressure. For example, a common Bourdon-tube gage (see Sec. 3.3) indicates only the difference between the pressure in the fluid to which it is tapped and the pressure in the atmosphere. In this case, then, the reference pressure is actually the atmospheric pressure at the gage. This type of pressure reading is called *gage pressure.*

The fundamental unit of pressure in the SI system is the pascal (Pa), which is one newton per square meter (N/m²). Gage and absolute pressures are usually identified after the unit.* For example, if a pressure of 50 kPa is measured with a gage referenced to the atmosphere and the absolute atmospheric pressure is 100 kPa, then the pressure can be expressed as either

$$p = 50 \text{ kPa gage} \qquad \text{or} \qquad p = 150 \text{ kPa absolute}$$

Whenever atmospheric pressure is used as a reference (or, in other words, when gage pressure is being measured), the possibility exists that the pressure thus measured can be either positive or negative. Negative gage pressures are also termed *vacuum pressures.* Hence, if a gage tapped into a tank indicates a vacuum pressure of 31.0 kPa, this can also be stated as 70.0 kPa absolute, or −31.0 kPa gage, assuming that the atmospheric pressure is 101 kPa absolute. An example of this reference system is depicted in Fig. 3.3 for arbitrary pressures of $p_A = 200$ kPa gage and $p_B = 51$ kPa absolute with an atmospheric pressure of 101 kPa absolute.

In the traditional foot–pound–second system of units, the gage or absolute designations are usually included as part of the abbreviated unit. For example, a gage pressure of 10 pounds per square foot is designated as psfg. Other combinations are psfa, psig, psia. The latter two designations are for pounds per square inch gage and pounds per square inch absolute.

FIGURE 3.3

Example of pressure relations.

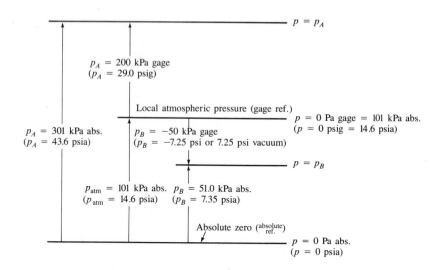

$p = p_A$

$p_A = 200$ kPa gage
($p_A = 29.0$ psig)

Local atmospheric pressure (gage ref.)

$p = 0$ Pa gage $= 101$ kPa abs.
($p = 0$ psig $= 14.6$ psia)

$p_A = 301$ kPa abs.
($p_A = 43.6$ psia)

$p_B = -50$ kPa gage
($p_B = -7.25$ psi or 7.25 psi vacuum)

$p = p_B$

$p_{atm} = 101$ kPa abs. $p_B = 51.0$ kPa abs.
($p_{atm} = 14.6$ psia) ($p_B = 7.35$ psia)

Absolute zero ($^{absolute}_{ref.}$)

$p = 0$ Pa abs.
($p = 0$ psia)

3.2 Pressure Variation with Elevation

Basic Differential Equation

For a static fluid, pressure varies only with the elevation within the fluid. This may be shown by isolating a cylindrical element of fluid and applying the equation of equilibrium to the element. Consider the element shown in Fig. 3.4. Here the element is oriented so that its longitudinal axis is parallel to an arbitrary ℓ direction. The element is $\Delta\ell$ long, ΔA in cross-sectional area, and inclined at an angle α with the horizontal. The equation of equilibrium for the ℓ direction, considering the pressure forces and gravitational force acting on the element in this direction, is

$$\sum F_\ell = 0$$

$$p\,\Delta A - (p + \Delta p)\,\Delta A - \gamma\,\Delta A\,\Delta\ell\,\sin\alpha = 0$$

FIGURE 3.4

Variation in pressure with elevation.

Upon simplifying, and dividing by the volume of the element, $\Delta\ell\,\Delta A$, this reduces to

$$\frac{\Delta p}{\Delta\ell} = -\gamma\sin\alpha$$

However, if we let the length of the element approach zero, then in the limit $\Delta p/\Delta\ell = dp/d\ell$. Also one notes that $\sin\alpha = dz/d\ell$. Therefore,

$$\frac{dp}{d\ell} = -\gamma\frac{dz}{d\ell} \qquad (3.5)$$

This can also be written as

$$\frac{dp}{dz} = -\gamma \qquad (3.6)$$

which is the basic equation for hydrostatic pressure variation with elevation. Equation (3.5) states that for static fluids a change of pressure in the ℓ direction, $dp/d\ell$, occurs only when there is a change of elevation in the ℓ direction, $dz/d\ell$. In other words, if one considers a path through the fluid that lies in a horizontal plane, the pressure everywhere along this path is constant. On the other hand, the greatest possible change in hydrostatic pressure occurs along a vertical path through the fluid. Furthermore, Eqs. (3.5) and (3.6) state that the pressure changes inversely with elevation. If one travels upward in the fluid (positive z direction), the pressure decreases; and if one goes downward (negative z), the pressure increases. Of course, a pressure increase is exactly what a diver experiences when descending in a lake or pool.

EXAMPLE 3.2 Compare the rate of change of pressure with elevation for air at sea level, 101.3 kPa absolute, at a temperature of 15.5°C, and for fresh water at the same pressure and temperature. Assuming constant specific weights for air and water, determine also the total pressure change that occurs for both with a 4-m decrease in elevation.

Solution First, determine specific weights of water and air:

$$\rho_{air} = \frac{p}{RT} = \frac{101.3 \times 10^3 \text{ N/m}^2}{287 \text{ J/kg K} \times (15.5 + 273) \text{ K}}$$

Then

$$\rho_{air} = 1.22 \text{ kg/m}^3$$

$$\gamma_{air} = \rho g = 1.22 \text{ kg/m}^3 \times 9.81 \text{ m/s}^2$$

$$= 11.97 \text{ kg/m}^2\text{s}^2 = 11.97 \text{ N/m}^3$$

and

$$\gamma_{water} = 9799 \text{ N/m}^3 \quad \text{(interpolated from Table A.5)}$$

$$\frac{dp}{dz} = -\gamma$$

Then $\left(\dfrac{dp}{dz}\right)_{air} = -11.97 \text{ N/m}^3$ $\left(\dfrac{dp}{dz}\right)_{water} = -9799 \text{ N/m}^3$

$$\text{Total pressure change for air} = (-11.97 \text{ N/m}^3) \times (-4 \text{ m})$$
$$= 47.9 \text{ N/m}^2 = 47.9 \text{ Pa} \quad \blacktriangleleft$$
$$\text{Total pressure change for water} = (-9799 \text{ N/m}^3) \times (-4 \text{ m})$$
$$= 39.2 \text{ kN/m}^2 = 39.2 \text{ kPa} \quad \blacktriangleleft$$

Pressure Variation for a Uniform-Density Fluid

Equations (3.5) and (3.6) are completely general in the sense that they describe the rate of change of pressure for all fluids in static equilibrium. However, much simplification accrues in practical applications of the equations if it can be assumed that the density, and thus the specific weight, of the fluid are the same throughout. Then γ is a constant in Eqs. (3.5) and (3.6). The reason for the simplification is that the integration of Eq. (3.5) or (3.6) becomes easier and the resulting equation simpler than if γ were a function of z. With constant specific weight, the following equation results from integration of Eq. (3.6):

$$p = -\gamma z + \text{constant} \qquad (3.7)$$

or

$$\left(\frac{p}{\gamma} + z\right) = \text{constant} \qquad (3.8)$$

The sum of the terms p/γ and z on the left-hand side of Eq. (3.8) is called the *piezometric head*. As shown by the equation, this is constant throughout an incompressible static fluid. Therefore, one can relate the pressure and elevation at one point to the pressure and elevation at another point in the fluid in the following manner:

$$\frac{p_1}{\gamma} + z_1 = \frac{p_2}{\gamma} + z_2 \qquad (3.9)$$

or

$$\Delta p = -\gamma \Delta z \qquad (3.10)$$

Note, however, that Eqs. (3.7) through (3.10) are applicable only in fluids with constant specific weights. In other words, Eqs. (3.9) and (3.10) can be applied between two points in a given fluid but not across an interface between two fluids having different specific weights.

EXAMPLE 3.3   What is the water pressure at a depth of 35 ft in the tank shown?

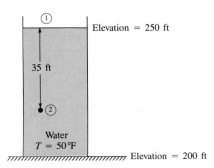

Solution At an elevation of 250 ft the gage pressure is zero, so

$$\frac{p_1}{\gamma} + z_1 = \frac{p_2}{\gamma} + z_2$$

$$0 + 250 = \frac{p_2}{\gamma} + 215$$

$$\frac{p_2}{\gamma} = 35 \text{ ft}$$

$$p_2 = 35 \times 62.4 = 2180 \text{ psfg} = 15.2 \text{ psig}$$ ◀

EXAMPLE 3.4 Oil with a specific gravity of 0.80 forms a layer 0.90 m deep in an open tank that is otherwise filled with water. The total depth of water and oil is 3 m. What is the gage pressure at the bottom of the tank?

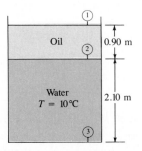

Solution First determine the pressure at the oil–water interface, staying within the oil, and then calculate the pressure at the bottom.

$$\frac{p_1}{\gamma} + z_1 = \frac{p_2}{\gamma} + z_2$$

where p_1 is the pressure at free surface of oil, z_1 is the elevation of free surface of oil, p_2 is the pressure at interface between oil and water, and z_2 is the elevation at interface between oil and water. For this example, $p_1 = 0$, $\gamma = 0.80 \times$ 9810 N/m^3, $z_1 = 3$ m, and $z_2 = 2.10$ m. Therefore,

$$p_2 = 0.90 \text{ m} \times 0.80 \times 9810 \text{ N/m}^3 = 7.06 \text{ kPa gage}$$

Now obtain p_3 from

$$\frac{p_2}{\gamma} + z_2 = \frac{p_3}{\gamma} + z_3$$

where p_2 has already been calculated and $\gamma = 9810$ N/m^3.

$$p_3 = 9810\left(\frac{7060}{9810} + 2.10\right) = 27.7 \text{ kPa gage} \qquad \blacktriangleleft$$

Pressure Variation for Compressible Fluids

The preceding section dealt with pressure variation in fluids for which the specific weight is constant. However, when the specific weight varies significantly, it must be expressed in such a form that Eq. (3.6) can be integrated. For the case of an ideal gas, this is accomplished through the equation of state, which relates the density of the gas to pressure and temperature:

$$\frac{p}{\rho} = RT$$

or

$$\rho = \frac{p}{RT} \qquad (3.11)$$

This can be expressed as follows when both sides of Eq. (3.11) are multiplied by g:

$$\gamma = \frac{pg}{RT} \qquad (3.12)$$

where R is the gas constant, 287 J/kg $\cdot$ K, for dry air, T is the absolute temperature K, and p is the absolute pressure Pa.

Equation (3.12) introduces another variable, temperature, so it becomes necessary to have additional data relating temperature and elevation. If one is interested in the pressure variation in the atmosphere, and if temperature-versus-elevation data for a local area at a given time are available, then one can quite accurately compute pressure versus elevation. Lacking such data, one can resort to the so-called *U.S. standard atmosphere* (2). This is a set of data compiled by the U.S. National Weather Service that represents average conditions over the United States at 40° N latitude. At sea level the standard atmospheric pressure is 101.3 kPa and the temperature is 288 K. Also, the atmosphere is divided into two layers, the *troposphere* and the *stratosphere.* In the troposphere, defined as the layer between sea level and 10,769 m, the temperature decreases linearly with increasing elevation at a *lapse rate* of 6.50 K/km. The stratosphere begins at the top of the troposphere and extends to an elevation of 32.3 km. In the stratosphere the temperature is constant at −55°C.

We now have sufficient information to calculate the pressure and density at any elevation. Let us first consider the troposphere.

Pressure Variation in the Troposphere

Let the temperature T be given by

$$T = T_0 - \alpha(z - z_0) \tag{3.13}$$

In this equation T_0 is the temperature at a reference level where the pressure is known and α is the lapse rate. If we use the specific weight of a gas from Eq. (3.12) in the basic hydrostatic equation, we obtain

$$\frac{dp}{dz} = -\frac{pg}{RT} \tag{3.14}$$

Substituting Eq. (3.13) for T, we get

$$\frac{dp}{dz} = -\frac{pg}{R[T_0 - \alpha(z - z_0)]}$$

Now we must separate the variables and integrate to obtain

$$\frac{p}{p_0} = \left[\frac{T_0 - \alpha(z - z_0)}{T_0}\right]^{g/\alpha R}$$

$$p = p_0\left[\frac{T_0 - \alpha(z - z_0)}{T_0}\right]^{g/\alpha R} \tag{3.15}$$

EXAMPLE 3.5   If at sea level the absolute pressure and temperature are 101.3 kPa and 15°C, what is the pressure at an elevation of 2000 m, assuming that standard atmospheric conditions prevail?

Solution   Use Eq. (3.15):

$$p = p_0\left[\frac{T_0 - \alpha(z - z_0)}{T_0}\right]^{g/\alpha R}$$

where $p_0 = 101{,}300 \text{ N/m}^2$, $T_0 = 273 + 15 = 288$ K, $\alpha = 6.50 \times 10^{-3}$ K/m, $z - z_0 = 2000$ m, and $g/\alpha R = 5.259$. Then

$$p = 101.3\left(\frac{288 - 6.50 \times 10^{-3} \times 2000}{288}\right)^{5.259} = 79.5 \text{ kPa} \quad \blacktriangleleft$$

Pressure Variation in the Stratosphere

In the stratosphere the temperature is assumed to be constant. Therefore, when Eq. (3.14) is integrated, we obtain

$$\ln p = -\frac{zg}{RT} + C$$

At $z = z_0$, $p = p_0$, so the foregoing equation reduces to

$$\frac{p}{p_0} = e^{-(z-z_0)g/RT}$$

or

$$p = p_0 e^{-(z-z_0)g/RT} \tag{3.16}$$

EXAMPLE 3.6   If the pressure and temperature are 3.28 psia ($p = 22.6$ kPa absolute) and $-67°F$ ($-55°C$) at an elevation of 36,000 ft (10,973 m), what is the pressure at 56,000 ft (17,069 m), assuming isothermal conditions over this range of elevation?

Solution   For isothermal conditions,

$$T = -67 + 460 = 393°R$$

$$p = p_0 e^{-(z-z_0)g/RT} = 3.28 e^{-(20,000)(32.2)/(1716 \times 393)} = 3.28 e^{-0.955}$$

Therefore the pressure at 56,000 ft is

$$p = 1.26 \text{ psia} \qquad \blacktriangleleft$$

SI units $p = 8.69$ kPa absolute $\blacktriangleleft$

3.3 Pressure Measurements

Numerous instruments have been devised to indicate the magnitude of pressure intensity, and most of these operate on either the principle of manometry or the principle of flexing of an elastic member whose deflection is directly proportional to the applied pressure. These principles and representative pressure gages are described in the following sections.

Manometry

Basically, this method utilizes the change in pressure with elevation to evaluate pressure. Consider the *piezometer,* or simple manometer, attached to a pipe as shown in Fig. 3.5. It is easy to compute the gage pressure at the center of the pipe; here the pressure is simply $p = \gamma h$, which follows directly from Eq. (3.10).

FIGURE 3.5

*Piezometer attached to
a pipe.*

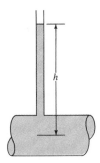

This type of pressure-indicating device is accurate and simple. However, the student may visualize how impractical it might become for measuring high pressures, and of course it is useless in its present form for pressure measurement in gases. For both these cases, a U-tube (a *differential manometer*) such as that shown in Fig. 3.6 can be employed. In this case, a knowledge of the specific weights of the fluids involved and of the linear measurements ℓ and Δh is needed to calculate the pressure in the pipe. Here the procedure is to calculate the pressure changes, step by step, from one level to the next in each fluid and to apply these changes finally to evaluate the unknown pressure. The following example illustrates the procedure for the case shown in Fig. 3.6.

EXAMPLE 3.7 Water is the liquid in the pipe of Fig. 3.6, and mercury is the manometer fluid. If the deflection Δh is 60 cm and ℓ is 180 cm, what is the gage pressure at the center of the pipe?

Solution Since the manometer is open to the atmosphere, we know that the gage pressure at point 1, the mercury surface, is zero. Then the pressure at 2 will be

$$p_2 = p_1 + \text{change in pressure between 1 and 2} = 0 - \gamma_m \Delta h$$

$$= 0 - \gamma_m(-0.60) \qquad \text{where } \gamma_m = 133 \text{ kN/m}^3$$

$$= 79.8 \text{ kPa}$$

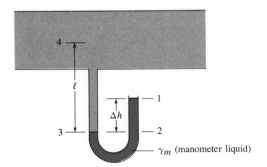

FIGURE 3.6

U-tube manometer.

Point 3 is at the same elevation as point 2 and in the same fluid; therefore, $p_3 = p_2$. The next step is to evaluate Δp from 3 to 4 and to apply this to the pressure at 3:

$$\Delta p_{3 \to 4} = -\gamma \times 1.80 \qquad \text{where } \gamma = 9810 \text{ N/m}^3$$

$$= -17.66 \text{ kPa}$$

Then

$$p_4 = p_3 - 17.66 \text{ kPa} = 62.1 \text{ kPa gage}$$ ◀

Once one is familiar with the basic principle of manometry, it should be easy to write a single equation rather than separate equations for each step in Example 3.7. The single equation for evaluation of the pressure in the pipe of Fig. 3.6 is

$$0 + \gamma_m \Delta h - \gamma \ell = p_p$$

One can read the equation in this way: zero pressure at the open end plus the change in pressure from 1 to 2 minus the change in pressure from 3 to 4 equals the pressure in the pipe. The main point that the student must remember in this process is that when one travels downward in the fluid, the pressure increases, and when one travels upward, the pressure decreases.

Up to this point we have considered only liquid-filled manometers. Let us consider Fig. 3.6 again with a gas-filled pipe. This is illustrated in the next example.

EXAMPLE 3.8 Air at 20°C is the fluid in the pipe of Fig. 3.6, and water is the manometer fluid. If the deflection Δh is 70 cm and ℓ is 140 cm, what is the gage pressure in the pipe? Also compute this pressure by neglecting the pressure change due to the 140-cm column of air. Assume standard atmospheric pressure.

Solution The specific weight of air is found from Eq. (3.12), which requires that the air pressure be known. Therefore, the air pressure at the bottom of the 70-cm column is first calculated:

$$p_{air} = 9790 \text{ N/m}^3 \times 0.70 \text{ m} = 6853 \text{ Pa gage}$$

Then the absolute air pressure is given as

$$p_{air} = 6853 \text{ Pa} + 101,300 \text{ Pa} = 108.15 \text{ kPa}$$

Then

$$\rho_{air} = \frac{p}{RT} = \frac{108,150 \text{ N/m}^2}{287 \text{ J/kg K} \times (20 + 273) \text{ K}} = 1.286 \text{ kg/m}^3$$

or

$$\gamma_{air} = 1.286 \text{ kg/m}^3 \times 9.81 \text{ m/s}^2 = 12.62 \text{ N/m}^3$$

Now compute the gage pressure in the pipe:

$$p_{pipe} = 6853 \text{ Pa} - 1.4 \text{ m} \times 12.62 \text{ N/m}^3 = 6835 \text{ Pa} \quad \blacktriangleleft$$

If the effect of the air column is neglected, the gage pressure in the pipe is

$$p_{pipe} = 9790 \text{ N/m}^3 \times 0.70 \text{ m} = 6853 \text{ Pa} \quad \blacktriangleleft$$

Results of the foregoing example show that when liquids and gases are both involved in a manometer problem, it is well within engineering accuracy to neglect the pressure changes due to the columns of gas.

EXAMPLE 3.9 What is the pressure of the air in the tank shown in the accompanying figure if $\ell_1 = 40$ cm (1.31 ft), $\ell_2 = 100$ cm (3.28 ft), and $\ell_3 = 80$ cm (2.62 ft)?

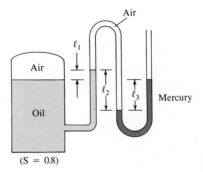

Solution

SI units

$$0 + 0.80 \text{ m} \times 133,000 \text{ N/m}^3 + 0.4 \text{ m} \times 9810 \text{ N/m}^3 \times 0.8 = p_{air}$$

$$p_{air} = 109.5 \text{ kPa gage} \quad \blacktriangleleft$$

Traditional units

$$0 + 2.62 \text{ ft} \times 846 \text{ lbf/ft}^3 + 1.31 \text{ ft} \times 62.4 \text{ lbf/ft}^3 \times 0.8 = p_{air}$$

$$p_{air} = 2282 \text{ psfg} = 15.85 \text{ psig} \quad \blacktriangleleft$$

Differential Manometer

It is often desirable to measure the difference in pressure between two points in a pipe. For this application a manometer is connected to the two points between which the pressure difference is to be measured. Such a setup is shown in

FIGURE 3.7

Differential manometer.

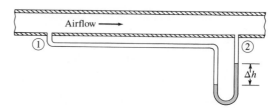

Fig. 3.7. In this case a gas is flowing, and the pressure difference between points 1 and 2 is given by $\Delta p = \gamma_m \Delta h$, where γ_m is the specific weight of the manometer liquid and Δh is the deflection of this liquid.

Bourdon-Tube Gage

This type of gage consists of a tube having an elliptical cross section and bent into a circular arc, as shown in Fig. 3.8*b*. When atmospheric pressure (zero gage pressure) prevails in the gage, the tube is undeflected, and for this condition the gage pointer is calibrated to read zero pressure. When pressure is applied to the gage, the curved tube tends to straighten (much like the party favors that straighten out when one blows into them), thereby actuating the pointer to read correspondingly higher pressure. The Bourdon-tube gage is a very common type that is reliable if not subjected to excessive pressure pulsations or undue external shock. However, because both these conditions sometimes prevail in engineering applications, pulsation dampers should be installed in the line leading to such gages and the gages should be periodically calibrated to check their accuracy.

Pressure Transducers

Modern factories and systems that involve flow processes are controlled automatically, and much of their operation involves sensing of pressure at critical points of the system. Therefore, pressure-sensing devices, such as pressure transducers, are designed to produce electronic signals that can be transmitted to oscillographs or digital devices for record-keeping and/or to control other devices

FIGURE 3.8

Bourdon-tube gage.
(a) View of typical gage.
(b) Internal mechanism
(schematic).

(a)

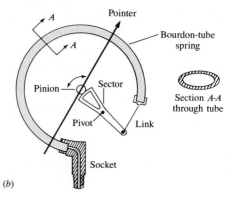

(b)

for process operation. Basically, most transducers are tapped into the system with one side of a small diaphragm exposed to the active pressure of the system. When the pressure changes, the diaphragm flexes and a sensing element connected to the other side of the diaphragm produces a signal that is usually linear with the change in pressure in the system. There are many types of sensing elements; one common type is the resistance-wire strain gage attached to a flexible diaphragm. As the diaphragm flexes, the wires of the strain gage change length, thereby changing the resistance of the wire. This change in resistance is utilized electronically to produce a voltage change that can then be used in various ways. Figure 3.9 shows a schematic arrangement for a pressure transducer and associated electronic equipment used to record pressure data.

Another type of pressure transducer used for measuring rapidly changing high pressures, such as the pressure in the cylinder head of an internal combustion engine, is the piezoelectric transducer (1). These transducers operate with a quartz crystal that generates a charge when subjected to a pressure. Sensitive electronic circuitry is required to convert the charge to a measurable voltage signal.

Computer data acquisition systems are used widely with pressure transducers. The analog signal from the transducer is converted (through an A/D converter) to a digital signal that can be processed by a computer. This expedites the data acquisition process and facilitates storing data on magnetic tapes or floppy disks.

FIGURE 3.9

Schematic of pressure transducer and associated equipment.

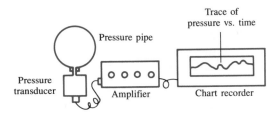

3.4   Hydrostatic Forces on Plane Surfaces

Surfaces that are horizontal or are subjected to gas pressure have essentially constant pressure over their entire surface. Therefore, the total force resulting from the pressure is equal to the product of the pressure and the area of the surface. For this case the resultant force acts at the centroid of the area, and its line of action is normal to the area.

If a plane surface is not horizontal and if it is acted on by a hydrostatic force such as that produced by static liquids, then the pressure is linearly distributed over the surface, and a more general type of analysis must be made to evaluate the magnitude of the resultant force and the location of its line of action. The following derivations assume atmospheric pressure at the liquid surface.

Magnitude of Resultant Hydrostatic Force

Consider the force on the top side of the plane surface AB in Fig. 3.10. Line AB is the edge view of a surface entirely submerged in the liquid. The plane of this surface intersects the horizontal liquid surface at axis 0-0 with an angle α. The distance from the axis 0-0 to the horizontal axis through the centroid of the area is given by $\bar{y}$. The distance from 0-0 to the differential area dA is y. The pressure on the differential area can be computed if the y distance to the point is known; that is, $p = \gamma y \sin \alpha$. Then it follows that the differential force on the differential area is

$$dF = p \, dA \qquad \text{or} \qquad dF = \gamma y \sin \alpha \, dA$$

The total force on the area is obtained by integrating the differential force over the entire area:

$$F = \int_A p \, dA$$

or
$$F = \int_A \gamma y \sin \alpha \, dA \qquad (3.17)$$

In Eq. (3.17), γ and $\sin \alpha$ are constants. Therefore, we obtain

$$F = \gamma \sin \alpha \int_A y \, dA \qquad (3.18)$$

Now, the integral in Eq. (3.18) is the first moment of the area. Consequently, this is replaced by its equivalent, $\bar{y}A$. Therefore, we obtain

$$F = \gamma \bar{y} A \sin \alpha$$

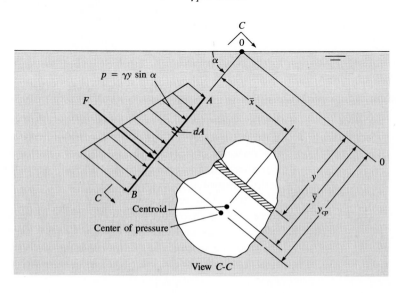

FIGURE 3.10

Distribution of hydrostatic pressure on a plane surface.

which can be rewritten in the following form:

$$F = (\gamma \bar{y} \sin \alpha)A \tag{3.19}$$

Reference to Fig. 3.10 will show that the product of the variables within the parentheses of Eq. (3.19) is the pressure at the centroid of the area. Consequently, we arrive at the conclusion that the magnitude of the resultant hydrostatic force on a plane surface is the product of the pressure at the centroid of the surface and the area of the surface:

$$F = \bar{p}A \tag{3.20}$$

For most hydrostatic problems we are interested only in the forces created in excess of the ambient atmospheric pressures, because atmospheric pressure usually acts on the opposite side of the area in question. Therefore, unless otherwise specified, the pressures used in the following section will be gage pressures.

EXAMPLE 3.10   Assuming that freshly poured concrete exerts a hydrostatic force similar to that exerted by a liquid of equal specific weight, determine the force acting on one side of a concrete form 2.44 m high and 1.22 m wide (8 ft by 4 ft) that is used for pouring a basement wall. *Note:* The specific weight of concrete may be taken as 23.6 kN/m³ (150 lbf/ft³).

Solution

$$F = \bar{p}A$$
$$\bar{p} = 1.22 \text{ m} \times 23.6 \times 10^3 \text{ N/m}^3 = 28.79 \text{ kPa}$$
$$A = 1.22 \times 2.44 = 2.98 \text{ m}^2$$

Then

$$F = 28.79 \times 10^3 \text{ N/m}^2 \times 2.98 \text{ m}^2 = 85.8 \text{ kN} \qquad \blacktriangleleft$$

Vertical Location of Line of Action of Resultant Hydrostatic Force

In general, the location of the line of action of the resultant hydrostatic force lies below the centroid because pressure increases with depth. We can derive an equation for this location by taking moments of the pressure forces about the horizontal axis 0-0. We call the point where the resultant force intersects the surface the *center of pressure* and identify the slant distance from 0-0 to this point by y_{cp} (Fig. 3.10). Then, by definition of the location of a resultant force, the following moment equation can be written:

$$y_{cp}F = \int y\, dF$$

But dF is given by $dF = p\,dA$; therefore,

$$y_{cp}F = \int_A yp\,dA$$

Also,

$$p = \gamma y \sin \alpha$$

so

$$y_{cp}F = \int_A \gamma y^2 \sin \alpha\,dA \tag{3.21}$$

Again, as in Eq. (3.17), γ and $\sin \alpha$ are constants, so we obtain

$$y_{cp}F = \gamma \sin \alpha \int_A y^2\,dA \tag{3.22}$$

The integral on the right-hand side of Eq. (3.22) is the second moment of the area (often called the area moment of inertia). This shall be identified as I_0. However, for engineering applications it is convenient to express the second moment with respect to the horizontal centroidal axis of the area. Hence by the transfer equation we have

$$I_0 = \bar{I} + \bar{y}^2A \tag{3.23}$$

When this is substituted into Eq. (3.22), we obtain

$$y_{cp}F = \gamma \sin \alpha(\bar{I} + \bar{y}^2A)$$

However, from Eq. (3.19), $F = \gamma\bar{y} \sin \alpha A$. Therefore,

$$y_{cp}(\gamma\bar{y} \sin \alpha A) = \gamma \sin \alpha(\bar{I} + \bar{y}^2A)$$

This reduces to

$$y_{cp} = \bar{y} + \frac{\bar{I}}{\bar{y}A} \tag{3.24}$$

or

$$y_{cp} - \bar{y} = \frac{\bar{I}}{\bar{y}A} \tag{3.25}$$

It can be seen from Eq. (3.25) that for a given area the center of pressure comes closer to the centroid as the area is lowered deeper into the liquid. Equation (3.25) is valid only when one liquid is involved. In addition, it is restricted to the case where $p = 0$ gage at the liquid surface. If the pressure is not zero at the surface, then an equivalent problem can be found that satisfies the restriction. That is, y must be measured from an equivalent free surface located above the centroid of the area a distance $\bar{p}/\gamma$.

EXAMPLE 3.11   An elliptical gate covers the end of a pipe 4 m in diameter. If the gate is hinged at the top, what normal force F is required to open the gate when water is 8 m deep above the top of the pipe and the pipe is open to the atmosphere on the other side? Neglect the weight of the gate.

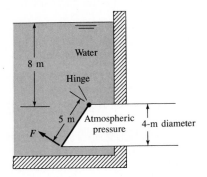

Solution First evaluate the magnitude of the hydrostatic force:

$$F = \bar{p}A$$

The area in question is an ellipse with major and minor axes of 5 m and 4 m. The area is given by the formula $A = \pi ab$ (from Fig. A.1 in the Appendix). Then

$$F = 10 \text{ m} \times 9810 \text{ N/m}^3 \times \pi \times 2 \text{ m} \times 2.5 \text{ m} = 1.541 \text{ MN}$$

Now calculate the slant distance between the centroid of the elliptical area and the center of pressure:

$$y_{cp} - \bar{y} = \frac{\bar{I}}{\bar{y}A} = \frac{\frac{1}{4}\pi a^3 b}{\bar{y}\pi ab} = \frac{\frac{1}{4}a^2}{\bar{y}}$$

Here $\bar{y} = 12.5$ m (slant distance from the water surface to the centroid). Thus

$$y_{cp} - \bar{y} = \frac{1}{4} \times \frac{6.25 \text{ m}^2}{12.5 \text{ m}} = 0.125 \text{ m}$$

Now take moments about the hinge at the top of the gate to obtain F:

$$\sum M_{\text{hinge}} = 0$$

$$1.541 \times 10^6 \text{ N} \times 2.625 \text{ m} - F \times 5 \text{ m} = 0$$

$$F = 809 \text{ kN} \qquad \blacktriangleleft$$

Note: Students are sometimes uncertain which axis to take the moment of inertia about when computing the distance to the center of pressure. A check of the derivation will reveal that the area moment of inertia as used in Eqs. (3.24) and (3.25) is always taken about the *horizontal-centroidal axis*. (Formulas for moments of inertia of selected areas are given in Fig. A.1 in the Appendix.)

Lateral Location of Line of Action of Resultant Hydrostatic Force

The same principles used for the vertical location of the line of action may be used for the lateral location—that is, by taking moments about a line normal to line 0-0 in Fig. 3.10. Areas that are symmetrical about an axis normal to 0-0 always yield a position for the center of pressure that is along the axis of symmetry and below the centroid. However, for asymmetrical areas it is necessary to carry out the analysis to evaluate the location.

EXAMPLE 3.12 Determine the magnitude of the hydrostatic force acting on one side of the submerged vertical plate shown in the figure and determine the location of the center of pressure.

Solution   The centroid of the plate is at a depth of 4 m. Therefore $F = 4$ m $\times$ 9810 N/m^3 $\times \frac{1}{2} \times$ 60 m^2 = 1.177 MN. The vertical location of the center of pressure is obtained from Eq. (3.25):

$$y_{cp} - \bar{y} = \frac{\bar{I}}{\bar{y}A} = \frac{bh^3/36}{\bar{y}\frac{1}{2}bh} = \frac{h^2}{18\bar{y}} = \frac{36}{72}$$

$$y_{cp} = 4 + \frac{1}{2} = 4.50 \text{ m}$$

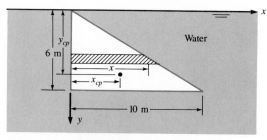

Obtain the lateral location of the center of pressure by summing moments of forces acting on the elemental strips and then dividing by F. Moments are taken about the vertical edge:

$$dM = \tfrac{1}{2}x\,dF = \tfrac{1}{2}x\,\gamma y x\,dy$$

But
$$x = \frac{10}{6}y \qquad \text{so} \qquad M = \frac{50}{36}\gamma \int_0^6 y^3\,dy$$

Then
$$M = \frac{50}{36}(9810 \text{ N/m}^3)\frac{y^4}{4}\Big|_0^6 = 4.414 \text{ MN} \cdot \text{m}$$

But
$$Fx_{cp} = M$$

so
$$x_{cp} = \frac{M}{F} = \frac{4.414 \text{ N} \cdot \text{m}}{1.177 \text{ N}} = 3.75 \text{ m}$$

◀

3.5 Hydrostatic Forces on Curved Surfaces

To determine the magnitude and line of action of the resultant hydrostatic force acting on a curved surface, it is convenient first to determine the horizontal and vertical components of the hydrostatic force and then to add these vectorially to get the total hydrostatic force on the surface.

Consider the two-dimensional curved surface AB in Fig. 3.11a. The arrows normal to the surface represent differential pressure forces acting on differential areas of the surface. If we focus on a representative differential area, Fig. 3.11b, we can derive formulas for the component forces. Assume that the surface in question has a dimension ℓ normal to the plane of the page. Then the differential area shown in Fig. 3.11b has a magnitude $\ell\,ds$, and the differential hydrostatic force is $\bar{p}\,dA$, or

$$dF = \gamma y \ell\, ds$$

Then the horizontal component of this differential force is

$$dF_x = \gamma y \ell\, ds \sin \alpha \qquad (3.26)$$

However, $\ell\,ds \sin \alpha$ is the vertical projection of the differential area dA. Therefore we can write Eq. (3.26) as

$$dF_x = \gamma y\, dA_v \qquad (3.27)$$

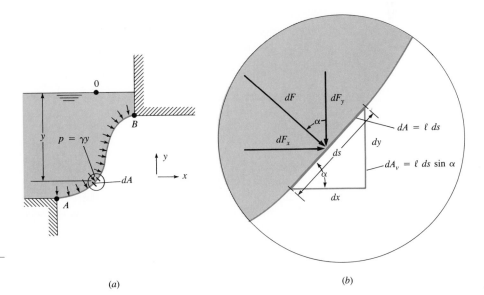

FIGURE 3.11

Hydrostatic force on a curved surface.

(a)

(b)

To obtain the total horizontal component of force, we sum dF_x over the entire surface to get

$$F_x = \int \gamma y \, dA_v \qquad (3.28)$$

Or, because γ is constant, we have

$$F_x = \gamma \int y \, dA_v = \gamma \bar{y}_v A_v$$

Thus $\qquad\qquad\qquad F_x = \bar{p}_v A_v \qquad\qquad\qquad\qquad (3.29)$

Equation (3.29) states that the magnitude of the horizontal component of hydrostatic force on a curved surface is equal to the product of the pressure at the centroid of the vertical projection of the surface and the area of the vertical projection.

To determine the location of the line of action of the horizontal component, we sum moments of the differential horizontal components and equate that sum to the moment of the horizontal component of force. Let us take moments about a horizontal axis that is normal to the hydrostatic forces in question (that passes through point 0 in Fig. 3.11a). Thus we have

$$\int y \, dF_x = y_{cp} F_x$$

$$\int y \, \gamma y \, dA_v = y_{cp} \gamma \bar{y}_v A_v \qquad (3.30)$$

We can divide both sides of Eq. (3.30) by γ. Then the left side of Eq. (3.30) will be the moment of inertia of the vertical projection of the curved surface about the axis passing through 0. Thus we can write

$$y_{cp} \bar{y}_v A_v = I_{0,v} \qquad (3.31)$$

By the transfer formula, $I_{0,v}$ is equal to $\bar{I}_v + \bar{y}_v^2 A_v$. By substituting this into Eq. (3.31) and simplifying and solving for y_{cp}, we obtain

$$y_{cp} = \bar{y}_v + \frac{\bar{I}_v}{\bar{y}_v A_v} \qquad (3.32)$$

Note that Eq. (3.32) is the same as Eq. (3.24) except that in Eq. (3.32) we are working with the vertical projection of the curved surface. Here $\bar{y}_v$ is the depth to the centroid of the vertical projection, $\bar{I}_v$ is the second moment of the vertical projection about its horizontal-centroidal axis, and A_v is the magnitude of the area of the vertical projection.

The vertical component of the differential force in Fig. 3.11 is

$$dF_y = \gamma y \ell \, ds \cos \alpha \qquad (3.33)$$

Note that $\ell\,ds\,\cos\alpha$ is the cross-sectional area of the prism of liquid vertically above the differential area dA. Thus the differential volume of the prism of liquid is $\gamma\ell\,ds\,\cos\alpha$, or

$$dF_y = \gamma\,d\Psi$$

Then
$$F_y = \int_\Psi \gamma\,d\Psi = \gamma\Psi \qquad (3.34)$$

Equation (3.34) states that the vertical component of force acting on the curved surface is equal to the weight of liquid located vertically above the surface. By taking moments, we can easily show that the line of action of the vertical component of force acts through the centroid of the volume of liquid lying above the curved surface.

In summary, the horizontal component of force acting on a curved surface is equal to the force acting on a vertical projection of that surface—which includes both magnitude and line of action. The magnitude of the vertical force acting on a curved surface is equal to the sum of all vertical forces acting on it. The lines of action of these vertical components are used to determine the line of action of the resultant vertical force.

EXAMPLE 3.13　In the figure the surface AB is a circular arc with a radius of 2 m. The distance DB is 4 m. If water is the liquid supported by the surface and if atmospheric pressure prevails on the other side of AB, determine the magnitude and line of action of the resultant hydrostatic force on AB per unit length.

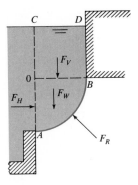

Solution　The vertical component is equal to the weight of water in volume $AOCDB$:

$$W_{OCDB} = \gamma\Psi_{OCDB} = 9810 \times 4 \times 2 \times 1 = 78.5 \text{ kN}$$

$$W_{AOB} = \gamma\Psi_{AOB} = \gamma\tfrac{1}{4}\pi r^2 \times 1 = 30.8 \text{ kN}$$

Therefore, the vertical component is $F_{Ry} = 109.3$ kN. The line of action of the vertical component acts through the centroid of the volume of water considered above, and this is calculated by taking moments about a horizontal axis through D and normal to plane ODB.

$$\bar{x}F_{Ry} = 1 \times 78.5 + r\left(1 - \frac{4}{3\pi}\right)30.8 = 114.0 \text{ kNm}$$

Note: The quantity $4/3\pi$ is the distance to the centroid of the quadrant and is obtained from Fig. A.1 in the Appendix.

$$\bar{x} = \frac{114.0}{109.3} = 1.04 \text{ m}$$

The magnitude of the horizontal component is given by the force on OA:

$$F_H = \bar{p}A = 5 \times 9810 \times 2 \times 1 = 98.1 \text{ kN}$$

The location of the line of action of the horizontal component is given by

$$y_{cp} - \bar{y} = \frac{\bar{I}}{\bar{y}A} = \frac{1 \times 2^3/12}{5 \times 2} = 0.0667 \text{ m}$$

$$y_{cp} = 5.067 \text{ m}$$

The resultant force is as shown in the figure. ◄

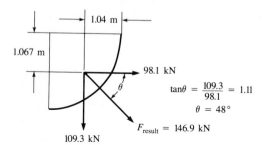

Up to this point we have considered a curved surface for which the external vertical forces act downward on the surface. However, it is easy to visualize the opposite sense for the hydrostatic forces if, for instance, the liquid and atmosphere are reversed from what is shown in Fig. 3.11. Such is the case in Fig. 3.12. If we consider the pressure point by point on the surface, we find that it is exactly the same as when the liquid is to the upper left. However, the pressure forces on the curved surface in the second case act in the opposite sense to the forces in the first. Therefore, the hydrostatic force acting on the curved surface in Fig. 3.12 is reversed from that for Fig. 3.11 but has the same line of action.

FIGURE 3.12

FIGURE 3.12

*Upward hydrostatic force on
a curved surface.*

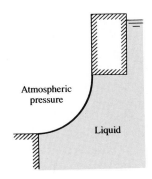

3.6 Buoyancy

Basic Development of the Principles of Buoyancy

The basic principles of buoyancy are readily grasped by referring to the principles of forces on curved surfaces that were introduced in Sec. 3.5. Let us consider a submerged body *ABCDEF* as shown in Fig. 3.13. We first examine the horizontal forces acting on the body in the *y* direction. We may think of these forces as the force that acts on the left end of the body, *AEBFD*, and the force that acts on the right end, *CEBFD*. But to evaluate the force on the left end is to evaluate the force on the curved surface *AEBFD*. This is the force acting on the vertical projection of the surface, the plane surface *EBFD*, and the force will act

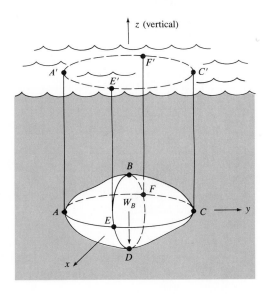

FIGURE 3.13

Analysis of forces on a submerged body.

in the positive y direction. Similarly, the horizontal force acting on the right end will be the force that acts on the same vertical projection *EBFD*. However, the sense of the right-end force will be reversed from that of the left-end force. Therefore, in the y direction we have two forces of equal and opposite magnitude acting on the body; consequently, they cancel one another. The result will be the same if we carry out a similar analysis for horizontal forces acting in the x direction. Thus it is apparent that the net horizontal force acting on a submerged body in a static fluid is zero.

Consider next the vertical forces acting on the submerged body. We must take into account the vertical force acting on the top part of the body and then that acting on the bottom part of the body. On the top the vertical force acts downward. Let us call this force F_{BV}; it equals the weight of the fluid above the top surface of the body. To visualize the force that acts on the bottom, we should think about the equilibrium condition if the body were replaced by an equal volume of fluid having the same properties as the surrounding fluid. It seems obvious that equilibrium would prevail. Therefore, the force acting on the bottom surface is equal to F_{BV} plus the weight of fluid having the same volume as the submerged body. The difference in forces on the bottom and top surfaces of the body, the buoyant force, is thus *equal to the weight of the displaced fluid and acts vertically upward*. The line of action of the buoyant force acts through the centroid of the displaced volume.

EXAMPLE 3.14   A 50-gal oil drum filled with air is to be used to help a diver raise an ancient ship anchor from the bottom of the ocean. The anchor weighs 400 lbf in sea water, and the empty barrel weighs 50 lbf in air. How much weight will the diver be required to lift when the submerged air-filled barrel is attached to the anchor?

Solution
$$\sum F_z = 0$$

$$F_D - W_B + F_B - W_A = 0$$

where F_D is the force applied by diver, W_B is the weight of barrel, or 50 lbf, F_B is the buoyant force of barrel, and W_A is the weight of anchor. Here

$$F_B = \frac{50 \text{ gal}}{7.48 \text{ gal/ft}^3} \times 64 \text{ lbf/ft}^3 = 427.8 \text{ lbf}$$

Therefore

$$F_D = W_B + W_A - F_B = 50 + 400 - 427.8 = 22.2 \text{ lbf} \quad \blacktriangleleft$$

The preceding discussion pertains to totally submerged bodies. However, the buoyant force on a floating body is also equal to the weight of the displaced liquid, and the force also acts vertically upward through the centroid of the volume displaced. This buoyant force just balances the weight of the body itself for equilibrium to prevail, which is why the gross tonnage of a ship is often referred to as the *displacement* of the ship.

Hydrometry

Precise measurement of the specific weight of a liquid is done by utilizing the principle of buoyancy. The device used for this, the *hydrometer,* is a glass bulb that is weighted on one end to make the hydrometer float in a vertical position and has a stem of constant diameter extending from the other end (Fig. 3.14). The hydrometer is so designed that only the stem end extends above the liquid surface. Therefore, appreciable vertical movement of the hydrometer is required to change the buoyant force or displaced volume of the device. Because the buoyant force (equal to the weight of the hydrometer) must be constant, the hydrometer will float deeper or shallower depending on the specific weight of the liquid. Consequently, graduations on the stem corresponding to different depths of submergence of the hydrometer can be made to indicate directly the specific weight or specific gravity of the liquid being measured.

FIGURE 3.14

Hydrometer.

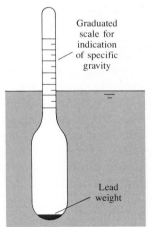

The figure shows a hydrometer with: Graduated scale for indication of specific gravity, and Lead weight.

3.7   Stability of Immersed and Floating Bodies

Immersed Bodies

The stability of an immersed body depends on the relative positions of the *center of gravity* of the body and the centroid of the displaced volume of fluid, which is called the *center of buoyancy.* If the center of buoyancy is above the center of gravity, such as in Fig. 3.15a, any tipping of the body produces a righting couple, and consequently, the body is stable. However, if the center of gravity is above the

FIGURE 3.15

Conditions of stability for immersed bodies. (a) *Stable.* (b) *Neutral.* (c) *Unstable.*

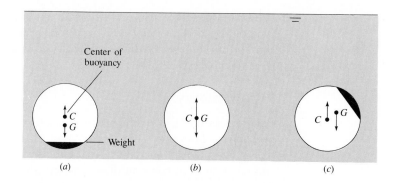

center of buoyancy, any tipping produces an increasing overturning moment, thus causing the body to turn through 180°. This is the condition shown in Fig. 3.15c. Finally, if the center of buoyancy and center of gravity are coincident, the body is neutrally stable—that is, it has the tendency for neither righting nor overturning.

Floating Bodies

The question of stability is more involved for floating bodies than for immersed bodies because the center of buoyancy may take different positions with respect to the center of gravity, depending on the shape of the body and the position in which it is floating. For example, consider the cross section of a ship shown in Fig. 3.16a. Here the center of gravity G is above the center of buoyancy C. Therefore, at first glance it would appear that the ship is unstable and could flip over. However, if we observe the position of C and G after the ship has taken a small angle of heel, as shown in Fig. 3.16b, we see that the center of gravity is in the same position but the center of buoyancy has moved outward of the center of gravity, thus producing a righting moment. A ship having such characteristics is stable.

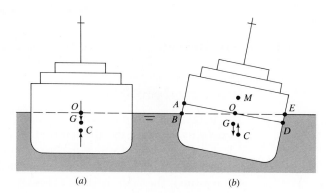

FIGURE 3.16

Ship stability relations.

The reason for the change in the center of buoyancy for the ship is that part of the original buoyant volume, as shown by the wedge shape *AOB*, is transferred to a new buoyant volume *EOD*. Because the buoyant center is at the centroid of the displaced volume, it follows that for this case the buoyant center must move laterally to the right. The point of intersection of the lines of action of the buoyant force before and after heel is called the *metacenter M*, and the distance *GM* is called the *metacentric height*. If *GM* is positive—that is, if *M* is above *G*—the ship is stable; however, if *GM* is negative, the ship is unstable. Quantitative relations involving these basic principles of stability are presented in the next paragraph.

Consider the ship shown in Fig. 3.17, which has taken a small angle of heel α. First we evaluate the lateral displacement of the center of buoyancy, *CC′*; then it will be easy by simple trigonometry to solve for the metacentric height *GM* or to evaluate the righting moment. Recall that the center of buoyancy is at the centroid of the displaced volume. Therefore, we must resort to the basic fundamentals of centroids to evaluate the displacement *CC′*. From the basic definition of the centroid of a volume, we can write the following equation:

$$\bar{x}\,\forall = \sum x_i \,\Delta\forall_i \tag{3.35}$$

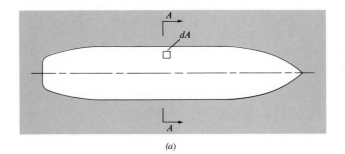

(a)

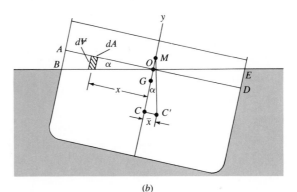

(b)

FIGURE 3.17

(a) *Plan view of ship at waterline.* (b) *Section* A-A *of ship.*

where $\bar{x} = CC'$, which is the distance from the plane about which moments are taken to the centroid of V; V is the total volume displaced; ΔV_i is the volume increment; and x_i is the moment arm of the increment of volume.

Here we take moments about the plane of symmetry of the ship. Recall from mechanics that when we apply this equation, volumes to the left produce negative moments and volumes to the right produce positive moments. For the right side of Eq. (3.35) we write terms for the moment of the submerged volume about the plane of symmetry. A convenient way to do this is to consider the moment of the volume before heel, subtract the moment of the volume represented by the wedge AOB, and then add the moment represented by the wedge EOD. In a general way this is given by the following equation:

$$\bar{x}\,V = \text{moment of } V \text{ before heel} - \text{moment of } V_{AOB} + \text{moment of } V_{EOD}$$

$$(3.36)$$

Because the original buoyant volume is symmetrical with y-y, the moment for the first term on the right is zero. Also, the sign of the moment of V_{AOB} is negative; therefore, when this negative moment is subtracted from the right-hand side of Eq. (3.36), we arrive at the following equation:

$$\bar{x}\,V = \sum x_i \Delta V_{i_{AOB}} + \sum x_i \Delta V_{i_{EOD}} \qquad (3.37)$$

Now, expressing Eq. (3.37) in integral form yields

$$\bar{x}\,V = \int_{AOB} x\,dV + \int_{EOD} x\,dV \qquad (3.38)$$

But it may be seen from Fig. 3.17b that dV can be given as the product of the length of the differential volume, $x \tan \alpha$, and the differential area, dA. Consequently, Eq. (3.38) can be written as

$$\bar{x}\,V = \int_{AOB} x^2 \tan \alpha\,dA + \int_{EOD} x^2 \tan \alpha\,dA$$

Here $\tan \alpha$ is a constant with respect to the integration. Also, since the two terms on the right-hand side are identical except for the area over which integration is to be performed, we combine them as follows:

$$\bar{x}\,V = \tan \alpha \int_{A_{\text{waterline}}} x^2\,dA \qquad (3.39)$$

The second moment or moment of inertia of the area defined by the waterline is given the symbol I_{00}, and the following is obtained:

$$\bar{x}\,V = I_{00} \tan \alpha$$

Next, replacing $\bar{x}$ by CC' and solving for CC', we get

$$CC' = \frac{I_{00} \tan \alpha}{\forall}$$

From Fig. 3.17, $$CC' = CM \tan \alpha$$

Thus eliminating CC' and $\tan \alpha$ yields

$$CM = \frac{I_{00}}{\forall}$$

However, $$GM = CM - CG$$

Therefore, $$GM = \frac{I_{00}}{\forall} - CG \qquad (3.40)$$

Equation (3.40) is used to determine the stability of floating bodies. As already noted, if GM is positive, the body is stable, and if GM is negative, it is unstable.

EXAMPLE 3.15   A block of wood 30 cm square in cross section and 60 cm long weighs 318 N. Will the block float with sides vertical as shown?

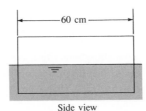

Side view   End view

Solution   First determine the depth of submergence of the block. This is calculated by applying the equation of equilibrium in the vertical direction.

$$\sum F_y = 0$$

$$-\text{weight} + \text{buoyant force} = 0$$

$$-318 \text{ N} + 9810 \text{ N/m}^3 \times 0.30 \text{ m} \times 0.60 \text{ m} \times d = 0$$

$$d = 0.18 \text{ m} = 18 \text{ cm}$$

Determine whether the block is stable about the longitudinal axis:

$$GM = \frac{I_{00}}{\forall} - CG = \frac{\frac{1}{12} \times 60 \times 30^3}{18 \times 60 \times 30} - (15 - 9)$$

$$= 4.167 - 6 = -1.833 \text{ cm}$$

Because the metacentric height is negative, the block is not stable about the longitudinal axis. Thus a slight disturbance will make it tip. Next, check to see if the block is stable about the transverse axis:

$$GM = \frac{\frac{1}{12} \times 30 \times 60^3}{18 \times 30 \times 60} - 6 = 10.67 \text{ cm}$$

The block is stable about the transverse axis and will float with the short sides vertical. ◄

Note that for small angles of heel α the righting moment or overturning moment is given as follows:

$$\text{R.M.} = \gamma \forall GM\alpha \qquad (3.41)$$

However, for large angles of heel, direct methods of calculation based on these same principles would have to be employed to evaluate the righting or overturning moment.

Problems

3.1 The Crosby gage tester shown in the figure is used to calibrate or to test pressure gages. When the weights and the piston together weigh 20 lbf (89.0 N), the gage being tested indicates 26.0 psi (179 kPa). If the piston diameter is 1 in., what percent error exists in the gage?

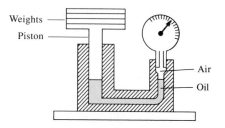

Weights
Piston
Air
Oil

PROBLEM 3.1

3.2 Two hemispheric shells are perfectly sealed together and the internal pressure is reduced to 30 kPa. The inner radius is 15 cm, and the outer radius is 15.5 cm. If the atmospheric pressure is 100 kPa, what force is required to pull the shells apart?

3.3 If the cabin of a jet liner is pressurized to 102 kPa absolute and the outside pressure is 20 kPa absolute when the plane is cruising at 12 km, what net force is exerted by the air on the cabin door, which is 2.2 m × 1.0 m in size?

3.4   If exactly 20 bolts of 2-cm diameter are needed to hold the air chamber together at *A-A* as a result of the high pressure within, how many bolts will be needed at *B-B*? Here $D = 50$ cm and $d = 25$ cm.

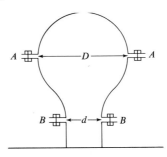

PROBLEM 3.4

3.5   The reservoir shown in the figure contains two immiscible liquids of specific weights γ_A and γ_B, respectively, one above the other. $\gamma_A > \gamma_B$. Which graph depicts the correct distribution of gage pressure along a vertical line through the liquids?

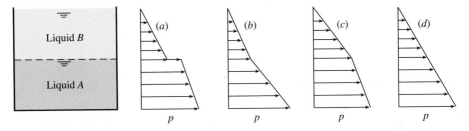

PROBLEM 3.5

3.6   This manometer contains water at room temperature. The glass tube on the left has an inside diameter of 1 mm ($d = 1.0$ mm). The glass tube on the right is three times as large. For these conditions, the water surface level in the left tube will be a) higher than the water surface level in the right tube, b) equal to the water surface level in the right tube, c) less than the water surface level in the right tube. State your main reason or assumption for making your choice.

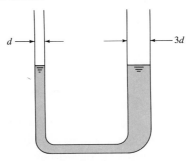

PROBLEM 3.6

3.7 If a 100-N force F_1 is applied to the piston with the 5-cm diameter, what is the magnitude of the force F_2 that can be resisted by the piston with the 10-cm diameter? Neglect the weights of the pistons.

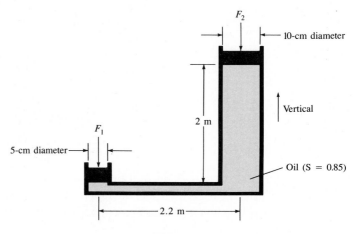

PROBLEM 3.7

3.8 Some skin divers go as deep as 40 m. What is the gage pressure at this depth in fresh water, and what is the ratio of the absolute pressure at this depth to normal atmospheric pressure? Assume $T = 20°C$.

3.9 Given: absolute atmospheric pressure of 98 kPa at a lake's surface. At a depth of 10 m the absolute pressure will be approximately a) 2 b) 3 c) 4 times the atmospheric pressure.

3.10 Water occupies the bottom 90 cm of a cylindrical tank. On top of the water is 1.0 m of kerosene, which is open to the atmosphere. If the temperature is 20°C, what is the gage pressure at the bottom of the tank?

3.11 What is the pressure at a depth of 10 m in an open tank of crude oil?

3.12 The density of air at sea level is 0.00247 slugs/ft³. If we assume air to be incompressible (a hypothetical condition), how thick an atmospheric air layer is necessary to give a sea level pressure of 14.7 psia?

3.13 The gage pressure at a depth of 5 m in an open tank of liquid is 73.6 kPa. What are the specific weight and the specific gravity of the liquid?

3.14 A liquid has the peculiar property that its mass density increases linearly with depth according to the expression $\rho = \rho_{water}(1 + 0.01d)$, where d is the depth below the liquid surface in meters. At a depth of 10 m, what is the gage pressure?

3.15 Consider the conditions given for Problem 3.14. The gage pressure at a particular point in the liquid is found by measurement to be 50 kPa. At what depth is that point?

3.16 What is the maximum gage pressure in the odd tank shown in the figure? Where will the maximum pressure occur? What is the hydrostatic force acting on the top (CD) of the last chamber on the right-hand side of the tank? Assume $T = 10°C$.

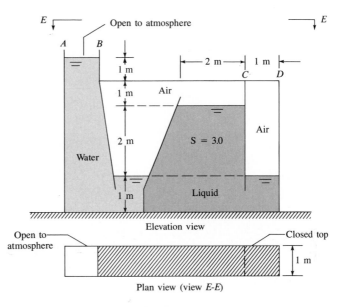

Elevation view

PROBLEM 3.16

3.17 Usually water is assumed to be incompressible for hydrostatic computations; however, extreme pressures may cause significant changes in density. Estimate the percentage difference in density of sea water between the surface and a point 6 km deep. Assume a constant temperature of 10°C and a bulk modulus of elasticity of 2.2 GPa.

3.18 The piston shown weighs 10 lbf. In its initial position, the piston is restrained from moving to the bottom of the cylinder by means of the metal stop. Assuming there is neither friction nor leakage between piston and cylinder, what volume of oil (S = 0.85) would have to be added to the 1-in. tube to cause the piston to rise 1 in. from its initial position?

3.19 A 1-cm layer of oil with a specific gravity of 0.80 lies on the surface of the water (T = 10°C) in a tank. Find the gage pressure (in Pa) at a point 10 cm below the oil surface.

3.20 Consider an air bubble rising from the bottom of a lake. Neglecting surface tension, determine approximately what will be the ratio of the density of the air in the bubble at a depth of 20 ft to its density at a depth of 10 ft.

3.21 What is the reading on a mercury barometer at a place where the atmospheric pressure is 98.0 kPa?

3.22 The figure shows a water barometer. If the temperature of the water is 40°C, what is the true atmospheric pressure for h = 10.00 m?

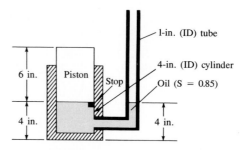

PROBLEM 3.18

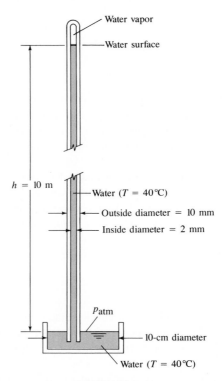

PROBLEM 3.22

3.23 The gage pressure at the center of the pipe has a value that is a) negative, b) zero, c) positive. (Neglect surface tension effects.)

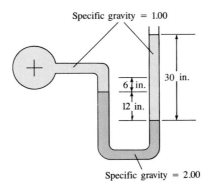

Specific gravity = 1.00

30 in.

6 in.

12 in.

Specific gravity = 2.00

PROBLEM 3.23

3.24 Determine the gage pressure at the center of pipe A in pounds per square inch when the temperature is 70°F.

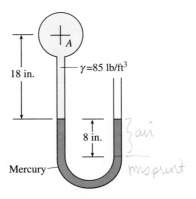

18 in.

$\gamma=85$ lb/ft^3

8 in.

Mercury

PROBLEM 3.24

3.25 Determine the gage pressure in pipe A.

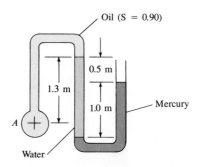

Oil (S = 0.90)

0.5 m

1.3 m

1.0 m

Mercury

A

Water

PROBLEM 3.25

3.26 Considering the effects of surface tension, estimate the gage pressure at the center of pipe A.

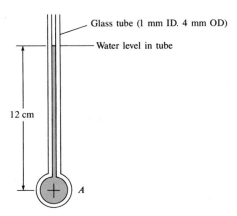

Glass tube (1 mm ID. 4 mm OD)

Water level in tube

12 cm

A

PROBLEM 3.26

3.27 What is the pressure at the center of pipe B?

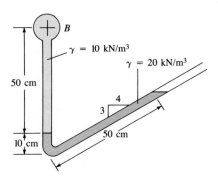

B

$\gamma = 10 \text{ kN/m}^3$

$\gamma = 20 \text{ kN/m}^3$

50 cm

4

3

50 cm

10 cm

PROBLEM 3.27

3.28 The ratio of cistern diameter to tube diameter is 8. When air in the cistern is at atmospheric pressure, the free surface in the tube is at position 1. When the cistern is pressurized, the liquid in the tube moves 50 cm up the tube from position 1 to position 2. What is the cistern pressure that causes this deflection? The liquid density is 800 kg/m^3.

3.29 The ratio of cistern diameter to tube diameter is 10. When air in the cistern is at atmospheric pressure, the free surface in the tube is at position 1. When the cistern is pressurized, the liquid in the tube moves 3 ft up the tube from position 1 to position 2. What is the cistern pressure that causes this deflection? The specific weight of the liquid is 50 lbf/ft^3.

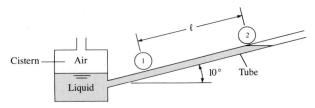

PROBLEMS 3.28, 3.29

3.30 The inclined manometer is filled with oil that has a specific gravity of 0.85. What angle α will yield a deflection of 20 cm in the inclined tube when the air pressure in the cistern is increased by 500 Pa?

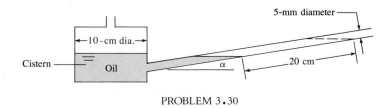

PROBLEM 3.30

3.31 Determine the gage pressure at the center of pipe A in pounds per square inch and in kilopascals.

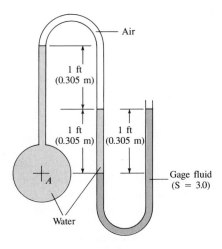

PROBLEM 3.31

3.32 Determine the gage pressure at the center of pipe A in pounds per square inch and in kilopascals.

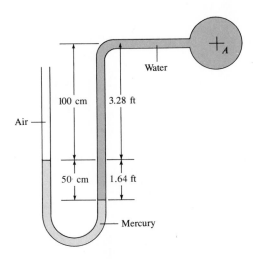

PROBLEM 3.32

3.33 A device for measuring the specific weight of a liquid consists of a U-tube manometer as shown. The manometer tube has an internal diameter of 0.5 cm and originally has water in it. Exactly 2 cm^3 of unknown liquid is then poured into one leg of the manometer, and a displacement of 5 cm is measured between the surfaces as shown. What is the specific weight of the unknown liquid?

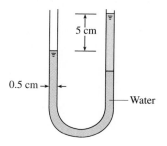

PROBLEM 3.33

3.34 Mercury is poured into the tube in the figure until the mercury occupies 1 ft of the tube's length. An equal volume of water is then poured into the left leg. Locate the water and mercury surfaces. Also determine the maximum pressure in the tube.

3.35 What is the specific gravity of the liquid in the left leg of the manometer tube?

3.36 A U-tube manometer is needed that will measure the difference in pressure between two points 100 m apart in a horizontal 6-cm pipe. The pipe carries water, and the maximum pressure difference is expected to be 60 kPa. Design the manometer and predict the probable degree of accuracy of measurement of Δp for your design.

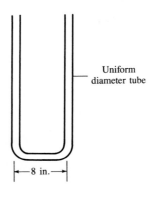

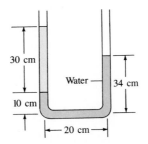

PROBLEM 3.35

3.37 Find $p_A - p_B$ if $x = 4.0$ ft, $y = 3.0$ ft, $z = 2.0$ ft, and fluids 1, 2, and 3 are kerosene, water, and kerosene, respectively. Assume S = 0.8 for kerosene and $T = 60°F$.

3.38 Find $p_A - p_B$ if $x = 3.0$ m, $y = 1.0$ m, $z = 2.0$ m, and the fluids 1, 2, and 3 are kerosene, water, and kerosene, respectively. Assume S = 0.8 for kerosene and $T = 20°C$.

3.39 Find z if $p_B - p_A = 2.0$ psi, fluid 1 is kerosene, $\gamma = 180$ lbf/ft^3 for fluid 2, and fluid 3 is water; $x = 1$ ft and $y = 3$ ft. Assume $T = 60°F$.

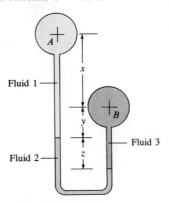

PROBLEMS 3.37, 3.38, 3.39, 3.40

3.40   Find z if $p_B - p_A = 3.0$ psi, fluid 1 is kerosene, fluid 2 is mercury, and fluid 3 is water; $x = 0$, $y = 3$ ft. Assume $T = 60°F$.

3.41   Find the pressure at the center of pipe A. Assume $T = 20°C$.

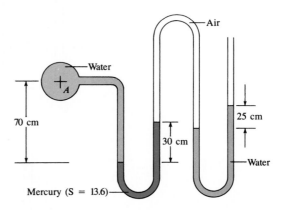

PROBLEM 3.41

3.42   Find the pressure at the center of pipe A. Assume $T = 10°C$.

3.43   Find the pressure at the center of pipe A if all dimensions given are in inches instead of centimeters.

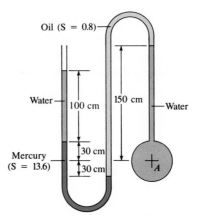

PROBLEM 3.42, 3.43

3.44   Determine (a) the difference in pressure and (b) the difference in piezometric head between points A and B. The elevations z_A and z_B are 10 m and 11 m, respectively, $\ell_1 = 1$ m, and the manometer deflection ℓ_2 is 50 cm.

3.45   The deflection on the mercury manometer is h meters when the pressure in the tank is 150 kPa absolute. If the absolute pressure in the tank is increased 50%, what will the deflection on the manometer be?

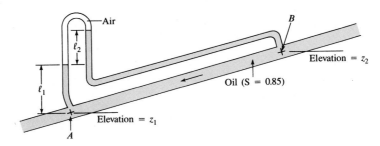

PROBLEM 3.44

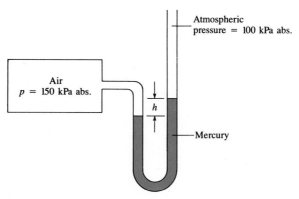

PROBLEM 3.45

3.46 A vertical conduit is carrying oil (S = 0.85). A differential mercury manometer is tapped into the conduit at points A and B. Determine the difference in pressure between A and B when $h = 3$ in. What is the difference in piezometric head between A and B?

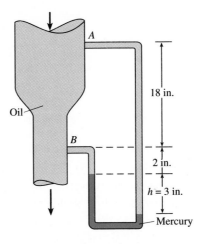

PROBLEM 3.46

3.47 For the closed tank with Bourdon-tube gages tapped into it, what are the specific gravity of the oil and the pressure reading on gage C?

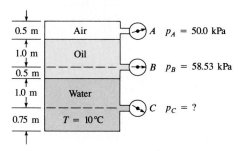

PROBLEM 3.47

3.48 One means of determining the surface level of liquid in a tank is by discharging a small amount of air through a small tube, the end of which is submerged in the tank, and reading the pressure on the gage that is tapped into the tube. Then the level of the liquid surface in the tank can be calculated. If the pressure on the gage is 30 kPa, what is the depth d of liquid in the tank?

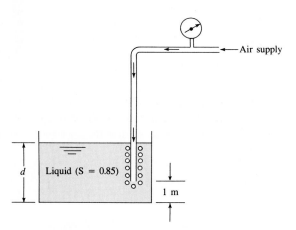

PROBLEM 3.48

3.49 Consider the ratio $(dp/dz)_0/(dp/dz)_{2000}$, where dp/dz is the pressure gradient in the atmosphere (z is positive up) and the subscripts 0 and 2000 refer to elevation in feet above sea level. Choose the correct statement: a) The ratio has a value less than 1. b) The ratio has a value equal to 1. c) The ratio has a value greater than 1.

3.50 Consider the two rectangular gates shown in the figure. They are both the same size, but one (Gate A) is held in place by a horizontal shaft through its midpoint and the other (Gate B) is cantilevered to a shaft at its top. Now consider the torque T required to hold the gates in place as H is increased. Choose the valid statement(s): a) T_A increases with H. b) T_B increases with H. c) T_A does not change with H. d) T_B does not change with H.

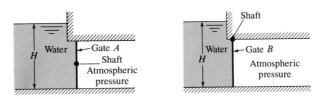

PROBLEM 3.50

3.51 For gate A (above), choose the statements that are valid: a) The hydrostatic force acting on the gate increases as H increases. b) The distance between the center of pressure on the gate and the centroid of the gate decreases as H increases. c) The distance between the center of pressure on the gate and the centroid of the gate remains constant as H increases. d) The torque applied to the shaft to prevent the gate from turning must be increased as H increases. e) The torque applied to the shaft to prevent the gate from turning remains constant as H increases.

3.52 The boiling point of water decreases with elevation because of the pressure change. What is the boiling point of water at an elevation of 1500 m and at an elevation of 3000 m for standard atmospheric conditions?

3.53 What is the atmospheric pressure at an elevation of 6 km if the pressure and temperature at sea level are 101 kPa absolute and 25°C? Assume that the standard lapse rate prevails.

3.54 From a depth of 10 m in a lake to an elevation of 4000 m in the atmosphere, plot the variation of absolute pressure. Assume that the lake water surface elevation is at mean sea level and assume standard atmospheric conditions.

3.55 What is the atmospheric pressure at an elevation of 20,000 ft (6096 m) if the pressure and temperature at sea level are 14.7 psia (101 kPa) and 60°F (15°C)? Assume standard atmospheric conditions.

3.56 Assume that a man must breathe a constant mass rate of air to maintain his metabolic processes. If he inhales and exhales 16 times per minute at sea level where the temperature is 59°F (15°C) and the pressure is 14.7 psia (101 kPa), what would you expect his rate of breathing at 18,000 ft (5486 m) to be? Use standard atmospheric conditions.

3.57 A pressure gage in an airplane indicates a pressure of 95 kPa at takeoff where the airport elevation is 1 km and the temperature is 10°C. If the standard lapse rate of 6.5°C/km is assumed, at what elevation is the plane when a pressure of 75 kPa is read? What is the temperature for that condition?

3.58 A pressure gage in an airplane indicates a pressure of 13.6 psia at takeoff where the airport elevation is 2000 ft and the temperature is 70°F. If the standard lapse rate of 0.003566°F/ft is assumed, at what elevation is the plane when a pressure of 10 psia is read?

3.59 Denver, Colorado, is called the "mile-high" city. What are the pressure, temperature, and density of the air when standard atmospheric conditions prevail? Give your answer in traditional and SI units.

3.60   A submerged square gate (pivoted about its vertical centroidal axis) is set between two reservoirs of equal depth as shown. What is the net hydrostatic force on the gate? What moment about the pivot axis is required to keep the gate closed?

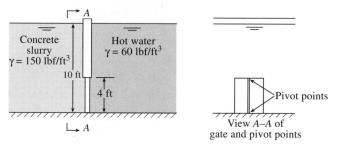

PROBLEM 3.60

3.61   A tank of liquid has two gates on the side walls pivoted on their horizontal centroidal axes. One gate is a circle, and the other is an ellipse. The circle and the ellipse have the same area, and the pivot lines (centroid axes) are at the same depth below the liquid surface. The major axis of the ellipse is in the vertical direction. The two gates are connected by a mechanism as shown, and a force on a handle is required to keep them closed. If the handle is released, in which direction will it move (to the right or the left)? Explain the reasons for your choice.

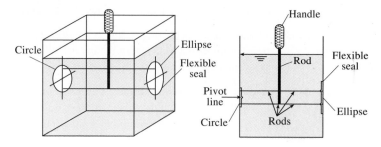

PROBLEM 3.61

3.62   Find the force of the gate on the block.

3.63   Assuming that concrete behaves as a liquid ($\gamma = 150$ lbf/ft^3) just after it is poured, determine the force per foot of length exerted on a form by the concrete if it is poured into forms for a wall that is to be 10 ft high. If the forms are held in place as shown, with ties between vertical braces spaced every 2 ft, what force is exerted on the bottom tie?

3.64   Neglecting the weight of the gate, determine the force acting on the hinge of the gate.

3.65   The gate shown is rectangular and has dimensions 6 m × 5 m. What is the reaction at point A? Neglect the weight of the gate.

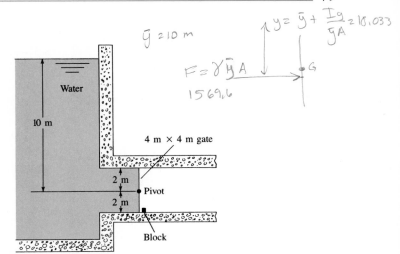

$\bar{y} = 10\,m$

$y = \bar{y} + \dfrac{I_g}{\bar{y}A} = 10.033$

$F = \gamma \bar{y} A$
156916

PROBLEM 3.62

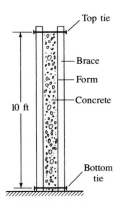

PROBLEM 3.63

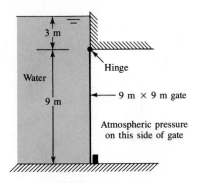

PROBLEM 3.64

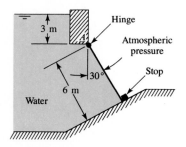

PROBLEM 3.65

3.66 The rectangular gate measures 10 ft by 6 ft ($\ell = 10$ ft) and is pin-connected at point B. If the surface on which the gate rests at A is frictionless, and if the water surface is 7 ft above point B, what is the reaction at A? Neglect the weight of the gate.

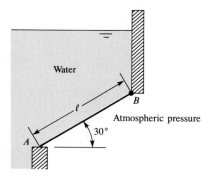

PROBLEM 3.66

3.67 In the figure, the gate holding back the oil is 80 cm high and 120 cm long. If it is held in place only along the bottom edge, what is the necessary resisting moment at that edge?

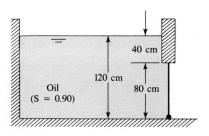

PROBLEM 3.67

3.68 If the rectangular gate shown is attached to a horizontal shaft at its midpoint, what torque would have to be applied to the shaft to open the gate? The rectangular conduit and gate are both 4 m wide, and $\ell = 5$ m.

3.69 If the rectangular gate shown is attached to a horizontal shaft at its midpoint, what torque would have to be applied to the shaft to open the gate? Here $\ell = 12$ ft and the rectangular conduit and gate are both 5 ft wide.

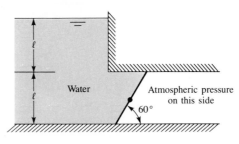

PROBLEMS 3.68, 3.69

3.70 If gate AB is rectangular and is 2 m wide (normal to the page), what force F is needed to open the gate if region C contains air at atmospheric pressure? The water surface lies 2 m above the hinge at B, and $\ell = 5$ m. Neglect the weight of the gate.

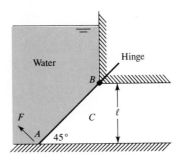

PROBLEM 3.70

3.71 Design a gate for the spillway of a dam that is to hold water at a normal high-water reservoir level of 2110.0 ft. The spillway crest elevation is 2090.0 ft, as shown. The gate is to be positioned between piers with a clear span of 30.0 ft. In case of flood, operators must be able to open the gate so that water may flow freely over the spillway. Obtain guidelines and the desired degree of design sophistication from your instructor.

3.72 The square gate shown is eccentrically pivoted so that it automatically opens at a certain value of h. What is that value in terms of ℓ?

3.73 The gate ABC is hinged at A. If the gate is 8 ft long (normal to the page), what will be the reaction at C, where the gate bears on the smooth (frictionless) surface?

3.74 This 10-ft-diameter butterfly valve is used to control the flow in a 10-ft-diameter outlet pipe in a dam. In the position shown, it is closed. The valve is supported by a horizontal shaft through its center. What torque would have to be applied to the shaft to hold the valve in the position shown?

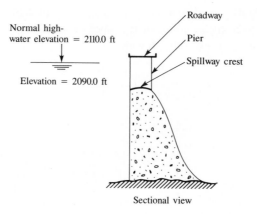

Normal high-
water elevation = 2110.0 ft

Elevation = 2090.0 ft

Roadway

Pier

Spillway crest

Sectional view

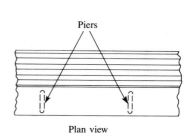

Piers

Plan view

PROBLEM 3.71

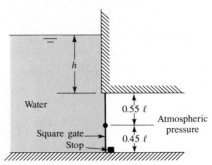

h

Water

0.55 ℓ

Atmospheric
pressure

Square gate

0.45 ℓ

Stop

PROBLEM 3.72

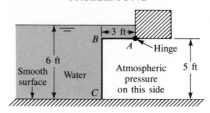

3 ft

B   A   Hinge

6 ft

Smooth
surface

Water

Atmospheric
pressure
on this side

5 ft

C

PROBLEM 3.73

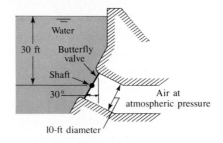

Water

30 ft

Butterfly
valve

Shaft

30°

Air at
atmospheric pressure

10-ft diameter

PROBLEM 3.74

3.75 For this gate, $\alpha = 45°$, $y_1 = 1$ m, and $y_2 = 3$ m. Will the gate fall or stay in position under the action of the hydrostatic and gravity forces if the gate itself weighs 90 kN and is 1.0 m wide? Assume $T = 10°C$.

3.76 For this gate, $\alpha = 45°$, $y_1 = 4$ ft, and $y_2 = 7.07$ ft. Will the gate fall or stay in position under the action of the hydrostatic and gravity forces if the gate itself weighs 18,000 lb and is 3 ft wide? Assume $T = 50°F$.

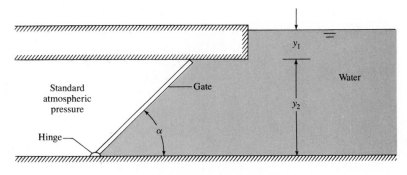

PROBLEMS 3.75, 3.76

3.77 Determine the hydrostatic force F on the triangular gate, which is hinged at the bottom edge and held by the reaction R_T at the upper corner. Express F in terms of γ, h, and W. Also determine the ratio R_T/F. Neglect the weight of the gate.

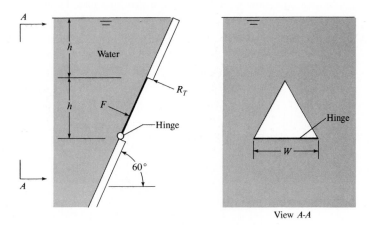

View A-A

PROBLEM 3.77

3.78 The triangular gate ABC is pivoted at the bottom edge AC and closes a triangular opening ABC in the wall of the tank. The opening is 4 m wide ($W = 4$ m) and 9 m high ($H = 9$ m). The depth d of water in the tank is 10 m. Determine the hydrostatic force on the gate and the horizontal force P required at B to hold the gate closed.

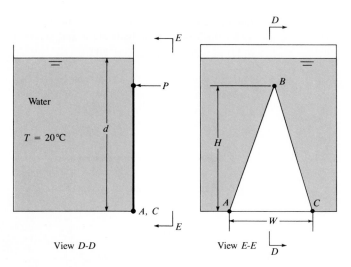

View D-D   View E-E

PROBLEM 3.78

3.79   A plate that has dimensions W (width), H (height), and t (thickness) is completely submerged with vertical orientation in a liquid. The liquid has a variable specific weight
given by $\gamma = \gamma_0(1 + kd/d_0)$, where k is a positive constant, d is depth below the liquid
surface, and γ_0 is the specific weight at reference depth d_0. Derive a formula for the magnitude of hydrostatic force on one side of the plate. Will the location of the center of pressure be below or above that for a plate located similarly in a liquid of constant density?

3.80   For the plane rectangular gate ($\ell \times w$ in size), Figure (a), what is the magnitude of the
reaction at A in terms of γ_w and the dimensions ℓ and w? For the cylindrical gate, Figure (b), will the magnitude of the reaction of A be greater than, less than, or the same as
that for the plane gate? Neglect the weight of the gates.

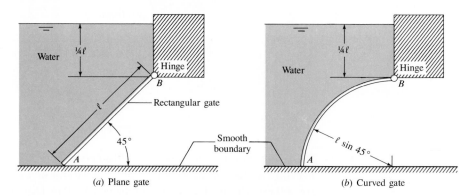

(a) Plane gate   (b) Curved gate

PROBLEM 3.80

3.81 The air above the liquid is under a pressure of 3 psig, and the specific gravity of the liquid in the tank is 0.80. If the rectangular gate is 5 ft wide and if $y_1 = 2$ ft and $y_2 = 10$ ft, what force P is required to hold the gate in place?

3.82 The air above the liquid is under a pressure of 40 kPa gage, and the specific gravity of the liquid in the tank is 0.80. If the rectangular gate is 1.0 m wide and if $y_1 = 1.0$ m and $y_2 = 3$ m, what force P is required to hold the gate in place?

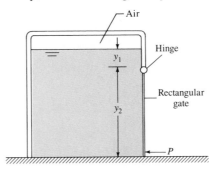

PROBLEMS 3.81, 3.82

3.83 If KH is large enough, the gate will be on the verge of opening when the water level is even with the hinge. What is K for this condition? Neglect the weight of the gate.

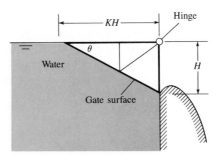

PROBLEM 3.83

3.84 In constructing dams, the concrete is poured in lifts of approximately 1.5 m ($y_1 = 1.5$ m). The forms for the face of the dam are reused from one lift to the next. The figure shows one such form, which is bolted to the already cured concrete. For the new pour, what moment will occur at the base of the form per meter of length (normal to the page)? Assume that concrete acts as a liquid when it is first poured and has a specific weight of 24 kN/m^3.

3.85 At what depth d will the rectangular gate automatically open if $W = 60$ kN? The gate is 4 m high and 2 m wide. Neglect the weight of the gate.

3.86 What weight W is needed for the gate to be on the verge of opening (no force from the gate on the stop) if the water depth d is 5 m (the water surface is above the top of the gate)? The gate is 4 m high and 2 m wide. Neglect the weight of the gate.

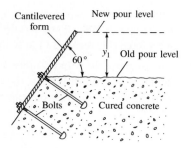

PROBLEM 3.84

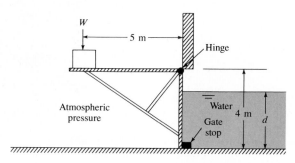

PROBLEMS 3.85, 3.86

3.87 The plane rectangular gate can pivot about the support at B. For the conditions given, is it stable or unstable? Neglect the weight of the gate.

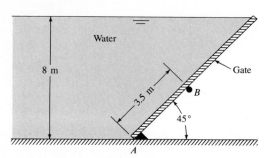

PROBLEM 3.87

3.88 The minimum water level on gate surface AB is 1 ft. At what depth h will the gate automatically open? Neglect the weight of the gate. *Note.* Both surfaces AB and BC pivot as a unit about the hinge at B.

3.89 Determine the minimum volume of concrete ($\gamma = 23.6 \text{ kN/m}^3$) needed to keep the gate (1 m wide) in a closed position; $\ell = 2$ m. Note the hinge at the bottom of the gate.

3.90 Determine the minimum volume of concrete ($\gamma = 150 \text{ lbf/ft}^3$) needed to keep the gate (2 ft wide) in a closed position; $\ell = 5$ ft.

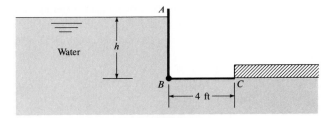

PROBLEM 3.88

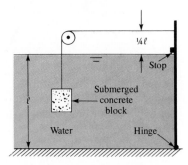

PROBLEMS 3.89, 3.90

3.91 The gate and mechanism shown are designed so that the gate will open when the water level h reaches a certain value. Neglecting the weights of the gate, chain, and cylindrical tank, determine at what value of h the gate will be on the verge of opening. *Note:* The beam is cantilevered to the gate.

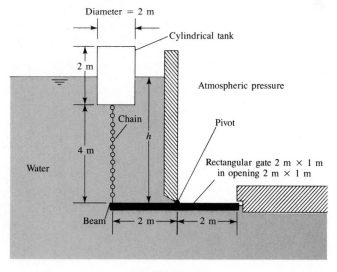

PROBLEM 3.91

3.92   The three walls are holding back water as shown. Per unit of width (normal to the page), which wall requires the greatest resisting moment (at A', B', or C') to resist the hydrostatic force, or are the moments equal? Explain. Neglect the weights of the walls.

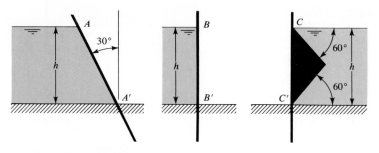

PROBLEM 3.92

3.93   A block of material of unknown volume is submerged in water and found to weigh 500 N (in water). The same block weighs 650 N in air. Determine the specific weight and volume of the material.

3.94   An object weighs 55 N when submerged in oil (S = 0.80), and a force of 45 N is required to hold it submerged in mercury. Determine its weight, volume, specific weight, and specific gravity.

3.95   A weather balloon is constructed of a flexible material such that the internal pressure of the balloon is always 10 kPa higher than the local atmospheric pressure. At sea level the diameter of the balloon is 1 m, and it is filled with helium. The balloon material, structure, and instruments have a mass of 100 g. This does not include the mass of the helium.

As the balloon rises, it will expand. The temperature of the helium is always equal to the local atmospheric temperature, so it decreases as the balloon gains altitude. Calculate the maximum altitude of the balloon in a standard atmosphere.

3.96   A rock weighs 912 N in air and 609 N in water. Find its volume.

3.97   This uniform-diameter rod is weighted at one end and is floating in the liquid as shown. The liquid a) is lighter than water, b) must be water, c) is heavier than water.

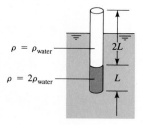

PROBLEM 3.97

3.98   A person is floating in a boat with an aluminum anchor. The anchor has a specific gravity of 2.2 and a volume of 0.5 ft³. The surface area of the water in the pond is 500 ft². The person throws the anchor out of the boat into the water. By how much (if any) will the water level in the pond change?

3.99   An inverted cone contains water as shown. The volume of the water in the cone is given by $\Psi = (\pi/3)h^3$. The original depth of the water is 10 cm. A block with a volume of 200 cm³ and a specific gravity of 0.5 is floated in the water. What will be the change (in cm) in water surface height in the cone?

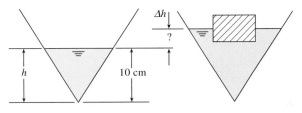

PROBLEM 3.99

3.100 Large concrete cylindrical shells sealed at each end are to be used in the construction of an offshore oil drilling rig. These cylinders are floated out to the site and then uprighted to form part of the structure. The cylinders are 20 m in diameter and, when floating in the horizontal position (axis horizontal), sink to a depth of 15 m. What will be their height above the water when erected?

3.101 A 1-ft-diameter cylindrical tank is filled with water to a depth of 2 ft. A cylinder of wood 6 in. in diameter and 3 in. long is set afloat on the water. The weight of the wood cylinder is 2 lbf. Determine the change (if any) in the depth of the water in the tank.

3.102 The floating platform shown is supported at each corner by a hollow sealed cylinder 1 m in diameter. The platform itself weighs 30 kN in air, and each cylinder weighs 1.0 kN per meter of length. What total cylinder length L is required for the platform to float 1 m above the water surface? Assume that the specific weight of the water (brackish) is 10,000 N/m³. The platform is square in plan view.

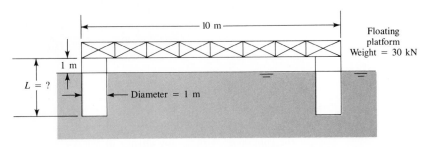

PROBLEM 3.102

3.103 A cylinder that is made of two materials having different densities is dropped into the liquid of Prob. 3.14. How deep will this cylinder float in the liquid?

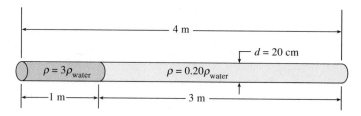

PROBLEM 3.103

3.104 Consider the cylinder of Prob. 3.103. How deep will that cylinder *float* in this water/oil bath?

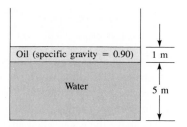

PROBLEM 3.104

3.105 To what depth d will this square block (it has a density 0.8 times that of water) float in the two-liquid reservoir?

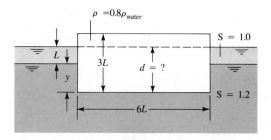

PROBLEM 3.105

3.106 A cylindrical container 4 ft high and 2 ft in diameter holds water to a depth of 2 ft. How much does the level of the water in the tank change when a 5-lb block of ice is placed in the container? Is there any change in the water level in the tank when the block of ice melts? Does it depend on the specific gravity of the ice? Explain all the processes.

3.107 The partially submerged wood pole is attached to the wall by a hinge as shown. The pole is in equilibrium under the action of the weight and buoyant forces. Determine the density of the wood.

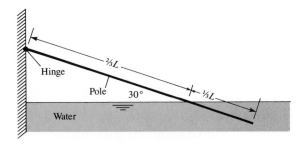

PROBLEM 3.107

3.108 This pole is pinned (hinged) at A and B, and the pole itself has a specific weight of 160 lb/ft^3. The liquid has a specific weight of 200 lb/ft^3. Upon release of the pin at B, will the end B rise, fall, or remain in its initial position?

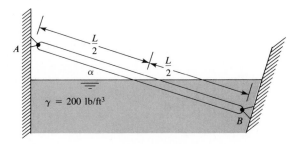

PROBLEM 3.108

3.109 An 800-ft ship has a displacement of 40,000 tons, and the area defined by the waterline is 40,000 ft^2. Will the ship take more or less draft when steaming from salt water to fresh water? How much will it settle or rise?

3.110 A submerged spherical steel buoy that is 1.2 m in diameter and weighs 1600 N is to be anchored in salt water 20 m below the surface. Find the weight of scrap iron that should be sealed inside the buoy in order that the force on its anchor chain will not exceed 4.5 kN.

3.111 A balloon is to be used to carry meteorological instruments to an elevation of 15,000 ft where the air pressure is 8.3 psia. The balloon is to be filled with helium, and the material from which it is to be fabricated weighs 0.01 lbf/ft^2. If the instruments weigh 10 lbf, what diameter should the spherical balloon have?

3.112 Water is held back by this radial gate. Does the resultant of the pressure forces acting on the gate pass above the pin, through the pin, or below the pin?

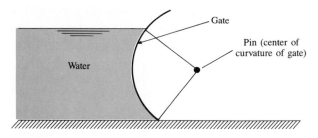

PROBLEM 3.112

3.113 For the curved surface AB:

    a. Determine the magnitude, direction, and line of action of the vertical component of hydrostatic force acting on the surface. Here $\ell = 1$ m.

    b. Determine the magnitude, direction, and line of action of the horizontal component of hydrostatic force acting on the surface.

    c. Determine the resultant hydrostatic force acting on the surface.

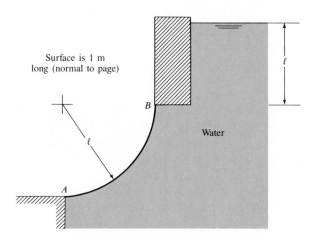

PROBLEM 3.113

3.114 Determine the hydrostatic force acting on the radial gate if the gate is 40 ft long (normal to the page). Show the line of action of the hydrostatic force acting on the gate.

3.115 A curved gate is to be constructed to conform to the curve $y = kx^3$ between $0 < x < a$ and $0 < y < b$. The liquid surface is along the line $y = b$. Find the total hydrostatic force on the gate in terms of γ on an element of gate 1 m wide ($z = 1$ m). Also find the lines of action of the vertical and horizontal components.

3.116 Determine the magnitude and direction of the horizontal and vertical components of the hydrostatic force acting on the two-dimensional curved metal surface per foot of width. Determine the location of the horizontal component of hydrostatic force.

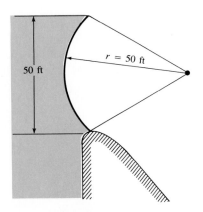

PROBLEM 3.114

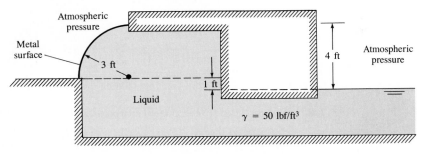

PROBLEM 3.116

3.117 The steel pipe and steel chamber together weigh 600 lbf. What force will have to be exerted on the chamber by all the bolts to hold it in place? The dimension ℓ is equal to 2 ft. *Note:* There is no bottom on the chamber—only a flange bolted to the floor.

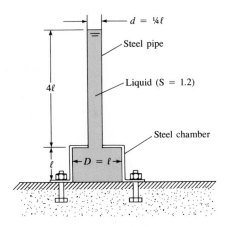

PROBLEM 3.117

3.118 What force must be exerted through the bolts to hold the dome in place? The metal dome and pipe weigh 1300 lbf. The dome has no bottom. Here $\ell = 2.0$ ft.

3.119 What force must be exerted through the bolts to hold the dome in place? The metal dome and pipe weigh 6 kN. The dome has no bottom. Here $\ell = 80$ cm.

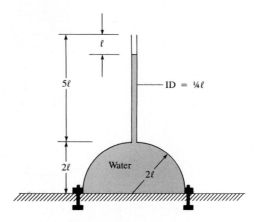

PROBLEMS 3.118, 3.119

3.120 Find the vertical component of force in the metal at the base of the spherical dome shown when gage A reads 10 psig. Indicate whether the metal is in compression or tension. The specific gravity of the enclosed fluid is 1.5. The dimension L is 3 ft. Assume the dome weighs 1000 lbf.

3.121 Find the vertical component of force in the metal at the base of the spherical dome shown when gage A reads 70 kPa. Indicate whether the metal is in compression or tension. The specific gravity of the enclosed fluid is 1.5. The dimension L is equal to 1 m. Assume the dome weighs 5000 N.

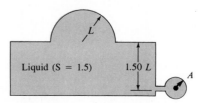

PROBLEMS 3.120, 3.121

3.122 This dome (hemisphere) is located below the water surface as shown. Determine the magnitude and sign of the force components needed to hold the dome in place and the line of action of the horizontal component of force. Here $y_1 = 1$ m and $y_2 = 2$ m. Assume $T = 10°C$.

3.123 Consider the dome of Prob. 3.122. This dome is 10 ft in diameter, but now the dome is not submerged. The water surface is at the level of the center of curvature of the dome.

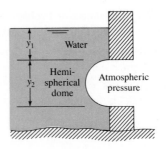

PROBLEM 3.122

For these conditions, determine the magnitude and direction of the resultant hydrostatic force acting on the dome.

3.124 For the hydrometer shown, the stem sinks 2 in. and the bulb displaces 1 in.^{3} of water. Find the weight of the hydrometer.

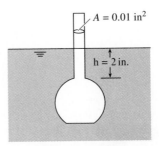

PROBLEM 3.124

3.125 For the hydrometer of Problem 3.124, how much would it sink ($h = ?$) if it were placed in oil ($S = 0.95$)?

3.126 The hydrometer shown sinks 5.3 cm in water (15°C). The bulb displaces 1.0 cm^3, and the stem area is 0.1 cm^2. Find the weight of the hydrometer.

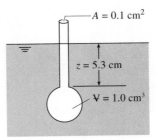

PROBLEM 3.126

3.127 The hydrometer of Problem 3.126 weighs 0.015 N. If the stem sinks 6.3 cm in oil ($z = 6.3$ cm), what is the specific gravity of the oil?

3.128 The hydrometer of Problem 3.126 is placed in a brine solution and sinks 3.3 cm (z = 3.3 cm). What is the specific gravity of the brine?

3.129 A hydrometer with the configuration shown has a bulb diameter of 2 cm, a bulb length of 8 cm, a stem diameter of 1 cm, a length of 8 cm, and a mass of 35 g. What is the range of specific gravities that can be measured with this hydrometer? (*Hint:* Liquid levels range between bottom and top of stem.)

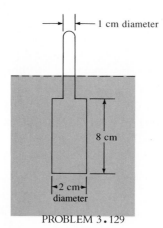

PROBLEM 3.129

3.130 A barge 20 ft wide and 50 ft long is loaded with rock as shown. Assume that the center of gravity of the rock and barge is located along the centerline at the top surface of the barge. If the rock and the barge weigh 400,000 lbf, will the barge float upright or tip over?

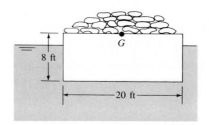

PROBLEM 3.130

3.131 A cylindrical block of wood 1 m in diameter and 1 m long has a specific weight of 8000 N/m³. Will it float in water with its axis vertical?

3.132 A cylindrical block of wood 1 m in diameter and 1 m long has a specific weight of 5000 N/m³. Will it float in water with the ends horizontal?

3.133 Consider the stability of the block given in Ex. 3.15 and show (1) that it is stable in the orientation shown only if the specific gravity of the block is quite small or quite large and (2) that it is unstable for intermediate values of specific gravity. Determine the limits of the specific gravity of the block for which it is stable.

3.134   Is the block in this figure stable floating in the position shown? Show your calculations.

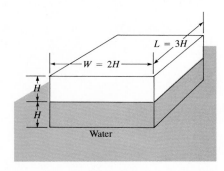

PROBLEM 3.134

3.135   A cylindrical wooden block of density 500 kg/m³ floats in water at 20°C. The length of the block is 20 cm. When the block is submerged from its equilibrium position and released, it oscillates about the equilibrium position. Neglecting the effect of hydrodynamic drag on the moving block, calculate the frequency of oscillation.

References

1. Holman, J. P., and Gajda, W. J., Jr. *Experimental Methods for Engineers*, McGraw-Hill, New York, 1984.

2. "Standard Atmosphere, Tables and Data for Altitudes to 65,800 Feet." *NACA Rept.*, 1235 (1955).

Fluids in Motion

This tripolar flow is developed from an unstable cyclonic vortex. The streaklines reveal the basic flow pattern and the lengths of the streaks are proportional to the flow velocities. This experimental study helps to advance the understanding of flow in the oceans. (Courtesy R.C. Kloosterzeil and G.J.F. van Heijst)

This chapter discusses the basic concepts paramount in the study of the kinematics of fluids in motion. If you develop a good understanding of velocity, acceleration, and flow visualization, you can more easily grasp the nature of pressure variation in flowing fluids when it is discussed in Chapter 5. Moreover, the control-volume approach introduced here will be used extensively in derivation and application of the momentum and energy equations that appear in Chapters 6 and 7, respectively. In addition, the control-volume approach is employed to derive the continuity equation, which is basic to all flow problems. This chapter is undoubtedly the most important chapter in the text; students who master the concepts presented here will indeed have invested their time wisely.

4.1   Velocity and Flow Visualization

The velocity of flow is of primary concern for most engineering problems in fluid mechanics. If the application involves flow past structural or machine parts, a knowledge of the velocity allows pressures and forces acting on the structure to be calculated. The information is then used in the design of the structure. In other cases, such as canal design or bridge-pier design, the engineer may be interested in velocity of flow from the point of view of its scouring action on the channel bottom itself rather than in the context of the overall pressure or force. In any case, the engineer involved with flow problems must be able to determine the velocity by either experimental or analytical means. The following sections describe methods for making such determinations.

Velocity; Lagrangian and Eulerian Viewpoints

There are two ways to express the equations for fluids in motion. One way, the *Lagrangian viewpoint,* has to do with considering an individual fluid particle for all time, which is the familiar approach in dynamics. In this case, the particle velocity is obtained by differentiating the particle's position vector with respect to time. Using the cartesian coordinate system, the position vector is expressed as

$$\mathbf{r}(t) = x\mathbf{i} + y\mathbf{j} + z\mathbf{k} \qquad (4.1)$$

where $\mathbf{i}$, $\mathbf{j}$, and $\mathbf{k}$ are the unit vectors in the x, y, and z directions. Differentiating Eq. (4.1) with respect to time, we obtain the velocity of the fluid particle:

$$\mathbf{V}(t) = \frac{dx}{dt}\mathbf{i} + \frac{dy}{dt}\mathbf{j} + \frac{dz}{dt}\mathbf{k} \qquad (4.2)$$

or
$$\mathbf{V}(t) = u\mathbf{i} + v\mathbf{j} + w\mathbf{k} \qquad (4.3)$$

where u, v, and w are the component velocities in their respective coordinate directions. Of course, the motion of one fluid particle is inadequate to describe an entire flow field, so the motion of all fluid particles must be considered simultaneously. The motion in the flow field is obtained by solving the equation of motion ($\mathbf{F} = m\mathbf{a}$) for each and every fluid particle in the field of flow.

The other way to express fluid velocity is to focus on a certain point in space and consider the motion of fluid particles that pass that point as time goes on. This is known as the *Eulerian approach*. In this case, the fluid particle velocity depends on the point in space and time:

$$u = f_1(x,y,z,t) \qquad v = f_2(x,y,z,t) \qquad w = f_3(x,y,z,t) \qquad (4.4)$$

In this method we observe the motion of particles passing a specific point in space, as opposed to the Lagrangian method in which we track the position of a specific particle as time passes. In order to describe the entire flow field, we must know the fluid motion at all points in the field.

It is an enormous task to keep track of the positions of all particles in a flow field, because unlike the movement of solid bodies, their relative positions continuously change with time. For this reason, the Eulerian approach is generally favored over the Lagrangian approach. In some areas—such as rarefied gas dynamics, where we are concerned with the motion of individual molecules—the Lagrangian approach is used, but the overwhelming majority of fluid dynamic analyses are based on the Eulerian approach.

Equations (4.4) give the component velocities as a function of space and time in the cartesian coordinate system, but there is also another useful way of expressing velocity by using the Eulerian viewpoint to describe the total velocity as a function of position along a streamline and time. This is given as

$$\mathbf{V} = \mathbf{V}(s,t) \qquad (4.5)$$

The streamline will be defined in the next section.

Streamlines and Flow Patterns

Often it is desirable to construct lines in the flow field to indicate the speed and direction of flow. Such a construction is called a *flow pattern,* and the lines, called *streamlines,* are defined as lines drawn through the flow field in such a manner that the velocity vector of the fluid at each and every point on the streamline is tangent to the streamline at that instant. Consequently, a tangent to the curve at any point along the streamline gives us the direction of the velocity vector at that point in the flow field. For example, consider a flow of water from a slot in the side of a tank, as shown in Fig. 4.1. The velocity vectors have been sketched at three different positions, *a*, *b*, and *c*. One can see that the flow pattern is a very effective way of illustrating the geometry of fluid flow. In Fig. 4.1 it may be noted that the two outer streamlines that bound the free jet also follow the walls inside the tank. This tangency of streamlines with fixed boundaries follows directly from the definition of a streamline. That is, since there is no flow through an impervious boundary, all velocity vectors of the flow adjacent to the boundaries must be parallel to the boundary. Therefore, all streamlines directly adjacent to the wall are also parallel and actually follow the contour of the wall. Later, in the section on the continuity equation, we will see that the speed of flow is also indicated by the flow pattern; it will become apparent that the speed of flow, *V*, is inversely proportional to the spacing of the streamlines in two-dimensional flow.

Whenever flow occurs around a body, part of it will go to one side and part to the other side. The streamline that follows the flow division (that divides on the upstream side of the body and joins again on the downstream side) is called the *dividing streamline*. Also, at the point of division the velocity will be zero. This point is called the *stagnation point*.

Now that the general concept of the flow pattern has been introduced, it is convenient to make distinctions between different types of flow. First, we shall consider whether a flow is *uniform* or *nonuniform*. In uniform flow the velocity does not change from point to point along any of the streamlines in the flow

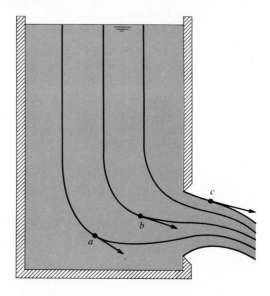

FIGURE 4.1

Flow from a slot.

field. Therefore, it follows that the streamlines depicting such flow must be straight and parallel. If they are not straight, there will be a directional change of velocity. If they are not parallel, there will be a change of speed along the streamlines. These conditions of uniform and nonuniform flow are expressed mathematically as follows:

$$\frac{\partial \mathbf{V}}{\partial s} = 0 \quad \text{(uniform flow)} \qquad \frac{\partial \mathbf{V}}{\partial s} \neq 0 \quad \text{(nonuniform flow)}$$

Here $\mathbf{V}$ is the total velocity at a given point on a streamline, and s is the distance along the streamline measured from an arbitrary point on the streamline. Flow patterns for uniform flow in an open channel and between parallel plates are shown in Fig. 4.2. In nonuniform flow the velocity changes from point to point along the streamline. Therefore, the flow pattern consists of streamlines that are either curving in space or converging or diverging. Flow patterns for cases of nonuniform flow are shown in Figs. 4.1 and 4.3.

Another flow classification is based on the variation of velocity with respect to time at a given point in a flow field. If at any given point the velocity does not vary in magnitude or direction with time, then the flow is *steady.* Steady and *unsteady* flow conditions are mathematically defined as follows:

$$\frac{\partial \mathbf{V}}{\partial t} = 0 \quad \text{(steady flow)} \qquad \frac{\partial \mathbf{V}}{\partial t} \neq 0 \quad \text{(unsteady flow)}$$

Here velocity $\mathbf{V}$ is at a given point in the flow field. An example of steady flow is a constant rate of flow through a pipe. However, if the discharge is being increased (by the opening of a valve, for example), then the flow is unsteady. And

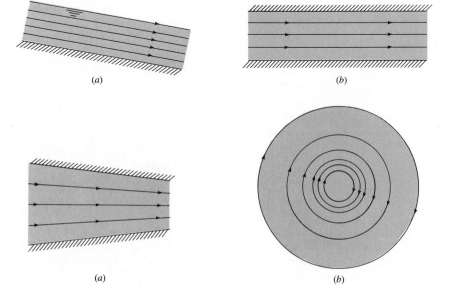

FIGURE 4.2

Uniform flow patterns.
(a) Open-channel flow.
(b) Flow between parallel plates.

(a)   (b)

FIGURE 4.3

Flow patterns for nonuniform flow.
(a) Converging flow.
(b) Vortex flow.

(a)   (b)

if the pipe is straight, one classifies the flow as uniform and unsteady. A simple case of nonuniform-unsteady flow is flow through a nozzle at a changing rate of discharge. Here the flow is nonuniform because of the converging flow passage and the consequently converging streamlines, and it is unsteady because of the time change of velocity associated with the change of flow rate.

If the student understands the fundamentals of flow patterns, he or she will appreciate the fact that by simply studying the flow pattern one can ascertain whether or not the flow is uniform. However, the flow pattern itself says nothing with regard to its steady or unsteady state. In general, a flow pattern for unsteady flow is only an instantaneous representation of flow geometry.

Laminar and Turbulent Flow

Turbulent flow is characterized by a mixing action throughout the flow field, and this mixing is caused by *eddies* of varying size within the flow. Simple observations will reveal this type of flow in rivers and in the atmosphere. Gusts of wind are the result of large-scale eddies that at times reinforce and at other times subtract from the mean wind velocity. Turbulent-flow phenomena can also be observed when smoke from a large stack discharges into the surrounding air.

Laminar flow, on the other hand, is devoid of the intense mixing phenomena and eddies common to turbulent flow. Thus this flow has a very smooth appearance. A typical example is the flow of honey or thick syrup from a pitcher.

To help in understanding the role of turbulence in the flow process, one can first consider laminar flow in a given situation and then visualize what would happen to the flow if in some hypothetical way one could superimpose a set of eddies onto this flow. The velocity distributions for these two situations are presented in Fig. 4.4 for flow in a straight pipe. For the laminar-flow case, the velocity distribution is parabolic across the section, and at any given distance from the pipe wall, the velocity is constant with respect to time. In the turbulent-flow case, two effects are readily apparent. First, because the eddies cause the flow to be mixed rather thoroughly, the velocity distribution over most of the section is more uniform than is the case in laminar flow. This results because the turbulent mixing process transports the low-velocity fluid near the wall toward the center and the higher-velocity fluid in the central region toward the wall.

The second effect of turbulence is to add continuously fluctuating components of velocity to the flow. At a given instant the distribution of the velocity component in the direction of flow is consequently irregular, as shown by the solid line in Fig. 4.4b. However, if we average the velocity over a sufficiently long period of time at various points across the section, the broken line shown in the same figure is obtained. It is thus apparent that at a given point the velocity varies with time. Therefore, according to our previous definition, the flow is called unsteady. However, if we consider the temporal mean (average with respect to time) velocity at a given point taken over a relatively long period of time,

FIGURE 4.4

Laminar and turbulent flow
in a straight pipe.
(a) Laminar flow.
(b) Turbulent flow.

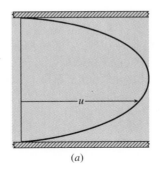

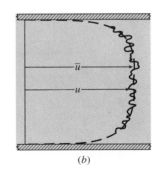

(a)   (b)

the velocity is virtually constant, and the flow is therefore termed steady. For most discussions, this latter point of view (based on the temporal mean velocity) is used to characterize the flow as steady or unsteady. An index relating to turbulence is the Reynolds number, defined as $\text{Re} = VD\rho/\mu$. If the Reynolds number is large ($\text{Re} > 2000$), the flow in a pipe will generally be turbulent; but if the Reynolds number is smaller ($\text{Re} < 2000$), the flow will be laminar. The significance of the Reynolds number will be considered in more detail in later chapters.

The student may wonder why flow is classified in so many different ways. Of the many reasons for the various classifications, two should serve as valid examples. First, certain types of problems (for example, steady-flow problems and unsteady-flow problems) require different methods of solution. The classifications help the engineer to organize his or her thoughts with regard to problem analysis. Sections on acceleration later in this chapter and discussions of pressure distribution in Chapter 5 include examples that illustrate this point. Second, the various classifications make it much easier for fellow engineers and scientists to communicate—which is why a language exists and what keeps it evolving.

Methods for Developing Flow Patterns

We have defined streamlines and have discussed how a number of streamlines make up the flow patterns, but we have not stated how these streamlines and flow patterns are obtained. The three basic methods used to predict flow patterns are analytical, analog, and experimental.

Analytical and Numerical Methods

The oldest and best known of the analytic methods derive from ideal-flow theory—that is, theory concerning the flow of incompressible, nonviscous fluids. The differential equation fundamental to ideal-flow theory is Laplace's equation, which is encountered frequently in science and engineering. Solutions of Laplace's equation can be applied to many flow problems, such as low-speed flows past airfoils and free-surface flows such as wave motion.

Ideal-flow theory is not applicable to flows where viscous and/or compressibility effects are important. In these cases, few closed-form analytic solutions exist, and these few are restricted to simple flow configurations.

The computational capability of the computer has led to new and incredible advances in the analysis of flow fields in recent years. Turbulent two-dimensional flows, unsteady-nonuniform flows, and rarefied gas dynamics are only a few examples of numerical solutions that have become available. Computational fluid mechanics is a field in which new developments are rapidly taking place in step with improved computer capability and capacity.

Analog Method

Where boundary conditions are complicated, it is often possible to develop flow patterns by analog methods, the most common of which is electrical analogy. In this we take advantage of the fact that the ideal-flow theory serves to describe electrical fields as well as certain fluid-flow fields. To apply the method, an electrical conductor is made in the same shape as the flow passage it represents, and a voltage potential is applied between the ends of the conductor, thus establishing a flow of electrons. By utilizing the data obtained by measuring the voltage at various points of the conductor, it is possible to construct the flow pattern and to predict the pressure distribution in the conduit. The method is valid for irrotational flow cases, a concept that will be covered in more detail in Sec. 4.6.

Experimental Method

For certain types of turbulent flow, mathematical or analog methods have not been developed to the point where they can reliably predict the flow pattern; therefore, it is necessary to resort to physical models. In other cases, such as basic research studies, experimental methods are also employed to define the pattern of flow. In these experimental setups, dye streams and floating or immersed particles are often used to define the flow pattern. When a photograph is taken of floating or neutrally buoyant particles in a flowing fluid (as in Fig. 4.5, which shows the approach channel to a spillway of a model dam), the particles produce light marks, which indicate the paths of the particles for the period of time of exposure. Each one of these light marks is a segment of a pathline for a given particle. By definition, then, a *pathline* is a line drawn through the flow field in such a way that it defines the path that a given particle of fluid has taken.

Another technique used to visualize flow patterns is to inject dye or smoke at a given point in the flow field and to observe the dye or smoke trace as it travels downstream. Such a trace is called a *streakline* (see Fig. 4.6; note the separation zone downstream of the airfoil where turbulence causes the smoke to be diffused).

We have discussed *streamline, pathline,* and *streakline,* all of which in one way or another are associated with the pattern of flow. For steady flow, all three

FIGURE 4.5

Pathlines of floating particles.

of these lines are coincident if they originate at the same point. Because streaklines and pathlines define streamlines for steady flow, it is often thought that they always define streamlines. This may be misleading, as the following paragraph explains.

Consider a body of fluid that has a velocity of $u = 1.0$ ft/s, $v = 0$, and $w = 0$ for a period of 4 s. Hypothetically assume that, at the end of 4 s, the fluid will take on a new velocity of $u = 0.707$ ft/s, $v = 0.707$ ft/s, and $w = 0$, and that the latter velocity will continue indefinitely. Figure 4.7 shows streamlines, a streakline, and a pathline for the flow at time $t = 4$ s, at time $t = 7$ s and at time $t = 10$ s. Note that injection of a single stream of dye was started at point A at $t = 0$ and that the particle of fluid for which the path is drawn passed by point A at time $t = 0$. For the first period, Fig. 4.7a, the pathline and streakline coincide. However, for the entire period there are widely differing traces (see Fig. 4.7b). One can conclude from this illustration that streaklines and pathlines *do not* in general define streamlines for *unsteady* flow.

One-, Two-, and Three-Dimensional Flow

In general, three coordinate directions are needed to describe the velocity and property changes in a flow field. Such flows are three-dimensional.

For some flow situations there are no changes in one coordinate direction; two dimensions suffice to describe such flow. For example, the flow between two parallel walls in which there is no velocity component in the direction normal to the wall is a case of two-dimensional flow.

FIGURE 4.6

Smoke traces about an airfoil with a large angle of attack. (Courtesy of Education Development Center, Inc., Newton, MA.)

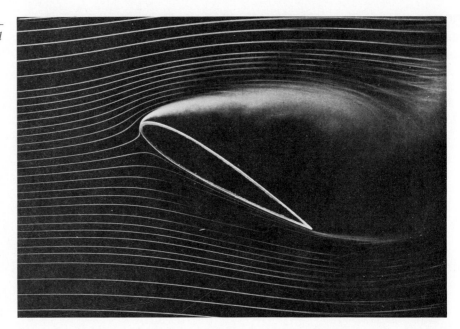

The simplest flow field is the one-dimensional case, in which only one co-ordinate is needed to relate velocity and property changes. An example is flow in a duct, or conduit, in which the velocity is constant across each section but varies with distance along the duct. Actually, the velocity is never completely constant across a conduit section. However, for problems in which we are primarily interested in the average velocity parallel to the conduit, this type of flow is called one-dimensional flow. In this case we are interested in the changes of average velocity and pressure occurring along the length of the conduit rather than across it. Such a concept is used extensively in applications of the momentum and energy equations to be introduced in Chapters 6 and 7. Throughout the remainder of the text the term *one-dimensional flow* will generally pertain to cases in which we are primarily interested in the mean, or average, velocity in a conduit.

FIGURE 4.7

Streamlines, streaklines, and pathline. (a) $t = 4$ s. *(b)* $t = 7$ s. *(c)* $t = 10$ s.

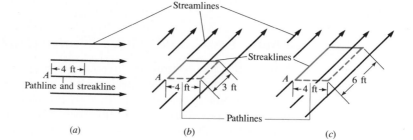

Rate of Flow

Volume Rate of Flow

Often simply called *flow rate* or *discharge,* the volume rate of flow refers to the rate at which the volume of fluid passes a given section in the flow stream. Consider the flow of a fluid in a pipe as shown in Fig. 4.8, where the velocity is constant throughout the pipe. Now consider the volume rate of flow past section *A–A*. First, assume that at time $t = 0$ we can instantaneously mark the fluid across section *A–A* (for example, by injecting dye into the stream). Thus the dashed line at section *A–A* shows the marked fluid at time zero. Then, if the fluid is allowed to flow for a time Δt, the marked fluid will travel a distance $V \Delta t$, as shown in Fig. 4.8. By referring to Fig. 4.8, one can conclude that the volume of fluid ($\Delta \math{V}$) that has passed through section *A–A* in the time Δt is given by the product of $AV \Delta t$, where A is the cross-sectional area of the pipe. However, we are more interested in the volume *rate* of flow past *A–A,* which is $\Delta \mathrm{V}/\Delta t = V \Delta t\, A/\Delta t$. If we let Q be the symbol for volume rate of flow, then we have $Q = V \Delta t\, A/\Delta t$ or $Q = VA$.

Flow Rate with Variable Velocity

The velocity, in general, is variable across the section through which it flows, such as in Fig. 4.9. Then the rate of flow past a differential area of the section is $V dA$, and the total volume rate of flow Q is obtained by integration over the entire flow section:

$$Q = \int_A V dA$$

In a similar manner, the mass rate of flow past a section is given by

$$\dot{m} = \int_A \rho V dA$$

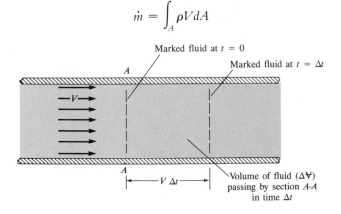

FIGURE 4.8

Flow of fluid in a pipe; constant velocity.

FIGURE 4.9

*Flow between parallel
boundaries.*

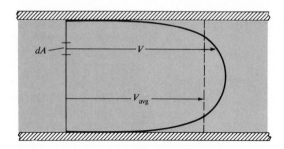

In the foregoing developments, the cross-sectional area was always oriented normal to the velocity vector. If other orientations are considered, such as in Fig. 4.10, where flow occurs past section A-A, it can be seen that only the normal component of velocity (the x component in this case) contributes to flow through the section. Therefore, to evaluate flow rate, one must always consider either the area of a section normal to the total velocity or the velocity component normal to the given area. Thus the discharge for the case of Fig. 4.10 is given by

$$Q = \int_A u \, dA \qquad \text{or} \qquad Q = \int_A V \cos \theta \, dA$$

If we define an area vector as one that has the magnitude of the area in question and that is oriented normal to the area, then by definition, $V \cos \theta \, dA = \mathbf{V} \cdot d\mathbf{A}$, and the discharge can be written as

$$Q = \int_A \mathbf{V} \cdot d\mathbf{A}$$

If the velocity is constant over the area, then the discharge is given as

$$Q = \mathbf{V} \cdot \mathbf{A}$$

In the foregoing equations, because of the scalar, or dot, product, only the normal component of velocity is multiplied by the area to obtain the discharge.

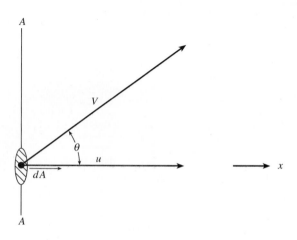

FIGURE 4.10

*Velocity not normal to
the section.*

Mean, or Average, Velocity

In many problems—for example, those concerning one-dimensional flow in pipes—one may be given the rate of flow and may need to find the mean (average) velocity without knowing the actual velocity distribution across the pipe. The mean velocity, by definition, is the discharge divided by the total cross-sectional area:

$$\overline{V} = \frac{Q}{A}$$

For turbulent flow in pipes, the mean velocity may be a fairly close approximation to the actual distribution of velocity over most of the section, as shown in Fig. 4.9. However, for laminar flow the mean velocity differs considerably from the velocity over most of the flow section. It is customary to leave the bar off the velocity symbol and simply to indicate the mean velocity with a V.

EXAMPLE 4.1   Air that has a mass density of 1.24 kg/m^3 (0.00241 slugs/ft^3) flows in a pipe with a diameter of 30 cm (0.984 ft) at a mass rate of flow of 3 kg/s (0.206 slugs/s). What are the mean velocity of flow in this pipe and the volume rate of flow?

Solution

$$\dot{m} = \rho Q \qquad \text{or} \qquad Q = \frac{\dot{m}}{\rho} = 2.42 \text{ m}^3/\text{s} \text{ (85.5 cfs)} \qquad \blacktriangleleft$$

$$V = \frac{Q}{A} = \frac{2.42}{(\frac{1}{4}\pi) \times (0.30)^2} = 34.2 \text{ m/s} \text{ (112 ft/s)} \qquad \blacktriangleleft$$

EXAMPLE 4.2   Water flows in a channel that has a slope of 30°. If the velocity is assumed to be constant, 12 m/s, and if the flow is uniform with a depth of 60 cm measured along a vertical line, what is the discharge per meter of width of the channel?

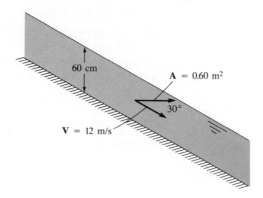

60 cm

A = 0.60 m²

30°

V = 12 m/s

Solution   The discharge per unit width is often designated by q, and in this case it can be solved for as follows:

$$q = \mathbf{V} \cdot \mathbf{A}$$

$$= 12 \text{ m/s} \times \cos 30° \times 0.60 \text{ m} \times 1.0 \text{ m} = 62.4 \text{ m}^3/\text{s per meter} \quad \blacktriangleleft$$

EXAMPLE 4.3   The water velocity in the channel shown in the accompanying figure has a distribution across the vertical section equal to $u/u_{max} = (y/d)^{1/2}$. What is the discharge in the channel if the channel is 2 m deep ($d = 2$ m) and 5 m wide and the maximum velocity is 3 m/s?

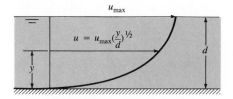

Solution

$$Q = \int_A u \, dA = \int_0^2 u_{max}(y/d)^{1/2}(5) \, dy$$

$$= \frac{5u_{max}}{d^{1/2}} \int_0^2 y^{1/2} \, dy = \frac{5 \times 3}{2^{1/2}} \times \frac{y^{3/2}}{3/2} \bigg|_0^2 = 20.0 \text{ m}^3/\text{s} \quad \blacktriangleleft$$

4.3   Acceleration

Basic Operations

The acceleration of a fluid particle is the rate of change of the particle's velocity with time. In the Lagrangian formulation, each component of velocity is a function of time only, so the differentiation of each component involves the differentiation of a single-variable function. By the Eulerian approach, the velocity components are functions of both space and time, as given by Eqs. (4.4). Thus

$$\mathbf{V} = u\mathbf{i} + v\mathbf{j} + w\mathbf{k}$$

where $u = f_1(x, y, z, t)$, $v = f_2(x, y, z, t)$, and $w = f_3(x, y, z, t)$.

The acceleration of a fluid particle in the x direction is given by

$$a_x = \frac{du}{dt}$$

where u is the x component of velocity as we follow the particle. By using the chain rule for differentiation of a multivariable function, we can express this as

$$a_x = \frac{\partial u}{\partial x}\frac{dx}{dt} + \frac{\partial u}{\partial y}\frac{dy}{dt} + \frac{\partial u}{\partial z}\frac{dz}{dt} + \frac{\partial u}{\partial t} \tag{4.6}$$

In a time dt, the fluid particle moves in the x direction a distance $dx = u\,dt$, so

$$u = \frac{dx}{dt}$$

and similarly $$v = \frac{dy}{dt} \quad \text{and} \quad w = \frac{dz}{dt}$$

Thus the acceleration component a_x of the particle is given by

$$a_x = u\frac{\partial u}{\partial x} + v\frac{\partial u}{\partial y} + w\frac{\partial u}{\partial z} + \frac{\partial u}{\partial t} \tag{4.7}$$

Similarly, for the y and z components, we obtain

$$a_y = u\frac{\partial v}{\partial x} + v\frac{\partial v}{\partial y} + w\frac{\partial v}{\partial z} + \frac{\partial v}{\partial t} \tag{4.8}$$

$$a_z = u\frac{\partial w}{\partial x} + v\frac{\partial w}{\partial y} + w\frac{\partial w}{\partial z} + \frac{\partial w}{\partial t} \tag{4.9}$$

Tangential and Normal Acceleration

If we solve for the acceleration using the velocity as given by Eq. (4.5), we obtain

$$a_t = \frac{\partial V_s}{\partial s}\frac{ds}{dt} + \frac{\partial V_s}{\partial t} = V_s\frac{\partial V_s}{\partial s} + \frac{\partial V_s}{\partial t} \tag{4.10}$$

This is only the acceleration component tangential to the streamline. There will also be a component normal to the streamline, which is given by

$$a_n = \frac{V_s^2}{r} \tag{4.11}$$

The term on the right-hand side of Eq. (4.11) follows from normal acceleration for curvilinear motions. Here r is the radius of curvature of the streamline, and the acceleration is toward the center of curvature of the streamline. Equations (4.10) and (4.11) represent the acceleration of a fluid particle with respect to a coordinate system within which the streamlines neither rotate nor accelerate.

Convective and Local Acceleration

Inspection of Eqs. (4.7) through (4.11) will reveal that the terms on the right-hand sides are of two different types: those that include changes of velocity with respect to position, $u\,\partial u/\partial x$, $v\,\partial v/\partial y$, and so on; and those that are changes of velocity with respect to time, $\partial u/\partial t$, $\partial v/\partial t$, $\partial w/\partial t$, and $\partial V_s/\partial t$. Terms of the first type are called *convective* accelerations, because they are associated with velocity changes that occur because of changes in position in the flow field. However, for the second type, acceleration results because the velocity changes with respect to time at a given point. These are called *local* accelerations. Obviously, local acceleration results when the flow is unsteady, and convective acceleration occurs when the flow is nonuniform—that is, when the velocity changes along a streamline.

EXAMPLE 4.4   A nozzle is designed so that its cross-sectional area changes linearly from the base to the tip. The inside diameters of the nozzle at the base and tip are 9 cm and 3 cm, respectively, and the nozzle is 36 cm long. What is the convective acceleration at a section midway between base and tip? Assume one-dimensional flow of water at a constant discharge of 0.02 m^3/s.

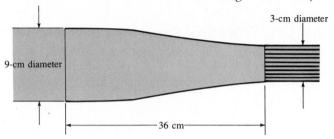

Solution   Using Eq. (4.10), we have the convective acceleration given by

$$a = V_s \frac{\partial V_s}{\partial s} \qquad \text{where } V_s = \frac{Q}{A}$$

However, because the area is linearly distributed, we can write

$$A = A_0 - \frac{\Delta A}{\Delta s} s$$

where A_0 = area of nozzle at the base = 6.36×10^{-3} m^2, ΔA = difference in area between base and tip of nozzle = 5.66×10^{-3} m^2, and Δs = distance between base and tip = 0.36 m. Thus

$$A = 0.00636 \text{ m}^2 - \frac{0.00566 \text{ m}^2}{0.36 \text{ m}} s = 0.00636 - 0.0157 s$$

The area at midsection is given by

$$A_{\text{mid}} = 0.00636 - 0.0157 \times 0.18 = 0.00353 \text{ m}^2$$

and

$$V_{\text{mid}} = \frac{Q}{A_{\text{mid}}} = \frac{0.02 \text{ m}^3/\text{s}}{0.00353 \text{ m}^2} = 5.66 \text{ m/s}$$

The velocity gradient is given by

$$\frac{\partial V_s}{\partial s} = \frac{\partial}{\partial s} \frac{Q}{A} = \frac{\partial}{\partial s} \frac{Q}{0.00636 - 0.0157 s} = \frac{0.0157 Q}{(0.00636 - 0.0157 s)^2}$$

Then, at midsection, where $s = 0.18$ m,

$$\frac{\partial V_s}{\partial s} = 25.1 \text{ s}^{-1}$$

Therefore $\qquad\qquad\qquad a_{\text{convec}} = V_s \frac{\partial V_s}{\partial s} = 142 \text{ m/s}^2 \qquad\qquad$ ◀

4.4 Basic Control-Volume Approach

As defined in Sec. 2.2, a *system* is a given quantity of matter. Therefore, because a given body in solid mechanics is easily identified, the system approach is generally used to solve dynamic problems for solids. For flow of fluids, however, individual particles are not easily identified one from another, so a method is needed that will enable us to solve problems by focusing on a given space through which the fluid flows rather than by directing attention solely to a given mass of fluid. Such a method is available. It is called the Reynolds transport theorem or, more descriptively, the control-volume approach, and it has wide application to a variety of problems in fluid mechanics. The next few paragraphs include preliminary developments that will be used in the derivation of the control-volume equation. Then the control-volume equation will be derived, and applications of it will be given.

Review of Extensive and Intensive Properties

In the derivation of the control-volume approach, we will be concerned with extensive and intensive properties. The extensive properties are mass M, momentum $M\mathbf{V}$, and energy E. The corresponding intensive properties are mass per unit mass, which is unity; momentum per unit mass, which is $\mathbf{V}$; and energy per unit of mass, which is e. Because we will be considering different intensive and extensive properties at different times when we actually apply the control-volume approach, it is desirable for this derivation that we use the special symbol B to represent a general extensive property. We will let β (beta) be the symbol for the corresponding intensive property. Because the intensive property is given as the extensive property per unit of mass, the relationship between the intensive and extensive properties for a given system is denoted by

$$B = \int \beta \, dm = \int \beta \rho \, d\Psi \qquad\qquad (4.12)$$

Here dm and $d\Psi$ are differential mass and differential volume, respectively, and the integral is over the volume occupied by the system at a given instant.

Control Volume and Control Surface

The control volume is a region in space that one establishes to aid in the solution of flow problems, and the control surface is the surface surrounding the control volume. In most problems, part of the control surface coincides with some physical boundary, such as the wall of a pipe. The remaining part of the control surface in such problems is a hypothetical surface through which fluid can flow, such as is shown in Fig. 4.11 for one-dimensional flow in a conduit.

FIGURE 4.11

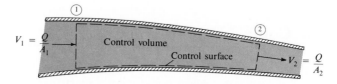

Flow into and out of a Control Volume

It has already been shown that the volume rate of flow past a given area A can be written as

$$Q = \mathbf{V} \cdot \mathbf{A}$$

Here $\mathbf{V}$ is the velocity vector of the flow and $\mathbf{A}$ is the area vector that has the magnitude of the area and is directed normal to the area in question. It is obvious that all surface areas have two sides, so we must establish a rule to govern whether the area vector is pointing in one direction or the other. The convention that we use with the control-volume approach is that the area vector always points outward from the control volume. Thus in Fig. 4.12 we show one-dimensional flow in a passage with proper orientation of the area vectors. The cosine of the angle between $\mathbf{V}_1$ and $\mathbf{A}_1$ is -1, and that of the angle between $\mathbf{V}_2$ and $\mathbf{A}_2$ is $+1$, so the flow rate out of the control volume minus the flow rate into the control volume can be given as

$$\text{Flow rate out} - \text{flow rate in} = V_2 A_2 - V_1 A_1$$

$$= \mathbf{V}_2 \cdot \mathbf{A}_2 + \mathbf{V}_1 \cdot \mathbf{A}_1$$

$$= \sum_{cs} \mathbf{V} \cdot \mathbf{A} \qquad (4.13)$$

Equation (4.13) states that if we sum up the dot products $\mathbf{V} \cdot \mathbf{A}$ for all flows into and out of a control volume, we will get the net rate of outflow. If the resulting summation is a positive number, the net rate of flow is out of the control volume. If the summation is negative, the net rate of flow is into the control volume. Of course, if the outflows and inflows are equal, then $\sum_{cs} \mathbf{V} \cdot \mathbf{A}$ will be zero.

FIGURE 4.12

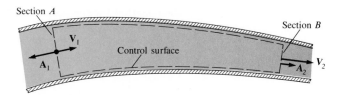

If we want to evaluate the mass rate of flow out of the control volume, we simply multiply the volume rate by ρ, or

$$\dot{m} = \sum_{cs} \rho \mathbf{V} \cdot \mathbf{A} \qquad (4.14)$$

In a similar manner, if we want the rate of flow of an extensive property B out of the control volume, we multiply the mass rate by the intensive property β:

$$\dot{B} = \sum_{cs} \beta \overbrace{\rho \mathbf{V} \cdot \mathbf{A}}^{\dot{m}} \qquad (4.15)$$

To reinforce the validity of Eq. (4.15), one may consider the dimensions involved. Equation (4.15) states that the flow rate of B is given by

$$\overset{\beta}{\searrow} \qquad \overset{\dot{m}}{\downarrow}$$

$$\left(\frac{B}{\text{mass}}\right)\left(\frac{\text{mass}}{\text{time}}\right) = \left(\frac{B}{\text{time}}\right) = \dot{B}$$

Equations (4.13), (4.14), and (4.15) are applicable for all one-dimensional flows. If the velocity varies across a flow section, then it becomes necessary to integrate the velocity across the section to obtain the rate of flow. A more general expression for the rate of flow of the extensive property from the control volume is thus

$$\dot{B} = \int_{cs} \beta \rho \mathbf{V} \cdot d\mathbf{A} \qquad (4.15a)$$

Equation (4.15a) will be used in the most general form of the control-volume equation.

Derivation of the Control-Volume Equation

To derive the basic control-volume equation, we first focus on a *system* that moves through space; however, in the process of derivation, the control volume becomes significant.

The basic equation for the control-volume approach is derived by considering the rate of change of an extensive property of the *system* of fluid that is flowing through the control volume. In Fig. 4.13 the solid line identifies the control surface that encloses the control volume, and this same surface serves to define the system, a given mass of fluid, at time t. At time $t + \Delta t$, this system, or mass of fluid, is identified by the dashed line in Fig. 4.13, and it has moved with respect to the control surface; it has moved downstream. The rate of change of the extensive property B of the system, dB_{sys}/dt, can be stated according to the fundamental definition of a derivative as

$$\frac{dB_{sys}}{dt} = \lim_{\Delta t \to 0} \left[\frac{B_{t+\Delta t} - B_t}{\Delta t} \right] \tag{4.16}$$

The mass of this system at time $t + \Delta t$ is the mass of the fluid within the control volume at time $t + \Delta t$ plus the mass that has moved out of the control volume in time Δt minus the mass of fluid that has moved into the control volume in time Δt (see Fig. 4.13). Let us call ΔM_{out} the mass that has moved out of the control volume in time Δt and ΔM_{in} the mass that has moved into the control volume in time Δt. Likewise, let the extensive property of the system that has moved out of the control volume in time Δt be ΔB_{out}, and let the extensive property of the system that has moved into the control volume in time Δt be ΔB_{in}. Thus the extensive property B of the system at time $t + \Delta t$ can be written $B_{cv,t+\Delta t} + \Delta B_{out} - \Delta B_{in}$, and the rate of change of the extensive property of the system with respect to time can be expressed as

$$\frac{dB_{sys}}{dt} = \lim_{\Delta t \to 0} \left[\frac{(B_{cv,t+\Delta t} + \Delta B_{out} - \Delta B_{in}) - (B_{cv,t})}{\Delta t} \right] \tag{4.17}$$

Rearranging terms yields

$$\frac{dB_{sys}}{dt} = \lim_{\Delta t \to 0} \left[\frac{B_{cv,t+\Delta t} - B_{cv,t}}{\Delta t} \right] + \lim_{\Delta t \to 0} \left[\frac{\Delta B_{out} - \Delta B_{in}}{\Delta t} \right] \tag{4.18}$$

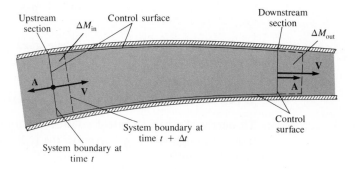

FIGURE 4.13

The first term on the right-hand side of Eq. (4.18) is simply the rate of change with respect to time of the extensive property B of the fluid inside the control volume at time t. That is,

$$\lim_{\Delta t \to 0} \left[\frac{B_{cv,t+\Delta t} - B_{cv,t}}{\Delta t} \right] = \frac{dB_{cv}}{dt} \qquad (4.19)$$

Substituting from Eq. (4.12), we get

$$\lim_{\Delta t \to 0} \left[\frac{B_{cv,t+\Delta t} - B_{cv,t}}{\Delta t} \right] = \frac{dB_{cv}}{dt} = \frac{d}{dt} \int_{cv} \beta \rho \, d\Psi \qquad (4.20)$$

The second term on the right-hand side of Eq. (4.18) can be analyzed in the following manner. The quantity $\Delta B_{out} - \Delta B_{in}$ represents the amount of the property B that has passed out of the control volume minus the amount of the property B that has passed into the control volume in time Δt. Thus, when this is divided by Δt and when we take the limit as $\Delta t \to 0$, we obtain the rate of flow of B out of the control volume minus the rate of flow of B into the control volume or, in other words, the net flow of B from the control volume. But Eq. (4.15), $\dot{B} = \Sigma_{cs} \beta \rho \mathbf{V} \cdot \mathbf{A}$, is precisely equal to this quantity, so we can make Eq. (4.18) more compact by substituting Eq. (4.15) for the second term on the right-hand side and substituting $d/dt \int_{cv} \beta \rho \, d\Psi$ for the first term on the right-hand side:

$$\frac{dB_{sys}}{dt} = \frac{d}{dt} \int_{cv} \beta \rho \, d\Psi + \sum_{cs} \beta \rho \mathbf{V} \cdot \mathbf{A} \qquad (4.21a)$$

The subscript on the summation sign of the second term on the right side of Eq. (4.21a) indicates that we are summing flows across the entire control surface. In the derivation of Eq. (4.21a), we first considered the rate of change of the extensive property B of the system, dB_{sys}/dt; then we showed that this could be expressed as the sum of the rate of change of B within the control volume plus the net rate of flow of B out of the control volume. Thus the left-hand side of Eq. (4.21a) is still the rate of change of the extensive property of the system (rate of change of the extensive property of the given quantity of matter), but the right-hand side refers to conditions within the control volume and to flow across the control surface for the first and second terms, respectively. In the derivation of Eq. (4.21a), one-dimensional flow was assumed; thus the rate of flow of B at each section was given as $\beta \rho \mathbf{V} \cdot \mathbf{A}$. However, when the velocity is variable across a section, the more general form for the rate of flow of the extensive property, Eq. (4.15a), must be used. Thus the control-volume equation is given as

$$\frac{dB_{sys}}{dt} = \frac{d}{dt} \int_{cv} \beta \rho \, d\Psi + \int_{cs} \beta \rho \mathbf{V} \cdot d\mathbf{A} \qquad (4.21b)$$

Control-Volume Equation for Steady Flow

In much of our work we are concerned only with steady-flow problems, and then the first term on the right-hand side of Eqs. (4.21) drops out. This occurs because the differentiation is taken with respect to conditions within the control volume; if the conditions do not change with time, then the $d(\ )/dt$ term is zero. For steady flow our basic control-volume equation thus reduces to

$$\frac{dB_{\text{sys}}}{dt} = \sum_{\text{cs}} \beta \rho \mathbf{V} \cdot \mathbf{A} \qquad (4.22)$$

Depending on the equation's application, β and B may be either scalar or vector quantities. For example, if we want to derive the general form of the continuity equation, we let B be the mass of the system (which is constant because a system is a given quantity of matter), and then β will be equal to unity. Both of these quantities are scalars. However, if we let $\mathbf{B} =$ momentum and $\beta = \mathbf{V}$, then we have vector quantities, whose use leads to the momentum equation. In Sec. 4.5, we will derive the continuity equation by utilizing the control-volume equation (Eqs. 4.21). Then in Chapters 6 and 7 the control-volume equation will be used to derive the momentum and energy equations, respectively.

Motion of the Control Volume and Steady or Unsteady Flow

The concepts of steady and unsteady flow were discussed earlier. It is interesting to see how the terms of Eqs. (4.21) can be interpreted in light of these concepts. Let us consider the first term on the right-hand side of Eqs. (4.21), which is the rate of change of B_{cv} with respect to time. If the flow is steady, there will be no change with respect to time, and the term will be zero. Conversely, if the flow process is unsteady, the term must be considered in any control-volume analysis. In this regard, it is most interesting to see how the motion of the control volume itself can, in some cases, dictate whether the flow condition is considered steady or unsteady. Consider the case in which a ship is traveling at a constant speed in shallow water, as shown in Fig. 4.14. If the control surface is fixed to the earth, the velocities within the control volume will continuously change with time, and by definition the flow will be unsteady. However, if the control surface moves at the same speed as the ship, then none of the flow characteristics will change with time, and this flow will be classified as steady. If the control surface is assumed to move in space, the reference system is likewise assumed to have the same motion, which means simply that the velocities considered in the basic equations are measured relative to the control surface—a requirement the student must be careful to observe in all problems involving control-volume analysis. Although moving the control surface with the body appears to be a simple way of converting an unsteady-flow process to a steady-flow process, a word of

FIGURE 4.14

Change from unsteady to steady flow by change of the velocity of the control volume. (a) Unsteady flow. (b) Steady flow.

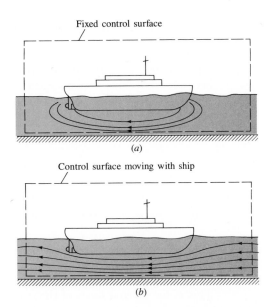

caution is necessary. If the body itself (and also the control surface) is accelerating and if the control-volume equation is applied to the momentum principle, then it is necessary to reference the velocity to the inertial frame rather than to the control surface itself. Because Eqs. (4.21) are involved with the rate of change of a flow property with time, it should not be surprising that this basic equation can serve as the starting point for many expressions involving the dynamics of a flow as well as for the continuity equation, which is simply kinematic in nature. In the next section we shall see how the control-volume approach can be applied to the continuity principle.

4.5   Continuity Equation

The Basic Concept

The continuity principle is based on the conservation of mass as it applies to the flow of fluids. In words, the continuity equation states that the mass rate of flow out of a region of space, such as a control volume, minus the mass rate of flow into the region is equal to the rate at which the fluid mass is being evacuated from the region. When the intensive property β of the basic control-volume equation is set equal to unity, the equation reduces to a general form of the continuity equation. This general equation can then be applied to different types of control volumes and different coordinate systems to yield special forms of the continuity equation. The general case and several special forms of the continuity equation will be considered in the paragraphs that follow.

General Form of the Continuity Equation

First we shall write the general control-volume equation with B = mass of the system, which means that β is equal to unity ($\beta = 1$). Then Eq. (4.21b) becomes

$$\frac{d(\text{mass})}{dt} = \frac{d}{dt} \int_{cv} \rho \, d\Psi + \int_{cs} \rho \mathbf{V} \cdot d\mathbf{A} \tag{4.23}$$

This is sometimes called the integral form of the continuity equation. The term on the left is the rate of change of the mass of the system. However, by definition, the mass of a system is constant. Therefore the left-hand side of the equation is zero, and the equation can be written as

$$\int_{cs} \rho \mathbf{V} \cdot d\mathbf{A} = -\frac{d}{dt} \int_{cv} \rho \, d\Psi \tag{4.24}$$

This is the general form of the continuity equation. It states that the net rate of outflow of mass from the control volume is equal to the rate of decrease of mass within the control volume.

The continuity equation involving flow streams having a uniform velocity across the flow sections is given as

$$\sum_{cs} \rho \mathbf{V} \cdot \mathbf{A} = -\frac{d}{dt} \int_{cv} \rho \, d\Psi \tag{4.25}$$

The following example illustrates the use of the continuity equation.

EXAMPLE 4.5   A jet of water discharges into an open tank, and water leaves the tank through an orifice in the bottom at a rate of 0.003 m³/s. If the cross-sectional area of the jet is 0.0025 m² where the velocity of water is 7 m/s, at what rate is water accumulating in (or evacuating from) the tank?

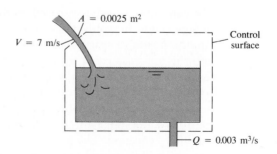

Solution   By drawing the control surface to enclose the entire tank, we can see from the control-volume equation that the mass rate at which the tank is emptying ($-d/dt \int \rho \, d\Psi$) is equal to $\Sigma_{cs} \rho \mathbf{V} \cdot \mathbf{A}$. Thus

$$\text{Emptying rate} = \sum_{cs} \rho \mathbf{V} \cdot \mathbf{A}$$

$$\text{Net rate of mass outflow} = \sum_{cs} \rho \mathbf{V} \cdot \mathbf{A}$$

$$= \rho Q_{\text{outflow}} + \rho \mathbf{V}_1 \cdot \mathbf{A}_1$$

Positive because of outflow

$$= (1000)(0.003) + 1000(-7.0 \times 0.0025)$$

Negative because area vector
opposes velocity vector

$$= 3.00 - 17.5 \text{ kg/s}$$

$$\text{Net rate of outflow} = -14.5 \text{ kg/s}$$

Since the net rate of outflow is -14.5 kg/s, we can also say the tank is filling at 14.5 kg/s. Or, if we want the volume rate at which the water is filling the tank, we divide by ρ to obtain a filling rate of

$$Q = 0.0145 \text{ m}^3/\text{s} \qquad \blacktriangleleft$$

The foregoing example is an application of the control-volume approach for a case of unsteady flow. In many problems, however, the flow is steady, and we are left with the equation

$$\int_{cs} \rho \mathbf{V} \cdot d\mathbf{A} = 0$$

The following example shows how this can be applied.

EXAMPLE 4.6   As shown in the accompanying figure, water flows steadily into a tank through pipes 1 and 2 and discharges at a steady rate out of the tank through pipes 3 and 4. The mean velocity of inflow and outflow in pipes 1, 2, and 3 is 50 ft/s, and the hypothetical outflow velocity in pipe 4 varies linearly from zero at the wall to a maximum at the center of the pipe. What are the mass rate of flow and the discharge from pipe 4, and what is the maximum velocity in pipe 4?

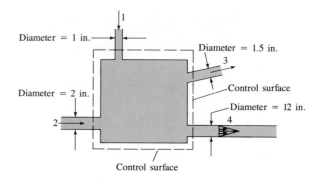

Solution   Since this is steady flow, we use

$$\int_{cs} \rho \mathbf{V} \cdot d\mathbf{A} = 0$$

$$-\rho_1 V_1 A_1 - \rho_2 V_2 A_2 + \rho_3 V_3 A_3 + \int_{A_4} \rho_4 V_4 dA_4 = 0$$

First let us solve for the mass rate of flow from pipe 4.

$$\dot{m}_4 = \int_{A_4} \rho_4 V_4 dA_4 = \rho_1 V_1 A_1 + \rho_2 V_2 A_2 - \rho_3 V_3 A_3$$

For water $\rho = 1.94$ slugs/ft^3. Therefore,

$$\dot{m}_4 = \frac{1.94 \times 50}{144} \frac{\pi}{4} (1 + 4 - 2.25) = 1.45 \text{ slugs/s} \qquad \blacktriangleleft$$

and $$Q_4 = \dot{m}_4/\rho = 0.750 \text{ ft}^3/\text{s} \qquad \blacktriangleleft$$

Now we can solve for V_{max} in pipe 4. The velocity is linearly distributed from zero at the wall to maximum at the center as shown below:

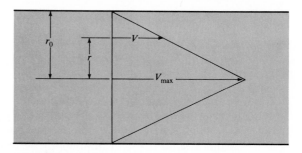

Therefore, by proportions we have

$$\frac{V}{r_0 - r} = \frac{V_{max}}{r_0} \qquad \text{or} \qquad V = V_{max} \left(1 - \frac{r}{r_0} \right)$$

Also
$$Q = \int_A V \, dA = \int_0^{r_0} V_{max}\left(1 - \frac{r}{r_0}\right) 2\pi r \, dr$$

$$= 2\pi V_{max} \int_0^{r_0}\left(1 - \frac{r}{r_0}\right) r \, dr = 2\pi V_{max} r_0^2 (\tfrac{1}{2} - \tfrac{1}{3}) = \tfrac{1}{3}\pi r_0^2 V_{max}$$

Thus
$$V_{max} = \frac{Q}{\tfrac{1}{3}\pi r_0^2} = \frac{0.75 \text{ ft}^3/\text{s}}{\tfrac{1}{3}\pi \times (\tfrac{1}{2})^2 \text{ ft}^2} = 2.86 \text{ ft/s}$$ ◀

Continuity Equation for Steady One-Dimensional Flow in a Conduit

Consider the steady-flow case where we have flow in a conduit, Fig. 4.15, in which we are interested in establishing a relationship between the mean velocities at two sections of the conduit. A control volume is drawn such that its two ends are the two sections 1 and 2. For flow through this control volume, we apply the continuity equation

$$\sum \rho \mathbf{V} \cdot \mathbf{A} = 0$$

When this is expanded for the case under consideration, we obtain

$$-\rho_1 V_1 A_1 + \rho_2 V_2 A_2 = 0$$
$$\rho_1 V_1 A_1 = \rho_2 V_2 A_2 \qquad (4.26)$$

Furthermore, if we have constant-density flow, the ρ's cancel, leaving

$$V_1 A_1 = V_2 A_2 \qquad (4.27)$$

This equation states that the volume rate of flow at section 1 is equal to the volume rate of flow at section 2. Therefore, it can also be written as

$$Q_1 = Q_2$$

Equation (4.27) is a very common form of the continuity equation and is used in numerous applications where the flow can be considered one dimensional and incompressible.

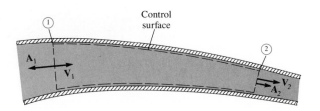

FIGURE 4.15

EXAMPLE 4.7   A 120-cm pipe is in series with a 60-cm pipe. The rate of flow of water in the system of pipes is 2 m³/s. What is the velocity of flow in each pipe?

Solution
$$Q = V_{120}A_{120}$$

Here
$$Q = 2 \text{ m}^3/\text{s} \quad \text{and} \quad A_{120} = \frac{\pi}{4} \times (1.20)^2 = 1.13 \text{ m}^2$$

Therefore
$$V_{120} = \frac{Q}{A_{120}} = \frac{2 \text{ m}^3/\text{s}}{1.13 \text{ m}^2} = 1.77 \text{ m/s} \quad \blacktriangleleft$$

Also
$$V_{120}A_{120} = V_{60}A_{60}$$

So
$$V_{60} = V_{120}\left(\frac{A_{120}}{A_{60}}\right) = V_{120}\left(\frac{120}{60}\right)^2 = 7.08 \text{ m/s} \quad \blacktriangleleft$$

EXAMPLE 4.8   The river discharges into the reservoir shown at a rate of 400,000 ft³/s (cfs), and the outflow rate from the reservoir through the flow passages in the dam is 250,000 cfs. If the reservoir surface area is 40 mi², what is the rate of rise of water in the reservoir?

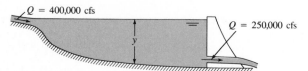

Solution   Apply the continuity equation to the control volume, as shown in the figure. Here flow past the control surface is occurring at three sections, 1, 2, and 3. The continuity equation is

$$\sum_{\text{cs}} \rho \mathbf{V} \cdot \mathbf{A} = 0$$

and because ρ is constant,

$$\sum_{\text{cs}} \mathbf{V} \cdot \mathbf{A} = 0$$

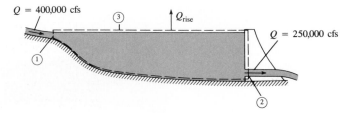

Applied to this example, the foregoing equation becomes

$$-400{,}000 \text{ ft}^3/\text{s} + 250{,}000 \text{ ft}^3/\text{s} + Q_{\text{rise}} = 0$$

or
$$V_{\text{rise}} A_R = 150{,}000 \text{ ft}^3/\text{s}$$

Here A_R is the area of the reservoir. Then we have

$$V_{\text{rise}} = \frac{150{,}000 \text{ ft}^3/\text{s}}{40 \text{ mi}^2 \times (5280)^2 \text{ ft}^2/\text{mi}^2}$$

$$\text{Rate of rise} = 1.34 \times 10^{-4} \text{ ft/s}$$

or
$$V_{\text{rise}} = 0.484 \text{ ft/hr} \qquad \blacktriangleleft$$

EXAMPLE 4.9 A 10-cm jet of water issues from a 1-m diameter tank, as shown. Assume that the velocity in the jet is $\sqrt{2gh}$ m/s. How long will it take for the water surface in the tank to drop from $h_0 = 2$ m to $h_f = 0.50$ m?

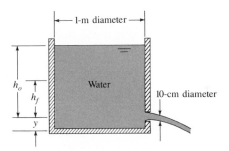

Solution In Example 4.8 we arbitrarily established a fixed control volume. In this example, however, we will approach the problem differently and let the control surface surround the volume of liquid within the tank at all times, so that the control volume will decrease in size as time passes. In other words, for this example, the control surface coincident with the liquid surface in the tank will drop right along with the liquid surface. Water is discharged from the control volume at section 1. We write the continuity equation as

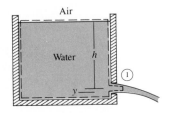

$$\sum_{\text{cs}} \rho \mathbf{V} \cdot \mathbf{A} = -\frac{d}{dt} \int_{\text{cv}} \rho \, d\Psi$$

$$\rho_1 V_1 A_1 = -\frac{d}{dt} \int_{\text{cv}} \rho \, d\Psi$$

The term $\int_{\text{cv}} \rho \, d\Psi$ represents the total mass of water in the control volume, so we write it as $\rho A_T(h + y)$. The continuity equation then becomes

$$\rho_1 V_1 A_1 = -\frac{d}{dt}[\rho A_T(h + y)]$$

where $\rho_1 = \rho = \rho_{\text{water}}$, A_T is the cross-sectional area of tank, h is the depth of water in tank above outlet, and y is the depth of water in tank below outlet. Since ρ in the foregoing equation is constant, we cancel it out, and when the differentiation is carried out on the right-hand side (remember that A_T and y are constant), the equation reduces to

$$V_1 A_1 = -A_T \frac{dh}{dt}$$

It was already noted that $V_1 = \sqrt{2gh}$, so substitution for V_1 yields

$$\sqrt{2gh}\, A_1 = -A_T \frac{dh}{dt}$$

Separating variables, we obtain

$$dt = \frac{-A_T}{\sqrt{2g}\, A_1} \frac{dh}{\sqrt{h}} \qquad \text{or} \qquad dt = \frac{-A_T}{\sqrt{2g}\, A_1} h^{-1/2} \, dh$$

Noting now that $A_T/\sqrt{2g}\, A_1$ is constant, we integrate the differential equation and get

$$t = \frac{-2A_T}{\sqrt{2g}\, A_1} h^{1/2} + C$$

The constant of integration is evaluated by arbitrarily letting $t = 0$ when $h = h_0$. Then

$$C = +\frac{2A_T}{\sqrt{2g}\, A_1} h_0^{1/2}$$

So we have

$$t = \frac{2A_T}{\sqrt{2g}\, A_1} (h_0^{1/2} - h^{1/2})$$

Thus, for this particular example, the elapsed time for the water level to drop from $h_0 = 2$ m to $h_f = 0.50$ m will be

$$t = \frac{2A_T}{\sqrt{2g}\,A_1}(2^{1/2} - 0.5^{1/2})$$

But

$$A_T = \frac{\pi}{4}D^2 = \frac{\pi}{4} \times 1^2 = \frac{\pi}{4}\ \text{m}^2$$

$$A_1 = \frac{\pi}{4}(0.10)^2 = 0.01\left(\frac{\pi}{4}\right)\ \text{m}^2$$

Hence

$$t = \frac{2\pi/4}{\sqrt{2g}\,(\pi/4 \times 0.01)}(1.414 - 0.707) = 31.9\ \text{s} \qquad \blacktriangleleft$$

EXAMPLE 4.10   Methane escapes through a small (10^{-7} m^2) hole in a 10-m^3 tank. The methane escapes so slowly that the temperature in the tank remains constant at 23°C. The mass flow rate of methane through the hole is given by $\dot{m} = 0.66pA/\sqrt{RT}$, where p is the pressure in the tank, A is the area of the hole, R is the gas constant, and T is the temperature in the tank. Calculate the time required for the absolute pressure in the tank to decrease from 500 to 400 kPa.

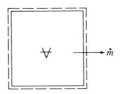

Solution   Consider a control surface that just encloses the tank. Writing the continuity equation for the control volume, we have

$$\frac{d}{dt}\int_{cv}\rho\,d\Psi + \sum \rho\mathbf{V}\cdot\mathbf{A} = 0$$

Since ρ is uniform throughout the tank and mass crosses the control surface only at the location of the leak,

$$\frac{d}{dt}(\rho\Psi) + \dot{m} = 0$$

The tank volume is constant, so

$$d\rho/dt = -\dot{m}/\Psi$$

The ideal-gas law provides the following relationship between pressure and density: $\rho = p/RT$. Since T is constant, $d\rho = dp/RT$. Substituting this and the mass-flow expression into the equation above gives

$$\frac{dp}{p} = -0.66\frac{A\sqrt{RT}\,dt}{\Psi}$$

Integrating, we obtain

$$t = \frac{1.52\Psi}{A\sqrt{RT}}\ln\frac{p_0}{p}$$

where p_0 is the initial pressure. Substituting in the appropriate values, we calculate

$$t = \frac{1.52(10\text{ m}^3)}{(10^{-7}\text{ m}^2)\left(518\dfrac{\text{J}}{\text{kg}\cdot\text{K}}300\text{ K}\right)^{1/2}}\ln\frac{500}{400} = 8.6\times10^4\text{ s}$$

or approximately 1 day.   ◀

Continuity at a Point

In the analysis of fluid flow, one of the fundamental independent equations needed for the solution of problems is a differential equation that is a statement of continuity at a point. This is derived by applying the basic continuity equation, Eq. (4.25), to a control volume of infinitesimal size for which flow is assumed to pass through all surfaces of the element of volume. Consider Fig. 4.16, which shows a cubical element oriented so that its sides are normal to the x, y, and z coordinate directions.

The general formulation of the continuity equation is

$$\sum \rho\mathbf{V}\cdot\mathbf{A} = -\frac{d}{dt}\int_{\text{cv}}\rho\,d\Psi$$

where $\mathbf{V}$ is measured with respect to the local control surface. Applying the Leibnitz theorem for differentiation of an integral allows the unsteady term to be written as

$$\frac{d}{dt}\int_{\text{cv}}\rho\,d\Psi = \int_{\text{cv}}\frac{\partial\rho}{\partial t}\,d\Psi + \sum\rho\mathbf{V}_s\cdot\mathbf{A}$$

FIGURE 4.16

*Continuity for elemental
control volume.*

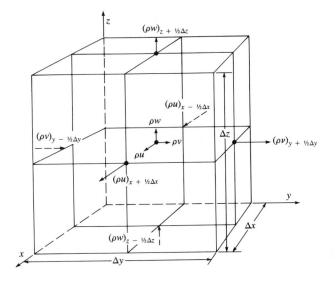

where $\mathbf{V}_s$ is the local velocity of the control surface with respect to the reference frame. For the cubical element with stationary sides (control surfaces) shown in Fig. 4.16, the velocity $\mathbf{V}_s = 0$, so the continuity equation can be written as

$$\sum \rho \mathbf{V} \cdot \mathbf{A} = -\int_{cv} \frac{\partial \rho}{\partial t} d\mathbb{V}$$

Considering the flow rates through the six faces of the cubical element and applying those to the foregoing form of the continuity equation, we obtain

$$[(\rho u)_{x+(1/2)\Delta x}] \Delta y \, \Delta z - [(\rho u)_{x-(1/2)\Delta x}] \Delta y \, \Delta z$$
$$+ [(\rho v)_{y+(1/2)\Delta y}] \Delta z \, \Delta x - [(\rho v)_{y-(1/2)\Delta y}] \Delta z \, \Delta x$$
$$+ [(\rho w)_{z+(1/2)\Delta z}] \Delta x \, \Delta y - [(\rho w)_{z-(1/2)\Delta z}] \Delta x \, \Delta y$$
$$+ \frac{\partial \rho}{\partial t} \Delta x \, \Delta y \, \Delta z = 0 \tag{4.28}$$

Dividing Eq. (4.28) by the volume of the element ($\Delta x \, \Delta y \, \Delta z$) yields

$$\frac{[(\rho u)_{x+(1/2)\Delta x}] - [(\rho u)_{x-(1/2)\Delta x}]}{\Delta x}$$

$$+ \frac{[(\rho v)_{y+(1/2)\Delta y}] - [(\rho v)_{y-(1/2)\Delta y}]}{\Delta y}$$

$$+ \frac{[(\rho w)_{z+(1/2)\Delta z}] - [(\rho w)_{z-(1/2)\Delta z}]}{\Delta z}$$

$$+ \frac{\partial \rho}{\partial t} = 0$$

Taking the limit as the volume approaches zero (that is, as Δx, Δy, and Δz uniformly approach zero) yields the differential form of the continuity equation, Eq. (4.29).

$$\frac{\partial}{\partial x}(\rho u) + \frac{\partial}{\partial y}(\rho v) + \frac{\partial}{\partial z}(\rho w) = -\frac{\partial \rho}{\partial t} \qquad (4.29)$$

If the flow is steady, we obtain

$$\frac{\partial}{\partial x}(\rho u) + \frac{\partial}{\partial y}(\rho v) + \frac{\partial}{\partial z}(\rho w) = 0 \qquad (4.30)$$

And if the fluid is incompressible, we have

$$\frac{\partial u}{\partial x} + \frac{\partial v}{\partial y} + \frac{\partial w}{\partial z} = 0 \qquad (4.31a)$$

for both steady and unsteady flow.

In vector notation Eq. (4.31a) is given as

$$\nabla \cdot \mathbf{V} = 0 \qquad (4.31b)$$

where ∇ is the del operator, defined as

$$\nabla = \frac{\partial}{\partial x}\mathbf{i} + \frac{\partial}{\partial y}\mathbf{j} + \frac{\partial}{\partial z}\mathbf{k}$$

EXAMPLE 4.11   The expression $\mathbf{V} = 10x\mathbf{i} - 10y\mathbf{j}$ is said to represent the velocity for a two-dimensional incompressible flow. Check it to see whether it satisfies continuity.

Solution

$$u = 10x \qquad \text{so} \qquad \frac{\partial u}{\partial x} = 10$$

$$v = -10y \qquad \text{so} \qquad \frac{\partial v}{\partial y} = -10$$

$$\frac{\partial u}{\partial x} + \frac{\partial v}{\partial y} = 10 - 10 = 0$$

Continuity is satisfied.   ◄

4.6 Rotation and Vorticity

Concept of Rotation

Consider a tank of liquid that is being rotated about a vertical axis. A plan view of such a tank is given in Fig. 4.17. If we focus on a given element, it can be seen that this element will rotate but not deform as time passes. In this process,

FIGURE 4.17

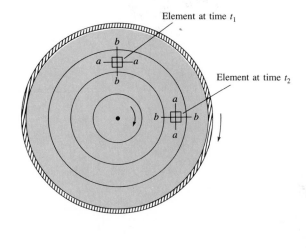

Element at time t_1

Element at time t_2

all lines drawn through the element, such as *a-a* and *b-b* in Figure 4.17, will rotate at the same rate. This is unquestionably a case of fluid rotation. Now consider fluid flow between two horizontal plates, Fig. 4.18, where the bottom plate is stationary and the top is moving to the right with a velocity *V.* The velocity distribution is linear; therefore, an element of fluid will deform as shown. Here we see that the element faces that were initially vertical rotate clockwise, whereas the horizontal faces do not. It is not so clear whether this is a case of rotational motion or not. The test is to determine the average rate of rotation of two initially mutually perpendicular lines that follow the motion of the faces of the fluid element. If the average rate of rotation of these lines is zero, then the flow is irrotational; if it is not zero, then by definition rotational flow exists. Thus the flow between the parallel boundaries is also a case of rotational flow. Now we shall derive an expression that indicates mathematically the degree of rotation in terms of velocity gradients of flow.

Consider a cubical element of fluid, one face of which is shown in Fig. 4.19. For this element we shall derive an expression that indicates the average rate of rotation of lines *AB* and *AC*, which are mutually perpendicular at time *t.* Here rotation is about the *z* axis. Therefore, this rotation is given as

$$\omega_z = \frac{\omega_{AB} + \omega_{AC}}{2} \tag{4.32}$$

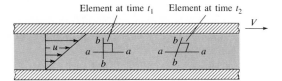

Element at time t_1 Element at time t_2

V

FIGURE 4.18

FIGURE 4.19

*Deformation of an element
of fluid.*

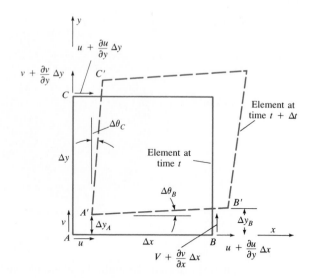

The rate of rotation of AB may be given as the rate of change of θ_B, which can be expressed as a function of the velocity distribution in the following manner:

$$\omega_{AB} = \lim_{\Delta t \to 0} \frac{\Delta \theta_B}{\Delta t}$$

Referring again to Fig. 4.19 above, we can see that $\Delta\theta_B$ may be expressed as $(\Delta y_B - \Delta y_A)/\Delta x$, where $\Delta y_B - \Delta y_A$ is given as

$$\Delta y_B - \Delta y_A = \left(v + \frac{\partial v}{\partial x}\Delta x - v \right) \Delta t$$

$$\Delta \theta_B = \frac{\partial v}{\partial x} \frac{\Delta x \, \Delta t}{\Delta x}$$

Then
$$\omega_{AB} = \lim_{\Delta t \to 0} \frac{(\partial v/\partial x)\,\Delta x\,\Delta t}{\Delta x\,\Delta t} = \frac{\partial v}{\partial x} \qquad (4.33)$$

In a similar manner it can be shown that the negative rate of rotation, clockwise, of line AC is given as

$$-\omega_{AC} = \frac{\partial u}{\partial y}$$

or
$$\omega_{AC} = -\frac{\partial u}{\partial y} \qquad (4.34)$$

When Eqs. (4.33) and (4.34) are substituted into Eq. (4.32), we obtain

$$\omega_z = \frac{1}{2}\left(\frac{\partial v}{\partial x} - \frac{\partial u}{\partial y} \right)$$

Likewise, the rates of rotation about the x and y axes are found to be

$$\omega_x = \frac{1}{2}\left(\frac{\partial w}{\partial y} - \frac{\partial v}{\partial z}\right) \quad \text{and} \quad \omega_y = \frac{1}{2}\left(\frac{\partial u}{\partial z} - \frac{\partial w}{\partial x}\right)$$

Vorticity

The vorticity is defined as twice the average rate of rotation. Consequently, the vorticity vector may be written in the following form:

$$\mathbf{\Omega} = \left(\frac{\partial w}{\partial y} - \frac{\partial v}{\partial z}\right)\mathbf{i} + \left(\frac{\partial u}{\partial z} - \frac{\partial w}{\partial x}\right)\mathbf{j} + \left(\frac{\partial v}{\partial x} - \frac{\partial u}{\partial y}\right)\mathbf{k} \qquad (4.35)$$

Remember that irrotational flow exists only when the average rates of rotation are zero; thus for irrotational flow each term inside the parentheses of Eq. (4.35) must have a zero value. It follows that for irrotational flow,

$$\frac{\partial w}{\partial y} = \frac{\partial v}{\partial z} \qquad (4.36)$$

$$\frac{\partial u}{\partial z} = \frac{\partial w}{\partial x} \qquad (4.37)$$

$$\frac{\partial v}{\partial x} = \frac{\partial u}{\partial y} \qquad (4.38)$$

The most extensive application of these equations is in the area of ideal-flow theory.

EXAMPLE 4.12   Is the flow in Example 4.11 irrotational?

Solution $u = 10x \qquad$ so $\qquad \dfrac{\partial u}{\partial y} = 0$

$\qquad\qquad\qquad\qquad\; v = -10y \qquad$ so $\qquad \dfrac{\partial v}{\partial x} = 0$

Then $\dfrac{\partial v}{\partial x} = \dfrac{\partial u}{\partial y}$

The flow is irrotational.   ◄

Vortices

A flow for which the streamlines are concentric circles is called a *vortex.* Fluid dynamicists have identified two types of vortices that can be easily described mathematically. One is called the *forced vortex,* in which the velocity increases linearly from the center of rotation, as described at the beginning of this section and depicted in Fig. 4.17. The other type of vortex is the *free,* or *potential, vortex.* In the free vortex, the product of the velocity at a point and the radial distance from the vortex center to that point is a constant ($Vr = C$). Thus, for the free vortex, the velocity increases toward the vortex center rather than decreases as it does in the forced-vortex case. Vortices in the real world (for example, tornadoes and whirlpools in rivers) are often a combination of free and forced vortex. Forced-vortex flow occurs at and near the center of the vortex, whereas free-vortex conditions are approximated outside this region. Figure 4.20 depicts a velocity distribution for such a vortex. It can be shown that flow for a *free* vortex is irrotational.

FIGURE 4.20

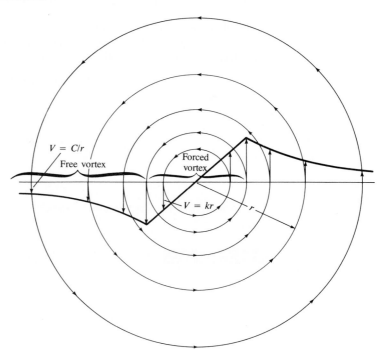

4.7   Separation, Vortices, and Turbulence

Flow patterns that are developed by ideal-flow theory, as already indicated, are for irrotational flow. In some applications irrotational flow is closely approximated; for example, flow of air or water in a region where the streamlines are converging usually approximates irrotational flow quite closely. However, in regions where boundaries turn away from the flow so as to cause the streamlines to diverge, the flow usually "separates" from the boundary and a recirculation pattern is generated in the region. This phenomenon is called *separation,* a typical case of which is shown schematically in Fig. 4.21*b*. Also shown, in Fig. 4.21*a*, is the pattern for ideal flow past a similar plate. The two flow patterns are markedly different, especially downstream of the plate. Therefore, the engineer must be cautious about applying ideal-flow theory to engineering problems. For some problems it can be very useful, but for others it gives erroneous results. The cause of separation and its effects are explained in the next chapter (Sec. 5.5). In the region between the high-velocity flow outside the zone of separation and the low-velocity zone inside it (see Fig. 4.21*b*), vortices are formed. These vortices are often called eddies. In rivers the larger eddies are often called whirlpools. These vortices or eddies lead to the phenomenon called turbulence. For the case of flow of water in rivers or wind in the atmosphere, the flow is almost always turbulent. The eddies that are initially developed are relatively large, but in the process of turbulent mixing, they break down into smaller and smaller eddies. Viscous resistance in the smallest eddies eventually dissipates virtually all of the kinetic energy that initially existed in the larger eddies. This process of vortex generation and decay is typical of all turbulent flows and is one of the most significant aspects of fluid mechanics. Research in this area will be especially challenging and rewarding for engineers and scientists for years to come. Only in recent years have advanced computational procedures and high-speed computers led to the development of good approximations for predicting the velocity and energy distribution for certain turbulent-flow problems. For additional information on vortices, see Lugt (3) and Hassaini and Salas (1).

In the foregoing discussion, separation induced by a solid boundary was the phenomenon that triggered the development of vortices; however, there are other processes that generate vortices. For example, the Coriolis effect associated with low-pressure storm centers in the atmosphere develop vortices (cyclonic storms) having diameters measured in hundreds of miles. Large-scale vortices also develop when a river discharges into a bay or ocean and when a smoke stack discharges its effluent into the atmosphere. But the fate of even these large-scale

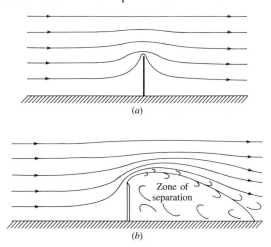

FIGURE 4.21

Flow past a plate. (a) Ideal flow past a plate. (b) Real flow past a plate.

(a)

Zone of separation

(b)

vortices is always the same; they eventually end up as very small eddies (of the order of 1 mm or less in diameter), and then they finally lose all kinetic energy by viscous action. See references (2), (4), (5), and (8) for a discussion of the size of the smallest eddies and other basic information about turbulence.

Problems

4.1 In the system in the figure, the valve at C is gradually opened in such a way that a constant rate of increase in discharge is produced. How would you classify the flow at B while the valve is being opened? How would you classify the flow at A?

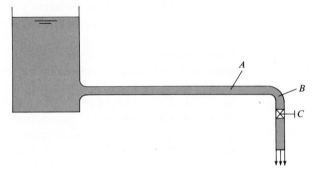

PROBLEM 4.1

4.2 a. Water flows in the passage shown. If the flow rate is decreasing with time, the flow is classified as a) steady, b) unsteady, c) uniform, d) nonuniform. b. In the flow there is a) local acceleration, b) convective acceleration, c) no acceleration.

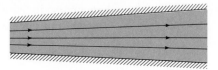

PROBLEM 4.2

4.3 If a flow pattern has converging streamlines, how would you classify the flow?

4.4 Consider flow in a straight-fixed conduit. The conduit is circular in cross section. Part of the conduit has a constant diameter, and part has a diameter that changes with distance. Then, relative to flow in that conduit, correctly match the items in column A with those in column B.

A	B
Steady flow	$V_s \, \partial V_s / \partial s = 0$
Unsteady flow	$V_s \, \partial V_s / \partial s \neq 0$
Uniform flow	$\partial V_s / \partial t = 0$
Nonuniform flow	$\partial V_s / \partial t \neq 0$

4.5   Refer to the flow as indicated by the pathlines in Fig. 4.5. Assume that the discharge is constant and the flow is nonturbulent. Which of the following statement(s) are true? a) There is convective acceleration in the flow. b) There is local acceleration in the flow. c) The pathlines are also streamlines. d) There is separation in this flow field. If you identified statement d) as true, indicate where separation is occurring.

4.6   At time $t = 0$, dye was injected at point A in a flow field of a liquid. When the dye had been injected for 4 s, a pathline for a particle of dye that was emitted at the 4-s instant was started. The streakline at the end of 10 s is shown below. Assume that the speed (but not the velocity) of flow is the same throughout the 10-s period. Draw the pathline of the particle that was emitted at $t = 4$ s. Make your own assumptions for any missing information.

PROBLEM 4.6

4.7   For a given hypothetical flow, the velocity from time $t = 0$ to $t = 10$ s was $u = 2$ m/s, $v = 0$. Then, from time $t = 10$ s to $t = 15$ s, the velocity was $u = +3$ m/s, $v = -4$ m/s. A dye streak was started at a point in the flow field at time $t = 0$, and the path of a particle in the fluid was also traced from that same point starting at the same time. Draw to scale the streakline, pathline of the particle, and streamlines at time $t = 15$ s.

4.8   At time $t = 0$, a dye streak was started at point A in a flow field of liquid. The speed of the flow is constant over a 10-s period, but the flow direction is not necessarily constant. At any particular instant the velocity in the entire field of flow is the same. The streakline produced by the dye is shown below. Draw (and label) a streamline for the flow field at $t = 8$ s.

Draw (and label) a pathline that one would see at $t = 10$ s for a particle of dye that was emitted from point A at $t = 2$ s.

PROBLEM 4.8

4.9   The figure shows the pathline of a fluid particle that was released from point A at time $t = 0$. Also, the sketch shows the streakline that was developed by dye that was emitted from point A over the time period from $t = 0$ to $t = 5$ s. For the flow associated with this pathline and streakline, carefully sketch a streamline for $t = 0$.

For the flow associated with this pathline and streakline, one can conclude that the flow is a) steady, b) unsteady.

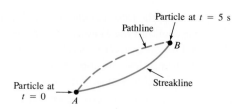

PROBLEM 4.9

4.10 For a given fluid the velocity for the entire field of flow is $u = 5$ m/s, $v = -2t$ m/s. Here t is the time in seconds. A particle is released at time $t = 0$ and a dye stream is started from the same point at the same time. Draw the streakline, the pathline of the particle, and the streamlines at time $t = 5$ s.

4.11 A flow has circular streamlines as shown. At time $t = 0$ the flow is in the clockwise direction and the angular velocity is π rad/s. This flow exists for 1 s, then reverses to the counterclockwise direction with an angular velocity of $-\pi$ rad/s and exists for 1 s. For the 2 s of flow, draw the pathline for a particle of fluid that started at the point shown at time $t = 0$. Also draw the streakline for a dye stream that was emitted from the same point for the 2-s period. Carefully identify by labels the pathline and streakline.

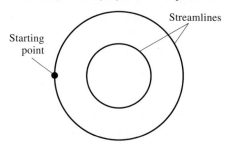

PROBLEM 4.11

4.12 Classify each of the following as a one-dimensional, two-dimensional, or three-dimensional flow.

   a. Water flow over the crest of a long spillway of a dam.

   b. Flow in a straight horizontal pipe.

   c. Flow in a constant-diameter pipeline that follows the contour of the ground in hilly country.

   d. Air flow from a slit in a plate at the end of a large rectangular duct.

   e. Air flow past an automobile.

   f. Air flow past a house.

   g. Water flow past a pipe that is laid normal to the flow across the bottom of a wide rectangular channel.

4.13 The discharge of water in a 30-cm pipe is 0.50 m³/s. What is the mean velocity?

4.14 A pipe with a 24-in. diameter carries water having a velocity of 3 ft/s. What is the discharge in cubic feet per second and in gallons per minute (1 cfs is equivalent to 449 gpm)?

4.15   A pipe with a 2-m diameter carries water having a velocity of 8 m/s. What is the discharge in cubic meters per second and in cubic feet per second?

4.16   A pipe whose diameter is 8 cm transports air with a temperature of 30°C and pressure of 200 kPa absolute at 20 m/s. Determine the mass-flow rate.

4.17   Natural gas (methane) flows at 10 m/s through a pipe with a 1-m diameter. The temperature of the methane is 15°C, and the pressure is 150 kPa gage. Determine the mass-flow rate.

4.18   An aircraft engine test pipe is capable of providing a flow rate of 200 kg/s at altitude conditions corresponding to an absolute pressure of 50 kPa and a temperature of $-18°C$. The velocity of air through the duct attached to the engine is 240 m/s. Calculate the diameter of the duct.

4.19   A heating and air-conditioning engineer is designing a system to move 1000 m³ of air per hour at 100 kPa and 30°C. The duct is rectangular with cross-sectional dimensions of 1 m by 20 cm. What will be the air velocity in the duct?

4.20   The hypothetical velocity distribution in a circular duct is $v/V_0 = 1 - r/R$, where r is the radial location in the duct, R is the duct radius, and V_0 is the velocity on the axis. Find the ratio of the mean velocity to the velocity on the axis.

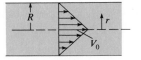

PROBLEM 4.20

4.21   Water flows in a two-dimensional channel of width W and depth D as shown in the diagram. The hypothetical velocity profile for the water is

$$V(x,y) = V_s\left(1 - \frac{4x^2}{W^2}\right)\left(1 - \frac{y^2}{D^2}\right)$$

where V_s is the velocity at the water surface midway between the channel walls. The coordinate system is as shown; x is measured from the center plane of the channel and y downward from the water surface. Find the discharge in the channel in terms of V_s, D, and W.

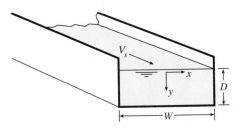

PROBLEM 4.21

4.22 Water flows in a pipe that has a 4-ft diameter and the following hypothetical velocity distribution: The velocity is maximum at the centerline and decreases linearly with r to a minimum at the pipe wall. If $V_{max} = 15$ ft/s and $V_{min} = 12$ ft/s, what is the discharge in cubic feet per second and in gallons per minute?

4.23 In Prob. 4.22, if $V_{max} = 8$ m/s, $V_{min} = 5$ m/s, and $D = 2$ m, what is the discharge in cubic meters per second and the mean velocity?

4.24 Air enters this square duct at section 1 with the velocity distribution as shown. Note that the velocity varies in the y direction only (for a given value of y, the velocity is the same for all values of z).

    a. What is the volume rate of flow?

    b. What is the mean velocity in the duct?

    c. What is the mass rate of flow if the mass density of the air is 1.3 kg/m³?

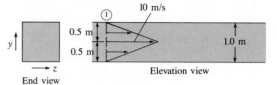

PROBLEM 4.24

4.25 The velocity at section A–A is 15 ft/s, and the vertical depth y at the same section is 4 ft. If the width of the channel is 25 ft, what is the discharge in cubic feet per second?

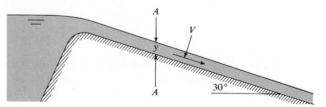

PROBLEM 4.25

4.26 The rectangular channel shown is 3 m wide. What is the discharge in the channel?

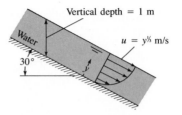

PROBLEM 4.26

4.27 If the velocity in the channel of Prob. 4.26 is given as $u = 10[\exp(y) - 1]$ m/s, what is the discharge in the channel and what is the mean velocity?

4.28 Water from a pipe is diverted into a weigh tank for exactly 13 min. The increased weight in the tank is 20 kN. What is the discharge in cubic meters per second? Assume $T = 20°C$.

4.29 Water enters the lock of a ship canal through 200 ports, each having an outlet that is 2 ft by 2 ft. The lock is 900 ft long and 85 ft wide. The lock is designed so that the water surface in it will rise at a maximum rate of 6 ft/min. For this condition, what will be the mean velocity of efflux from the ports?

4.30 An empirical equation for the velocity distribution in a horizontal, rectangular, open channel is given by $u = u_{max}(y/d)^n$, where u is the velocity at a distance y feet above the floor of the channel. If the depth d of flow is 1.2 m, $u_{max} = 3$ m/s, and $n = 1/6$, what is the discharge in cubic meters per second per meter of width of channel? What is the mean velocity?

4.31 The hypothetical water velocity in a V-shaped channel (see the accompanying figure) varies linearly with depth from zero at the bottom to maximum at the water surface. Determine the discharge if the maximum velocity is 6 ft/s.

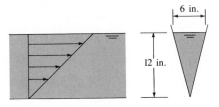

6 in.

12 in.

PROBLEM 4.31

4.32 The velocity of flow in a circular pipe varies according to the equation $V/V_C = (1 - r^2/r_0^2)^n$, where V_C is the centerline velocity, r_0 is the pipe radius, and r is the radial distance from the centerline. The exponent n is general and is chosen to fit a given profile ($n = 1$ for laminar flow). Determine the mean velocity as a function of V_C and n.

4.33 Plot the velocity distribution across the pipe, and determine the discharge of a fluid flowing through a pipe 1 m in diameter that has a velocity distribution given by $V = 12(1 - r^2/r_0^2)$ m/s. Here r_0 is the radius of the pipe and r is the radial distance from the centerline. What is the mean velocity?

4.34 Water flows through a 2-in. pipeline at 200 lbm/min. Calculate the mean velocity. Assume $T = 60°F$.

4.35 Water flows through a 20-cm pipeline at 1000 kg/min. Calculate the mean velocity in meters per second if $T = 20°C$.

4.36 Water from a pipeline is diverted into a weigh tank for exactly 9 min. The increased weight in the tank is 4765 lbf. What is the average flow rate in gallons per minute and in cubic feet per second? Assume $T = 60°F$.

4.37 The mean velocity of water in a 4-in. pipe is 9 ft/s. Determine the flow in slugs per second, gallons per minute, and cubic feet per second if $T = 60°F$.

4.38 Figure 5.9 on p. 179 shows the flow pattern for flow past a circular cylinder. Assume that the approach velocity at A is constant (does not vary with time).

   a. Is the flow past the cylinder steady or unsteady?

   b. Is this a case of one-dimensional, two-dimensional, or three-dimensional flow?

   c. Are there any regions of the flow where local acceleration is present? If so, show where they are and show vectors representing the local acceleration in the regions where it occurs.

   d. Are there any regions of flow where convective acceleration is present? If so, show vectors representing the convective acceleration in the regions where it occurs.

   e. If the velocity at point A is 10 ft/s, estimate the velocity at point C.

4.39 Given:

$$u = xt + 2y \qquad v = xt^2 - yt \qquad w = 0$$

What is the total acceleration at a point $x = 1$ m, $y = 1$ m, and at time $t = 2$ s?

4.40 Tests on a sphere are conducted in a wind tunnel at an airspeed of U_0. The velocity of flow toward the sphere along the longitudinal axis is found to be $u = -U_0(1 - r_0^3/x^3)$, where r_0 is the radius of the sphere and x the distance from its center. Determine the acceleration of the air as it passes a point ahead of the center of the sphere in terms of x, r_0, and U_0.

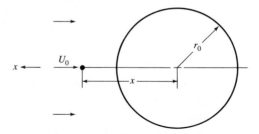

PROBLEM 4.40

4.41 Two parallel disks of diameter D are brought together, each with a normal speed of V. When their spacing is h, what is the radial component of convective acceleration at the section just inside the edge of the disk (section A) in terms of V, h, and D? Assume uniform velocity distribution across the section.

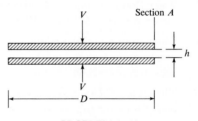

PROBLEM 4.41

4.42   For the conditions given in Prob. 4.41, find the radial component of local acceleration at A in terms of D, V, and h.

4.43   Two streams discharge into a pipe as shown. The flows are incompressible. The volume flow rate of stream A into the pipe is given by $Q_A = 0.01t$ m³/s and that of stream B by $Q_B = 0.005t^2$ m³/s, where t is in seconds. The exit area of the pipe is 0.01 m². Find the velocity and acceleration of the flow at the exit at $t = 1$ s.

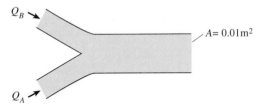

PROBLEM 4.43

4.44   Consider the flow field given in Prob. 4.11. The velocity at a radius of 10 m is given as $V_\theta = 10t$ m/s, where t is in seconds. Find the magnitude of the acceleration of a particle of fluid at $r = 10$ m when $t = 1$ s.

4.45   In this flow passage the discharge is varying with time according to the following expression:

$$Q \text{ in m}^3/\text{s} = Q_0 - Q_1 \frac{t}{t_0}$$

At time $t = 0.50$ s, it is known that at section A–A the velocity gradient in the S direction is $+2$ m/s per meter. Given that Q_0, Q_1, and t_0 are constants with values of 0.985 m³/s, 0.5 m³/s, and 1 s, respectively, and assuming that one-dimensional flow applies, answer the following questions for time $t = 0.5$ s.

    a. What is the velocity at A–A?
    b. What is the local acceleration at A–A?
    c. What is the convective acceleration at A–A?

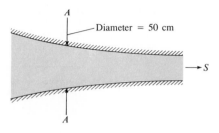

PROBLEM 4.45

4.46   The nozzle in the figure is shaped such that the velocity of flow varies linearly from the base of the nozzle to its tip. Assuming one-dimensional flow, what is the convective acceleration midway between the base and the tip if the diameters D and d are 3 in. and

1 in., respectively, the nozzle length L is 18 in., and the discharge is 0.40 cfs? Also, what is the local acceleration midway between the base and the tip?

4.47 In Prob. 4.46 the discharge varies with time according to $Q = 2t$, where Q is in cubic feet per second and t is in seconds. What is the local acceleration midway along the nozzle when $t = 2$ s?

4.48 Liquid flows through this two-dimensional slot at a rate of $q = 2q_0 t/t_0$, where q is the discharge per unit of width of slot and q_0 and t_0 are reference values. Assume that along the line $y = 0$ the velocity is given by $u = q/b$. Then what will be the local acceleration at $x = 2B$ and $y = 0$ in terms of B, t, t_0, and q_0?

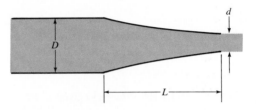

PROBLEMS 4.46, 4.47

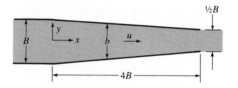

PROBLEMS 4.48, 4.49

4.49 What will be the convective acceleration for the conditions of Prob. 4.48?

4.50 The velocity of water flow in the nozzle shown is given by the following expression: $V = 2t/(1 - 0.5x/L)^2$, where V = velocity in feet per second, t = time in seconds, x = distance along the nozzle, and L = length of nozzle = 4 ft. When $x = 0.5L$ and $t = 3$ s, what is the local acceleration along the centerline? What is the convective acceleration? Assume one-dimensional flow prevails.

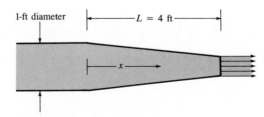

PROBLEM 4.50

4.51 Air discharges downward in the pipe and then outward between the parallel disks. Assuming negligible density change in the air, derive a formula for the acceleration of air at point A, which is a distance r from the center of the disks. Express the acceleration in terms of the constant air discharge Q, the radial distance r, and the disk spacing h. If $D = 10$ cm, $h = 1$ cm, and $Q = 0.380$ m^3/s, what are the velocity in the pipe and the acceleration at point A where $r = 20$ cm?

4.52 All the conditions of Prob. 4.51 are the same except that the discharge is given as $Q = Q_0(t/t_0)$, where $Q_0 = 0.1$ m^3/s and $t_0 = 1$ s. For the additional condition, what will be the acceleration at point A when $t = 2$ s and $t = 3$ s?

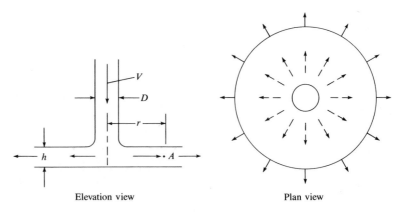

Elevation view   Plan view

PROBLEMS 4.51, 4.52

4.53 A mechanical pump is used to pressurize a bicycle tire. The discharge through the pump is 1 cfm. The density of the air entering the pump is 0.075 lbm/ft^3. The inflated volume of a bicycle tire is 0.04 ft^3. The density of air in the inflated tire is 0.4 lbm/ft^3. How many seconds does it take to pressurize the tire if there initially was no air in the tire?

4.54 For the conditions in each of the flow cases (a and b) shown, respond to the following questions and statements concerning the application of the control-volume equation to the continuity principle.

a. What is the value of β?

b. Determine the value of dB_{sys}/dt.

c. Determine the value of $\Sigma \beta\rho\mathbf{V} \cdot \mathbf{A}$.

d. Determine the value of $d/dt \int_{cv} \beta\rho \, d\Psi$.

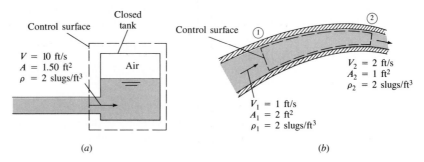

(a)   (b)

PROBLEM 4.54

4.55   Gas flows into and out of the chamber as shown. For the conditions shown, which of the following statement(s) are true of the application of the control-volume equation to the continuity principle?

a. $B_{sys} = 0$

d. $\dfrac{d}{dt} \displaystyle\int \beta \rho \, d\Psi = 0$

b. $dB_{sys}/dt = 0$

e. $\beta = 0$

c. $\displaystyle\sum_{cs} \beta \rho \mathbf{V} \cdot \mathbf{A} = 0$

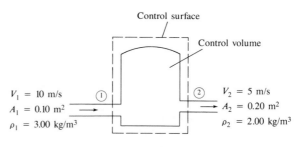

PROBLEM 4.55

4.56   Both pistons are moving to the left, but piston A has a speed twice as great as that of piston B. Then the water level in the tank is a) rising, b) not moving up or down, c) falling.

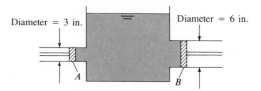

PROBLEM 4.56

4.57   The piston in the cylinder is moving up. Assume that the control volume is the volume inside the cylinder above the piston (the control volume changes in size as the piston moves). A gaseous mixture exists in the control volume. For the given conditions, indicate which of the following statements are true.

a. $\displaystyle\sum_{cs} \rho \mathbf{V} \cdot \mathbf{A}$ is equal to zero.

b. $\dfrac{d}{dt} \displaystyle\int_{cv} \rho \, d\Psi$ is equal to zero.

c. The mass density of the gas in the control volume is increasing with time.

d. The temperature of the gas in the control volume is increasing with time.

e. The flow inside the control volume is unsteady.

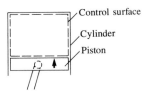

4.58 For the conditions shown, respond to the following questions and statements concerning application of the control-volume equation to the continuity principle.

a. What is the value of β?

b. Determine the value of dB_{sys}/dt.

c. Determine the value of $\Sigma \, \beta\rho\mathbf{V} \cdot \mathbf{A}$.

d. Determine the value of $d/dt \int_{cv} \beta\rho \; d\forall$.

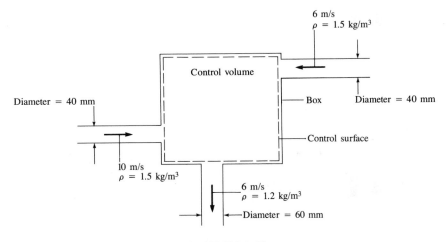

PROBLEM 4.58

4.59 This plunger moves downward in the conical receptacle, which is filled with oil. At what level (y in terms of d) above the bottom of the receptacle will the mean upward velocity of the oil (between the plunger and the receptacle wall) be exactly the same magnitude as the downward velocity of the plunger?

4.60 A 6-in.-diameter cylinder falls at a rate of 1 ft/s in an 8-in.-diameter tube containing an incompressible liquid. What is the mean velocity of the liquid (with respect to the tube) in the space between the cylinder and the tube wall?

4.61 This circular tank of water is being filled from a pipe as shown. The velocity of flow of water from the pipe is 10 ft/s. What will be the rate of rise of the water surface in the tank?

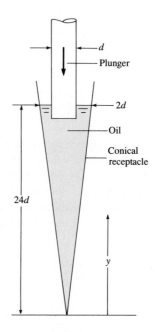

PROBLEM 4.59

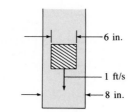

PROBLEM 4.60

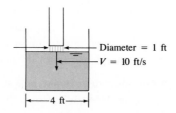

PROBLEM 4.61

4.62 A sphere 1 ft in diameter falls axially through a closed cylinder 1.05 ft in diameter at a speed of 0.5 ft/s. What is the mean upward velocity of the surrounding fluid at the mid-section of the sphere?

4.63   A rectangular air duct 30 cm by 50 cm carries a flow of 1.8 m³/s. Determine the velocity in the duct. If the duct tapers to 15 cm by 40 cm, what is the velocity in the latter section? Assume constant air density.

4.64   A 30-cm pipe divides into a 20-cm branch and a 15-cm branch. If the total discharge is 0.30 m³/s and if the same mean velocity occurs in each branch, what is the discharge in each branch?

4.65   The conditions are the same as in Prob. 4.64 except that the discharge in the 20-cm branch is twice that in the 15-cm branch. What is the mean velocity in each branch?

4.66   Water flows in an 8-in. pipe that is connected in series with a 6-in. pipe. If the rate of flow is 449 gpm (gallons per minute), what is the mean velocity in each pipe?

4.67   What is the velocity of the flow of water in leg *B* of the tee shown in the figure?

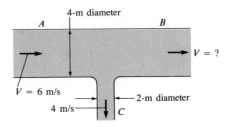

PROBLEM 4.67

4.68   For a steady flow of gas in the conduit shown, what is the mean velocity at section 2?

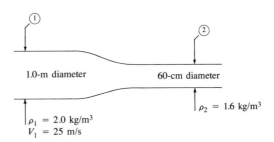

PROBLEM 4.68

4.69   Two pipes are connected to an open water tank. The water is entering the bottom of the tank from pipe *A* at 10 cfm. The water level in the tank is rising at 0.8 in./min, and the surface area of the tank is 100 ft². Calculate the discharge in a second pipe, pipe *B*, that is also connected to the bottom of the tank. Is the flow entering or leaving the tank from pipe *B*?

4.70   Is the tank in the figure filling or emptying? At what rate is the water level rising or falling in the tank?

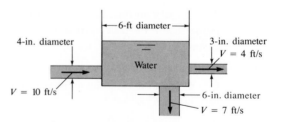

PROBLEM 4.70

4.71   Given: Flow velocities as shown in the figure and water surface elevation (as shown) at $t = 0$ s. At the end of 22 s, will the water surface in the tank be rising or falling, and at what speed?

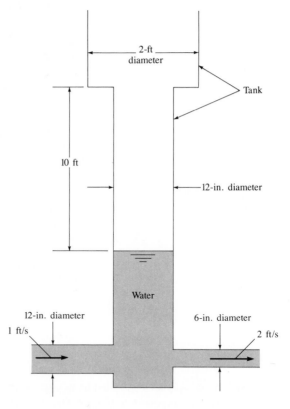

PROBLEM 4.71

4.72   A lake with no outlet is fed by a river with a constant flow of 1000 ft³/s. Water evaporates from the surface at a constant rate of 15 ft³/s per square mile surface area. The area varies with depth h (feet) as A (square miles) = 4.5 + 5.5h. What is the equilibrium elevation of the lake? Below what river discharge will the lake dry up?

4.73 A stationary nozzle discharges water against a plate moving toward the nozzle at half the jet velocity. When the discharge from the nozzle is 8 cfs, at what rate will the plate deflect water?

4.74 The open tank shown has a constant inflow discharge of 20 ft³/s. A 1.0-ft-diameter drain provides a variable outflow velocity V_{out} equal to $\sqrt{(2gh)}$ ft/s. What is the equilibrium height h_{eq} of the liquid in the tank?

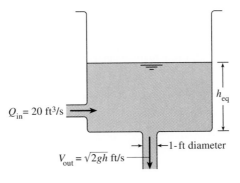

PROBLEM 4.74

4.75 Assuming that complete mixing occurs between the two inflows before the mixture discharges from the pipe at C, find the mass rate of flow, the velocity, and the specific gravity of the mixture in the pipe at C.

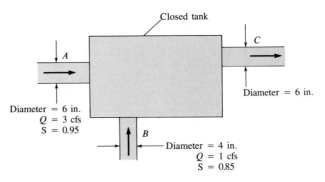

PROBLEM 4.75

4.76 Oxygen and methane are mixed at 200 kPa absolute pressure and 100°C. The velocity of the gases into the mixer is 5 m/s. The density of the gas leaving the mixer is 2.2 kg/m³. Determine the exit velocity of the gas mixture.

4.77 A slow leak develops in a tire (assume constant volume), in which it takes 3 hr for the pressure to decrease from 30 psig to 25 psig. The air volume in the tire is 0.5 ft³, and the temperature remains constant at 60°F. The mass-flow rate of air is given by $\dot{m} = 0.68pA/\sqrt{RT}$. Calculate the area of the hole in the tire. Atmospheric pressure is 14 psia.

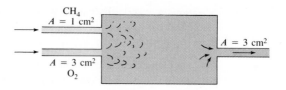

PROBLEM 4.76

4.78 Oxygen leaks slowly through a small orifice in an oxygen bottle. The volume of the bottle is 0.1 m³, and the diameter of the orifice is 0.15 mm. The temperature in the tank remains constant at 18°C, and the mass-flow rate is given by $\dot{m} = 0.68pA/\sqrt{RT}$. How long will it take the absolute pressure to decrease from 10 to 5 MPa?

4.79 How long will it take the water surface in the tank shown to drop from $h = 3$ m to $h = 30$ cm?

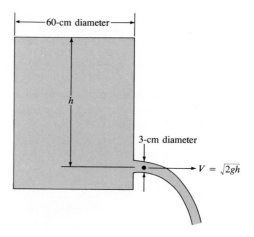

PROBLEM 4.79

4.80 For the type of tank shown, the tank diameter is given as $D = d + C_1h$, where d is the bottom diameter and C_1 is a constant. Derive a formula for the time of fall of liquid surface from $h = h_0$ to $h = h$ in terms of d_j, d, h_0, h, and C_1. Solve for t if $h_0 = 1$ m, $h = 20$ cm, $d = 20$ cm, $C_1 = 0.4$, and $d_j = 5$ cm.

4.81 Water drains out of a trough as shown. The angle with the vertical of the sloping sides is α, and the distance between the parallel sides is B. The width of the trough is $W_0 + 2h \tan \alpha$, where h is the distance from the trough bottom. The velocity of the water issuing from the opening in the bottom of the trough is equal to $V_e = \sqrt{2gh}$. The area of the water stream at the bottom is A_e. Derive an expression for the time to drain to depth h in terms of h/h_0, W_0/h_0, $\tan \alpha$, and $A_e g^{0.5}/(h_0^{1.5}B)$, where h_0 is the original depth. Find the time to drain to one-half the original depth for $W_0/h_0 = 0.2$, $\alpha = 30°$, $A_e g^{0.5}/(h_0^{1.5}B) = 0.01$ s^{-1}.

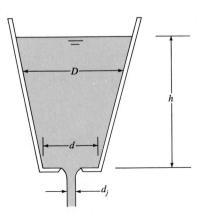

PROBLEM 4.80

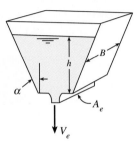

PROBLEM 4.81

4.82 The velocity components for a two-dimensional flow are

$$u = \frac{C(y^2 - x^2)}{(y^2 + x^2)^2} \qquad v = \frac{-2Cxy}{(x^2 + y^2)^2}$$

where C is a constant. Does this description of a flow field satisfy continuity? Is the flow irrotational?

4.83 Consider the flow shown in Fig. 4.18. Write an equation for the velocity, and then answer the following questions: Is the differential form of the continuity equation satisfied for the given flow field? Is the flow rotational or irrotational?

4.84 It is predicted that a flow field will have the following velocity components:

$$u = V(x^3 + xy^2) \qquad v = V(y^3 + yx^2) \qquad w = 0$$

V is a constant. Is such a flow field possible? (Does it satisfy continuity?)

4.85 The velocity variation for flow near a wall is given by $u = Vy$, $v = Cy$, and $w = 0$, where V and C are constants. Is the flow irrotational?

4.86 A two-dimensional flow field is defined by $u = xt + 2y$ and $v = xt^2 - yt$. What is the total acceleration at a point $x = 1$ m, $y = 1$ m, and at time $t = 1$ s? Is the flow rotational or irrotational?

4.87 The velocity components of a flow field are given by

$$u = \frac{y}{(x^2 + y^2)^{3/2}} \qquad v = \frac{-x}{(x^2 + y^2)^{3/2}}$$

Is continuity satisfied? Is the flow irrotational?

4.88 A u component of velocity is given by $u = Axy$, where A is a constant. What is a possible v component? What must the v component be if the flow is irrotational?

4.89 An end-burning rocket motor has a chamber diameter of 10 cm and a nozzle exit diameter of 8 cm. The density of the propellant is 1800 kg/m^3 and regresses at the rate of 1 cm/s. The gases crossing the nozzle exit plane have a pressure of 10 kPa and a temperature of 2000°C. The gas constant of the exhaust gases is 415 J/kg K. Calculate the gas velocity at the nozzle exit plane.

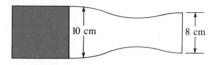

PROBLEM 4.89

4.90 A cylindrical-port rocket motor has a grain design consisting of a cylindrical shape as shown. The curved internal surface and both ends burn. The propellant surface regresses uniformly at 1.2 cm/s. The propellant density is 2000 kg/m^3. The inside diameter of the motor is 20 cm. The propellant grain is 40 cm long and has an inside diameter of 12 cm. The diameter of the nozzle exit plane is 20 cm. The gas velocity at the exit plane is 2000 m/s. Determine the gas density at the exit plane.

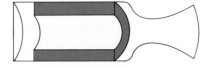

PROBLEM 4.90

4.91 The mass-flow rate through a nozzle is given by

$$\dot{m} = 0.65 \frac{p_c A_t}{\sqrt{RT_c}}$$

where p_c and T_c are pressure and temperature in the rocket chamber and R is the gas constant of the gases in the chamber. The propellant burning rate (surface regression rate) can be expressed as $\dot{r} = ap_c^n$, where a and n are two empirical constants. Show, by application of the continuity equation, that the chamber pressure can be expressed as

$$p_c = \left(\frac{a\rho_p}{0.65}\right)^{1/(1-n)}\left(\frac{A_g}{A_t}\right)^{1/(1-n)}(RT_c)^{1/[2(1-n)]}$$

where ρ_p is the propellant density and A_g is the grain surface burning area. If the operating chamber pressure of a rocket motor is 3.5 MPa and $n = 0.3$, how much will the chamber pressure increase if a crack develops in the grain, increasing the burning area by 20%?

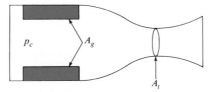

PROBLEM 4.91

4.92 A piston is moving up during the exhaust stroke of a four-cycle engine. Mass escapes through the exhaust port at a rate given by

$$\dot{m} = 0.65\frac{p_c A_v}{\sqrt{RT_c}}$$

where p_c and T_c are the cylinder pressure and temperature, A_v is the valve opening area, and R is the gas constant of the exhaust gases. The bore of the cylinder is 10 cm, and the piston is moving upward at 30 m/s. The distance between the piston and the head is 10 cm. The valve opening area is 1 cm^2, the chamber pressure is 300 kPa, the chamber temperature is 600°C, and the gas constant is 350 J/kg K. Applying the continuity equation, determine the rate at which the gas density is changing in the cylinder. Assume the density and pressure are uniform in the cylinder and the gas is ideal.

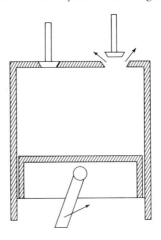

PROBLEM 4.92

4.93   Assume that the air flow in the outer portion ($r > 50$ mi) of a cyclonic storm is approximated by irrotational vortex flow. If you observed a tangential wind speed of 15 mph at a radial distance of 200 mi from the center of such a storm, what wind speeds would you predict at radial distances of 100 mi and 50 mi?

References

1. Hussaini, M.Y., and M.D. Salas, eds. *Studies of Vortex Dominated Flows,* Springer-Verlag, New York, 1987.

2. Landahl, M.T., and E. Mollo-Christensen. *Turbulence and Random Processes in Fluid Mechanics,* Cambridge University Press, New York, 1986.

3. Lugt, H.J. *Vortex Flows in Nature and Technology,* John Wiley, New York, 1983.

4. Panofsky, H.A., and J.A. Dutton. *Atmospheric Turbulence,* John Wiley, New York, 1984.

5. Vinnichenko, N.K., et al. *Turbulence in the Free Atmosphere,* 2nd ed., Consultants Bureau, a Division of Plenum Publishing Co., New York, 1980.

Films

6. Kline, S. J. *Flow Visualization.* National Committee for Fluid Mechanics Films, distributed by Encyclopaedia Britannica Educational Corporation.

7. Lumley, John L. *Eulerian and Lagrangian Descriptions in Fluid Mechanics.* National Committee for Fluid Mechanics Films, distributed by Encyclopaedia Britannica Educational Corporation.

8. Stewart, R.W. *Turbulence.* National Committee for Fluid Mechanics Films, distributed by Encyclopaedia Britannica Educational Corporation.

9. Shapiro, Ascher H. *Vorticity.* National Committee for Fluid Mechanics Films, distributed by Encyclopaedia Britannica Educational Corporation.

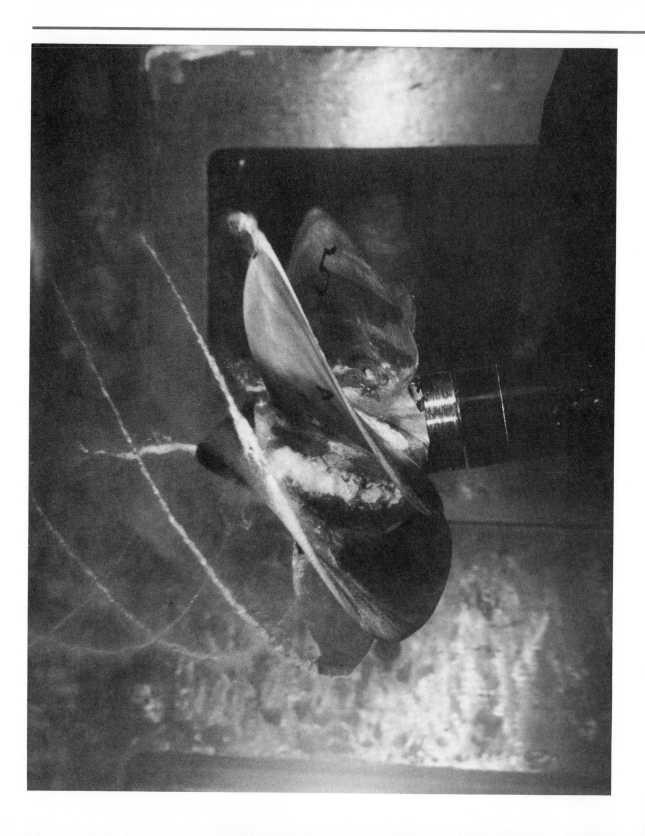

Pressure Variation in Flowing Fluids

This ship propeller being tested in a water tunnel shows cavitation developing on the front sides of blades as well as at the tips of the blades. The bubbles from the tips form a helical pattern as they follow the tip vortices that are shed from the blades. (Courtesy Carderock Division, Naval Surface Warfare Center [formerly David Taylor Research Center])

P ressure variation is important to the engineer for several reasons. In certain cases, as in the design of tall structures, the pressure variation resulting from the wind must be considered in the design of individual parts, such as windows, as well as in the design of the basic structure so it can resist the overall wind load. In the design of pump impellers, hydrofoils, and even pipelines, the engineer must avoid the possibility of *cavitation,* the phenomenon of boiling in a flowing liquid at normal temperatures, which results from low pressure. In this case, pressure variation is the most significant aspect of the problem. In aircraft design, the pressure variation around the wing produces lift, but it also contributes to the drag of the aircraft. Thus pressure plays a major role in many areas of engineering design and analysis.

Even in our daily lives, many phenomena that affect us are related to pressures caused by flowing fluids. For example, one indicator of our health, blood pressure, is directly related to the flow of blood in veins and arteries. The atmospheric pressure readings reported in weather forecasts are related to the vortex strength (tangential velocity of the wind) in cyclonic storms and the position of the observer relative to the storm center. The drag that we experience with gusts of wind is a pressure-related phenomenon. Even the vortex motion as we stir a cup of coffee and the rise velocity of bubbles in our favorite beverage result from pressure variations in flowing fluids. The basic principles introduced in this chapter will help to explain natural phenomena such as these and will also set the stage for a more thorough treatment of pressure-related topics later in the text.

5.1   Basic Causes of Pressure Variation in a Flowing Fluid

In Chapter 3 we saw that gravity causes pressure to vary with elevation; now we shall consider other causes of pressure variation. In fluid flow there·are basically two causes of pressure variation in addition to the weight effect—these are acceleration and viscous resistance. To accelerate a mass of fluid in a given direction, there must be a net force in the direction of acceleration. Therefore, the pressure must decrease in the direction of acceleration. The fact that a net force on the fluid acts in the direction of decreasing pressure is illustrated in Fig. 5.1 for the case of flow in a pipe. When we isolate a mass of fluid such as this, we see that the greater pressure on the left end acts to the right in the direction of decreasing pressure, and the smaller pressure at the right end acts in the opposite direction. Since the areas are the same, the net pressure force on the fluid acts to the right in the direction of decreasing pressure. In addition to acceleration, pressure variation is needed to overcome the viscous resistance, which, like friction in solids, acts in opposition to the motion of the fluid. In the foregoing discussion, it has been assumed that the net force that produces acceleration and overcomes viscous resistance is a result of the pressure distribution. However, gravity may also enter the problem. Therefore, fluid mechanics is an extension of basic mechanics with added variations in the way that forces are applied.

Pressure Variation due to Weight and Acceleration

Consider the cylindrical element of fluid shown in Fig. 5.2. Here the element is being accelerated in the ℓ direction and is acted on by pressure and weight forces only. Note also that the coordinate axis z is vertically upward and that it is assumed that the pressure varies along the length of the element. Applying Newton's second law in the ℓ direction and using the system approach, we have

$$\sum F_\ell = M a_\ell$$

or
$$p \Delta A - (p + \Delta p) \Delta A - \Delta W \sin \alpha = \rho \Delta \ell \Delta A \, a_\ell \qquad (5.1)$$

However, $\Delta W = \gamma \Delta \ell \Delta A$, so Eq. (5.1) reduces to

$$-\frac{\Delta p}{\Delta \ell} - \gamma \sin \alpha = \rho a_\ell \qquad (5.2)$$

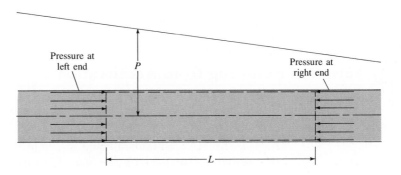

FIGURE 5.1

Variation of pressure in a pipe.

FIGURE 5.2

Pressure and weight forces
acting on an accelerating
fluid element.
(a) Fluid element.
(b) Trigonometric relation.

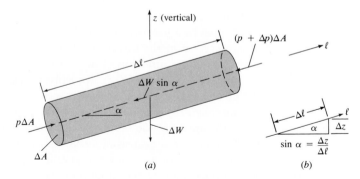

Pressure is a function of both position and time. Taking the limit of $\Delta p/\Delta \ell$ at a given time as $\Delta \ell$ approaches zero yields the partial derivative:

$$\lim_{\Delta \ell \to 0} \frac{\Delta p}{\Delta \ell}\bigg|_t = \frac{\partial p}{\partial \ell}$$

Fig. 5.2b also shows that $\sin \alpha$ is equal to $\Delta z/\Delta \ell$. Taking the limit as $\Delta \ell$ approaches zero at a given time yields

$$\sin \alpha = \lim_{\Delta \ell \to 0} \frac{\Delta z}{\Delta \ell} = \frac{\partial z}{\partial \ell}$$

Thus the limiting form of Eq. (5.2) when $\Delta \ell$ approaches zero is

$$-\frac{\partial p}{\partial \ell} - \gamma \frac{\partial z}{\partial \ell} = \rho a_\ell$$

or, taking γ as a constant,

$$-\frac{\partial}{\partial \ell}(p + \gamma z) = \rho a_\ell \tag{5.3}$$

Equation (5.3) is Euler's equation of motion for a fluid. It is of interest to note that when the acceleration is zero, Eq. (5.3) reduces to $\partial/\partial \ell(p + \gamma z) = 0$, which, when integrated, gives the familiar expression for hydrostatics, $p + \gamma z = C$. In other words, along a path of zero acceleration the pressure distribution must be hydrostatic. Again, this assumes that gravity and pressure forces are the only forces acting.

5.2 Examples of Pressure Variation Resulting from Acceleration

Uniform Acceleration of a Tank of Liquid

Assume that the open tank of liquid shown in Fig. 5.3 is accelerated to the right, the positive x direction, at a rate of a_x. For this to occur, a net force must act on

FIGURE 5.3

*Uniform acceleration of a
tank of liquid.*

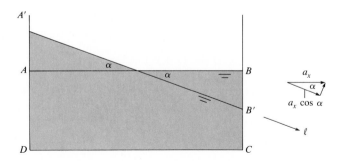

the liquid in the x direction; this is accomplished when the liquid redistributes itself in the tank as shown by $A'B'CD$. Under this condition the hydrostatic force at the left end is greater than the hydrostatic force at the right, which is consistent with the requirement of $F = Ma$.

Further quantitative analysis of the acceleration of the tank of liquid is made with Eq. (5.3). First consider application of the equation along the liquid surface $A'B'$. Here the pressure is constant, $p = p_{atm}$. Consequently, $\partial p/\partial \ell = 0$. The acceleration along $A'B'$ is given by $a_\ell = a_x \cos \alpha$. Hence, Eq. (5.3) reduces to

$$\frac{d}{d\ell}(\gamma z) = -\rho a_x \cos \alpha \qquad (5.4)$$

where the total derivative is used because the variables do not change with time. The specific weight in Eq. (5.4) is constant. Therefore, Eq. (5.4) becomes

$$\frac{dz}{d\ell} = -\frac{a_x \cos \alpha}{g}$$

But $dz/d\ell = -\sin \alpha$. Thus we obtain

$$\sin \alpha = \frac{a_x \cos \alpha}{g}$$

or $$\tan \alpha = \frac{a_x}{g} \qquad (5.5)$$

Still further analysis can be made if Eq. (5.3) is applied along a horizontal plane in the liquid, such as at the bottom of the tank. Now z is constant and Eq. (5.3) reduces to $\partial p/\partial x = -\rho a_x$, which shows that the pressure must decrease in the direction of acceleration. The change in pressure is consistent with the change in depth of liquid because hydrostatic pressure variation prevails in the vertical direction, since there is no component of acceleration in that direction. Thus as the depth decreases in the direction of acceleration, the pressure along the bottom of the tank must also decrease. Another case of uniform acceleration is given in the following example.

EXAMPLE 5.1 The tank on a tank truck is filled completely with gasoline, which has a specific weight of 42 lbf/ft^3 (6.60 kN/m^3).

a. If the tank on the trailer is 20 ft (6.1 m) long and if the pressure at the top rear end of the tank is atmospheric, what is the pressure at the top front when the truck decelerates at a rate of 10 ft/s^2 (3.05 m/s^2)?

b. If the tank is 6 ft (1.83 m) high, what is the maximum pressure in the tank?

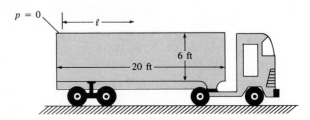

Solution Apply Eq. (5.3) along the top of the tank. Here z is constant and the pressure does not vary with time during this phase of deceleration. Therefore, one may write

$$\frac{dp}{d\ell} = -\rho a_\ell$$

Integrating, one obtains

$$p = -\rho \ell a_\ell + C$$

When $\ell = 0$, $p = 0$; hence, $C = 0$ and $p = -\rho \ell a_\ell$.

Now substituting -10 ft/s^2 (-3.05 m/s^2) for a_ℓ, 20 ft (6.1 m) for ℓ, and 1.30 slugs/ft^3 (672 kg/m^3) for ρ, which is equal to γ/g, one obtains

$$p = -1.30 \text{ slugs/ft}^3 \times (-10 \text{ ft/s}^2) \times 20 \text{ ft} = 260 \text{ psfg} \qquad \blacktriangleleft$$

SI units $\qquad p = -673 \text{ kg/m}^3 \times (-3.05 \text{ m/s}^2) \times 6.1 \text{ m}$

$$= 12,500 \text{ N/m}^2 = 12,500 \text{ Pa gage} \qquad \blacktriangleleft$$

The maximum pressure in the tank will occur at the front end of the tank bottom. Since the pressure variation is hydrostatic in the vertical direction, one obtains $p + \gamma z = $ constant, or

$$p_{\text{bottom}} + \gamma z_{\text{bottom}} = p_{\text{top}} + \gamma z_{\text{top}}$$

Solving yields $\qquad p_{\text{bottom}} = 260 + (42)(6)$

$$p_{\max} = p_{\text{bottom}} = 512 \text{ psfg} \qquad \blacktriangleleft$$

SI units $\qquad p_{\max} = p_{\text{bottom}} = 12,500 \text{ N/m}^2 + 6.6 \text{ kN/m}^3 \times 1.83 \text{ m}$

$$= 24.6 \text{ kPa gage} \qquad \blacktriangleleft$$

Rotation of Tank of Liquid

In Sec. 5.3 it will be shown that the variation in piezometric head ($p/\gamma + z$) can be easily determined by use of Bernoulli's equation if the flow is irrotational. However, for the case of pure rotation, such as a tank of liquid in rotation, one must use Eq. (5.3) at the outset.

Consider a cylindrical tank of liquid rotating at a constant rate ω, as shown in Fig. 5.4. Here AA depicts the liquid surface before rotation, and surface $A'A'$ shows how it appears after a period of time when a steady state has been established. Applying Eq. (5.3) in a radial direction for this tank, we obtain

$$\frac{d}{dr}(p + \gamma z) = -\left(-\rho\frac{V^2}{r}\right) \tag{5.6}$$

Here, Eq. (5.3) is written for the r direction, so r is used as the space variable instead of ℓ. We know from mechanics that the acceleration in the r direction is negative (acceleration is toward the center of rotation). Hence, a negative sign is given to the acceleration term. Substituting $r\omega$ for V and clearing the negative signs then yield

$$\frac{d}{dr}(p + \gamma z) = \rho r\omega^2 \tag{5.7}$$

Integrating Eq. (5.7) with respect to r then gives us

$$p + \gamma z = \frac{\rho r^2 \omega^2}{2} + \text{constant}$$

but $V = r\omega$ and $\rho = \gamma/g$, so we have

$$\frac{p}{\gamma} + z - \frac{V^2}{2g} = \text{constant} \tag{5.8}$$

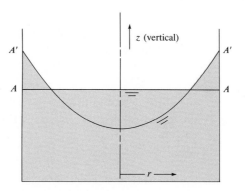

FIGURE 5.4

Rotating vessel of liquid.

Replacing V by $r\omega$ in Eq. (5.8) reveals that the liquid surface is in the form of a paraboloid of revolution. *Note:* Although Eq. (5.8) at first glance appears similar to Bernoulli's equation (see next section)—the terms p/γ, z, and $V^2/2g$ are the same as in Bernoulli's equation—the similarity is not complete because the sign on the $V^2/2g$ term of Eq. (5.8) is different from that on the other two terms. In Bernoulli's equation the signs are all the same.

EXAMPLE 5.2   When the U-tube is not rotated, the water stands in the tube as shown. If the tube is rotated about the eccentric axis at a rate of 8 rad/s, what are the new levels of water in the tube?

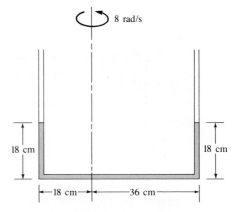

Solution   Solution of this problem is based on Eq. (5.8) and also on the fact that the water occupies a given volume of the tube, which may be expressed in terms of a given length of tube. Let the elevation reference be at the level of the horizontal part of the tube. Then, by considering a point at the water surface in the left tube where $p = 0$ and also a point at the surface in the right tube, one can write Eq. (5.8) as follows:

$$z_l - \frac{r_l^2 \omega^2}{2g} = z_r - \frac{r_r^2 \omega^2}{2g}$$

Another independent equation involving the volume of tube occupied by the liquid may be written as

$$z_l + z_r = 0.36 \text{ m}$$

Substituting $r_l = 0.18$ m, $r_r = 0.36$ m, and $\omega = 8$ rad/s into the first equation above and solving the equations for z_l and z_r yield

$$z_l = 2.1 \text{ cm} \qquad \text{and} \qquad z_r = 33.9 \text{ cm} \qquad \blacktriangleleft$$

5.3 Bernoulli's Equation

Bernoulli's Equation Along a Streamline

Bernoulli's equation can be derived by integrating Euler's equation along a streamline for steady-incompressible flow. However, limiting its application to a given streamline is often too restrictive; therefore, a more useful range of application is sought. It can be shown that if the flow is steady, incompressible, and also irrotational, Bernoulli's equation can be applied throughout the entire flow field. The derivation of Bernoulli's equation for this more useful application is given in the next section.

Bernoulli's Equation for Irrotational Flow

To derive Bernoulli's equation for irrotational-incompressible-steady flow, we utilize the cartesian coordinate system in two dimensions.* First, we write Euler's equation for the x coordinate direction.

$$-\frac{\partial}{\partial x}(p + \gamma z) = \rho a_x$$

If the fluid is incompressible (constant γ), then we can divide through by γ to obtain

$$-\frac{\partial}{\partial x}\left(\frac{p}{\gamma} + z\right) = \frac{a_x}{g}$$

By definition $(p/\gamma) + z = h$. Therefore, we obtain

$$-\frac{\partial h}{\partial x} = \frac{a_x}{g} \tag{5.9}$$

From Sec. 4.3 the acceleration in the x direction for steady flow is given as

$$a_x = u\frac{\partial u}{\partial x} + v\frac{\partial u}{\partial y} \tag{5.10}$$

When Eq. (5.10) is substituted into Eq. (5.9) we get

$$-\frac{\partial h}{\partial x} = \frac{1}{g}\left(u\frac{\partial u}{\partial x} + v\frac{\partial u}{\partial y}\right) \tag{5.11}$$

---

*The same type of derivation can easily be used to derive Bernoulli's equation for three-dimensional flow. The two-dimensional case is presented here for brevity.

Similarly, for the y direction we have

$$-\frac{\partial h}{\partial y} = \frac{1}{g}\left(u\frac{\partial v}{\partial x} + v\frac{\partial v}{\partial y}\right) \tag{5.12}$$

The condition for irrotationality of flow in the x-y plane was given by Eq. (4.38), which is

$$\frac{\partial v}{\partial x} = \frac{\partial u}{\partial y}$$

Therefore, when $\partial u/\partial y$ in Eq. (5.11) is replaced by $\partial v/\partial x$, we obtain

$$-\frac{\partial h}{\partial x} = \frac{1}{g}\left(u\frac{\partial u}{\partial x} + v\frac{\partial v}{\partial x}\right) \tag{5.13}$$

However, note that $u\,\partial u/\partial x = \partial/\partial x(u^2/2)$ and $v\,\partial v/\partial y = \partial/\partial y(v^2/2)$. Therefore, we can write Eq. (5.13) as

$$\frac{\partial}{\partial x}\left(h + \frac{u^2 + v^2}{2g}\right) = 0$$

The sum $u^2 + v^2$ is the total velocity squared, V_s^2, so we have

$$\frac{\partial}{\partial x}\left(h + \frac{V_s^2}{2g}\right) = 0 \tag{5.14}$$

Thus when the flow is steady, incompressible, and irrotational, there is no change in $h + V_s^2/2g$ in the x direction according to Eq. (5.14).

In a similar manner, if we take Eq. (5.12) and replace $\partial v/\partial x$ by $\partial u/\partial y$, because of the irrotationality condition it can be shown that

$$\frac{\partial}{\partial y}\left(h + \frac{V_s^2}{2g}\right) = 0 \tag{5.15}$$

This shows that there is no change in $h + V_s^2/2g$ in the y direction for the given conditions. Therefore, we have shown that $h + V_s^2/2g$ is constant throughout the flow field, or

$$h + \frac{V_s^2}{2g} = C \tag{5.16}$$

Also, $h = p/\gamma + z$, so we have

$$\frac{p}{\gamma} + z + \frac{V_s^2}{2g} = C \tag{5.17}$$

In most applications the subscript s is omitted from the velocity symbol, and the equation is written between two points in the field of flow as follows:

$$\frac{p_1}{\gamma} + \frac{V_1^2}{2g} + z_1 = \frac{p_2}{\gamma} + \frac{V_2^2}{2g} + z_2 \tag{5.18}$$

Bernoulli's equation relates the pressure, velocity, and elevation between any two points in the flow field for flow that is *steady, irrotational, nonviscous,* and *incompressible*. The nonviscous limitation follows from the inherent limitation of Euler's equation, from which Bernoulli's equation is derived. Note that V is the speed of the fluid; thus component velocities are not valid in Bernoulli's equation (it is a scalar equation). Bernoulli's equation can be used to predict the pressure distribution within the fluid or the pressure distribution on a body if the flow pattern about the body is known.

5.4 Application of Bernoulli's Equation

Stagnation Tube

Consider a curved tube such as that shown in Fig. 5.5. When Bernoulli's equation is written between points 1 and 2, we obtain

$$\frac{p_1}{\gamma} + \frac{V_1^2}{2g} + z_1 = \frac{p_2}{\gamma} + \frac{V_2^2}{2g} + z_2 \qquad (5.19)$$

However, $z_2 = z_1$ and the velocity at point 2 is zero (a *stagnation point*). Hence, Eq. (5.19) reduces to

$$\frac{V_1^2}{2g} = \frac{p_2}{\gamma} - \frac{p_1}{\gamma} \qquad (5.20)$$

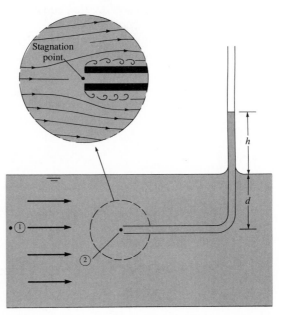

FIGURE 5.5

Stagnation tube.

By the equations of hydrostatics (there is no acceleration normal to the stream-lines where the streamlines are straight and parallel), $p_1 = \gamma d$ and $p_2 = \gamma(h + d)$. Therefore, Eq. (5.20) can be written as

$$\frac{V_1^2}{2g} = \frac{\gamma(h + d) - \gamma d}{\gamma}$$

which reduces to $\qquad\qquad V_1 = \sqrt{2gh} \qquad\qquad\qquad (5.21)$

Thus it is seen that a very simple device such as this curved tube can be used to measure the velocity of flow.

Pitot Tube

The Pitot tube, named after the eighteenth-century French hydraulic engineer who invented it, is based on the same principle as the stagnation tube, but it is much more versatile than the stagnation tube. The Pitot tube has a pressure tap at the upstream end of the tube for sensing the *stagnation pressure*. There are also ports located several tube diameters downstream of the front end of the tube for sensing the *static pressure* in the fluid where the velocity is essentially the same as the approach velocity. When Bernoulli's equation is applied between points 1 and 2 in Fig. 5.6, we have

$$\frac{p_1}{\gamma} + \frac{V_1^2}{2g} + z_1 = \frac{p_1}{\gamma} + \frac{V_2^2}{2g} + z_2$$

But $V_1 = 0$, so solving that equation for V_2 gives

$$V_2 = \left\{ 2g \left[\left(\frac{p_1}{\gamma} + z_1 \right) - \left(\frac{p_2}{\gamma} + z_2 \right) \right] \right\}^{1/2}$$

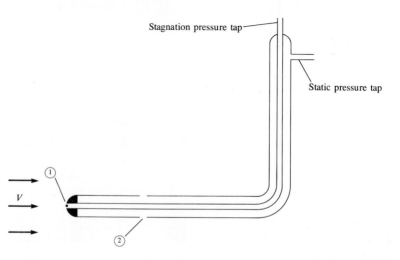

FIGURE 5.6

Pitot tube.

Here $V_2 = V$ and $p/\gamma + z = h$. Hence we obtain

$$V = \sqrt{2g(h_1 - h_2)} \qquad (5.22)$$

where V is the velocity of the stream and h_1 and h_2 are the piezometric heads at points 1 and 2, respectively.

By connecting a pressure gage or manometer between taps that lead to points 1 and 2, we can easily measure the flow velocity with the Pitot tube. A major advantage of the Pitot tube is that it can be used to measure velocity in a pressure pipe; a simple stagnation tube is not convenient to use in such a situation. In gas-flow measurement, where a single differential pressure gage is connected across the taps, Eq. (5.22) simplifies to $V = \sqrt{2\,\Delta p/\rho}$, where Δp is the pressure difference across the taps.

EXAMPLE 5.3   A mercury–kerosene manometer is connected to the Pitot tube as shown. If the deflection on the manometer is 7 in., what is the kerosene velocity in the pipe? Assume that the specific gravity of the kerosene is 0.81.

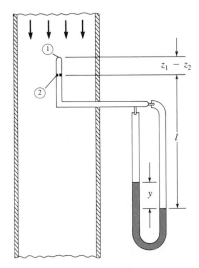

Solution   We need to know $h_1 - h_2$, the difference in piezometric head, between points 1 and 2. We evaluate this first by applying the principles of hydrostatics through the manometer:

$$p_1 + (z_1 - z_2)\gamma_{\text{kero}} + l\gamma_{\text{kero}} - y\,\gamma_{\text{Hg}} - (l - y)\gamma_{\text{kero}} = p_2$$

This reduces to

$$p_1 - p_2 = y(\gamma_{\text{Hg}} - \gamma_{\text{kero}}) - (z_1 - z_2)\gamma_{\text{kero}}$$

or

$$\frac{p_1 - p_2}{\gamma_{\text{kero}}} + z_1 - z_2 = \frac{y(\gamma_{\text{Hg}} - \gamma_{\text{kero}})}{\gamma_{\text{kero}}}$$

Then for
$$h_1 - h_2 = \left(\frac{p_1}{\gamma} + z_1\right) - \left(\frac{p_2}{\gamma} + z_2\right)$$

we have
$$h_1 - h_2 = y\left(\frac{\gamma_{Hg}}{\gamma_{kero}} - 1\right) = \frac{7}{12}(16.7 - 1)$$

The velocity is then

$$V = \sqrt{2g\left[\frac{7}{12}(16.7 - 1)\right]} = 24.3 \text{ ft/s} \quad \blacktriangleleft$$

Note: The -1 of the quantity $16.7 - 1$ reflects the effect of the column of kerosene in the right leg of the manometer, which tends to counterbalance the mercury in the left leg. Thus if we have a gas–liquid manometer, the counterbalancing effect is negligible.

EXAMPLE 5.4   A differential pressure gage is connected across the taps of a Pitot tube. When this Pitot tube is used in a wind tunnel test, the gage indicates a Δp of 730 Pa. What is the air velocity in the tunnel? The pressure and temperature in the tunnel are 98 kPa absolute and 20°C.

Solution
$$V = \sqrt{2\,\Delta p/\rho}$$

where
$$\rho = \frac{p}{RT} = \frac{98 \times 10^3 \text{ N/m}^2}{(287 \text{ J/kg K}) \times (20 + 273) \text{ K}} = 1.17 \text{ kg/m}^3$$

$$\Delta p = 730 \text{ Pa}$$

Therefore
$$V = \sqrt{(2 \times 730 \text{ N/m}^2)/(1.17 \text{ kg/m}^3)} = 35.3 \text{ m/s} \quad \blacktriangleleft$$

Pressure Variation near Curved Boundaries

If flow passages are converging, such as is shown in Fig. 5.7, then irrotational flow will be approximated for low-viscosity fluids such as water or air. Hence Bernoulli's equation can be used to obtain the pressure variation between points in the flow field, including points adjacent to the boundaries. A common procedure for such an application is to use one point as a reference (for example, point 0 far upstream in Fig. 5.7). When we write Bernoulli's equation between the reference point and any other point, we have

$$\frac{p}{\gamma} + \frac{V^2}{2g} + z = \frac{p_0}{\gamma} + \frac{V_0^2}{2g} + z_0 \tag{5.23}$$

FIGURE 5.7

Flow net for transition (half-section).

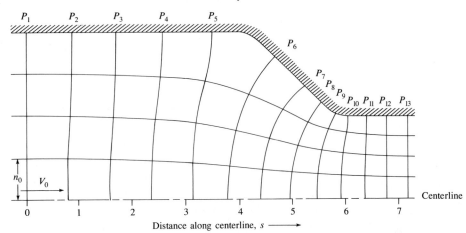

where p_0, V_0, and z_0 are pressure, velocity, and elevation, respectively, at the reference point; and p, V, and z are pressure, velocity, and elevation at any other given point.

By simple rearrangement, Eq. (5.23) is written as

$$p - p_0 = \gamma(z_0 - z) + \frac{\rho}{2}(V_0^2 - V^2) \qquad (5.24)$$

Equation (5.24) expresses the pressure change in terms of the change in hydrostatic pressure (the first term on the right) and the change in kinetic pressure (the second term on the right). Thus the dynamic effect is given by

$$\left(\frac{p}{\gamma} + z\right) - \left(\frac{p_0}{\gamma} + z_0\right) = \frac{\rho}{2\gamma}(V_0^2 - V^2) \qquad (5.25)$$

Equation (5.25) expresses the change in piezometric head as a function of the difference in the velocities squared, and it reduces to

$$h - h_0 = \frac{V_0^2 - V^2}{2g} \qquad (5.26)$$

where h is the piezometric head at a given point, and h_0 is the piezometric head at the reference point.

For gases in which hydrostatic effects are negligible, Eq. (5.24) can be expressed as

$$p - p_0 = \frac{\rho}{2}(V_0^2 - V^2) \qquad (5.27)$$

When we nondimensionalize Eqs. (5.26) and (5.27) by dividing through by $V_0^2/2g$ and $\rho V_0^2/2$, respectively,* we obtain the following equations:

$$\frac{h - h_0}{V_0^2/2g} = 1 - \left(\frac{V}{V_0}\right)^2 \tag{5.28}$$

$$\frac{p - p_0}{\rho V_0^2/2} = 1 - \left(\frac{V}{V_0}\right)^2 \tag{5.29}$$

These are dimensionless forms of Bernoulli's equation, the latter being for the case where the hydrostatic variation of pressure is negligible.

Because the velocity is inversely proportional to the cross-sectional area through which flow occurs in a flow passage (that is, $V/V_0 = A_0/A$), we can express the relative pressure distribution or piezometric-head distribution in terms of the dimensions of the flow passage. For two-dimensional flow, the streamline spacing is directly proportional to the flow area. Thus we have $V/V_0 = n_0/n$ for the relationship between the relative-velocity distribution and the relative-streamline spacing. Here n is the distance between two adjacent streamlines measured along the line (possibly curved) perpendicular to both streamlines. These perpendicular or normal lines are shown in Fig. 5.7. We can now express the relative-piezometric-head distribution or relative-pressure distribution for the case of gases, where the γz term is negligible, in terms of the streamline spacing for a two-dimensional-irrotational flow pattern:

$$\frac{h - h_0}{V_0^2/2g} = 1 - \left(\frac{n_0}{n}\right)^2 \tag{5.30a}$$

$$\frac{p - p_0}{\rho V_0^2/2} = 1 - \left(\frac{n_0}{n}\right)^2 \tag{5.30b}$$

Furthermore, since both sides of Eqs. (5.30) are dimensionless, application of the equations is not a function of the density of the fluid or the absolute size of the passage that controls the flow. Consequently, tests can be made on a small-scale structure (a model), and the results may be applied to a larger-scale structure. This is the principle of model testing, which will be covered in more detail in Chapter 8. The left side of either of Eqs. (5.30) is often called the *pressure coefficient* C_p. It is the change in piezometric head (or pressure) between two points in the flow field relative to the velocity head (kinetic pressure) of the reference velocity.

Because the terms $V_0^2/2g$ and $\rho V_0^2/2$ occur so often in hydraulics and fluid mechanics calculations, they have been given special names: velocity head and kinetic pressure, respectively. Another term, dynamic pressure, which is closely associated with kinetic pressure, is equal to the difference between the total pressure (pressure at a point of stagnation) and the static pressure. Under conditions where Bernoulli's equation applies, kinetic and dynamic pressure are equal. However, in high-speed gas flow, where compressibility effects are important, the two may have significantly different values.

For the conduit of Fig. 5.7, the relative pressure (pressure coefficient) along the centerline and along the boundary at various sections is plotted in Fig. 5.8. Because there are greater variations of velocity near the boundary, the pressure variations are also greater along the boundary than they are along the centerline.

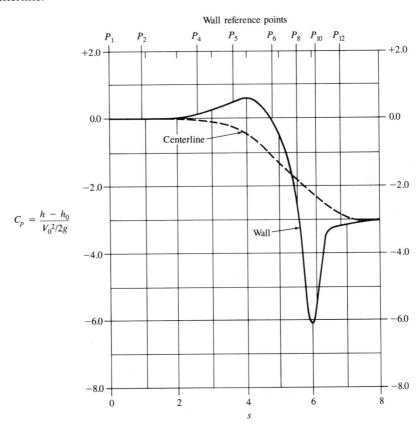

FIGURE 5.8

Relative piezometric head along the wall and centerline of transition in Fig. 5.7.

EXAMPLE 5.5 If air ($\rho = 1.2$ kg/m³) flows through the two-dimensional passage of Fig. 5.7, what is the difference in pressure between points P_5 and P_{10} when the flow rate is 60 m³/s per meter of width? If water flows through the passage, what is the difference in pressure between the same two points? Assume the half-size of the passage at the reference section is 2 m and the downstream size is 1 m. The view shown in Fig. 5.7 is an elevation view.

Solution To solve for the pressure difference, we use Fig. 5.8, which shows that at point P_5,

$$\frac{h_5 - h_0}{V_0^2/2g} = 0.5 \quad \text{or} \quad h_5 - h_0 = 0.5\frac{V_0^2}{2g}$$

whereas at point P_{10}

$$h_{10} - h_0 = -5.4 \frac{V_0^2}{2g}$$

Then

$$h_5 - h_{10} = \frac{V_0^2}{2g}[0.5 - (-5.4)] = 5.9 \frac{V_0^2}{2g}$$

But

$$h = \frac{p}{\gamma} + z$$

so

$$\frac{p_5}{\gamma} + z_5 - \frac{p_{10}}{\gamma} - z_{10} = 5.9 \frac{V_0^2}{2g}$$

$$p_5 - p_{10} = \gamma \left[(z_{10} - z_5) + 5.9 \frac{V_0^2}{2g} \right]$$

For air flow,

$$p_5 - p_{10} = (1.2 \times 9.81) \left[(1 - 2) + 5.9 \left(\frac{V_0^2}{2 \times 9.81} \right) \right]$$

But

$$V_0 = \frac{60}{4} = 15 \text{ m/s}$$

hence

$$p_5 - p_{10} = 11.77(-1.0 + 67.7) = 785 \text{ Pa} \qquad \blacktriangleleft$$

For water flow,

$$p_5 - p_{10} = 9810(1 - 2 + 67.7) = 654 \text{ kPa} \qquad \blacktriangleleft$$

Pressure Distribution Around a Circular Cylinder—Ideal Fluid

If a fluid is nonviscous (an *ideal* fluid) and if the flow of such a fluid is initially incompressible and irrotational, then the flow will be irrotational throughout the entire flow field.* Then, if the flow is also steady, Bernoulli's equation will apply because all the restrictions for Bernoulli's equation will have been satisfied. The flow pattern about a circular cylinder with such restrictions is shown in Fig. 5.9.

Because the flow pattern is symmetrical with either the vertical or the horizontal axis through the center of the cylinder, the pressure distribution

---

*This can be seen in a qualitative sense if one visualizes a small spherical mass of fluid within a non-viscous flow field. Since pressure forces act normal to the surface of the spherical mass, the mass deforms if the pressure is not of equal intensity over the entire surface. However, the mass cannot rotate (irrotational situation) because there is no shear stress (viscosity is zero) on the surface of the sphere to possibly cause rotation.

FIGURE 5.9

Irrotational flow past a cylinder.

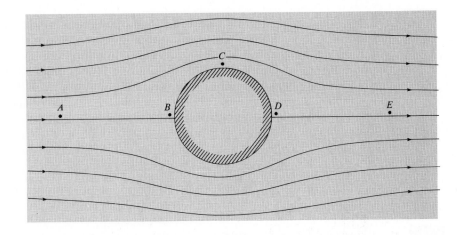

Note: Positive C_p
plotted inward from
cylinder surface;
negative C_p plotted
outward.

$$C_p = \frac{p - p_0}{\rho V_0^2 / 2}$$

$C_p = -3.0$

Negative C_p

$C_p = +1$

C_p

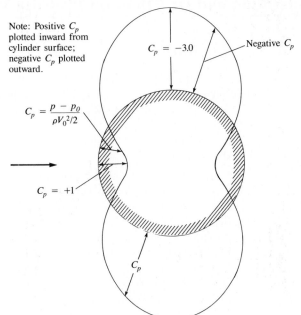

FIGURE 5.10

Pressure distribution on a cylinder — irrotational flow.

on the surface of the cylinder, obtained by application of Bernoulli's equation, is also symmetrical. In Fig. 5.10 the relative pressure C_p is plotted outward (negative) or inward (positive) from the surface of the cylinder, depending on the sign of the relative pressure and on a line normal to the surface of the cylinder. It should also be noted that p_0 and V_0 are the pressure and velocity of the free stream far upstream or downstream of the body. Thus we see that the points at the front and rear of the cylinder are points of stagnation ($C_p = +1.0$) and that the minimum pressure ($C_p = -3.0$) occurs at the midsection where the velocity is highest. If we visualize a fluid particle as it travels around the cylinder from *A* to *B* to *C* and finally to *D* in Fig. 5.9, we see that it first decelerates, which is consistent with the increase in pressure from *A* to *B*. Then as it passes from *B* to *C*, it is accelerated to its highest speed by the action of the pressure gradient; that is, the pressure decreases over the entire path from *B* to *C*. Next, as the particle travels from *C* to *D*, its momentum at *C* is sufficient to allow it to travel to *D* against the adverse pressure gradient (pressure increases in the direction of flow here). Finally, the particle accelerates to the freestream velocity in its passage from *D* to *E*. Understanding this qualitative description of how the fluid travels from one point to another will be helpful when the phenomenon of separation is explained in the next section.

5.5   Separation and Its Effect on Pressure Variation

Separation

In Chapter 4 we introduced the concept of separation and emphasized that separation usually occurs where the physical boundaries turn away from the main stream of flow. We will now consider the basic cause of separation and its consequences.

     Consider the flow of a real (viscous) fluid past a circular cylinder, as shown in Fig. 5.11. The flow pattern upstream of the midsection of the cylinder is quite similar to the flow pattern of irrotational flow about a cylinder, except very close to the boundary surface. Here, because of the viscous resistance, a thin layer of fluid has its velocity reduced from that predicted by irrotational theory. In fact, the fluid particles directly adjacent to the surface have zero velocity (this "no-slip" condition at a boundary is characteristic of all real fluids). The normal tendency is for the layer of reduced velocity (called the *boundary layer*) to grow in thickness in the direction of flow. However, because the main stream of fluid outside the boundary layer is accelerating in the same direction, the boundary layer remains quite thin up to approximately the midsection.

FIGURE 5.11 _____

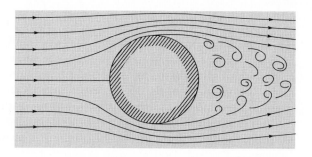

Flow of a real fluid past a circular cylinder.

Downstream of the midsection (from C to D in Fig. 5.9), the normal ir-rotational-flow pattern shows a significant deceleration of fluid next to the boundary, with a corresponding increase in pressure. For real flow, however, deceleration of the fluid next to the boundary is limited because its velocity is already small (much smaller than for irrotational flow) because of the viscous resistance. Therefore, the fluid near the boundary can proceed only a very short distance against the adverse pressure gradient before stopping completely. Once the motion of the fluid next to the boundary ceases, this causes the main stream of flow to be diverted away, or to be "separated" from the boundary. Thus the process of separation is produced. Downstream of this point of separation, the fluid outside the surface of separation has a high velocity and the fluid inside the surface of separation has a relatively low velocity. Because of the steep veloc-ity gradient along the surface of separation, eddies are generated, which through viscous action are finally dissipated into heat.

Since the location of the point of separation on a rounded body, such as the cylinder, depends on the character of the flow in the boundary layer, it is not surprising that roughness of the surface or turbulence in the approach flow has an effect on the location of the separation point. These effects will be considered in more detail in Chapter 11. For angular-type bodies, however, the point of separation occurs at the sharp break in boundary configuration. Thus in Fig. 5.12 we observe flow separation at the boundary discontinuity for flow past a square rod and a disk and through a sharp-edged orifice.

We have already indicated that the point of separation may be related to the shape and roughness of a body. Because separation is closely associated with the viscous resistance of the fluid, it is not surprising that the Reynolds number, the value of which is inversely proportional to the relative viscous resistance, is an indicator of the onset of separation. For example, in flow past a circular cylin-der, separation occurs for a Reynolds number ($VD\rho/\mu$) greater than 50. For Reynolds numbers less than 50, the entire flow field is dominated by relatively large viscous stresses that inhibit the onset of eddy motion in the fluid.

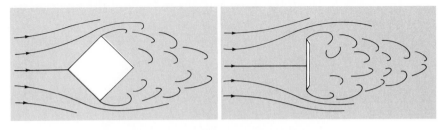

FIGURE 5.12

Flow past a square rod and a disk and through a sharp-edged orifice.

The Effect of Separation on Pressure Distribution

When separation occurs, the flow pattern is changed from that of irrotational flow. Therefore, one would expect corresponding changes in the pressure distribution, which do, in fact, occur. For flow past a blunt body, the slight change (due to viscosity) of the flow pattern next to the forward part of the body changes the pressure distribution only slightly. However, in the zone of separation, marked changes in pressure result. It is a rule of thumb that the pressure that prevails at the point of separation also prevails over the body within the zone of separation. This is borne out for flow past a cylinder and a disk, as shown in Fig. 5.13. Note that for both disk and cylinder, the pressure on the rear half of the body is much less than the pressure on the front half; consequently, a net force in the downstream direction is imposed on the body. This force is the drag of the body, which will be considered in more detail in Chapter 11.

5.6   Cavitation

Cavitation occurs in liquid systems when the pressure at any point in the system is reduced to the vapor pressure of the liquid. Under such conditions, vapor bubbles form (boiling occurs) and then collapse (condense), thereby producing dynamic effects that can often lead to decreased efficiency and/or equipment failure.

Consider water flow through the pipe restriction shown in Fig. 5.14. In Fig. 5.14a, the physical configuration and the plots of piezometric head along the wall of the conduit for different flows are shown. In Fig. 5.14b, the dimensionless plot of piezometric head along the wall is shown. Here, the reference point is taken at the center of the pipe. For low and medium rates of flow, there is a relatively small drop in pressure at the constriction. Therefore, the pressure in the water remains well above the vapor pressure, and cavitation does not

FIGURE 5.13

Pressure distribution on a circular cylinder and a disk. (a) Circular cylinder, $Re = 10^5$; after Fage and Warsap (1). (b) Disk, $Re = 10^5$; after Rouse (8).

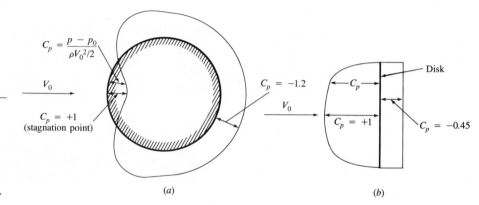

$$C_p = \frac{p - p_0}{\rho V_0^2/2}$$

V_0

$C_p = +1$ (stagnation point)

$C_p = -1.2$

V_0

Disk

C_p

$C_p = +1$

$C_p = -0.45$

(a)   (b)

FIGURE 5.14

Flow through pipe restriction. (a) Variation of piezometric head. (b) Relative piezometric head.

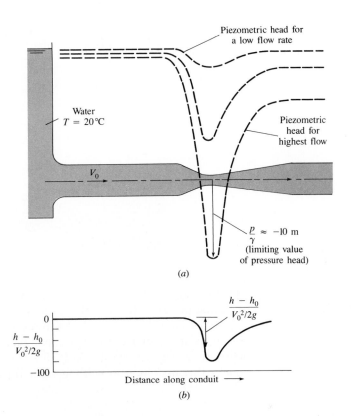

occur. This is indicated in Fig. 5.14*a* for the low rate where the piezometric head is everywhere above the conduit itself. If the rate of flow is high, however, the piezometric-head line actually drops below the pipe, thereby indicating a less-than-atmospheric pressure for the liquid in the constriction. Of course, the pressure can drop no lower than the vapor pressure of the liquid, because at this pressure the liquid boils. Such a condition is depicted in Fig. 5.15*a*, where vapor bubbles are shown forming at the restriction, growing in size, then collapsing as they move into a region of higher pressure as they are swept downstream with the flow. Experimental and theoretical studies reveal that very high intermittent pressures develop in the vicinity of the bubbles when they collapse. These pressures may exceed 800 MPa (115,000 psi) (4).

Therefore, if the bubbles collapse close to physical boundaries, such as pipe walls, pump impellers, ship propellers, valve casings, or dam-spillway floors, they can cause damage.* Usually this damage occurs in the form of a fatigue failure brought about by the action of millions of bubbles impacting (in effect,

---

Cavitation in an enclosed pipe or machine can often be detected by the characteristic sound generated. In large structures, it sounds like large rocks are being carried through the system and are hitting the sides of the conduit.

FIGURE 5.15

Formation of vapor bubbles in the process of cavitation. (a) Cavitation. (b) Cavitation—higher flow rate.

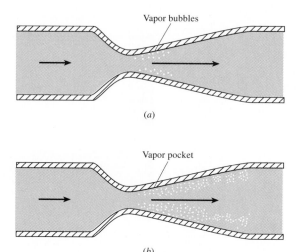

implooding) against the surface material over a long period of time, thus producing pitting of the material in the vicinity of the zone of cavitation.

If the flow is increased even more than indicated above, the minimum pressure is still restricted to the vapor pressure of the water, but the zone of vaporization increases, as shown in Fig. 5.15b. For such a condition the entire vapor pocket may intermittently grow and collapse, producing serious vibration problems. Fig. 5.16 shows severe cavitation damage that occurred on a centrifu-

FIGURE 5.16

Cavitation damage to impeller of a centrifugal pump.

*Cavitation damage to a
power dam spillway tunnel.*

gal pump impeller, and Fig. 5.17 shows very serious erosion produced by cavitation in a spillway tunnel of Hoover Dam. Needless to say, cavitation should be avoided or minimized by the proper design of equipment and structures and by their proper operation.

See Knapp, Daily, and Hammitt (4) and Hammitt (3) for background information along with detailed discussions of cavitation. Basic research in cavitation has focused on topics such as bubble dynamics, boundary layer effects, and surface roughness effects. Kuiper (5) summarizes past and current research on these topics. Other research has concentrated on such topics as erosion and noise reduction in cavitating fluid machinery, reduced efficiency of cavitating propellers, and prevention of cavitation in spillways and tunnels. Typical papers on such topics are given in references (2), (6), (7), and (9).

Problems

5.1 The hypothetical liquid in the tube shown in the figure has zero viscosity and a specific weight of 8 kN/m^3. If $p_B - p_A$ is equal to 7 kPa, one can conclude that the liquid in the tube is being accelerated a) upward, b) downward, c) neither: acceleration = 0.

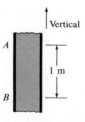

PROBLEM 5.1

5.2 If the piston and water are accelerated upward at a rate of 0.5g, what will be the pressure at a depth of 2 ft in the water column?

PROBLEM 5.2

5.3 Water stands at a depth of 10 ft in a vertical pipe that is open at the top and closed at the bottom by a piston. What upward acceleration of the piston is necessary to create a pressure of 8 psi immediately above the piston?

5.4 What pressure gradient is required to accelerate water in a horizontal pipe at a rate of 3 m/s²?

5.5 A pipe slopes upward in the direction of liquid flow at an angle of 30° with the horizontal. What is the pressure gradient in the flow direction along the pipe in terms of the specific weight of the liquid if the liquid is decelerating (accelerating opposite to flow direction) at a rate of 0.3g?

5.6 What pressure gradient is required to accelerate kerosene (S = 0.8) vertically upward in a vertical pipe at a rate of 0.2g?

5.7 Water is accelerated from rest in a horizontal pipe that is 100 m long and 30 cm in diameter. If the acceleration rate (toward the downstream end) is 6 m/s², what is the pressure at the upstream end if the pressure at the downstream end is 70 kPa gage?

5.8 Water stands at a depth of 13 ft in a vertical pipe that is closed at the bottom by a piston. Assuming that $p_v = 0$ psia, determine the maximum downward acceleration that can be given to the piston without causing the water immediately above it to vaporize.

5.9 A liquid with a specific weight of 100 lb/ft³ is in the conduit. This is a special kind of liquid that has zero viscosity. The pressure at points A and B is 170 psf and 100 psf, respectively. Which one (or more) of the following conclusions can one draw with certainty? a) The velocity is in the positive s direction. b) The velocity is in the negative s direction.

c) The acceleration is in the positive s direction. d) The acceleration is in the negative s direction.

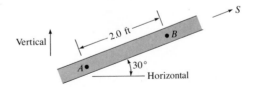

PROBLEM 5.9

5.10 If the velocity varies linearly with distance through this water nozzle, what is the pressure gradient, dp/dx, halfway through the nozzle?

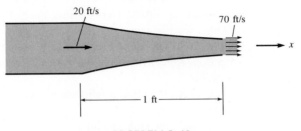

PROBLEM 5.10

5.11 Consider the flow of water over the surfaces shown below. For each case the depth of water at section D–D is the same (1 ft), and the mean velocity is the same and equal to 10 ft/s. Which of the following statements are valid?

    a. $p_B < p_C < p_A$
    b. $p_A < p_B < p_C$
    c. $p_A = p_B = p_C$
    d. $p_C > p_B > p_A$
    e. $p_B > p_C > p_A$

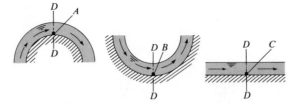

PROBLEM 5.11

5.12 In this two-dimensional conduit, which discharges water into the atmosphere, the discharge is given by $q = q_0 t/t_0$, where q_0 and t_0 are 0.2 m³/s per meter and 1 s, respectively. Also, $B = 30$ cm, $b = 20$ cm, $L = 1$ m, and $T = 10°C$. What is the rate of change of pressure with respect to x at point A when $t = 2$ s?

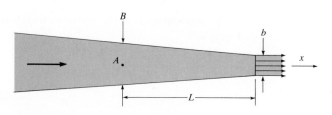

PROBLEM 5.12

5.13   In Prob. 4.48, assume that $B = 40$ cm, $q_0 = 0.2$ m^2/s, and $t_0 = 0.1$ s. For these conditions, what will be the pressure gradient at $x = 2B$ when $t = 0.5$ s if water is flowing?

5.14   In Prob. 4.45, what is the pressure gradient in terms of ρ at section A–A when $t = 0.5$ s?

5.15   The flow rate in cubic meters per second for the nozzle shown is given by the following equation: $Q = 0.03t$. Here the time t is given in seconds, the fluid is a nonviscous liquid with a specific gravity of 1.6, $D_0 = 10$ cm, $d_n = 4$ cm, $L_1 = 60$ cm, and $L_2 = 30$ cm.

   a. If the simplifying assumption is made that the velocity is uniform across any given section, what will be the convective acceleration of a fluid particle at point A when $t = 2$ s?

   b. With the same conditions as given above, what will be the local acceleration at point A?

   c. With the same conditions as above, what is the pressure gradient in the z direction at point A?

   d. What is the pressure at point A if the pressure surrounding the free jet is atmospheric?

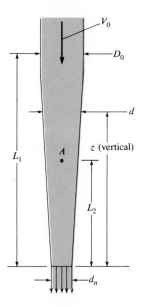

PROBLEMS 5.15, 5.16

5.16 A nozzle similar to that shown in the figure (discharging vertically downward) is to be designed so that the pressure between the base of the nozzle and the nozzle tip is zero gage for a given discharge of water. Derive a formula for d as a function of V_0, z_0, z, and D_0 to achieve the desired objective.

5.17 In Prob. 4.51, assume that the air density is 1.4 kg/m^3. What is the pressure gradient at point A? If the maximum disk radius is 50 cm, what is the pressure at point A? Neglect frictional effects (viscous effects) in this problem, and assume that the atmospheric pressure (pressure at outlet) is 100 kPa.

5.18 This tank is accelerated in the x direction in such a way that the liquid surface does not change slope. What is the acceleration of the tank?

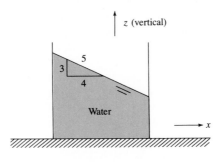

PROBLEM 5.18

5.19 The closed tank shown, which is full of liquid, is accelerated downward at $1.4g$ ft/s^2 and to the right at $0.9g$ ft/s^2. Here $L = 3$ ft, $H = 4$ ft, and the specific gravity of the liquid is 1.1. Determine $p_C - p_A$ and $p_B - p_A$.

5.20 The closed tank shown, which is full of liquid, is accelerated downward at $\frac{2}{3}g$ m/s^2 and to the right at g m/s^2. Here $L = 2$ m, $H = 3$ m, and the liquid has a specific gravity of 1.5. Determine $p_C - p_A$ and $p_B - p_A$.

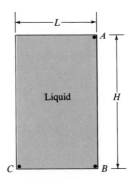

PROBLEMS 5.19, 5.20

5.21 A truck carries a tank that is open at the top. The tank is 18 ft long, 6 ft wide, and 7 ft high. Assuming that the driver will not accelerate or decelerate the truck at a rate greater than 8.02 ft/s², to what maximum depth may the tank be filled so that water will not be spilled?

5.22   A truck carries a cylindrical tank (axis vertical) of liquid that is open at the top. Assuming that the driver will not accelerate or decelerate the truck at a rate greater than $\frac{1}{3}g$, to what maximum depth may the tank be filled so that water will not be spilled? Also, if the truck goes around an unbanked curve ($r = 50$ m), what maximum speed can it go before water will be spilled? Assume that the tank's height is the same as its diameter and that the depth for the second part of the problem is the same as that for the first.

5.23   Given: Liquid orientation in a tank as shown. The conditions could be caused by a) constant acceleration of the tank to the right, b) the tank being placed on a vehicle that travels at a constant speed about a circular track (center of the circle to the left of the vehicle), c) the tank being placed on a vehicle that travels at a constant speed about a circular track (center of the circle to the right of the vehicle), d) none of the above apply.

PROBLEM 5.23

5.24   The tank shown is 4 m long, 3 m high, and 3 m wide, and it is closed except for a small opening at the right end. It contains oil (S = 0.85) to a depth of 2 m in a static situation. If the tank is uniformly accelerated to the right at a rate of 9.81 m/s², what will be the maximum pressure intensity in the tank during acceleration?

$$\tan \alpha = -\frac{a_x}{g}$$

$$V_1 = V_2 = 4(3) = \tfrac{1}{2} d^2 (3)$$

$$d = 2.83$$

$$\int dp = -\gamma \frac{a_x}{g} \int_c^A dx - \gamma \int_B^C dz$$

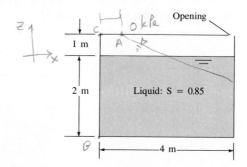

PROBLEM 5.24

5.25   This closed tank, which is 4 ft in diameter, is filled with water and is spun around its vertical centroidal axis at a rate of 15 rad/s. An open piezometer is connected to the tank as shown so that it is also rotating with the tank. For these conditions, what is the pressure at the center of the bottom of the tank?

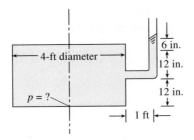

PROBLEM 5.25

5.26 A tank of liquid (S = 0.80) that is 1 ft in diameter and 1.2 ft high (h = 1.2 ft) is rigidly fixed (as shown) to a rotating arm having a 2-ft radius. The arm rotates such that the speed at point A is 20 ft/s. If the pressure at A is 30 psf, what is the pressure at B?

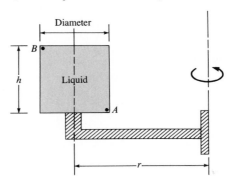

PROBLEM 5.26

5.27 A closed tank of water is rotated about a vertical axis (see the figure), and at the same time the entire tank is accelerated upward at 4 m/s². If the rate of rotation is 10 rad/s, what is the difference in pressure between points A and B ($p_B - p_A$)? Point B is at the bottom of the tank at a radius of 0.5 m from the axis of rotation, and point A is at the top on the axis of rotation.

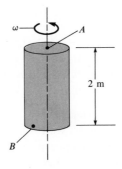

PROBLEM 5.27

5.28   A U-tube is rotated at 30 rev/min about one leg. The fluid at the bottom of the U-tube has a specific gravity of 3.0. The distance between the two legs of the U-tube is 1 ft. A 6-in. height of another fluid is in the outer leg of the U-tube. Both legs are open to the atmosphere. Calculate the specific gravity of the other fluid.

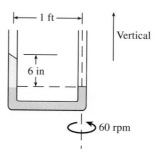

PROBLEM 5.28

5.29   If the U-tube is rotated about the axis 0–0 at a rate of 16.06 rad/s, determine the new position of the water surface in the outside leg during rotation. The values of L_1, L_2, and L_3 are 1 m, 18 cm, and 30 cm, respectively.

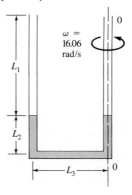

PROBLEM 5.29

5.30   The U-tube is attached to platform B, and the liquid levels in the U-tube are shown for conditions at rest. The platform and U-tube are then rotated about axis $A–A$ at a rate of 4 rad/s. What will be the elevation of the liquid in the smaller leg of the U-tube after rotation?

5.31   A manometer spins about one leg as shown. The leg about which the manometer rotates contains water with a height of 10 cm. The other leg, which is 1 m from the axis of rotation, contains mercury (specific gravity = 13.6) with a height of 1 cm. What is the rotational speed (in radians per second)?

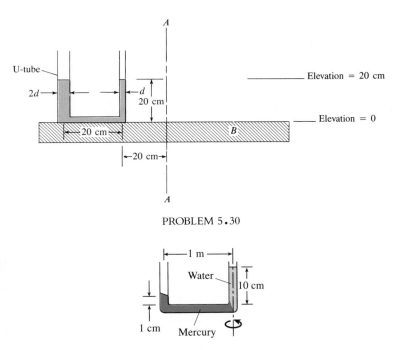

PROBLEM 5.30

PROBLEM 5.31

5.32 A fuel tank for a rocket in space under a zero-g environment is rotated to keep the fuel in one end of the tank. The system is rotated at 3 rev/min. The end of the tank (point A) is 1.5 m from the axis of rotation, and the fuel level is 1 m from the rotation axis. The pressure in the nonliquid end of the tank is 0.1 kPa, and the density of the fuel is 800 kg/m³. What is the pressure at the exit (point A)?

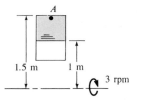

PROBLEM 5.32

5.33 Water stands in these tubes as shown when no rotation occurs. Derive a formula for the angular speed at which water will just begin to spill out of the small tube when the entire system is rotated about axis A–A.

5.34 Mercury is the liquid in the rotating U-tube. Determine the rate of rotation ω if $\ell = 8$ in. Then, if rotation is stopped, to what level z will the mercury level drop in the larger leg?

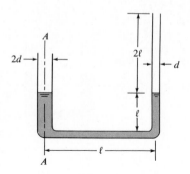

PROBLEM 5.33

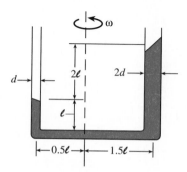

PROBLEM 5.34

5.35   Water fills a slender tube 1 cm in diameter, 50 cm long, and closed at one end. When the tube is rotated in the horizontal plane about its open end at a constant speed of 60.8 rad/s, what force is exerted on the closed end?

5.36   Mercury maintains the position shown in the rotating tube. What is the rate of rotation in terms of g and ℓ?

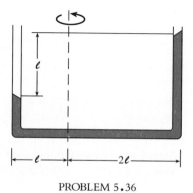

PROBLEM 5.36

5.37 The U-tube is shown with water standing in it when there is no rotation.

a. If the tube is rotated about the left leg at a rate of 5 rad/s, where will the water stand in the tube for $\ell = 30$ cm?

b. Where will the water surface be when the rate of rotation is increased to 15 rad/s for $\ell = 30$ cm?

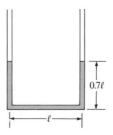

PROBLEM 5.37

5.38 This tube with liquid is rotated about a vertical axis as indicated. If the rate of rotation is 8 rad/s, what will be the pressure in the liquid at point A (at the axis of rotation) and at point B? Then, if the tube is rotated at a rate of 20 rad/s, what will be the pressure at points A and B?

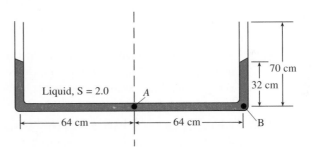

PROBLEM 5.38

5.39 Water stands in the closed-end U-tube when there is no rotation. If $\ell = 10$ cm and if the entire system is rotated about axis A–A, at what angular speed will water just begin to spill out of the small tube? Assume that the temperature for the system is the same before and after rotation.

5.40 A simple centrifugal pump consists of a 10-cm disk with radial ports as shown. Water is pumped from a reservoir through a central tube on the axis. The wheel spins at 2000 rev/min, and the liquid discharges to atmospheric pressure. To establish the maximum height for operation of the pump, assume that the flow rate is zero and the pressure at the pump intake is atmospheric pressure. Calculate the maximum operational height z for the pump.

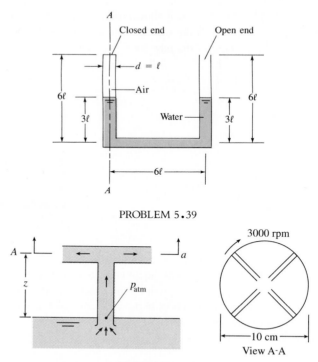

PROBLEM 5.39

PROBLEM 5.40

5.41 A closed cylindrical tank of water is rotated about its horizontal axis as shown. The water inside the tank rotates with the tank ($V = r\omega$). Derive an equation for dp/dz along a vertical-radial line through the center of rotation. What is dp/dz along this line for $z = -1$ m, $z = 0$, and $z = +1$ m when $\omega = 5$ rad/s? Here $z = 0$ at the axis.

5.42 For the conditions of Prob. 5.41, derive an equation for the maximum pressure difference in the tank as a function of the significant variables.

5.43 The tank shown is 5 ft in diameter and 12 ft long and is closed and filled with water. It is rotated about its horizontal-centroidal axis, and the water in the tank rotates with the tank ($V = r\omega$). The maximum velocity is 20 ft/s. What is the maximum difference in pressure in the tank? Where is the point of minimum pressure?

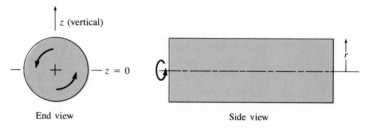

PROBLEMS 5.41, 5.42, 5.43

5.44 The velocity in the outlet pipe from this reservoir is 13 ft/s and $h = 15$ ft. Because of the rounded entrance to the pipe, the flow is assumed to be irrotational. Under these conditions, what is the pressure at A?

5.45 The velocity in the outlet pipe from this reservoir is 8 m/s and $h = 15$ m. Because of the rounded entrance to the pipe, the flow is assumed to be irrotational. Under these conditions, what is the pressure at A?

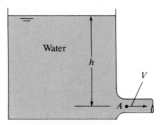

PROBLEMS 5.44, 5.45

5.46 Water flows from the large orifice at the bottom of the tank as shown. Assume that the flow is irrotational. Point B is at zero elevation, and point A is at 1 ft elevation. If $V_A = 9$ ft/s at an angle of 45° with the horizontal and if $V_B = 20$ ft/s vertically downward, what is the value of $p_A - p_B$?

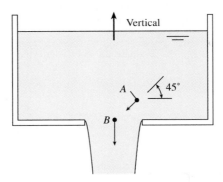

PROBLEM 5.46

5.47 Water is forced out of this cylinder by the piston. If the piston is driven at a speed of 5 ft/s, what will be the speed of efflux of the water from the nozzle if $d = 2$ in. and $D = 4$ in.? Neglecting friction and assuming irrotational flow, determine the force F that will be required to drive the piston. The exit pressure is atmospheric pressure.

5.48 Refer to the figure for Prob. 4.50 on p. 146. Assume that the jet issuing from the nozzle into the atmosphere has a diameter of 0.50 ft. Water is flowing at a constant rate of 20 ft³/s. Assuming irrotational flow, find the gage pressure in the pipe.

5.49 Refer to Prob. 4.40 on p. 144. The air density is 1.2 kg/m³, and $U_0 = 30$ m/s. Assume irrotational flow. Determine the pressure in the air at $x = r_0$, $x = 1.1r_0$, and $x = 2r_0$.

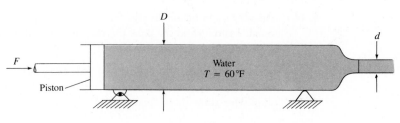

PROBLEM 5.47

5.50 Refer to Prob. 4.48 on p. 146. Water is discharging into the atmosphere from the slot, and $B = 20$ cm. Assume irrotational flow. If the pressure in the flow passage having the dimension B is 100 kPa gage, what is the discharge per unit of width of slot?

5.51 An engineer is designing an "elbow" meter for measuring the discharge of a liquid. The elbow meter, shown below, consists of an elbow with pressure taps mounted across the meter. The elbow has a rectangular cross section. The velocity across the elbow varies as $V = K/r$, where K is a constant and r is the radius of curvature of the flow. Assume this is an irrotational flow field, so Bernoulli's equation is valid across streamlines.

   a. Develop an equation for the discharge, Q, through the elbow as a function of pressure difference Δp between the taps in the form

$$Q = (2 \Delta p/\rho)^{0.5} A_c f(r_2/r_1)$$

where ρ is the fluid density, A_c is the cross-sectional area of the elbow, and r_2 and r_1 are the outer and inner radii of the elbow, respectively. Find the functional relationship $f(r_2/r_1)$.

   b. Evaluate the coefficient $f(r_2/r_1)$ for an elbow meter with a radius ratio of 1.5.

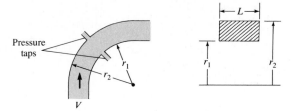

PROBLEM 5.51

5.52 Liquid flows with a free surface around a bend. The liquid is inviscid and incompressible, and the flow is steady and irrotational. The velocity varies with the radius across the flow as $V = 1/r$ m/s, where r is in meters. Find the difference in depth of the liquid from the inside to the outside radius. The inside radius of the bend is 1 m and the outside radius is 2 m.

5.53 The flow pattern through the pipe contraction is as indicated, and the discharge of water is 70 cfs. For $d = 2$ ft and $D = 6$ ft, what will be the pressure at point B if the pressure at point A is 3500 psf?

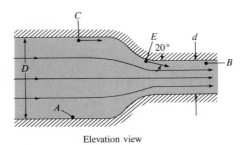

Elevation view

PROBLEMS 5.53, 5.54

5.54 Consider the flow of water in the pipe contraction. Assuming steady, irrotational, nonviscous flow, find the pressure at point E if the velocity at point E is 50 ft/s and the pressure and velocity at point C are 10 psi and 10 ft/s, respectively?

5.55 Refer to Prob. 4.51 on p. 146. Determine and plot the variation of the gage pressure on the bottom disk. Use your own good assumption(s) regarding the pressure variation from $r = 0$ to $r = D/2$. Assume $\rho = 1.2$ kg/m^3. What pressure force acts on the disk? Assume irrotational flow.

5.56 The pressure coefficient distribution on a cylinder in a crossflow is given by

$$C_p = 1 - 4 \sin^2\theta$$

where θ is the angular displacement from the forward stagnation point. Assume that two pressure taps are located at $\theta = \pm30°$ and connected to a water manometer. The cylinder is immersed in air with a velocity of 50 m/s and a density of 1.2 kg/m^3 and rotated 20° from the position shown in the diagram. What will be the deflection on the manometer, in centimeters?

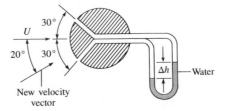

PROBLEM 5.56

5.57 The velocity and pressure are given at two points in the flow field. Assume that the two points lie in a horizontal plane and that the fluid density is uniform in the flow field and is equal to 1000 kg/m^3. Assume steady flow. Then, given these data, determine which of the following statements is true. a) The flow in the contraction is nonuniform and irrotational. b) The flow in the contraction is uniform and irrotational. c) The flow in the contraction is nonuniform and rotational. d) The flow in the contraction is uniform and rotational.

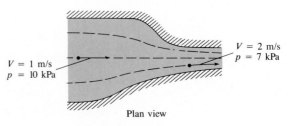

Plan view

PROBLEM 5.57

5.58   A Pitot static tube is mounted on an airplane to measure airspeed. At an altitude of 10,000 ft, where the temperature is 23°F and the pressure is 10.0 psia, a pressure difference corresponding to 10 in. of water is measured. What is the airspeed?

5.59   An arm with a Pitot tube on the end is rotated in a horizontal plane 10 cm below a liquid surface as shown. The arm is 20 cm long, and the tube at the center of rotation extends above the liquid surface. The liquid in the tube is the same as that in the tank and has a specific weight of 10,000 N/m³. Find the location of the liquid surface in the central tube.

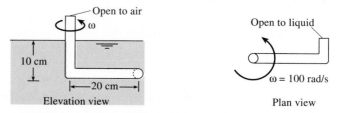

PROBLEM 5.59

5.60   A glass tube with a 90° bend is open at both ends. It is inserted into a flowing stream of water so that one opening is directed upstream and the other is vertical (see Fig. 5.5). If the water surface in the vertical tube is 9 in. higher than the stream surface, what is the velocity?

5.61   A glass tube like the one described in Prob. 5.60 is inserted into a flowing stream of water with one opening directed upstream and the other end vertical (see Fig. 5.5). If the water velocity is 2 m/s, how high will the water rise in the vertical leg relative to the level of the water surface of the stream?

5.62   A bourdon-tube gage is tapped into the center of a disk as shown. Then for a disk that is about 1 ft in diameter and for an approach velocity of air (V_0) of 40 ft/s, the gage would read a pressure intensity that is a) less than $\rho V_0^2/2$, b) equal to $\rho V_0^2/2$, c) greater than $\rho V_0^2/2$.

5.63   An air–water manometer is connected to a Pitot tube used to measure air velocity. If the manometer deflects 1.60 in., what is the velocity? Assume $T = 60°F$ and $p = 15$ psia.

5.64   A flow-metering device consists of a stagnation probe at station 2 and a static pressure tap at station 1. The cross-sectional area of the tube at station 2 is one-half that at station 1. Air with a density of 1.2 kg/m³ flows through the duct. A water manometer is connected

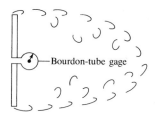

PROBLEM 5.62

between the stagnation probe and the pressure tap, and a deflection of 10 cm is measured. What is the velocity at station 2?

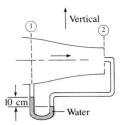

PROBLEM 5.64

5.65   A "spherical" Pitot probe is used to measure the flow velocity in water. Pressure taps are located at the forward stagnation point and at 90° from the forward stagnation point. The speed of fluid next to the surface of the sphere varies with θ ($V = 1.5\ V_0 \sin \theta$, where V_0 is the free-stream velocity and θ is measured from the forward stagnation point). The pressure taps are at the same level; that is, they are in the same horizontal plane. The pressure difference between the two taps is 2 kPa. What is the free-stream velocity V_0?

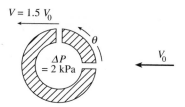

PROBLEM 5.65

5.66   An engineer is designing a velocity probe that consists of a sphere similar to the one in Prob. 5.65 except that the second pressure tap (the one not located at the stagnation point) is not located 90° from the forward one. The engineer wants to select the angle θ such that the free-stream velocity is related to the pressure difference by $V_0 = \sqrt{\Delta p/\rho}$. Find the value of θ to achieve this.

5.67   As shown in the figure, water flows in the two-dimensional bend, which turns in a horizontal plane. Here $V_1 = V_2 = 13$ m/s, $p_1 = 10$ kPa gage, and $p_2 = 0$ gage. Would you expect cavitation to occur anywhere within the bend if the atmospheric pressure is 100 kPa absolute? *Hint:* The streamlines are drawn to scale.

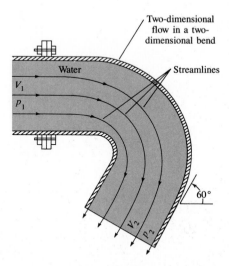

PROBLEM 5.67

5.68 When gage A indicates a pressure of 100 kPa, then cavitation just starts to occur in the venturi meter. If $D = 40$ cm and $d = 10$ cm, what is the water discharge in the system for this condition of incipient cavitation? The atmospheric pressure is 100 kPa, and the water temperature is 10°C. Neglect gravitational effects.

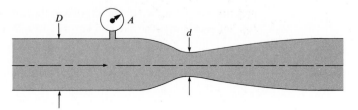

PROBLEM 5.68

5.69 Two Pitot tubes are shown. The one on the left is used to measure the velocity of air, and it is connected to an air–water manometer as shown. The one on the right is used to measure the velocity of water, and it too is connected to an air–water manometer as shown. If the deflection h is the same for both manometers, then one can conclude that a) $V_A = V_W$, b) $V_A > V_W$, c) $V_A < V_W$.

5.70 A Pitot tube is used to measure the velocity at the center of a 12-in. pipe. If kerosene at 33°F is flowing and the deflection on a mercury–kerosene manometer connected to the Pitot tube is 7 in., what is the velocity?

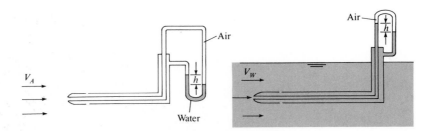

PROBLEM 5.69

5.71 A Pitot tube used to measure air velocity is connected to a differential pressure gage. If the air temperature is 20°C at standard atmospheric pressure at sea level, and if the differential gage reads a pressure difference of 2 kPa, what is the air velocity?

5.72 A Pitot tube used to measure air velocity is connected to a differential pressure gage. If the air temperature is 60°F at standard atmospheric pressure at sea level, and if the differential gage reads a pressure difference of 10 psf, what is the air velocity?

5.73 A Pitot tube is used to measure the gas velocity in a duct. A pressure transducer connected to the Pitot tube registers a pressure difference of 1 psi. The density of the gas in the duct is 0.1 lbm/ft^3. What is the gas velocity in the duct?

5.74 A sphere moves horizontally through still water at a speed of 10 ft/s. A short distance directly ahead of the sphere (call it point A), the velocity, with respect to the earth, induced by the sphere is 1 ft/s in the same direction as the motion of the sphere. If p_0 is the pressure in the undisturbed water at the same depth as the center of the sphere, then the value of the ratio p_A/p_0 will be a) less than unity, b) equal to unity, c) greater than unity.

5.75 Body A travels through water at a constant speed of 12 m/s. Velocities at points B and C are induced by the moving body and are observed to have magnitudes of 4 m/s and 2 m/s, respectively. What is $p_B - p_C$?

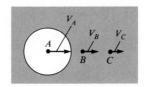

PROBLEM 5.75

5.76 Consider the flow field produced by the ship shown in Fig. 4.14a, p. 121. At a point A, 3 m ahead of the bow and 2 m below the keel elevation, the water velocity was measured as 0.20 m/s with a direction vertically downward. At a point B, at the same elevation but midway between bow and stern, the water velocity was measured as 0.8 m/s directed horizontally toward the stern. The ship is steaming in salt water ($\rho = 1050$ kg/m^3) at a speed of 7 m/s. Calculate $p_A - p_B$.

5.77 Water in a flume is shown for two conditions. If the depth d is the same for each case, will gage A read greater or less than gage B? Explain.

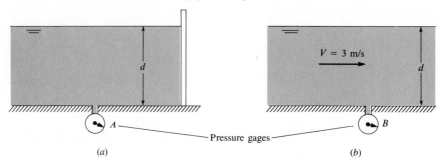

(a)   (b)

Pressure gages

PROBLEM 5.77

5.78 The apparatus shown in the figure is used to measure the velocity of air at the center of a duct having a 10-cm diameter. A tube mounted at the center of the duct has a 2-mm diameter and is attached to one leg of a slant-tube manometer. A pressure tap in the wall of the duct is connected to the other end of the slant-tube manometer. The well of the slant-tube manometer is sufficiently large that the elevation of the fluid in it does not change significantly when fluid moves up the leg of the manometer. The air in the duct is at a temperature of 20°C, and the pressure is 150 kPa. The manometer liquid has a specific gravity of 0.8, and the slope of the leg is 30°. When there is no flow in the duct, the liquid surface in the manometer lies at 2.3 cm on the slanted scale. When there is flow in the duct, the liquid moves up to 6.7 cm on the slanted scale. Find the velocity of the air in the duct. Assuming a uniform velocity profile in the duct, calculate the rate of flow of the air.

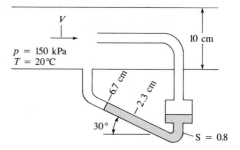

PROBLEM 5.78

5.79 A spherical probe is used for finding gas velocity by measuring the pressure difference between the upstream and downstream points A and B. The pressure coefficients at points A and B are 1.0 and −0.4. The pressure difference $p_A - p_B$ is 5 kPa, and the gas density is 1.50 kg/m^3. Calculate the gas velocity.

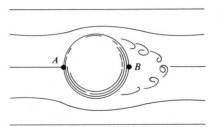

PROBLEM 5.79

5.80 A rugged instrument used frequently for monitoring gas velocity in smokestacks consists of two open tubes oriented to the flow direction, as shown, and connected to a manometer. The pressure coefficient is 1.0 at A and -0.3 at B. Assume that water, at 20°C, is used in the manometer and that a 0.5-cm deflection is noted. The pressure and temperature of the stack gases are 101 kPa and 250°C. The gas constant of the stack gases is 200 J/kg K. Determine the velocity of the stack gases.

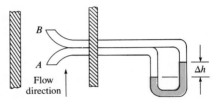

PROBLEM 5.80

5.81 A special spherical probe is used to measure the velocity of water. Pressure taps are located at the forward stagnation point and at the maximum width of the sphere. A mercury manometer is connected to the pressure taps as shown. A deflection of 5 cm is measured on the manometer. The velocity distribution around the sphere is given by

$$u = 1.5U \sin \theta$$

where U is the free-stream velocity. Find the free-stream velocity.

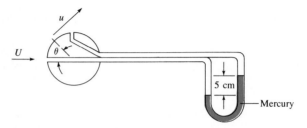

PROBLEM 5.81

5.82 A Pitot tube is used to measure the airspeed of an airplane. The Pitot tube is connected to a pressure-sensing device calibrated to indicate the correct airspeed when the temperature is 17°C and the pressure is 101 kPa. The airplane flies at an altitude of 3000 m, where the pressure and temperature are 70 kPa and −5°C. The indicated airspeed is 60 m/s. What is the true airspeed?

5.83 An atomizer consists of a constriction in an air duct that reduces the pressure and draws water from the reservoir into the airstream. The pressure on the water and at the exit of the atomizer is atmospheric. The velocity of the air at the exit is 5 m/s, and the air density is 1.2 kg/m³. Assuming the air flow is steady, incompressible, and nonviscous, calculate the maximum permissible area ratio (A_c/A_e) for which the atomizer will operate.

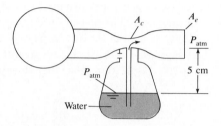

PROBLEM 5.83

5.84 Refer to Fig. 5.7. If in this flow field, air is flowing ($\rho = 1.4$ kg/m³) and the maximum gage pressure in the field is 10 kPa at a point where the velocity is 20 m/s, what is the minimum pressure in the field?

5.85 Air ($\rho = 1.20$ kg/m³) flows past a circular disk, and a pressure distribution as shown in Fig. 5.13*b* is developed. The free-stream air velocity is 50 m/s. What are the maximum and minimum gage pressures on the disk?

5.86 Ideal-flow theory will yield a flow pattern past an airfoil similar to that shown. If the approach air velocity V_0 is 50 m/s, what is the pressure difference between the bottom and the top of this airfoil at points where the velocities are $V_1 = 90$ m/s and $V_2 = 70$ m/s? Assume ρ_{air} is uniform at 1.1 kg/m³.

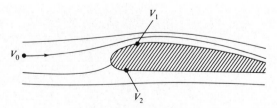

PROBLEM 5.86

5.87 When the hydrofoil shown was tested, the minimum pressure on the surface of the foil was found to be 70 kPa absolute when the foil was submerged 1.80 m and towed at a speed of 7 m/s. At the same depth, at what speed will cavitation first occur? Assume irrotational flow for both cases and $T = 10°C$.

5.88 For the hydrofoil of Prob. 5.87, at what speed will cavitation begin if the depth is increased to 3 m?

5.89 When the hydrofoil shown was tested, the minimum pressure on the surface of the foil was found to be 2.5 psi vacuum when the foil was submerged 4 ft and towed at a speed of 20 ft/s. At the same depth, at what speed will cavitation first occur? Assume irrotational flow for both cases and $T = 50°F$.

5.90 For the conditions of Prob. 5.89, at what speed will cavitation begin if the depth is increased to 10 ft?

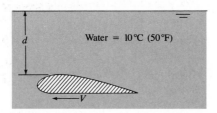

PROBLEMS 5.87, 5.88, 5.89, 5.90

5.91 A sphere is moving in water at a depth where the absolute pressure is 20 psia. The maximum velocity on a sphere occurs 90° from the forward stagnation point and is 1.5 times the free-stream velocity. The density of water is 62.4 lbm/ft³. Calculate the speed of the sphere at which cavitation will occur.

5.92 The minimum pressure on a cylinder moving horizontally in water ($T = 10°C$) at 5 m/s at a depth of 1 m is 90 kPa absolute. At what velocity will cavitation begin? Atmospheric pressure is 100 kPa absolute.

5.93 A disk like that shown in Fig. 5.13b, p. 182, is placed in a large pipe (see the accompanying figure). When the water velocity in the pipe is 32 ft/s and the pressure in the pipe is 100.0 psfg, what will be the minimum pressure on the disk? Is cavitation likely to occur near the disk under these conditions?

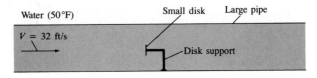

PROBLEM 5.93

5.94 Consider the conduit shown in Fig. 5.7, p. 175. Design a manometer setup that is to be tapped into the wall of the conduit such that it can be used to determine the velocity, V_0. Assume that water will be flowing in the conduit and that the range of V_0 will be from 0 to 5 m/s. Also develop an equation that gives the discharge as a function of the deflection on the manometer. The conduit width (normal to the page) is given by the symbol B.

5.95 Utilize a disk like that shown in Fig. 5.13b, p. 182, in designing a velocity-measuring device. Design the device for measuring air flow, and use a manometer as the velocity indicator. Make your own assumption(s) about range of airspeed.

References

1. Fage, A., and J. H. Warsap. "The Effects of Turbulence and Surface Roughness on the Drag of a Circular Cylinder." *Aero. Res. Comm., London, Rept. Mem.,* 1283 (1929).

2. Falvey, H. T. "Predicting Cavitation in Spillway Tunnels." *Water Power and Dam Construction* (August 1982).

3. Hammitt, F. G. *Cavitation and Multiphase Flow Phenomena.* McGraw-Hill, New York, 1980.

4. Knapp, R. T., J. W. Daily, and F. G. Hammitt. *Cavitation,* McGraw-Hill, New York, 1970.

5. Kuiper, G. "Reflections on Cavitation Inception," *Cavitation and Multiphase Flow Forum—1985,* ASME, 345 East 47th St., New York, 1985.

6. Izenson, M. G. "A New Scaling Parameter for Prediction of Cavitating Pump Performance," *Cavitation and Multiphase Flow—1988 Forum,* ASME, 345 East 47th St., New York, 1988.

7. Rayan, M. A., et al. "Cavitation Erosive Wear in Centrifugal Pump Impeller," pp. 92–95, *Cavitation and Multiphase Flow Forum—1987,* ASME, 345 East 47th St., New York, 1987.

8. Rouse, H. *Elementary Mechanics of Fluids.* John Wiley, New York, 1946.

9. Thomas, H. A., and E. P. Schuleen. "Cavitation in Outlet Conduits of High Dams." *Trans. Am. Soc. Civil Eng.,* 107 (1942).

Films

10. Eisenberg, Philip. *Cavitation*. National Committee for Fluid Mechanics Films, distributed by Encyclopaedia Britannica Educational Corporation.

11. Shapiro, A. J. *Pressure Fields and Fluid Acceleration*. National Committee for Fluid Mechanics Films, distributed by Encyclopaedia Britannica Educational Corporation.

Momentum Principle

*Penstock "anchors,"
Shasta Dam, California.
The massive concrete
structures at the upper
ends of these 15 ft
(4.57 m) diameter
penstock sections are
anchor blocks designed
to withstand the dynamic
forces (pressure and
momentum) acting on the
bend from a water flow
in each penstock of
3,800 cfs (108 m³/s)
under a pressure of
140 psi (965 kPa). The
penstocks' plate thickness
is 2 in. and each anchor
weighs more than
2 million lb (9 MN).
(Courtesy U.S. Bureau
of Reclamation)*

I n solid mechanics the impulse, $\int \mathbf{F}\,dt$, applied to a body is equal to the change of momentum, $M\mathbf{V}_2 - M\mathbf{V}_1$, for a given time interval. The directly analogous situation for fluid flow is the case where liquid in a pipe is accelerated by means of a pressure differential along the pipe. Both these cases may be termed simple unsteady-state cases, for which the basic equation of linear momentum is applicable. When we have nonuniform flow, which may be either steady or unsteady, the process becomes more complicated, and we turn to the basic control-volume approach to develop the momentum equations.

6.1 │ The Momentum Equation

Derivation of Basic Equation

First, let us consider Eq. (4.21b), which is rewritten here for convenience:

$$\frac{dB_{\text{sys}}}{dt} = \frac{d}{dt} \int_{\text{cv}} \beta\rho \, d\forall + \int_{\text{cs}} \beta\rho\mathbf{V} \cdot d\mathbf{A} \qquad (4.21b)$$

We let B equal the momentum of the system, or, in other words, the momentum of a given quantity of matter. Then dB/dt will be the rate of change of momentum of the system with respect to time, $d(\text{momentum})/dt$. Momentum, by definition, is the product of mass and velocity. Therefore β, the corresponding intensive property—or momentum per unit mass—is simply the velocity $\mathbf{v}$ referenced to an inertial reference frame. We use a small $\mathbf{v}$ here to distinguish it from the $\mathbf{V}$ in $\mathbf{V} \cdot \mathbf{A}$ of Eq. (4.21a). We now substitute $d(\text{momentum})/dt$ for dB/dt and $\mathbf{v}$ for β in Eq. (4.21a) to obtain

$$\frac{d(\text{momentum})}{dt} = \int_{\text{cs}} \mathbf{v}\rho\mathbf{V} \cdot d\mathbf{A} + \frac{d}{dt} \int_{\text{cv}} \mathbf{v}\rho \, d\forall \qquad (6.1)$$

According to Newton's second law, the summation of all external forces on a system is equal to the rate of change of momentum of that system, $\Sigma \mathbf{F} = d(\text{momentum})/dt$. Thus when the appropriate substitution is made in Eq. (6.1), we have

$$\sum \mathbf{F} = \int_{cs} \mathbf{v}\rho \mathbf{V} \cdot d\mathbf{A} + \frac{d}{dt} \int_{cv} \mathbf{v}\rho \, d\forall \qquad (6.2)$$

The force term on the left may include various types of forces. For example, the two types of forces usually considered are surface forces and body forces. When these are so designated, Eq. (6.2) can be written as

$$\sum \mathbf{F}_S + \sum \mathbf{F}_B = \int_{cs} \mathbf{v}\rho \mathbf{V} \cdot d\mathbf{A} + \frac{d}{dt} \int_{cv} \mathbf{v}\rho \, d\forall \qquad (6.3a)$$

A simplified form of the momentum equation results when there is uniform velocity in the streams crossing the control surface. It is

$$\sum \mathbf{F}_S + \sum \mathbf{F}_B = \sum_{cs} \mathbf{v}\rho \mathbf{V} \cdot \mathbf{A} + \frac{d}{dt} \int_{cv} \mathbf{v}\rho \, d\forall \qquad (6.3b)$$

There are no limitations on use of the momentum equation taken as a whole; however, there are certain limitations and assumptions for each term of the equation. These are considered in detail below.

External Forces

Let us first consider the terms on the left of Eqs. (6.3). These represent the external forces acting on the system, or, in other words, acting on the mass inside the control volume at the instant the equation is applied. The surface forces $\Sigma \mathbf{F}_S$ may be in the form of pressure forces transmitted through the fluid, such as those shown acting at section 1 in Fig. 6.1; or they may be forces transmitted through a solid, such as the force transmitted by the metal in the joint that joins the pipe to the nozzle of Fig. 6.1. The surface forces may also include the force of the physical boundary on the fluid if the control surface is so drawn. In other words, whenever forces are transmitted across the control surface, they must be

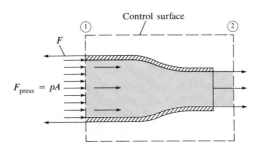

FIGURE 6.1

considered in $\Sigma \mathbf{F}_S$ of Eqs. (6.3). The body forces $\Sigma \mathbf{F}_B$ for most engineering applications consist of the forces of gravity acting on the mass inside the control volume. However, more advanced studies, such as magnetohydrodynamics, also include electromagnetic forces in this category.

It is important to recognize that the term $\Sigma \mathbf{F}$ in Eq. (6.2) can be considered in exactly the same manner as one considers the forces on a free body in basic engineering mechanics. In our case the body under consideration is the mass that at the instant is inside the control volume; therefore, the control surface delineates the body. One should keep this free-body concept in mind when using the momentum equation to solve flow problems.

Velocity Reference

Recall that in the basic derivation of the control-volume equation, the velocity in the $\mathbf{V} \cdot \mathbf{A}$ term is always referenced to the control volume itself because it represents the discharge across the control surface. In its application in the momentum equation, this term is subject to the same requirement.

The velocity $\mathbf{v}$ that is used for β (momentum per unit mass) must always be referenced to an inertial reference frame (nonaccelerating reference frame). *If the control volume itself is not accelerating (it can be stationary or moving with a constant velocity), then the velocity v that is used for β will also be referenced to the same reference frame to which the control volume is fixed. Most of your problems will be of this type.* However, if the problem is one for which it would be desirable to let the control volume accelerate—for example, to let the control volume accelerate with a body that is also accelerating—then the velocity used for β must be with respect to the inertial frame of reference. This is necessary to satisfy Newton's second law, which is used in the derivation of the momentum equation and is valid only in an inertial frame of reference.

Unsteadiness

When conditions at a point change with respect to time, we have unsteady flow. In Eqs. (6.3) this is taken into account by the last term on the right. A simple case of unsteady flow in a straight pipe illustrates the applicability of this term, as is shown in Example 6.1.

EXAMPLE 6.1   A straight, horizontal pipe with a 20-cm diameter discharges water into the air. If the rate of flow is 0.01 m^3/s and is increasing at the rate of 0.15 $\mathrm{m}^3/\mathrm{s}^2$, what will be the pressure 100 m from the outlet end? Neglect viscous effects.

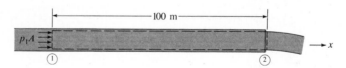

Solution   If we write Eq. (6.3b) for the x direction, we have

$$p_1 A_1 - p_2 A_2 = \sum_{cs} v_x \rho \mathbf{V} \cdot \mathbf{A} + \frac{d}{dt} \int_{cv} v_x \rho \, d\Psi$$

Here $\sum_{cs} v_x \rho \mathbf{V} \cdot \mathbf{A} = \rho v_x(-V_1 A_1 + V_2 A_2) = 0$. The last term on the right-hand side of the equation represents the rate of change of momentum inside the control volume, which can be expressed as

$$\frac{d}{dt} \int_{cv} v_x \rho \, d\Psi = \frac{d}{dt}\left(\frac{Q}{A} \rho \Psi\right) = \frac{d}{dt}\left(\frac{Q}{A} \rho L A\right) = \rho L \frac{dQ}{dt}$$

In the statement of the example, it was given that the discharge is increasing at a rate of 0.15 m³/s² ($dQ/dt = +0.15$). Hence this substitution is made along with values for ρ and L in the foregoing expression:

$$\frac{d}{dt} \int_{cv} v_x \rho \, d\Psi = 1000 \times 100 \times 0.15 = 15 \text{ kN}$$

Then $(p_1 - p_2)A = 15 \text{ kN}$

Since $p_2 = 0$ gage,

$$p_1 = p_{100} = \frac{15 \text{ kN}}{0.0314 \text{ m}^2} = 478 \text{ kPa}$$ ◀

In many problems the flow inside the control volume is steady, and the unsteadiness term drops out of the equation. However, the student should always be aware of this term and include it whenever it is significant.

Nonuniformity of Flow

The first term on the right-hand side of either of Eqs. (6.3) is the net flow of momentum out of the control volume. Therefore, if the flow through the control volume is uniform, as in Example 6.1, the flow of momentum entering the control volume is the same as the flow of momentum leaving the control volume; therefore, this term will be zero. However, if the flow is nonuniform, a change in the flow of momentum will undoubtedly exist between the inflow and outflow sections, and the magnitude of the term must be evaluated. Common applications involving nonuniform flow are flow in bends and flow through nozzles. These will be covered in detail later in this chapter.

Advantages of the Momentum Equation

The momentum equation can be used in the solution of numerous problems. A common application is determination of the force exerted on a piece of equipment, such as a nozzle or a bend in a pipe, given a certain discharge and pressure. One way to solve for such a force is to evaluate the pressure and shear stress, τ, over the surface of the device in question and to carry out the integration of $p\,dA$ and $\tau\,dA$ for the particular direction in question. In theory this can be done, but in practice it is usually too difficult to solve for the exact pressure and shear-stress distribution over the area. Therefore, we turn to the momentum equation for a simple means of evaluating the force. If the flow is steady, we need only know the pressure and velocity where the flow crosses the control surface to apply the momentum equation. This is of great significance, because for many problems we can draw the control surface such that it crosses one-dimensional flow zones where the velocity and piezometric head across each section are essentially uniform, thus avoiding complicated flow patterns. Indeed, making a judicious choice of control surface often justifies our use of the mean velocity in calculations instead of the more exact variable velocity, which would require integration. For some problems it is often advantageous to use both the momentum and Bernoulli's equations to obtain a solution.

Momentum Equation in the Cartesian Coordinate System

It is often convenient to use separate momentum equations in the x, y, and z coordinate directions instead of a single vector equation. Thus Eq. (6.2), modified for the case of uniform velocity across flow sections, is written as follows in these directions:

x direction:
$$\sum F_x = \sum_{cs} v_x(\rho \mathbf{V} \cdot \mathbf{A}) + \frac{d}{dt}\int_{cv} v_x \rho\, d\forall \qquad (6.4)$$

y direction:
$$\sum F_y = \sum_{cs} v_y(\rho \mathbf{V} \cdot \mathbf{A}) + \frac{d}{dt}\int_{cv} v_y \rho\, d\forall \qquad (6.5)$$

z direction:
$$\sum F_z = \sum_{cs} v_z(\rho \mathbf{V} \cdot \mathbf{A}) + \frac{d}{dt}\int_{cv} v_z \rho\, d\forall \qquad (6.6)$$

Several examples given in the next section will illustrate the methods of solution and show the wide range of application of the momentum equation.

6.2 Applications of the Momentum Equation

Initial Setup and Signs

When they apply the momentum principle, students often have difficulty initially in setting up the problem and then in getting the correct signs for the terms of Eqs. (6.4), (6.5), and (6.6). The following explanation may be helpful.

First, you must establish a coordinate system and indicate positive (and negative) coordinate directions. Next you must draw a control surface for your particular problem. Where to draw the control surface is somewhat arbitrary, but seeing what other people have done (by way of example problems) and practicing will help you to learn to choose a satisfactory control surface. (It is much like getting used to drawing free-body diagrams in your statics and dynamics courses.) In any case, the control surface is chosen such that the unknown(s) you are seeking in the solution are isolated by your choice. Once you have set up the coordinate system and have drawn your control surface, you must indicate on your sketch the forces and velocities relevant to your problem. The forces to include are the forces that would be included in a free-body diagram. If it is a steady-flow problem (most of the problems for this course are), the only velocities you will be concerned with are the velocities that occur at the control surface. That is, when flow occurs across a control surface, the velocity at that control surface should be shown and considered in your solution.

Once you have completed the initial set up of the problem you can start to work out the magnitudes and signs for the various terms of the equation(s). For example, a problem might involve the high-velocity steady flow of water in a vertical curve of an open channel, as shown below. Let this control surface and coordinate system be fixed. Therefore, the coordinate system will also be an inertial reference frame. Let it be the one we use. In this problem, assume you are interested in solving for the x and y components of force that the channel exerts on the water to effect the turning of the water. For the control surface we have drawn and for the force and velocity vectors of this problem, the momentum equation in the x direction (Eq. 6.4) is written as

$$F_1 - F_{2x} + F_x = V_1(-\rho V_1 A_1) + V_{2x}(+\rho V_2 A_2)$$

Let us briefly discuss the terms in the above equations and their signs.

1. The first term on the left, F_1, is the hydrostatic force acting on the water within the control volume at section 1. This force acts to the right (positive x direction), so it has a positive sign. One may calculate the magnitude of this

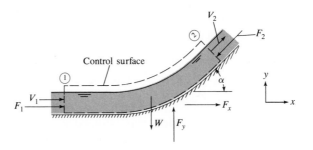

FIGURE 6.2

Flow in a vertical bend.

force from the formula given in Chapter 3, $F = \bar{p}_1 A_1 = \gamma(y_1/2)B_1 y_1$, where y_1 is the depth at section 1 and B is the width of the channel at that section.

2. The second term, F_{2x} is the component of hydrostatic force acting at section 2. This component acts to the left, so it has a negative sign. The magnitude of the force would be $F_2 \cos \alpha$ where $F_2 = \bar{p}_2 A_2$.

3. The force F_x (the third term on the left side of the equation) is the one being solved for. By letting it have a positive sign, we are assuming that it will act in the positive x direction. If that assumption turns out to be incorrect, the solution for F_x will yield a negative number, which will tell us that the force actually acts to the left, opposite to the direction we had assumed. By assuming a positive sign on the unknown quantity, then, we ensure that we will always have the correct sign when we solve for it.

4. In the first term on the right side of the equation, $V_1(-\rho_1 V_1 A_1)$, the V_1 is the velocity in the x direction at section 1 relative to the inertial reference frame. In this example the coordinate system is also our inertial reference frame. Because the V_1 is in the positive x direction, it carries a plus sign (understood when no sign appears). The $-\rho_1 V_1 A_1$ comes from the evaluation of $\rho \mathbf{V} \cdot \mathbf{A}$ in Eq. (6.4). The minus sign results because the velocity vector and the area vector are in opposite directions.

5. The second term on the right side of the equation involves flow across the control surface at section 2. The V_{2x} is the x component of the velocity at this section; it has a plus sign because this velocity component is in the positive x direction. The plus sign on the $\rho_2 V_2 A_2$ term results because the velocity and area vectors both point in the same direction, so $\mathbf{V}_2 \cdot \mathbf{A}_2 = V_2 A_2 \cos \theta = V_2 A_2(+1) = V_2 A_2$.

We have just described how to set up and interpret the momentum equation in the x direction for the example given. It should be obvious that if the velocities V_1 and V_2 were known and if the fluid density were given, one could easily solve for F_x. In a similar manner we could write the momentum equation for the y direction (using Eq. 6.5) to solve for the force F_y. In that case the forces would be the weight, W, which would have a negative sign, and F_y. The component velocity would be V_{2y} and it would have a positive sign.

The foregoing discussion should make it clear that once you have chosen your coordinate system and inertial reference frame and have drawn your control surface, you must *studiously* evaluate the magnitude and signs for each term of the equation(s). If you do not do so, your attempts to apply the momentum principle will be haphazard at best.

Jet Deflected by a Plate or a Vane

Consider a jet of water turned through a horizontal angle, such as is shown in Fig. 6.3. Here it is assumed that the pressure in the stream leaving the vane is the same as that in the stream entering the vane, and it is also assumed

FIGURE 6.3

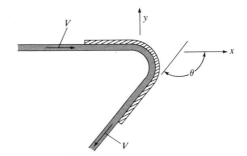

that the speed of the jet is not appreciably reduced by the surface resistance of the vane. Then, if we draw the control surface so that it includes only the jet, as shown in Fig. 6.4, we can easily apply the momentum equation to solve for the force components F_x and F_y, which are the forces exerted by the vane on the jet. We initially assume that these forces have a positive sense as shown. The momentum equation for the x direction is

$$F_x = \sum_{cs} v_x \rho \mathbf{V} \cdot \mathbf{A} + \frac{d}{dt} \int_{cv} v_x \rho \, d\forall \qquad (6.7)$$

If the flow is steady, the second term on the right of Eq. (6.7) is equal to zero, which leaves the following:

$$F_x = \sum_{cs} v_x \rho \mathbf{V} \cdot \mathbf{A} \qquad (6.8)$$

Also, because the vane is not accelerating, we can use the control surface as a reference frame for v_x. At section 1 the velocity is uniform over the section and the area vector $\mathbf{A}_1$ is in the reverse direction of the velocity vector $\mathbf{V}_1$. Therefore, for this part of the control surface we have

$$(v_x \rho \mathbf{V} \cdot \mathbf{A})_1 = V_{1x} \rho_1 (-V_1 A_1)$$

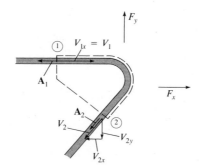

FIGURE 6.4

By a similar analysis for section 2 (in this case, velocity $\mathbf{V}_2$ and area $\mathbf{A}_2$ have the same sense), we get

$$v_x \rho \mathbf{V} \cdot \mathbf{A} = V_{2x} \rho_2 V_2 A_2$$

When these substitutions are made, Eq. (6.8) becomes

$$F_x = -V_{1x} \rho (V_1 A_1) + V_{2x} \rho (V_2 A_2) \tag{6.9}$$

In a similar manner the force in the y direction, F_y, is given by

$$F_y = -V_{1y} \rho (V_1 A_1) + V_{2y} \rho (V_2 A_2) \tag{6.10}$$

and for this particular case, Fig. 6.4, $F_y = V_{2y} \rho (V_2 A_2)$. Because the forces given by Eqs. (6.9) and (6.10) are the forces exerted by the vane on the system (on the jet, in this case), we obtain the forces of the jet on the vane by simply reversing the signs on F_x and F_y.

EXAMPLE 6.2   Consider a water jet that is deflected by a stationary vane, such as is shown in the figure. If the jet speed and diameter are 100 ft/s and 1 in., respectively, and the jet is deflected 60°, what force is exerted on the vane by the jet?

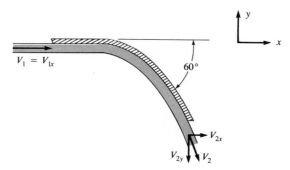

Solution   First draw the control surface as shown. Next solve for F_x, the component of the force of the vane on the jet, using Eq. (6.9).

$$F_x = -V_{1x}(\rho_1 V_1 A_1) + V_{2x}(\rho_2 V_2 A_2)$$

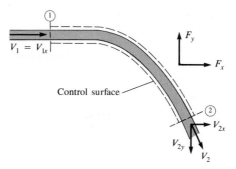

The velocity in the x direction at section 2 is given as

$$V_{2x} = 100 \cos 60° = 100 \times 0.500 = 50 \text{ ft/s}$$

Also
$$V_{1x} = 100 \text{ ft/s}$$

and
$$\rho_1 V_1 A_1 = \rho_2 V_2 A_2 = \dot{m} = 1.94 \times 100\left(\frac{0.785}{144}\right) = 1.06 \text{ slugs/s}$$

Therefore
$$F_x = \dot{m}(V_{2x} - V_{1x})$$

$$F_x = 1.06 \text{ slugs/s} \times (50 - 100) \text{ ft/s} = -53.0 \text{ lbf}$$

Similarly determined, the y component of force on the jet is

$$F_y = 1.06 \text{ slugs/s} \times (-86.6 \text{ ft/s} - 0) = -91.8 \text{ lbf}$$

Then the force on the vane will be the reactions to the forces of the vane on the jet, or

$$F_x = +53.0 \text{ lbf} \qquad F_y = +91.8 \text{ lbf} \qquad \blacktriangleleft$$

If the vane were moving, such as on a turbine runner, we could carry out the same type of analysis, except that it would be convenient to let the control volume move with the vane. For that case the flow is still steady; so the momentum equation becomes

$$\sum \mathbf{F} = \sum \mathbf{v}(\rho \mathbf{V} \cdot \mathbf{A})$$

In addition, because the vane is moving at a constant velocity, the vane represents an inertial reference frame for the velocities. The control volume for this application is shown in Fig. 6.5, where the velocities are referenced to the moving control volume. The force applied to the vane in the x direction is then

$$F_{vx} = -\rho(V - V_v)A[(V - V_v)_{2x} - (V - V_v)_{1x}] \qquad (6.11)$$

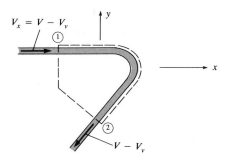

FIGURE 6.5

Control volume for jet deflected by moving vane.

The power delivered by the vane is equal to the product of the force on the vane and the speed of the vane. The resulting power is then $F_{vx}V_v$. Obviously, no power results unless the vane speed is greater than zero and less than the velocity of the approaching jet. Other interesting and useful relations can be developed, such as the optimum speed and optimum θ for the most efficient generation of power. Such topics are discussed in more detail in Chapter 14.

In the next example, we show a case where the inertial reference frame may be either fixed or moving.

EXAMPLE 6.3   A jet of water from a nozzle is deflected 90° by a vane on a moving cart. The velocity in the water jet with respect to the nozzle is V_j, and the area of the jet is A_j. The cart moves at a constant speed of V_c with respect to the nozzle. Find the resisting force, F, acting on the cart.

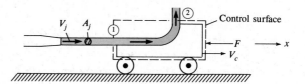

Solution   Draw a control surface around the cart as shown. Let the control surface move with the cart. Take the direction to the right as the positive x direction. Mass crosses the control surface at two sections, 1 and 2. A resisting force F acts on the cart to the left as shown. Assume the rolling resistance is negligible.

The general form of the momentum equation in the x direction is

$$\sum F_x = \frac{d}{dt}\int \rho v_x\,d\forall + \sum v_x \rho \mathbf{V}\cdot\mathbf{A}$$

where $\mathbf{V}$ is the velocity vector with respect to the control surface of the fluid crossing the control surface, and v_x is the velocity component of the fluid in the x direction at the control surface with respect to the inertial reference frame.

The flow inside the control volume is steady, so the first term on the right-hand side is zero. Thus the momentum equation simplifies to

$$-F = v_{1x}\rho\mathbf{V}\cdot\mathbf{A} + v_{2x}\rho\mathbf{V}\cdot\mathbf{A}$$

The speed of the water crossing the control surface with respect to the control surface at section 1 is

$$V = V_j - V_c$$

The mass flow at section 1 is

$$\rho\mathbf{V}_1\cdot\mathbf{A}_1 = -\rho(V_j - V_c)A_j$$

Because the flow inside the control volume is steady, the mass flow across the control surface at section 2 is

$$\rho \mathbf{V}_2 \bullet \mathbf{A}_2 = -\rho \mathbf{V}_1 \bullet \mathbf{A}_1 = \rho(V_j - V_c)A_j$$

Substituting the mass flows into the momentum equation gives

$$F = (v_{1x} - v_{2x})\rho(V_j - V_c)A_j$$

Let us take the nozzle as the inertial reference frame. Then the velocity component v_x of the fluid at section 1 is simply the jet velocity, V_j, and the corresponding component at 2 is the speed of the cart, V_c. Thus the resisting force acting on the cart is

$$F = (V_j - V_c)\rho(V_j - V_c)A_j = \rho A_j(V_j - V_c)^2$$

As an alternative approach, one can use the cart as an inertial reference frame because it is moving at a constant velocity (no acceleration). Thus the velocity component v_x at section 1 becomes $V_j - V_c$ and the corresponding velocity at section 2 is zero, because the velocity is measured with respect to the moving cart. Substituting these values into the momentum equation yields

$$F = (V_j - V_c - 0)\rho(V_j - V_c)A_j = \rho A_j(V_j - V_c)^2$$

which is the same expression that is obtained when the nozzle is used as the inertial reference frame. The purpose of this example is to show that one can choose an inertial reference frame arbitrarily, so long as it is not accelerating.

Flow Through a Nozzle

In a nozzle type of problem one often wants to solve for the force needed to hold the nozzle on the supply hose or pipe, given information about the physical features of the nozzle and the rate of flow of fluid through the nozzle or pressure in the supply line. If one is given the pressure in the supply line, then one can solve for the velocities and discharge by means of Bernoulli's equation and the continuity equation. If the discharge is given, then one can solve for the pressure in the line by means of the same equations. Once one has the pressure, velocities, and discharge, one can use them in the momentum equation to solve for the desired force. Example 6.4 illustrates the procedure.

EXAMPLE 6.4　Find the force needed to hold the nozzle on the pipe. Neglect head losses. The fluid is oil ($S = 0.85$) and $p_1 = 100$ psig. Assume that the nozzle and the oil in it weigh 10 lbf.

Solution   Equations (6.4) and (6.6) will be used to solve for the unknown force. For steady flow, the values for the last terms of the equations are zero. Thus we have

$$\sum F_x = \sum v_x \rho \mathbf{V} \cdot \mathbf{A} \qquad\qquad (6.4)$$

$$\sum F_z = \sum v_z \rho \mathbf{V} \cdot \mathbf{A} \qquad\qquad (6.6)$$

If we draw a control surface around the nozzle itself, we have

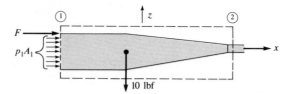

The pressure in the pipe is given; therefore, we need to solve for the velocity and the discharge by means of the continuity equation and Bernoulli's equation.

$$p_1 + \rho \frac{V_1^2}{2} = p_2 + \rho \frac{V_2^2}{2}$$

We will eliminate V_1 by means of the continuity equation.

$$V_1 A_1 = V_2 A_2$$
$$V_1 = V_2 A_2 / A_1 = V_2 (d_2/d_1)^2$$
$$= V_2 (1/3)^2 = (1/9) V_2$$

Substituting V_1 into Bernoulli's equation gives

$$V_2^2 - (1/81) V_2^2 = (2/\rho)(p_1 - p_2)$$
$$V_2^2 [1 - (1/81)] = (2/\rho)(p_1 - p_2)$$
$$V_2 = 0.994 \sqrt{(2/\rho)(p_1 - p_2)}$$

where
$$p_1 = 100 \text{ lbf/in}^2 = 100 \text{ lbf/in}^2 \times 144 \text{ in}^2/\text{ft}^2$$
$$= 14{,}400 \text{ lbf/ft}^2 \text{ gage}$$

$$p_2 = 0 \text{ gage}$$

$$\rho = 0.85 \times 1.94 \text{ slugs/ft}^3 = 1.65 \text{ lbf} \cdot \text{s}^2/\text{ft}^4$$

$$V_2 = 0.994 \sqrt{(2/1.65 \text{ lbf} \cdot \text{s}^2/\text{ft}^4)(14{,}400 \text{ lbf/ft}^2)}$$

$$V_2 = 131.3 \text{ ft/s and } V_1 = (1/9) V_2 = 14.59 \text{ ft/s}$$

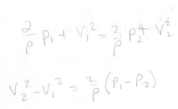

Also
$$Q = V_2 A_2 = (131.3 \text{ ft/s}) [(\pi/4) (1/12)^2 \text{ ft}^2]$$
$$= 0.716 \text{ ft}^3/\text{s}$$

Now we can solve the momentum equations for the unknown forces.

$$\sum F_x = \sum V_x \rho \mathbf{V} \cdot \mathbf{A}$$

$$F_{\text{nozzle},x} + p_1 A_1 = v_{1x} \rho(-V_1 A_1) + v_{2x} \rho(V_2 A_2)$$

where F_{nozzle} is the unknown force acting on the nozzle in the x direction to hold it in place. Also note that $V_1 A_1 = V_2 A_2 = Q$, so the foregoing equation reduces to

$$F_{\text{nozzle},x} = \rho Q(v_{2x} - v_{1x}) - p_1 A_1$$
$$F_{\text{nozzle},x} = (1.65 \text{ lbf} \cdot \text{s}^2/\text{ft}^4) (0.716 \text{ ft}^3/\text{s}) (131.3 \text{ ft/s} - 14.59 \text{ ft/s})$$
$$- (100 \text{ lbf/in}^2) (\pi/4) (3^2 \text{ in}^2)$$
$$= -569 \text{ lbf}$$

In the z direction we have

$$F_{\text{nozzle},z} - 10 \text{ lbf} = 0$$
$$F_{\text{nozzle},z} = 10 \text{ lbf}$$

Thus the total force that must be applied to the nozzle to hold it in place is

$$F_{\text{nozzle}} = -569\mathbf{i} + 10\mathbf{k} \text{ lbf} \qquad \blacktriangleleft$$

Forces on Bends

Consider the bend in the pipe shown in Fig. 6.6 on page 226. Here it is assumed that the flow is steady and that the velocity is uniform across sections 1 and 2. Assume also that the bend is oriented such that y is vertically upward. In the design of such a bend, one of the primary considerations is the force required to hold the bend in place. For small- or medium-sized pipes, this force is supplied by the flange bolts or by welds if it happens to be a welded connection; however, if the pipe is large, specially designed anchor blocks may be required to hold the bend. In the latter case, we are primarily interested in the reactions R_x and R_y needed to hold the bend in place.

The analysis is carried out by isolating the bend by means of a control surface, as shown in Fig. 6.6b, and the momentum equation is applied for the x, y, and z directions with due care to be certain that all relevant forces such as pressure and gravity are included in the solution.

FIGURE 6.6

Flow in a bend. (a) Bend.
(b) Control volume.

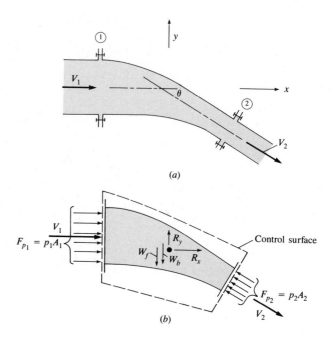

EXAMPLE 6.5   Water flows through a 180° vertical reducing bend, as shown. The discharge is 0.25 m³/s and the pressure at the center of the inlet of the bend is 150 kPa. If the bend volume is 0.10 m³ and it is assumed that Bernoulli's equation is valid, what reaction is required to hold the bend in place? Assume that the metal in the bend weighs 500 N.

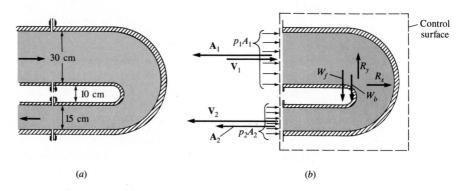

Solution   The momentum equation is applied with the control volume shown. For the *x* direction we have

$$\sum F_x = \sum_{cs} v_x \rho \mathbf{V} \cdot \mathbf{A}$$

which becomes the following for this example:

$$R_x + p_1 A_1 + p_2 A_2 = v_{1x}\rho(-V_1 A_1) + v_{2x}\rho(V_2 A_2)$$

Note that in the x direction the forces acting on the system are the pressure forces $p_1 A_1$ and $p_2 A_2$, both acting in the positive x direction for this example, and R_x. Here R_x is the net force acting across the metal at the flange. Note also at section 1 that the velocity is assumed to be uniform across the section and that the area vector $\mathbf{A}_1$ is pointing in the opposite direction to $\mathbf{V}_1$; therefore, $v_{1x}\rho\mathbf{V} \cdot \mathbf{A} = V_1\rho(-V_1 A_1)$. Note that we are using the control surface as the inertial reference frame. At section 2 the area and velocity vectors point in the same direction, so the sign on $\mathbf{V} \cdot \mathbf{A}$ is positive. However, $v_{2x} = -V_2$; consequently, there is a negative sign on the last term on the right. The reaction in the x direction is thus given as

$$R_x = -p_1 A_1 - p_2 A_2 - \rho V_1^2 A_1 - \rho V_2^2 A_2$$

However,
$$V_1 A_1 = V_2 A_2 = Q$$

so this reduces to

$$R_x = -p_1 A_1 - p_2 A_2 - \rho Q(V_1 + V_2)$$

The velocities in the foregoing equation are next determined.

$$V_1 = \frac{Q}{A_1} = \frac{0.25}{(\pi/4) \times 0.30^2} = 3.54 \text{ m/s}$$

$$V_2 = \frac{Q}{A_2} = \frac{0.25}{(\pi/4) \times 0.15^2} = 14.15 \text{ m/s}$$

The initial pressure p_1 is given, but p_2 has to be obtained using Bernoulli's equation written between the center of section 1 and the center of section 2.

$$\frac{p_1}{\gamma} + \frac{V_1^2}{2g} + z_1 = \frac{p_2}{\gamma} + \frac{V_2^2}{2g} + z_2$$

Here $z_2 - z_1 = -0.325$ m, $V_1 = 3.54$ m/s, and $V_2 = 14.15$ m/s. Thus

$$\frac{p_2}{\gamma} = \frac{150 \times 10^3}{9810} + \frac{3.54^2}{2 \times 9.81} + 0.325 - \frac{14.15^2}{2 \times 9.81}$$

and
$$p_2 = 59.3 \text{ kPa}$$

Now we can solve for R_x as follows:

$$R_x = -150 \times 10^3 \times (\pi/4) \times 0.3^2 - 59.3 \times 10^3 \times (\pi/4)$$
$$\times 0.15^2 - 1000 \times 0.25(14.15 + 3.54)$$
$$= -10{,}603 - 1048 - 4423 = -16.07 \text{ kN}$$

◀

Because there are no velocities in the y direction where the flow passes across the control surface, R_y is obtained by static equilibrium: $\Sigma F_y = 0$, or

$$R_y - 500 - 0.1 \times 9810 = 0$$

$$R_y = 500 + 981 = 1.48 \text{ kN} \qquad \blacktriangleleft$$

EXAMPLE 6.6   A pipe that is 1 m in diameter has a 30° horizontal bend in it, as shown in the figure, and carries crude oil (S = 0.94) at a rate of 2 m³/s. If the pressure in the bend is assumed to be essentially uniform at 75 kPa gage, if the volume of the bend is 1.2 m³, and if the metal in the bend weighs 4 kN, what forces must be applied to the bend to hold it in place?

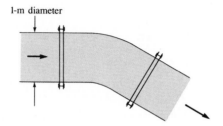

1-m diameter

Solution   Isolate the bend by a control surface as shown, and solve for the unknown forces R_x, R_y, and R_z using the momentum equation. First solve for the forces in the x direction. *Note:* Our coordinate system is set up so that x is positive to the left. The flow is steady, so the relevant equation is

$$\sum F_x = \sum v_x \rho \mathbf{V} \cdot \mathbf{A}$$

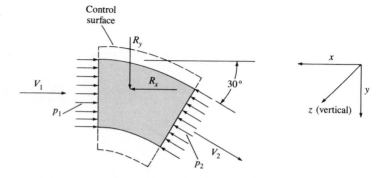

which for this example becomes

$$R_x - p_1 A_1 + 0.866 p_2 A_2 = v_{1x}\rho(-V_1 A_1) + v_{2x}\rho(V_2 A_2)$$

$$v_{1x} = -\frac{Q}{A_1} = -2.55 \text{ m/s} \qquad v_{2x} = -\frac{Q}{A_2}\cos 30° = -2.21 \text{ m/s}$$

$$R_x = +p_1 A_1 - 0.866\, p_2 A_2 + \rho Q(-v_{1x} + v_{2x})$$

$$= 59 \times 10^3 - 51 \times 10^3 + 0.94 \times 10^3 \times 2(2.55 - 2.21)$$

$$= 8 \times 10^3 + 0.64 \times 10^3 \text{ N} = 8.64 \text{ kN} \qquad \blacktriangleleft$$

For the y direction we have

$$\sum F_y = \sum v_y \rho \mathbf{V} \cdot \mathbf{A}$$

$$R_y - p_2 A_2 \sin 30° = v_{1y}\rho(-V_1 A_1) + v_{2y}\rho(V_2 A_2)$$

But $\qquad v_{1y} = 0 \qquad$ and $\qquad v_{2y} = +\dfrac{Q}{A_2}\sin 30° = +1.27 \text{ m/s}$

$$R_y = +p_2 A_2 \sin 30° + v_{2y}\rho Q$$

$$= 29.5 \text{ kN} + 1.27 \times 0.94 \times 10^3 \times 2 \text{ N} = 31.9 \text{ kN}$$

There are no components of velocity in the z direction (normal to the xy plane). Hence the momentum equation reduces to

$$\sum F_z = 0$$

$$R_z - W_B - W_f = 0$$

$$R_z = W_B + W_f$$

where W_B is the weight of the bend and W_f is the weight of the fluid in the bend. But

$$W_B = 4 \text{ kN} \qquad \text{and} \qquad W_f = \rho g \mathbb{V}$$

where $\mathbb{V}$ is the volume of fluid in bend. Thus

$$W_f = 0.94 \times 1000 \times 9.81 \times 1.2 = 11.1 \text{ kN}$$

and $\qquad R_z = 4 \text{ kN} + 11.1 \text{ kN} = 15.1 \text{ kN}$

$$\mathbf{R} = (8.64\mathbf{i} + 31.9\mathbf{j} + 15.1\mathbf{k}) \text{ kN} \qquad \blacktriangleleft$$

Problems Involving Nonuniform Velocity Distribution

All the foregoing examples were cases wherein it was assumed that the velocity across each flow section was uniform. Thus we used the simplified form of the momentum equation, Eq. (6.3b) or Eqs. (6.4), (6.5), and (6.6), to solve them.

The next example involves a nonuniform velocity across a flow section. Therefore, the more general form, Eq. (6.2), must be used.

EXAMPLE 6.7   A torpedo-like device is tested in a wind tunnel with an air density of 1 kg/m^3. The tunnel is 1 m in diameter, the upstream pressure is 1.5 kPa gage, and the downstream pressure is 1 kPa gage. If the mean air velocity, V, is 30 m/s, what are the mass rate of flow and the maximum velocity at the downstream section at C? If the pressure is assumed to be constant across the sections at A and C, what is the drag of the device and supporting vane? Assume viscous resistance at the wall is negligible.

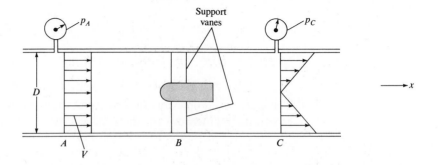

Solution   We draw the control surface so that it surrounds the air in the wind tunnel between sections A and C. Note that a separate part of the control surface then surrounds the torpedo device and its strut so that these will not be within the control volume. Because we are neglecting wall resistance, the forces acting on the system within the control volume in the x direction are the pressure forces acting at sections A and C and the force of the device on the air of the system. But the force of the device on the air is negative of the air drag on the device; that is,

$$F_{\text{device on air}} = -\text{Drag}$$

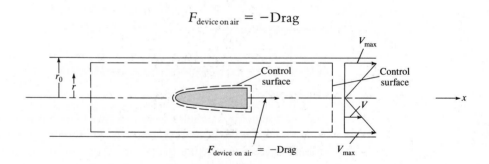

Writing the momentum equation using the control surface as the inertial reference frame (steady flow in the x direction) yields

$$\sum F_x = \int_{cs} v_x \rho \mathbf{V} \cdot d\mathbf{A}$$

$$p_1 A_1 - p_2 A_2 + F_{\text{device on air}} = V_A \rho(-V_A A_A) + \int_{A_C} \rho V_C^2 dA_C$$

$$= -V_A \dot{m} + \int_{A_C} \rho V_C^2 dA_C$$

The volume rate of flow, Q, is given by

$$Q = V_A A_A$$

$$= (30 \text{ m/s})\,(\pi/4)\,(1^2 \text{ m}^2) = 23.56 \text{ m}^3/\text{s}$$

Then $V_A \dot{m} = V_A \rho Q = (30 \text{ m/s})\,(1 \text{ kg/m}^3)\,(23.56 \text{ m}^3/\text{s}) = 706.8 \text{ N}$

To evaluate the integral of the second term on the right-hand side of the momentum equation, we need to write an equation for the velocity at section C as a function of r, where r is the radial distance from the axis of the tunnel. If $V_{\max}$ is the velocity next to the wall of the tunnel, then $V_C = (r/r_0)V_{\max}$, where r_0 is the radial distance to the wall of the tunnel. We can determine $V_{\max}$ from the continuity equation.

$$Q_A = Q_C$$

$$V_A A_A = \int_{A_C} V_C dA_C$$

$$(V_A \text{ m/s})\,(\pi/4)\,(1^2 \text{ m}) = \int_0^{r_0} (r/r_0)V_{\max}(2\pi r\, dr)$$

Letting $r_0 = 0.50$ m and solving the continuity equation for $V_{\max}$ yields $V_{\max} = 1.50 V_A$. Next, substitute this and given pressures in the momentum equation. Also substitute negative Drag for $F_{\text{device on air}}$. Then for the momentum equation, we get

$$(1{,}500 \text{ N/m}^2)\,(\pi/4)\,(1^2 \text{ m}^2) - (1{,}000 \text{ N/m}^2)\,(\pi/4)\,(1^2 \text{ m}^2) - \text{Drag}$$

$$= -706.8 \text{ N} + \int_0^{r_0} (1 \text{ N} \cdot \text{s}^2/\text{m}^4)\,\{[(r/r_0)\,(1.5V_A)]^2\,(\text{m/s})^2\}\,[(2\pi r\, dr)\text{ m}^2]$$

$$1{,}178 \text{ N} - 785 \text{ N} - \text{Drag} = -706.8 \text{ N} + (9\pi/8)V_A^2 r_0^2 \text{ N}$$

Then for $V_A = 30$ m/s and $r_0 = 0.5$ m, we get

$$\text{Drag} = 305 \text{ N}$$

◀

Motion of a Rocket

Our first example involving a rocket is a case where a rocket motor is tested (nonaccelerating condition).

EXAMPLE 6.8   A rocket motor is tested on a test stand (see the accompanying figure). During a test, fuel is fed in at the rate of 1 kg/s, and LOX (liquid oxygen) is fed in at 10 kg/s. The flexible couplings and supporting rollers offer negligible horizontal force. The gases exit at atmospheric pressure and with a velocity of 2000 m/s. Under these conditions, what force will be exerted on the thrust stand?

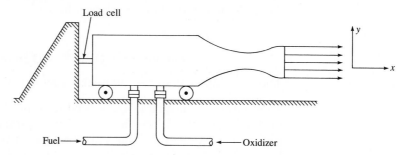

Solution   First, for computational purposes, we isolate the rocket with a control surface as shown and apply the steady-flow momentum equation in the x direction:

$$\sum F_x = \sum v_x \rho \mathbf{V} \cdot \mathbf{A}$$

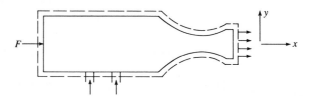

Since the pressure is uniform around the control surface except for the fuel and LOX lines, there is no net force due to pressure in the x direction. The flexible couplings and rollers offer no force in the x direction. The single surface force is F, the force of the test stand on the rocket motor. The body force is due to gravity, so there is no component in the x direction. Thus the momentum equation becomes

$$F = \sum v_x \rho \mathbf{V} \cdot \mathbf{A}$$

Mass enters the control surface through the fuel and oxidizer connections. However, the flow in these lines is vertical, so v_x is zero. Thus the momentum equation reduces to

$$F = v_{x,\text{exit}} \rho \mathbf{V} \bullet \mathbf{A} = v_{x,\text{exit}} \dot{m}$$

$$= (2000 \text{ m/s})[(1 + 10) \text{ kg/s}] = 22,000 \text{ N}$$

This force is the force of the thrust stand on the rocket motor. Thus the reaction to this force is the force on the thrust stand, which will be in the negative x direction:

$$F_{\text{on stand}} = -22,000 \text{ N} \qquad \blacktriangleleft$$

A further interesting application of the momentum equation is the derivation of the equation of motion for a rocket accelerating along a vertical rectilinear path. Consider the rocket illustrated in Fig. 6.7. The exhaust gases exit through a nozzle with area A_e with a velocity V_e with respect to the nozzle and a pressure in the exhaust stream of p_e. The pressure of the atmosphere is p_0. Let us take our control surface around the rocket and allow it to accelerate with the rocket. In regard to the accelerating control volume, it should be noted that our applications to this point have all been for nonaccelerating control volumes. If the control volume is accelerating, it simply means that we must use Eq. (6.2) and make certain that the velocity $\mathbf{v}$ is referenced to the inertial frame of reference. Thus,

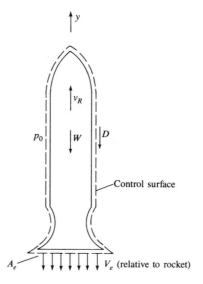

FIGURE 6.7

when we apply the momentum equation in the vertical direction to the accelerating control volume (assuming uniform velocity across the exit section), we have

$$\sum F_y = \sum_{cs} v_y \rho \mathbf{V} \cdot \mathbf{A} + \frac{d}{dt} \int_{cv} v_y \rho \, d\forall$$

where v_y refers to the velocity with respect to the inertial reference frame, and $\mathbf{V}$ represents the velocity measured with respect to the control volume—that is, the rocket. It is convenient to express the velocity v_y of elements of mass inside the control volume as a sum of the velocity of the rocket with respect to inertial space, v_R, and the velocity relative to the rocket, V:

$$v = v_R + V$$

Substituting into the momentum equation above, we have for the y direction

$$\sum F_y = (v_R - V_e)\rho_e V_e A_e + \frac{d}{dt} \int_{cv} v_R \rho \, d\forall + \frac{d}{dt} \int_{cv} V \rho \, d\forall \qquad (6.12)$$

where ρ_e is the exit density and V_e is the exit velocity of the exhaust gases relative to the rocket. Also, $v_R - V_e$ is the velocity of the exiting gases with respect to the inertial reference frame. Because v_R is uniform over the space of the control volume, we have

$$\frac{d}{dt} \int_{cv} v_R \rho \, d\forall = \frac{d}{dt} \left[v_R \int_{cv} \rho \, d\forall \right] = \frac{d}{dt}(v_R M)$$

where M is the instantaneous mass of the rocket. The last term in Eq. (6.12) is zero because the relative velocity of the solid elements in the rocket is zero and the density and relative velocity of the gas flow in the rocket do not vary with time. The momentum equation then reduces to

$$\sum F_y = -\rho_e V_e^2 A_e + v_R \left(\rho_e V_e A_e + \frac{dM}{dt} \right) + M \frac{dv_R}{dt}$$

Applying the continuity equation to the control volume, we have

$$\sum \rho \mathbf{V} \cdot \mathbf{A} + \frac{d}{dt} \int \rho \, d\forall = 0$$

$$\rho_e V_e A_e + \frac{dM}{dt} = 0$$

which further simplifies the momentum equation to

$$\sum F_y + \rho_e V_e^2 A_e = M \frac{dv_R}{dt}$$

The external forces on the rocket accrue from the pressure difference between the atmosphere and nozzle at the exit plane, the drag on the forward part of the rocket (upstream of the exit plane), and the force due to gravity.

$$-D - W + (p_e - p_0)A_e + \rho_e V_e^2 A_e = M\frac{dv_R}{dt}$$

The combination $(p_e - p_0)A_e + \rho_e V_e^2 A_e$ is the thrust T of the rocket engine, so the equation of motion becomes

$$-D - W + T = M\frac{dv_R}{dt} \tag{6.13}$$

EXAMPLE 6.9   Neglecting the drag force, calculate the velocity a rocket will achieve after a 100-s burn if the thrust is 200 kN and the initial and final masses are 10,000 and 1000 kg, respectively. Assume that the rocket's mass decreases linearly with time.

Solution   Using Eq. (6.13), we have

$$T - Mg = M\frac{dv_R}{dt} \tag{6.14}$$

$$\frac{T}{M} - g = \frac{dv_R}{dt}$$

The mass decreases linearly with time, so

$$M = M_0 - \lambda t$$

where λ has a value of 90 kg/s.
Substituting into Eq. (6.14) and integrating, we have

$$\int_0^{t_b} \frac{T\,dt}{M_0 - \lambda t} - \int_0^{t_b} g\,dt = v_R$$

or

$$v_R = \frac{T}{\lambda}\ln\frac{M_0}{M_f} - gt_b$$

The final result is

$$v_R = \frac{2 \times 10^5\ \text{N}}{90\ \text{kg/s}}\ln\frac{10{,}000\ \text{kg}}{1000\ \text{kg}} - (9.81\ \text{m/s}^2)(100\ \text{s}) = 4136\ \text{m/s} \quad \blacktriangleleft$$

Force on Rectangular Sluice Gate

A gate across a channel under which the water flows is called a sluice gate. In Fig. 6.8 the pressure distribution near the upper part of the sluice gate is

FIGURE 6.8

Flow under a sluice gate.

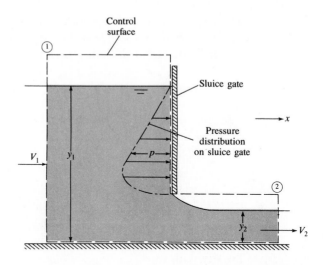

approximately hydrostatic because the velocity of flow in this region is very low. However, near the bottom of the sluice gate the velocity is quite large. Consequently, there is a large deviation from the hydrostatic pressure distribution. Because of this variable pressure on the sluice gate, it is not easy to determine the force on the gate by integration of the pressure over the gate surface. The simplest, although indirect, way to obtain the force is to apply the momentum equation to the liquid in the control volume shown in Fig. 6.8. Here the control surface is identified by the dashed line, and the flow across the control surface on the upstream side and downstream side is assumed to be uniform and steady. Hence the pressure distribution across both sections 1 and 2 will be hydrostatic. The upstream hydrostatic pressure will act against the water in the control volume in the positive x direction (to the right), and the downstream hydrostatic pressure will act to the left, or with a negative sign. Thus, for a sluice gate in a horizontal channel, the momentum equation for the x direction yields

$$\sum F_x = \sum_{cs} v_x \rho \mathbf{V} \cdot \mathbf{A}$$

$$\frac{\gamma y_1^2 b}{2} - \frac{\gamma y_2^2 b}{2} + F_{GW} - F_{\text{visc}} = V_1 \rho (-V_1 y_1 b) + V_2 \rho (V_2 y_2 b) \qquad (6.15)$$

where F_{GW} is the force of the gate acting on the water in the control volume, F_{visc} is the viscous force from the bottom of the channel, and b is the width of channel and gate.

For most applications of this type the viscous force is negligible. Therefore, given the discharge plus upstream and downstream depths, it is easy to solve for the gate force. The procedure is to solve for F_{GW} and then change the sign on F_{GW} to obtain the force on the gate F_G:

$$F_G = \rho Q(V_1 - V_2) + \tfrac{1}{2}b\gamma(y_1^2 - y_2^2) \qquad (6.16)$$

Water Hammer—Physical Description

Whenever a valve is closed in a pipe, a positive pressure wave is created upstream of the valve and travels up the pipe at the speed of sound. In this context a positive pressure wave is defined as one for which the pressure is greater than the steady-state pressure. This pressure wave may be great enough to cause pipe failure. Therefore, a basic understanding of this process, which is called *water hammer,* is necessary for the proper design and operation of such systems. The simplest case of water hammer will be considered here. For a more comprehensive treatment of the subject, the reader is referred to Chaudhry (1) and Streeter and Wylie (2).

Consider flow in the pipe shown in Fig. 6.9 on page 238. Initially the valve at the end of the pipe is only partially open (Fig. 6.9a); consequently, an initial velocity V and initial pressure p_0 exist in the pipe. At time $t = 0$ it is assumed that the valve is instantaneously closed, thus creating the pressure wave that travels toward the reservoir at the speed of sound, c. All the water between the pressure wave and the upper end of the pipe will have the initial velocity V, but all the water on the other side of the pressure wave (between the wave and the valve) will be at rest. This condition is shown in Fig. 6.9b. Once the pressure wave reaches the upper end of the pipe (after time $t = L/c$), it can be visualized that all of the water in the pipe will be under a pressure $p_0 + \Delta p$; however, the pressure in the reservoir at the end of the pipe is only p_0. This imbalance of pressure at the reservoir end causes the water to flow from the pipe back into the reservoir with a velocity V. Thus a new pressure wave is formed that travels toward the valve end of the pipe (Fig. 6.9c), and the pressure on the reservoir side of the wave is reduced to p_0. When this wave finally reaches the valve, all the water in the pipe is flowing toward the reservoir with a velocity V. This condition is only momentary, however, because the closed valve prevents any sustained flow.

Next, during time $2L/c < t < 3L/c$, a rarefied wave of pressure $(p < p_0)$ travels up to the reservoir, as shown in Fig. 6.9d. When the wave reaches the reservoir, all the water in the pipe has a pressure less than that in the reservoir. This imbalance of pressure causes flow to be established again in the entire pipe, as shown in Fig. 6.9f, and the condition is exactly the same as in the initial condition (Fig. 6.9a). Hence the process will repeat itself in a periodic manner.

From this description, it may be seen that the pressure in the pipe immediately upstream of the valve will be alternately high and low, as shown in

FIGURE 6.9

Water-hammer process.
(a) Initial condition.
(b) Condition during time
$0 < t < L/c.$
(c) Condition during time
$L/c < t < 2L/c.$
(d) Condition during time
$2L/c < t < 3L/c.$
(e) Condition during time
$3L/c < t < 4L/c.$
(f) Condition at time
$t = 4L/c.$

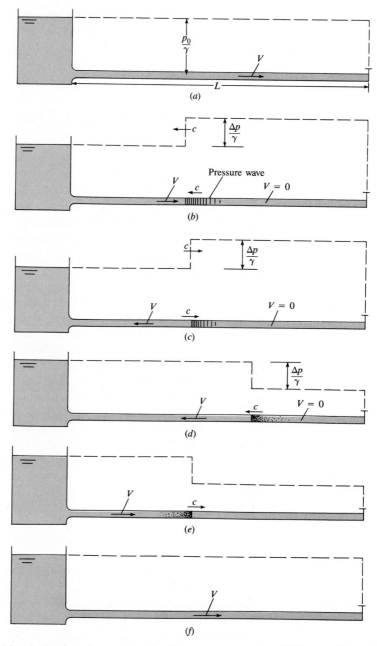

Fig. 6.10a. A similar observation for the pressure at the midpoint of the pipe reveals a more complex variation of pressure with time, as shown in Fig. 6.10b. Obviously, a valve cannot be closed instantaneously, and viscous effects, which were neglected here, will have a damping effect on the process. Therefore, a

more realistic pressure–time trace for the point just upstream of the valve is given in Fig. 6.10c. The finite time of closure erases the sharp discontinuities in the pressure trace that were present in Fig. 6.10a. However, it should be noted that the maximum pressure developed at the valve will be virtually the same as for instantaneous closure if the time of closure is less than $2L/c$. That is, the change in pressure will be the same for a given change in velocity unless the negative wave from the reservoir mitigates the positive pressure, and it takes a time $2L/c$ before this negative wave can reach the valve. The value $2L/c$ is called the *critical time of closure* and is given the symbol t_c.

Magnitude of Water-Hammer Pressure and Speed of Pressure Wave

We can analyze the quantitative relations for water hammer with the momentum equation by letting the control volume either move with the pressure wave, thus creating steady motion, or be fixed, thus retaining the inherently unsteady character of the process. To illustrate the use of the momentum equation with unsteady motion, let us take the latter approach. Consider a pressure wave in a rigid pipe, as shown in Fig. 6.11. The density, pressure, and velocity of the fluid on the reservoir side of the pressure wave are ρ, p, and V, respectively, and the similar quantities on the valve side of the wave are $\rho + \Delta\rho$, $p + \Delta p$, and 0. Because the wave in this case is traveling from the valve to the reservoir, its distance from the valve at any time t is given as ct. We can now apply the momentum

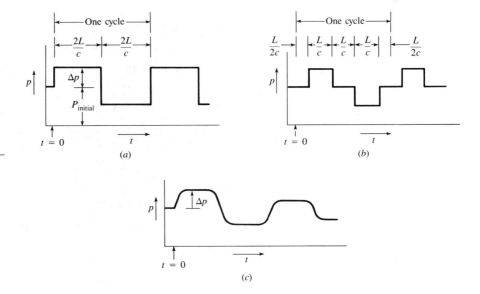

FIGURE 6.10

Variation of water-hammer pressure with time at two points in a pipe. (a) Location: adjacent to valve. (b) Location: at midpoint of pipe. (c) Actual variation of pressure near valve.

FIGURE 6.11

Pressure wave in a pipe.

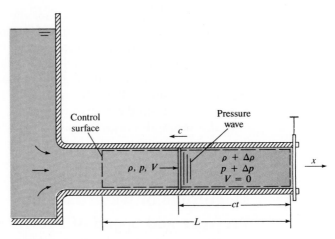

equation to the flow in the control volume. Let the x direction be along the pipe. Then the momentum equation in the x direction is

$$\sum F_x = \sum_{cs} v_x \rho \mathbf{V} \cdot \mathbf{A} + \frac{d}{dt} \int_{cv} v_x \rho \, d\forall$$

where

$$\sum F_x = pA - (p + \Delta p)A \tag{6.17}$$

$$\sum_{cs} v_x \rho \mathbf{V} \cdot \mathbf{A} = V\rho(-VA) \tag{6.18}$$

$$\frac{d}{dt} \int_{cv} v_x \rho \, d\forall = \frac{d}{dt}[V\rho(L - ct)A]$$

$$= -V\rho cA \tag{6.19}$$

When Eqs. (6.17), (6.18), and (6.19) are substituted into the basic momentum equation, we obtain

$$pA - (p + \Delta p)A = -\rho V^2 A - \rho V c A$$

This reduces to

$$\Delta p = \rho V^2 + \rho V c$$

In this equation the first term on the right-hand side is usually negligible with respect to the second term on the right, because for liquids c is much greater than V. Consequently, the equation reduces to

$$\Delta p = \rho V c \tag{6.20}$$

We will now determine the speed of the pressure wave by applying the continuity equation to the control volume in Fig. 6.11. The basic continuity equation is

$$0 = \sum_{cs} \rho \mathbf{V} \cdot \mathbf{A} + \frac{d}{dt} \int_{cv} \rho \, d\forall$$

Applying this to Fig. 6.11, we obtain

$$0 = \rho(-VA) + \frac{d}{dt}\left[\rho(L - ct)A + (\rho + \Delta\rho)ctA\right]$$

which reduces to

$$\frac{\Delta\rho}{\rho} = \frac{V}{c}$$

or

$$c = \frac{V}{\Delta\rho/\rho} \qquad (6.21)$$

However, by definition $E_v = \Delta p/(\Delta\rho/\rho)$. Therefore,

$$\frac{\Delta\rho}{\rho} = \frac{\Delta p}{E_v} \qquad (6.22)$$

Now when $\Delta\rho/\rho$ is eliminated between Eqs. (6.21) and (6.22), we have

$$c = \frac{VE_v}{\Delta p} \qquad (6.23)$$

From Eq. (6.20), $\Delta p = \rho Vc$. Therefore, Eq. (6.23) becomes

$$c = \sqrt{\frac{E_v}{\rho}} \qquad (6.24)$$

Thus, by application of the momentum and continuity equations, we have derived equations for both Δp and c.

EXAMPLE 6.10   A rigid pipe leading from a reservoir is 3000 ft long, and water is flowing through it with a velocity of 4 ft/s. If the initial pressure at the downstream end is 40 psig, what maximum pressure will develop at the downstream end when a rapid-acting valve at that end is closed in 1 s?

Solution

$$c = \sqrt{\frac{E_v}{\rho}} = \sqrt{\frac{320{,}000 \text{ lbf/in.}^2 \times 144 \text{ in}^2/\text{ft}^2}{1.94 \text{ slugs/ft}^3}} = 4874 \text{ ft/s}$$

Next determine whether the closure time is greater or less than the critical closure time, t_c.

$$t_c = 2L/c$$

$$= 2(3000 \text{ ft}/4874 \text{ ft/s}) = 1.23 \text{ s}$$

Since the closure time is less than t_c, the maximum value of Δp will be equal to ρVc.

$$\Delta p = 1.94 \text{ slugs/ft}^3 \times 4 \text{ ft/s} \times 4874 \text{ ft/s}$$

$$= 37{,}820 \text{ lbf/ft}^2 = 263 \text{ psi}$$

The maximum pressure is the initial pressure plus the pressure change, which is

$$p_{\max} = 40 + 263 = 303 \text{ psig}$$ ◄

As indicated by Example 6.10, water-hammer pressures can be quite large. Therefore, engineers must design piping systems to keep the pressure within acceptable limits. This is done by installing an accumulator near the valve and/or operating the valve in such a way that rapid closure is prevented. Accumulators may be in the form of air chambers for relatively small systems, or surge tanks (a surge tank is a large open tank connected by a branch pipe to the main pipe) for cases such as large hydropower systems. Another way to eliminate excessive water-hammer pressures is to install pressure-relief valves at critical points in the pipe system. These valves are pressure-activated so that water is automatically diverted out of the system when the water-hammer pressure reaches excessive levels.

6.3   Moment-of-Momentum Equation

The moment-of-momentum equation relates the moment applied to a system to the rate of change of angular momentum of the system and the flow of angular momentum across the control surface. Because the phenomenon involves the rate of change of a physical property, we can employ the basic control-volume approach. The basic control-volume equation is

$$\frac{dB_{\text{sys}}}{dt} = \sum_{\text{cs}} \beta \rho \mathbf{V} \cdot \mathbf{A} + \frac{d}{dt} \int_{\text{cv}} \beta \rho \, d\forall \qquad (4.21a)$$

Here, if we let $\beta = \mathbf{r} \times \mathbf{v}$ where $\times$ denotes the vector product of $\mathbf{r}$ and $\mathbf{v}$, then the extensive property B_{sys} will be

$$B_{\text{sys}} = \int (\mathbf{r} \times \mathbf{v}) \rho \, d\forall = \text{angular momentum of the system} \qquad (6.25)$$

When Eq. (6.25) and $\beta = \mathbf{r} \times \mathbf{v}$ are substituted back into Eq. (4.21a), we obtain

$$\frac{d\left[\int (\mathbf{r} \times \mathbf{v}) \rho \, d\forall\right]}{dt} = \sum_{\text{cs}} (\mathbf{r} \times \mathbf{v}) \rho \mathbf{V} \cdot \mathbf{A} + \frac{d}{dt} \int_{\text{cv}} (\mathbf{r} \times \mathbf{v}) \rho \, d\forall \qquad (6.26)$$

Equation (6.26) states that the rate of change with respect to time of the angular momentum of the system is equal to the net flow of angular momentum from the control volume plus the rate of change of angular momentum within the control volume. Here the origin for the angular momentum is the origin of the position vector $\mathbf{r}$. From basic mechanics it is known that the net moment, $\Sigma\,\mathbf{M}$, applied to a system is equal to the rate of change of angular momentum of the system. Therefore, the left side of Eq. (6.26) can be replaced by $\Sigma\,\mathbf{M}$. We then have

$$\sum \mathbf{M} = \sum_{cs} (\mathbf{r} \times \mathbf{v})\rho\mathbf{V} \cdot \mathbf{A} + \frac{d}{dt}\int_{cv} (\mathbf{r} \times \mathbf{v})\rho\,d\forall \qquad (6.27a)$$

Equation (6.27a) is the basic moment-of-momentum equation for cases where the velocity distribution is uniform across the flow section. When the velocity is variable across the flow section, we use the more general form:

$$\sum \mathbf{M} = \int_{cs} (\mathbf{r} \times \mathbf{v})\rho\mathbf{V} \cdot \mathbf{dA} + \frac{d}{dt}\int_{cv} (\mathbf{r} \times \mathbf{v})\rho\,d\forall \qquad (6.27b)$$

Equation 6.27 is often used in applications involving hydraulic machines (pumps and turbines), and in those cases the origin for $\mathbf{r}$ is taken at the axis of the rotating shaft of the machine. In other cases, such as in the example that follows, the origin for $\mathbf{r}$ is more or less arbitrary but is usually chosen such that the resulting moment is relevant to the design of whatever is being considered. Also, Eq. 6.27 is rarely applied for a case of an accelerating control volume; therefore, in most applications both $\mathbf{v}$ and $\mathbf{V}$ are referenced to the control volume.

EXAMPLE 6.11   If the bend of Example 6.5 is to be supported from above by an external system of supports, what moment about an axis that is level with the top of the pipe and 30 cm to the right of the flange must the support system be designed for? Assume that the center of mass of the bend and the water taken together is 30 cm below the top of the pipe and 50 cm to the right of the flanges of the bend.

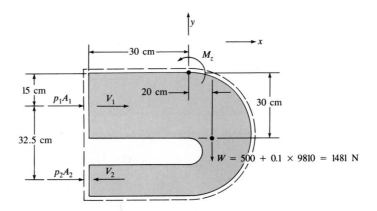

Solution   We draw the control surface as shown and apply Eq. (6.27a). We have steady flow, so the last term of Eq. (6.27a) drops out. Then we have

$$\sum M_{z\,axis} = \sum_{cs} rV\rho \mathbf{V} \cdot \mathbf{A}$$

Here the moments will be from the pressure forces, the weight of the bend, the weight of the water, and the external moment M_z applied at the axis in question. Also note that we use the right-hand rule for sign convention of moments and that the origin for the position vector $\mathbf{r}$ is at the moment center noted in the problem statement (30 cm to the right of the flange). Inserting these moments, the velocities, and their moment arms in the foregoing equation yields

$$M_z + 0.15p_1A_1 + 0.475p_2A_2 - 0.2 \times 1481 = (0.15 \times 3.54)(1000)$$
$$\times [-3.54 \times 0.3^2 \times (\pi/4)] + (-0.475 \times 14.15)(1000)$$
$$\times [14.15 \times 0.15^2 \times (\pi/4)] \text{ N} \cdot \text{m}$$

Here $p_1 = 150$ kPa and $p_2 = 59.3$ kPa, so we solve for M_z:

$$M_z = -0.15 \times 150 \times 10^3 \times 0.3^2 \times (\pi/4) - 0.475 \times 59.3 \times 10^3$$
$$\times 0.15^2 \times (\pi/4) + 0.2 \times 1481 - 133 - 1681$$
$$= -1590 - 498 + 296 - 133 - 1681 = -3.61 \text{ kN} \cdot \text{m} \quad \blacktriangleleft$$

Thus a moment of 3610 N · m applied in the clockwise sense about the axis in question is needed. Stated differently, the support system must be designed to withstand a counterclockwise moment of 3610 N · m.

EXAMPLE 6.12   Determine the power produced by a simple sprinkler-like turbine that rotates in a horizontal plane at 500 rpm. The radius of the turbine is 0.5 m. Water enters the turbine from a vertical pipe that is coaxial with the axis of rotation, and it exits through nozzles, each of which has a cross-sectional area of 10 cm². The exit velocity of the water is 50 m/s with respect to the nozzle. The water density is 1000 kg/m³, and the pressure at the exit is atmospheric.

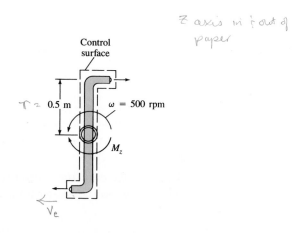

Solution We draw a control surface around the turbine as shown, taking the z axis as the axis of rotation. Mass enters the control volume at the hub and exits at each nozzle. Applying the moment-of-momentum equation, we have

$$M_z = \sum_{cs} (\mathbf{r} \times \mathbf{v})\rho \mathbf{V} \cdot \mathbf{A}$$

since the flow is steady. The moment of momentum associated with the flow into the hub is zero because $\mathbf{r} = 0$. Thus the moment-of-momentum equation reduces to

$$M_z = (\mathbf{r} \times \mathbf{v})\dot{m}$$

where $\dot{m}$ is the mass flow rate from the nozzles. The velocity $\mathbf{v}$ must be referenced to an inertial frame. Thus

$$\mathbf{r} \times \mathbf{v} = -r(V_e - \omega r)$$

where r is the radius and V_e is the exit velocity with respect to the nozzle. Thus the moment applied to the system (that is, to the arms) by the generator is $-r(V_e - \omega r)\dot{m}$, but the moment applied to the generator is the negative of the moment of the generator on the arms, or

$$M_z = (0.5)[50 - 500 \times 2\pi \times (0.50/60)](50)(10 \times 10^{-4})(1000) \times 2$$
$$= 1190 \text{ N} \cdot \text{m}$$

The power produced by the turbine-generator is the product of the moment and angular velocity:

$$P = M_z\omega = \frac{(500)(2\pi)(1190)}{60} = 62.4 \text{ kW} \qquad \blacktriangleleft$$

6.4   **Navier–Stokes Equations**

The foregoing momentum equations (Eqs. 6.3, 6.4, 6.5, and 6.6) and their applications were for cases of control volumes of macro size. In many cases, however, it is desirable to use the differential form of the momentum equation. The differential form is obtained by applying Newton's second law to an *element* of fluid. In this case the forces acting on the element are the pressure, gravitational force, and viscous forces, and in general the fluid is undergoing both local and convective acceleration. The resulting equations are called the Navier–Stokes equations after L. M. Navier (1785–1836) and G. G. Stokes (1819–1903), who are credited with their development. These equations for flow of an incompressible fluid* in a cartesian coordinate system are

---

* *When compressibility effects are significant or the viscosity is nonuniform, additional terms are included in the Navier–Stokes equations to account for these effects.*

$$\rho g_x - \frac{\partial p}{\partial x} + \mu\left(\frac{\partial^2 u}{\partial x^2} + \frac{\partial^2 u}{\partial y^2} + \frac{\partial^2 u}{\partial z^2}\right) = \rho\frac{du}{dt} \qquad (6.28a)$$

$$\rho g_y - \frac{\partial p}{\partial y} + \mu\left(\frac{\partial^2 v}{\partial x^2} + \frac{\partial^2 v}{\partial y^2} + \frac{\partial^2 v}{\partial z^2}\right) = \rho\frac{dv}{dt} \qquad (6.28b)$$

$$\rho g_z - \frac{\partial p}{\partial z} + \mu\left(\frac{\partial^2 w}{\partial x^2} + \frac{\partial^2 w}{\partial y^2} + \frac{\partial^2 w}{\partial z^2}\right) = \rho\frac{dw}{dt} \qquad (6.28c)$$

It is assumed that the coordinate system is arbitrarily defined. Therefore, in these equations ρg_x, ρg_y, and ρg_z are the gravitational forces per unit of volume in the x, y, and z direction, respectively. Equations 6.28, together with the continuity equation (4.31a) and the equation of state, can be used to solve, numerically or by computational techniques, for the unknowns ρ, u, v, w, and p.

The derivation of the Navier–Stokes equations is beyond the scope of this text. The reader is referred to Schlichting (3) for their derivation. These equations will be used in Chapter 16 in the introduction to computational fluid mechanics.

Problems

6.1 A horizontal water jet issues from a circular orifice in a large tank and strikes a vertical plate that is normal to the axis of the jet. The discharge in the jet is 0.3 m³/s. What horizontal force is required to hold the plate in position if the pressure in the tank is 50 kPa?

6.2 A horizontal water jet impinges on a vertical-perpendicular plate. The discharge is 1 cfs. If the external force required to hold the plate in place is 160 lbf, what is the velocity of the water?

6.3 A horizontal water jet issues from a circular orifice in a large tank. The jet strikes a vertical plate that is normal to the axis of the jet. A force of 200 lbf is needed to hold the plate in place against the action of the jet. If the pressure in the tank is 10 psig, what is the diameter of the jet just downstream of the orifice?

6.4 Solve Prob. 6.3 for a force of 3.0 kN and a pressure of 80 kPa.

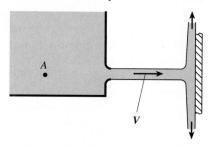

PROBLEMS 6.1, 6.2, 6.3, 6.4, 6.5

6.5   A horizontal circular jet of water is discharged from a tank and then strikes a vertical plate (see figure). The axis of the jet is normal to the plate. The cross-sectional area of the jet is 0.01 m². What velocity in the jet is needed to produce a force of 1000 N on the plate? What pressure at A in the tank is needed to produce this velocity?

6.6   A conveyor belt discharges gravel into a barge as shown at a rate of 50 yd³/min. If the gravel weighs 120 lb/ft³, what is the tension in the hawser that secures the barge to the dock?

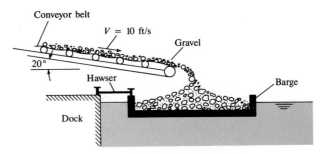

PROBLEM 6.6

6.7   A jet of water is discharging at a constant rate of 1.20 cfs from the upper tank. If the jet diameter at section 1 is 4 in., what forces will be measured by scales A and B? Assume the empty tank weighs 200 lbf, the cross-sectional area of the tank is 4 ft², $h = 9$ ft, and $H = 1$ ft.

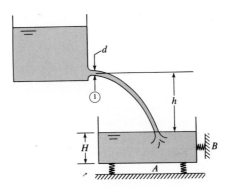

PROBLEM 6.7

6.8   An open boxcar is rolling down a slight incline (the slope is a 1-ft drop in 100 ft) at a speed of 5 ft/s. Then grain is poured into the car (it drops vertically downward) at a rate of 500 lbm/s, with a speed of 30 ft/s when it hits the car. Will the boxcar accelerate or decelerate when the grain is poured into it?

6.9   Determine the external reactions in the x and y directions needed to hold this fixed vane, which turns the oil jet in a horizontal plane. Here V_1 is 28 m/s, $V_2 = 27$ m/s, and $Q = 0.07$ m³/s.

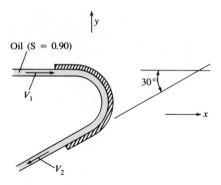

PROBLEMS 6.9, 6.10

6.10 Solve Prob. 6.9 for $V_1 = 90$ ft/s, $V_2 = 85$ ft/s, and $Q = 1$ cfs.

6.11 This horizontal two-dimensional water jet is deflected by the fixed vane. What force per foot of width is required to hold the vane in place?

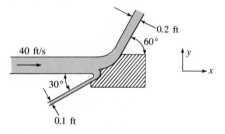

PROBLEM 6.11

6.12 Starting with the momentum equation, show all the steps leading to Eq. (6.11), p. 221.

6.13 A horizontal jet of water that is 6 cm in diameter and has a velocity of 20 m/s is deflected by the vane as shown. If the vane is moving at a rate of 7 m/s in the x direction, what components of force are exerted on the vane by the water in the x and y directions? Assume negligible friction between the water and the vane.

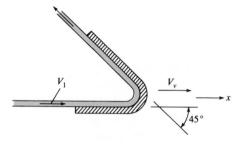

PROBLEM 6.13

6.14   A vane on this moving cart deflects a 10-cm water jet as shown. The initial speed of the water in the jet is 20 m/s, and the cart moves at a speed of 2 m/s. If the vane splits the jet so that half goes one way and half the other, what force is exerted on the vane by the jet?

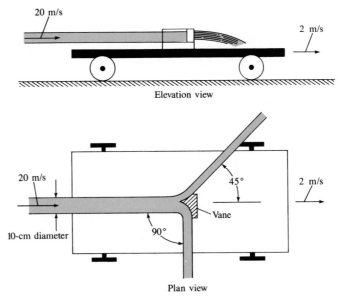

PROBLEM 6.14

6.15   Water flows through the blade passage of a turbine. The turbine is not rotating. The blade deflects the water 30°. The inlet water velocity is 30 ft/s. The ratio of inlet area to exit area of the blade passage is 2. Find the lateral force per unit mass flow produced by the water acting on the blade. The pressures at the inlet and exit are atmospheric.

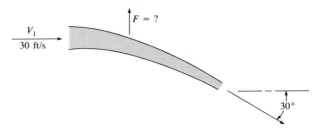

PROBLEM 6.15

6.16   The water in this jet has a speed of 30 m/s to the right and is deflected by a cone that is moving to the left with a speed of 15 m/s. The diameter of the jet is 10 cm. Determine the external horizontal force needed to move the cone. Assume negligible friction between the water and the vane.

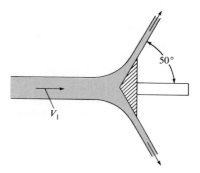

PROBLEMS 6.16, 6.17

6.17 This two-dimensional water jet is deflected by the two-dimensional vane, which is moving to the right with a speed of 60 ft/s. The initial jet is 0.30 ft thick (vertical dimension), and its speed is 100 ft/s. What power per foot of the jet (normal to the page) is transmitted to the vane?

6.18 Water strikes a block as shown and is deflected 30°. The flow rate of the water is 1 kg/s, and the velocity is 10 m/s. The mass of the block is 1 kg. The coefficient of static friction between the block and the surface is 0.1 (friction force/normal force). If the force parallel to the surface exceeds the frictional force, the block will move. Determine whether the block will move. Neglect the weight of the water.

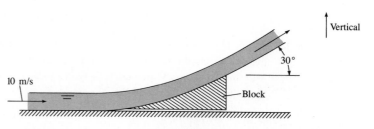

PROBLEM 6.18

6.19 A jet of water with velocity V_j is directed in the positive x direction, and it is deflected by a disk moving in the negative x direction with a velocity V_d. The absolute magnitudes (speeds) of V_j and V_d are equal. If the jet cross-sectional area is A, then the magnitude of the force F required to move the disk is a) $\rho V_j^2 A$, b) $2\rho V_j^2 A$, c) $3\rho V_j^2 A$, d) $4\rho V_j^2 A$.

6.20 Just after jet planes touch down, vanes are actuated to produce reverse thrust to cause the plane to decelerate faster. Each engine takes in air at a rate of 150 kg/s, and the combustion products are exhausted at a speed of 750 m/s relative to the engine. The fuel-to-air ratio is 1 to 40. What is the reverse thrust per engine produced by the thruster?

6.21 Plate A is 50 cm in diameter and has a sharp-edged orifice at its center. A water jet strikes the plate concentrically with a speed of 30 m/s. With the plate held stationary, what external force is needed to hold the plate in place if the jet issuing from the orifice also has a speed of 30 m/s? The diameters of the jets are $D = 10$ cm and $d = 4$ cm.

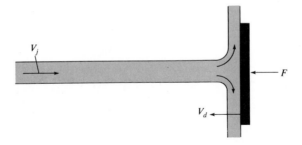

PROBLEM 6.19

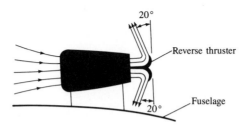

PROBLEM 6.20

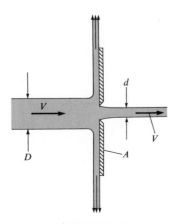

PROBLEM 6.21

6.22 This two-dimensional liquid jet impinges on the horizontal floor. Derive formulas for d_2 and d_3 as functions of b_1 and θ. Assume that the jet speed is large enough to cause the gravitational effects to be negligible. Also assume that the speed of the liquid is the same at sections 1, 2, and 3.

6.23 A two-dimensional liquid jet impinges on a vertical wall. Assuming that the incoming jet speed is the same as the exiting jet speed ($V_1 = V_2$), derive an expression for the force per unit width of jet exerted on the wall. What form do you think the upper liquid surface will take next to the wall? Sketch the shape you think it will take, and explain your reasons for drawing it that way.

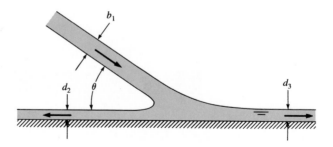

PROBLEM 6.22

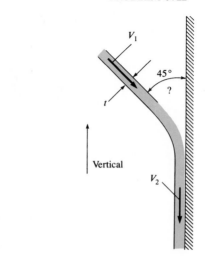

PROBLEM 6.23

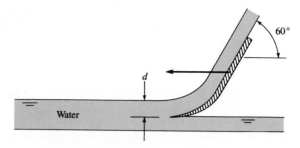

PROBLEM 6.24

6.24 Assume that the scoop shown, which is 20 cm wide, is used as a braking device for studying deceleration effects, such as those on space vehicles. If the scoop is attached to a 900-kg sled that is initially traveling horizontally at the rate of 100 m/s, what will be the initial deceleration of the sled? The scoop dips into the water 8 cm ($d = 8$ cm).

6.25 This snowplow "cleans" a swath of snow that is 3 in. deep ($d = 3$ in.) and 2 ft wide ($B = 2$ ft). The snow leaves the blade in the direction indicated in the sketches. Neglect-

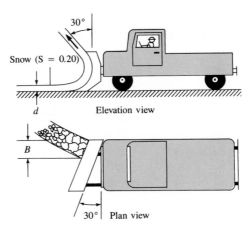

PROBLEM 6.25

ing friction between the snow and the blade, estimate the power required for just the snow removal if the speed of the snowplow is 40 ft/s.

6.26 A large tank of liquid is resting on a frictionless plane as shown. Explain in a qualitative way what will happen after the cap is removed from the short pipe.

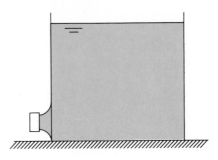

PROBLEM 6.26

6.27 A ramjet operates by taking in air at the inlet, providing fuel for combustion, and exhausting the hot air through the exit. The mass flow at the inlet and outlet of the ramjet is 50 kg/s (the mass-flow rate of fuel is negligible). The inlet velocity is 200 m/s. The density of the gases at the exit is 0.25 kg/m^3, and the exit area is 0.5 m^2. Calculate the thrust delivered by the ramjet. The ramjet is not accelerating and the flow within the ramjet is steady.

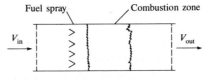

PROBLEM 6.27

6.28 A water jet with a 6-cm diameter and a velocity $V_j = 15$ m/s causes the cart to move at a constant speed of 10 m/s. What is the rolling resistance of the cart?

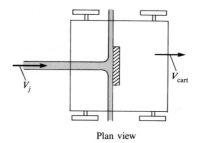

Plan view

PROBLEM 6.28

6.29 A discharge Q of a liquid of very low viscosity drops vertically into a short horizontal-rectangular channel of width B as shown. The depth at section 2 is y_2. Derive an equation that gives y_1 in terms of y_2, Q, B, and γ.

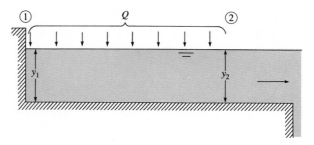

PROBLEM 6.29

6.30 The end section of a pipe has a slot cut in it so that liquid will discharge laterally as shown. The liquid is discharged from the slot at a rate given by

$$dQ = \Delta y \sqrt{2p/\rho}\, dx$$

where dQ is the discharge per differential length dx along the slot, Δy is the thickness of the jet issuing from the slot, and p is the gage pressure in the pipe. Tell qualitatively how the pressure will change in the pipe from $x = 0$ to $x = L$. Also devise a way of solving for the pressure distribution, given p_{end} (pressure at the end of the pipe) and the slot dimensions. Neglect viscous resistance in developing your solution procedure.

6.31 A cone that is held stable by a wire is free to move in the vertical direction and has a jet of water striking it from below. The cone weighs 50 N. The initial speed of the jet as it comes from the orifice is 15 m/s, and the initial jet diameter is 3 cm. Find the height to which the cone will rise and remain stationary. *Note:* The wire is only for stability and should not enter into your calculations.

6.32 A 6-in. horizontal pipe has a 180° bend in it. If the rate of flow of water in the bend is 6 cfs and the pressure therein is 20 psi, what external force in the original direction of flow is required to hold the bend in place?

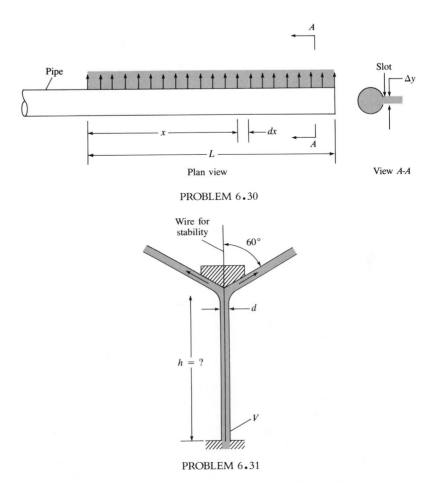

Pipe

Slot

Δy

x

dx

L

Plan view

View A-A

PROBLEM 6.30

Wire for
stability

60°

d

$h = ?$

V

PROBLEM 6.31

6.33 A hot gas stream enters a uniform-diameter return bend as shown. The entrance velocity is 100 ft/s, the gas density is 0.02 lbm/ft³, and the mass flow rate is 1 lbm/s. Water is sprayed into the duct to cool the gas down. The gas exits with a density of 0.06 lbm/ft³. The mass flow of water into the gas is negligible. The pressures at the entrance and exit are the same and equal to the atmospheric pressure. Find the force required to hold the bend.

6.34 Assume that the gage pressure p is the same at sections 1 and 2 in this horizontal bend. The fluid flowing in the bend has density ρ, discharge Q, and velocity V. The cross-sectional area of the pipe is A. Then the magnitude of the force (neglecting gravity) required at the flanges to hold the bend in place will be a) pA, b) $pA + \rho QV$, c) $2pA + \rho QV$, d) $2pA + 2\rho QV$.

6.35 This pipe has a 180° vertical bend in it. The diameter D is 1 ft, and the pressure at the center of the upper pipe is 10 psig. If the flow in the bend is 15 cfs, what external force will be required to hold the bend in place against the action of the water? The bend weighs 100 lbf, and the volume of the bend is 3 ft³. Neglect head losses.

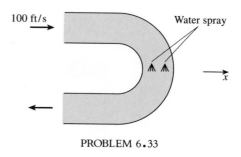

PROBLEM 6.33

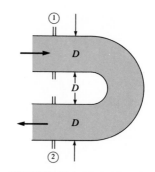

PROBLEMS 6.34, 6.35, 6.36

6.36 This pipe has a 180° horizontal bend in it as shown, and D is 30 cm. The discharge of water in the pipe and bend is 0.60 m³/s, and the pressure in the pipe and bend is 120 kPa gage. If the bend volume is 0.10 m³ and the bend itself weighs 500 N, what force must be applied at the flanges to hold the bend in place?

6.37 Water flows in the horizontal bend at a rate of 10 cfs and discharges into the atmosphere past the downstream flange. The pipe diameter is 1 ft. What force must be applied at the upstream flange to hold the bend in place? Assume that the volume of water downstream of the upstream flange is 4 ft³ and that the bend and pipe weigh 100 lbf.

6.38 The gage pressure in the horizontal 90° pipe bend is 200 kPa. If the pipe diameter is 1 m and the water flow rate is 10 m³/s, what x component of force must be applied to the bend to hold it in place against the water action?

6.39 This 30° vertical bend in a pipe with a 2-ft diameter carries water at a rate of 31.4 cfs. If the pressure p_1 is 10 psi at the lower end of the bend, where the elevation is 100 ft, and p_2 is 8 psi at the upper end, where the elevation is 103 ft, what will be the vertical component of force that must be exerted by the "anchor" on the bend to hold it in position? The bend itself weighs 300 lb, and the length L is 4 ft.

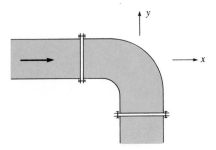

PROBLEMS 6.37, 6.38

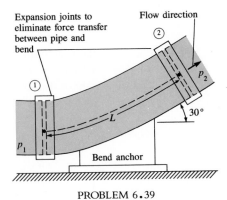

PROBLEM 6.39

6.40 A 90° horizontal bend narrows from a 2-ft diameter upstream to a 1-ft diameter downstream. If the bend is discharging water into the atmosphere and the pressure upstream is 25 psig, what is the magnitude of the component of external force exerted on the bend in the x direction (direction parallel to the initial flow direction) required to hold the bend in place?

6.41 A horizontal reducing bend turns the flow of water through 90°. The inlet diameter is 20 cm and the outlet diameter is 10 cm. The inlet gage pressure is 150 kPa and the outlet gage pressure is 74 kPa. The discharge is 0.1 m³/s. The interior volume of the bend is 0.02 m³, and the weight of the metal in the bend is 200 N. What external force is required to hold the bend in place?

6.42 A 180° horizontal bend narrows from a 14-in. diameter upstream to a 10-in. diameter downstream. If the bend discharges water into the atmosphere at a rate of 20 cfs, what force component in the x direction (direction parallel to the initial flow direction) is required to hold the bend in place?

6.43 This bend discharges water into the atmosphere. Determine the magnitude and direction of the external force components at the flange required to hold the bend in place. The bend lies in a horizontal plane, and water is flowing. Assume that the interior volume of the bend is 0.25 m³, $D_1 = 60$ cm, $D_2 = 30$ cm, and $V_2 = 10$ m/s.

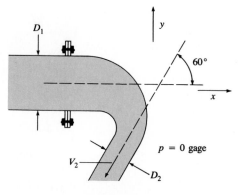

PROBLEM 6.43

6.44   This nozzle bends the flow from vertically upward to 30° with the horizontal and discharges water ($\gamma = 62.4$ lbf/ft^3) at a speed of 100 ft/s. The volume within the nozzle itself is 2 ft^3, and the weight of the nozzle is 100 lbf. For these conditions, what *vertical* force must be applied to the nozzle at the flange to hold it in place?

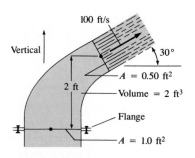

PROBLEM 6.44

6.45   A pipe 1 ft in diameter bends through an angle of 135°. The velocity of flow of gasoline ($S = 0.8$) is 15 ft/s, and the pressure is 12 psi in the bend. What external force is required to hold the bend against the action of the gasoline? Neglect the gravitational force.

6.46   A pipe 30 cm in diameter bends through 135°. The velocity of flow of gasoline ($S = 0.8$) is 8 m/s, and the pressure is 100 kPa gage in the bend. Neglecting gravitational force, determine the external force required to hold the bend against the action of the gasoline.

6.47   A pipe 40 cm in diameter has a 135° horizontal bend in it. The pipe carries water under a pressure of 90 kPa gage at a rate of 0.40 m^3/s. What external force component in a direction parallel to the initial flow direction is necessary to hold the bend in place under the action of the water? What horizontal force component normal to the initial direction of flow is required to hold the bend?

6.48   A horizontal reducing bend turns the flow of water through 60°. The inlet area is 0.001 m^2, and the outlet area is 0.0001 m^2. The water from the outlet discharges into the atmosphere with a velocity of 50 m/s. What horizontal force (parallel to the initial flow direction) acting through the metal of the bend at the inlet is required to hold the bend in place?

6.49 A horizontal reducing bend turns a flow of water through 90°. The inlet diameter is 20 cm, and the outlet diameter is 10 cm. The flow rate is 0.1 m^3/s, and the upstream pressure is 150 kPa gage. Neglecting body forces and assuming irrotational flow, calculate the external force required to hold the bend in place.

6.50 This 130-cm overflow pipe from a small hydroelectric plant conveys water from the 70-m elevation to the 40-m elevation. The pressures in the water at the bend entrance and exit are 20 kPa and 25 kPa, respectively. The bend interior volume is 3 m^3, and the bend itself weighs 10 kN. Determine the force that a thrust block must exert on the bend to secure it if the discharge is 15 m^3/s.

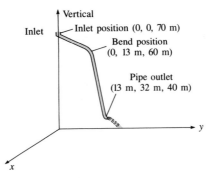

PROBLEM 6.50

6.51 Water flows in a duct as shown. The inlet water velocity is 10 m/s. The cross-sectional area of the duct is 0.2 m^2. Water is injected normal to the duct wall at the rate of 400 kg/s midway between stations 1 and 2. Neglect frictional forces on the duct wall. Calculate the pressure difference $(p_1 - p_2)$ between stations 1 and 2.

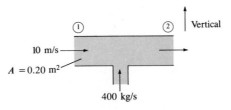

PROBLEM 6.51

6.52 Water flows vertically into a T section. Assuming an inviscid fluid, calculate the gage pressure at section 1.

6.53 For this wye fitting, which lies in a horizontal plane, the cross-sectional areas at sections 1, 2, and 3 are 1 ft^2, 1 ft^2, and 0.25 ft^2, respectively. At these same respective sections the pressures are 1000 psfg, 900 psfg, and 0 psfg, and the water discharges are 20 cfs to the right, 12 cfs to the right, and 8 cfs. What x component of force would have to be applied to the wye to hold it in place?

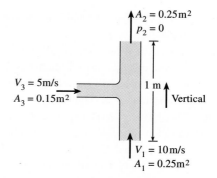

PROBLEM 6.52

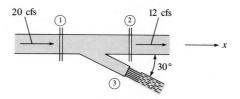

PROBLEM 6.53

6.54 Water flows through a horizontal bend and T section as shown. The mass-flow rate entering at section a is 10 lbm/s, and those exiting at sections b and c are 5 lbm/s each. The entering velocity is 6 ft/s, and the exit velocity at each outlet is 3 ft/s. The pressure at section a is 1 psig. The pressure at the two outlets is atmospheric. The cross-sectional areas of the pipes are the same: 5 in^2. Find the x component of force necessary to restrain the section.

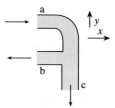

PROBLEMS 6.54, 6.55

6.55 Water flows through a horizontal bend and T section as shown. At section a the flow enters with a velocity of 6 m/s, and the pressure is 48 Pa. At both sections b and c the flow exits the device with a velocity of 3 m/s, and the pressure at these sections is atmospheric ($p = 0$). The cross-sectional areas at a, b, and c are all the same: 0.20 m^2. Find the x and y components of force necessary to restrain the section.

6.56 For this horizontal T through which water ($\rho = 1000$ kg/m^3) is flowing, the following data are given: $Q_1 = 0.25$ m^3/s, $Q_2 = 0.15$ m^3/s, $p_1 = 100$ kPa, $p_2 = 70$ kPa, $p_3 =$

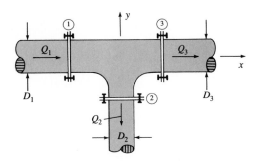

PROBLEM 6.56

80 kPa, $D_1 = 15$ cm, $D_2 = 10$ cm, and $D_3 = 15$ cm. For these conditions, what external force in the x-y plane (through the bolts or other supporting devices) is needed to hold the T in place?

6.57 For this weird nozzle, what force would have to be applied through the bolts in the flange to hold the nozzle in place? Assume irrotational flow. Water is flowing, and the nozzle itself weighs 200 N.

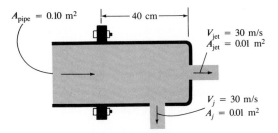

PROBLEM 6.57

6.58 Water flows through this nozzle at a rate of 15 cfs and discharges into the atmosphere. $D_1 = 12$ in. and $D_2 = 8$ in. Determine the force required at the flange to hold the nozzle in place. Assume irrotational flow. Neglect gravitational forces.

6.59 Solve Prob. 6.58 using the following values: $Q = 0.30$ m³/s, $D_1 = 30$ cm, and $D_2 = 15$ cm.

6.60 This "double" nozzle discharges water into the atmosphere at a rate of 15.7 cfs. If the nozzle is lying in a horizontal plane, what x component of force acting through the flange bolts is required to hold the nozzle in place? *Note:* Assume irrotational flow, and assume the water speed in each jet to be the same. Jet A is 4 in. in diameter, jet B is 4.5 in. in diameter, and the pipe is 1 ft in diameter.

6.61 This "double" nozzle discharges water into the atmosphere at a rate of 0.50 m³/s. If the nozzle is lying in a horizontal plane, what x component of force acting through the flange bolts is required to hold the nozzle in place? *Note:* Assume irrotational flow, and assume the water speed in each jet to be the same. Jet A is 10 cm in diameter, jet B is 12 cm in diameter, and the pipe is 30 cm in diameter.

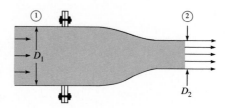

PROBLEMS 6.58, 6.59

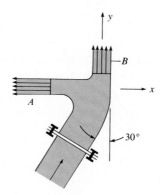

PROBLEMS 6.60, 6.61

6.62 A 15-cm nozzle is bolted with six bolts to the flange of a 30-cm pipe. If water discharges from the nozzle into the atmosphere, calculate the tension load in each bolt when the pressure in the pipe is 280 kPa. Assume irrotational flow.

6.63 A nozzle at the end of a 3-in. hose produces a jet 1½ in. in diameter. Determine the longitudinal force required in the joint at the base of the nozzle to hold the nozzle when it is discharging 300 gpm of water.

6.64 A nozzle on a garden hose 3 cm in diameter has an outlet diameter of 15 mm. Determine the longitudinal force required at the nozzle-hose joint to hold the nozzle for a discharge of 0.005 m^3/s.

6.65 Water is discharged from the two-dimensional slot shown at the rate of 4 cfs per foot of slot. Determine the pressure p at the gage and the water force per foot on the vertical end plates A and C. The slot and jet dimensions B and b are 8 in. and 3 in., respectively.

6.66 Water is discharged from the two-dimensional slot shown at the rate of 0.40 m^3/s per meter of slot. Determine the pressure p at the gage and the water force per meter on the vertical end plates A and C. The slot and jet dimensions B and b are 20 cm and 7 cm, respectively.

6.67 A 3-in. orifice is located in a plate at the end of a 6-in. pipe. Water flows through the pipe and orifice at a rate of 4 cfs. If the water jet downstream of the orifice is 2.5 in. in diameter, what is the external force required to hold the orifice in place?

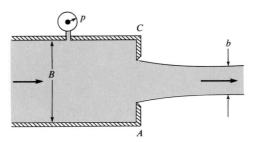

PROBLEMS 6.65, 6.66

6.68 A 4-cm orifice is located in a plate at the end of a 10-cm pipe. Water flows through the pipe and orifice at 0.05 m³/s. The diameter of the water jet downstream of the orifice is 3.5 cm. Calculate the external force required to hold the orifice in place.

6.69 Refer to the figure for Probs. 6.58 and 6.59. Water is flowing through the nozzle and is discharging into the atmosphere at section 2. The cross-sectional area at section 1 is 0.001 m², and the cross-sectional area at section 2 is 0.0001 m². If the velocity at section 1 is 3 m/s, what force is required at the flange to hold the nozzle in place? Neglect gravitational forces.

6.70 A 1-ft horizontal pipe carries water to three jets discharging into the atmosphere from the end of the pipe. The velocity of water in the jets is 40 ft/s. One jet discharges parallel to the pipe and has a 2-in. diameter. The second jet is inclined upward and in the direction of the flow in the pipe at an angle of 30° with respect to the pipe and has a 1-in. diameter. The third jet discharges vertically upward and has a 3-in. diameter. If an upstream section in the pipe is considered, what horizontal component of force must be transmitted through the metal at this section to hold the end in place?

6.71 This spray head discharges water at a rate of 2 ft³/s. Assuming irrotational flow and an efflux speed of 65 ft/s in the free jet, determine what force acting through the bolts of the flange is needed to keep the spray head on the 6-in. pipe. Neglect gravitational forces.

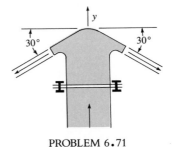

PROBLEM 6.71

6.72 Two circular water jets of 1-in. diameter (d = 1 in.) issue from this unusual nozzle. If the efflux speed is 80.2 ft/s, what force is required at the flange to hold the nozzle in place? The pressure in the 4-in. pipe (D = 4 in.) is 43 psig.

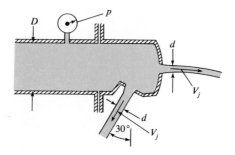

PROBLEM 6.72

6.73 Liquid (S = 1.5) enters the "black sphere" through a 2-in. pipe with velocity of 40 ft/s and a pressure of 60 psig. It leaves the sphere through two jets as shown. The velocity in the vertical jet is 100 ft/s, and its diameter is 1 in. The other jet's diameter is also 1 in. What force through the 2-in. pipe wall is required in the x and y directions to hold the sphere in place? Assume the sphere plus the liquid inside it weighs 200 lbf.

6.74 Liquid (S = 1.5) enters the "black sphere" through a 5-cm pipe with a velocity of 10 m/s and a pressure of 400 kPa. It leaves the sphere through two jets as shown. The velocity in the vertical jet is 30 m/s, and its diameter is 25 mm. The other jet's diameter is also 25 mm. What force through the 5-cm pipe wall is required in the x and y directions to hold the sphere in place? Assume the sphere plus the liquid inside it weighs 500 N.

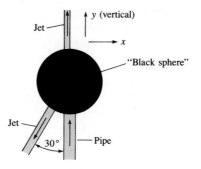

PROBLEMS 6.73, 6.74

6.75 What force is required to hold this "black box" in place if gravitational forces are neglected and if the liquid's specific gravity is 2.0?

6.76 This "black box" and associated equipment weigh 200 lbf and, when operating as shown, contain 3 ft³ of water. Assuming that steady flow prevails, what external force must be applied to the black box in the y direction (vertical) to keep it in equilibrium?

6.77 Neglecting viscous resistance, determine the force of the water per unit of width acting on a sluice gate for which the upstream depth is 3 ft and the downstream depth is 0.6 ft.

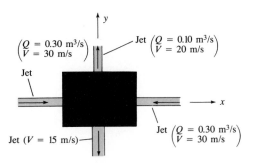

PROBLEM 6.75

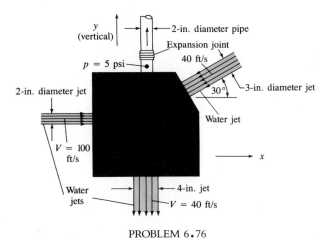

PROBLEM 6.76

6.78 For laminar flow in a pipe, the velocity distribution changes from uniform to parabolic as shown. At the fully developed section (section 2), the velocity is distributed as follows: $u = u_{max}[1 - (r/r_0)^2]$. Derive a formula for the resisting shear force F_τ as a function of U (the mean velocity in the pipe), ρ, p_1, p_2, and D (the pipe diameter).

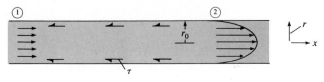

PROBLEM 6.78

6.79 The propeller on a swamp boat produces a slipstream 3 ft in diameter with a velocity relative to the boat of 80 ft/s. If the air temperature is 80°F, what is the propulsive force when the boat is not moving and also when its forward speed is 20 ft/s? *Hint:* Assume that the pressure, except in the immediate vicinity of the propeller, is atmospheric.

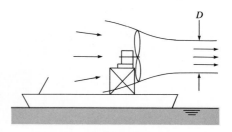

PROBLEM 6.79

6.80 A windmill is operating in a 10-m/s wind that has a density of 1.2 kg/m³. The diameter of the windmill is 4 m. The constant pressure (atmospheric) streamline has a diameter of 3 m upstream of the windmill and 4.5 downstream. Assume that the velocity distributions are uniform and the air is incompressible. Determine the thrust on the windmill.

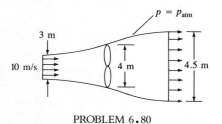

PROBLEM 6.80

6.81 The figure illustrates the principle of the jet pump. Derive a formula for $p_2 - p_1$ as a function of D_j, V_j, D_0, V_0, and ρ. Assume that the fluid from the jet and the fluid initially flowing in the pipe are the same, and assume that they are completely mixed at section 2, so that the velocity is uniform across that section. Also assume that the pressures are uniform across both sections 1 and 2. What is $p_2 - p_1$ if the fluid is water, $A_j/A_0 = \frac{1}{3}$, $V_j = 15$ m/s, and $V_0 = 2$ m/s?

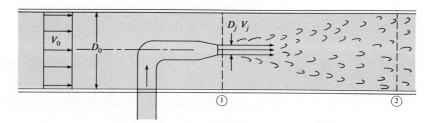

PROBLEM 6.81

6.82 Jet-type pumps are sometimes used for special purposes, such as to circulate the flow in basins in which fish are being reared. The use of a jet-type pump eliminates the need for mechanical machinery that might be injurious to the fish. The accompanying figure shows the basic concept for this type of application. For this type of basin the jets would have to increase the water surface elevation by an amount equal to $6V^2/2g$, where V is the

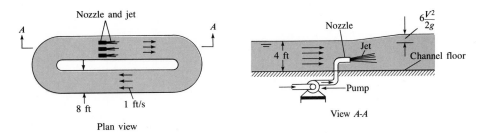

Plan view

View A-A

PROBLEM 6.82

average velocity in the basin (1 ft/s as shown in this example). Propose a basic design for a jet system that would make such a recirculating system work for a channel 8 ft wide and 4 ft deep. That is, determine the speed, size, and number of jets.

6.83 An engineer is measuring the lift and drag on an airfoil section mounted in a two-dimensional wind tunnel. The wind tunnel is 0.5 m high and 0.5 m deep (into the paper). The upstream wind velocity is uniform at 10 m/s, and the downstream velocity is 12 m/s and 8 m/s as shown. The vertical component of velocity is zero at both stations. The test section is 1 m long. The engineer measures the pressure distribution in the tunnel along the upper and lower walls and finds

$$p_u = 100 - 10x - 20x(1 - x) \text{ (Pa, gage)}$$

$$p_l = 100 - 10x - 20x(1 - x) \text{ (Pa, gage)}$$

where x is the distance in meters measured from the beginning of the test section. The gas density is homogeneous throughout and equal to 1.2 kg/m^3. The lift and drag are the vectors indicated on the figure. The forces acting on the fluid are in the opposite direction to these vectors. Find the lift and drag forces acting on the airfoil section.

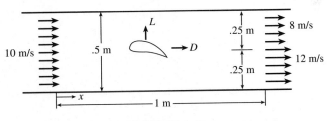

PROBLEM 6.83

6.84 A torpedo-like device is tested in a wind tunnel with an air density of 0.0026 slugs/ft^3. The tunnel is 3 ft in diameter, the upstream pressure is 0.24 psig, and the downstream pressure is 0.10 psig. If the mean air velocity V is 100 ft/s, what are the mass rate of flow and the maximum velocity at the downstream section at C? If the pressure is assumed to be uniform across the sections at A and C, what is the drag of the device and support vanes? Assume viscous resistance at the wall is negligible.

6.85 Consider a tank of water in a container that rests on a sled. A high pressure is maintained by a compressor so that a jet of water leaving the tank horizontally from an orifice does so

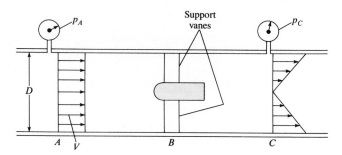

PROBLEM 6.84

at a constant speed of 25 m/s relative to the tank. If there is 0.10 m³ of water in the tank at time t and the diameter of the jet is 15 mm, what will be the acceleration of the sled at time t if the empty tank and compressor have a weight of 350 N and the coefficient of friction between the sled and the ice is 0.05?

6.86   The hemicircular nozzle sprays a sheet of liquid through 180° of arc as shown. The velocity is V at the efflux section where the sheet thickness is t. Derive a formula for the external force F (in the y direction) required to hold the nozzle system in place. This force should be a function of ρ, V, r, and t.

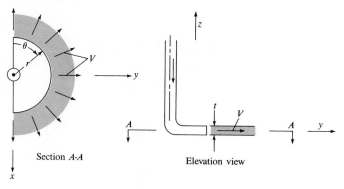

PROBLEM 6.86

6.87   A modern turbofan engine in a commercial jet takes in air, part of which passes through the compressors, combustion chambers, and turbine, and the rest of which bypasses the compressor and is accelerated by the fans. The mass-flow rate of bypass air to the mass-flow rate through the compressor-combustor-turbine path is called the "bypass ratio." The total flow rate of air entering a turbofan is 300 kg/s with a velocity of 300 m/s. The engine has a bypass ratio of 2. The bypass air exits at 600 m/s, whereas the air through the compressor-combustor-turbine path exits at 1000 m/s. What is the thrust of the turbofan engine? Clearly show your control-volume and application of momentum equation.

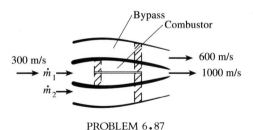

PROBLEM 6.87

6.88 It is common practice in rocket-trajectory analyses to neglect the body-force term and drag, so the velocity at burnout is given by

$$v_{bo} = \frac{T}{\lambda} \ln \frac{M_0}{M_f}$$

Assuming a thrust-to-mass-flow ratio of 3000 N · s/kg and a final mass of 50 kg, calculate the initial mass needed to establish the rocket in an earth orbit at a velocity of 7200 m/s.

6.89 A very popular toy on the market several years ago was the water rocket. Water was loaded into a plastic rocket and pressurized with a hand pump. The rocket was released and would travel a considerable distance in the air. Assume that a water rocket has a mass of 50 g and is charged with 100 g of water. The pressure inside the rocket is 100 kPa gage. The exit area is one-tenth of the chamber cross-sectional area. The inside diameter of the rocket is 5 cm. Assume that Bernoulli's equation is valid for the water flow inside the rocket. Neglecting air friction, calculate the maximum velocity it will attain.

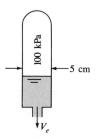

PROBLEM 6.89

6.90 A rocket is designed to have four nozzles, each canted at 30° with respect to the rocket's centerline. The gases exit at 2000 m/s through an exit area of 1 m². The density of the exhaust gases is 0.3 kg/m³, and exhaust pressure is 50 kPa. The atmospheric pressure is 10 kPa. Determine the thrust on the rocket in newtons.

6.91 A rocket-nozzle designer is concerned about the force required to hold the nozzle section on the body of a rocket. The nozzle section is shaped as shown in the figure. The pressure and velocity at the entrance to the nozzle are 1.5 MPa and 100 m/s. The exit pressure

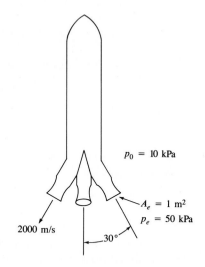

$p_0 = 10$ kPa

$A_e = 1$ m^2
$p_e = 50$ kPa

2000 m/s

30°

PROBLEM 6.90

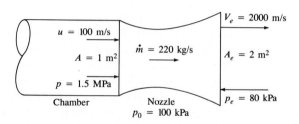

$u = 100$ m/s

$A = 1$ m^2

$\dot{m} = 220$ kg/s

$p = 1.5$ MPa

$V_e = 2000$ m/s

$A_e = 2$ m^2

$p_e = 80$ kPa

Chamber   Nozzle
$p_0 = 100$ kPa

PROBLEM 6.91

and velocity are 80 kPa and 2000 m/s. The mass flow through the nozzle is 220 kg/s. The atmospheric pressure is 100 kPa. The rocket is not accelerating. Calculate the force on the nozzle-chamber connection.

6.92 The expansion section of a rocket nozzle is often conical in shape; and because the flow diverges, the thrust derived from the nozzle is less than it would be if the exit velocity were everywhere parallel to the nozzle axis. By considering the flow through the spherical section subtended by the cone and assuming that the exit pressure is equal to the atmospheric pressure, show that the thrust is given by

$$T = \dot{m}V_e\frac{(1 + \cos \alpha)}{2}$$

where $\dot{m}$ is the mass flow through the nozzle, V_e is the exit velocity, and α is the nozzle half-angle.

6.93 A valve at the end of a gasoline pipeline is rapidly closed (assume it is closed instantaneously). If the gasoline velocity was initially 11 m/s, what will be the water-hammer pressure rise? The bulk modulus of elasticity of the gasoline is 715 MPa and the density of gasoline is 680 kg/m^3.

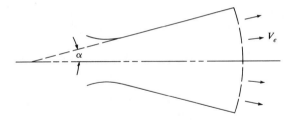

PROBLEM 6.92

6.94 Estimate the maximum water-hammer pressure that is generated in a rigid pipe if the initial water velocity is 2 m/s and the pipe is 10 km long with a valve at the downstream end that is closed in 10 s.

6.95 The length of a 20-cm rigid pipe carrying 0.15 m³/s of water is estimated by instantaneously closing a valve at the downstream end and noting the time required for the pressure fluctuation to complete a cycle. If the time interval is 3 s, what is the pipe length?

6.96 Estimate the maximum water-hammer pressure that is generated in a rigid pipe if the initial water velocity is 6 ft/s and the pipe is 5 mi long with a valve at the downstream end that is closed in 10 s.

6.97 A rigid pipe 4 km long and 12 cm in diameter discharges water at the rate of 0.025 m³/s. If a valve at the end of the pipe is closed in 3 s, what is the maximum force that will be exerted on the valve as a result of the pressure rise? Assume that the water temperature is 10°C.

6.98 By letting the control volume move with the water-hammer wave, steady-flow conditions are established. Using the momentum and continuity equations and the steady-flow approach, derive Eq. (6.20).

6.99 The 60-cm pipe carries water with an initial velocity, V_0, of 0.10 m/s. If the valve at C is instantaneously closed at time $t = 0$, what will the pressure-versus-time trace look like at point B for the next 5 s? Graph your results and indicate significant quantitative relations or values from $t = 0$ to $t = 5$ s. What does the pressure versus the position along the pipe look like at $t = 1.5$ s? Plot your results and indicate the velocity or velocities in the pipe.

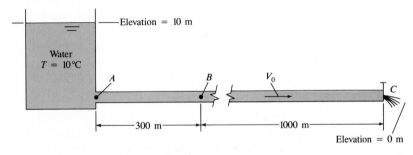

PROBLEM 6.99

6.100 Steady flow initially occurs in this 1-m steel pipe. There is a rapid-acting valve at the end of the pipe at point B, and there are pressure transducers at both points A and B. If the valve is closed at B and the p-versus-t traces are made as shown, estimate the initial discharge and the length L from A to B.

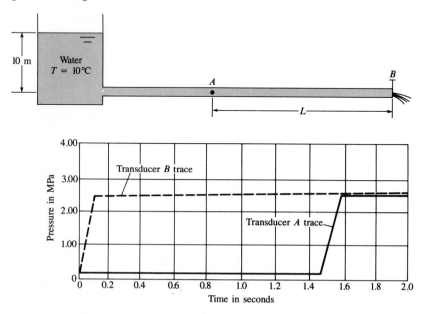

PROBLEM 6.100

6.101 Water is discharged from the slot in the pipe as shown. If the resulting two-dimensional jet is 100 cm long and 15 mm thick, and if the pressure at section A–A is 30 kPa, what is the reaction at section A–A? In this calculation, do not consider the weight of the pipe.

6.102 What is the reaction at section 1? Water is flowing in the system. Neglect gravitational forces.

6.103 What is the reaction at section 1? Water is flowing, and the axes of the two jets lie in a vertical plane. The pipe and nozzle system weighs 90 N.

6.104 A reducing pipe bend is held in place by a pedestal as shown. There are expansion joints at sections 1 and 2, so no force is transmitted through the pipe past these sections. If the pressure at section 1 is 20 psig and if the rate of flow of water is 2 cfs, what is the resultant force system that must be transmitted through the base of the pedestal at section 3? That is, determine the twisting and bending moments on the pedestal as well as the force components at section 3 that are due to the dynamic effect of the water. Assume the flow is irrotational.

6.105 Show how the momentum equation can be applied to derive Euler's equation for the flow of inviscid fluids, Eq. (5.3). *Hint:* Select an arbitrary control volume of length Δs enclosed by a stream tube in an unsteady, nonuniform flow as shown. The volume of the

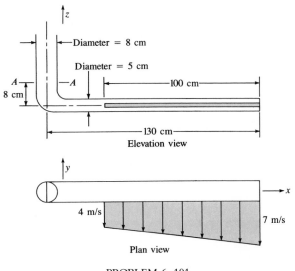

PROBLEM 6.101

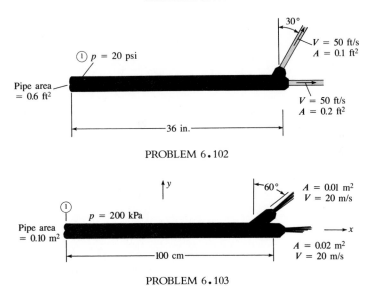

PROBLEM 6.102

PROBLEM 6.103

control volume is $[A + (\partial A/\partial s)(\Delta s/2)]\Delta s$. First derive the continuity equation by applying the continuity principle to the flow through the control volume. Then apply the momentum equation along the stream-tube direction, and use the continuity equation to reduce it to Euler's equation.

6.106 Two small liquid-propellant rocket motors are mounted at the tips of a helicopter rotor to augment power under emergency conditions. The diameter of the helicopter rotor is 8 m, and it rotates at 1 rev/s. The air enters at the tip speed of the rotor, and exhaust gases

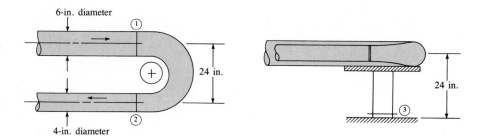

PROBLEM 6.104

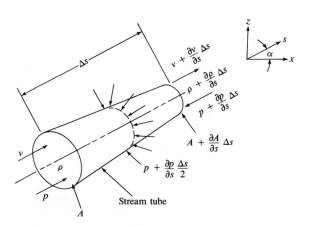

PROBLEM 6.105

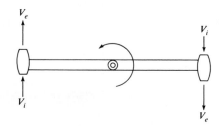

PROBLEM 6.106

exit at 500 m/s with respect to the rocket motor. The intake area of each motor is 20 cm², and the air density is 1.2 kg/m³. Calculate the power provided by the rocket motors. Neglect the mass rate of flow of fuel in this calculation.

6.107 Design a rotating lawn sprinkler to deliver 0.25 in. of water per hour over a circle of 50-ft radius. Make the simplifying assumptions that the pressure to the sprinkler is 50 psig and that frictional effects involving the flow of water through the sprinkler flow passages are negligible (Bernoulli's equation is applicable). However, do not neglect the friction between the rotating element and the fixed base of the sprinkler.

References

1. Chaudhry, M. H. *Applied Hydraulic Transients,* 2nd ed. Van Nostrand Reinhold, New York, 1987.

2. Streeter, V. L., and E. B. Wylie. *Fluid Transients.* FEB Press, Ann Arbor, Mich., 1983.

3. Schlichting, H. *Boundary Layer Theory.* McGraw-Hill, New York, 1979.

4. Sutton, G. P. *Rocket Propulsion Elements.* John Wiley, New York, 1963.

Energy Principle

Grand Coulee Dam in Washington State has a power-generating capacity of about 5600 MW. At this generating level the turbines discharge water at 250,000 cfs (7,080 m³/s) and operate under a head of about 308 ft (94 m). To generate significant power during dry seasons, water is stored in Lake Roosevelt (shown in this photo) as well as in three other reservoirs in Canada. (Courtesy U.S. Bureau of Reclamation)

To this point, we have been concerned with the mechanical forces (pressure, gravity, shear stress) on a fluid. The energy equation allows us to incorporate the thermal energies as well. In the early nineteenth century, J. P. Joule carried out a number of tests that verified the general principle of the conservation of energy that had been previously hypothesized. From this was developed the *first law of thermodynamics,* which can be written for a given system (given quantity of matter) as follows:

$$\Delta E = Q - W$$

Here Q is the heat transferred to the system in a given time t, and W is the work done by the system on its surroundings in this same interval of time.* The energy E of a system can take a variety of forms, such as kinetic and potential energy of the system as a whole and energy associated with motion of the molecules. The latter includes energy involved with the structure of the atom, chemical energy, and electrical energy. It is convenient to consider kinetic energy E_k and potential energy E_p separately and to lump all other energies into a single term called *internal energy* E_u. Thus the total energy of the system is

$$E = E_u + E_k + E_p$$

Now, if we want the rate of change of E with time, the differential form of the first law of thermodynamics is given as follows:

$$\frac{dE}{dt} = \dot{Q} - \dot{W}$$

---

*Heat transferred to the system and work done by the system are defined, by convention, to be positive quantities. Heat transferred from the system and work done on the system are negative quantities.

In the next section the first law of thermodynamics, along with the control-volume approach, will be used to develop the energy equation for fluid flow.

7.1   Derivation of the Energy Equation

Control-Volume Approach Applied to the First Law of Thermodynamics

The energy E introduced above refers to the total energy of the system; thus E is an extensive property of the system. Then the corresponding intensive property (energy per unit of mass) is given by e, which is made up of e_k, e_p, and u. In applying the control-volume equation, Eq. (4.21a), we let $B_{sys} = E$ and $\beta = e$ to obtain

$$\frac{dE}{dt} = \frac{d}{dt} \int_{cv} e\rho \, d\forall + \sum_{cs} e\rho \mathbf{V} \cdot \mathbf{A} \tag{7.1}$$

However, from the first law of thermodynamics, $dE/dt = \dot{Q} - \dot{W}$. Consequently, substitution is made for dE/dt in Eq. (7.1) to yield

$$\dot{Q} - \dot{W} = \frac{d}{dt} \int_{cv} e\rho \, d\forall + \sum_{cs} e\rho \mathbf{V} \cdot \mathbf{A} \tag{7.2}$$

When we replace e by its equivalent, $e_k + e_p + u$, we obtain

$$\dot{Q} - \dot{W} = \frac{d}{dt} \int_{cv} (e_k + e_p + u)\rho \, d\forall + \sum_{cs} (e_k + e_p + u)\rho \mathbf{V} \cdot \mathbf{A} \tag{7.3}$$

Several terms on the right of Eq. (7.3) are too general for practical application. Therefore, let us examine these carefully and express them in terms of variables associated with the flow of fluids.

The kinetic energy per unit of mass, e_k, is given by the total kinetic energy of mass having velocity V divided by the mass itself, or

$$e_k = \frac{\Delta M V^2 / 2}{\Delta M} = \frac{V^2}{2} \tag{7.4}$$

The potential energy per unit of mass, e_p, is given by $E_p/\Delta M$, where E_p is the product of weight and the elevation of the centroid of the incremental mass. In this case, the potential energy is referenced to the datum from which elevation is measured. Then

$$e_p = \frac{\gamma \Delta \forall z}{\Delta M} = \frac{\gamma \Delta \forall z}{\rho \Delta \forall} = gz \tag{7.5}$$

When Eqs. (7.4) and (7.5) are substituted into Eq. (7.3), we obtain

$$\dot{Q} - \dot{W} = \frac{d}{dt} \int_{cv} \left(\frac{V^2}{2} + gz + u \right) \rho \, d\forall + \sum_{cs} \left(\frac{V^2}{2} + gz + u \right) \rho \mathbf{V} \cdot \mathbf{A}$$

$$(7.6)$$

On the left side of this equation are the two terms $\dot{Q}$ and $\dot{W}$, which are the rate of flow of heat into the system and the rate of work done by the system on its surroundings, respectively. For convenience of analysis, work is divided into shaft work W_s and flow work W_f. These are discussed next.

Flow Work

Flow work is the work done by pressure forces as the system moves through space. Let us consider the basic figure (Fig. 7.1) depicting the system and control volume to get a better understanding of flow work. Consider the area A_2, which is the right end of the fluid system. The force acting on the surrounding fluid will be $p_2 A_2$, and the distance traveled by the area in the time Δt will be $\Delta \ell_2 = V_2 \Delta t$. The work done on the surrounding fluid because of this force in time Δt will be the product of the component of force in the direction of motion ($p_2 A_2$) and the distance traveled by the area ($V_2 \Delta t$). Hence the flow work done by the system on the surrounding fluid in time Δt by the downstream end of the system will be

$$\Delta W_{f2} = V_2 p_2 A_2 \Delta t$$

The rate at which flow work is done on the area is obtained by dividing through by Δt:

$$\dot{W}_{f2} = V_2 p_2 A_2 \qquad (7.7)$$

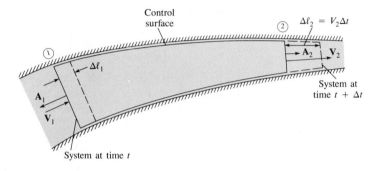

FIGURE 7.1

It can also be seen from Fig. 7.1 that $V_2 A_2 = \mathbf{V}_2 \cdot \mathbf{A}_2$. Consequently, Eq. (7.7) reduces to

$$\dot{W}_{f2} = p_2 \mathbf{V}_2 \cdot \mathbf{A}_2$$

In a similar manner it can be shown that the flow work done on the surrounding fluid by the upstream end of the system will be $-p_1 V_1 A_1$. A negative sign occurs here because the pressure force on the surrounding fluid acts in a direction opposite to the motion of the system boundary. The rate at which work is done on the surrounding fluid by the upstream end of the system can also be given in terms of the scalar product:

$$\dot{W}_{f1} = p_1 \mathbf{V}_1 \cdot \mathbf{A}_1$$

A negative rate of work results from this product because the velocity vector $\mathbf{V}_1$ and the area vector $\mathbf{A}_1$ have opposite sense; thus the scalar product has a negative sign. Hence all system surfaces that are moving (represented by streams of fluid passing across the control surface) do work on the surrounding fluid according to the expression

$$\dot{W}_f = p \mathbf{V} \cdot \mathbf{A} \qquad (7.8)$$

Then the rate at which flow work is done on the system's surroundings is obtained by summing Eq. (7.8) for all streams passing through the control surface:

$$\dot{W}_f = \sum_{cs} p \mathbf{V} \cdot \mathbf{A} \qquad (7.9)$$

Multiplying and dividing the quantity following the summation symbol in Eq. (7.9) by ρ will yield simplification later in this derivation. Hence we obtain

$$\dot{W}_f = \sum_{cs} \frac{p}{\rho} \rho \mathbf{V} \cdot \mathbf{A} \qquad (7.10)$$

Shaft Work

Shaft work is defined as any work other than flow work. It is usually the form of work done through a shaft (from which the term originates) that either takes energy out of the system or puts energy into the system. In the first case, the system does work on a mechanism (such as a turbine blade) attached to the shaft. In the second case, the shaft is attached to a mechanism (such as a pump impeller) that does work on the system. In this latter case, to retain our original terminology for W, we say that the fluid system is doing negative work on its surroundings.

Basic Form of the Energy Equation

If we substitute for $\dot{W}$ in Eq. (7.6) the sum of the shaft-work rate $\dot{W}_s$ and the flow-work rate, Eq. (7.10), the following form of the energy equation for fluid flow results:

$$\dot{Q} - \dot{W}_s - \sum_{cs} \frac{p}{\rho} \rho \mathbf{V} \cdot \mathbf{A}$$

$$= \frac{d}{dt} \int_{cv} \left(\frac{V^2}{2} + gz + u \right) \rho \, d\Psi + \sum_{cs} \left(\frac{V^2}{2} + gz + u \right) \rho \mathbf{V} \cdot \mathbf{A} \quad (7.11)$$

The last terms on both sides of Eq. (7.11) are the same form. Hence these two terms may be combined as follows:

$$\dot{Q} - \dot{W}_s = \frac{d}{dt} \int_{cv} \left(\frac{V^2}{2} + gz + u \right) \rho \, d\Psi + \sum_{cs} \left(\frac{p}{\rho} + \frac{V_2}{2} + gz + u \right) \rho \mathbf{V} \cdot \mathbf{A}$$

$$(7.12a)$$

Equation (7.12a) is the basic form of the energy equation.

More General Form of the Energy Equation

If the velocity distribution across a flow section is not uniform, then it is necessary to use the more general form of the control-volume equation, Eq. (4.21b), in the derivation. The steps in the derivation are essentially the same as given above, except that integration of the flow past the control surface results in the following more general form of the energy equation:

$$\dot{Q} - \dot{W}_s = \frac{d}{dt} \int_{cv} \left(\frac{V^2}{2} + gz + u \right) \rho \, d\Psi + \int_{cs} \left(\frac{p}{\rho} + \frac{V^2}{2} + gz + u \right) \rho \mathbf{V} \cdot d\mathbf{A}$$

$$(7.12b)$$

7.2   Simplified Forms of the Energy Equation

Steady-Flow Energy Equation

When Eq. (7.12a) is written for steady flow, we have

$$\dot{Q} - \dot{W}_s = \sum_{cs} \left(\frac{p}{\rho} + \frac{V^2}{2} + gz + u \right) \rho \mathbf{V} \cdot \mathbf{A} \quad (7.13)$$

An alternative form of Eq. (7.13) is

$$\dot{Q} - \dot{W}_s = \sum_{cs} \left(\frac{V^2}{2} + gz + h \right) \rho \mathbf{V} \cdot \mathbf{A} \quad (7.14)$$

where $h = p/\rho + u$ is defined as the specific enthalpy of the fluid.

Equations (7.13) and (7.14) are much simpler than Eqs. (7.12) because in their application we are no longer concerned with the fluid mass inside the control surface. Equations (7.13) and (7.14) pertain only to the rate of transfer of heat and energy across the control surface plus the rate at which work is done by the system on its surroundings.

EXAMPLE 7.1   A steam turbine receives superheated steam at 1.4 MPa absolute and 400°C, which corresponds to a specific enthalpy of 3121 kJ/kg. The steam leaves the turbine at 101 kPa absolute and 100°C, for which the specific enthalpy is 2676 kJ/kg. The steam enters the turbine at 15 m/s and exits at 60 m/s. The elevation difference between entry and exit ports is negligible. The heat lost through the turbine wall is 7600 kJ/h. Calculate the power output if the mass flow through the turbine is 0.5 kg/s.

Solution   First sketch the general layout of the turbine, indicating the inlet and outlet velocities, shaft work, and heat transfer, as shown.

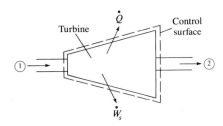

Writing down the energy equation for this turbine, neglecting the elevation terms, gives

$$\dot{Q} - \dot{W}_s = \left(\frac{V_2^2}{2} + h_2\right)\rho_2 V_2 A_2 - \left(\frac{V_1^2}{2} + h_1\right)\rho_1 V_1 A_1$$

However, $\rho_1 V_1 A_1 = \rho_2 V_2 A_2 = \dot{m}$

from the conservation of mass, so we have

$$\dot{W}_s = \dot{Q} + \dot{m}\left(\frac{V_1^2}{2} - \frac{V_2^2}{2} + h_1 - h_2\right)$$

We must be careful to check units before substituting numbers into the equation and evaluating $\dot{W}_s$. The units for enthalpy are joules per kilogram; for velocity squared, they are meters squared per second squared. One finds that meters squared per second squared is equivalent to joules per kilogram:

$$\frac{m^2}{s^2} = \frac{kg \cdot m^2}{kg \cdot s^2} = \frac{kg \cdot m}{s^2}\frac{m}{kg} = \frac{N \cdot m}{kg} = \frac{J}{kg}$$

Substituting numbers into the energy equation, while realizing that $\dot{Q}$ (the heat transfer) has a negative value and using the correct units, gives

$$\dot{W}_s = \frac{-7600}{3600} \frac{\text{kJ}}{\text{h}} \frac{\text{h}}{\text{s}} + 0.5 \frac{\text{kg}}{\text{s}} \left[\frac{15^2 - 60^2}{2 \times 10^3} \frac{\text{kJ}}{\text{kg}} + (3121 - 2676) \frac{\text{kJ}}{\text{kg}} \right]$$

$$= -2.11 + 0.5(-1.69 + 445) = 220 \frac{\text{kJ}}{\text{s}} = 220 \text{ kW} \quad \blacktriangleleft$$

The kinetic-energy term in this type of problem is usually negligible compared with the enthalpy difference, which is the case in this example.

Energy Equation for Steady, One-Dimensional Flow of an Incompressible Fluid in a Pipe

Consider flow through the pipe system shown in Fig. 7.2. Here the magnitude of the velocity is variable across the flow sections; thus we use the more general form of the energy equation, Eq. (7.12b). When Eq. (7.12b) is written between sections 1 and 2 and when the flow quantities relating to section 1 are transferred to the left side of the equation, we obtain

$$\dot{Q} - \dot{W}_s + \int_{A_1} \left(\frac{p_1}{\rho} + gz_1 + u_1 \right) \rho V_1 \, dA_1 + \int_{A_1} \frac{\rho V_1^3}{2} dA_1$$

$$= \int_{A_2} \left(\frac{p_2}{\rho} + gz_2 + u_2 \right) \rho V_2 \, dA_2 + \int_{A_2} \frac{\rho V_2^3}{2} dA_2 \quad (7.15)$$

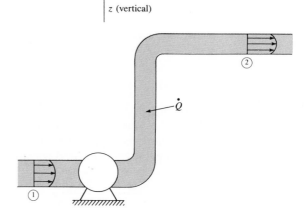

FIGURE 7.2

At sections 1 and 2, where the flow is uniform, hydrostatic conditions prevail across the section. Therefore, $p/\rho + gz$ is constant across the section.* At any section the internal energy is also virtually constant across the section. Therefore, $p/\rho + gz + u$ can be taken outside the integral to yield

$$\dot{Q} - \dot{W}_s + \left(\frac{p_1}{\rho} + gz_1 + u_1\right)\int_{A_1} \rho V_1 \, dA_1 + \int_{A_1} \rho \frac{V_1^3}{2} dA_1$$

$$= \left(\frac{p_2}{\rho} + gz_2 + u_2\right)\int_{A_2} \rho V_2 \, dA_2 + \int_{A_2} \rho \frac{V_2^3}{2} dA_2 \quad (7.16)$$

It can be seen that $\int \rho V dA = \rho \overline{V} A = \dot{m}$, the mass rate of flow. Consequently, some simplification will result if we divide through by $\dot{m}$. However, $\dot{m}$ does not appear as a factor of $\int (\rho V^3/2)\, dA$; so it is common to express $\int (\rho V^3/2)\, dA$ as $\alpha(\rho \overline{V}^3/2)A$. Then when we factor out $\dot{m}$ from each of these terms, Eq. (7.16) becomes

$$\dot{Q} - \dot{W}_s + \left(\frac{p_1}{\rho} + gz_1 + u_1 + \alpha_1 \frac{\overline{V}_1^2}{2}\right)\dot{m} = \left(\frac{p_2}{\rho} + gz_2 + u_2 + \alpha_2 \frac{\overline{V}_2^2}{2}\right)\dot{m}$$

$$(7.17)$$

When we divide through by $\dot{m}$, we get

$$\frac{1}{\dot{m}}(\dot{Q} - \dot{W}_s) + \frac{p_1}{\rho} + gz_1 + u_1 + \alpha_1 \frac{\overline{V}_1^2}{2} = \frac{p_2}{\rho} + gz_2 + u_2 + \alpha_2 \frac{\overline{V}_2^2}{2}$$

$$(7.18)$$

The coefficients α_1 and α_2 are kinetic-energy correction factors and are evaluated by the original expressions in which they were first introduced:

$$\alpha \frac{\rho \overline{V}^3 A}{2} = \int_A \frac{\rho V^3 \, dA}{2} \quad (7.19)$$

or

$$\alpha = \frac{1}{A} \int_A \left(\frac{V}{\overline{V}}\right)^3 dA \quad (7.20)$$

Thus $\alpha = 1$ when the velocity is uniform across the section, and $\alpha > 1$ for nonuniform velocity distributions. Computations show that $\alpha = 2$ for laminar flow in a pipe where the velocity has a parabolic distribution across the section. For most cases of turbulent flow, $\alpha \approx 1.05$. Because this is quite close to unity, it is common practice in engineering applications to let $\alpha_1 = \alpha_2 = 1$. A similar correction factor could be used with the momentum flux terms in the momentum equation for one-dimensional flow. However, it deviates even less from unity than does α for a given velocity distribution.

* The pressure is hydrostatically distributed across the section because the flow is assumed to be uniform at each section. For uniform flow, the streamlines are straight and parallel. Hence there is no acceleration normal to the streamlines, and the pressure is thus hydrostatically distributed. For hydrostatic conditions, p/γ + z = constant; see p. 38. Therefore, if we multiply p/γ + z by the constant g, we get p/ρ + gz, which is also constant.

EXAMPLE 7.2   The velocity distribution for laminar flow in a pipe is given by the equation

$$V = V_{max}\left[1 - \left(\frac{r}{r_0}\right)^2\right]$$

Here r_0 is the radius of the pipe and r is the radial distance from the center. Determine $\overline{V}$ in terms of V_{max} and evaluate the kinetic-energy correction factor α.

Solution   A sketch for the velocity distribution is shown in the figure.

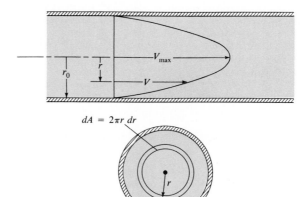

$$dA = 2\pi r\,dr$$

The discharge is given by $Q = \int V dA$, or

$$Q = \overline{V}A = \int_0^{r_0} V_{max}\left(1 - \frac{r^2}{r_0^2}\right)2\pi r\,dr$$

or

$$\overline{V} = \frac{-\pi}{A}r_0^2 V_{max}\frac{(1 - r^2/r_0^2)^2}{2}\bigg|_0^{r_0} = \frac{1}{2}V_{max} \qquad \blacktriangleleft$$

The mean velocity is one-half the maximum velocity. To evaluate α we apply Eq. (7.20):

$$\alpha = \frac{1}{\pi r_0^2}\int_0^{r_0}\frac{V_{max}^3(1 - r^2/r_0^2)^3}{(\frac{1}{2})^3 V_{max}^3}2\pi r\,dr = 2 \qquad \blacktriangleleft$$

The shaft-work term in Eq. (7.18) is usually the result of a turbine or pump in the flow system. When fluid passes through a turbine, the fluid system is doing shaft work on the surroundings; on the other hand, a pump does work on the fluid. It is convenient, then, to represent the shaft-work term as

$$\dot{W}_s = \dot{W}_t - \dot{W}_p \tag{7.21}$$

where $\dot{W}_t$ and $\dot{W}_p$ are magnitudes of power (work per unit time) delivered by a turbine or supplied to a pump. Substituting this expression for shaft work into Eq. (7.18) and dividing by g results in

$$\frac{\dot{W}_p}{\dot{m}g} + \frac{p_1}{\gamma} + z_1 + \alpha_1 \frac{\overline{V}_1^2}{2g} = \frac{\dot{W}_t}{\dot{m}g} + \frac{p_2}{\gamma} + z_2 + \alpha_2 \frac{\overline{V}_2^2}{2g} + \frac{u_2 - u_1}{g} - \frac{\dot{Q}}{\dot{m}g} \tag{7.22}$$

All of the terms of Eq. (7.22) have one dimension, length. Hence we designate the first term involving shaft work as h_p, head supplied by a pump, and the second term involving shaft work as h_t, head given up to a turbine. Equation (7.22) is then written as

$$\frac{p_1}{\gamma} + \alpha_1 \frac{\overline{V}_1^2}{2g} + z_1 + h_p = \frac{p_2}{\gamma} + \alpha_2 \frac{\overline{V}_2^2}{2g} + z_2 + h_t + \left[\frac{1}{g}(u_2 - u_1) - \frac{\dot{Q}}{\dot{m}g} \right] \tag{7.23}$$

At this point we have separated the energy equation into a mechanical part and a thermal part, the latter of which is represented by the last term enclosed in the brackets.

In the flow process, some of the system's mechanical energy is converted to thermal energy through viscous action between fluid particles. Consequently, there may be a finite increase in internal energy of the fluid between sections 1 and 2, or some of the heat that is generated through energy dissipation may escape from the pipe $(-\dot{Q})$. In any event, the bracketed terms in Eq. (7.23) represent a loss of mechanical energy due to viscous stresses. This lost energy is usually simply lumped together in a single term called *head loss* and symbolized by h_L.* Thus Eq. (7.23) becomes

$$\frac{p_1}{\gamma} + \alpha_1 \frac{\overline{V}_1^2}{2g} + z_1 + h_p = \frac{p_2}{\gamma} + \alpha_2 \frac{\overline{V}_2^2}{2g} + z_2 + h_t + h_L \tag{7.24}$$

In Chapter 10 we shall consider specific practical ways to estimate h_L. However, for the present we shall use it only in a general way.

In most applications of the energy equation it is understood that the kinetic-energy term $\alpha \overline{V}^2/2g$ involves the mean velocity $\overline{V}$. Hence the bar over the V is usually omitted. The coefficient α is often also omitted; it usually has a value

---

* *It might appear from Eq. (7.23) that with sufficient heat transfer to the conduit (positive $\dot{Q}$), h_L could be eliminated or even made negative. However, increasing $\dot{Q}$ also increases $u_2 - u_1$, maintaining positive head loss. The fact that h_L is always positive can be shown by application of the entropy principle of thermodynamics.*

very near unity because most flows are turbulent. In addition, it is often necessary to convert the total power of a pump or turbine to h_p or h_t, respectively, or
vice versa. This is done using the original definition of h_p and h_t. That is,
$h_p = \dot{W}_p/\dot{m}g = \dot{W}_p/Q\rho g = \dot{W}_p/Q\gamma$. Similarly, $h_t = \dot{W}_t/Q\gamma$.

It must be cautioned that Eq. (7.24) is not valid when compressibility effects are significant. The energy equation for compressible flow will be covered
in Chapter 12.

EXAMPLE 7.3   A horizontal pipe carries cooling water for a thermal power
plant from a reservoir as shown. The head loss in the pipe is given as

$$\frac{0.02(L/D)V^2}{2g}$$

where L is the length of the pipe from the reservoir to the point in question, V is
the mean velocity in the pipe, and D is the diameter of the pipe. If the pipe diameter is 20 cm and the rate of flow is 0.06 m³/s, what is the pressure in the pipe
at $L = 2000$ m?

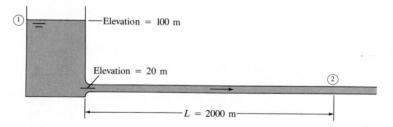

Solution   Write the one-dimensional energy equation, Eq. (7.24), between the
water surface in the reservoir and a section in the pipe:

$$\frac{p_1}{\gamma} + \frac{V_1^2}{2g} + z_1 = \frac{p_2}{\gamma} + \frac{V_2^2}{2g} + z_2 + h_L$$

where $p_1 = 0$, $V_1 = 0$, $z_1 = 100$ m, $z_2 = 20$ m, $V_2 = Q/A = 0.06/[(\pi/4) \times
0.2^2] = 1.91$ m/s, and $h_L = 0.02(L/0.2) \times 1.91^2/(2 \times 9.81) = 0.0186L$ m.
Then

$$\frac{p_2}{\gamma} = 100 - 20 - 0.186 - 0.0186L = 79.8 - 0.0186L$$

At $L = 2000$ m

$$\frac{p_2}{\gamma} = 79.8 - 37.2 = 42.6 \text{ m}$$

$$p_2 = 417.9 \text{ kPa} \qquad \blacktriangleleft$$

EXAMPLE 7.4   The pipe in Fig. 7.2 is 50 cm in diameter and carries water at a rate of 0.5 m³/s. Also, $z_2 = 40$ m, $z_1 = 30$ m, and $p_1 = 70$ kPa gage. What power in kilowatts and in horsepower must be supplied to the flow by the pump if the gage pressure at section 2 is to be 350 kPa? Assume $h_L = 3$ m of water and $\alpha_1 = \alpha_2 = 1$.

Solution   Write the one-dimensional energy equation, Eq. (7.24), between sections 1 and 2:

$$\frac{p_1}{\gamma} + \frac{V_1^2}{2g} + z_1 + h_p = \frac{p_2}{\gamma} + \frac{V_2^2}{2g} + z_2 + h_L$$

where $p_1 = 70$ kPa, $\gamma = 9.81$ kN/m³, $V_1 = Q/A_1 = (0.50 \text{ m}^3/\text{s})/[\pi(0.25^2 \text{ m}^2)] = 2.55$ m/s, $V_2 = 2.55$ m/s, $z_1 = 30$ m, $z_2 = 40$ m, $p_2 = 350$ kPa, $h_L = 3$ m, and $g = 9.81$ m/s². Then

$$h_p = \frac{p_2 - p_1}{\gamma} + \frac{V_2^2 - V_1^2}{2g} + z_2 - z_1 + h_L$$

$$= \frac{(350 - 70) \text{ kN/m}^2}{9.81 \text{ kN/m}^3} + \frac{(2.55^2 - 2.55^2) \text{ m}^2/\text{s}^2}{9.81 \text{ m/s}^2} + 40 \text{ m} - 30 \text{ m} + 3 \text{ m}$$

$$= 28.5 + 0 + 10 + 3 = 41.5 \text{ m} = 41.5 \text{ m} \cdot \text{N/N}$$

The energy supplied by the pump is 41.6 m · N/N of fluid that is flowing. Therefore, we obtain the total power supplied by the pump by taking the product of h_p and the weight rate of flow.

$$P = \dot{W}_p = Q\gamma h_p = 0.5 \text{ m}^3/\text{s} \times 9.81 \text{ kN/m}^3 \times 41.5 \text{ m} = 204 \text{ m} \cdot \text{kN/s}$$

$$= 204 \text{ kJ/s} = 204 \text{ kW} \qquad \blacktriangleleft$$

$$= 273 \text{ hp} \qquad \blacktriangleleft$$

EXAMPLE 7.5   At the maximum rate of power generation this hydroelectric power plant takes a discharge of 141 m³/s. If the head loss through the intakes, penstock, and outlet works is 1.52 m, what is the rate of power generation?

Solution   Assume $\alpha_1 = \alpha_2 = 1$. Then, for this application, Eq. (7.24) reduces to

$$\frac{p_1}{\gamma} + \frac{V_1^2}{2g} + z_1 = \frac{p_2}{\gamma} + \frac{V_2^2}{2g} + h_L + h_t$$

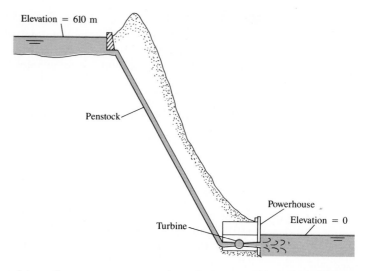

If section 1 is at the upstream reservoir and section 2 is at sea level, then $p_1 = 0$, $V_1 = 0$, $z_1 = 610$ m, $p_2 = 0$, $V_2 = 0$, and $h_L = 1.52$ m. Then

$$h_t = 610 - 1.52 = 608.5 \text{ m}$$
$$P = \dot{W}_t = Q\gamma h_t = 141 \times 9810 \times 608.5 = 842 \text{ MW} \qquad \blacktriangleleft$$

Energy Equation for Nonviscous, One-Dimensional, Incompressible Flow

Nonviscous flow means that there will be no head loss in the system. Consequently, Eq. (7.24) reduces to

$$\frac{p_1}{\gamma} + \alpha_1 \frac{V_1^2}{2g} + z_1 + h_p = \frac{p_2}{\gamma} + \alpha_2 \frac{V_2^2}{2g} + z_2 + h_t \qquad (7.25)$$

In addition, if we apply the equation to a stream tube (a filament of fluid of infinitesimal cross section bounded by streamlines) where no shaft work is involved, then Eq. (7.25) becomes

$$\frac{p_1}{\gamma} + \frac{V_1^2}{2g} + z_1 = \frac{p_2}{\gamma} + \frac{V_2^2}{2g} + z_2 \qquad (7.26)$$

Note here that α_1 and α_2 have been dropped. For a stream tube the velocities at the end sections are point velocities; hence α has its limiting value of unity. Equation (7.26) is the same as Bernoulli's equation derived in Chapter 5. Indeed it should be, because the limitations are exactly the same.

7.3 Application of the Energy, Momentum, and Continuity Equations in Combination

The energy, momentum, and continuity equations are independent equations. They can therefore be used together to solve a variety of problems. To illustrate the use of these equations in combination, we shall consider flow through an abrupt expansion and forces on transitions.

Abrupt Expansion

Consider the flow from a small pipe into a larger pipe, as shown in Fig. 7.3. We want to solve for the head loss due to the expansion as a function of the flow velocities in the two pipes. Normally such a problem is not amenable to analytic solution; however, one can solve this problem with a reasonable assumption about the pressure distribution at the change in section. Experience tells us that when flow occurs past an abrupt enlargement such as this, the flow separates from the boundary. Hence, in effect, a jet of fluid from the smaller pipe discharges into the larger pipe. Because the streamlines in the jet are initially straight and parallel, the pressure distribution across the jet will be simply hydrostatic. Because the fluid in the zone of separation at section 1 has such low velocity, it can be likewise assumed that the pressure in this zone is the same as that in the jet except for the hydrostatic variation. With this basic assumption for the pressure at section 1, we can apply the momentum and energy equations to solve

FIGURE 7.3

Flow through an abrupt expansion.

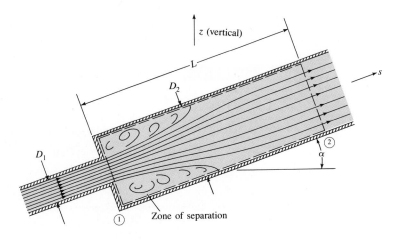

for the head loss due to the expansion. The one-dimensional energy equation written between sections 1 and 2 is

$$\frac{p_1}{\gamma} + \alpha_1 \frac{V_1^2}{2g} + z_1 = \frac{p_2}{\gamma} + \alpha_2 \frac{V_2^2}{2g} + z_2 + h_L \tag{7.27}$$

We are assuming turbulent flow conditions here, so we may assume $\alpha_1 = \alpha_2 = 1$. The momentum equation for the fluid in the large pipe between section 1 and section 2, written for the s direction, is

$$\sum F_s = \sum_{cs} V_s \rho \mathbf{V} \cdot \mathbf{A}$$

$$p_1 A_2 - p_2 A_2 - \gamma A_2 L \sin \alpha = \rho V_2^2 A_2 - \rho V_1^2 A_1$$

or
$$\frac{p_1}{\gamma} - \frac{p_2}{\gamma} - (z_2 - z_1) = \frac{V_2^2}{g} - \frac{V_1^2}{g} \frac{A_1}{A_2} \tag{7.28}$$

The continuity equation, $V_1 A_1 = V_2 A_2$, is also a valid independent equation for this problem. Therefore, when Eqs. (7.27) and (7.28), along with the continuity equation, are used to solve for h_L, the following is obtained:

$$h_L = \frac{(V_1 - V_2)^2}{2g} \tag{7.29}$$

Hence we see that the head loss for an abrupt expansion is given by the square of the difference in mean velocities in the two pipes divided by $2g$. *Note:* The head loss for an abrupt expansion is *not* $(V_1^2 - V_2^2)/2g$.

Forces on Transitions

The method for determining the forces required to hold a transition or any other flow passage in place is presented in the form of an example.

EXAMPLE 7.6   Water flows through the contraction at a rate of 0.707 m³/s.

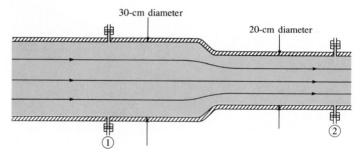

The head loss due to this particular transition is given by the empirical equation

$$h_L = 0.2\frac{V_2^2}{2g}$$

Here V_2 is the velocity in the 20-cm pipe. What horizontal force is required to hold the transition in place if $p_1 = 70$ kPa?

Solution   Write the momentum equation for the transition between sections 1 and 2. The control surface is drawn so that it encloses the fluid and the transition itself; consequently, the control volume is as shown. The pressures are gage pressures, so the pressure on the exterior of the transition is zero. Hence the

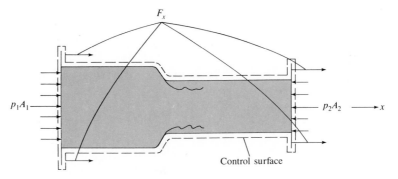

force on the exterior is zero. Then, for this control volume, the momentum equation is

$$p_1 A_1 - p_2 A_2 + F_x = \sum_{cs} V_x \rho \mathbf{V} \cdot \mathbf{A}$$

$$p_1 A_1 - p_2 A_2 + F_x = \rho V_2^2 A_2 - \rho V_1^2 A_1$$

$$F_x = \rho Q(V_2 - V_1) + p_2 A_2 - p_1 A_1$$

Here Q and the velocities are known, and F_x is the unknown. Also, p_2 is still unknown. We obtain p_2 by applying the energy equation between sections 1 and 2:

$$\frac{p_1}{\gamma} + \frac{V_1^2}{2g} + z_1 = \frac{p_2}{\gamma} + \frac{V_2^2}{2g} + z_2 + h_L$$

Here we have assumed $\alpha_1 = \alpha_2 = 1$. For our problem, $z_1 = z_2$, so the foregoing equation reduces to

$$\frac{p_1}{\gamma} + \frac{V_1^2}{2g} = \frac{p_2}{\gamma} + \frac{V_2^2}{2g} + h_L$$

$$p_2 = p_1 - \gamma\left(\frac{V_2^2}{2g} - \frac{V_1^2}{2g} + h_L\right)$$

Substituting this expression for p_2 into the equation for the force on the transition yields

$$F_x = \rho Q(V_2 - V_1) + A_2\left[p_1 - \gamma\left(\frac{V_2^2}{2g} - \frac{V_1^2}{2g} + h_L\right)\right] - p_1 A_1$$

$$V_1 = \frac{Q}{A_1} = \frac{0.707}{(\pi/4) \times 0.3^2} = 10 \text{ m/s}$$

$$V_2 = \frac{Q}{A_2} = \frac{0.707}{(\pi/4) \times 0.2^2} = 22.5 \text{ m/s}$$

$$h_L = \frac{0.2V_2^2}{2g} = \frac{0.2 \times 22.5^2}{2 \times 9.81} = 5.16 \text{ m}$$

Then

$$F_x = 1000 \times (0.707)(22.5 - 10) + \frac{\pi}{4} \times 0.2^2$$

$$\times \left[70,000 - 9810\left(\frac{22.5^2}{2 \times 9.81} - \frac{10^2}{2 \times 9.81} + 5.16\right)\right]$$

$$- 70,000 \times 0.785 \times 0.3^2$$

$$= -1.88 \text{ kN} \qquad \blacktriangleleft$$

Thus a force of 1.88 kN must be applied in the negative x direction to hold the transition in place for the given conditions.

7.4 Concept of the Hydraulic and Energy Grade Lines

The units of Eq. (7.24) are meters or feet, and we can attach a useful physical relationship to them. Consider the flow in the pipe shown in Fig. 7.4. The sum of the terms on the left side of Eq. (7.24) represents the total mechanical energy (stated in energy per unit weight of flowing liquid, N · m/N) plus the flow-work term in the fluid at the upstream section plus the energy supplied by a pump. The sum on the right-hand side is the total energy per unit weight at the downstream section plus the head loss and energy per unit weight given up to a turbine between the two sections. For the case shown in Fig. 7.4, the velocity of flow in the reservoir is negligible. Hence the total energy at the surface is potential energy and, in terms of energy per unit weight, is simply z. At the downstream section the liquid is at a different elevation than in the reservoir and it has significant velocity. Therefore, the "energy per unit weight" here is given in terms of p_2/γ, z_2, and $\alpha_2 V_2^2/2g$. It is common practice to lump all these terms together as *total* head in feet or meters. Note that the sum of these terms plus the

FIGURE 7.4

EGL and HGL in a
straight pipe.

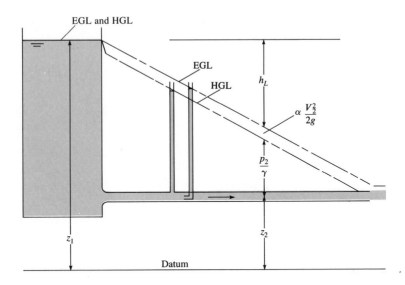

head loss between sections 1 and 2 is equal to the total head at section 1. By analyzing a sketch such as Fig. 7.4 it is possible at a glance to ascertain the pressure
at any point in the pipeline: one simply picks off values of p/γ and multiplies by
γ to obtain p.

   If one were to tap a piezometer into the pipe in Fig. 7.4, the liquid would
rise to a height p/γ above the pipe. Hence the height of the p/γ line is called the
hydraulic grade line (HGL). The total head ($p/\gamma + \alpha V^2/2g + z$) in the system is
greater than $p/\gamma + z$ by an amount $\alpha V^2/2g$. Consequently, the *energy grade line*
(EGL) is above the HGL by a distance $\alpha V^2/2g$. The engineer who develops a vi-
sual concept of the energy equation as explained above will find it much easier to
sense trouble spots in the system (usually points of low pressure) and to devise
ways of solving the problem.

   Here are some other helpful hints for drawing hydraulic grade lines and
energy grade lines.

   1. By definition, the EGL is positioned above the HGL an amount $\alpha V^2/2g$.
Thus if the velocity is zero, as in a lake or reservoir, the HGL and EGL will coincide (see Fig. 7.4) with the liquid surface.

   2. Head loss for flow in a pipe or channel always means that the EGL will
slope downward in the direction of flow (see Fig. 7.4). The only exception to
this rule occurs when a pump supplies energy (and pressure) to the flow. Then
an abrupt rise in the EGL (and the HGL) occurs from the upstream side to the
downstream side of the pump (see Fig. 7.5).

   3. In point 2 above, it was noted that a pump can cause an abrupt rise in
the EGL (and the HGL) because energy is introduced into the flow by the
pump. Similarly, if energy is abruptly taken out of the flow—by a turbine, for

FIGURE 7.5

*Rise in EGL and HGL
due to pump.*

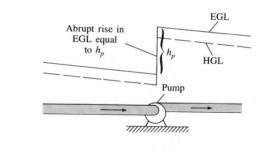

FIGURE 7.6

*Drop in EGL and HGL
due to turbine.*

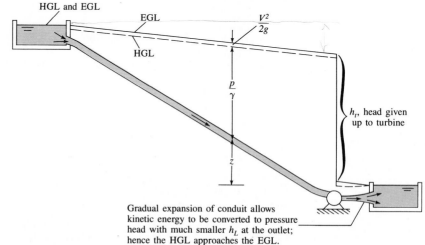

Gradual expansion of conduit allows
kinetic energy to be converted to pressure
head with much smaller h_L at the outlet;
hence the HGL approaches the EGL.

example—then the EGL and the HGL will drop abruptly, as in Fig. 7.6. In Fig. 7.6 and other figures (Fig. 7.7 through Fig. 7.9), it is assumed that $\alpha = 1.0$. Figure 7.6 also shows that much of the kinetic energy can be converted to pressure if there is a gradual expansion such as at the outlet. Thus the head loss at the outlet is reduced, making the turbine installation more efficient. If the outlet to a reservoir is an abrupt expansion as in Fig. 7.8, all the kinetic energy is lost. Thus the EGL drops an amount $\alpha V^2/2g$ at the outlet.

4. In a pipe or channel where the pressure is zero, the HGL is coincident with the system because $p/\gamma = 0$ at these points. This fact can be used to locate the

FIGURE 7.7

*Change in HGL and
EGL due to flow through
a nozzle.*

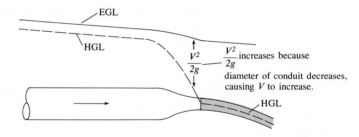

FIGURE 7.8

*Change in EGL and HGL
due to change in diameter
of pipe.*

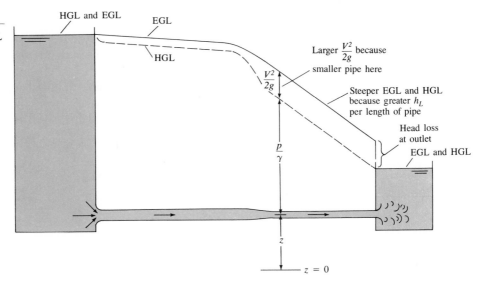

HGL at certain points in a physical system, such as at the outlet end of a pipe where the liquid discharges into the atmosphere or at the upstream end where the gage pressure is zero at the reservoir surface (see Fig. 7.4).

5. For steady flow in a pipe that has uniform physical characteristics (diameter, roughness, shape, and so on) along its length, the head loss per unit length will be constant. Thus the slope ($\Delta h_L/\Delta L$) of the EGL and the HGL will be constant along the length of pipe (see Fig. 7.4).

6. If a flow passage changes diameter, such as in a nozzle or by means of a change in pipe size, the velocity therein will also change. Hence, the distance between the EGL and the HGL will change (see Fig. 7.7). In addition, the slope on the EGL will change because the head loss per length will be larger in the conduit with the larger velocity (see Fig. 7.8). You will learn more about this latter point in Chapter 10 when head loss in pipes is considered in more detail.

7. If the HGL falls below the pipe, then p/γ is negative, indicating subatmospheric pressure (see Fig. 7.9).

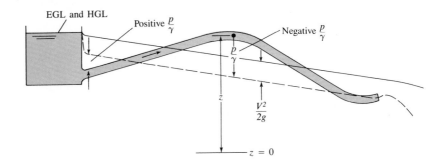

FIGURE 7.9

*Subatmospheric pressure
when pipe is above HGL.*

EXAMPLE 7.7   A pump draws water from a reservoir, where the water-surface elevation is 520 ft, and forces the water through a pipe 5000 ft long and 1 ft in diameter. This pipe then discharges the water into a reservoir with water-surface elevation of 620 ft. The flow rate is 7.85 cfs, and the head loss in the pipe is given by $0.01(L/D)(V^2/2g)$. Determine the head supplied by the pump, h_p, and the power supplied to the flow, and draw the HGL and the EGL for the system. Assume that the pipe is horizontal and is 510 ft in elevation.

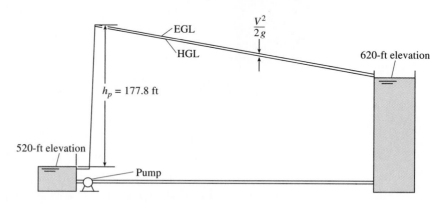

Solution   First solve for h_p by using the energy equation (one-dimensional flow assumed) written from the water surface in the lower reservoir to the water surface in the upper reservoir.

$$\frac{p_1}{\gamma} + \frac{V_1^2}{2g} + z_1 + h_p = \frac{p_2}{\gamma} + \frac{V_2^2}{2g} + z_2 + h_L$$

In this example, p_1/γ, p_2/γ, V_1, and V_2 are all zero, but $z_1 = 520$ ft, and $z_2 = 620$ ft.

$$h_L = 0.01 \times \frac{5000}{1} \frac{V^2}{2g} \qquad \text{and} \qquad V = \frac{Q}{A_p} = 10 \text{ ft/s}$$

Then $h_p = 620 - 520 + 0.01 \times \dfrac{5000}{1} \dfrac{100}{64.4}$ ft-lbf/lbf $= 178$ ft   ◀

However, the product of the flow rate Q and specific weight will give us the weight rate of flow. Then the power supplied by the pump will be $h_p \times$ weight rate of flow, or

$$P = h_p Q \gamma \text{ ft-lbf/s} = \frac{h_p Q \gamma}{550} = 158 \text{ hp} \qquad ◀$$

Problems

7.1 A turbine receives steam at 1.80 MPa, 500°C (enthalpy = 3470 kJ/kg) at a velocity of 5 m/s. The steam exits at an enthalpy of 2630 kJ/kg with a velocity of 70 m/s. The steam flows through at a rate of 1 kg/s, and the turbine develops 830 kW. Calculate the rate of heat loss from the turbine. Neglect the potential energy due to the elevation difference.

7.2 A turbine receives steam at 300 psia, $T = 700°F$ (enthalpy = 1268 Btu/lbm). The entrance velocity to the turbine is 50 ft/s. The steam exits the turbine at 10 psia, $T = 193°F$, as a gas–liquid mixture with an enthalpy of 1098 Btu/lbm. The exit velocity is 200 ft/s. The heat loss from the turbine is 3500 Btu/hr, and the mass flow rate of steam through the turbine is 5800 lbm/hr. Calculate the power output. The elevation difference between inlet and exit ports is negligible.

7.3 A turbine receives steam at 2.0 MPa, 500°C (enthalpy = 3062 kJ/kg) at a velocity of 10 m/s. The steam leaves the turbine as a gas–liquid mixture at 101 kPa, 373 K, with an enthalpy of 2621 kJ/kg. The exit velocity is 50 m/s. Thermal energy is lost through the turbine walls at a rate of 5 kJ/h. The potential energy due to the elevation difference between inlet and exit ports can be neglected. Calculate the power if 4000 kg/h of steam pass through the turbine.

7.4 A compressor is used to supply high-pressure air for a supersonic wind tunnel. The enthalpy of the air is 300 kJ/kg at the entrance and 450 kJ/kg at the compressor outlet. The outlet velocity is 200 m/s, and the inlet velocity is negligible. The mass flow through the compressor is 1.5 kg/s. The system is adiabatic. Find the power (in kilowatts) required to operate the compressor.

7.5 Air flows in a pipe 1000 m long and 8 cm in diameter at a rate of 0.5 kg/s. The pressure and temperature at the upstream end of the pipe are 50 kPa gage and 20°C. The pipe is well insulated, so that a negligible amount of heat is transferred to or from the air in the pipe. What are the velocity and temperature in the stream of air at the outlet as it discharges into the atmosphere? Assume that atmospheric pressure is 100 kPa. Assume that air is an ideal gas and that the specific enthalpy is given by h (kJ/kg) $= 1.004T$ (K).

7.6 A hypothetical velocity distribution in a pipe has a maximum velocity of V_{max} at the center and a minimum velocity of $0.7V_{max}$ at the wall. If the velocity is linearly distributed from the center to the wall, what is α? What is the mean velocity V in terms of V_{max}?

7.7 For this hypothetical velocity distribution in a wide rectangular channel, evaluate the kinetic-energy correction factor α.

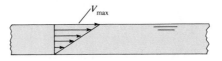

PROBLEM 7.7

7.8 For these velocity distributions in a pipe, indicate whether the kinetic-energy correction factor α is greater than, equal to, or less than unity.

7.9   Calculate α for case (c) in Prob. 7.8.

7.10 Calculate α for case (d) in Prob. 7.8.

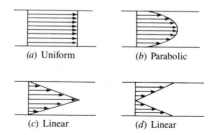

(a) Uniform   (b) Parabolic

(c) Linear   (d) Linear

PROBLEMS 7.8, 7.9, 7.10

7.11 If the value of α (kinetic-energy correction factor) for flow in a pipe is 1.06, then the flow must be a) laminar, b) turbulent.

7.12 An approximate equation for the velocity distribution in a pipe with turbulent flow is

$$\frac{V}{V_{max}} = \left(\frac{y}{r_0}\right)^n$$

where V_{max} is the centerline velocity, y is the distance from the wall of the pipe, r_0 is the radius of the pipe, and n is an exponent that depends on the Reynolds number and varies between $\frac{1}{6}$ and $\frac{1}{8}$ for most applications. Derive a formula for α as a function of n. What is α if $n = \frac{1}{6}$?

7.13 An approximate equation for the velocity distribution in a rectangular channel with turbulent flow is

$$\frac{V}{V_{max}} = \left(\frac{y}{d}\right)^n$$

where V_{max} is the velocity at the surface, y is the distance from the floor of the channel, d is the depth of flow, and n is an exponent that varies from about $\frac{1}{6}$ to $\frac{1}{8}$ depending on the Reynolds number. Derive a formula for α as a function of n. What is the value of α for $n = \frac{1}{7}$?

7.14 Water flows from a pressurized tank as shown. The pressure in the tank above the water surface is 100 kPa gage, and the water surface level is 10 m above the outlet. The water exit velocity is 9 m/s. The head loss in the system varies as

$$h_L = K_L V^2/2g$$

where K_L is the head-loss coefficient. Find the value for K_L.

7.15 A reservoir with water is pressurized as shown. The pipe diameter is 1 in. The head loss in the system is given by $h_L = 6V^2/2g$. The height between the water surface and the pipe outlet is 10 ft. A discharge of 0.10 ft^3/s is needed. What must the pressure in the tank be to achieve such a flow rate?

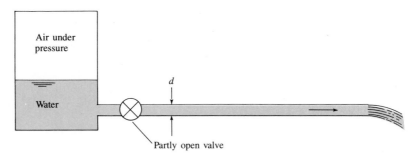

PROBLEMS 7.14, 7.15

7.16 A pipe drains a tank as shown. If $x = 10$ ft, $y = 5$ ft, and head losses are neglected, what is the pressure intensity at point A?

7.17 A pipe drains a tank as shown. If $x = 3$ m, $y = 2$ m, and head losses are neglected, what is the pressure intensity at point A?

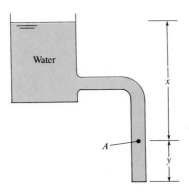

PROBLEM 7.16, 7.17

7.18 If $D_A = 20$ cm, $D_B = 10$ cm, and $L = 2$ m, and if crude oil ($S = 0.90$) is flowing at a rate of 0.06 m³/s, determine the difference in pressure between sections A and B. Neglect head losses.

7.19 In the figure for Probs. 7.14 and 7.15, suppose that the reservoir is open to the atmosphere at the top. The valve is used to control the flow rate from the reservoir. The head loss across the valve is given as $h_L = 10V^2/2g$, where V is the velocity in the pipe. The cross-sectional area of the pipe is 5 cm². The head loss due to friction in the pipe is negligible. The elevation of the water level in the reservoir above the pipe outlet is 10 m. Find the discharge in the pipe.

7.20 With steady viscous flow of a liquid in a constant-diameter pipe, the pressure along the pipe is observed to increase in the direction of the flow. What sort of situation can cause the increase in pressure along the pipe?

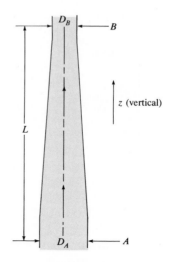

PROBLEM 7.18

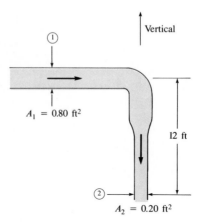

PROBLEM 7.21

7.21 Gasoline having a specific gravity of 0.8 is flowing in the pipe shown at a rate of 5 cfs. What is the pressure at section 2 when the pressure at section 1 is 10 psig and the head loss is 5 ft between the two sections?

7.22 Determine the discharge in the pipe and the pressure at point B. Neglect head losses.

7.23 The discharge in the siphon is 2.80 cfs, $D = 8$ in., $L_1 = 5$ ft, and $L_2 = 3$ ft. Determine the head loss between the reservoir surface and point C. Determine the pressure at point B if $\frac{2}{3}$ of the head loss (as found above) occurs between the reservoir surface and point B.

7.24 For this siphon the elevations at A, B, C, and D are 30 m, 32 m, 27 m, and 26 m, respectively. The head loss between the inlet and point B is $\frac{3}{4}$ of the velocity head, and the head

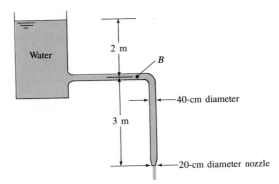

PROBLEM 7.22

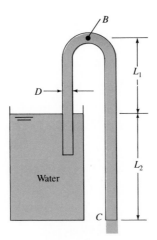

PROBLEM 7.23

loss in the pipe itself between point B and the end of the pipe is $\frac{1}{4}$ of the velocity head. For these conditions, what is the discharge and what is the pressure at point B? The pipe diameter = 30 cm.

7.25 For this system, point B is 10 m above the bottom of the upper reservoir. The head loss from A to B is $2V^2/2g$, and the pipe area is 10^{-4} m^2. Assume a constant discharge of 7×10^{-4} m^3/s. For these conditions, what will be the depth of water in the upper reservoir for which cavitation will begin at point B? Vapor pressure = 1.23 kPa and atmospheric pressure = 100 kPa.

7.26 Water flows at a steady rate in this vertical pipe. The pressure at A is 10 kPa, and at B it is 115 kPa. Then the flow in the pipe is a) upward, b) downward, c) no flow.

7.27 In this system, d = 6 in., D = 12 in., Δz_1 = 4 ft, and Δz_2 = 12 ft. The discharge of water in the system is 10 cfs. Is the machine a pump or a turbine? What are the pressures at points A and B? Neglect head losses.

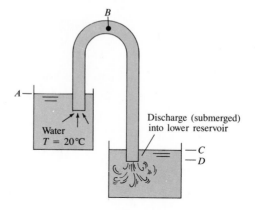

PROBLEMS 7.24, 7.25

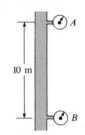

PROBLEM 7.26

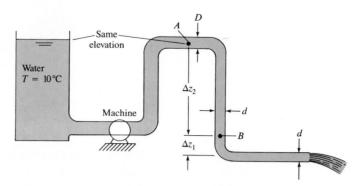

PROBLEM 7.27

7.28 For this system, the discharge of water is 0.20 m³/s, $x = 1.0$ m, $y = 2.0$ m, $z = 8.0$ m, and the pipe diameter is 30 cm. Neglecting head losses, what is the pressure head at point 2 if the jet from the nozzle is 15 cm in diameter?

7.29 A pump draws water through an 8-in. suction pipe and discharges it through a 6-in. pipe in which the velocity is 12 ft/s. The 6-in. pipe discharges horizontally into air at C. To what height h above the water surface at A can the water be raised if 30 hp is delivered to

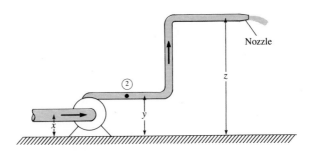

PROBLEM 7.28

the pump? Assume that the pump operates at 70% efficiency and that the head loss in the pipe between A and C is equal to $1.8V_C^2/2g$.

7.30   A pump draws water through a 20-cm suction pipe and discharges it through a 15-cm pipe in which the velocity is 4 m/s. The 15-cm pipe discharges horizontally into air at point C. To what height h above the water surface at A can the water be raised if 35 kW is delivered to the pump? Assume that the pump operates at 70% efficiency and that the head loss in the pipe between A and C is equal to $1.8V_C^2/2g$.

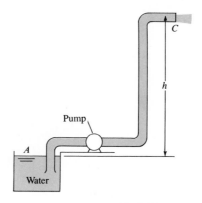

PROBLEMS 7.29, 7.30

7.31   As shown in the figure, the pump supplies energy to the flow such that the upstream pressure (12-in. pipe) is 10 psi and the downstream pressure (6-in. pipe) is 30 psi when the flow of water is 3.92 cfs. What horsepower is delivered by the pump to the flow?

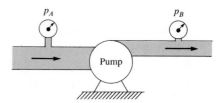

7.32 A water discharge of 7.85 m³/s is to flow through this horizontal pipe, which is 1 m in diameter. If the head loss is given as $6.0V^2/2g$ (V is velocity in the pipe), how much power will have to be supplied to the flow by the pump to produce this discharge?

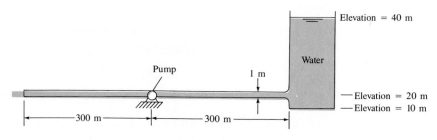

PROBLEM 7.32

7.33 Water is flowing at a rate of 0.25 m³/s, and it is assumed that $h_L = 1.5V^2/2g$ from the reservoir to the gage, where V is the velocity in the 30-cm pipe. What power must the pump supply?

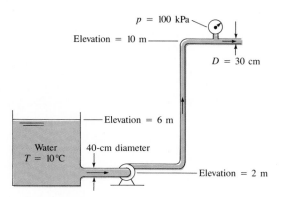

PROBLEM 7.33

7.34 In the pump test shown, the rate of flow is 5 cfs of oil (S = 0.88). Calculate the horsepower that the pump supplies to the oil if there is a differential reading of 36 in. of mercury in the U-tube manometer.

7.35 If the discharge is 250 cfs, what power output may be expected from the turbine? Assume that the turbine efficiency is 80% and that the overall head loss is $V^2/2g$, where V is the velocity in the 6-ft penstock.

7.36 A small-scale hydraulic power system is shown. The elevation difference between the reservoir water surface and the pond water surface downstream of the reservoir, H, is 10 m. The velocity of the water exhausting into the pond is 5 m/s, and the discharge through the system is 1 m³/s. The head loss due to friction in the penstock is negligible. Find the power produced by the turbine in kilowatts.

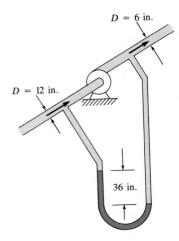

PROBLEM 7.34

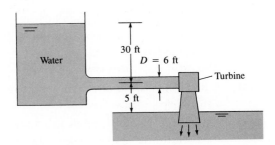

PROBLEM 7.35

7.37 The discharge of water through this turbine is 1000 cfs. What power is generated if the turbine efficiency is 85% and the total head loss is 3 ft? $H = 100$ ft. Also carefully sketch the energy grade line (EGL) and the hydraulic grade line (HGL).

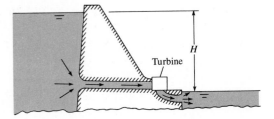

PROBLEMS 7.36, 7.37

7.38 Neglecting head losses, determine what power the pump must deliver to produce the flow as shown. Here the elevations at points A, B, C, and D are 40 m, 65 m, 35 m, and 30 m, respectively. The nozzle area is 30 cm².

7.39 Neglecting head losses, determine what horsepower the pump must deliver to produce the flow as shown. Here the elevations at points A, B, C, and D are 110 ft, 200 ft, 110 ft, and 90 ft, respectively. The nozzle area is 0.10 ft^2.

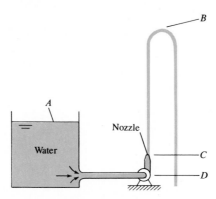

PROBLEMS 7.38, 7.39

7.40 A 40-cm pipe abruptly expands to a 60-cm size. These pipes are horizontal, and the discharge of water from the smaller size to the larger is 0.8 m^3/s. What horizontal force is required to hold the transition in place if the pressure in the 40-cm pipe is 70 kPa gage? Also, what is the head loss?

7.41 A 10-cm pipe carries water with a mean velocity of 6 m/s. If this pipe abruptly expands to a 15-cm pipe, what will be the head loss due to the abrupt expansion?

7.42 A 6-in. pipe abruptly expands to a 12-in. size. If the discharge of water in the pipes is 4 cfs, what is the head loss due to abrupt expansion?

7.43 This abrupt expansion is to be used to dissipate the high-energy flow of water in the 5-ft-diameter penstock.

a. What power (in horsepower) is lost through the expansion?
b. If the pressure at section 1 is 5 psig, what is the pressure at section 2?
c. What force is needed to hold the expansion in place?

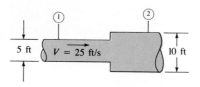

PROBLEM 7.43

7.44 This rough aluminum pipe is 6 in. in diameter. It weighs 2 lb per foot of length, and the length L is 50 ft. If the discharge of water is 4 cfs and the head loss due to friction from

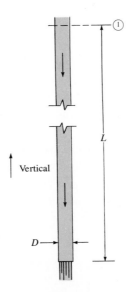

PROBLEM 7.44

section 1 to the end of the pipe is 10 ft, what is the longitudinal force transmitted across section 1 through the pipe wall?

7.45 Water flows in this bend at a rate of 5 m³/s, and the pressure at the inlet is 650 kPa. If the head loss in the bend is 10 m, what will the pressure be at the outlet of the bend? Also estimate the force of the anchor block on the bend in the x direction required to hold the bend in place.

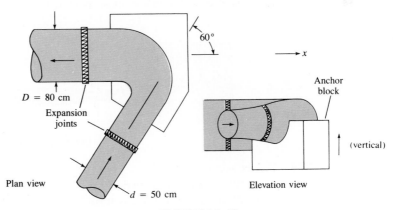

PROBLEM 7.45

7.46 What is the head loss at the outlet of the pipe that discharges water into the reservoir at a rate of 12 cfs if the diameter of the pipe is 12 in.?

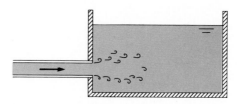

PROBLEMS 7.46, 7.47

7.47 What is the head loss at the outlet of the pipe that discharges water into the reservoir at a rate of 0.8 m³/s if the diameter of the pipe is 50 cm?

7.48 The pipe diameter D is 30 cm, d is 15 cm, and the atmospheric pressure is 100 kPa. What is the maximum allowable discharge before cavitation occurs at the throat of the venturi meter if $H = 5$ m?

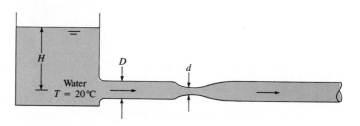

PROBLEM 7.48

7.49 In this system $d = 20$ cm, $D = 40$ cm, and the head loss from the venturi meter to the end of the pipe is given by $h_L = 0.9V^2/2g$, where V is the velocity in the pipe. Neglecting all other head losses, determine what head H will first initiate cavitation if the atmospheric pressure is 100 kPa absolute. What will be the discharge at incipient cavitation?

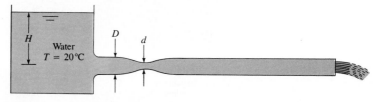

PROBLEM 7.49

7.50 For the system shown,
    a. What is the flow direction?
    b. What kind of machine is at A?
    c. Do you think both pipes, AB and CA, are the same diameter?
    d. Sketch in the EGL for the system.
    e. Is there a vacuum at any point or region of the pipes? If so, identify the location.

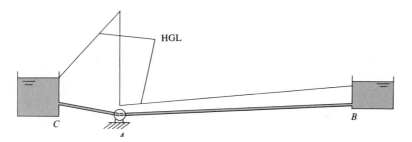

PROBLEM 7.50

7.51 Water flows from the reservoir through a pipe and then discharges from a nozzle as shown. The head loss in the pipe itself is given as $h_L = 0.02(L/D)V^2/2g$, where L and D are the length and diameter of the pipe and V is the velocity in the pipe. What is the discharge of water? Also draw the HGL and EGL for the system.

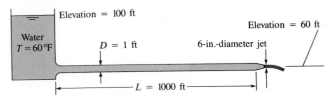

PROBLEM 7.51

7.52 Sketch the HGL and the EGL for the system of Example 7.5.

7.53 Carefully sketch the HGL and the EGL for the flow system of Prob. 7.49.

7.54 Sketch the HGL and the EGL for the reservoir and pipe of Example 7.3.

7.55 The energy grade line for steady flow in a uniform-diameter pipe is shown. Which of the following could be in the "black box"? a) A pump, b) a partially closed valve, c) an abrupt expansion, d) a turbine. Choose valid answer(s).

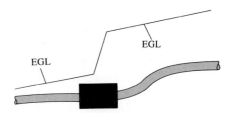

PROBLEM 7.55

7.56 If the pipe shown has constant diameter, is this type of HGL possible? If so, under what additional conditions? If not, why not?

7.57 Sketch the HGL and the EGL for this conduit, which tapers uniformly from the left end to the right end.

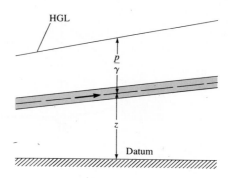

PROBLEM 7.56

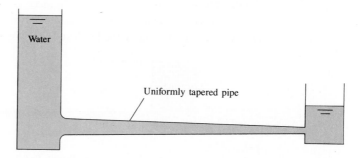

PROBLEM 7.57

7.58 The HGL and the EGL for a pipeline are shown in the figure.

    a. Indicate which is the HGL and which is the EGL.

    b. Are all pipes the same size? If not, which is the smallest? *A-B*

    c. Is there any region in the pipes where the pressure is below atmospheric pressure? If so, where?

    d. Where is the point of maximum pressure in the system? *A =bot oftank*

    e. Where is the point of minimum pressure in the system? *e*

    f. What do you think is located at the end of the pipe at point E?

    g. Is the pressure in the air in the tank above or below atmospheric pressure?

    h. What do you think is located at point B?

7.59 Refer to Fig. 7.8 on p. 297. Assume that the head loss in the pipes is given by $h_L = 0.02(L/D)(V^2/2g)$, where V is the mean velocity in the pipe, D is the pipe diameter, and L is the pipe length. The water surface elevations of the upper and lower reservoirs are 100 m and 70 m, respectively. The respective dimensions for upstream and downstream pipes are $D_u = 30$ cm, $L_u = 200$ m, and $D_d = 20$ cm, $L_d = 100$ m. Determine the discharge of water in the system.

7.60 What horsepower must be supplied to the water to pump 2.5 cfs at 68°F from the lower to the upper reservoir? Assume that the head loss in the pipes is given by $h_L = 0.015(L/D)(V^2/2g)$, where L is the length of pipe in feet and D is the pipe diameter in feet. Sketch the HGL and the EGL.

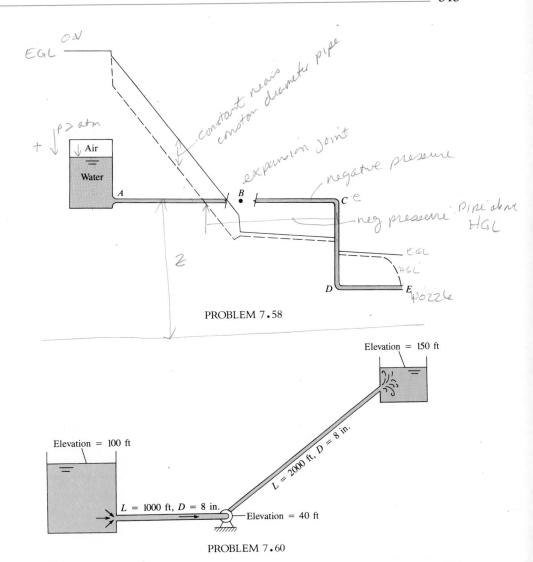

PROBLEM 7.58

PROBLEM 7.60

7.61 Water flows from reservoir A to reservoir B. The water temperature in the system is 10°C, the pipe diameter D is 1 m, and the pipe length L is 300 m. If $H = 16$ m, $h = 2$ m, and the pipe head loss is given by $h_L = 0.01(L/D)(V^2/2g)$, where V is the velocity in the pipe, what will be the discharge in the pipe? In your solution, include the head loss at the pipe outlet, and also sketch the HGL and the EGL. What will be the pressure at point P halfway between the two reservoirs?

7.62 Water flows from the reservoir on the left to the reservoir on the right at a rate of 16 cfs. The formula for the head losses in the pipes is $h_L = 0.02(L/D)(V^2/2g)$. What elevation in the left reservoir is required to produce this flow? Also carefully sketch the HGL and the EGL for the system. *Note:* Assume the head-loss formula can be used for the smaller pipe and also for the larger pipe.

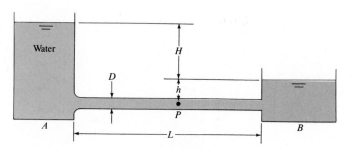

PROBLEM 7.61

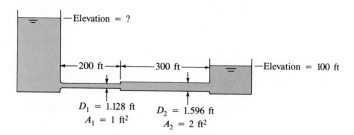

PROBLEM 7.62

7.63 What power is required to pump water at a rate of 2 m³/s from the lower to the upper reservoir? Assume the pipe head loss is given by $h_L = 0.014(L/D)(V^2/2g)$, where L is the length of pipe, D is the pipe diameter, and V is the velocity in the pipe. The water temperature is 10°C, the water surface elevation in the lower reservoir is 150 m, and the surface elevation in the upper reservoir is 250 m. The pump elevation is 100 m, $L_1 = 100$ m, $L_2 = 1000$ m, $D_1 = 1$ m, and $D_2 = 50$ cm. Assume the pump and motor efficiency is 74%. In your solution, include the head loss at the pipe outlet and sketch the HGL and the EGL.

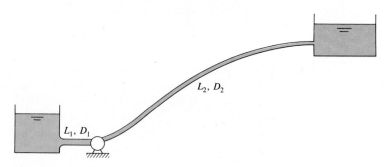

PROBLEM 7.63

7.64 Refer to Fig. 7.9 on p. 297. Assume that the head loss in the pipe is given by $h_L = 0.02(L/D)(V^2/2g)$, where V is the mean velocity in the pipe, D is the pipe diameter, and

L is the pipe length. The elevations of the reservoir water surface, the highest point in the pipe, and the pipe outlet are 200 m, 200 m, and 185 m, respectively. The pipe diameter is 30 cm, and the pipe length is 200 m. Determine the water discharge in the pipe, and, assuming that the highest point in the pipe is halfway along the pipe, determine the pressure in the pipe at that point.

7.65 A pump is used to fill a tank 5 m in diameter from a river as shown. The water surface in the river is 2 m below the bottom of the tank. The pipe diameter is 5 cm, and the head loss in the pipe is given by

$$h_L = 10V^2/2g$$

where V is the mean velocity in the pipe. The flow in the pipe is turbulent, so $\alpha = 1$. The head provided by the pump varies with discharge through the pump as

$$h_p = 20 - 5 \times 10^4 Q^2$$

where the discharge is given in cubic meters per second (m³/s) and h_p is in meters. How long will it take to fill the tank to a depth of 10 m?

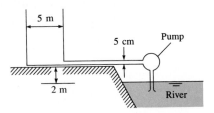

PROBLEM 7.65

7.66 The HGL and the EGL are as shown for a certain flow system.
  a. Is flow from A to E or from E to A?
  b. Does it appear that a reservoir exists in the system?
  c. Does the pipe at E have a uniform or a variable diameter?
  d. Is there a pump in the system?
  e. Sketch the physical setup that could yield the conditions shown between C and D.
  f. Is anything else revealed by the sketch?

7.67 Assume that the head loss in the pipe is given by $h_L = 0.014(L/D) (V^2/2g)$, where L is the length of pipe in feet and D is the pipe diameter in feet.
  a. Determine the discharge of water through this system.
  b. Draw the HGL and the EGL for the system.
  c. Locate the point of maximum pressure.
  d. Locate the point of minimum pressure.
  e. Calculate the maximum and minimum pressures in the system.

7.68 In Prob. 6.80, what power is developed by the windmill?

7.69 An engineer is designing a subsonic wind tunnel. The test section is to have a cross-sectional area of 4 m² and an airspeed of 50 m/s. The air density is 1.2 kg/m³. The area

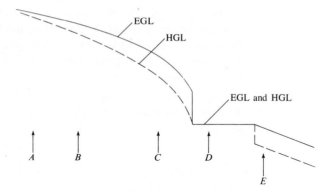

PROBLEM 7.66

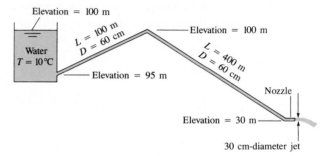

PROBLEM 7.67

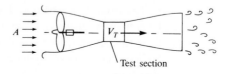

PROBLEM 7.69

of the tunnel exit is 10 m². The head loss through the tunnel is given by $h_L = (0.02)(V_T^2/2g)$, where V_T is the airspeed in the test section. Calculate the power needed to operate the wind tunnel. *Hint:* Assume negligible energy loss for the flow approaching the tunnel in region A, and assume atmospheric pressure at the outlet section of the tunnel.

7.70 Fluid flowing along a pipe of diameter D accelerates around a disk of diameter d as shown in the figure. The velocity far upstream of the disk is U, and the fluid density is ρ. Assuming incompressible flow and that the pressure downstream of the disk is the same as that at the plane of separation, develop an expression for the force required to hold the disk in place in terms of U, D, d, and ρ. Using the expression you developed, determine the force when $U = 10$ m/s, $D = 5$ cm, $d = 4$ cm, and $\rho = 1.2$ kg/m³.

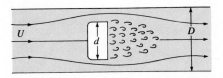

PROBLEM 7.70

7.71 If the fluid in Prob. 4.62 is water, estimate the specific gravity of the sphere necessary to produce the 0.5-ft/s fall velocity. *Hint:* Approach the solution by means of the energy equation, and assume that the head loss from the smallest-flow section to the open-tube section (downstream of the smallest section) is like that for an abrupt expansion.

References

1. Van Wylen, G. J., and R. E. Sonntag. *Fundamentals of Classical Thermodynamics.* John Wiley, New York, 1965.

2. Wark, K. *Thermodynamics.* McGraw-Hill, New York, 1966.

Dimensional Analysis and Similitude

Wind tunnel tests are made to help predict wind forces as well as to show flow patterns around the building. The turntable allows different wind directions to be set, and the blocks on the wind tunnel floor are used to generate the proper turbulence and velocity distribution. This model (1 to 400 scale) is for the Standard Oil Company (SOHIO) Headquarters Building in Cleveland, Ohio. (Courtesy J. E. Cermak and J. A. Peterka, Colorado State University)

The solutions of most engineering problems involving fluid mechanics rely on data acquired by experimental means. In many cases the empirical data are general enough that engineers have need for them in their normal design practice. Consequently, they are made available by publication in journals and textbooks. Examples of such data are the resistance coefficients for pipes and the drag coefficients for blunt bodies.* For many problems, however, either the geometry of the structure that guides the flow or the flow conditions themselves are so unique that special tests on a replica of the structure at a different scale are required to predict the flow patterns and pressure variation. When such tests are performed, the replica of the structure on which the tests are made is called the *model,* and the full-scale structure employed in the actual engineering design is called the *prototype.* The model is usually made much smaller than the prototype for economic reasons.

8.1 The Need for Dimensional Analysis

Fluid mechanics is more heavily involved with empirical work than is structural engineering, machine design, or electrical engineering because the analytical tools presently available are not capable of yielding exact solutions to many of the problems in fluid mechanics. It is true that exact solutions are obtainable for all hydrostatic problems and for many laminar-flow problems. However, the most general equations solved on the largest computers yield only fair approximations for turbulent-flow problems—hence the need for experimental evaluation and verification.

For analyzing model studies and for correlating the results of experimental research, it is essential that researchers employ *dimensionless parameters.* To appreciate the advantages of using dimensionless parameters, let us consider the flow of water through the unusual orifice illustrated in Fig. 8.1. Actually, this is much like a flow nozzle, which will be presented in Chapter 13, except that the flow is

These coefficients will be presented in Chapters 10 and 11.

FIGURE 8.1

*Flow through inverted flow
nozzle.*

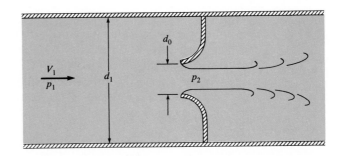

in the opposite direction to that in a nozzle operating in normal fashion. It is
true that an orifice operating in such a manner will have a much different performance than a flow nozzle. However, it is not unlikely that a firm or city
water department might have such a situation where the flow may occur the
"right way" most of the time and the "wrong way" part of the time. Hence the
need for such knowledge. The test procedure involves testing several orifices,
each with a different throat diameter d_0. We want to measure the pressure difference $p_1 - p_2$ as a function of the velocity V_1, density ρ, and diameter d_0. We
think that by carrying out numerous measurements at different values of V_1, d_0,
and ρ, we can plot our data as shown in Fig. 8.2a. Soon, however, we realize
that our first plan will involve a terrific amount of work, so we look for a better
scheme. We then think that Bernoulli's equation must have relevance here in a
manner similar to that shown in Chapter 5. It was noted in Chapter 5 that the
conditions for application of Bernoulli's equation are closely approximated if the
fluid has fairly low viscosity, as does water, if the streamlines converge in the direction of flow, and if the flow is steady. These conditions prevail between
sections 1 and 2 in our problem. In Sec. 5.4 we knew by the character of the
flow passages how V_1 and V_2 were related. Thus we could write Bernoulli's equation between two points in the flow field and solve for Δp directly. However, in

FIGURE 8.2

*Relations of pressure,
velocity, and diameter.*

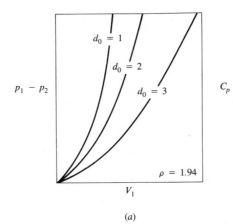

(a)

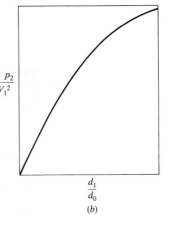

(b)

the orifice in Fig. 8.1, it should be expected that separation will occur downstream of the throat. Thus the smallest flow section is smaller than the throat section. Therefore, it is necessary to determine experimentally the relationship between Δp and the other variables.

We use a general form of Bernoulli's equation, much as we did in Chapter 5, but we insert coefficients for unknown relations. In other words, we write Bernoulli's equation as

$$p_1 + \rho \frac{V_1^2}{2} = p_2 + \rho \frac{V_2^2}{2}$$

or, in a dimensionless form, as

$$\frac{p_1 - p_2}{\rho V_1^2/2} = \frac{V_2^2}{V_1^2} - 1 \tag{8.1}$$

In Eq. (8.1) we know that $V_2/V_1 = A_1/A_2$. However, we do not know the value of A_2, and we are most interested in relating the velocity ratio to the ratio of A_1 to the orifice area A_0. Therefore, let us simply express A_1/A_0 in functional form: $V_2/V_1 = f(A_1/A_0) = f_1(d_1/d_0)^2$. Then Eq. (8.1) can be rewritten as

$$\frac{p_1 - p_2}{\rho V_1^2/2} = [f_1(d_1/d_0)^2]^2 - 1 \tag{8.2}$$

The left-hand side of Eq. (8.2) is a form of pressure coefficient C_p, and since the right-hand side is solely a function of d_1/d_0, we can write

$$C_p = f_2\left(\frac{d_1}{d_0}\right) \tag{8.3}$$

If we plot the pressure coefficient C_p (instead of the pressure difference) versus d_1/d_0, we will get a single curve, as shown in Fig. 8.2b. Inspection will convince the reader that everything given on Fig. 8.2a is included on Fig. 8.2b.

Thus by considering a nondimensional form of Bernoulli's equation, we have made a tremendous reduction in the experimental work that would have been required had we not considered the nondimensional form. The process of nondimensionalizing the equation reduced the correlating parameters from five $(\Delta p, \rho, V, d_1, d_0)$ to two $[\Delta p/(\rho V^2/2), d_1/d_0]$. To collect the data for Fig. 8.2a would require, say, 20 data points for each d_0 value. If five d_0's were tested, then 100 data points would be needed. On the other hand, only about 20 would be required to yield the curve in Fig. 8.2b, so a considerable amount of time is saved.

In the foregoing development we had a clue about the governing equation. By considering a dimensionless form of that equation, we were able to obtain a set of dimensionless parameters with which to correlate our data. In many cases, however, the governing equation is not available, and we must seek the dimensionless parameters by using a formal procedure called *dimensional analysis*. Sections 8.2 and 8.3 present some basic material leading up to dimensional analysis, and Sec. 8.4 contains an explanation of the actual procedure of dimensional analysis.

8.2   Dimensions and Equations

All variables used in science or engineering are expressed in terms of a limited number of basic dimensions. For most engineering problems the basic dimensions are either force, length, and time; or mass, length, and time—either group being equally as acceptable as the other. Thus we can designate the dimensions of pressure as follows:

$$[p] = \frac{F}{L^2} \qquad (8.4)$$

Here the brackets mean "dimension of." Therefore, Eq. (8.4) reads as follows: "The dimensions of p equal force per length squared." In this case, L^2 represents the dimension of area.

It goes without saying that all equations must balance in magnitude. However, all rational equations (those developed from basic laws of physics) must also be dimensionally homogeneous. That is, the left-hand side of the equation must have the same dimensions as the right-hand side. Moreover, every term in the equation must have the same dimensions.

8.3   The Buckingham II Theorem

In 1915 Buckingham (4) showed that the number of independent dimensionless groups of variables (dimensionless parameters) needed to correlate the variables in a given process is equal to $n - m$, where n is the number of variables involved and m is the number of basic dimensions included in the variables. Thus if the drag force F of a fluid flowing past a sphere is known to be a function of the velocity V, mass density ρ, viscosity μ, and diameter D, then five variables ($F, V, \rho, \mu,$ and D) and three basic dimensions ($L, F,$ and T) are involved. We will have $5 - 3 = 2$ basic groupings of variables that can be used to correlate experimental results. Note that the same reduction in correlating parameters occurred in our discussion concerning the reverse-flow nozzle in Sec. 8.1. In the next section we will show in a step-by-step process how variables can be combined to form the dimensionless parameters.

8.4   Dimensional Analysis

The Step-by-Step Method

Several methods may be used to carry out the process of dimensional analysis, but the step-by-step approach, very clearly presented by Ipsen (6), is one of the easiest and reveals much about the process. The basic objective in dimensional analysis, as already noted, is to reduce the number of separate variables involved in a problem to a smaller number of independent dimensionless groups of variables (dimensionless parameters). The rationale for this is that all rational equations can be nondimensionalized, as illustrated in Sec. 8.1, and that all rational equations have a certain number of independent terms in them. Thus, by the procedure of dimensional analysis, we will be simply arranging variables into a dimensionless equation. In the process of combining the variables to form dimensionless groups or terms, the number of independent groups thus obtained is less than the number of original variables.

We start the process by identifying only those variables that are significant to the problem. Then we include these variables in a functional equation and note their dimensions. As an example, let us consider a simple problem we are already familiar with so that we can get a feel for the process.

EXAMPLE 8.1   Suppose that we want to consider the fall velocity of a body in a vacuum and we know that this velocity V is a function of the acceleration due to gravity, g, and the distance through which the body falls, h. However, let us say that we do not know how to derive the equation for the fall velocity but want to obtain the proper relationship by dimensional analysis and experimentation. Thus in this example we want to determine the dimensionless parameter(s) that apply.

Solution   We include the significant variables in a functional equation:

$$V = f(g, h)$$

where

$$[V] = L/T$$

$$[g] = L/T^2$$

$$[h] = L$$

By referring to this list of variables and their dimensions, we note that the velocity V has the dimension of time in the denominator. The only variable on the right-hand side of the equation with the dimension of time is g, which has T^2 in its denominator. Thus we know that $\sqrt{g}$ will have T to the first power in the denominator. Now, if we divide the left-hand side by $\sqrt{g}$, that side will be dimensionless in T. The only way we can make the right-hand side dimensionless in T is to combine g with itself—that is, to divide g by g—thereby deleting it from the right-hand side. We then have

$$\frac{V}{\sqrt{g}} = f_1(h)$$

Note that we have reduced the number of variables on the right-hand side by one—namely, g—which is the essence of the technique of dimensional analysis.

In a similar manner, we combine h with $V/\sqrt{g}$ to make the entire combination dimensionless in L as well as T, which gives

$$\frac{V}{\sqrt{gh}} = C$$

Of course, our experience in basic physics tells us that $C = \sqrt{2}$.   ◀

In the foregoing illustration of the step-by-step method of dimensional analysis, it should be noted that the remaining single dimensionless parameter was consistent with the number of dimensionless parameters, $n - m$, that the Buckingham Π theorem predicts. That is, we had three variables ($n = 3$) and two dimensions [L and T ($m = 2$)]. Hence we should have one ($3 - 2$) dimensionless parameter, and we do. In fact, the step-by-step method of dimensional analysis will always yield the correct number of parameters.*

We will take another example that is somewhat more involved than the first. However, the procedure is the same as before.

EXAMPLE 8.2   If the drag F_D of a sphere in a fluid flowing past the sphere is a function of the viscosity μ, the mass density ρ, the velocity of flow V, and the diameter of the sphere D, what dimensionless parameters are applicable to the flow process?

Solution $F_D = f_1(V, \rho, \mu, D)$ (8.5)

where $[F_D] = F$

$$[V] = L/T$$

$$[\rho] = FT^2/L^4$$

$$[\mu] = FT/L^2$$

$$[D] = L$$

---

* Note that, in rare instances, the number of parameters may be one more than predicted by the Buckingham Π theorem. This anomaly can occur because it is possible that two-dimensional categories can be eliminated when dividing (or multiplying) by a given variable. See Ipsen (7, page 172) for an example of this.

Eliminate F as a dimension by combining ρ with all variables that have the force dimension in them in such a way that F is canceled:

$$\frac{F_D}{\rho} = f_2\left(V, \frac{\mu}{\rho}, D\right)$$

where

$$[F_D/\rho] = FL^4/FT^2 = L^4/T^2$$

$$[V] = L/T$$

$$[\mu/\rho] = FTL^4/L^2FT^2 = L^2/T$$

$$[D] = L$$

Now eliminate the time dimension T by combining V with the remaining groups of variables that include the time dimension:

$$\frac{F_D}{\rho V^2} = f_3\left(\frac{\mu}{\rho V}, D\right)$$

where

$$\left[\frac{F_D}{\rho V^2}\right] = L^2$$

$$\left[\frac{\mu}{\rho V}\right] = L$$

$$[D] = L$$

We now have groupings of variables that have only the length dimension left. We can make these dimensionless by combining the length variable D with them. We finally obtain

$$\frac{F_D}{\rho V^2 D^2} = f_4\left(\frac{\mu}{\rho V D}\right) \tag{8.6}$$

Although Eq. (8.6) still has as many variables as Eq. (8.5), for purposes of correlation the number has been reduced from five to two. We can treat the two dimensionless parameters as separate variables when determining the functional relationship.

The actual functional relationship between the parameters is determined by running a series of tests to observe values of $F_D/\rho V^2 D^2$ for different values of the ratio $\mu/\rho VD$. The results of these tests are plotted to yield a usable functional relationship between the two parameters. Since the relationship is experimentally determined, it is just as valid to express Eq. (8.6) with the right-hand side inverted:

$$\frac{F_D}{\rho V^2 D^2} = f_5\left(\frac{VD\rho}{\mu}\right) \tag{8.7}$$

Now the combination on the right is the Reynolds number, which was first introduced in Chapter 4. Hence the parameter $F_D/\rho V^2 D^2$ is a function of the Reynolds number, $VD\rho/\mu$.

The solution of Example 8.2 was found by combining ρ, then V, and finally D with other variables to produce the dimensionless groupings. The question arises whether we must use these variables or proceed in this order to produce dimensionless groupings. The answer is that this is not the only way it can be done. For example, let us solve Example 8.2 again with a different order of attack:

We start by noting as before that

$$F_D = f_1(V, \rho, \mu, D)$$

where

$$[F_D] = F$$

$$[V] = L/T$$

$$[\rho] = FT^2/L^4$$

$$[\mu] = FT/L^2$$

$$[D] = L$$

Now we eliminate the length dimension first by combining D with the other variables. Therefore, we have

$$F_D = f_2\left(\frac{V}{D}, \rho D^4, \mu D^2\right)$$

where

$$[F_D] = F$$

$$[V/D] = T^{-1}$$

$$[\rho D^4] = FT^2$$

$$[\mu D^2] = FT$$

Next we eliminate T by the combination D/V, which itself has the dimension of T. We get

$$F_D = f_3\left(\frac{VD}{DV}, \rho D^4 \frac{V^2}{D^2}, \mu D^2 \frac{V}{D}\right)$$

The combination VD/DV is a pure number (unity); it is not a variable and hence is dropped from the groups of variables. We then have

$$F_D = f_3(\rho D^2 V^2, \mu DV)$$

where $[\rho D^2 V^2] = F$ and $[\mu DV] = F$. Finally, we use the combination $\rho V^2 D^2$, the dimension of which is F, to eliminate F as a dimension. We obtain

$$\frac{F_D}{\rho V^2 D^2} = f_4\left(\frac{\rho D^2 V^2}{\rho D^2 V^2}, \frac{\mu DV}{\rho V^2 D^2}\right)$$

or
$$\frac{F_D}{\rho V^2 D^2} = f_5\left(\frac{\mu}{\rho V D}\right) = f_6\left(\frac{V D \rho}{\mu}\right)$$

We end up with the same result we had before. We could have gotten different combinations if we had, for example, used μ, V, and D as combining variables instead of ρ, V, and D or combinations thereof.* However, it is the usual practice in fluid mechanics to use ρ, V, and a length variable for combining variables. Note that when the variables are combined in this order (first ρ, then V, and finally a length variable) to eliminate F, T, and L, respectively, the process of dimensional analysis becomes quite simple. After the force dimension is eliminated, it is not reintroduced when V is combined with the group(s) of variables to eliminate T, because the F dimension is not included in V. This is also the case when the length variable is combined with the group(s), because neither F nor T is included in a length variable. If different choices of variables (other than ρ, V, and L) are used for combining variables, it is still desirable to combine them in order from the more complex (the one with the greatest number of dimensions) to the simplest (the one with merely a single dimension).

The Exponent Method

An alternative method for finding the dimensionless parameters, once the significant variables have been identified, is to solve a set of algebraic simultaneous equations that derive from the requirement of dimensional homogeneity for equations describing physical systems. This will be referred to as the exponent method. The method is best explained by example.

The problem described in Example 8.2 is solved here using the exponent method. For this application, mass, length, and time will be used as the basic dimensions instead of force, length, and time. The dimensions of the variables are

$$[F] = \frac{ML}{T^2}$$

$$[V] = \frac{L}{T}$$

$$[\rho] = \frac{M}{L^3}$$

$$[\mu] = \frac{M}{LT}$$

$$[D] = L$$

---

*For example, if μ had first been combined with F_D and ρ to cancel the F dimension, we would have obtained $F_D/\mu = f_1(\rho/\mu, V, D)$. Then, upon combining V and D to cancel the T and L dimensions, respectively, we would have obtained $F_D/\mu DV = f_4(VD\rho/\mu)$.

Dimensional homogeneity requires that each term making up the function f_1 in Eq. (8.5) have the same dimensions as force. Thus the variables in each term must combine in such a way that the combination has the dimensions of force (ML/T^2). We assume that the variables in f_1 combine as the product

$$V^a \rho^b \mu^c D^d$$

where a, b, c, and d are exponents to be selected that yield the dimensions of force. Dimensional homogeneity requires

$$[F] = [V^a \rho^b \mu^c D^d]$$

or

$$\frac{ML}{T^2} = \left(\frac{L}{T}\right)^a \left(\frac{M}{L^3}\right)^b \left(\frac{M}{LT}\right)^c L^d$$

$$= \frac{L^{a-3b-c+d} M^{b+c}}{T^{a+c}}$$

Equating the powers of M, L, and T on each side of the equation results in three algebraic equations,

$$L: \quad a - 3b - c + d = 1$$

$$M: \quad b + c = 1$$

$$T: \quad a + c = 2$$

We have three equations and four unknowns. However, we can solve for three of the unknowns in terms of the fourth unknown. The solution procedure is simplified if we select as the fourth unknown the exponent that appears most frequently in the equations. Let us take a, b, and d as the exponents to be solved for in terms of c. The equations can be written in matrix form as

$$\begin{pmatrix} 1 & -3 & 1 \\ 0 & 1 & 0 \\ 1 & 0 & 0 \end{pmatrix} \begin{pmatrix} a \\ b \\ d \end{pmatrix} = \begin{pmatrix} 1+c \\ 1-c \\ 2-c \end{pmatrix}$$

If the determinant of the matrix is not zero, a unique solution for a, b, and d is obtainable.* The result is

$$a = 2 - c$$

$$b = 1 - c$$

$$d = 2 - c$$

* *The choice of the three unknowns is not necessarily arbitrary. If the determinant is zero, then a different set of the three unknowns must be chosen. If all the possible combinations of the three unknowns yield a zero determinant, then two unknowns must be chosen and solved for in terms of the remaining unknowns. In this situation, the number of dimensionless parameters will be two (not three) less than the number of variables.*

These exponents are now substituted back into the combination of the physical variables, and the result is

$$V^a \rho^b \mu^c D^d = \rho V^2 D^2 \left(\frac{\mu}{\rho V D}\right)^c$$

The factor $\rho V^2 D^2$ has the dimensions of force and will be common to every term in f_1. The combination $(\mu/\rho V D)$ is dimensionless. Thus Eq. (8.5) can be rewritten as

$$F = \rho V^2 D^2 f_1 \left(\frac{\mu}{\rho V D}\right)$$

Dividing through by $\rho V^2 D^2$ yields the same dimensionless equation as Eq. (8.6).

$$\frac{F}{\rho V^2 D^2} = f_1 \left(\frac{\mu}{\rho V D}\right)$$

Note that in the process of dimensional analysis, we did not need to evaluate the exponent c. We sought only the form of the dimensionless parameter. The functional form of f_1 must be obtained from experiment.

Recapitulation of the Process of Dimensional Analysis

In the foregoing paragraphs, the concept and procedure of dimensional analysis have been presented. For review purposes and handy reference, the procedure of dimensional analysis can be summarized as follows:

1. Identify all significant variables associated with the problem and include these in a functional equation, such as

$$Z = f(U, V, W, X, Y)$$

2. First combine one variable that includes a given dimension with other variables having that same dimension in such a way as to eliminate that dimension. Then choose a second variable to combine with other variables and groups of variables to eliminate a second dimension. Repeat the process until the entire equation consists of dimensionless groups or parameters. If the exponent method is used, a set of algebraic equations for exponents of the physical variables that satisfy dimensionless homogeneity is developed. Three exponents (for systems with three dimensions) are selected and solved for in terms of the remaining exponents. Substituting the three exponents back into the combination of physical variables yields the dimensionless parameters.

3. The final form will appear something like the following:

$$\pi_1 = f(\pi_2, \pi_3, \ldots, \pi_{n-m})$$

where $\pi_1, \pi_2, \ldots, \pi_{n-m}$ represent the dimensionless parameters obtained, n is the number of variables, and m is the number of dimensions included in the list of variables.

Limitations of Dimensional Analysis

All the foregoing procedures deal with straightforward situations. However, some problems do occur. In order to apply dimensional analysis we must first decide which variables are significant. If we do not understand the problem well enough to make a good initial choice of variables, dimensional analysis seldom provides clarification.

One error might be inclusion of variables whose influence is already accounted for. For example, one might tend to include two or three length variables in a scale-model test where only one may be sufficient.

Another serious error might be the omission of a significant variable. If this is done, one of the significant dimensionless parameters will likewise be missing. In this regard, we may sometimes identify a list of variables that we think are significant to a problem and find that one dimensional category (such as F or L or T) is included in only one variable. When this occurs, we know that an error in choice of variables has been made because it will not be possible to combine two variables to eliminate the lone dimension. Either the variable with the lone dimension should not have been included in the first place (it is not significant), or another one should have been included (a significant variable was omitted).

How do we know whether a variable is significant for a given problem? Probably the truest answer is by experience. After working in the field of fluid mechanics for several years, one develops a feel for the significance of variables to certain kinds of applications. However, even the inexperienced engineer will appreciate the fact that free-surface effects have no significance in closed-conduit flow; consequently, surface tension, σ, would not be included as a variable. In closed-conduit flow, if the velocity is less than approximately one-third the speed of sound, compressibility effects are usually negligible. Such guidelines, which have been observed by previous experimenters, help the beginning engineer develop confidence in her or his application of dimensional analysis and similitude.

8.5 Common Dimensionless Numbers

In Example 8.2, one of the dimensionless parameters obtained was a form of the Reynolds number. This parameter and others recur repeatedly in fluid-flow studies. By considering all variables that might be significant in a general flow situation, we can derive these parameters or *dimensionless numbers* by means of dimensional analysis.

Visualize a hypothetical flow condition in which the pressure difference between two points in the flow field is expected to be a function of V, L, ρ, μ, E_v, σ, and $\Delta\gamma$. Respectively, these variables are velocity, characteristic length involved in the fluid flow, mass density of the fluid, viscosity of the fluid, bulk modulus of elasticity of the fluid, surface tension of the fluid, and the difference in specific weight between the flowing fluid and the fluid above the free surface. The latter variable, $\Delta\gamma$, is significant when free-surface phenomena such as waves exist. It would never be applied to flow in completely enclosed conduits with no interface between two fluids. The functional relationship for these variables is given as

$$\Delta p = f(V, L, \rho, \mu, E_v, \sigma, \Delta\gamma)$$

where
$$[\Delta p] = F/L^2$$
$$[V] = L/T$$
$$[L] = L$$
$$[\rho] = FT^2/L^4$$
$$[\mu] = FT/L^2$$
$$[E_v] = F/L^2$$
$$[\sigma] = F/L$$
$$[\Delta\gamma] = F/L^3$$

When we make a dimensional analysis on these variables using ρ, V, and L as combining variables, we obtain the following functional relationship between the dimensionless parameters:

$$\frac{\Delta p}{\rho V^2} = f\left(\frac{VL\rho}{\mu}, \frac{V}{\sqrt{E_v/\rho}}, \frac{\rho LV^2}{\sigma}, \frac{V^2}{L\,\Delta\gamma/\rho}\right)$$

Without affecting the results with respect to dimensional considerations, we can rewrite the left-hand side of this equation as $\Delta p/\frac{1}{2}\rho V^2$. We do this to put the term in the same form as the pressure coefficient that was introduced in Chapter 5. The speed of sound c is given by $\sqrt{E_v/\rho}$. We then have

Re, $Mach$, $Weber$, Fr $\Rightarrow$

$$\frac{\Delta p}{\frac{1}{2}\rho V^2} = f_1\left(\frac{VL\rho}{\mu}, \frac{V}{c}, \frac{\rho LV^2}{\sigma}, \frac{V^2}{L\,\Delta\gamma/\rho}\right) \tag{8.8}$$

The parameter on the left-hand side of Eq. (8.8) is the pressure coefficient, and it is seen to be dependent on the dimensionless parameters on the right-hand side of the equation. On the right-hand side, the first parameter is the Reynolds number already referred to, the second parameter is the Mach number, the third is the Weber number, and the last is the Froude (rhymes with food) number squared. The Froude number is defined as the square root of this last parameter, and it is customary to replace ρ by γ/g:

$$\text{Fr} = \frac{V}{\sqrt{(\Delta\gamma/\gamma)gL}}$$

In this particular form it is called the densimetric Froude number, and it is applied in studying the motion of fluids in which there is density stratification, such as between salt water and fresh water in an estuary. It also has application in the study of thermal plumes from stacks or from heated-water effluents associated with thermal power plants. However, in most applications of the Froude number, the flowing fluid is usually a liquid such as water and the adjacent fluid at the interface is a gas such as air. Thus for all practical purposes, $\Delta\gamma = \gamma_{\text{liquid}}$. Then $\Delta\gamma/\gamma = 1$. Hence the Froude number is most often given as

$$\text{Fr} = \frac{V}{\sqrt{gL}} \tag{8.9}$$

We can then write Eq. (8.8) as

$$\frac{\Delta p}{\frac{1}{2}\rho V^2} = f_2\left(\frac{VL\rho}{\mu}, \frac{V}{c}, \frac{\rho L V^2}{\sigma}, \frac{V}{\sqrt{gL}}\right) \tag{8.10}$$

or $$C_p = f_2(\text{Re}, \text{M}, \text{W}, \text{Fr}) \tag{8.11}$$

In Chapter 5 it was noted that C_p is a dimensionless parameter derived from Bernoulli's equation. One can view it as a ratio of a pressure force relative to an inertial reaction (mass times acceleration) of the fluid. The pressure force, F_p, is simply $\Delta p A$, or, from a dimensional point of view, $F_p \propto \Delta p L^2$. The inertial, or Ma, force is expressed as

$$F_i = Ma$$

$$F_i \propto \rho L^3 V \frac{dV}{ds} \propto \frac{1}{2}\rho V^2 L^2$$

where M represents the mass of the fluid, and $V dV/ds$ is acceleration in the form of convective acceleration. The ratio of the pressure force to inertial reaction is then given as

$$\frac{F_p}{F_i} \propto \frac{\Delta p L^2}{\frac{1}{2}\rho V^2 L^2} = \frac{\Delta p}{\frac{1}{2}\rho V^2} = C_p$$

Similarly, the other dimensionless numbers can be viewed as relative quantities — that is, inertial-force reaction relative to forces relating to fluid characteristics. The next paragraphs demonstrate these relations.

The Reynolds number can be viewed as a ratio of inertial to viscous forces. The viscous force acting on a surface is $F_v = \tau A = (\mu\, du/dy)A$. However, we are primarily interested in dimensional considerations; hence F_v can be written as

$$F_v \propto \mu \frac{V}{L} A \propto \mu \frac{V}{L} L^2$$

$$F_v \propto \mu V L$$

Thus the ratio of inertial to viscous force in terms of the basic variables is

$$\frac{F_i}{F_v} \propto \frac{\rho L^2 V^2}{\mu V L} = \frac{V L \rho}{\mu} = \text{Re}$$

Because, as we have shown, the magnitude of the Reynolds number is inversely proportional to the shear force, very low Reynolds numbers imply relatively large viscous shear forces and that very large Reynolds numbers imply relatively low viscous shear forces. This concept of the Reynolds number and its relationship to viscous stresses is often very helpful in understanding certain flow phenomena and in scoping solutions to them. For example, in Sec. 8.9 it will be shown that because the viscous effects are relatively insignificant at very high Reynolds numbers, one can assume that C_p is virtually constant for problems involving convective acceleration.

Expressed in terms of the basic variables involved, the other forces (surface tension, gravity, and elastic) that influence the flow are as follows:

$$F_\sigma \propto \sigma L \qquad \text{(surface-tension force)}$$

$$F_g \propto \Delta \gamma L^3 \qquad \text{(gravity force)}$$

$$F_c \propto \rho V c L^2 \qquad \text{(elastic or compressibility force)}$$

The elastic-force relation follows from $F_c = \Delta p_c L^2$, where $\Delta p_c = \rho V c$ (the pressure change that occurs in water hammer or shock waves in supersonic flow). Thus the Mach number is a ratio of inertial to elastic force,

$$M = \frac{\rho L^2 V^2}{\rho V c L^2} = \frac{V}{c}$$

the Weber number is a ratio of inertial to surface-tension force,

$$W = \frac{\rho L^2 V^2}{\sigma L} = \frac{\rho V^2 L}{\sigma}$$

and the Froude number squared is the ratio of inertial to gravity force,

$$(\text{Fr})^2 = \frac{\rho L^2 V^2}{\Delta \gamma L^3} = \frac{V^2}{L \, \Delta \gamma / \rho}$$

With respect to the Froude number, it should be noted that when gravity causes only a hydrostatic pressure distribution, such as in a closed conduit, its effect is of no significance in controlling the pattern of flow. However, if the gravitational force influences the pattern of flow, such as in flow over a spillway or in the formation of waves created by a ship as it cruises over the sea, the Froude number is a most significant parameter.

The Mach number is an indicator of how important compressibility effects are in a fluid flow. If the Mach number is small, then the inertial force associated with the fluid motion does not cause a significant density change, and the flow can be treated as incompressible (constant density). On the other hand, if the Mach number is large, there are often appreciable density changes that must be considered in model studies. The significance of the Mach number in compressible flow studies will be discussed in more detail in Chapter 12.

The Weber number is an important parameter in liquid atomization. The surface tension of the liquid at the surface of a droplet is responsible for maintaining the droplet's shape. If a droplet is subjected to an air jet and there is a relative velocity between the droplet and the gas, inertial forces due to this relative velocity cause the droplet to deform. If the Weber number is too large, the inertial force overcomes the surface-tension force to the point that the droplet shatters into even smaller droplets. Thus a Weber-number criterion can be useful in predicting the droplet size to be expected in liquid atomization. The size of the droplets resulting from liquid atomization is a very significant parameter in gas-turbine and rocket combustion.

8.6 Similitude

Scope of Similitude

Whenever it is necessary to perform tests on a model to obtain information that cannot be obtained by analytical means alone, the rules of similitude must be applied. *Similitude* is the theory and art of predicting prototype performance from model observations. We shall see that the theory of similitude involves the application of dimensionless numbers, such as the Reynolds number or the Froude number, to predict prototype performance from model tests. The art of similitude enters the problem when the engineer must make decisions about model design, model construction, performance of tests, or analysis of results that are not included in the basic theory.

Present engineering practice makes use of model tests more frequently than most people realize. For example, whenever a new airplane is being designed, tests are made not only on the general scale model of the prototype airplane but also on various components of the plane. Numerous tests are made on individual wing sections as well as on the engine pods and tail sections.

Models of automobiles and high-speed trains are also tested in wind tunnels to predict the drag and flow patterns for the prototype. Information derived from these model studies often indicates potential problems that can be corrected before the prototype is built, thereby saving considerable time and expense in development of the prototype.

In civil engineering, model tests are always used to predict flow conditions for the spillways of large dams. In addition, river models assist the engineer in the design of flood-control structures as well as in the analysis of sediment movement in the river. Marine engineers make extensive tests on model ship hulls to predict the drag of the ships. Much of this type of testing is done at the Naval Ship Research and Development Center near Washington, D.C. (see Fig. 8.3). Tests are also regularly performed on models of tall buildings to help predict the wind loads on the buildings, the stability characteristics of the buildings, and the air-flow patterns in their vicinity. The latter information is used by the architects to design walkways and passageways that are safer and more comfortable for pedestrians to use.

Geometric Similitude

The basic and perhaps the most obvious requirement of similitude is that the model be an exact geometric replica of the prototype.* Consequently, if a

FIGURE 8.3

Ship-model test at the Naval Ship Research and Development Center, Carderock, Maryland.

For most model studies this is a basic requirement. However, for certain types of problems, such as river models, distortion of the vertical scale is often necessary to obtain meaningful results.

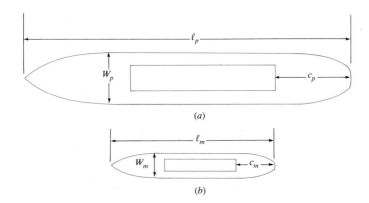

(a)

(b)

1:10 scale model is specified, all linear dimensions of the model must be $\frac{1}{10}$ of those of the prototype. In Fig. 8.4 if the model and prototype are geometrically similar, the following equalities hold:

$$\frac{\ell_m}{\ell_p} = \frac{w_m}{w_p} = \frac{c_m}{c_p} = L_r \qquad (8.12)$$

Here ℓ, w, and c are specific linear dimensions associated with the model and prototype, and L_r is the scale ratio between model and prototype. It follows that the ratio of corresponding areas between model and prototype will be the square of the length ratio: $A_r = L_r^2$. The ratio of corresponding volumes will be given by $\Psi_m/\Psi_p = L_r^3$.

Dynamic Similitude

The basic requirement for dynamic similitude is that the forces that act on corresponding masses in the model and prototype must be in the same ratio ($F_m/F_p =$ constant) throughout the entire flow field. Since the forces acting on the fluid elements will thus control the motion of these elements, it follows that dynamic similarity will yield similarity of flow patterns. Consequently, the flow patterns will be the same in the model as in the prototype if geometric similarity is satisfied and if the relative forces acting on the fluid are the same in the model as in the prototype. This latter condition requires that we have equality of the appropriate dimensionless numbers introduced in Sec. 8.4, because these dimensionless numbers are, in fact, indicators of relative forces within the fluid.

Now we shall give a more physical interpretation to the foregoing developments. Consider flow over the spillway shown in Fig. 8.5a. Here corresponding masses of fluid in the model and prototype are acted on by corresponding forces. These forces are the force of gravity F_g, the pressure force F_p, and the viscous resistance force F_v. These forces add vectorially as shown in Fig. 8.5 to yield a resultant force F_R, which will in turn produce an acceleration of the volume of

FIGURE 8.5

*Model–prototype relations:
view (a) and view (b).*

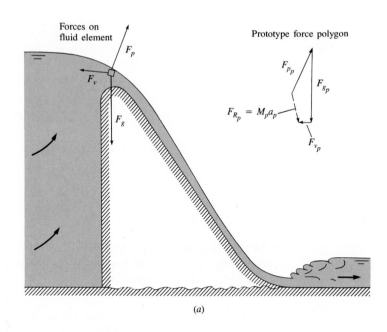

(a)

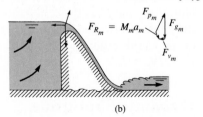

(b)

fluid in accordance with Newton's second law. Hence, because the force polygons in the prototype and model are similar, the magnitudes of the forces in the prototype and model will be in the same ratio as the magnitude of the *Ma* vectors:

$$\frac{M_m a_m}{M_p a_p} = \frac{F_{g_m}}{F_{g_p}}$$

or

$$\frac{\rho_m L_m^3 (V_m/t_m)}{\rho_p L_p^3 (V_p/t_p)} = \frac{\gamma_m L_m^3}{\gamma_p L_p^3}$$

which reduces to

$$\frac{V_m}{g_m t_m} = \frac{V_p}{g_p t_p}$$

But

$$\frac{t_m}{t_p} \propto \frac{L_m/V_m}{L_p/V_p}$$

so

$$\frac{V_m^2}{g_m L_m} = \frac{V_p^2}{g_p L_p} \tag{8.13}$$

Taking the square root of each side of Eq. (8.13), we obtain

$$\frac{V_m}{\sqrt{g_m L_m}} = \frac{V_p}{\sqrt{g_p L_p}} \tag{8.14}$$

or

$$\mathrm{Fr}_m = \mathrm{Fr}_p$$

Thus it has been shown that the Froude number in the model must be equal to the Froude number in the prototype. However, we have looked only at the Ma, or inertial, forces and the gravity forces to derive Eq. (8.14). If we equate the ratio of inertial forces to the ratio of viscous forces, we obtain

$$\frac{M_m a_m}{M_p a_p} = \frac{F_{v_m}}{F_{v_p}} \tag{8.15}$$

Here $F_v \propto \mu V L$, so with a little algebra, Eq. (8.15) reduces to

$$\mathrm{Re}_m = \mathrm{Re}_p$$

And finally

$$\frac{M_m a_m}{M_p a_p} = \frac{F_{p_m}}{F_{p_p}} \tag{8.16}$$

where

$$F_p \propto \Delta p L^2$$

which yields

$$C_{p_m} = C_{p_p}$$

The foregoing development then means that dynamic similitude (similarity of pressure coefficients) will be completely achieved for flow over a spillway if the Froude number and the Reynolds number are the same in the model as in the prototype. As will be shown in Sec. 8.7, the latter part of this requirement can be relaxed if the model is reasonably large. In the force polygon shown in Fig. 8.5, it can be seen that the polygon can be completed with only three of the forces; hence one of them is dependent on the others. Thus, if we think of the pressure force as a dependent one, it follows that the pressure coefficient is dependent on the other parameters. In other words, if we have equality of Fr and Re, then we automatically have equality of C_p between model and prototype. This is exactly the conclusion arrived at in Sec. 8.5, reached here in a different manner. If other forces, such as surface-tension or elastic forces, were significant in establishing the flow pattern, then equality of additional dimensionless parameters, such as the Weber number and the Mach number, respectively, would be required.

We conclude from the foregoing developments that *the requirement for similarity of flow between model and prototype is that the significant dimensionless parameters must be equal for model and prototype.*

For example, if we were model-testing a valve (enclosed flow), the only forces that might affect the flow pattern would be the shear forces produced by viscous effects. Consequently, dynamic similarity would prevail, with equality of Reynolds numbers. That is, the model Reynolds number would have to be the same as the prototype Reynolds number. For flow over a spillway, the gravitational force is predominant in establishing the flow pattern. Therefore, the Froude number in the model must be the same as the Froude number in the prototype for dynamic similarity in this type of application. The significance of these requirements and their actual use in predicting prototype performance will be considered in the next two sections.

8.7 Model Studies for Flows Without Free-Surface Effects

Free-surface effects are absent in the flow of liquids or gases in closed conduits, including control devices such as valves, or in the flow about bodies (for example, aircraft) that travel through air or are deeply submerged in a liquid such as water (submarines). Free-surface effects are also absent where a structure such as a building is stationary and wind flows past it. In all these cases, given relatively low Mach numbers, it is the Reynolds-number criterion that must be used for dynamic similarity. That is, the Reynolds number in the model must equal the Reynolds number in the prototype. The following examples illustrate the application.

EXAMPLE 8.3   The drag characteristics of a blimp 5 m in diameter and 60 m long are to be studied in a wind tunnel. If the speed of the blimp through still air is 10 m/s, and if a $\frac{1}{10}$ scale model is to be tested, what airspeed in the wind tunnel is needed for dynamically similar conditions? Assume the same air pressure and temperature for both model and prototype.

Solution   For dynamic similarity, the Reynolds number of the model must equal the Reynolds number of the prototype, or

$$\mathrm{Re}_m = \mathrm{Re}_p$$

Hence

$$\frac{V_m L_m \rho_m}{\mu_m} = \frac{V_p L_p \rho_p}{\mu_p}$$

From this we can solve for V_m:

$$V_m = V_p \frac{L_p}{L_m} \frac{\rho_p}{\rho_m} \frac{\mu_m}{\mu_p} = 10 \times 10 \times 1 \times 1 = 100 \text{ m/s} \qquad \blacktriangleleft$$

Here, we get the result that the air speed in the wind tunnel must be 100 m/s for true Reynolds-number similitude. This speed is quite large, and in fact Mach-number effects may start to become important at such a speed. However, we will see in Sec. 8.9 that it is not always necessary to operate models at true Reynolds-number criteria to obtain useful results.

However, if the engineer thinks that it is essential to test models at the same Reynolds number as the prototype, then only a few alternatives are available. One way to produce high Reynolds numbers at nominal air speeds is to increase the density of the air. A pressurized wind tunnel achieves this. At one time NASA at Langley Field in Virginia had a variable-density wind tunnel that could be pressurized up to 20 atmospheres (13). There are several problems that are peculiar to a pressurized tunnel. First, a shell (essentially a pressurized bottle) must surround the entire tunnel and its components, and this adds to the cost of the tunnel. Second, it takes a long time to pressurize the tunnel in preparation for operation, increasing the time from the start to the finish of runs.

Another method of obtaining high Reynolds numbers is to build a tunnel in which the test medium (gas) is at a very low temperature, thus producing a relatively high-density–low-viscosity fluid. NASA has built such a tunnel and operates it at the Langley Research Center. This tunnel, called the National Transonic Facility, can be pressurized up to 9 atmospheres. The test medium is nitrogen, which is cooled by injecting liquid nitrogen into the system. In this wind tunnel it is possible to reach Reynolds numbers of 10^8 based upon a model size of 0.25 m (9). Because of its sophisticated design, its initial cost of approximately $100,000,000 (2), and its operating expenses of about $10 per second (9) are both high.

Another modern approach in wind-tunnel technology is the development of magnetic or electrostatic suspension of models. The use of the magnetic suspension with model airplanes is being researched (2), and the electrostatic suspension for the study of single-particle aerodynamics has already been reported (8).

The use of wind tunnels for aircraft design has grown significantly as the size and sophistication of aircraft have increased. For example, in the 1930s the DC-3 and B-17 each had about 100 hours of wind-tunnel tests at a rate of $100 per hour of run time. By contrast the F-15 fighter required about 20,000 hours of tests at a cost of $20,000 per hour (2). The latter test time is even more staggering when one realizes that a much greater volume of data per hour at higher accuracy is obtained from the modern wind tunnels because of the high-speed data acquisition made possible by computers.

EXAMPLE 8.4   The valve shown is the type used in the control of water in large conduits. Model tests are to be done, using water as the fluid, to determine how the valve will operate under wide-open conditions. The prototype size is 6 ft in diameter at the inlet. What flow rate is required for the model if the prototype flow is 700 cfs? Assume that temperature for model and prototype is 60°F and that the model inlet diameter is 1 ft.

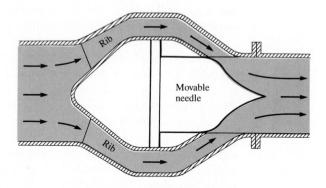

Solution   Again we use the Reynolds-number criterion. Therefore

$$Re_m = Re_p$$

$$\frac{V_m L_m}{\nu_m} = \frac{V_p L_p}{\nu_p}$$

$$\frac{V_m}{V_\rho} = \frac{L_p}{L_m} \frac{\nu_m}{\nu_p}$$

However, we are interested in the total rate of flow. Therefore, if we multiply both sides of the foregoing equation by the area ratio A_m/A_p, the resulting product on the left-hand side yields the desired discharge ratio:

$$\frac{A_m}{A_p} \frac{V_m}{V_p} = \frac{A_m}{A_p} \frac{L_p}{L_m} \frac{\nu_m}{\nu_p}$$

or

$$\frac{Q_m}{Q_p} = \frac{A_m}{A_p} \frac{L_p}{L_m} \frac{\nu_m}{\nu_p}$$

Since $A_m/A_p = L_r^2 = (L_m/L_p)^2$, we obtain

$$\frac{Q_m}{Q_p} = \frac{L_m^2}{L_p^2} \frac{L_p}{L_m} \frac{\nu_m}{\nu_p}$$

But $\nu_m = \nu_p$, or $\nu_m/\nu_p = 1$, because water is used for both model and prototype, so

$$\frac{Q_m}{Q_p} = \frac{L_m}{L_p}$$

Finally, $Q_m = \frac{1}{6} Q_p = \frac{1}{6} \times 700 = 116.7$ cfs   ◀

Note: This discharge is very large indeed, and in fact it serves to emphasize that very few model studies are made that completely satisfy the Reynolds-number criterion. This subject will be discussed further in the next sections.

8.8 Significance of the Pressure Coefficient

In the foregoing examples it was demonstrated that dynamic similarity between model and prototype exists when the significant dimensionless parameters are the same in the model and the prototype. Since none of the parameters we have considered (Re, Fr, M, or W) explicitly includes Δp, one may wonder how Δp in the model is related to Δp in the prototype. It is related by means of the pressure coefficient. If we refer to Eq. (8.10),

$$\frac{\Delta p}{\frac{1}{2}\rho V^2} = f\left(\frac{VL\rho}{\mu}, \frac{V}{c}, \frac{\rho L V^2}{\sigma}, \frac{V}{\sqrt{gL}}\right)$$

we see that the pressure coefficient, $\Delta p/\frac{1}{2}\rho V^2$, is a function of the basic parameters of similitude. Consequently, when dynamic similarity exists—that is, when the significant dimensionless numbers are the same in the model and the prototype—then it follows that the pressure coefficient will also be the same in the model and the prototype. Thus, when we have dynamic similarity,

$$C_{p,\text{model}} = C_{p,\text{prototype}} \tag{8.17}$$

or
$$\frac{\Delta p_m}{\frac{1}{2}\rho_m V_m^2} = \frac{\Delta p_p}{\frac{1}{2}\rho_p V_p^2} \tag{8.18}$$

The pressure coefficient can be used like any of the other basic parameters for model analysis. It is useful not only in relating pressure changes in the model to those in the prototype but also in relating total forces in model and prototype. The latter is accomplished by multiplying the pressure ratio by the area ratio.

EXAMPLE 8.5   For the given conditions of Example 8.3, if the pressure difference between two points on the surface of the model blimp is measured to be 17.8 kPa, what will be the pressure difference in the prototype for dynamically similar conditions?

Solution   From Eq. (8.18), the prototype pressure difference can be given as

$$\Delta p_p = \Delta p_m \frac{\rho_p}{\rho_m} \frac{V_p^2}{V_m^2}$$

However, $\qquad\qquad\qquad\qquad\rho_p = \rho_m$

and using the Reynolds similarity criterion we get

$$\frac{V_p}{V_m} = \frac{L_m}{L_p}$$

Therefore, $\qquad\Delta p_p = \Delta p_m \left(\frac{1}{10}\right)^2 = \frac{1}{100}\Delta p_m = \frac{17{,}800}{100} = 178 \text{ Pa}$ ◄

EXAMPLE 8.6 For the given conditions of Example 8.3, if the drag force on the model blimp is measured to be 1530 N, what corresponding force could be expected in the prototype?

Solution Again using Eq. (8.18), we can write

$$\frac{\Delta p_p}{\Delta p_m} = \frac{\rho_p}{\rho_m}\frac{V_p^2}{V_m^2}$$

Also, $\qquad\qquad\qquad\qquad \text{Re}_p = \text{Re}_m$

$$\frac{V_p L_p}{\nu_p} = \frac{V_m L_m}{\nu_m}$$

or

$$\frac{V_p}{V_m} = \frac{\nu_p}{\nu_m}\frac{L_m}{L_p}$$

However, for this example, $\rho_p = \rho_m$ and $\nu_p = \nu_m$. Thus when we let $\rho_p/\rho_m = 1$, we obtain from the first equation in this example the following:

$$\frac{\Delta p_p}{\Delta p_m} = \frac{V_p^2}{V_m^2}$$

From the Reynolds-number relationship, when we set $\nu_p/\nu_m = 1$, we get

$$\frac{V_p}{V_m} = \frac{L_m}{L_p} \qquad \frac{V_p^2}{V_m^2} = \frac{L_m^2}{L_p^2}$$

Thus when we substitute L_m^2/L_p^2 for V_p^2/V_m^2, we get

$$\frac{\Delta p_p}{\Delta p_m} = \frac{L_m^2}{L_p^2}$$

Now, multiplying both sides of the foregoing equation by the area ratio A_p/A_m, which is the same as L_p^2/L_m^2, we obtain

$$\frac{\Delta p_p}{\Delta p_m}\frac{L_p^2}{L_m^2} = \frac{L_m^2}{L_p^2}\frac{L_p^2}{L_m^2}$$

The left-hand side of this equation is the ratio of the prototype force to the model force, and the right-hand side is unity. Hence we arrive at the interesting result that the model force is the same as the prototype force. When we use the Reynolds-number criterion and use the same fluid for model and prototype, the forces on the model will always be the same as the forces on the prototype. Consequently, for this example the force on the prototype blimp would be 1530 N. ◄

8.9 Approximate Similitude at High Reynolds Numbers

The primary justification for model tests is that it is more economical to get answers needed for engineering design by such tests than by any other means. However, as revealed by Examples 8.3, 8.4, and 8.6, true similarity according to the Reynolds-number criterion yields quantities for the model that would require very costly model setups. Consider the size and power required for wind-tunnel tests of the blimp in Example 8.3. The wind tunnel would probably require a section at least 2 m by 2 m to accommodate the model blimp. With a 100 m/s airspeed in the tunnel, the power required for producing continuously a stream of air of this size and velocity is in the order of 4 MW. Such a test is not prohibitive, but it is very expensive. It is also conceivable that the 100 m/s airspeed would introduce Mach-number effects not encountered with the prototype, thus generating concern over the validity of the model data. Furthermore, a force of 1530 N is indeed larger than that usually associated with model tests. Therefore, especially in the study of problems involving non-free-surface flows, it is desirable to perform model tests in such a way that large magnitudes of forces or pressures are not encountered.

For many cases, it is possible to obtain all the needed information from abbreviated tests. Often the Reynolds-number effect (relative viscous effect) either becomes insignificant at high Reynolds numbers or becomes independent of the Reynolds number. The point where testing can be stopped often can be detected by inspection of a graph of the pressure coefficient C_p versus the Reynolds number Re. Such a graph for a venturi meter in a pipe is shown in Fig. 8.6. In this meter, Δp is the pressure difference between the points shown, and V is the velocity in the restricted section of the venturi meter. Here it is seen that viscous forces affect the value of C_p below a Reynolds number of approximately 50,000. However, for higher Reynolds numbers, C_p is virtually constant. Physically this means that at low Reynolds numbers (relatively high viscous forces), a significant part of the change in pressure comes from viscous resistance, and the remainder comes from the inertial reaction (change in kinetic energy) of the fluid as it passes through the venturi meter. However, with high Reynolds numbers (equivalent to either small viscosity or a large product of V, D, and ρ), the viscous resistance is negligible compared with that required to overcome the inertial

FIGURE 8.6

Cₚ for a venturi meter as a function of the Reynolds number.

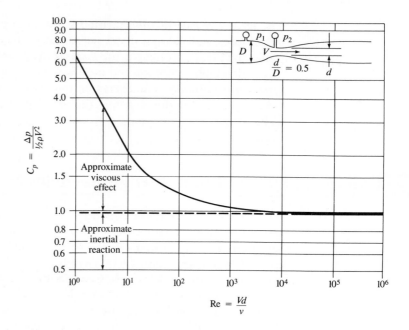

reaction of the fluid. Since the ratio of Δp to the inertial reaction does not change (constant C_p) for high Reynolds numbers, there is no need to carry out tests at higher Reynolds numbers. This is true in general, so long as the flow pattern does not change with the Reynolds number.

In a practical sense, whoever is in charge of the model test will try to predict from previous works approximately what maximum Reynolds number will be needed to reach the point of insignificant Reynolds-number effect and then will design the model accordingly. After a series of tests has been made on the model, C_p versus Re will be plotted to see whether the range of constant C_p has indeed been reached. If so, then no more data are needed to predict the prototype performance. However, if C_p has not reached a constant value, the test program has to be expanded or results extrapolated. Thus the results of some model tests can be used to predict prototype performance even though the Reynolds numbers are not the same for the model and the prototype. This is especially valid for angular-shaped bodies, such as model buildings, tested in wind tunnels.

EXAMPLE 8.7 Tests were made by Roberson and Crowe (14) on model-building shapes such as that shown in part (*a*) in the figure to determine the effects of free-stream turbulence and angle of incidence on the surface pressure distribution. Part (*b*) in the figure is an example of the temporal-mean pressure distribution obtained at a relative elevation on the building of. $z/H = 0.7$ for an angle of incidence of α of 8° and a wind speed V of 20 m/s. The air temperature in the wind tunnel was 20°C. If it is assumed that C_p does not change with higher Re's, what would be the maximum and the minimum temporal-mean pressures on a similar building 100 m high from a 50-m/s wind (10°C) with the same angle of incidence? If the instantaneous magnitude of the lowest pressure is three times more negative than the lowest temporal-mean pressure (this is due to turbulence), and if the pressure inside the building is the same as the mean pressure on the leeward wall, then what will be the total wind force in the lowest pressure zone on a window that is 2 m by 2 m?

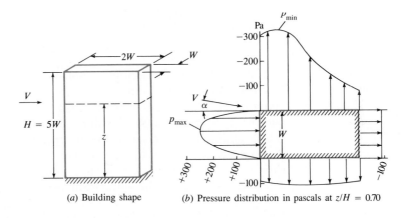

(a) Building shape   (b) Pressure distribution in pascals at $z/H = 0.70$

Solution   We are assuming the same C_p for the model as for the full-scale prototype building. Therefore, we have

$$C_{p_m} = C_{p_p}$$

$$\frac{\Delta p_m}{\rho_m V_m^2/2} = \frac{\Delta p_p}{\rho_p V_p^2/2}$$

$$\frac{p_m - p_0}{\rho_m V_m^2/2} = \frac{p_p - p_0}{\rho_p V_p^2/2}$$

However, our reference pressure p_0 is zero gage, so we can solve for p_p:

$$p_p = \frac{\rho_p}{\rho_m} \frac{V_p^2}{V_m^2} p_m$$

We obtain

$$p_{p_{max}} = \frac{1.25}{1.20} \left(\frac{50}{20}\right)^2 p_{m_{max}}$$

$$= \frac{1.25}{1.20} \times 6.25 \times 250 = 1.63 \text{ kPa}$$

In a similar manner $p_{p_{\min}} = -1.95$ kPa, and the pressure on the leeward wall (and inside the building) is

$$p_{p_{\text{inside}}} = -0.58 \text{ kPa}$$

The force on the window is the product of the difference of pressure across the window and the area. Here the lowest outside pressure is assumed to be $3 \times p_{\min}$, so

$$F_{\text{on window}} = (p_{\text{inside}} - p_{\text{outside}}) \times A$$
$$= \{-0.58 \text{ kPa} - [3 \times (-1.95 \text{ kPa})]\} \times 4 \text{ m}^2 = 21.1 \text{ kN} \quad \blacktriangleleft$$

Before leaving the section on non-free-surface flows, we should note that Mach-number effects (compressibility) usually become significant for Mach numbers exceeding 0.3. It is not always possible to tell which parameter—Mach number or Reynolds number—will be the most significant in a certain situation. Which similitude parameter is chosen depends a great deal on what information the engineer is seeking. If the engineer is interested in the viscous motion of fluid near a wall in shock-free supersonic flow, then the Reynolds number should be selected as the significant similitude parameter. However, if the shock-wave pattern over a body is of interest, then the Mach number should be selected as the similitude parameter. The Mach number and its significance are discussed in more detail in Chapter 12.

8.10   Free-Surface Model Studies

Spillway Models

The major influence, besides the spillway geometry itself, on the flow of water over a spillway is the action of gravity. Hence the Froude-similarity criterion is used for such model studies. It can be appreciated that for large spillways with depths of water in the order of 3 or 4 m and velocities in the order of 10 m/s or more, the Reynolds number is very large. At high values of the Reynolds number, the relative viscous forces are often independent of the Reynolds number, as already noted in the foregoing section (Sec. 8.9). However, if the reduced-scale model is made too small, the viscous forces as well as the surface-tension forces would have a larger relative effect on the flow in the model than in the prototype. Therefore, in practice, spillway models are made large enough so that the viscous effects have about the same relative effect in the model as in the prototype (the viscous effects are nearly independent of the Reynolds number). Then the Froude number is the significant similarity parameter. Most model spillways

FIGURE 8.7

Comprehensive model for Hell's Canyon Dam. Tests were made at the Albrook Hydraulic Laboratory, Washington State University.

are made at least 1 m high, and for precise studies, such as calibration of individual spillway bays, it is not uncommon to design and construct model spillway sections that are 2 or 3 m high. Figures 8.7 and 8.8 show a comprehensive model and spillway model for Hell's Canyon Dam in Idaho.

FIGURE 8.8

Spillway model for Hell's Canyon Dam. Tests were made at the Albrook Hydraulic Laboratory, Washington State University.

EXAMPLE 8.8 A $\frac{1}{49}$ scale model of a proposed dam is used to predict proto-type flow conditions. If the design flood discharge over the spillway is 15,000 m³/s, what water flow rate should be established in the model to simulate this flow? If a velocity of 1.2 m/s is measured at a point in the model, what is the velocity at a corresponding point in the prototype?

Solution   The Froude-number criterion will be used. Therefore,

$$\text{Fr}_m = \text{Fr}_p$$

$$\frac{V_m}{\sqrt{g_m L_m}} = \frac{V_p}{\sqrt{g_p L_p}}$$

However,

$$g_m = g_p$$

Thus we have

$$\frac{V_m}{V_p} = \sqrt{\frac{L_m}{L_p}}$$

We multiply both sides of this equation by the area ratio A_m/A_p:

$$\frac{V_m}{V_p}\frac{A_m}{A_p} = \frac{A_m}{A_p}\sqrt{\frac{L_m}{L_p}}$$

The left-hand side of the equation is the discharge ratio, and $A_m/A_p = L_m^2/L_p^2$. Hence we obtain

$$\frac{Q_m}{Q_p} = \left(\frac{L_m}{L_p}\right)^{5/2}$$

$$Q_m = Q_p\left(\frac{1}{49}\right)^{5/2} = 15,000\,\frac{1}{16,800} = 0.89 \text{ m}^3/\text{s} \qquad \blacktriangleleft$$

From the fourth equation in this example,

$$\frac{V_m}{V_p} = \sqrt{\frac{L_m}{L_p}}$$

Consequently, $V_p = V_m\sqrt{\dfrac{L_p}{L_m}} = 1.2 \times 7 = 8.4 \text{ m/s} \qquad \blacktriangleleft$

At the given point in the prototype, we would have a velocity of 8.4 m/s.

Ship Model Tests

The resistance that the propulsion system of the ship must overcome is the sum of the wave resistance and the surface resistance of the hull. The wave resistance is a free-surface, or Froude-number, phenomenon, and the hull resistance is a viscous, or Reynolds-number, phenomenon. Because both wave and viscous effects contribute significantly to the overall resistance, it would appear that both the Froude and Reynolds criteria should be used. However, it is impossible to satisfy both if the model liquid is water (the only practical test liquid), because the Reynolds-number criterion dictates a higher velocity for the model than for the prototype [equal to $V_p(L_p/L_m)$], whereas the Froude-number criterion dictates a lower velocity for the model [equal to $V_p(\sqrt{L_m}/\sqrt{L_p})$]. To circumvent such a dilemma, the procedure is to model for the phenomenon that is the most difficult to predict analytically and to account for the other resistance by analytical means. Since the wave resistance is the most difficult problem, the model is operated according to the Froude-number criterion and the hull resistance is accounted for analytically.

To illustrate how the test results and the analytical solutions for surface resistance are merged to yield design data, we note the necessary sequential steps:

1. First make model tests according to the Froude-number criterion, and the total model resistance is measured. This total model resistance will be equal to the wave resistance plus the surface resistance of the hull of the model.

2. Then estimate the surface resistance of the model by analytical calculations.

3. Subtract the surface resistance calculated in step 2 from the total model resistance of step 1 to yield the wave resistance of the model.

4. Using the Froude-number criterion, scale the wave resistance of the model up to yield the wave resistance of the prototype.

5. Then estimate the surface resistance of the hull of the prototype by analytical means.

6. The sum of the wave resistance of the prototype from step 4 and the surface resistance of the prototype from step 5 yields the total prototype resistance, or drag.

Problems

8.1 Determine which of the following equations are dimensionally homogeneous:

a.
$$Q = \tfrac{2}{3} C L \sqrt{2g}\, H^{3/2}$$

where Q is discharge, C is pure number, L is length, g is acceleration due to gravity, and H is head.

b.
$$V = \frac{1.49}{n} R^{2/3} S^{1/2}$$

where V is velocity, n is length to the one-sixth power, R is length, and S is slope.

c.
$$h_f = f \frac{L}{D} \frac{V^2}{2g}$$

where h_f is head loss, f is a dimensionless resistance coefficient, L is length, D is diameter, V is velocity, and g is acceleration due to gravity.

d.
$$D = \frac{0.074}{\text{Re}^{0.2}} \frac{Bx\rho V^2}{2}$$

where D is drag force, Re is Vx/ν, B is width, x is length, ρ is mass density, and V is velocity.

8.2 Determine the dimensions of the following variables and combinations of variables in terms of the length, force, and time system of units and in terms of the length, mass, and time system of units. (*Hint:* Convert F to M by Newton's second law.)

a. T (torque)
b. $\rho V^2/2$, where V is velocity and ρ is mass density
c. $\sqrt{\tau/\rho}$, where τ is shear stress
d. Q/ND^3, where Q is discharge, D is diameter, and N is angular speed of a pump

8.3 It takes a certain length of time for the liquid level in a tank of diameter D to drop from position h_1 to position h_2 as the tank is being drained through an orifice of diameter d at the bottom. Determine the dimensionless parameters that apply to this problem. Assume that the fluid is nonviscous.

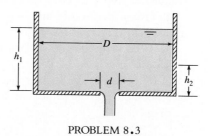

PROBLEM 8.3

8.4 The motion of small-amplitude surface waves depends on the wave amplitude h, the surface tension of the fluid σ, the specific weight of the fluid γ, and the acceleration due to gravity g. Use dimensional analysis to find a nondimensional functional form for the wave celerity (velocity) V.

8.5 The maximum rise of a liquid in a small capillary tube is a function of the diameter of the tube, the surface tension, and the specific weight of the liquid. What are the significant dimensionless parameters for the problem?

8.6 For very low velocities it is known that the drag force F_D of a small sphere is a function solely of the velocity V of flow past the sphere, the diameter d of the sphere, and the viscosity μ of the fluid. Determine the dimensionless relationship involving these variables.

8.7 The drag force F_D on a very rough sphere held inside a pipe in which liquid is flowing is a function of D, ρ, μ, V, and k. D is the diameter of the sphere, ρ is mass density, μ is viscosity, V is the velocity of the liquid, and k is the height of the roughness elements on the sphere. By dimensional analysis, determine the relevant dimensionless parameters for this problem.

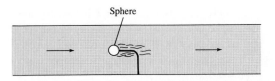

PROBLEM 8.7

8.8 Consider steady viscous flow through a small horizontal tube. For this type of flow, the pressure gradient along the tube, $\Delta p/\Delta \ell$, should be a function of the viscosity μ, the mean velocity V, and the diameter D. By dimensional analysis, derive a formula relating these variables.

8.9 It is known that the pressure developed by a centrifugal pump, Δp, is a function of the diameter D of the impeller, the speed of rotation n, the discharge Q, and the fluid density ρ. By dimensional analysis, determine the dimensionless parameters relating these variables.

8.10 The velocity V of very small ripples on a liquid surface is a function of the ripple length L, density ρ, and surface tension σ of the liquid. By dimensional analysis, derive an expression for V.

8.11 By dimensional analysis, develop the relationship between the pressure gradient in the radial direction (dp/dr) for a rotating tank of liquid (vertical axis of rotation) and the significant variables ρ, ω, and r, where ρ is mass density, ω is angular velocity, and r is the radial distance from the axis of rotation.

8.12 A smooth circular plate is positioned a distance S away from a smooth boundary as shown in the figure. Oil with viscosity μ fills the space between the plate and the boundary. It should be obvious that the torque required to rotate the plate as shown will be a function of μ, ω, S, and D, where ω is the angular velocity in radians per second. Determine the dimensionless parameters that would be involved if you were to correlate results by experimental means only.

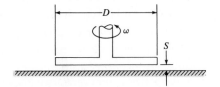

PROBLEM 8.12

8.13 A general study is to be made of the height of rise of liquid in a capillary tube as a function of time after the start of a test. Other significant variables include surface tension, mass density, specific weight, viscosity, and diameter of the tube. Determine the dimensionless parameters that apply to the problem.

8.14 The flow of a gas–particle mixture in a tube gives rise to erosion of the tube wall. That is, the collisions of particles with the wall result in removal of material from the wall. Material properties that may be of significance in this problem are modulus of elasticity E, ultimate strength σ, and Brinell hardness number Br. Particle and flow properties that may be important are velocity V, particle diameter d, particle mass flow rate $\dot{M}_p$, and tube diameter D. The Brinell hardness number is dimensionless. Perform a dimensional analysis for the erosion rate e, which has units of kg/m^2 · s.

8.15 By dimensional analysis, determine the dimensionless relationship for the change in pressure that occurs when water or oil flows through a horizontal pipe with an abrupt contraction as shown.

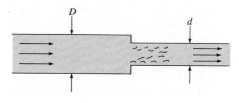

PROBLEM 8.15

8.16 For flow through a transition section from a large pipe to a small pipe, are the viscous forces relatively *large* or *small* compared to the inertial forces when the Reynolds number is very large?

8.17 Water with a kinematic viscosity of 10^{-6} m^2/s flows through a 20-cm pipe. What would the velocity of water have to be for the water flow to be dynamically similar to oil ($\nu = 10^{-5}$ m^2/s) flowing through the same pipe at a velocity of 3 m/s?

8.18 Oil with a kinematic viscosity of 4×10^{-6} m^2/s flows through a smooth pipe 20 cm in diameter at 5 m/s. What velocity should water have at 20°C in a smooth pipe 5 cm in diameter to be dynamically similar?

8.19 A large venturi meter is calibrated by means of a $\frac{1}{10}$ scale model using the prototype liquid. What is the discharge ratio Q_m/Q_p for dynamic similarity? If a pressure difference of 300 kPa is measured across ports in the model for a given discharge, what pressure difference will occur between similar ports in the prototype for dynamically similar conditions?

8.20 It is known that for flow past cylinders, vortices are shed alternately from one side of the cylinder and from the other. If the frequency of shedding n is a function of the approach velocity V, the diameter d of the cylinder, the mass density ρ, and the viscosity μ, what are the dimensionless parameters for this phenomenon?

8.21 A $\frac{1}{4}$ scale model of an experimental bathosphere that will operate at great depths is to be tested to determine its drag characteristic by towing it behind a submarine. For true similitude, what should be the towing speed relative to the speed of the prototype?

8.22 A spherical balloon that is to be used in air at 60°F is tested by towing a $\frac{1}{3}$ scale model in a lake. The model is 1 ft in diameter, and a drag of 20 lbf is measured when the model is being towed in deep water at 5 ft/s. What drag (in pounds force and newtons) can be expected for the prototype in air under dynamically similar conditions? Assume that the water temperature is 60°F.

8.23 A water tunnel is used to model the flow of high-pressure, high-temperature steam past a heat exchanger element (a circular cylinder). The actual element has a diameter of 1 in., and the steam velocity is 100 ft/s. The kinematic viscosity of the steam is $1.5 \cdot 10^{-3}$ ft²/s. The water in the tunnel is 40°F, and the velocity in the tunnel is 5 ft/s. What should the diameter of the model be to ensure dynamic similitude?

8.24 An airplane travels in air ($p = 100$ kPa, $T = 10°C$) at 150 m/s. If a $\frac{1}{5}$ scale model of the plane is tested in a wind tunnel at 25°C, what must be the density of the air in the tunnel so that both the Reynolds-number and the Mach-number criteria are satisfied? For E_v take the adiabatic relation as given in Chapter 2. *Note:* The dynamic viscosity is independent of pressure.

8.25 Flow in a given pipe is to be tested with air and then with water. Assume that the velocities (V_A and V_W) are such that the flow with air is dynamically similar to the flow with water. Then for this condition, the magnitude of the ratio of the velocities, V_A/V_W, will be a) less than unity, b) equal to unity, c) greater than unity.

8.26 A smooth pipe designed to carry crude oil (diameter = 48″, $\rho = 1.75$ slugs/ft³, and $\mu = 4 \times 10^{-4}$ lb-s/ft²) is to be modeled with a smooth pipe 4 in. in diameter carrying water ($T = 60°F$). If the mean velocity in the prototype is 20 ft/s, what should be the mean velocity of water in the model to ensure dynamically similar conditions?

8.27 Water at 20°C flows at 3 m/s in a 20-cm smooth pipe. What must be the velocity of air (standard atmospheric pressure) at 30°C in a 10-cm smooth pipe for the two flows to be dynamically similar? If the pressure change for the water flow was measured as 2.0 kPa over a given length of pipe, what pressure change should you get for the air flow with the same length of pipe?

8.28 Using the Reynolds-number criterion, a 1:1 scale model of a torpedo is tested in a wind tunnel. If the velocity of the torpedo in water is 6 m/s, what should be the air velocity (standard atmospheric pressure) in the wind tunnel? The temperature for both tests is 10°C.

8.29 Colonization of the moon will require an improved understanding of fluid flow under reduced gravitational forces. The gravitational force on the moon is $\frac{1}{5}$ that on the surface of the earth. An engineer is designing a model experiment for flow in a conduit on the moon. The important scaling parameters are the Froude number and the Reynolds number. The model will be full-scale. The kinematic viscosity of the fluid to be used on the moon is 10^{-5} m²/s. What should be the kinematic viscosity of the fluid to be used for the model on earth?

8.30 A drying tower at an industrial site is 10 m in diameter. The air inside the tower has a kinematic viscosity of 4×10^{-5} m²/s and enters at 10 m/s. A $\frac{1}{8}$ scale model of this tower is fabricated to operate with water that has a kinematic viscosity of 10^{-6} m²/s. What should the entry velocity of the water be to achieve Reynolds-number scaling?

8.31 A discharge meter to be used in a 40-cm pipeline carrying oil ($\nu = 10^{-5}$ m²/s, $\rho = 860$ kg/m³) is to be calibrated by means of a model ($\frac{1}{4}$ scale) carrying water ($T = 20°C$ and standard atmospheric pressure). If the model is operated with a velocity of 1 m/s and a meter coefficient is determined, at what velocity in the prototype may we be certain that the prototype meter coefficient will have the same magnitude as that in the model? For the given conditions, if the pressure difference in the model was measured as 3.0 kPa, what pressure difference would you expect for the discharge meter in the oil pipeline?

8.32 Water at 10°C flowing through a rough pipe 10 cm in diameter is to be simulated by air (20°C) flowing through the same pipe. If the velocity of the water is 1.5 m/s, what will the air velocity have to be to achieve dynamic similarity? Assume the absolute air pressure in the pipe to be 150 kPa. If the pressure difference between two sections of the pipe during air flow was measured as 780 Pa, what pressure difference occurs between these two sections when water is flowing under dynamically similar conditions?

8.33 The "noisemaker" B is towed behind the minesweeper A to set off enemy acoustic mines such as that shown at C. The drag force of the "noisemaker" is to be studied in a water tunnel at a $\frac{1}{4}$ scale (the model is $\frac{1}{4}$ the size of the full scale). If the full-scale towing speed is 3 m/s, what should be the water velocity in the water tunnel for the two tests to be exactly similar? What will be the prototype drag force if the model drag force is found to be 868 N? Assume that sea water at the same temperature is used in both the full-scale and the model tests.

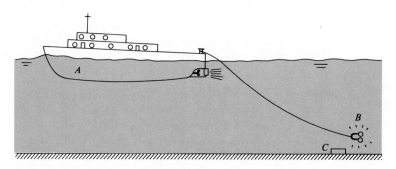

PROBLEM 8.33

8.34 An experiment is being designed to measure aerodynamic forces on a building. The model is a $\frac{1}{100}$ scale replica of the prototype. The wind velocity on the prototype is 30 ft/s, and the density is 0.0024 slugs/ft^3. The maximum velocity in the wind tunnel is 300 ft/s. The viscosity of the air flowing for the model and the prototype is the same. Find the density needed in the wind tunnel for dynamic similarity. A force of 50 lbf is measured on the model. What will the force be on the prototype?

8.35 A 60-cm valve is designed for control of flow in a petroleum pipeline. A $\frac{1}{2}$ scale model of the full-size valve is to be tested with water in the laboratory. If the prototype flow rate is to be 0.5 m³/s, what flow rate should be established in the laboratory test for dynamic similitude to be established? Also, if the pressure coefficient C_p in the model is found to be 1.07, what will be the corresponding C_p in the full-scale valve? The relevant fluid properties for the petroleum are S = 0.82 and $\mu = 3 \times 10^{-3}$ N · s/m². The viscosity of water is 10^{-3} N · s/m².

8.36 The moment acting on a submarine rudder is studied by a $\frac{1}{60}$ scale model. If the test is made in a water tunnel and if the moment measured on the model is 3 m · N when the fresh-water speed in the tunnel is 10 m/s, what are the corresponding moment and speed for the prototype? Assume the prototype operates in sea water. Assume $T = 10°C$ for both the fresh water and the sea water.

8.37 A model hydrofoil is tested in a water tunnel. For a given angle of attack, the lift of the hydrofoil is measured to be 20 kN when the water velocity is 20 m/s in the tunnel. If the prototype hydrofoil is to be twice the size of the model, what lift force would be expected for the prototype for dynamically similar conditions? Assume a water temperature of 20°C for both model and prototype.

8.38 A $\frac{1}{10}$ scale model of an automobile is tested in a pressurized wind tunnel. The test is to simulate the automobile traveling at 80 km/h in air at atmospheric pressure and 25°C. The wind tunnel operates with air at 25°C. At what pressure in the test section must the tunnel operate to have the same Mach and Reynolds numbers? The speed of sound in air at 25°C is 345 m/s.

8.39 If the tunnel in Prob. 8.38 were to operate at atmospheric pressure and 25°C, what speed would be needed to achieve the same Reynolds number for the prototype? At this speed, would you conclude that Mach-number effects were important?

8.40 An important parameter in rarefied gas dynamics is the ratio M/Re. If this ratio exceeds unity, the flow is rarefied. A spherical satellite 2 ft in diameter reenters the earth's atmosphere at 24,000 mph, where the pressure and temperature of the air are 22 psfa and −67°F. Can the flow around the satellite be classified as rarefied? The dynamic viscosity at −67°F is 3.0×10^{-7} lbf-s/ft², and the speed of sound is 975 ft/s.

8.41 Experimental studies have shown that the condition for breakup of a droplet in a gas stream is

$$W/Re^{1/2} = 0.5$$

where Re is the Reynolds number and W is the Weber number based on the droplet diameter. What diameter water droplet would break up in a 30-m/s air stream at 20°C and standard atmospheric pressure? The surface tension of water is $7.3 \cdot 10^{-2}$ N/m.

8.42 The critical Weber number for breakup of a liquid droplet is 6.0 based on the droplet diameter. The surface tension of heptane is 0.02 N/m. When a spray of heptane is atomized by discharging at 30 m/s into air at atmospheric pressure and 100°C, what is the expected size of the droplets?

8.43 Water is sprayed from a nozzle at 15 m/s into air at atmospheric pressure and 20°C. Estimate the size of the droplets produced if the Weber number for breakup is 6.0 based on the droplet diameter.

8.44 Determine the relationship between the kinematic viscosity ratio v_m/v_p and the scale ratio if both the Reynolds-number and the Froude-number criteria are to be satisfied in a given model test.

8.45 A hydraulic model, $\frac{1}{10}$ scale, is built to simulate the flow conditions of a spillway of a dam. For a particular run, the waves downstream were observed to be 10 cm high. How high would be similar waves on the full-scale dam operating under the same conditions? If the wave period in the model is 1 s, what would the wave period in the prototype be?

8.46 The scale ratio between a model dam and its prototype is $\frac{1}{16}$. In the model test, the velocity of flow near the crest of the spillway was measured to be 2.0 m/s. What is the corresponding prototype velocity? If the model discharge is 0.10 m³/s, what is the prototype discharge?

8.47 A seaplane model is built at a $\frac{1}{10}$ scale. To simulate takeoff conditions at 110 km/h, what should be the corresponding model speed?

8.48 If the scale ratio between a model spillway and its prototype is $\frac{1}{25}$, what velocity and discharge ratio will prevail between model and prototype? If the prototype discharge is 4000 m³/s, what is the model discharge?

8.49 The depth and velocity at a point in a river are measured to be 25 ft and 15 ft/s, respectively. If a $\frac{1}{64}$ scale model of this river is constructed and the model is operated under dynamically similar conditions to simulate the free-surface conditions, then what velocity and depth can be expected in the model at the corresponding point?

8.50   A $\frac{1}{25}$ scale model of a spillway is tested in a laboratory. If the model velocity and discharge are 7.87 ft/s and 3.53 cfs, respectively, what are the corresponding values for the prototype?

8.51   Flow around a bridge pier is studied using a model at $\frac{1}{10}$ scale. When the velocity in the model is 0.9 m/s, the standing wave at the pier nose is observed to be 2.5 cm in height. What are the corresponding values of velocity and wave height in the prototype?

8.52   A $\frac{1}{25}$ scale model of a spillway is tested. The discharge in the model is 0.1 m³/s. To what prototype discharge does this correspond? If it takes 2 min for a particle to float from one point to another in the model, how long would it take a similar particle to traverse the corresponding path in the prototype?

8.53   A tidal estuary is to be modeled at $\frac{1}{300}$ scale. In the actual estuary, the maximum water velocity is expected to be 3 m/s and the tidal period is approximately 12.5 h. What corresponding velocity and period would be observed in the model?

8.54   The maximum wave force on a $\frac{1}{36}$ model sea wall was found to be 80 N. For a corresponding wave in the full-scale wall, what full-scale force would you expect? Assume fresh water is used in the model study. Assume $T = 10°C$ for both model and prototype water.

8.55   A model of a spillway is to be built at $\frac{1}{20}$ scale. If the prototype has a discharge of 200 m³/s, what must be the water discharge in the model to ensure dynamic similarity? The total force on part of the model is found to be 20 N. To what prototype force does this correspond?

8.56   A newly designed dam is to be modeled in the laboratory. The prime objective of the general model study is to determine the adequacy of the spillway design and to observe the water velocities, elevations, and pressures at critical points of the structure. The reach of the river to be modeled is 1200 m long, the width of the dam (also the maximum width of the reservoir upstream) is to be 300 m, and the maximum flood discharge to be modeled is 5000 m³/s. The maximum laboratory discharge is limited to 0.90 m³/s, and the floor space available for the model construction is 50 m long and 20 m wide. Determine the largest feasible scale ratio (model/prototype) for such a study.

8.57   A ship model 4 ft long is tested in a towing tank at a speed that will produce waves that are dynamically similar to those observed around the prototype. The test speed is 3 ft/s. What should be the prototype speed, given that the prototype length is 150 ft? Assume both the model and the prototype are to operate in fresh water.

8.58   The wave resistance of a model of a ship at $\frac{1}{25}$ scale is 2 lbf at a model speed of 5 ft/s. What are the corresponding velocity and wave resistance of the prototype?

8.59   The wave resistance of a model of a ship at $\frac{1}{16}$ scale ratio is 10 N at a speed of 3 m/s. What are the corresponding velocity and wave resistance of the prototype? Assume both the model and the prototype are to operate in fresh water.

8.60   A $\frac{1}{20}$ scale model building that is rectangular in plan view and is three times as high as it is wide is tested in a wind tunnel. If the drag of the model in the wind tunnel is measured to be 200 N for a wind speed of 20 m/s, then the prototype building in a 40-m/s wind (same temperature) should have a drag of about a) 40 kN, b) 80 kN, c) 230 kN, d) 320 kN.

8.61   A model of a high-rise office building at $\frac{1}{250}$ scale is tested in a wind tunnel to estimate the pressures and forces on the full-scale structure. The wind-tunnel air speed is 20 m/s at 20°C, and the full-scale structure is expected to withstand winds of 160 km/h (10°C). If the extreme values of the pressure coefficient are found to be 1.0, -2.7, and -0.8 on the windward wall, side wall, and leeward wall of the model, respectively, what corresponding full-scale pressures could be expected for the design wind? If the lateral wind force (wind force on building normal to wind direction) was measured as 20 N in the model, what lateral force might be expected in the prototype in the 160 km/h wind?

References

1. Allen, J. *Scale Models in Hydraulic Engineering.* Longmans, Green & Co., London, 1952.

2. Baals, D. D., and W. R. Corliss. *Wind Tunnels of NASA,* U.S. Govt. Printing Office, Washington, D.C., 1981.

3. Bain, D. C., P. J. Baker, and M. J. Rowat. *Wind Tunnels, An Aid to Engineering Structure Design.* British Hydromechanics Research Association, Cranfield, England, 1971.

4. Buckingham, E. "Model Experiments and the Forms of Empirical Equations." *Trans. ASME,* 37 (1915), 263.

5. Freeman, John R. (ed.). *Hydraulic Laboratory Practice.* American Society of Mechanical Engineers, 1929.

6. Hickox, G. H. "Hydraulic Models." In C.V. Davis (ed.), *Handbook of Applied Hydraulics,* 2nd ed. McGraw-Hill, New York, 1952.

7. Ipsen, D. C. *Units, Dimensions and Dimensionless Numbers.* McGraw-Hill, New York, 1960.

8. Kale, S., et al. "An Experimental Study of Single-Particle Aerodynamics." *Proc. of First Nat. Congress on Fluid Dynamics,* Cincinnati, Ohio, July 1988.

9. Kilgore, R. A., and D. A. Dress. "The Application of Cryogenics to High Reynolds-Number Testing in Wind Tunnels," Part 2. Development and application of the cryogenic wind tunnel concept, *Cryogenics,* Vol. 24, no. 9, September 1984.

10. Langhaar, Henry L. *Dimensional Analysis and Theory of Models.* John Wiley, New York, 1951.

11. "Manual of Engineering Practice No. 25." In *Hydraulic Models.* American Society of Civil Engineers, 1942.

12. Potthoff, J. "Luftwiderstand und Auftrieb Moderner Kraftfahrzeuge." Paper No. 12, *Proc. 1st Symp. Road Vehicle Aerodyn.,* London, 1969.

13. Rae, W. H., Jr., and A. Pope. *Low-Speed Wind Tunnel Testing,* John Wiley, New York, 1984.

14. Roberson, J. A., and C. T. Crowe. "Pressure Distribution on Model Buildings at Small Angles of Attack in Turbulent Flow." In *Proc. 3rd U.S. Natl. Conf. on Wind Engineering Research.* University of Florida, Gainesville, 1978.

15. Sharp, J. J. *Hydraulic Modelling.* Butterworth's, London, 1981.

16. U.S. Department of the Interior (Water and Power Resources Service). *Hydraulic Laboratory Techniques.* U.S. Government Printing Office, Washington, D.C., 1980.

Surface Resistance

The high speed attained by surfers is owing to the fact that "planing" of the board virtually eliminates wave drag, thus leaving only surface resistance, which is relatively small. (Courtesy Australian Tourist Commission)

A fluid medium through which bodies such as aircraft or ships move exerts a resistance to motion on the bodies; this resistance is called *drag*. Aeronautical engineers and naval architects are vitally interested in the drag of an airplane or ship because the success or failure of the craft is directly related to its resistance to motion. If the drag is too large, the craft may be an economic failure because of the excessive costs (initial and operational) of the propulsion system. The drag of a body results from two types of forces acting on the body: shear forces and pressure forces. The drag, or surface resistance, from shear forces is often called *skin-friction drag,* whereas the drag resulting from pressure forces is called *form drag.* In this chapter we shall concentrate on the mechanics of skin-friction drag. In Chapter 11, we will cover form drag.

9.1 Introduction

In general, the shear stress on a smooth plane surface is variable over the surface. Hence the total shear force in a given direction is obtained by integrating the component of shear stress in that direction over the total area of the surface. The shear stress on a smooth plane is a direct function of the velocity gradient next to the plane, as given by Eq. (2.5). Therefore, any problem involving shear stress also involves the flow pattern in the vicinity of the surface. The layer of fluid near the surface that has undergone a change in velocity because of the shear stress at the surface is called the *boundary layer,* and the general area of study of the flow pattern in the boundary layer, as well as of the associated shear stress at the boundary, is called *boundary-layer theory.*

We will first consider the simplest cases of surface resistance resulting from uniform laminar flow in which the velocity gradient and also the shear stress are constant. We will then examine a laminar boundary layer that develops from the leading edge of a smooth, flat plate. Finally, we will consider the characteristics of a turbulent boundary layer.

9.2   Surface Resistance with Uniform Laminar Flow

Uniform Flow Produced by Relative Movement of Parallel Plates

Consider the motion of a plate as shown in Fig. 9.1. The moving plate causes the fluid in contact with it to have the same velocity as the plate. If the pressure gradient along the plate (x direction) is zero, then the flow produced is called *Couette flow* after the French scientist M. Couette, a pioneer in the analysis of shear flows between parallel plates and rotating cylinders. We now consider fluid within the control volume *abcd* of Fig. 9.1, having unit width normal to the plane of the page, and apply the momentum equation in the x direction to obtain

$$p\,\Delta y - \left(p + \frac{dp}{dx}\Delta x\right)\Delta y - \tau\Delta x + \left(\tau + \frac{d\tau}{dy}\Delta y\right)\Delta x = \int_{cs} u\rho \mathbf{u} \cdot d\mathbf{A} \qquad (9.1)$$

The flow is uniform because of the parallel boundaries, so the right-hand side of Eq. (9.1) is zero. Because $dp/dx = 0$, the equation reduces to $d\tau/dy = 0$, which means τ is constant. Because $\tau = \mu\,du/dy$, we conclude that the velocity gradient is also constant, as shown in Fig. 9.1. In other words, the velocity varies linearly from zero at the fixed boundary to maximum at the moving surface.

EXAMPLE 9.1   If the fluid between the plates of Fig. 9.1 is SAE 30 lubricating oil at $T = 38°C$, if the plates are spaced 0.3 mm apart, and if the upper plate is moved at a velocity of 1.0 m/s, what is the surface resistance for 1.0 m^2 of the upper plate?

Solution   From the Appendix (Table A.4), $\mu = 1.0 \times 10^{-1}$ N $\cdot$ s/m^2. Then

$$\tau = \mu\frac{du}{dy} = \mu\frac{\Delta u}{\Delta y}$$

$$= (1.0 \times 10^{-1} \text{ N} \cdot \text{s/m}^2)(1.0 \text{ m/s})/(3 \times 10^{-4} \text{ m}) = 333 \text{ N/m}^2$$

$$F_s = \tau A = 333 \text{ N/m}^2 \times 1.0 \text{ m}^2 = 333 \text{ N} \qquad \blacktriangleleft$$

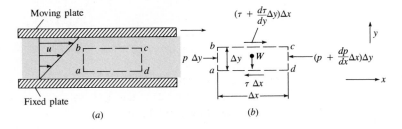

FIGURE 9.1

Flow between a fixed plate and a moving plate.

Uniform Liquid Flow Down an Inclined Plane

Consider the case shown in Fig. 9.2, where the plane is inclined at an angle θ to the horizontal and the flow is assumed to be uniform—that is, the depth and velocity distribution do not change with distance along the plane. Here we apply the momentum equation to an element of fluid, the top of which is the liquid surface. The element has unit width normal to the plane of the page. The pressure is hydrostatic across any section normal to the inclined plane because uniform flow is being assumed. In addition, the depth is constant, so the pressure forces on the end sections of the element cancel. Furthermore, the shear stress on the liquid surface at the liquid–air interface is assumed to be negligible. Therefore, the momentum equation reduces to

$$W \sin \theta - \tau \Delta s = 0$$

Consequently,
$$\gamma(d - y)\, \Delta s \sin \theta = \tau \Delta s$$

$$\tau = \gamma \sin \theta (d - y)$$

This shows that the shear stress across the section varies linearly from zero at the liquid surface, where $y = d$, to maximum at the plate, which means that physically the shear stress is proportional to the weight of the element, and the weight of the element increases directly with the thickness of the element $(d - y)$. Experiments have shown that if the Reynolds number based on the depth of flow, $\mathrm{Re} = Vd/\nu$, is less than 500, one can expect laminar flow in an open channel such as this. For laminar flow we can substitute $\mu\, du/dy$ for τ and then solve for the velocity distribution:

$$\mu\frac{du}{dy} = \gamma \sin \theta (d - y) \qquad \frac{du}{dy} = \frac{\gamma}{\mu} \sin \theta (d - y)$$

Integrating the latter equation yields

$$u = \frac{\gamma}{\mu} \sin \theta \left(yd - \frac{y^2}{2} \right) + C$$

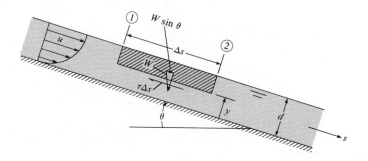

FIGURE 9.2

Free-surface flow down an inclined plane.

To evaluate the constant of integration, we observe that $u = 0$ when $y = 0$; therefore, $C = 0$. Then, after a little algebraic manipulation, the velocity distribution is given as

$$u = \frac{\gamma}{2\mu}(\sin \theta)y(2d - y) = \frac{g \sin \theta}{2\nu} y(2d - y) \qquad (9.2)$$

We now obtain the discharge per unit width by integrating the velocity u over the depth of flow:

$$q = \int_0^d u \, dy = \frac{\gamma}{2\mu} \sin \theta \left[dy^2 - \frac{y^3}{3} \right]_0^d \qquad (9.3)$$

$$= \frac{1}{3} \frac{\gamma}{\mu} d^3 \sin \theta \qquad (9.3)$$

The average velocity is now obtained by dividing Eq. (9.3) by the cross-sectional area d:

$$V = \frac{q}{d} = \frac{1}{3} \frac{\gamma}{\mu} d^2 \sin \theta$$

This reduces to

$$V = \frac{gd^2}{3\nu} \sin \theta \qquad (9.4a)$$

The slope, $S_0 = \tan \theta$, is approximately equal to $\sin \theta$ for small slopes. Thus Eq. (9.4a) can be given as

$$V = \frac{gS_0 d^2}{3\nu} \qquad (9.4b)$$

EXAMPLE 9.2 Crude oil, $\nu = 9.3 \times 10^{-5}$ m²/s, S = 0.92, flows over a flat plate that has a slope of $S_0 = 0.02$. If the depth of flow is 6 mm, what is the maximum velocity and what is the flow rate per meter of width of plate? Also determine the Reynolds number for this flow.

Solution First, we assume that the flow is laminar. Since $S_0 \approx \sin \theta$,

$$u = \frac{gS_0}{2\nu} y(2d - y)$$

Therefore, the maximum velocity will occur when y is maximum ($y = d$), or

$$u_{max} = \frac{(9.81 \text{ m/s}^2)(0.02)(0.006)^2 \text{ m}^2}{2 \times 9.3 \times 10^{-5} \text{ m}^2/\text{s}} = 0.038 \text{ m/s} \qquad \blacktriangleleft$$

The discharge per meter of width is given by Eq. (9.3), and for a small slope we can replace $\sin \theta$ by S_0:

$$q = \frac{1}{3} \frac{\gamma}{\mu} S_0 d^3 = \frac{1}{3} \frac{\gamma}{\rho \nu} S_0 d^3 = \frac{1}{3} \frac{g}{\nu} S_0 d^3$$

Then $q = \dfrac{1}{3} \dfrac{(9.81 \text{ m/s}^2)}{(9.3 \times 10^{-5} \text{ m}^2/\text{s})} (0.02)(0.006)^3 \text{ m}^3 = 1.52 \times 10^{-4} \text{ m}^2/\text{s}$ ◀

Now we check the Reynolds number to see if our original assumption of laminar flow (Re < 500) was correct.

$$\text{Re} = \frac{Vd}{\nu} = \frac{q}{\nu} = \frac{1.52 \times 10^{-4} \text{ m}^2/\text{s}}{9.3 \times 10^{-5} \text{ m}^2/\text{s}} = 1.63$$ ◀

The Reynolds number is less than 500. Therefore, our assumption of laminar flow was valid and the use of Eqs. (9.2) and (9.3) is justified.

Flow Between Parallel Plates with a Pressure Gradient Along the Plates

Here we consider the general two-dimensional case where two parallel plates are inclined to the horizontal at an angle θ and a pressure gradient exists in the fluid along the length of the plates. The schematic arrangement for these conditions is shown in Fig. 9.3. When the momentum equation is applied to the element of fluid (having dimensions $\Delta s \times \Delta y \times$ unity) shown in Fig. 9.3, the following equation results:

$$\sum F_s = 0$$

$$-\frac{dp}{ds} \Delta s \, \Delta y + \frac{d\tau}{dy} \Delta s \, \Delta y + W \sin \theta = 0$$

FIGURE 9.3

Flow between parallel boundaries with a pressure gradient.

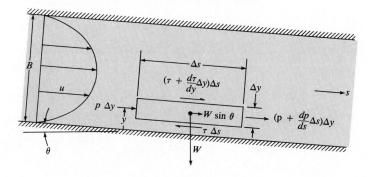

Here $\sin \theta = -dz/ds$ and $W = \gamma \Delta s\,\Delta y$. Then, after making these substitutions and dividing by $\Delta s\,\Delta y$, we have

$$\frac{d\tau}{dy} = \frac{d}{ds}(p + \gamma z)$$

$$\frac{d\tau}{dy} = \gamma \frac{d}{ds}\left(\frac{p}{\gamma} + z\right)$$

or
$$\frac{d\tau}{dy} = \gamma \frac{dh}{ds} \tag{9.5}$$

However,
$$\tau = \mu \frac{du}{dy}$$

so we get
$$\frac{d}{dy}\left(\mu \frac{du}{dy}\right) = \gamma \frac{dh}{ds}$$

Here μ is a constant, so we can write the foregoing equation as

$$\frac{d^2u}{dy^2} = \frac{\gamma}{\mu}\frac{dh}{ds}$$

Integrating this equation twice yields

$$u = \frac{\gamma}{\mu}\frac{dh}{ds}\frac{y^2}{2} + C_1 y + C_2$$

The constants of integration C_1 and C_2 can be evaluated from the boundary conditions $u = 0$ at $y = 0$ and $y = B$. Thus we obtain

$$C_2 = 0 \quad\text{and}\quad C_1 = -\frac{\gamma}{\mu}\frac{dh}{ds}\frac{B}{2}$$

Therefore
$$u = -\frac{\gamma}{2\mu}(By - y^2)\frac{dh}{ds} \tag{9.6a}$$

where
$$h = p/\gamma + z$$

As was the case with unconfined laminar-liquid flow over a plane surface, the velocity distribution is parabolic. However, in this case the maximum velocity occurs midway between the two plates. It can also be shown, by integrating the velocity over the section and dividing by the section area, that the mean velocity V is two-thirds of u_{max}. Note furthermore that flow is the result of a change of the piezometric head, not just a change of p or z alone. Experiments reveal that if the Reynolds number (VB/ν) is less than 1000, the flow is laminar.

EXAMPLE 9.3   Oil having a specific gravity of 0.8 and a viscosity of 2×10^{-2} N · s/m² flows downward between two vertical smooth plates spaced 10 mm apart. If the discharge per meter of width is 0.01 m³/s, what is the pressure gradient dp/ds for this flow?

Solution   First we check to see if the flow is laminar or turbulent:

$$\text{Re} = \frac{VB}{\nu} = \frac{VB\rho}{\mu} = \frac{q\rho}{\mu}$$

$$= \frac{(0.01 \text{ m}^2/\text{s}) \times 800 \text{ N} \cdot \text{s}^2/\text{m}^4}{0.02 \text{ N} \cdot \text{s/m}^2} = 400$$

The Reynolds number is less than 1000; therefore, the flow is laminar. The velocity distribution is given by Eq. (9.6a):

$$u = -\frac{\gamma}{2\mu}(By - y^2)\frac{dh}{ds}$$

For this example, $B = 0.01$ m and

$$\nu = \mu/\rho = \frac{2 \times 10^{-2} \text{ N} \cdot \text{s/m}^2}{0.8 \times 1000 \text{ N} \cdot \text{s}^2/\text{m}^4} = 2.5 \times 10^{-5} \text{ m}^2/\text{s}$$

Then we have $u_{\max}$, where $du/dy = 0$ at $y = B/2$:

$$u_{\max} = \frac{-0.80 \times 9810 \text{ N/m}^3}{2 \times 2 \times 10^{-2} \text{ N} \cdot \text{s/m}^2}\left(\frac{0.01^2}{2} \text{ m}^2 - \frac{0.01^2}{4} \text{ m}^2\right)\frac{dh}{ds}$$

$$= -4.905 \, dh/ds \qquad\qquad\qquad (9.6b)$$

But $V = \frac{2}{3}u_{\max}$ and $q = VB$

Therefore $q = \frac{2}{3}u_{\max}B$

or $u_{\max} = \frac{3}{2}q/B$

Here B is the plate spacing and $q = 0.01$ m³/s, so

$$u_{\max} = \frac{3}{2} \times 0.01/0.01 = 1.50 \text{ m/s} \qquad\qquad (9.6c)$$

Now we can solve for dh/ds from Eqs. (9.6b) and (9.6c) to obtain

$$dh/ds = -0.306$$

However, $\dfrac{dh}{ds} = \dfrac{d}{ds}\left(\dfrac{p}{\gamma} + z\right)$

Therefore $\dfrac{d}{ds}\left(\dfrac{p}{\gamma} + z\right) = -0.306$

Because the plates are vertically oriented and s is positive downward, $dz/ds = -1$. Thus

$$\frac{d(p/\gamma)}{ds} = 1 - 0.306$$

or $\qquad \dfrac{dp}{ds} = (0.8 \times 9810 \text{ N/m}^3) \times 0.694 = 5447 \text{ N/m}^2 \text{ per meter}$ ◄

In other words, the pressure is increasing downward at a rate of 5.45 kPa per meter of length of plate.

9.3 ## Qualitative Description of the Boundary Layer

Flow Pattern in a Boundary Layer

As we noted in Sec. 9.1, the *boundary layer* is the region next to a boundary of an object in which the fluid has had its velocity changed because of the shearing resistance created by the boundary. Outside the boundary layer the velocity is essentially the same as if an ideal (nonviscous) fluid were flowing past the object. In Sec. 9.2 we considered liquid flow over a plane surface and flow between parallel plates. In a narrow sense, each of these is boundary-layer flow—that is, uniform flow for a fully developed laminar boundary layer. Now we will look at a boundary layer that is still developing. That is, we will look at boundary layers that grow in thickness and have significant changes of velocity with distance along the boundary. To visualize the flow pattern associated with the boundary layer, we will qualitatively analyze the interaction between a fluid and the surface of a thin, flat plate as the fluid (for example, air) passes by the plate. Figure 9.4 illustrates such a plate. Fluid passes over the top and underneath the plate, so two boundary layers are depicted in Fig. 9.4 (one above and one below the plate). In Fig. 9.4 the fluid has a uniform velocity U_0 before it reaches the vicinity of the plate. However, the fluid touching the plate has zero velocity (the velocity of the plate), because of the no-slip condition characterizing continuum flows. Therefore, a velocity gradient must exist between the fluid in the free stream and the fluid next to the plate. Consistent with this gradient, there is a shear stress at the plate surface. As the fluid particles next to the plate pass the leading edge of the plate, a retarding force (from the shear stress) begins to act on them. As these particles progress downstream they continue to be subjected to shear stress from the plate, so they continue to decelerate. In addition, these particles (because of their lower velocity) retard other particles adjacent to them but farther out from the plate. Thus the boundary layer becomes thicker, or

FIGURE 9.4

Development of boundary layer and distribution of shear stress along a thin, flat plate. (a) Flow pattern in boundary layers above and below the plate. (b) Shear-stress distribution on either side of the plate.

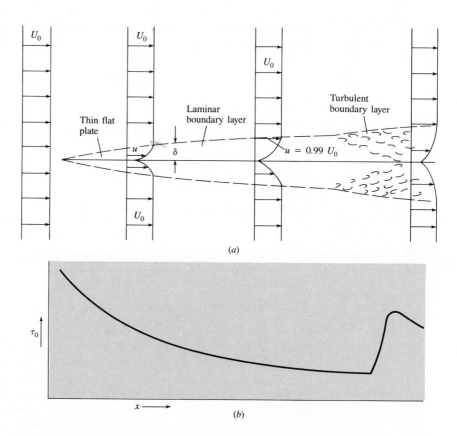

"grows," in the downstream direction. The broken line in Fig. 9.4 identifies the outer limit of the boundary layer. Because the boundary layer becomes thicker, the velocity gradient becomes less steep as one proceeds along the plate.

Thickening of the laminar boundary layer continues smoothly in the downstream direction until the thickness is so great that the flow becomes unstable and the boundary layer becomes turbulent. In the turbulent boundary layer, eddies mix higher-velocity fluid into the region close to the wall so that the velocity gradient du/dy at the wall becomes greater than it is at the wall in the laminar boundary layer just upstream of the transition point.

If one considers an element of fluid in a boundary layer (similar to the element shown in Fig. 4.18), it becomes obvious that the side normal to the boundary will rotate while the side parallel to the boundary will have virtually no rotation. Thus boundary layers, whether laminar or turbulent, are cases of rotational flow.

Shear-Stress Distribution along the Boundary

In the foregoing section we noted that it is the shearing force of the plate that decelerates the fluid to produce the boundary layer, but no mention was made of the way the shear stress changes along the boundary. Because the shear stress is given by $\tau = \mu \, du/dy$, it is easy to visualize that the shear stress must be relatively large near the leading edge of the plate where the velocity gradient is steep, and that it becomes smaller as the velocity gradient becomes smaller in the downstream direction. However, where the boundary layer becomes turbulent, the shear stress at the boundary again becomes larger, consistent with the greater velocity gradient next to the wall in the turbulent boundary layer. Figure 9.4*b* depicts such a distribution of shear stress. These qualitative aspects of the boundary layer serve as a foundation for the quantitative relations presented in the next section.

9.4   Quantitative Relations for the Laminar Boundary Layer

Boundary-Layer Equations

In 1904 Prandtl (9) first stated the essence of the boundary-layer hypothesis, which is that viscous effects are concentrated in a thin layer of fluid (the boundary layer) next to solid boundaries. Along with his discussion of the qualitative aspects of the boundary layer, he also simplified the general equations of motion of a fluid (Navier–Stokes equations) for application to the boundary layer. Then in 1908, Blasius, one of Prandtl's students, obtained a solution for the flow in a laminar boundary layer (3). The solution was for the case of zero pressure gradient along the plate, $dp/dx = 0$, and one of Blasius' key assumptions was that the shape of the nondimensional velocity distribution did not vary from section to section along the plate. That is, he assumed that a plot of the relative velocity u/U_0 versus the relative distance from the boundary, y/δ, would be the same at each section. Here δ is the thickness of the boundary layer, defined as the distance from the boundary to the point in the fluid where the velocity is 99% of the free-stream velocity. With this assumption and with Prandtl's equations of motion for boundary layers, Blasius obtained a solution for the relative velocity distribution, shown in Fig. 9.5. In this plot, x is the distance from the leading edge of the plate, and Re_x is the Reynolds number based on the free-stream velocity and the length along the plate ($\mathrm{Re}_x = U_0 x/\nu$). In Fig. 9.5 the outer limit of the boundary layer ($u/U_0 = 0.99$) occurs at approximately $y\mathrm{Re}_x^{1/2}/x = 5$. Since $y = \delta$ at this point, we have a relationship for the thickness of the boundary layer:

$$\frac{\delta}{x} \, \mathrm{Re}_x^{1/2} = 5 \qquad \text{or} \qquad \delta = \frac{5x}{\mathrm{Re}_x^{1/2}} \tag{9.7}$$

We can also obtain from Fig. 9.5 the inverse of the slope of the curve at the boundary, which is equal to 0.332, or

$$\left. \frac{d(u/U_0)}{d[(y/x)\mathrm{Re}_x^{1/2}]} \right|_{y=0} = 0.332$$

FIGURE 9.5

Velocity distribution in laminar boundary layer. [After Blasius (3)].

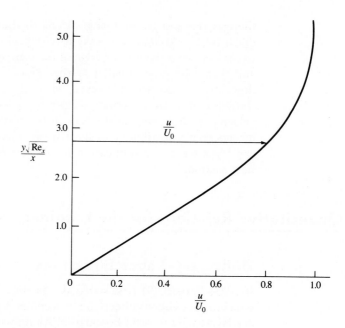

However, at any given section, x, Re_x, and U_0 are constants. Therefore, we express the velocity gradient at the boundary as follows:

$$\left.\frac{du}{dy}\right|_{y=0} = 0.332\frac{U_0}{x}\,\mathrm{Re}_x^{1/2} \tag{9.8}$$

$$\left.\frac{du}{dy}\right|_{y=0} = 0.332\frac{U_0}{x}\left(\frac{U_0 x}{\nu}\right)^{1/2}$$

$$\left.\frac{du}{dy}\right|_{y=0} = 0.332\frac{U_0^{3/2}}{x^{1/2}\nu^{1/2}} \tag{9.9}$$

Equation (9.9) shows that the velocity gradient decreases as the distance x along the boundary increases.

Shear Stress

The shear stress at the boundary is obtained by multiplying the velocity gradient at the wall, Eq. (9.8), by the absolute viscosity:

$$\tau_0 = 0.332\mu\frac{U_0}{x}\,\mathrm{Re}_x^{1/2} \tag{9.10}$$

Equation (9.10) is used to obtain the local shear stress at any given section (any given value of x) for the laminar boundary layer, as shown in the following example.

EXAMPLE 9.4 Crude oil at 70°F ($\nu = 10^{-4}$ ft²/s, S = 0.86) with a free-stream velocity of 10 ft/s flows past a thin, flat plate that is 4 ft wide and 6 ft long in a direction parallel to the flow. Determine and plot the boundary-layer thickness and the shear-stress distribution along the plate.

Solution
$$\text{Re}_x = \frac{U_0 x}{\nu} = \frac{10x}{10^{-4}} = 10^5 x$$

$$\text{Re}_x^{1/2} = 3.16(10^2 x^{1/2})$$

The shear stress is given by

$$\tau_0 = 0.332 \mu \frac{U_0}{x} \text{Re}_x^{1/2}$$

where $\mu = \rho\nu = 1.94 \times 0.86 \times 10^{-4} = 1.67 \times 10^{-4}$ lbf-s/ft²

Then $\tau_0 = 0.332(1.67 \times 10^{-4})\dfrac{10}{x}(3.16)(10^2 x^{1/2}) = \dfrac{0.175}{x^{1/2}}$ psf

The thickness of the boundary layer is

$$\delta = \frac{5x}{\text{Re}_x^{1/2}} = \frac{5x}{3.16(10^2 x^{1/2})} = 1.58(10^{-2} x^{1/2}) \text{ ft}$$

$$= 1.58(12)(10^{-2} x^{1/2}) = 0.190 x^{1/2} \text{ in.}$$

The results for Example 9.4 are plotted in the accompanying figure and listed in Table 9.1.

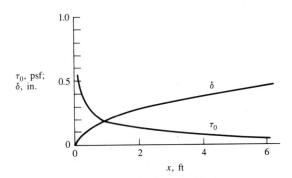

	$x = 0.1$ ft	$x = 1.0$ ft	$x = 2$ ft	$x = 4$ ft	$x = 6$ ft
TABLE 9.1	**RESULTS—δ AND τ_0 FOR DIFFERENT VALUES OF x**				
$x^{1/2}$	0.316	1.00	1.414	2.00	2.45
τ_0, psf	0.552	0.174	0.123	0.087	0.071
δ, ft	0.005	0.016	0.022	0.031	0.039
δ, in.	0.060	0.189	0.270	0.380	0.466

Shearing Resistance for a Surface of Given Size

Because the shear stress at the boundary, τ_0, varies along the plate, it is necessary to integrate this stress over the entire surface to obtain the total shearing force on the surface:

$$F_s = \int_0^L \tau_0 B \, dx \tag{9.11}$$

where F_s is the surface resistance produced by viscous stresses on one side of the plate, B is the width of the plate, and L is the length. When Eq. (9.10) is substituted into Eq. (9.11), we get

$$F_s = \int_0^L 0.332 B\mu \frac{U_0 U_0^{1/2} x^{1/2}}{x \nu^{1/2}} \, dx$$

$$= 0.664 B\mu U_0 \frac{U_0^{1/2} L^{1/2}}{\nu^{1/2}}$$

$$= 0.664 B\mu U_0 \mathrm{Re}_L^{1/2} \tag{9.12}$$

In Equation (9.12) Re_L is the Reynolds number based on the approach velocity and the length of the plate.

Shear-Stress Coefficients

It is convenient to express the shear stress at the boundary, τ_0, and the total shearing force F_s in terms of dimensionless resistance coefficients and the dynamic pressure of the free stream, $\rho U_0^2 / 2$. The coefficients c_f and C_f are defined as follows:

$$c_f = \frac{\tau_0}{\rho U_0^2 / 2} \quad \underleftarrow{\text{shear stress at boundary}} \tag{9.13}$$

and

$$C_f = \frac{F_s}{BL\rho U_0^2 / 2} \quad \underleftarrow{\text{total shearing force}} \tag{9.14}$$

Combining Eq. (9.10) with Eq. (9.13) and Eq. (9.12) with Eq. (9.14) produces the relationship between these coefficients and the corresponding Reynolds numbers for each case:

$$c_f = \frac{0.664}{\mathrm{Re}_x^{1/2}} \tag{9.15}$$

$$C_f = \frac{1.33}{\mathrm{Re}_L^{1/2}} \tag{9.16}$$

EXAMPLE 9.5   For the conditions of Example 9.4, determine the resistance of one side of the plate.

Solution
$$F_s = \frac{C_f B L \rho U_0^2}{2}$$

Here
$$C_f = \frac{1.33}{\mathrm{Re}_L^{1/2}} = \frac{1.33}{(3.16)\,(10^2)\,(6^{1/2})} = 0.0017$$

Then
$$F_s = 0.0017(4)\,(6)\,(0.86)\,(1.94)\left(\frac{10^2}{2}\right) = 3.40 \text{ lbf} \qquad \blacktriangleleft$$

Experiment versus Theory for the Laminar Boundary Layer

Experimental evidence indicates that the Blasius solution is valid except very near the leading edge of the plate. In the vicinity of the leading edge, an error results because of certain simplifying assumptions. However, the discrepancy is not significant for most engineering problems. For very thin, smooth plates the laminar boundary layer can be expected to change to a turbulent boundary layer at a Reynolds number of approximately 500,000. However, if the approach flow is turbulent and/or if the plate is rough, the laminar boundary layer can be expected to become turbulent at a smaller Reynolds number.

9.5 Quantitative Relations for the Turbulent Boundary Layer

Velocity Distribution in the Turbulent Boundary Layer along a Smooth Wall

The turbulent boundary layer is more complex than the laminar boundary layer. The former has three zones of flow that require different equations for the velocity distribution, as opposed to the single relationship of the laminar case. Figure 9.6 shows a portion of a turbulent boundary layer in which the three different zones of flow are identified. Each of these zones will be discussed separately.

The zone immediately adjacent to the wall is a layer of fluid that, because of the damping effect of the wall, remains relatively smooth even though most of the flow in the boundary layer is turbulent. This very thin layer is called the *viscous sublayer*. The velocity distribution in this layer is related to the shear stress and viscosity by Newton's viscosity law, which was introduced in Chapter 2. For convenience, it is repeated here as follows: $\tau = \mu\, du/dy$. In the viscous sublayer, τ

FIGURE 9.6

Pattern of flow in turbulent boundary layer.

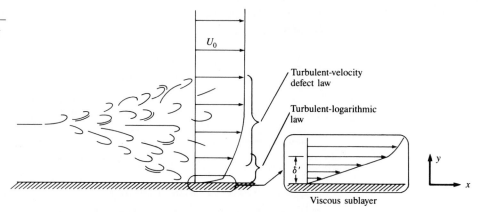

is virtually constant and equal to the shear stress at the wall, τ_0. Thus $du/dy = \tau_0/\mu$, which on integration yields

$$u = \frac{\tau_0 y}{\mu} \tag{9.17}$$

If we multiply and divide the right side of Eq. (9.17) by ρ, we obtain the following:

$$u = \frac{\tau_0/\rho}{\mu/\rho} y$$

$$\frac{u}{\sqrt{\tau_0/\rho}} = \frac{\sqrt{\tau_0/\rho}}{\nu} y \tag{9.18}$$

The combination of variables $\sqrt{\tau_0/\rho}$ recurs again and again in derivations involving boundary-layer theory and has been given the special name *shear velocity*. The name is appropriate because the variable τ_0 relates to the shear stress and the units of the combination are those of velocity. The shear velocity (which is also sometimes called *friction velocity*) is symbolized as u_*. Thus, by definition,

$$u_* = \sqrt{\frac{\tau_0}{\rho}} \tag{9.19}$$

Now, when we substitute u_* for $\sqrt{\tau_0/\rho}$ in Eq. (9.18), we have the customary form for expressing the velocity distribution in the viscous sublayer:

$$\frac{u}{u_*} = \frac{y}{\nu/u_*} \tag{9.20}$$

Equation (9.20) shows that the relative velocity (velocity relative to the shear velocity) in the viscous sublayer is equal to a dimensionless distance from the wall. Experimental results show that the viscous sublayer occurs only in a film of fluid for which yu_*/ν is less than approximately 5. Consequently, the thickness of the viscous sublayer, identified by δ', is given as

$$\delta' = \frac{5\nu}{u_*} \tag{9.21}$$

It follows that the thickness of the viscous sublayer will be small for flows in which the shear stress is large, because u_* will also be large for these cases. Furthermore, the viscous sublayer will become larger along the wall in the direction of flow because the shear stress decreases along the wall in the downstream direction.

The flow zone outside the viscous sublayer is turbulent; therefore, a completely different type of flow is involved. In fact, the turbulence alters the flow regime so much that the shear stress as given by $\tau = \mu\,du/dy$ is not significant. What happens is that the mixing action of turbulence causes small fluid masses to be swept back and forth in a direction transverse to the mean flow direction. Thus, as a small mass of fluid is swept from a low-velocity zone next to the viscous sublayer into a higher-velocity zone farther out in the stream, the mass has a retarding effect on the higher-velocity stream. Through an exchange of momentum, this mass of fluid creates the effect of a retarding shear stress applied to the higher-velocity stream, much like the "conveyor belt" analogy introduced on page 17. Similarly, a small mass of fluid that originates farther out in the boundary layer in a high-velocity flow zone and is swept into a region of low velocity has an effect on the low-velocity fluid much like shear stress augmenting the flow velocity. In other words, the mass of fluid with relatively higher momentum tends to accelerate the lower-velocity fluid in the region into which it moves. Although the process just described is primarily a momentum-exchange phenomenon, it has the same effect as a shear stress applied to the fluid; thus in turbulent flow these "stresses" are termed *apparent shear stresses,* or *Reynolds stresses* after the British scientist–engineer who first did extensive research in turbulent flow in the late 1800s.

The mixing action of turbulence causes the velocities at a given point in a flow to fluctuate with time. If one places a velocity-sensing device, such as a hot-wire anemometer, in a turbulent flow, one can measure a fluctuating velocity, as illustrated in Fig. 9.7. It is convenient to think of the velocity as composed of two parts: a mean value, $\bar{u}$, plus a fluctuating part, u'. The fluctuating part of the velocity is responsible for the mixing action and the momentum exchange, which manifests itself as an apparent shear stress as we have noted. In fact, the apparent shear stress is related to the fluctuating part of the velocity by

$$\tau_{\text{app}} = -\rho\overline{u'v'} \tag{9.22}$$

where u' and v' refer to the x and y components of the velocity fluctuations, respectively, and the bar over these terms denotes the product of $u'v'$ averaged over

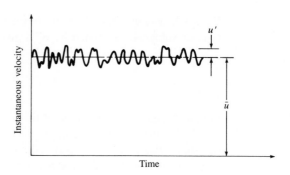

a period of time.* The expression for apparent shear stress is not very useful in this form, so Prandtl developed a theory to relate the apparent shear stress to the temporal mean velocity distribution.

Prandtl's Mixing-Length Theory

In the turbulent boundary layer, the principal flow is parallel to the boundary. However, because of turbulent eddies, there are fluctuating components transverse to the principal flow direction. By focusing on a small mass of fluid that is moving transverse to the principal flow under the action of turbulence, one can explain Prandtl's mixing-length theory. Assume that the small fluid mass is initially in a zone of relatively low mean velocity near the wall (see Fig. 9.8). Then, under the action of turbulence, it is transported to a new position farther from the wall (in direction of increasing y). This mass of fluid in its new location will

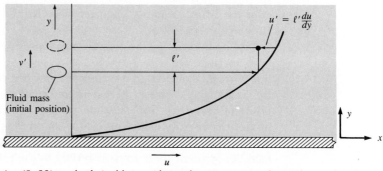

FIGURE 9.8

Concept of mixing length.

---

* *Equation (9.22) can be derived by considering the momentum exchange that results when the transverse component of turbulent flow passes through an area parallel to the x-z plane. Or, by including the fluctuating velocity components in the Navier–Stokes equations, one can obtain the apparent shear-stress terms, one of which is Eq. (9.22). Details of these derivations appear in Chapter 18 of Schlichting (11).*

have a lower velocity (in the principal flow direction) than the surrounding fluid (it tends to retain its initial momentum), and the difference in velocity between the surrounding fluid and the small mass of fluid will be given by $\ell' du/dy$, where du/dy is the mean velocity gradient and ℓ' is the distance the small fluid mass traveled in the transverse direction in going from the initial position to the position farther out in the flow field. Note also that, at the point farther out in the flow field, the velocity has changed from an initial value of u to a value of $u - u'$. Therefore, Prandtl assumed that the magnitude of the fluctuating velocity component in the principal flow direction is equal to $\ell' du/dy$ or $|u'| = \ell' du/dy$. Furthermore, Prandtl assumed that the magnitude of the transverse fluctuating-velocity component is proportional to the magnitude of the fluctuating component in the principal flow direction: $|v'| = K \times |u'|$, where K is a constant. This would seem to be a reasonable assumption because both components arise from the same set of eddies. Also, it should be noted that a positive v' will be associated with a negative u' (when a small mass of fluid moves away from the boundary $+v'$, it produces a negative u' in the vicinity into which it moves, as explained above). Thus $\overline{u'v'}$ will be negative, and when $\ell' du/dy$ is substituted for u' and $K\ell' du/dy$ is substituted for v' in Eq. (9.22), one obtains

$$\tau_{\text{app}} = K\rho\ell'^2 \left(\frac{du}{dy}\right)^2$$

If we let $K\ell'^2$ in the foregoing equation be equal to ℓ^2, then we may write it as

$$\tau_{\text{app}} = \rho\ell^2 \left(\frac{du}{dy}\right)^2 \tag{9.23}$$

A more general form of Eq. (9.23) is

$$\tau_{\text{app}} = \rho\ell^2 \left|\frac{du}{dy}\right| \frac{du}{dy}$$

However, this form is not needed in this text.

The theory leading to Eq. (9.23) is called Prandtl's mixing-length theory and is used extensively in analyses involving turbulent flow.* The advantage of Eq. (9.23) over Eq. (9.22) is that Eq. (9.23) is expressed in terms of the temporal mean velocity distribution and can thus be integrated to obtain formulas for the velocity variation. In Eq. (9.23), ℓ is the mixing length, which was shown by Prandtl to be essentially proportional to the distance from the wall ($\ell = \kappa y$) for the region close to the wall. If we consider the velocity distribution in a

---

*Prandtl published an account of his mixing-length concept in 1925. G. I. Taylor published a similar concept in 1915, but the idea has been traditionally attributed to Prandtl.

boundary layer where du/dy is positive, such as is shown in Fig. 9.6, and if we substitute κy for ℓ, then Eq. (9.23) reduces to

$$\tau_{\text{app}} = \rho\kappa^2 y^2 \left(\frac{du}{dy}\right)^2$$

For the zone of flow near the boundary, it is assumed that the shear stress is uniform and approximately equal to the shear stress at the wall. Thus the foregoing equation becomes

$$\tau_0 = \rho\kappa^2 y^2 \left(\frac{du}{dy}\right)^2 \tag{9.24}$$

Taking the square root of each side of Eq. (9.24) and rearranging yield

$$du = \frac{\sqrt{\tau_0/\rho}}{\kappa}\frac{dy}{y}$$

Integrating the above equation and substituting u_* for $\sqrt{\tau_0/\rho}$ give

$$\frac{u}{u_*} = \frac{1}{\kappa}\ln y + C \tag{9.25}$$

Experiments on smooth boundaries indicate that the constant of integration C can be given in terms of u_*, ν, and a pure number as

$$C = 5.56 - \frac{1}{\kappa}\ln\frac{\nu}{u_*}$$

When this expression for C is substituted into Eq. (9.25), we have

$$\frac{u}{u_*} = \frac{1}{\kappa}\ln\frac{yu_*}{\nu} + 5.56 \tag{9.26}$$

In Eq. (9.26), κ has sometimes been called the universal turbulence constant, or Karman's constant. Experiments show this constant to have a value of approximately 0.40 for the turbulent zone next to the viscous sublayer. Introducing this into Eq. (9.26) and expressing the equation in terms of the logarithm to the base 10, we get

$$\frac{u}{u_*} = 5.75 \log\frac{yu_*}{\nu} + 5.56 \tag{9.27}$$

This logarithmic velocity distribution is valid for values of yu_*/ν ranging from approximately 30 to 500. Thus we have a form of velocity distribution for this zone that is far different from that for the viscous sublayer. However, the relative velocity is still a function of yu_*/ν. Thus, for the range of yu_*/ν from 0 to approximately 500 (Fig. 9.9), the velocity distribution is called the *law of the wall*.

Making a semilogarithmic plot of the velocity distribution in a turbulent boundary layer, as shown in Fig. 9.9, makes it easy to identify the velocity

FIGURE 9.9

Velocity distribution in a turbulent boundary layer. [Adapted from Schlichting (11) and Daily and Harleman (5)].

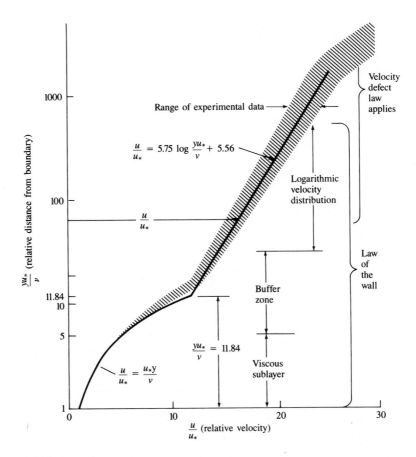

distribution in the viscous sublayer and in the region where the logarithmic equation applies. However, note that this form of plot accentuates the nondimensional distance yu_*/ν near the wall. So that the student may view this plot in better perspective, the graph shown in Fig. 9.9 is repeated in Fig. 9.10, except that in the latter, both the relative distance yu_*/ν and the relative velocity are plotted on linear scales. Figure 9.10 properly indicates that the laminar sublayer and the buffer zone, which is defined below, are a small part of the thickness of the turbulent boundary layer.

For $y/\delta > 0.15$ the law of the wall is no longer valid. Therefore, in this outer region we have a third zone given by the *velocity-defect law,* Fig. 9.11. In fact, the velocity-defect law applies not only to this outer region but extends well into the logarithmic zone, Fig. 9.9. The velocity-defect law relates the relative defect of velocity $(U_0 - u)/u_*$ to y/δ. Another important point about this law is that it applies to rough as well as smooth surfaces.

The foregoing discussion about the three zones of flow (the viscous sublayer, the logarithmic velocity distribution, and the outer zone where the

FIGURE 9.10

*Velocity distribution in a
turbulent boundary
layer–linear scales.*

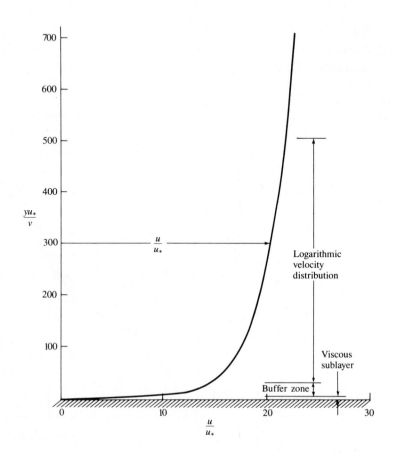

velocity-defect law applies) may seem to imply that there is a sharp demarcation between zones. This is definitely not the case, as can be seen in Fig. 9.9, which shows a smooth transition of velocity between the viscous sublayer and the zone of logarithmic velocity distribution. The band of experimental data shows that there is a range of distance over which neither law applies. This region has been aptly called the *buffer zone*. In some cases it may be desirable to define the velocity in the buffer zone by a separate equation, for which the student is referred to Rouse (10). However, for some boundary-layer calculations, it is convenient to ignore the precise form of velocity distribution in the buffer zone and simply let the distribution be given by the extension of the equations already cited. In this case we refer to the nominal thickness of the viscous sublayer, δ'_N, which is the distance from the boundary to the point where the velocity-distribution curves for the viscous sublayer and the logarithmic velocity distribution intersect. This point of intersection occurs (see Fig. 9.9) at

$$\frac{yu_*}{\nu} = 11.84 \qquad (9.28)$$

FIGURE 9.11

Velocity-defect law for boundary layers. [After Schlichting (11)].

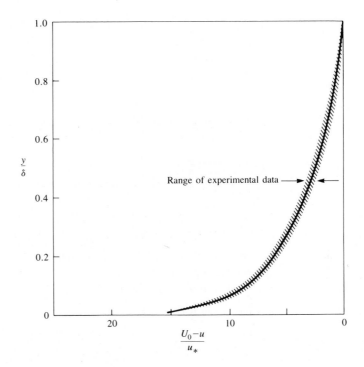

In other words, for this value of $y u_*/\nu$ we can substitute δ'_N for y:

$$\delta'_N = \frac{11.84\nu}{u_*} \tag{9.29}$$

As noted in the preceding section, the actual limit of the viscous sublayer occurs at about $\delta' = 5\nu/u_*$. That is, turbulence affects the velocity distribution in the boundary layer from $\delta' = 5\nu/u_*$ outward. However, the effect is not appreciable up to the point of the intersection of the velocity-distribution curve for the zone of logarithmic distribution (see Fig. 9.9). Thus, in derivations and analyses involving turbulent boundary layers, it is not uncommon to assume that the velocity distribution in the viscous sublayer extends out to $\delta'_N = 11.84\nu/u_*$.

Power-Law Formula for Velocity Distribution

Analyses have shown that for a wide range of Reynolds numbers ($10^5 < \mathrm{Re} < 10^7$), the velocity profile in the turbulent boundary layer is reasonably approximated by the equation

$$\frac{u}{U_0} = \left(\frac{y}{\delta}\right)^{1/7} \tag{9.30}$$

Comparisons with experimental results show that this formula conforms to those results very closely over about 90% of the boundary layer ($0.1 < y/\delta < 1$). For the inner 10% of the boundary layer, one must resort to equations for the law of the wall (see Fig. 9.9) to obtain a more precise indication of velocity. Because Eq. (9.30) is valid over the major portion of the boundary layer, it is used to advantage in deriving the overall thickness of the boundary layer as well as other relations for the turbulent boundary layer. These will be considered in the next sections.

EXAMPLE 9.6   Water (60°F) flows with a velocity of 20 ft/s past a flat plate. The plate is oriented parallel to the flow. At a particular section downstream of the leading edge of the plate, the shear stress next to the plate is 0.896 lbf/ft^2 and the boundary layer thickness is 0.0880 ft. Find the velocity of the water at a distance of 0.0088 ft from the plate as determined by

    a. The logarithmic velocity distribution.
    b. The velocity-defect law.
    c. The power-law formula.

What is the nominal thickness of the viscous sublayer?

Solution   Logarithmic velocity distribution:

$$u/u_\star = 5.75 \log(y u_\star/\nu) + 5.56$$

where $u_\star = (\tau_0/\rho)^{1/2}$, $\nu = 1.22 \times 10^{-5}$ ft^2/s, and $y = 0.0088$ ft.

$$\tau_0 = 0.896 \text{ lbf/ft}^2 \quad \text{and} \quad \rho = 1.94 \text{ slugs/ft}^3 = 1.94 \text{ lbf} \cdot \text{s}^2/\text{ft}^4$$

Thus

$$(\tau_0/\rho)^{1/2} = [(0.896 \text{ lbf/ft}^2)/(1.94 \text{ lbf} \cdot \text{s}^2/\text{ft}^4)]^{1/2} = 0.680 \text{ ft/s}$$

Also,

$$y u_\star/\nu = (0.0088 \text{ ft})(0.680 \text{ ft/s})/(1.22 \times 10^{-5} \text{ ft}^2/\text{s})$$
$$= 490$$

Then $u/u_\star = 5.75 \log(490) + 5.56$

              $u/u_\star = 21.03 \quad \text{or} \quad u = 21.03(u_\star) = 21.03 \times 0.680$ ft/s

              $u = 14.3$ ft/s   ◀

Velocity-defect law:

$$y/\delta = 0.0088 \text{ ft}/0.088 \text{ ft} = 0.10$$

Then, from Fig. 9.11, we find that

$$(U_0 - u)/u_\star \approx 8.2 \quad \text{for } y/\delta = 0.10$$
$$U_0 - u = 8.2 u_\star$$

or
$$u = U_0 - 8.2u_*$$
$$= 20 \text{ ft/s} - (8.2)(0.68) \text{ ft/s}$$
$$= 14.42 \text{ ft/s} \qquad \blacktriangleleft$$

Power-law formula:
$$u/U_0 = (y/\delta)^{1/7}$$
$$u = (U_0)(0.10)^{1/7}$$
$$= (20 \text{ ft/s})(0.7197)$$
$$= 14.40 \text{ ft/s} \qquad \blacktriangleleft$$

Nominal viscous sublayer thickness:
$$\delta'_N = 11.84\nu/u_* = (11.84)(1.22 \times 10^{-5} \text{ ft}^2/\text{s})/(0.68 \text{ ft/s})$$
$$= 2.12 \times 10^{-4} \text{ ft} = 2.54 \times 10^{-3} \text{ in.} \qquad \blacktriangleleft$$

Momentum Equation
Applied to the Boundary Layer

When the form of velocity distribution in the boundary layer is known, it is possible to derive equations for the shear stress and the thickness of the boundary layer by utilizing the momentum equation. These basic relationships involved with the momentum equation will be introduced in this section.

Consider the control volume in Fig. 9.12 for the boundary layer over a flat plate with zero pressure gradient along the plate. Flow comes into the control

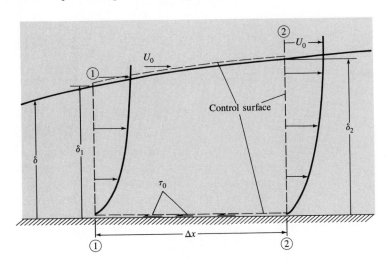

FIGURE 9.12

Control volume applied to boundary layer.

volume at section 1-1 and through the top of the boundary layer; it leaves the control volume at section 2-2. When we write the momentum equation for the x direction for the fluid in this control volume, we have

$$\sum F_x = \int_{cs} u\rho \mathbf{V} \cdot d\mathbf{A} \tag{9.31}$$

We assume a unit width normal to the page and designate the mass inflow through the top of the boundary layer as $\dot{m}_\delta$. Therefore, Eq. (9.31) becomes

$$-\tau_0 \Delta x = \int_0^{\delta_2} \rho u_2^2 \, dy - \int_0^{\delta_1} \rho u_1^2 \, dy - U_0 \dot{m}_\delta \tag{9.32}$$

The mass-flow rate into the control volume from the top of the boundary layer can be expressed as the difference in the mass-flow rate past sections 1-1 and 2-2,

$$\dot{m}_\delta = \int_0^{\delta_2} \rho u_2 \, dy - \int_0^{\delta_1} \rho u_1 \, dy \tag{9.33}$$

When Eq. (9.33) is substituted into Eq. (9.32), we obtain

$$-\tau_0 \Delta x = \int_0^{\delta_2} \rho u_2^2 \, dy - \int_0^{\delta_1} \rho u_1^2 \, dy - U_0 \left[\int_0^{\delta_2} \rho u_2 \, dy - \int_0^{\delta_2} \rho u_1 \, dy \right] \tag{9.34}$$

Assuming ρ is constant, we can rearrange this equation to yield

$$
\begin{aligned}
-\tau_0 \Delta x &= \rho \int_0^{\delta_2} (u_2^2 - U_0 u_2) \, dy - \rho \int_0^{\delta_1} (u_1^2 - U_0 u_1) \, dy \\
&= \rho U_0^2 \left\{ \int_0^{\delta_2} \left[\left(\frac{u_2}{U_0} \right)^2 - \frac{u_2}{U_0} \right] dy - \int_0^{\delta_1} \left[\left(\frac{u_1}{U_0} \right)^2 - \frac{u_1}{U_0} \right] dy \right\} \\
&= \rho U_0^2 \Delta \left\{ \int_0^{\delta} \left[\left(\frac{u}{U_0} \right)^2 - \frac{u}{U_0} \right] dy \right\}
\end{aligned}
$$

If we let Δx approach zero in the limit, then the foregoing equation reduces to

$$\tau_0 = \rho U_0^2 \frac{d}{dx} \int_0^{\delta} \frac{u}{U_0} \left(1 - \frac{u}{U_0} \right) dy \tag{9.35}$$

Equation (9.35) states that the shear stress at the wall is equal to ρU_0^2 times the rate of change with respect to x of an integral that is a function of the velocity distribution across the section. This equation can be used to evaluate the shear stress on a boundary indirectly by measuring the velocity distribution at various sections or, as will be shown in the next section, to derive other boundary-layer equations.

Thickness of the Turbulent Boundary Layer on a Flat Plate

Using the power-law formula for velocity distribution, Eq. (9.30), in Eq. (9.35), we have

$$\frac{\tau_0}{\rho} = U_0^2 \frac{d}{dx} \int_0^\delta \left(\frac{y}{\delta}\right)^{1/7} \left[1 - \left(\frac{y}{\delta}\right)^{1/7}\right] dy \tag{9.36}$$

The integration in Eq. (9.36) yields

$$\frac{\tau_0}{\rho} = \frac{7}{72} U_0^2 \frac{d\delta}{dx} \tag{9.37}$$

Here we have an equation involving δ. However, it is not very useful because τ_0 and δ are both unknowns. We can get another equation by referring to the velocity distribution in Fig. 9.9. By fitting a power-law distribution to the curve for $100 < yu_*/\nu < 1000$, we get

$$\frac{u}{u_*} = 8.74 \left(\frac{yu_*}{\nu}\right)^{1/7} \tag{9.38}$$

Now, if we consider the outer limit of the boundary layer where $u = U_0$ and $y = \delta$ and if we recall that $u_* = \sqrt{\tau_0/\rho}$ and substitute all of these into Eq. (9.38) and solve for τ_0/ρ, we obtain

$$\frac{\tau_0}{\rho} = 0.0225 U_0^2 \left(\frac{\nu}{U_0\delta}\right)^{1/4} \tag{9.39}$$

This is valid for smooth surfaces up to a Reynolds number, based on the length of the boundary, of about 10^7. When Eqs. (9.37) and (9.39) are combined, the resulting equation is

$$\frac{0.0225\nu^{1/4}}{U_0^{1/4}} = \frac{7}{72} \delta^{1/4} \frac{d\delta}{dx}$$

When the variables are separated and the integration is performed, we have

$$\delta^{5/4} = \frac{5}{4} \left[\frac{(0.0225)(72)}{7}\right] \frac{x\nu^{1/4}}{U_0^{1/4}} + C \tag{9.40}$$

We are assuming that the boundary layer starts from the leading edge, so we evaluate C by taking $x = 0$ and $\delta = 0$; therefore, $C = 0$. When Eq. (9.40) is solved for δ and simplified, we obtain

$$\delta = \frac{0.37x}{\text{Re}_x^{1/5}} \tag{9.41}$$

Thus we have the thickness of the turbulent boundary layer as a function of both the distance along the boundary and the Reynolds number based on the distance along the boundary.

Shearing Resistance of the
Turbulent Boundary Layer on a Flat Plate

When δ of Eq. (9.41) is substituted back into Eq. (9.39), we can express τ_0 in terms of the Reynolds number based on the distance along the boundary:

$$\tau_0 = \rho \frac{U_0^2}{2} \frac{0.058}{\mathrm{Re}_x^{1/5}} \tag{9.42}$$

Since $c_f = \tau_0 / \frac{1}{2}\rho U_0^2$, we can solve for c_f from Eq. (9.42):

$$c_f = \frac{\tau_0}{\rho U_0^2 / 2} = \frac{0.058}{\mathrm{Re}_x^{1/5}} \tag{9.43}$$

When τ_0 from Eq. (9.42) is integrated over the area of the boundary, we obtain the overall shearing resistance, which is

$$F_s = \frac{0.072BL}{\mathrm{Re}_L^{1/5}} \rho \frac{U_0^2}{2}$$

From the definition of C_f, Eq. (9.14), we conclude that C_f for the turbulent boundary layer along a smooth plate is $C_f = 0.072/\mathrm{Re}_L^{1/5}$. However, Schlichting (11) indicates that better agreement with experimental results is obtained when the numerical coefficient is given a value of 0.074. Then C_f is

$$C_f = \frac{0.074}{\mathrm{Re}_L^{1/5}} \tag{9.44}$$

Summary of Relations for the
Turbulent Boundary Layer on a Flat Plate

To summarize, the equations for the boundary-layer thickness and resistance for the turbulent boundary layer are

Boundary-layer thickness: $\qquad \delta = \dfrac{0.37x}{\mathrm{Re}_x^{1/5}} \qquad$ for $\mathrm{Re} < 10^7$

Local shear stress: $\qquad \tau_0 = c_f \rho \dfrac{U_0^2}{2}$

where $\qquad c_f = \dfrac{0.058}{\mathrm{Re}_x^{1/5}} \qquad$ for $\mathrm{Re} < 10^7$

Overall shearing resistance:
$$F_s = C_f B L \rho \frac{U_0^2}{2}$$

where
$$C_f = \frac{0.074}{\mathrm{Re}_L^{1/5}} \quad \text{for Re} < 10^7$$

In the foregoing developments we have applied the integral momentum equation over the boundary layer in order to derive useful equations for local shear stress and overall plate resistance. As a point of interest, the integral momentum equation, Eq. (9.31), is also used in a variety of other applications to relate the drag of a body to the difference in pressure distribution and momentum flux between an upstream and a downstream section. Applications of this type include the determination of the drag of an airfoil section, the analysis of boundary layers over rough surfaces, the analysis of boundary layers over evaporating surfaces [see Crowe et al. (4)], and the prediction of points of separation on curved surfaces.

EXAMPLE 9.7   Air at a temperature of 20°C and with a free-stream velocity of 30 m/s flows past a smooth, thin plate that is 3 m wide and 6 m long in the direction of flow. Assuming that the boundary layer is forced to be turbulent from the leading edge, determine the shear stress, the thickness of the viscous sublayer, and the thickness of the boundary layer 5 m downstream of the leading edge.

Solution   First compute Re_x at a distance 5 m from the leading edge:

$$\mathrm{Re}_x = U_0 \frac{x}{\nu} = \frac{(30 \text{ m/s})(5 \text{ m})}{1.49(10^{-5}) \text{ m}^2/\text{s}} = 10^7$$

$$\mathrm{Re}_x^{1/5} = (10^7)^{1/5} = 25$$

Compute τ_0, where $\tau_0 = c_f \rho U_0^2/2$. Here

$$c_f = 0.058 \mathrm{Re}_x^{-1/5} = \frac{0.058}{25}$$

Also
$$\rho = 1.20 \text{ kg/m}^3 \qquad \text{(from Appendix)}$$

Then
$$\tau_0 = \frac{0.058}{25}(1.20 \text{ kg/m}^3)\frac{30^2}{2}\text{m}^2/\text{s}^2$$

$$= 1.25 \text{ kg/m} \cdot \text{s}^2 = 1.25 \text{ N/m}^2 \qquad \blacktriangleleft$$

Now compute $u_* = \sqrt{\tau_0/\rho}$ and the thickness of the viscous sublayer and the boundary layer.

$$u_* = \left(\frac{\tau_0}{\rho}\right)^{1/2} = \left(\frac{1.25 \text{ N/m}^2}{1.20 \text{ kg/m}^3}\right)^{1/2} = 1.02 \text{ m/s}$$

The thickness of the viscous sublayer is given by

$$\delta' = \frac{5\nu}{u_*}$$

But $\nu = 1.49 \times 10^{-5}$ m^2/s, so

$$\delta' = \frac{5(1.49)(10^{-5}) \text{ m}^2/\text{s}}{1.02 \text{ m/s}} = 7.30 \times 10^{-5} \text{ m} = 0.07 \text{ mm} \qquad \blacktriangleleft$$

Compute the thickness of the boundary layer:

$$\delta = 0.37 \frac{x}{\text{Re}_x^{1/5}} = \frac{(0.37)(5) \text{ m}}{25} = 0.074 \text{ m} = 74 \text{ mm} \qquad \blacktriangleleft$$

The foregoing equations for the completely turbulent boundary layer on a smooth boundary are valid up to a Reynolds number of about 10^7. For higher Reynolds numbers, the $\frac{1}{7}$ power law is not precise and more refined analyses are required. See Schlichting (11) for the local shear stress and C_f at higher Reynolds numbers. One formula for C_f at high Reynolds numbers is

$$C_f = \frac{0.455}{(\log_{10} \text{Re}_L)^{2.58}} - \frac{1700}{\text{Re}_L} \qquad \text{for Re} > 10^7 \qquad (9.45)$$

Analysis of Schlichting's data shows that the magnitude of the relative boundary-layer thickness, δ/x, is related to Re_x by the following expression:

$$\frac{\delta}{x} = 0.079 \text{Re}_x^{-0.085} \qquad (9.46)$$

for Reynolds numbers greater than 10^7. Also, Schlichting showed that the local shear-stress coefficient is a function of Re_x at these high Reynolds numbers according to the equation

$$c_f = (2 \log \text{Re}_x - 0.65)^{-2.3} \qquad (9.47)$$

A plot of the average shear-stress coefficient C_f is shown in Fig. 9.13. Here the curve marked "completely turbulent boundary layer" is for the case where, by artificial roughening at the upstream edge of the boundary, the boundary layer is forced to be turbulent from the outset. It is of interest to note that marine engineers utilize this technique on ship models to produce a boundary layer that can be predicted more precisely than the combination of laminar and turbulent boundary layers.

We have derived formulas for the velocity distribution and shearing resistance of the laminar and turbulent boundary layers. Now we shall discuss the existence of laminar and turbulent boundary layers together on a smooth, flat plate and examine how one determines the total resistance when they both occur

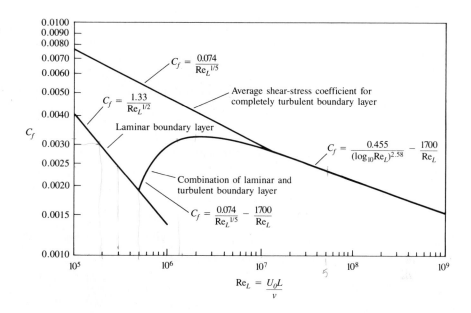

FIGURE 9.13

Average shear-stress coefficients. [After Schlichting (11)].

on the same plate. As noted in Sec. 9.3, the laminar boundary layer first develops on the upstream end of the plate. Then, as this layer grows in thickness, it becomes unstable and turbulence sets in. A turbulent boundary layer develops over the remainder of the plate. The onset of turbulence depends to a certain extent on the degree of smoothness of the plate and on the degree of turbulence. However, when the approach flow is nonturbulent, the transition or critical region on a smooth plate occurs at a Reynolds number, Re_x, of about 500,000. Then, when the turbulent boundary layer develops downstream of the laminar layer, we wonder whether the turbulent boundary layer will have the characteristics of one with the origin at the leading edge of the plate (same as the origin for the laminar layer) or of one with the origin at the downstream end of the laminar layer. Experiments reveal that the former is the most valid model. Thus, when calculating the overall resistance of a plate with a laminar and a turbulent boundary layer, we must evaluate the two resistances separately and sum them to obtain the total resistance. The calculation of the resistance of the laminar part is straightforward, as given in Example 9.4. However, in calculating the resistance of the turbulent part of the boundary layer, we must compute the resistance as though the entire boundary layer were turbulent and then subtract from that the resistance that would have occurred on the plate up to the transition zone.

When the foregoing model for resistance is stated mathematically, we have

$$F_s = \left(\frac{1.33}{\mathrm{Re}_{cr}^{1/2}} Bx_{cr} + \frac{0.074}{\mathrm{Re}_L^{1/5}} BL - \frac{0.074}{\mathrm{Re}_{cr}^{1/5}} Bx_{cr} \right) \rho \frac{U_0^2}{2} \qquad (9.48)$$

where Re_{cr} is the Reynolds number at the transition, Re_L is the Reynolds number at the end of the plate, and x_{cr} is the distance from the leading edge of the plate to the critical or transition zone.

Then if we define the average resistance coefficient as $F_s = C_f BL\rho U_0^2/2$, we can solve for C_f utilizing Eq. (9.48). Doing so yields

$$C_f = \frac{1.33}{\mathrm{Re}_{cr}^{1/2}}\frac{x_{cr}}{L} + \frac{0.074}{\mathrm{Re}_L^{1/5}} - \frac{0.074}{\mathrm{Re}_{cr}^{1/5}}\frac{x_{cr}}{L}$$

Here $x_{cr}/L = \mathrm{Re}_{cr}/\mathrm{Re}_L$. Therefore, we get

$$C_f = \frac{1.33}{\mathrm{Re}_{cr}^{1/2}}\frac{\mathrm{Re}_{cr}}{\mathrm{Re}_L} + \frac{0.074}{\mathrm{Re}_L^{1/5}} - \frac{0.074}{\mathrm{Re}_{cr}^{1/5}}\frac{\mathrm{Re}_{cr}}{\mathrm{Re}_L}$$

or

$$C_f = \frac{0.074}{\mathrm{Re}_L^{1/5}} - \frac{\mathrm{Re}_{cr}}{\mathrm{Re}_L}\left(\frac{0.074}{\mathrm{Re}_{cr}^{1/5}} - \frac{1.33}{\mathrm{Re}_{cr}^{1/2}}\right)$$

For $\mathrm{Re}_{cr} = 500,000$, we have

$$C_f = \frac{0.074}{\mathrm{Re}_L^{1/5}} - \frac{1700}{\mathrm{Re}_L} \tag{9.49}$$

When Eq. (9.49) is applied for various values of Re_L from 5×10^5 to 10^7, we obtain the laminar-turbulent curve for C_f shown in Fig. 9.13.

Even though the equations in this chapter have been developed for flat plates, they are useful for engineering estimates for some surfaces that are not truly flat plates. For example, the skin-friction drag of the submerged part of the hull of a ship can be estimated with Eq. (9.45).

EXAMPLE 9.8 Assume that a boundary layer over a smooth, flat plate is laminar at first and then becomes turbulent at a critical Reynolds number of 5×10^5. If we have a plate 3 m long and 1 m wide and if air, at 20°C and normal atmospheric pressure, flows past this plate with a velocity of 30 m/s, what will be the average resistance coefficient C_f for the plate? Also, what will be the total shearing resistance of one side of the plate and what will be the resistance due to the turbulent part and the laminar part of the boundary layer?

Solution The total resistance is $F_s = C_f BL\rho(U_0^2/2)$, where from Eq. (9.49) we get C_f:

$$C_f = \frac{0.074}{\mathrm{Re}_L^{1/5}} - \frac{1700}{\mathrm{Re}_L}$$

In addition, $\mathrm{Re}_L = UL/\nu$, or

$$\mathrm{Re}_L = \frac{30 \text{ m/s} \times 3 \text{ m}}{(1.49)(10^{-5}) \text{ m}^2/\text{s}} = 6.04 \times 10^6$$

Then, solving Eq. (9.49), we have

$$C_f = 0.00326 - 0.000281 = 0.00298$$

◀

The total resistance is calculated:

$$F_s = C_f B L \rho \frac{U_0^2}{2} = 0.00298 \times 1 \times 3 \times 1.2 \times \frac{30^2}{2} = 4.83 \text{ N} \quad \blacktriangleleft$$

Then x_{cr} is determined:

$$\frac{U x_{cr}}{\nu} = 500{,}000$$

or

$$x_{cr} = \frac{500{,}000 \times 1.49 \times 10^{-5}}{30} = 0.248 \text{ m}$$

Thus the laminar resistance will be

$$F_{s,\text{lam}} = \frac{1.33}{(5 \times 10^5)^{1/2}} \times 1 \times 0.248 \times 1.2 \times \frac{30^2}{2} = 0.252 \text{ N} \quad \blacktriangleleft$$

Then $\quad F_{s,\text{turb}} = 4.83 \text{ N} - 0.25 \text{ N} = 4.58 \text{ N} \quad \blacktriangleleft$

EXAMPLE 9.9 Determine the total drag of the plate given in Example 9.7.

Solution The shearing resistance of one side is given as $F_s = C_f B L \rho (U_0^2 / 2)$. Therefore, the total drag will be twice this for two sides of the plate:

$$F_s = C_f B L \rho U_0^2$$

Since C_f is a function of Re_L, we compute that as

$$\text{Re}_L = U_0 \frac{L}{\nu} = \frac{(30 \text{ m/s}) (6 \text{ m})}{(1.49)(10^{-5} \text{ m}^2/\text{s})} = 1.21(10^7)$$

From Eq. 9.49, $C_f = 0.0027$. Then

$$F_s = 0.0027(3 \text{ m}) (6 \text{ m}) (1.20 \text{ N} \cdot \text{s}^2/\text{m}^4) (30^2 \text{ m}^2/\text{s}^2) = 52.5 \text{ N} \quad \blacktriangleleft$$

9.6 Boundary-Layer Control

The preceding sections show that the boundary layer on a flat plate is initially laminar and then becomes turbulent at a distance along the plate where the Reynolds number is approximately 5×10^5. Also it was shown (Figs. 9.4 and 9.13) that the skin friction, for a given Reynolds number, is much greater for the turbulent part of the boundary layer than for the laminar part. Thus if there were some way to prevent the boundary layer from becoming turbulent, the skin-friction drag could be reduced. This is an active area of research today, and it has been for the past several decades. Procedures for producing this desirable effect are called *boundary-layer control*.

One way to control the boundary layer, for example on an airfoil, is to shape the airfoil in such a way that the pressure distribution on its surface delays the onset of turbulence. Several such airfoils have been developed, but the amount of drag reduction that it is possible to achieve in this manner is limited because alteration of the airfoil's basic shape so often decreases its lift.

Another means of boundary-layer control is to make the surface of the boundary, such as an airfoil surface, somewhat porous and to apply a reduced pressure to the surface (reduced pressure or suction is maintained inside the airfoil) so that part of the boundary layer is drawn away through the porous surface (part of the air of the boundary layer is sucked into the interior of the airfoil). This keeps the boundary layer thin, so it remains stable in its laminar state and a reduced skin-friction drag results. Wind-tunnel tests prove that boundary-layer control achieved by applying suction to a porous surface is effective in reducing drag; studies indicate that a commercial aircraft with such control would burn 20% less fuel (13)! The initial cost of the aircraft would be increased, but it is estimated that the fuel savings would pay for the additional cost within 6 months (13).

Some of the severe practical problems associated with boundary-layer control by suction involve the surface finish. For example, the perforations made in the airfoil to produce the porous surface must be very small; otherwise, flow disturbances in the vicinity of each perforation will induce turbulence. One suction surface that was successfully tested at the NASA Langley Research Center in Virginia was made of titanium sheet bonded to fiberglass sandwich panel. The perforations on the sheet were 0.0025 in. in diameter with a spacing of 0.025 in. (13). These perforations were produced by electron beams.

Once such an airfoil is put in service, it must be "groomed" in such a way as to preserve its smoothness in order to prevent the development of turbulence from undesirable roughness. Insects and/or dirt cannot be allowed to build up on the surface, so an effective cleansing system is essential. For these and other details please refer to Wagner and Fisher (13).

Aircraft are not the only bodies that could benefit from boundary-layer control. Bar-Hain et al. (1) studied submerged bodies such as submarines and torpedoes. Their study focused on the application of a distributed suction technique over a portion of an axisymmetric body.

Another boundary-layer control measure involves the use of a dilute polymer solution for drag reduction in liquids. This method is discussed in more detail in Chapter 10 (see page 434) in connection with the reduction of head loss in pipes. Still another drag-reduction phenomenon associated with boundary layers is that of the compliant surface. Studies reveal that porpoises have a special kind of skin (its outer surface is very compliant) that delays the onset of turbulence in the boundary layer. Thus they can swim faster than any other animal of their size. Considerable research has been directed to the development of artificial surfaces to yield reduced drag, but only limited success has been achieved. See Kramer (8) and Benjamin (2) for more information on this boundary-layer phenomenon.

Problems

9.1 A flat plate is pulled to the right at a speed of 30 cm/s. Oil with a viscosity of 3 N · s/m²
fills the space between the plate and the solid boundary. The plate is 1 m long ($L = 1$ m)
by 30 cm wide, and the spacing between the plate and boundary is 2.0 mm.

   a. Express the velocity mathematically in terms of the coordinate system shown.
   b. By mathematical means, determine whether this flow is rotational or irrotational.
   c. Determine whether continuity is satisfied, using the differential form of the continuity equation.
   d. Calculate the force required to produce this plate motion.

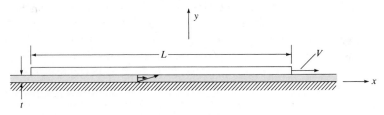

PROBLEM 9.1

9.2 Refer to the figure that accompanies Problem 2.18 in your text and assume that the figure is for the velocity distribution in a liquid such as oil. Indicate whether each of the following statements is true or false.

   a. The velocity gradient at the boundary is infinitely large.
   b. The maximum shear stress in the liquid occurs next to the boundary.
   c. The maximum shear stress in the liquid occurs midway between the walls.
   d. The flow is rotational.
   e. The flow is irrotational.

9.3 A cube weighing 100 N and measuring 30 cm on a side is allowed to slide down an inclined surface on which there is a film of oil having a viscosity of 10^{-2} N · s/m². What is
the terminal velocity of the block if the oil has a thickness of 0.1 mm?

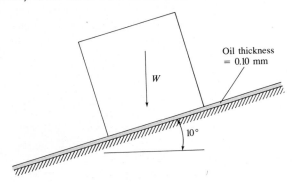

PROBLEM 9.3

9.4 A board 3 ft by 3 ft that weighs 20 lbf slides down an inclined ramp with a velocity of 0.5 fps. The board is separated from the ramp by a layer of oil 0.02 in. thick. Neglecting the edge effects of the board, calculate the approximate dynamic viscosity μ of the oil.

9.5 A board 1 m by 1 m that weighs 13 N slides down an inclined ramp with a velocity of 12 cm/s. The board is separated from the ramp by a layer of oil 0.5 mm thick. Neglecting the edge effects of the board, calculate the approximate dynamic viscosity μ of the oil.

$$\tau = \mu \frac{\Delta u}{\Delta y}$$

$A = 3 \times 3 = 9\ ft^2$

$W \sin \theta = \tau A$

$\dfrac{W \sin \theta\ \Delta y}{A\ \Delta u} = \mu = \dfrac{20\ lbf \left(\frac{5}{13}\right) 0.02^{in}}{9\ ft^2\ 0.5\ \frac{f}{s}\ 12^{in/ft}} = 0.00285$

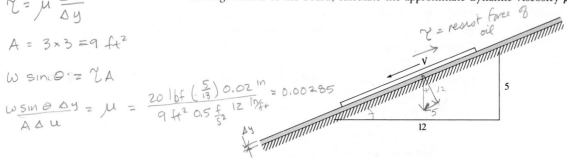

$\tau = $ resist force of oil

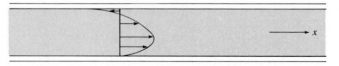

PROBLEMS 9.4, 9.5

9.6 Uniform, steady flow is occurring between horizontal parallel plates as shown.

    a. In a few words, tell what other condition must be present to cause the odd velocity distribution.

    b. Where is the minimum shear stress located?

PROBLEM 9.6

9.7 Under certain conditions (pressure decreasing in the x direction and a moving plate), the laminar velocity distribution will be as shown here. For such a condition, indicate whether each of the following statements is true or false.

    a. The greatest shear stress in the liquid occurs next to the fixed plate.
    b. The shear stress midway between the plates is zero.
    c. The minimum shear stress in the liquid occurs next to the moving plate.
    d. The shear stress is greatest where the velocity is greatest.
    e. The minimum shear stress occurs where the velocity is the greatest.

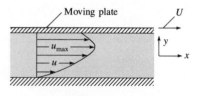

PROBLEM 9.7

9.8 A tube and a wire positioned concentrically within the tube are submerged in oil. If the wire is drawn through the tube at a constant rate, will the viscous shear stress on the wire be greater than, equal to, or less than the shear stress on the tube wall?

9.9 The upper plate shown is moving to the right with a velocity V, and the lower plate is free to move laterally under the action of the viscous forces applied to it. For steady-state conditions, derive an equation for the velocity of the lower plate. Assume that the area of oil contact is the same for the upper plate, each side of the lower plate, and the fixed boundary.

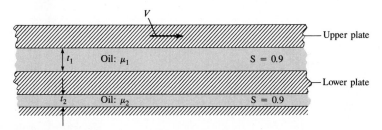

PROBLEM 9.9

9.10 A circular horizontal disk with a 12-in. diameter has a clearance of 0.001 ft from a horizontal plate. What torque is required to rotate the disk about its center at an angular velocity of 120 rpm when the clearance space contains oil ($\mu = 0.10$ lbf-s/ft^2)?

9.11 A circular horizontal disk with a 20-cm diameter has a clearance of 2.0 mm from a horizontal plate. What torque is required to rotate the disk about its center at an angular speed of 100 rad/s when the clearance space contains oil ($\mu = 6$ N $\cdot$ s/m^2)?

9.12 A movable cone fits inside a stationary conical depression as shown. When a torque is applied to the cone, the cone rotates at a speed depending on the angles θ and β, the radius r_0, and the viscosity μ of the liquid. Derive an equation for the torque in terms of the other variables, including only the viscous resistance. Assume that θ is very small.

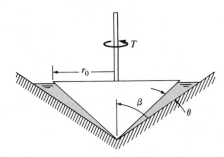

PROBLEM 9.12

9.13 A plate 2 mm thick and 1 m wide (normal to the page) is pulled between the walls shown in the figure at a speed of 0.20 m/s. Note that the space that is not occupied by the plate is filled with glycerine at a temperature of 20°C. Also, the plate is positioned midway between

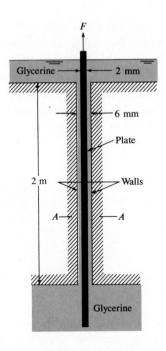

PROBLEM 9.13

the walls. Sketch the velocity distribution of the glycerine at section *A–A*. Neglecting the weight of the plate, estimate the force required to pull the plate at the speed given.

9.14   A bearing uses SAE 30 oil with a viscosity of 0.1 N · s/m². The bearing is 20 mm in diameter, and the gap between the shaft and the casing is 1 mm. The bearing has a length of 1 cm. The shaft turns at 100 rad/s. Assuming that the flow between the shaft and the casing is a Couette flow, find the torque required to turn the bearing.

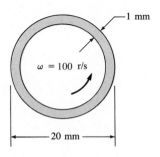

PROBLEM 9.14

9.15   An important application of surface resistance is found in lubrication theory. Consider a shaft that turns inside a stationary cylinder, with a lubricating fluid in the annular region. By considering a ring of fluid of radius r and width Δr and realizing that under steady-state operation the net torque on this ring is zero, show that $d(r^2\tau)/dr = 0$, where τ is the

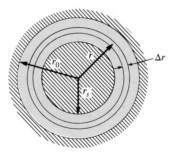

PROBLEM 9.15

viscous shear stress. For a fluid that has a tangential component of velocity only, the shear stress is related to the velocity by $\tau = \mu r\, d(v/r)/dr$. Show that the torque per unit length acting on the inner cylinder is given by $T = 4\pi\mu\omega r_s^2/(1 - r_s^2/r_0^2)$, where ω is the angular velocity of the shaft.

9.16 Using the equation developed in Prob. 9.15, find the power necessary to rotate a 2-cm shaft at 400 rad/s if the inside diameter of the casing is 2.2 cm, the bearing is 4 cm long, and SAE 30 oil at 38°C is the lubricating fluid.

9.17 The analysis developed in Prob. 9.15 applies to a device used to measure the viscosity of a fluid. By applying a known torque to the inner cylinder and measuring the angular velocity achieved, one can calculate the viscosity of the fluid. Assume you have a 4-cm inner cylinder and a 4.5-cm outer cylinder. The cylinders are 10 cm long. When a force of 0.6 N is applied to the tangent of the inner cylinder, it rotates at 30 rpm. Calculate the viscosity of the fluid.

9.18 If a thin film of oil ($\nu = 10^{-3}$ m²/s) 3.5 mm thick flows down a surface inclined at 30° to the horizontal, what will be the maximum and mean velocity of flow?

9.19 What are the depth and discharge per unit width of oil (SAE 30 at 100°F) flowing down a 45° incline at a Reynolds number of 200?

9.20 Rain falls on a smooth roof 15 ft by 40 ft at a rate of 0.4 in./hr, and the roof has a slope, in the 15-ft direction, of 10°. Estimate the depth and average velocity of flow at the lower end of the roof, assuming a temperature of 50°F.

9.21 Show that the mean velocity for laminar flow over a plate is $2u_{max}/3$.

9.22 Two horizontal parallel plates are spaced 2 mm apart. If the pressure decreases at a rate of 1.80 kPa/m in the horizontal x direction in the fluid between the plates, what is the maximum fluid velocity in the x direction? The fluid has a dynamic viscosity of 10^{-1} N · s/m² and a specific gravity of 0.80. What is the magnitude of the shearing force on the upper plate if it is 2 m long (in the direction of flow) and 1.5 m wide?

9.23 Two horizontal parallel plates are spaced 0.01 ft apart. The pressure decreases at a rate of 14 psf/ft in the horizontal x direction in the fluid between the plates. What is the maximum fluid velocity in the x direction? The fluid has a dynamic viscosity of 10^{-3} lbf-s/ft² and a specific gravity of 0.80.

9.24 A viscous fluid fills the space between these two plates, and the pressures at A and B are 160 psf and 100 psf, respectively. The fluid is not accelerating. If the specific weight of

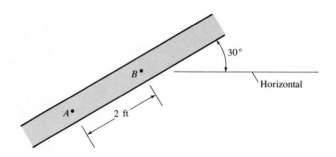

PROBLEM 9.24

the fluid is 100 lbf/ft^3, then one must conclude that a) flow is downward, b) flow is upward, c) there is no flow.

9.25 Gylcerine at 20°C flows downward between two vertical parallel plates separated by a distance of 1 cm. The ends are open, so there is no pressure gradient. Calculate the discharge per unit width, q, in m^2/s.

9.26 Two vertical parallel plates are spaced 0.01 ft apart. If the pressure decreases at a rate of 9 psf/ft in the vertical z direction in the fluid between the plates, what is the maximum fluid velocity in the z direction? The fluid has a viscosity of 10^{-3} lbf-s/ft^2 and a specific gravity of 0.80.

9.27 Two vertical parallel plates are spaced 2 mm apart. If the pressure decreases at a rate of 10 kPa/m in the positive z direction (vertically upward) in the fluid between the plates, what is the maximum fluid velocity in the z direction? The fluid has a viscosity of 10^{-1} N · s/m^2 and a specific gravity of 0.85.

9.28 Two vertical parallel plates are spaced 0.01 ft apart. If the pressure decreases at a rate of 70 psf/ft in the vertical z direction in the fluid between the plates, what is the maximum fluid velocity in the z direction? The fluid has a viscosity of 10^{-3} lbf-s/ft^2 and a specific gravity of 0.80.

9.29 Two parallel plates are spaced 0.10 in. apart, and motor oil (SAE 30) with a temperature of 100°F flows at a rate of 0.0083 cfs per foot of width between the plates. What is the pressure gradient in the direction of flow if the plates are inclined at 60° with the horizontal and if the flow is downward between the plates?

9.30 Two parallel plates are spaced 2 mm apart, and oil ($\mu = 10^{-1}$ N · s/m^2, S = 0.80) flows at a rate of 24×10^{-4} m^3/s per meter of width between the plates. What is the pressure gradient in the direction of flow if the plates are inclined at 60° with the horizontal and if the flow is downward between the plates?

9.31 Glycerine at 20°C flows downward in the annular region between two cylinders. The internal diameter of the outer cylinder is 2 cm, and the external diameter of the inner cylinder is 1.8 cm. The pressure is uniform along the flow direction. The flow is laminar. Calculate the discharge. (*Hint:* The flow between the two cylinders can be treated as the flow between two flat plates.)

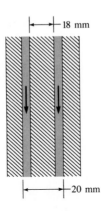

PROBLEM 9.31

9.32 One type of bearing that can be used to support very large structures is shown in the accompanying figure. Here fluid under pressure is forced from the bearing midpoint (slot A) to the exterior zone B. Thus a pressure distribution occurs as shown. For this bearing, which is 30 cm wide, what discharge of oil from slot A per meter of length of bearing is required to support a 50-kN load per meter of bearing length with a clearance space t between the floor and the bearing surface of 0.60 mm? Assume an oil viscosity of 10^{-1} N · s/m². How much oil per hour would have to be pumped per meter of bearing length for the given conditions?

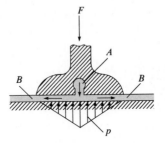

PROBLEM 9.32

9.33 Refer to the figure for Probs. 4.51 and 4.52 on p. 147. Instead of air, oil (SAE 10W at 100°F) flows in the system. The flow is laminar and the disks lie in the horizontal plane, so the velocity distribution is parabolic. The discharge per unit width is given by $q = -(B^3/12\mu)\, dp/dr$, where B is the distance between disks, p is the pressure, and μ is the viscosity. Continuity requires that $Q = 2\pi r q = $ constant. Derive an equation for the pressure change, Δp, between radii r_1, and r_2 in terms of Q, μ, B, and r_1 and r_2. Calculate the pressure change for $r_1 = 1$ cm, $r_2 = 10$ cm, and $B = 5$ mm for a discharge of 0.0003 m²/s. Neglect velocity head changes.

9.34 A thin plate 5 ft long and 3 ft wide is submerged and held stationary in a stream of water ($T = 60°F$) that has a velocity of 6 ft/s. What is the thickness of the boundary layer on the plate for $\mathrm{Re}_x = 500,000$ (assume the boundary layer is still laminar), and at what distance downstream of the leading edge does this Reynolds number occur? What is the shear stress on the plate at this point?

9.35 Determine the shearing force due to the laminar part of the boundary layer on one side of the plate of Prob. 9.34.

9.36 Oil ($\mu = 10^{-2}$ N · s/m²; $\rho = 900$ kg/m³) flows past a plate in a tangential direction so that a boundary layer develops. If the velocity of approach is 4 m/s, then at a section 30 cm downstream of the leading edge the ratio of τ_δ (shear stress at the edge of the boundary layer) to τ_0 (shear stress at the plate surface) is approximately a) 0, b) 0.24, c) 2.4, d) 24.

9.37 Verify Eqs. (9.15) and (9.16).

9.38 You want to use the integral technique to determine the thickness of a laminar boundary layer. Assume that the velocity profile can be approximated by $u/U_0 = (y/\delta)^{1/2}$. Experimental data show that $\tau_0 = 1.66 U \mu/\delta$. Use Eq. (9.35) and the foregoing relations to obtain a differential equation for δ and solve for $\delta = f(x)$. Compare your results with those of Blasius, Eq. (9.7).

9.39 A liquid ($\rho = 1000$ kg/m³, $\mu = 10^{-2}$ N · s/m², $\nu = 10^{-5}$ m²/s) flows tangentially past a flat plate. If the approach velocity is 1 m/s, what is the liquid velocity 1 m downstream from the leading edge of the plate and 2 mm away from the plate?

9.40 The plate of Prob. 9.39 has a total length of 3 m (parallel to the flow direction), and it is 2 m wide. What is the skin-friction drag (shear force) on one side of the plate?

9.41 Oil ($\nu = 10^{-4}$ m²/s) flows tangentially past a thin plate. If the free-stream velocity is 6 m/s, what is the velocity 1 m downstream from the leading edge and 8 mm away from the plate?

9.42 Oil ($\nu = 10^{-4}$ m²/s, S = 0.9) flows past a plate in a tangential direction so that a boundary layer develops. If the velocity of approach is 1 m/s, what is the oil velocity 1 m downstream from the leading edge of the plate and 11 cm away from the plate?

9.43 A thin plate 1 m long and 1 m wide is submerged and held stationary in a stream of water ($T = 10°C$) that has a velocity of 2 m/s. What is the thickness of the boundary layer on the plate for $\mathrm{Re}_x = 500,000$ (assume the boundary layer is still laminar), and at what distance downstream of the leading edge does this Reynolds number occur? What is the shear stress on the plate on this point?

9.44 An airplane wing of 2 m chord (leading edge to trailing edge distance) and 10 m span flies at 200 km/h in air at 30°C. Assume that the resistance of the wing surfaces is like that of a flat plate.

    a. What is the friction drag on the wing?
    b. What power is required to overcome this?
    c. How much of the chord is laminar?
    d. What would be the change in drag if the turbulent boundary layer were tripped at the leading edge?

$$Re_L = \frac{V_0 L}{\nu} = \frac{1 \frac{m}{s} \, 3m}{10^{-5} \, m^2/s} = 3 \times 10^5$$

$$C_f = \frac{1.33}{Re_L^{1/2}} \approx 0.0024$$

$$F_s = \frac{1}{2}\left(C_f \, \rho \, A \, U_0^2\right)$$

$$\delta(x) = \delta(1m) = \frac{5x}{\sqrt{Re_L}} = 5 \times 10^{-2}$$
$$(L = 1m)$$

$$Re_L = \frac{V_0 L}{\nu} = 10^4$$

9.45 A turbulent boundary layer exists in the flow of water at 20°C over a flat plate. The local shear stress measured at the surface of the plate is 0.1 N/m². What is the velocity at a point 1 cm from the plate surface?

9.46 A flat plate 1.5 m long and 1.5 m wide is towed in water at 20°C in the direction of its length at a speed of 20 cm/s. Determine the resistance of the plate and the boundary-layer thickness at its aft end.

9.47 A liquid flows tangentially past a flat plate. The fluid properties are $\mu = 10^{-5}$ N · s/m² and $\rho = 1.5$ kg/m³. Find the skin-friction drag on the plate per unit width if the plate is 2 m long and the approach velocity is 20 m/s. Also, what is the velocity gradient at a point that is 1 m downstream of the leading edge and just next to the plate ($y = 0$)?

9.48 Starting with Eq. (9.38), carry out the steps leading to Eq. (9.39).

9.49 A model airplane has a wing span of 3 ft and a chord (leading edge–trailing edge distance) of 6 in. The model flies in air at 60°F and atmospheric pressure. The wing can be regarded as a flat plate so far as drag is concerned. At what speed will a turbulent boundary layer start to develop on the wing? What will be the total drag force on the wing just before turbulence appears?

9.50 For the hypothetical boundary layer on the flat plate shown, what are the skin-friction drag on the top side per meter of width and the shear stress on the plate at the downstream end? Here $\rho = 1.2$ kg/m³ and $\mu = 1.8 \times 10^{-5}$ N · s/m².

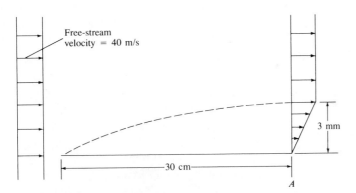

PROBLEMS 9.50, 9.53

9.51 Starting with Eq. (9.36), perform the integration and simplify to obtain Eq. (9.37).

9.52 Assume that the velocity profile in a boundary layer is replaced by a step profile, as shown in the figure, where the velocity is zero adjacent to the surface and equal to the free-stream velocity (U) at a distance greater than δ_* from the surface. Assume also that the density is uniform and equal to the free-stream density (ρ_∞). The distance δ_* (displacement thickness) is so chosen that the mass flux corresponding to the step profile is equal to the mass flux through the actual boundary layer. Derive an integral expression for the displacement thickness as a function of ρ, ρ_∞, u, U, y, and δ.

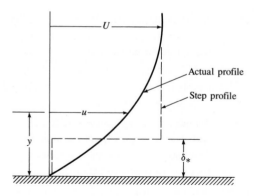

PROBLEM 9.52

9.53 Because of the reduction of velocity associated with the boundary layer, the streamlines outside the boundary layer are shifted away from the boundary. This amount of displacement of the streamlines is defined as the displacement thickness δ_*. For the boundary layer of section A of Prob. 9.50, what is the magnitude of the displacement thickness?

9.54 What is the ratio of the skin-friction drag of a plate 20 m long and 5 m wide to that of a plate 10 m long and 5 m wide if both plates are towed lengthwise through water ($T = 20°C$) at 10 m/s?

9.55 Estimate the power required to pull the sign shown if it is towed at 30 m/s and if it is assumed that the sign has the same resistance characteristics as a flat plate. Assume standard atmospheric pressure and a temperature of 10°C.

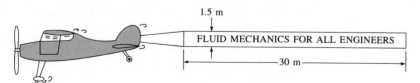

PROBLEM 9.55

9.56 A thin plastic panel (3 mm thick) is lowered from a ship to a construction site on the ocean floor. The plastic panel weighs 200 N in air and is lowered at a rate of 2 m/s. Assuming that the panel remains vertically oriented, calculate the tension in the cable.

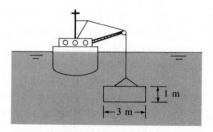

PROBLEM 9.56

9.57 A turbulent boundary layer develops from the leading edge of a flat plate with water at 20°C flowing tangentially past the plate with a free-stream velocity of 5 m/s. Determine the thickness of the viscous sublayer, δ', at a distance 1 m downstream from the leading edge. Would a roughness element 100 μm high affect the local skin friction coefficient? Why?

9.58 The plate shown in the figure is weighted at the bottom so that it will fall stably and steadily in a liquid. If it weighs 19 N in air and has a volume of 0.002 m³, estimate its speed of fall in a liquid with the following properties: $\nu = 10^{-6}$ m²/s and $\rho = 815$ kg/m³.

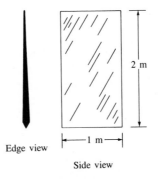

Edge view

Side view

PROBLEM 9.58

9.59 A model airplane descends in a vertical dive through air at standard conditions (1 atmosphere and 20°). The majority of the drag is due to skin friction on the wing (like that on a flat plate). The wing has a span of 1 m (tip to tip) and a chord (leading edge to trailing edge distance) of 10 cm. The leading edge is rough, so the turbulent boundary layer is "tripped." The model weighs 3 N. Determine the speed (in meters per second) at which the model will fall.

9.60 An outboard racing boat "planes" at 50 mph over water at 60°F. The hull has an average width of 5 ft and a length of 8 ft. Estimate the power required to overcome its surface resistance.

9.61 A javelin is approximately 265 cm long, has an average diameter of approximately 25 mm, and weighs 8.0 N. With a straight throw (javelin oriented parallel to the line of flight so that the relative air speed is parallel to the javelin), what will be the air drag at a speed of 30 m/s? What will be the javelin's deceleration (parallel to the line of flight) at its trajectory azimuth for the 30-m/s speed? What will be the corresponding drag and acceleration if the javelin is thrown into a 5-m/s wind and then thrown with a 5-m/s tailwind? Estimate the maximum distance of the throw if the thrower releases the javelin with a speed of 32 m/s. Assume the air temperature is 20°C.

9.62 A motor boat pulls a long, smooth, water-soaked log (1 m in diameter and 50 m long) at a speed of 1.7 m/s. Assuming total submergence, estimate the force required to overcome the surface resistance of the log. Assume a water temperature of 10°C.

9.63 Modern high-speed passenger trains are streamlined to reduce surface resistance. The cross section of a passenger car of one such train is shown. For a train 150 m long, estimate the surface resistance for a speed of 100 km/h and for one of 200 km/h. What power is required for just the surface resistance at these speeds? Assume $T = 10°C$.

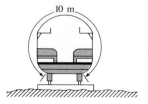

PROBLEM 9.63

9.64 Consider the boundary layer next to the smooth hull of a ship. The ship is cruising at a speed of 30 ft/s in 60°F fresh water. Assuming that the boundary layer on the ship hull develops the same as on a flat plate, determine

a. The thickness of the boundary layer at a distance of 100 ft downstream from the bow.
b. The velocity of the water at a point in the boundary layer at $y/\delta = 0.50$.
c. The shear stress, τ_0, adjacent to the hull at this position.

9.65 A ship 600 ft long steams at a rate of 30 ft/s through still fresh water ($T = 50°$). If the submerged area of the ship is 60,000 ft^2, what is the skin-friction drag of this ship?

9.66 A river barge has the dimensions shown below. It draws 2 ft of water when empty. Estimate the skin-friction drag of the barge when it is being towed at a speed of 10 ft/s through still fresh water.

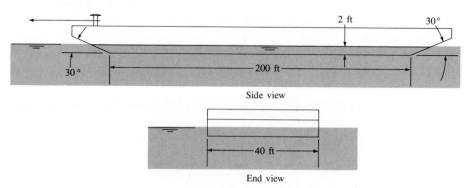

Side view

End view

PROBLEM 9.66

9.67 A supertanker has length, breadth, and draught (fully loaded) dimensions of 325 m, 48 m, and 19 m, respectively. In open seas the tanker normally operates at a speed of 15 kts (1 kt = 0.515 m/s). For these conditions, and assuming that flat-plate boundary-layer conditions are approximated, estimate the skin-friction drag of such a ship steaming in

10°C water. What power is required to overcome the skin-friction drag? What is the boundary-layer thickness at 300 m from the bow?

9.68 A model test is to be done to predict the wave drag on a ship. The ship is 500 ft long and operates at 30 ft/s in sea water at 10°C. The wetted area of the prototype is 25,000 ft². The model : prototype scale ratio is $\frac{1}{100}$. Modeling is done in fresh water at 60°F to match the Froude number. The viscous drag can be calculated by assuming a flat plate with the wetted area of the model and a length corresponding to the length of the model. A total drag of 0.1 lbf is measured in the model tests. Calculate the wave drag on the actual ship.

9.69 A ship is designed so that it is 250 m long, its beam measures 30 m, and its draft is 12 m. The surface area of the ship below the water line is 8800 m². A $\frac{1}{30}$ scale model of the ship is tested and is found to have a total drag of 38.0 N when towed at a speed of 1.45 m/s. Using the methods outlined in Sec. 8.10, answer the following questions, assuming that model tests are made in fresh water (20°C) and that prototype conditions are sea water (10°C).

a. To what speed in the prototype does the 1.45 m/s correspond?

b. What are the model skin-friction drag and wave drag?

c. What would the ship drag be in salt water corresponding to the model-test conditions in fresh water?

9.70 A hydroplane 4 m long skims across a very calm lake ($T = 20$°C) at a speed of 20 m/s. For this condition, what will be the minimum shear stress along the smooth bottom?

9.71 Estimate the power required to overcome the surface resistance of a water skier if he is towed at 30 mph and each ski is 4 ft by 6 in. Assume the water temperature is 60°F.

9.72 If the wetted area of an 80-m ship is 1500 m², approximately how great is the surface drag when the ship is traveling at a speed of 10 m/s? What is the thickness of the boundary layer at the stern? Assume $T = 10$°C.

References

1. Bar-Hain, B., and D. Weihs. "Boundary-Layer Control as a Means of Reducing Drag on Fully Submerged Bodies of Revolution." *Trans. of the ASME Fluids Engineering Division,* vol. 107 (September 1985), p. 107.

2. Benjamin, T. B. "Fluid Flow with Flexible Boundaries." *Proc. Intern. Congress of Applied Math,* 11th ed., p. 109. Springer-Verlag, 1964.

3. Blasius, H. "Grenzschichten in Flüssigkeiten mit kleiner Reibung." *Z. Mat. Physik.* (1908), p. 1. Summarized in Durand (6).

4. Crowe, C. T., J. A. Nicholls, and R. B. Morrison. "Drag Coefficients of Inert and Burning Particles Accelerating in Gas Streams." In *Ninth International Symposium on Combustion,* pp. 395–406. Academic Press, New York, 1963.

5. Daily, J. W., and D. J. R. Harleman. *Fluid Dynamics.* Addison-Wesley, Reading, Mass., 1966.

6. Durand, William Frederick (ed.). *Aerodynamic Theory.* Dover Publications, New York, 1963.

7. Kline, S. J., M. V. Morkovin, G. Sarron, and D. J. Cockrell (eds.). "Procedures of Computation of Turbulent Boundary Layers." Thermosciences Division, Department of Mechanical Engineers, Stanford University, Palo Alto, Calif., 1968.

8. Kramer, M. O. "Boundary Layer Stabilization by Distributed Damping." *Jour. of Aeron. Sci.,* Vol. 24, p. 459 (1957).

9. Prandtl, L. "Über Flussigkeitsbewegung bei sehr kleiner Reibung." *Verhandlungen des III. Internationalen Mathematiker-Kongresses.* Leipzig, 1905.

10. Rouse, H., et al. *Advanced Mechanics of Fluids.* John Wiley, New York, 1959.

11. Schlichting, H. *Boundary Layer Theory.* McGraw-Hill, New York, 1979.

12. Todd, F. H. "Viscous Resistance of Ships." In *Advances in Hydroscience,* vol. 3. Academic Press, New York, 1966.

13. Wagner, R. D., and M. C. Fisher. "Fresh Attack on Laminar Flow." *Aerospace America* (March 1984).

14. White, Frank M. *Viscous Fluid Flow.* McGraw-Hill, New York, 1974.

Flow in Conduits

This trapezoidal canal is part of the Colorado River Aqueduct that conveys water to Southern California. The canal is 55 feet across the top, 20 feet wide at the bottom, 11 feet deep and delivers water at a maximum rate of 1,800 cfs. (Courtesy of Metropolitan Water District of Southern California)

W hen one considers the conveniences and necessities of everyday life, it is truly amazing to note the role played by conduits. For example, all of the water that we use in our homes is pumped through pipes so that it will be available when and where we want it. In addition, virtually all of this water leaves our homes as dilute wastes through sewers, another type of conduit. In addition to domestic use, the consumption of water by industry is enormous, from the processing of agricultural products to the manufacturing of steel and paper. All of the water used in these manufacturing processes is transported by means of piping systems. In the United States the petroleum industry alone transports approximately 20 million barrels of liquid petroleum per day in addition to the billions of cubic feet of gas transported by pipeline.

In the foregoing examples it is the transportation of the fluid that is the primary objective. However, there are numerous applications in which flow is a necessary but secondary part of the process. For example, heating and ventilating systems, as well as electric generating stations, utilize conduit flow to circulate fluids in order to transport energy from one location to another. Piping systems are also used extensively for controlling the operation of machinery.

Thus the application of flow in conduits cuts across all fields of engineering. Consequently, every engineer should understand the basic fluid mechanics involved with such flow. In this chapter we shall introduce the fundamental theory of flow in conduits as well as basic design procedures.

10.1   Shear-Stress Distribution Across a Pipe Section

The velocity distribution in a pipe is directly linked to the shear-stress distribution; hence it is important to understand the latter. To determine the shear-stress distribution, we start with the equation of equilibrium applied to a cylindrical element of fluid that is oriented coaxially with the pipe, as shown in Fig. 10.1.

FIGURE 10.1

Variation of shear stress in a pipe.

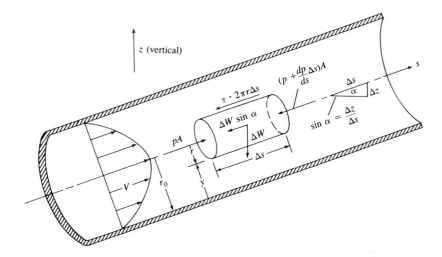

For the conditions shown in Fig. 10.1, it is assumed that the flow is uniform (streamlines are straight and parallel). Therefore, the pressure across any section of the pipe will be hydrostatically distributed. Thus the pressure force acting on an end face of the fluid element will be the product of the pressure at the center of the element (also at the center of the pipe) and the area of the face of the element. With steady uniform flow, equilibrium between the pressure, gravity, and shearing forces acting on the fluid will prevail. Consequently, the equilibrium equation yields the following:

$$\sum F_s = 0$$

$$pA - \left(p + \frac{dp}{ds}\Delta s\right)A - \Delta W \sin \alpha - \tau(2\pi r)\,\Delta s = 0 \qquad (10.1)$$

In Eq. (10.1) $\Delta W = \gamma A\,\Delta s$ and $\sin \alpha = dz/ds$. Therefore, Eq. (10.1) reduces to

$$-\frac{dp}{ds}\Delta s\,A - \gamma A\,\Delta s\frac{dz}{ds} - \tau(2\pi r)\,\Delta s = 0 \qquad (10.2)$$

Then, when we divide Eq. (10.2) through by $\Delta s\,A$ and simplify, we obtain

$$\tau = \frac{r}{2}\left[-\frac{d}{ds}(p + \gamma z)\right] \qquad (10.3)$$

Since the gradient itself, $d/ds(p + \gamma z)$, is negative (see Sec. 7.5) and constant across the section for uniform flow,* it follows that $-d/ds(p + \gamma z)$ will be positive and constant across the pipe section. Thus τ in Eq. (10.3) will be zero at

---

* *The combination p + γz is constant across the section because the streamlines are straight and parallel in uniform flow, and for this condition there will be no acceleration of the fluid normal to the streamline. Thus hydrostatic conditions prevail across the flow section. For a hydrostatic condition, p/γ + z = constant or p + γz = constant, as shown in Chapter 3.*

the center of the pipe and will increase linearly to a maximum at the pipe wall. We will use Eq. (10.3) in the following section to derive the velocity distribution for laminar flow.

10.2 Laminar Flow in Pipes

We determine how the velocity varies across the pipe by substituting for τ in Eq. (10.3) its equivalent $\mu\, dV/dy$ and integrating. First, making the substitution, we have

$$\mu\frac{dV}{dy} = \frac{r}{2}\left[-\frac{d}{ds}(p + \gamma z)\right] \tag{10.4}$$

Because $dV/dy = -dV/dr$, Eq. (10.4) becomes

$$\frac{dV}{dr} = -\frac{r}{2\mu}\left[-\frac{d}{ds}(p + \gamma z)\right] \tag{10.5}$$

When we separate variables and integrate across the section, we obtain

$$V = -\frac{r^2}{4\mu}\left[-\frac{d}{ds}(p + \gamma z)\right] + C \tag{10.6}$$

We can evaluate the constant of integration in Eq. (10.6) by noting that when $r = r_0$, the velocity $V = 0$. Therefore, the constant of integration is given by $C = (r_0^2/4\mu)\,[-d/ds(p + \gamma z)]$, and Eq. (10.6) then becomes

$$V = \frac{r_0^2 - r^2}{4\mu}\left[-\frac{d}{ds}(p + \gamma z)\right] \tag{10.7}$$

Equation (10.7) indicates that the velocity distribution for laminar flow in a pipe is parabolic across the section with the maximum velocity at the center of the pipe. Figure 10.2 shows the variation of the shear stress and velocity in the pipe.

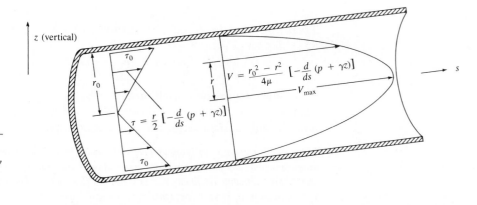

FIGURE 10.2

Distribution of shear stress and velocity for laminar flow in a pipe.

EXAMPLE 10.1 Oil ($S = 0.90$; $\mu = 5 \times 10^{-1}$ N · s/m²) flows steadily in a 3-cm pipe. The pipe is vertical, and the pressure at an elevation of 100 m is 200 kPa. If the pressure at an elevation of 85 m is 250 kPa, is the flow direction up or down? What is the velocity at the center of the pipe and at 6 mm from the center, assuming that the flow is laminar?

Solution First determine the rate of change of $p + \gamma z$. Taking s in the z direction,

$$\frac{d}{ds}(p + \gamma z) = \frac{(p_{100} + \gamma z_{100}) - (p_{85} + \gamma z_{85})}{15}$$

$$= \frac{[200 \times 10^3 + 8830(100)] - [250 \times 10^3 + 8830(85)]}{15}$$

$$= \frac{(1.083 \times 10^6 - 1.00 \times 10^6) \text{ N/m}^2}{15 \text{ m}} = 5.53 \text{ kN/m}^3$$

The quantity $p + \gamma z$ is not constant with elevation—it increases upward (decreases downward). Therefore, the direction of flow is downward. This can be seen by substituting $d(p + \gamma z)/ds = 5.53$ kN/m³ into Eq. (10.7). When this is done, V is negative for all values of r in the flow. When $r = 0$ (center of the pipe), the velocity will be maximum. Thus

$$V_{\text{center}} = V_{\text{max}} = \frac{r_0^2}{4\mu}(-5.53 \text{ kN/m}^3)$$

$$= \frac{0.015^2 \text{ m}^2}{4(5 \times 10^{-1} \text{ N} \cdot \text{s/m}^2)}(-5.53 \times 10^3 \text{ N/m}^3) = -0.622 \text{ m/s} \quad \blacktriangleleft$$

At first it may seem strange that the velocity is in a direction opposite to the direction of decreasing pressure. However, it may not seem so peculiar if one realizes that in this example the pipe is vertical, so the gravitational force as well as pressure helps to establish the flow. What counts when flow is other than in the horizontal direction is how the combination $p + \gamma z$ changes with s. If $p + \gamma z$ is constant, then we have the equation of hydrostatics and no flow occurs. However, if $p + \gamma z$ is not constant, flow will occur in the direction of decreasing $p + \gamma z$.

Next determine the velocity at $r = 6$ mm $= 0.006$ m. Using Eq. (10.7), we find that

$$V = \frac{0.015^2 \text{ m}^2 - 0.006^2 \text{ m}^2}{4(5 \times 10^{-1} \text{ N} \cdot \text{s/m}^2)}(-5.53 \times 10^3 \text{ N/m}^3) = -0.522 \text{ m/s} \quad \blacktriangleleft$$

For many problems we wish to relate the pressure change to the rate of flow or mean velocity $\overline{V}$ in the conduit. Therefore, it is necessary to integrate $dQ = V dA$ over the cross-sectional area of flow. That is,

$$Q = \int V dA$$

$$= \int_0^{r_0} \frac{(r_0^2 - r^2)}{4\mu} \left[-\frac{d}{ds}(p + \gamma z) \right] (2\pi r\, dr) \qquad (10.8)$$

The factor $\pi [d(p + \gamma z)/ds]/4\mu$ is constant across the pipe section. Therefore, upon integration, we obtain

$$Q = \frac{\pi}{4\mu} \left[\frac{d}{ds}(p + \gamma z) \right] \frac{(r^2 - r_0^2)^2}{2} \bigg|_0^{r_0} \qquad (10.9)$$

which reduces to

$$Q = \frac{\pi r_0^4}{8\mu} \left[-\frac{d}{ds}(p + \gamma z) \right] \qquad (10.10)$$

If we divide through by the cross-sectional area of the pipe, we have an expression for the mean velocity:

$$\overline{V} = \frac{r_0^2}{8\mu} \left[-\frac{d}{ds}(p + \gamma z) \right] \qquad (10.11)$$

Comparing Eqs. (10.11) and (10.7) reveals that $\overline{V} = V_{max}/2$. Also, by substituting $D/2$ for r_0, we have

$$\overline{V} = \frac{D^2}{32\mu} \left[-\frac{d}{ds}(p + \gamma z) \right] \qquad (10.12)$$

or

$$\frac{d}{ds}(p + \gamma z) = -\frac{32\mu\overline{V}}{D^2} \qquad (10.13)$$

Integrating Eq. (10.13) along the pipe between sections 1 and 2, we obtain

$$p_2 - p_1 + \gamma(z_2 - z_1) = -\frac{32\mu\overline{V}}{D^2}(s_2 - s_1) \qquad (10.14)$$

Here $s_2 - s_1$ is the length L of pipe between the two sections. Therefore, Eq. (10.14) can be rewritten as

$$\frac{p_1}{\gamma} + z_1 = \frac{p_2}{\gamma} + z_2 + \frac{32\mu L\overline{V}}{\gamma D^2} \qquad (10.15)$$

It can be seen that when the general energy equation for incompressible flow in conduits, Eq. (7.26), is reduced to one for uniform flow in a constant-diameter pipe where $V_1 = V_2$, the result is

$$\frac{p_1}{\gamma} + z_1 = \frac{p_2}{\gamma} + z_2 + h_f \qquad (10.16)$$

Here h_f is used instead of h_L to signify head loss due to frictional resistance of the pipe. Comparison of Eqs. (10.15) and (10.16) then shows that the head loss is given by

$$h_f = \frac{32\mu LV}{\gamma D^2} \qquad (10.17)$$

Here the bar over the V has been omitted to conform to the standard practice of denoting the mean velocity in one-dimensional flow analyses by V without the bar.

10.3   Criterion for Laminar or Turbulent Flow in a Pipe

To predict whether flow will be laminar or turbulent, it is necessary to explore the characteristics of flow in both laminar and turbulent states. Although other scientists before him had sensed the marked physical difference between laminar and turbulent flow, it was Osborne Reynolds (31) who first developed the basic laws of turbulent flow. With his analytical and experimental work he showed that the Reynolds number was a basic parameter relating to laminar as well as turbulent flow. For example, using an experimental apparatus such as that shown in Fig. 10.3, he found that the onset of turbulence in a smooth pipe was related to the Reynolds number ($VD\rho/\mu$) in a very interesting way. If the fluid in the upstream reservoir was not completely still or if the pipe had some vibration in it, the flow in the pipe as it was gradually increased from a low rate to higher rates was initially laminar but then changed from laminar to turbulent flow at a Reynolds number in the neighborhood of 2100. However, Reynolds found that if the fluid was initially completely motionless and if there was no vibration in the equipment while the flow was increased, it was possible to reach a much higher Reynolds number before the flow became turbulent. He also found that, when going from high-velocity turbulent flow to low-velocity flow, the change from turbulent flow always occurred at a Reynolds number of about 2000.

FIGURE 10.3

Schematic diagram of apparatus used by Reynolds to study laminar and turbulent flow.

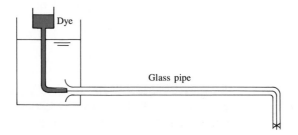

These experiments of Reynolds indicate that under carefully controlled conditions it is possible to have laminar flow in pipes at Reynolds numbers much higher than 2000. However, the slightest disturbances will trigger the onset of turbulence at high values of Re. Because most engineering applications involve some vibration or flow disturbance, it is reasonable to expect that pipe flow will be laminar for Reynolds numbers less than 2000 and turbulent for Reynolds numbers greater than 3000. When Re is between 2000 and 3000, the type of flow is very unpredictable and often changes back and forth between laminar and turbulent states. Fortunately, however, most engineering applications either are not in this range or are not significantly affected by the unstable flow.

EXAMPLE 10.2   Oil (S = 0.85) with a kinematic viscosity of $6 \times 10^{-4} \text{ m}^2/\text{s}$ flows in a 15-cm pipe at a rate of $0.020 \text{ m}^3/\text{s}$. What is the head loss per 100 m of length of pipe?

Solution   First we determine whether the flow is laminar or turbulent by checking to see if the Reynolds number is below 2000 or above 3000.

$$V = \frac{Q}{A} = \frac{0.020 \text{ m}^3/\text{s}}{(\pi/4)D^2 \text{ m}^2} = \frac{0.020 \text{ m}^3/\text{s}}{0.785(0.15^2 \text{ m}^2)} = 1.13 \text{ m/s}$$

Then
$$\text{Re} = \frac{VD}{\nu} = \frac{(1.13 \text{ m/s})(0.15 \text{ m})}{6 \times 10^{-4} \text{ m}^2/\text{s}} = 283$$

Since the Reynolds number is less than 2000, the flow is laminar. The head loss per 100 m is obtained from Eq. (10.17):

$$h_f = \frac{32\mu LV}{\gamma D^2}$$

Here $\mu/\gamma = \nu/g$; hence

$$h_f = \frac{32\nu LV}{gD^2}$$

Then
$$h_f = \frac{32(6)(10^{-4} \text{ m}^2/\text{s})(100 \text{ m})(1.13 \text{ m/s})}{(9.81 \text{ m/s}^2)(0.15^2 \text{ m}^2)} = 9.83 \text{ m}$$

The head loss is 9.83 m/100 m of length.   ◀

EXAMPLE 10.3   Kerosene (0°C) flows under the action of gravity in the pipe shown, which is 6 mm in diameter and 100 m long. Determine the rate of flow in the pipe.

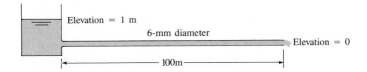

Solution   Because the pipe diameter is small and because the head producing flow is also quite small, it is expected that the velocity in the pipe will be small. Hence it will be initially assumed that the flow is laminar and $V^2/2g$ is negligible. Then, to solve for the velocity, we apply the energy equation to the problem. We write this equation between a section at the upstream water surface and the outlet of the pipe. Thus we have

$$\frac{p_1}{\gamma} + \frac{\alpha_1 V_1^2}{2g} + z_1 = \frac{p_2}{\gamma} + \frac{\alpha_2 V_2^2}{2g} + z_2 + \frac{32\mu LV}{\gamma D^2}$$

With the assumption we have noted, this equation reduces to

$$0 + 0 + 1 = 0 + 0 + 0 + \frac{32\mu LV}{\gamma D^2}$$

or
$$\frac{32\mu LV}{\gamma D^2} = 1$$

For 0°C the viscosity (from Figs. A.1 and A.2 in the Appendix) is

$$\mu = 3.2 \times 10^{-3} \text{ N} \cdot \text{s/m}^2 \qquad \nu = 3.9 \times 10^{-6} \text{ m}^2/\text{s}$$

Then $V = \dfrac{1 \times \gamma D^2}{32\mu L}$

$$= \frac{1(814 \text{ N/m}^3)(0.006^2 \text{ m}^2)}{32(3.2 \times 10^{-3} \text{ N} \cdot \text{s/m}^2)(100 \text{ m})} = 0.00286 \text{ m/s} = 2.86 \text{ mm/s}$$

Now check Re to see if the flow is laminar, and check $V^2/2g$ to see if it is indeed negligible:

$$\text{Re} = \frac{VD}{\nu} = \frac{(0.00286 \text{ m/s})(0.006 \text{ m})}{3.9 \times 10^{-6} \text{ m}^2/\text{s}} = 4.40$$

$$\frac{V^2}{2g} = \frac{(0.00286 \text{ m/s})^2}{(2)(9.81 \text{ m/s})} = 4.17 \times 10^{-7} \text{ m} \quad \text{(negligible)}$$

Therefore, the flow is laminar and the velocity is valid. The discharge is then calculated as follows:

$$Q = VA = (0.00286 \text{ m/s})\left(\frac{\pi}{4}\right)(0.006 \text{ m})^2 = 8.09 \times 10^{-8} \text{ m}^3/\text{s} \qquad \blacktriangleleft$$

10.4 | Turbulent Flow in Pipes

Turbulence and its Influence in Pipe Flow

In the preceding section it was pointed out that pipe flow is turbulent when the Reynolds number is larger than approximately 3000. However, to say that the flow is turbulent is only a gross description of it. We can obtain a better "feel" for the flow by exploring the similarities between turbulent flow in a pipe and flow in a turbulent boundary layer and by relating the shear stress in the pipe to the level of turbulence. Once we understand these basic physical relationships, we will be better equipped to proceed to the development of equations for the velocity distribution and the resistance to turbulent flow in pipes.

The similarities between turbulent-boundary-layer flow and turbulent flow in pipes are many. In fact, it is valid to think of turbulent flow in a pipe as a turbulent boundary layer that has become as thick as the radius of the pipe. With this perspective we realize that flow in a smooth pipe has a viscous sublayer just as a flat-plate boundary layer does. In addition, the velocity gradient in the viscous sublayer will be consistent with the shear stress, as given by $\tau = \mu\, du/dy$. However, outside the viscous sublayer the viscous shear stress is negligible compared with the resistance resulting from turbulence. We have already referred in Chapter 9 to the *apparent shear stress,* $\tau_{\text{app}} = -\rho\overline{u'v'}$, which involves an exchange of momentum, but its effect is like that of a true shear stress. It is zero at the pipe center and increases to a maximum near the wall, as shown in Fig. 10.4. Here it is seen that the apparent shear stress increases linearly almost to the edge of the pipe. This linear change in τ_{app} is in accordance with Eq. (10.3), which was developed in Sec. 10.1. Near the wall, in the viscous sublayer, τ_{app} reduces to zero because all of the shear stress there is in the form of viscous shear stress.

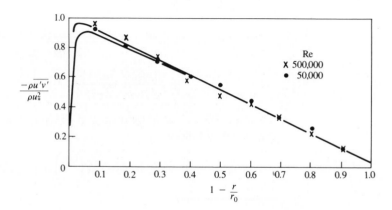

FIGURE 10.4

Apparent shear stress in a pipe. [After Laufer (22)]

We have shown that there are indeed many analogies between turbulent-boundary-layer flow and turbulent flow in pipes. The primary difference is that pipe flow is uniform and boundary-layer flow is not. Of course, this difference does not apply near the inlet of the pipe, where the flow is nonuniform.

Velocity Distribution and Resistance in Smooth Pipes

Experiments have shown that, in the viscous sublayer and in the turbulent zone near the wall, the velocity-distribution equations are of the same form as those for the turbulent boundary layer. That is, for a smooth pipe,

$$\frac{u}{u_*} = \frac{u_* y}{\nu} \qquad \text{for } 0 < \frac{y u_*}{\nu} < 5 \qquad (10.18)$$

$$\frac{u}{u_*} = 5.75 \log \frac{u_* y}{\nu} + 5.5 \qquad \text{for } 20 < \frac{y u_*}{\nu} \leq 10^5 \qquad (10.19)$$

Figure 10.5 is a plot of Eqs. (10.18) and (10.19) as well as an indication of the range of experimental data from various sources. For flow near the center of the

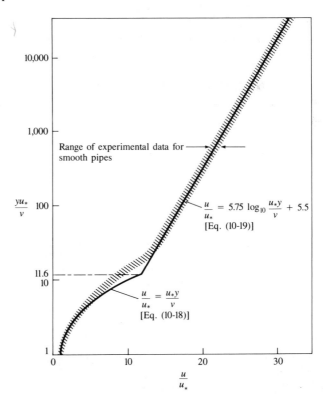

FIGURE 10.5

Velocity distribution for smooth pipes. [After Schlichting (35)]

pipe, as for flow near the outer limit of the boundary layer, the velocity-defect law is applicable, as shown in Fig. 10.6. Figure 10.6 also includes the range of experimental velocity data obtained from flow in rough conduits. Again, a power-law formula like that for the turbulent boundary layer is applicable everywhere except close to the wall. This formula is

$$\frac{u}{u_{\max}} = \left(\frac{y}{r_0}\right)^m \tag{10.20}$$

Here y is the distance from the wall and m is an empirically determined quantity. Some references indicate that m has a value of $\frac{1}{7}$ for turbulent flow. However, Schlichting (35) shows that m varies from $\frac{1}{6}$ to $\frac{1}{10}$ depending on the Reynolds number. His values for m are given in Table 10.1.

TABLE 10.1 EXPONENTS FOR POWER-LAW EQUATION AND RATIO OF MEAN TO MAXIMUM VELOCITY					
Re $\rightarrow$	4×10^3	2.3×10^4	1.1×10^5	1.1×10^6	3.2×10^6
$m \rightarrow$	$\dfrac{1}{6.0}$	$\dfrac{1}{6.6}$	$\dfrac{1}{7.0}$	$\dfrac{1}{8.8}$	$\dfrac{1}{10.0}$
$\overline{V}/V_{\max} \rightarrow$	0.791	0.807	0.817	0.850	0.865

SOURCE: Schlichting (35).

In Chapter 9 the shear stress on a flat plate was expressed as

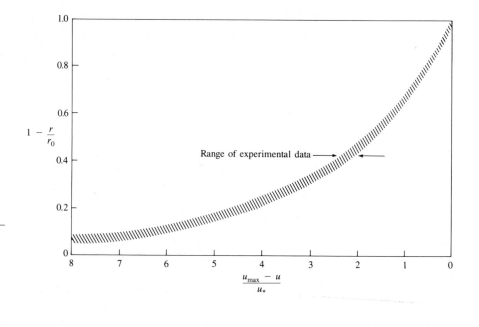

FIGURE 10.6

Velocity-defect law for turbulent flow in smooth and rough pipes. [After Schlichting (35)]

$$\tau_0 = c_f \rho \frac{V_0^2}{2}$$

where c_f is a function of the character of flow (laminar or turbulent) and the Reynolds number. For pipe flow it is customary to express τ_0 in a similar manner; however, we use the mean velocity as the reference velocity, and the coefficient of proportionality is given as $f/4$ instead of c_f. Here f is called the *resistance coefficient* or friction factor of the pipe. Thus we have

$$\tau_0 = \frac{f}{4} \rho \frac{V^2}{2} \tag{10.21}$$

Or, because $\sqrt{\tau_0/\rho} = u_*$, we have

$$\frac{u_*}{V} = \sqrt{\frac{f}{8}}$$

Noting that $\tau = \tau_0$ when $r = r_0$ in Eq. (10.3), we can eliminate τ_0 between Eqs. (10.3) and (10.21). Then, by integrating between two sections along the pipe, we obtain

$$h_1 - h_2 = f \frac{L}{D} \frac{V^2}{2g}$$

$$h_f = f \frac{L}{D} \frac{V^2}{2g} \tag{10.22}$$

where h_f is the head loss created by viscous effects and is equal to the change in piezometric head along the pipe. Equation (10.22) is called the *Darcy–Weisbach equation.* It is named after Henry Darcy, a French engineer of the nineteenth century, and Julius Weisbach, a German engineer and scientist of the same era. Weisbach first proposed the use of the nondimensional resistance coefficient, and Darcy carried out numerous tests on water pipes. Brief accounts of their work are given by Rouse and Ince (33). For laminar flow, it can be easily shown by a simultaneous solution of Eqs. (10.17) and (10.22) that the resistance coefficient is given by

$$f = \frac{64}{\text{Re}} \tag{10.23}$$

For turbulent flow, analytical and empirical results on smooth pipes yield the following approximate relation for f:

$$\frac{1}{\sqrt{f}} = 2 \log(\text{Re}\sqrt{f}) - 0.8 \qquad \text{for Re} > 3000 \tag{10.24}$$

Equation (10.24) was first developed by Prandtl.

Velocity Distribution and
Resistance—Rough Pipes

Numerous tests on flow in rough pipes all show that a semilogarithmic velocity distribution is valid over most of the pipe section (29, 34). This relationship is given in the following form:

$$\frac{u}{u_*} = 5.75 \log \frac{y}{k} + B \tag{10.25}$$

Here y is the distance from the rough wall, k is a measure of the height of the roughness elements, and B is a function of the character of roughness. That is, B is a function of the type, concentration, and size variation of the roughness. Research by Roberson and Chen (32) shows that B can be analytically determined for artificially roughened boundaries. More recent works by Wright (43), Calhoun (7), Kumar (21), and Eldridge (13) show that using a numerical approach with the same theory yields solutions for B and f for natural roughness, such as found in rock-bedded streams and commercial pipes when the roughness characteristics are known.

In 1933 Nikuradse (29) carried out a number of tests on flow in pipes that were roughened with uniform-sized sand grains. From these tests he found that the value of B with this kind of roughness was 8.5. Thus, for his tests, Eq. (10.25) becomes

$$\frac{u}{u_*} = 5.75 \log \frac{y}{k_s} + 8.5$$

In Nikuradse's tests the distance y was measured from the geometric mean of the wall surface, and k_s was the size of the sand grains.

Nikuradse's tests revealed two very important characteristics of rough-pipe flow. First, with low Reynolds numbers and with small-sized sand grains, the flow resistance is virtually the same as that for a smooth pipe. Second, for high values of the Reynolds number, the resistance coefficient is solely a function of the relative roughness k_s/D. These characteristics are shown in Fig. 10.7, where f is plotted as a function of Re for various values of the relative roughness k_s/D. The reason the resistance is like that of a smooth pipe for low values of k_s/D and Re is that for these conditions the roughness elements become submerged in the viscous sublayer and hence have negligible influence on the main flow in the pipe. However, at high values of the Reynolds number, the viscous sublayer is so thin that the roughness elements project into the main stream of flow and the flow resistance is determined by the drag of the individual roughness elements. Hence, for relatively large values of k_s/D and for large Reynolds numbers, the resistance to flow is proportional to V^2; thus f becomes constant for these conditions.

FIGURE 10.7
_____

Resistance coefficient f versus Re for sand-roughened pipe. [After Nikuradse (29)]

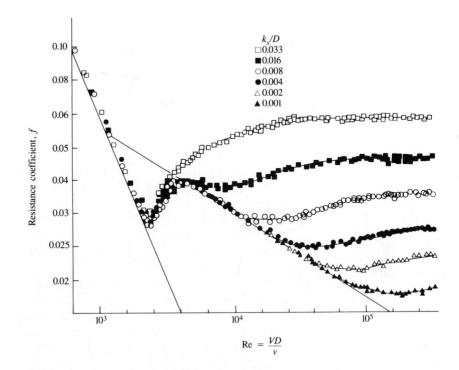

$$\text{Re} = \frac{VD}{v}$$

The uniform character of the sand grains used in Nikuradse's tests produced a dip in the f-versus-Re curve (Fig. 10.7) before the curve reached a constant value of f. However, tests on commercial pipes where the roughness is somewhat random reveal that no such dip occurs. By plotting data for commercial pipes from a number of sources, Moody (28) developed a design chart similar to that shown in Fig. 10.8 on page 428.

In Fig. 10.8 the variable k_s is the symbol used to denote the *equivalent sand roughness.* That is, a pipe that has the same resistance characteristics at high Re values as a sand-roughened pipe of the same size is said to have a size of roughness equivalent to that of the sand-roughened pipe. Figure 10.9 gives approximate values of k_s and k_s/D for various kinds of pipes. Both of these figures are used to solve certain kinds of pipe-flow problems.

In Fig. 10.8 the abscissa (labeled at the bottom) is the Reynolds number Re, and the ordinate (labeled at the left) is the resistance coefficient f. Each blue curve is for a constant relative roughness k_s/D and the values of k_s/D are given on the right at the end of each curve. To find f, given Re and k_s/D, one goes to the right to find the correct relative roughness curve. Then one looks at the bottom of the chart to find the given value of Re and, with this value of Re, moves vertically upward until the given k_s/D curve is reached. Finally, from this point one moves horizontally to the left scale to read the value of f. If the curve for the

FIGURE 10.8

Resistance coefficient f versus Re. Reprinted with minor variations. [After Moody (28). Reprinted with permission from the A.S.M.E.]

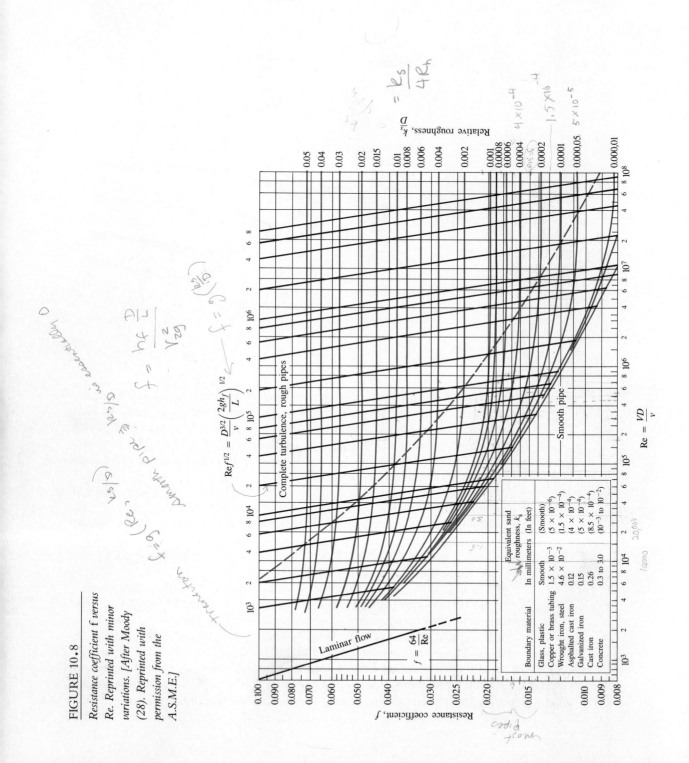

FIGURE 10.9

Relative roughness for various kinds of pipe. [After Moody (28). Reprinted with permission from the A.S.M.E.]

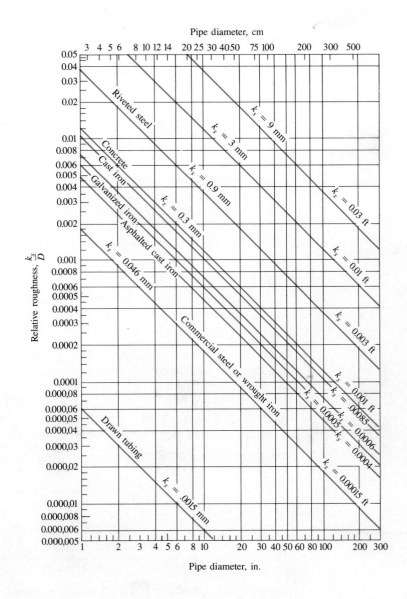

given value of k_s/D is not plotted in Fig. 10.8, then one simply finds the proper position on the graph by interpolation between the k_s/D curves that bracket the given k_s/D.

For some problems it is convenient to enter Fig. 10.8 using a value of the parameter $\mathrm{Re}\,f^{1/2}$. This parameter is useful when h_f and k_s/D are known but the velocity V is not. Without V the Reynolds number cannot be computed, so f cannot be read by entering the chart with Re and k_s/D. But from

$h_f = f(L/D)V^2/2g$ [or $V = (2gh_f/L)^{1/2}(D/f)^{1/2}$] and Re $= VD/\nu$, one can see that Re can be given as

$$\text{Re} = \frac{D^{3/2}}{f^{1/2}}\left(\frac{2gh_f}{L}\right)^{1/2}$$

Upon multiplying both sides of the above equation by $f^{1/2}$, we get

$$\text{Re}f^{1/2} = \frac{D^{3/2}}{\nu}(2gh_f/L)^{1/2}$$

Thus a value of $\text{Re}f^{1/2}$ can be calculated for this type of flow problem, which, in turn, enables us to determine f directly, using Fig. 10.8, where curves of constant $\text{Re}f^{1/2}$ are plotted slanting from the upper left to lower right and the values of $\text{Re}f^{1/2}$ for each line are given at the top of the chart.

There are basically three types of problems involved with uniform flow in a single pipe. These are

1. Determine the head loss, given the kind and size of pipe and the flow rate.
2. Determine the flow rate, given the head, kind, and size of pipe.
3. Determine the size of pipe needed to carry the flow, given the kind of pipe, head, and flow rate.

In the first type of problem, the Reynolds number and k_s/D are first computed and then f is read from Fig. 10.8, after which the head loss is obtained by the use of Eq. (10.22).

EXAMPLE 10.4   Water ($T = 20°C$) flows at a rate of 0.05 m³/s in a 20-cm asphalted cast-iron pipe. What is the head loss per kilometer of pipe?

Solution   First compute the Reynolds number, VD/ν. Here $V = Q/A$. Thus

$$V = \frac{0.05 \text{ m}^3/\text{s}}{(\pi/4)(0.20^2 \text{ m}^2)} = 1.59 \text{ m/s}$$

$$\nu = 1.0 \times 10^{-6} \text{ m}^2/\text{s} \qquad \text{(from Table A.5)}$$

Then

$$\text{Re} = \frac{VD}{\nu} = \frac{(1.59 \text{ m/s})(0.20 \text{ m})}{10^{-6} \text{ m}^2/\text{s}} = 3.18 \times 10^5$$

From Fig. 10.9, $k_s/D = 0.0006$. Then from Fig. 10.8, using the values obtained for k_s/D and Re, we find $f = 0.019$. Finally, the head loss is computed from the Darcy–Weisbach equation:

$$h_f = f\frac{L}{D}\frac{V^2}{2g} = 0.019\left(\frac{1000 \text{ m}}{0.20 \text{ m}}\right)\left(\frac{1.59^2 \text{ m}^2/\text{s}^2}{2(9.81 \text{ m/s}^2)}\right) = 12.2 \text{ m}$$

The head loss per kilometer is 12.2 m.

◀

In the second type of problem, k_s/D and the value of $(D^{3/2}/\nu)\sqrt{2gh_f/L}$ are computed so that the top scale can be used to enter the chart of Fig. 10.8. Then, once f is read from the chart, the velocity from Eq. (10.22) is solved for and the discharge is computed from $Q = VA$.

EXAMPLE 10.5   The head loss per kilometer of 20-cm asphalted cast-iron pipe is 12.2 m. What is the discharge of water?

Solution   First compute the parameter $D^{3/2}\sqrt{2gh_f/L}/\nu$. Assume $T = 20°C$, so that

$$D^{3/2}\frac{\sqrt{2gh_f/L}}{\nu} = (0.20\text{ m})^{3/2}\frac{[2(9.81\text{ m/s}^2)(12.2\text{ m}/1000\text{ m})]^{1/2}}{1.0 \times 10^{-6}\text{ m}^2/\text{s}} = 4.38 \times 10^4$$

From Fig. 10.9, $k_s/D = 0.0006$. Using Fig. 10.8, we read $f = 0.019$. We use this f in the Darcy–Weisbach equation to solve for V:

$$h_f = f\frac{L}{D}\frac{V^2}{2g}$$

$$12.2\text{ m} = \frac{0.019(1000\text{ m})}{0.20\text{ m}}\frac{V^2}{2(9.81\text{ m/s}^2)}$$

$$V^2 = 2.52\text{ m}^2/\text{s}^2$$

$$V = 1.59\text{ m/s}$$

Finally we compute the discharge:

$$Q = VA = V\frac{\pi}{4}D^2$$

$$= (1.59\text{ m/s})(0.785)(0.20^2\text{ m}^2) = 0.050\text{ m}^3/\text{s} \qquad \blacktriangleleft$$

Examples 10.4 and 10.5 are good checks on the validity of the methods of solution because their basic data are exactly the same—in one case, the head loss is unknown; in the other, the discharge is unknown.

In Example 10.5 the head loss in the pipe was given. Therefore, it was possible to obtain a direct solution by entering Fig. 10.8 with a value of $\text{Re}f^{1/2}$. However, many problems for which the discharge Q is desired cannot be solved directly. For example, a problem in which water flows from a reservoir through a pipe and into the atmosphere cannot be solved directly. Here part of the available head is lost to friction in the pipe, and part of the head remains as kinetic energy in the jet as it leaves the pipe. Therefore, at the outset one does not know how much head loss occurs in the pipe itself. To effect a solution, one must iterate on f. The energy equation is written and an initial value for f is guessed. Because f tends to a constant value at high values of Re, an "educated" first guess is to use this limiting value of f. Next one solves for the velocity V. With this value of V, one then computes a Reynolds number that makes it possible to determine a better value of f using Fig. 10.8, and so on. This type of solution usually converges quite rapidly because f changes more slowly than Re. Once f and V have been determined, one calculates the discharge by using the continuity equation.

EXAMPLE 10.6   Determine the discharge of water through the 50-cm steel pipe shown in the figure.

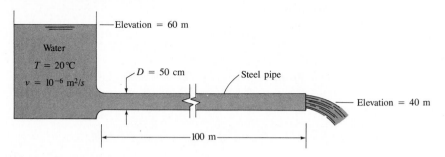

Solution   From Fig. 10.8 the value of $k_s = 4.6 \times 10^{-5}$ m. Therefore, $k_s/D = 9.2 \times 10^{-5}$ m. Now write the energy equation from the reservoir water surface to the free jet at the end of the pipe:

$$\frac{p_1}{\gamma} + \frac{V_1^2}{2g} + z_1 = \frac{p_2}{\gamma} + \frac{V_2^2}{2g} + z_2 + h_L$$

$$0 + 0 + 60 = 0 + \frac{V_2^2}{2g} + 40 + f\frac{L}{D}\frac{V_2^2}{2g}$$

or

$$V = \left(\frac{2g \times 20}{1 + 200f}\right)^{1/2}$$

First trial: Assume $f = 0.020$; then $V = 8.86$ m/s and Re $= 4.43 \times 10^6$. With Re $= 4.43 \times 10^6$ and $k_s/D = 9.2 \times 10^{-5}$, then $f = 0.012$ (from Fig. 10.8). This f then yields $V = 10.7$ m/s.

Second trial: For $V = 10.7$ m/s, Re $= 5.35 \times 10^6$ and $f = 0.012$,

$$Q = VA = 10.7 \text{ m/s} \times (\pi/4) \times (0.50)^2 \text{ m}^2 = 2.10 \text{ m}^3/\text{s} \quad \blacktriangleleft$$

In the third type of problem, it is usually best first to assume a value of f and then to solve for D, after which a better value of f is computed based on the first estimate of D. This iterative procedure is continued until a valid solution is obtained. A trial-and-error procedure is necessary because without D one cannot compute k_s/D or Re to enter Fig. 10.8.

EXAMPLE 10.7   What size of asphalted cast-iron pipe is required to carry water at a discharge of 3 cfs and with a head loss of 4 ft per 1000 ft of pipe?

Solution   First assume $f = 0.015$. Then

$$h_f = \frac{fL}{D}\frac{V^2}{2g} = \frac{fL}{D}\frac{Q^2/A^2}{2g} = \frac{fLQ^2}{2g(\pi/4)^2 D^5}$$

or

$$D^5 = \frac{fLQ^2}{0.785^2(2gh_f)}$$

For this example,

$$D^5 = \frac{0.015(1000 \text{ ft})(3 \text{ ft}^3/\text{s})^2}{0.615(64.4 \text{ ft/s}^2)(4 \text{ ft})} = 0.852 \text{ ft}^5$$

$$D = 0.97 \text{ ft}$$

Now compute a more accurate value of f:

$$\frac{k_s}{D} = 0.0004 \qquad V = \frac{Q}{A} = \frac{3 \text{ ft}^3/\text{s}}{0.785(0.94 \text{ ft}^2)} = 4.07 \text{ ft/s}$$

Then

$$\text{Re} = \frac{VD}{\nu} = \frac{(4.07 \text{ ft/s})(0.97 \text{ ft})}{1.21(10^{-5} \text{ ft}^2/\text{s})} = 3.26 \times 10^5$$

From Fig. 10.8, $f = 0.0175$. Now recompute D by applying the ratio of f's to previous calculations for D^5:

$$D^5 = \frac{0.0175}{0.015}(0.85 \text{ ft}^5) = 0.992 \text{ ft}^5$$

$$D = 0.998 \text{ ft}$$

Use a pipe with a 12-in. diameter.   ◀

Note: If a size that is not available commercially is calculated during design, it is customary practice to choose the next larger available size. The cost will be less than that for odd-sized pipe, and the pipe will be more than large enough to carry the flow.

Explicit Equations for h_f, Q, and D

In the foregoing discussion, methods were presented by which h_f, Q, and D can be calculated. All of these methods involve the use of the Moody diagram (Fig. 10.8). With the advent of computers and programmable calculators, it is most desirable to be able to solve similar problems without having to resort to the Moody diagram. By utilizing the Colebrook–White formula, from which the Moody diagram was developed, Swamee and Jain (38) developed explicit formulas relating f, h_f, Q, and D. It is reported that their formulas yield results that deviate no more than 3% from those obtained from the Moody diagram for the following ranges of k_s/D and Re: $10^{-5} < k_s/D < 2 \times 10^{-2}$ and $4 \times 10^3 < \text{Re} < 10^8$. The formulas for f and Q are

$$f = \frac{0.25}{\left[\log\left(\dfrac{k_s}{3.7D} + \dfrac{5.74}{\text{Re}^{0.9}} \right) \right]^2} \tag{10.26}$$

$$Q = -2.22 D^{5/2} \sqrt{gh_f/L} \, \log\left(\frac{k_s}{3.7D} + \frac{1.78\nu}{D^{3/2}\sqrt{gh_f/L}} \right) \tag{10.27}$$

They also developed a formula for the explicit determination of D. A modified version of that formula, given by Streeter and Wylie (37), is

$$D = 0.66 \left[k_s^{1.25} \left(\frac{LQ^2}{gh_f} \right)^{4.75} + \nu Q^{9.4} \left(\frac{L}{gh_f} \right)^{5.2} \right]^{0.04} \tag{10.28}$$

   If you want to solve for head loss given Q, L, D, k_s, and ν, simply solve for f by Eq. (10.26) and compute h_f with the Darcy–Weisbach equation, Eq. (10.22). Straightforward calculations for Q and D can also be made if h_f is known. However, for problems involving head losses in addition to h_f, an iterative solution is required. For computing Q, you can assume an f and solve for Q from the energy equation after substituting Q/A in that equation. Then compute Re and use the result in Eq. (10.26) to get a better estimate of f, and so on, until Q converges analogous to the procedure for determining Q using the Moody diagram. In this case, however, Eq. (10.26) is substituted for the Moody diagram. Similarly, you can determine D if you are given Q, ν, the change in pressure or head, and the geometric configuration.

EXAMPLE 10.8   Solve Example 10.5 using Eq. (10.27).

Solution   From Fig. 10.8, k_s for asphalted cast-iron pipe is given as 1.2×10^{-4} m. From the given conditions, $h_f/L = 0.0122$. Assume $T = 20°C$, so $\nu = 10^{-6}$ m²/s. Then, using Eq. (10.27), we have

$$Q = -2.22(0.20 \text{ m})^{5/2} \sqrt{9.81 \text{ m/s}^2 \times 0.0122}$$

$$\times \log\left(\frac{1.2 \times 10^{-4} \text{ m}}{3.7 \times 0.20 \text{ m}} + \frac{1.78 \times 10^{-6} \text{ m}^2/\text{s}}{(0.20 \text{ m})^{3/2} \sqrt{9.81 \text{ m/s}^2 \times 0.0122}} \right)$$

$$= 0.050 \text{ m}^3/\text{s} \qquad \blacktriangleleft$$

Reduction of Head Loss with Polymers

In 1948 Toms (39) reported on the effect on head loss of minute amounts of soluble polymer added to water. These solutions are often called drag-reducing agents. Since then hundreds of experiments have confirmed his initial findings. The essential features of this phenomenon are as follows:

1. The head loss of water flow in pipes can be reduced as much as 70% with the addition of a polymer solution in concentrations as low as 5 ppm (27).

2. Significant reductions in head loss occur only for turbulent flow. No reduction in head loss occurs in laminar flow.

3. The fluid mechanics of this phenomenon is not yet well understood; however, most experiments indicate that it is related to the turbulence associated with flow very near the wall of the pipe. Some researchers think it is associated with the thickness of the viscous sublayer (40), whereas at least one study indicates that it is primarily associated with the buffer zone of the boundary layer (6).

4. Polymer solutions are more effective on small pipes than on large ones (6).

Because of the potential for achieving significant economic advantages with reduced head loss, research in this area will undoubtedly be extensive for the foreseeable future. For background information on the subject, see references (17) and (25).

A practical application of the head-loss reduction phenomenon is the use of polymer solutions in the Trans-Alaska Pipeline. For more details see reference (4). Other details about the use of drag-reducing agents in oil pipelines is given by Lester (23).

10.5 Flow at Pipe Inlets and Losses from Fittings

In the preceding section, formulas were presented that are used to determine the head loss for uniform flow in a pipe. However, pipe systems also include inlets, outlets, bends, and other appurtenances that create additional head losses. Flow separation and the generation of additional turbulence therefrom usually cause these head losses. In this section we will consider the flow patterns and resulting head losses for some of these flow transitions.

Flow in a Pipe Inlet

If the inlet to a pipe is well rounded, as shown in Fig. 10.10, the boundary layer will develop from the inlet and grow in thickness until it extends to the center of the pipe. After that point, the flow in the pipe will be uniform. The length L_e of the developing region at the entrance is equal to approximately $0.05D$ Re for laminar flow and approximately $50D$ for turbulent flow. Velocity and pressure distribution for the inlet region of a pipe with turbulent flow are shown in

FIGURE 10.10

Flow characteristics at a pipe inlet (not to scale).

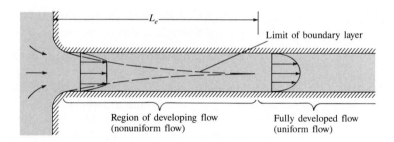

Fig. 10.11. The head loss that is produced by inlets, outlets, or fittings is expressed by the equation

$$h_L = K\frac{V^2}{2g} \tag{10.29}$$

In Eq. (10.29) V is the mean velocity in the pipe and K is the loss coefficient for the particular fitting that is involved. For example, K for a well-rounded inlet with flow at high values of Re is approximately 0.10. Hence the head loss for such a transition is quite small compared with that for an abrupt pipe outlet for which K is 1.0.

If the pipe inlet is abrupt, or sharp-edged, as in Fig. 10.12, separation occurs just downstream of the entrance. Hence the streamlines converge and then diverge with consequent turbulence and relatively high head loss. The loss coefficient for the abrupt inlet is approximately 0.5.

Flow through an Elbow

Although the cross-sectional area of an elbow may not change from section to section, considerable head loss is produced because separation occurs near the inside of the bend and downstream of the midsection. Thus when the flow leaves the elbow, the eddies produced by separation create considerable head loss. The approximate flow pattern for an elbow is shown in Fig. 10.13.

FIGURE 10.11

Distribution of velocity and pressure in the inlet region of a pipe [Barbin and Jones (3)]. (a) Velocity distribution. (b) Pressure distribution.

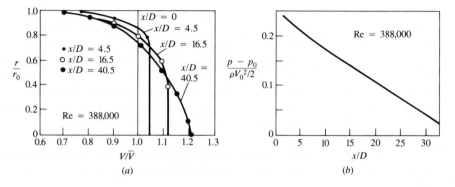

FIGURE 10.12

Flow at a sharp-edged inlet.

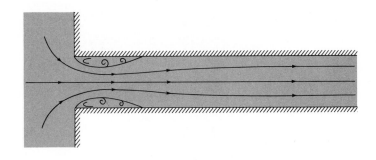

FIGURE 10.13

Flow pattern in an elbow.

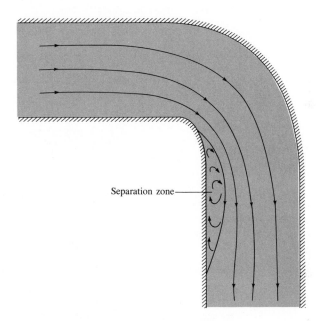

Separation zone

The loss coefficient for an elbow at high Reynolds numbers depends primarily on the shape of the elbow. If it is a very short-radius elbow, the loss coefficient is quite high. For larger-radius elbows, the coefficient decreases until a minimum value is found at an *r/d* value of about 4 (see Table 10.2). However, for still larger values of *r/d*, an increase in loss coefficient occurs because, for larger *r/d* values, the elbow itself is significantly longer than elbows with small *r/d* values. The greater length creates an additional head loss. The loss coefficients for various types of elbows, and for a number of other fittings and flow transitions, are given in Table 10.2.*

* *Engineering handbooks usually include extensive tables of loss coefficients. References (2), (10), (18), and (36) are particularly useful in this respect.*

	TABLE 10.2 LOSS COEFFICIENTS FOR VARIOUS TRANSITIONS AND FITTINGS			
Description	**Sketch**	**Additional Data**	**K**	**Source**
Pipe entrance $h_L = K_e V^2/2g$		r/d 0.0 0.1 >0.2	K_e 0.50 0.12 0.03	(2)
Contraction $h_L = K_C V_2^2/2g$		D_2/D_1 0.0 0.20 0.40 0.60 0.80 0.90	K_C $\theta = 60°$ K_C $\theta = 180°$ 0.08 0.50 0.08 0.49 0.07 0.42 0.06 0.32 0.05 0.18 0.04 0.10	(2)
Expansion $h_L = K_E V_1^2/2g$		D_1/D_2 0.0 0.20 0.40 0.60 0.80	K_E $\theta = 10°$ K_E $\theta = 180°$ 1.00 0.13 0.92 0.11 0.72 0.06 0.42 0.03 0.16	(2)
90° miter bend		Without vanes	$K_b = 1.1$	(36)
		With vanes	$K_b = 0.2$	(36)
90° smooth bend		r/d 1 2 4 6 8 10	$K_b = 0.35$ 0.19 0.16 0.21 0.28 0.32	(5) and (19)

Threaded pipe fittings	Globe valve—wide open
	Angle valve—wide open
	Gate valve—wide open
	Gate valve—half open
	Return bend
	Tee
	  straight-through flow
	  side-outlet flow
	90° elbow
	45° elbow
		$K_\nu = 10.0$
		$K_\nu = 5.0$
		$K_\nu = 0.2$
		$K_\nu = 5.6$
		$K_b = 2.2$
		
		$K_t = 0.4$
		$K_t = 1.8$
		$K_b = 0.9$
		$K_b = 0.4$
				(36)

EXAMPLE 10.9   If oil ($\nu = 4 \times 10^{-5}$ m²/s, S = 0.9) flows from the upper to the lower reservoir at a rate of 0.028 m³/s in the 15-cm smooth pipe, what is the elevation of the oil surface in the upper reservoir?

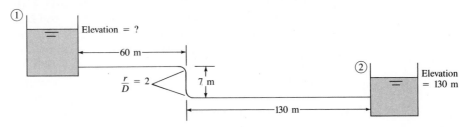

Solution   Apply the energy equation between the surfaces of the upper and lower reservoirs:

$$\frac{p_1}{\gamma} + \frac{V_1^2}{2g} + z_1 = \frac{p_2}{\gamma} + \frac{V_2^2}{2g} + z_2 + \sum h_L$$

$$0 + 0 + z_1 = 0 + 0 + 130 \text{ m} + \frac{fL}{D}\frac{V^2}{2g} + 2K_b\frac{V^2}{2g} + K_e\frac{V^2}{2g} + K_E\frac{V^2}{2g}$$

Here K_b, K_e, and K_E are loss coefficients for bend, entrance, and outlet, respectively. These have values of 0.19, 0.5, and 1.0 (Table 10.2). To determine f, we get Re in order to enter Fig. 10.8:

$$\text{Re} = \frac{VD}{\nu}$$

But

$$V = \frac{Q}{A} = \frac{(0.028 \text{ m}^3/\text{s})}{0.785(0.15 \text{ m})^2} = 1.58 \text{ m/s}$$

Then

$$\text{Re} = \frac{1.58 \text{ m/s }(0.15 \text{ m})}{4 \times 10^{-5} \text{ m}^2/\text{s}} = 5.93 \times 10^3$$

Now we read f from Fig. 10.8 (smooth pipe curve): $f = 0.035$. Then

$$z_1 = 130 \text{ m} + \frac{V^2}{2g}\left[\frac{0.035(197 \text{ m})}{0.15 \text{ m}} + 2(0.19) + 0.5 + 1\right]$$

$$= 130 \text{ m} + \left[\frac{(1.58 \text{ m/s})^2}{2(9.81 \text{ m/s}^2)}\right](46 + 0.38 + 0.5 + 1)$$

$$= 130 \text{ m} + 6.1 \text{ m} = 136.1 \text{ m} \qquad \blacktriangleleft$$

Transition Losses and Grade Lines

In Chapter 7 the effect on the energy and hydraulic grade lines (EGL and HGL) of the pipe head loss and head loss at an abrupt expansion were discussed in some detail. However, no mention was made of head loss due to entrances, bends, and other flow transitions. The primary effect, just as for abrupt expansions, is to cause the EGL to drop an amount equal to the head loss produced by that transition. Generally, this drop occurs over a distance of several diameters downstream of the transition, as seen in Fig. 10.14. The HGL also drops sharply immediately downstream of the entrance because of the high-velocity flow in the contracted portion of the stream. Then, as the turbulent mixing occurs even farther downstream, energy is lost owing to viscous action occurring in the mixing process. Thus the EGL at the entrance is steeper than it is farther downstream where the flow has become uniform. As the additional energy loss from the transition subsides, the EGL takes on the slope created by the head loss of the pipe itself.

Even though many transitions produce grade lines that have interesting local details, such as those noted for a sharp-edged entrance, it is common as a gross indication to simply show abrupt changes in the EGL and to neglect local departures between the EGL and the HGL owing to local changes in $V^2/2g$. Thus, Fig. 10.15 is a simplified plot of the EGL and the HGL for a pipe with several transitions in it.

FIGURE 10.14

EGL and HGL at a sharp-edged pipe entrance.

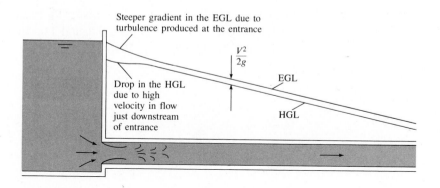

FIGURE 10.15

Head losses in a pipe.

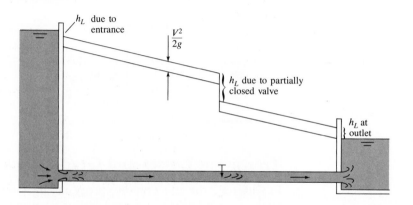

EXAMPLE 10.10   Refer to Fig. 10.15. The difference in water-surface elevation between the reservoirs is 5.0 m, and the horizontal distance between them is 300 m. Using the explicit formula for f, Eq. (10.26), determine the size of steel pipe needed for a discharge of 2 m³/s when the gate valve is wide open.

Solution   First write the energy equation:

$$\frac{p_1}{\gamma} + \frac{V_1^2}{2g} + z_1 = \frac{p_2}{\gamma} + \frac{V_2^2}{2g} + z_2 + \sum h_L$$

$$5 \text{ m} = \sum h_L$$

$$= \frac{V^2}{2g}\left(\frac{fL}{D} + K_e + K_v + K_E\right)$$

$$= \frac{Q^2}{2gA^2}(f \times 300/D + 0.5 + 0.2 + 1.0)$$

$$= \frac{Q^2}{2g(\pi/4)^2 D^4}\left[300\left(\frac{f}{D}\right) + 1.7\right]$$

$$= \frac{(2 \text{ m}^3/\text{s})^2}{2 \times 9.81 \text{ m/s}^2 \times (\pi/4)^2 \times D^4}\left[300\left(\frac{f}{D}\right) + 1.7\right]$$

$$= \frac{0.33 \text{ m}^5}{D^4}\left(300\frac{f}{D} + 1.7\right)$$

The other equations for solving this problem are Eq. (10.26), $V = Q/A$, and $\text{Re} = VD/\nu$. As we did when using the Moody diagram, we make an initial assumption for D. Next we compute V and Re, after which we compute f from Eq. (10.26). Then we compute D from the energy equation (above). With this calculated value of D we go through the process again to get a better estimate of D, and so on, until the change in D is negligibly small. In this example, $\nu = 10^{-6}$ m²/s and $k_s = 4.6 \times 10^{-5}$ m (from Fig. 10.9). We assume an initial value for D of 1 m. Then

$$V = Q/A = \frac{2 \text{ m}^3/\text{s}}{(\pi/4) \times (1 \text{ m})^2} = 2.55 \text{ m/s}$$

$$\text{Re} = \frac{VD}{\nu} = 2.55 \times \frac{1}{10^{-6}} = 2.55 \times 10^6$$

$$f = 0.25\left[\log\left(\frac{k_s}{3.70} + \frac{5.74}{\text{Re}^{0.9}}\right)\right]^{-2}$$

$$= 0.25\left[\log\left(\frac{4.6 \times 10^{-5} \text{ m}}{3.7 \times 1 \text{ m}} + \frac{5.74}{(2.55 \times 10^6)^{0.9}}\right)\right]^{-2} = 0.0116$$

With this value of f, we solve the energy equation for D to obtain

$$D = 0.79 \text{ m}$$

With $D = 0.79$ m we repeat the computational procedure again and again, until convergence. The next iteration yields $D = 0.79$ m. Since there is no significant change, we have a solution: $D = 0.79$ m.   ◀

10.6 Pipe Systems

Simple Pump in a Pipeline

We have considered a number of pipe-flow problems in which the head for producing the flow was explicitly given. Now we shall consider flow in which the head is developed by a pump. However, the head produced by a centrifugal pump is a function of the discharge. Hence a direct solution is usually not immediately available. The solution (that is, the flow rate for a given system) is obtained when the system equation (or curve) of head versus discharge is solved simultaneously with the pump equation (or curve) of head versus discharge. The solution of these two equations (or the point where the two curves intersect) yields the operating condition for the system. Consider flow of water in the system of Fig. 10.16. When the energy equation is written from the reservoir water surface to the outlet stream, we obtain the following equation:

$$\frac{p_1}{\gamma} + \frac{V_1^2}{2g} + z_1 + h_p = \frac{p_2}{\gamma} + \frac{V_2^2}{2g} + z_2 + \sum K_L \frac{V^2}{2g} + \sum \frac{fL}{D}\frac{V^2}{2g}$$

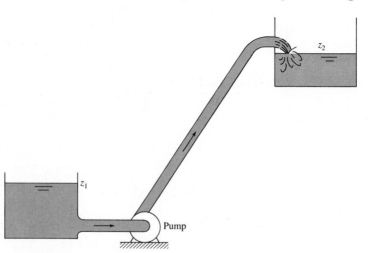

FIGURE 10.16

Pump and pipe combination.

For a system with one size of pipe, this simplifies to

$$h_p = (z_2 - z_1) + \frac{V^2}{2g}\left(1 + \sum K_L + \frac{fL}{D}\right) \qquad (10.30)$$

Hence, for any given discharge, a certain head h_p must be supplied to maintain that flow. Thus we can construct a head-versus-discharge curve, as shown in Fig. 10.17. Such a curve is called the *system curve*. Any given centrifugal pump has a head-versus-discharge curve that is characteristic of that pump at a given pump speed. Such curves are supplied by the pump manufacturer; a typical one for a centrifugal pump is shown in Fig. 10.17.

Figure 10.17 reveals that, as the discharge increases in a pipe, the head required for flow also increases. However, the head that is produced by the pump decreases as the discharge increases. Consequently, the two curves intersect, and the operating point is at the point of intersection—that point where the head produced by the pump is just the amount needed to overcome the head loss in the pipe.

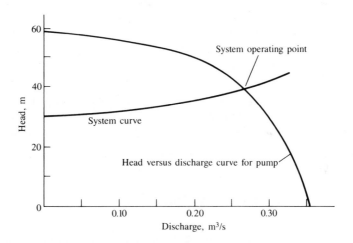

FIGURE 10.17

Pump and system curves.

EXAMPLE 10.11   What will be the discharge in this water system if the pump has the characteristics shown in Fig. 10.17? Assume $f = 0.015$.

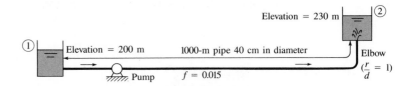

Solution   First we write the energy equation from water surface to water surface:

$$\frac{p_1}{\gamma} + \frac{V_1^2}{2g} + z_1 + h_p = \frac{p_2}{\gamma} + \frac{V_2^2}{2g} + z_2 + \sum h_L$$

$$0 + 0 + 200 + h_p = 0 + 0 + 230 + \left(\frac{fL}{D} + K_e + K_b + K_E\right)\frac{V^2}{2g}$$

Here $K_e = 0.5$, $K_b = 0.35$, and $K_E = 1.0$. Hence

$$h_p = 30 + \frac{Q^2}{2gA^2}\left[\frac{0.015(1000)}{0.40} + 0.5 + 0.35 + 1\right]$$

$$= 30 + \frac{Q^2}{2 \times 9.81 \times [(\pi/4) \times 0.4^2]^2}(39.3) = 30\text{ m} + 127Q^2\text{ m}$$

Now we make a table of Q versus h_p (see below) to give values to produce a

Q, m³/s	Q², m⁶/s²	127Q²	$h_p = 30\text{ m} + 127Q^2$ m
0	0	0	30
0.1	1×10^{-2}	1.3	31.3
0.2	4×10^{-2}	5.1	35.1
0.3	9×10^{-2}	11.4	41.4

system curve that will be plotted with the pump curve. When the system curve is plotted on the same graph as the pump curve, it is seen (Fig. 10.17) that the operating condition occurs at $Q = 0.27$ m³/s.   ◀

Pipes in Parallel

Consider a pipe that branches into two parallel pipes and then rejoins, as shown in Fig. 10.18. A problem involving this configuration might be to determine the division of flow in each pipe, given the total flow rate.

No matter which pipe is involved, the pressure difference between the two junction points is the same. Also, the elevation difference between the two junction points is the same. Because $h_L = (p_1/\gamma + z_1) - (p_2/\gamma + z_2)$, it follows that

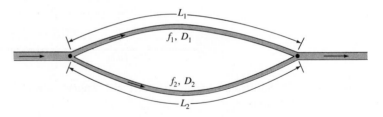

FIGURE 10.18

Flow in parallel pipes.

h_L between the two junction points is the same in both of the pipes of the parallel pipe system. Thus we can write

$$h_{L_1} = h_{L_2}$$

$$f_1 \frac{L_1}{D_1} \frac{V_1^2}{2g} = f_2 \frac{L_2}{D_2} \frac{V_2^2}{2g}$$

Then $\qquad \left(\frac{V_1}{V_2}\right)^2 = \frac{f_2}{f_1} \frac{L_2}{L_1} \frac{D_1}{D_2} \qquad$ or $\qquad \frac{V_1}{V_2} = \left(\frac{f_2}{f_1} \frac{L_2}{L_1} \frac{D_1}{D_2}\right)^{1/2}$

If f_1 and f_2 are known, the division of flow can be easily determined. However, some trial-and-error analysis may be required if f_1 and f_2 are in the range where they are functions of the Reynolds number.

Pipe Networks

The most common pipe networks are the water-distribution systems for municipalities. These systems have one or more sources (discharges of water into the system) and numerous loads: one for each household and commercial establishment. For purposes of simplification, the loads are usually lumped throughout the system. Figure 10.19 shows a simplified distribution system with two sources and seven loads.

The engineer is often engaged to design the original system or to recommend an economical expansion to the network. An expansion may involve additional housing or commercial developments, or it may be to handle increased loads within the existing area.

In the design of such a system, the engineer will have to estimate the future loads for the system and will need to have sources (wells or direct pumping from streams or lakes) to satisfy the loads. Also, the layout of the pipe network must be made (usually parallel to streets), and pipe sizes will have to be determined. The object of the design is to arrive at a network of pipes that will deliver the design flow at the design pressure for minimum cost. The cost will include

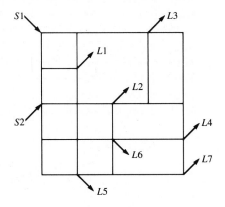

FIGURE 10.19

Pipe network.

first costs (materials and construction) as well as maintenance and operating costs. The design process usually involves a number of iterations on pipe sizes and layouts before the optimum design (minimum cost) is achieved.

So far as the fluid mechanics of the problem is concerned, the engineer must determine pressures throughout the network for various conditions —that is, for various combinations of pipe sizes, sources, and loads. The solution of a problem for a given layout and a given set of sources and loads requires that three conditions be satisfied:

1. Continuity must be satisfied. That is, the flow into a junction of the network must equal the flow out of the junction. This must be satisfied for all junctions.

2. The head loss between any two junctions must be the same regardless of the path in the series of pipes taken to get from one junction point to the other. This requirement results because pressure must be continuous throughout the network (pressure cannot have two values at a given point). This condition leads to the conclusion that the algebraic sum of head losses around a given loop must be equal to zero. Here the sign (positive or negative) for the head loss in a given pipe is given by the sense of the flow with respect to the loop, that is, whether the flow has a clockwise or counterclockwise sense.

3. The flow and head loss must be consistent with the appropriate velocity-head-loss equation.

Only a few years ago, these solutions were made by trial-and-error hand computation, but modern applications using digital computers have made the older methods obsolete. Even with these advances, however, the engineer charged with the design or analysis of such a system must understand the basic fluid mechanics of the system to be able to interpret the results properly and to make good engineering decisions based on the results. Therefore, an understanding of the original method of solution by Hardy Cross (11) may help you to gain this basic insight. The Hardy Cross method is as follows.

The engineer first distributes the flow throughout the network so that loads at various nodes are satisfied. In the process of distributing the flow through the pipes of the network, the engineer must be certain that continuity is satisfied at all junctions (flow into a junction = flow out of the junction), thus satisfying requirement 1. The first guess at the flow distribution obviously will not satisfy requirement 2 regarding head loss; therefore, corrections are applied. For each loop of the network, a discharge correction is applied to yield a zero net head loss around the loop. For example, consider the isolated loop in Fig. 10.20. In this loop, the loss of head in the clockwise sense will be given by

$$\sum h_{L_c} = h_{L_{AB}} + h_{L_{BC}}$$

$$= \sum k Q_c^n \tag{10.31}$$

FIGURE 10.20

*A typical loop of a
pipe network.*

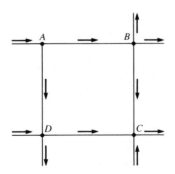

FIGURE 10.20

*A typical loop of a
pipe network.*

The loss of head for the loop in the counterclockwise sense is

$$\sum h_{L_{cc}} = \sum_{cc} k Q_{cc}^n \qquad (10.32)$$

For a solution, the clockwise and counterclockwise head losses have to be equal, or

$$\sum h_{L_c} = \sum h_{L_{cc}}$$

$$\sum k Q_c^n = \sum k Q_{cc}^n$$

As we noted, the first guess for flow in the network is undoubtedly in error; therefore, a correction in discharge, ΔQ, will have to be applied to satisfy the head loss requirement. If the clockwise head loss is greater than the counterclockwise head loss, ΔQ will have to be applied in the counterclockwise sense. That is, subtract ΔQ from the clockwise flows and add it to the counterclockwise flows:

$$\sum k(Q_c - \Delta Q)^n = \sum k(Q_{cc} + \Delta Q)^n \qquad (10.33)$$

Expanding the summation on either side of Eq. (10.33) and including only two terms of the expansion, we obtain

$$\sum k(Q_c^n - nQ_c^{n-1}\Delta Q) = \sum k(Q_{cc}^n + nQ_{cc}^{n-1}\Delta Q)$$

Then solving for ΔQ, we get

$$\Delta Q = \frac{\sum k Q_c^n - \sum k Q_{cc}^n}{\sum nk Q_c^{n-1} + \sum nk Q_{cc}^{n-1}} \qquad (10.34)$$

Thus if ΔQ as computed from Eq. (10.34) is positive, the correction is applied in a counterclockwise sense (add ΔQ to counterclockwise flows and subtract it from clockwise flows).

A different ΔQ is computed for each loop of the network and applied to the pipes. Some pipes will have two ΔQ's applied because they will be common to two loops. The first set of corrections usually will not yield the final desired result because the solution is approached only by successive approximations. Thus the corrections are applied successively until the corrections are negligible. Experience has shown that for most loop configurations, applying ΔQ as computed by Eq. (10.34) produces too large a correction. Fewer trials are required to solve for Q's if approximately 0.6 of the computed ΔQ is used.

For information on more modern methods of solution of pipe network problems, see references (16) and (20).

10.7 Turbulent Flow in Noncircular Conduits

Basic Development

Earlier in this chapter (Sec. 10.4), τ_0 was eliminated between Eqs. (10.3) and (10.21) to yield the Darcy–Weisbach equation, Eq. (10.22). It should be noted that Eq. (10.3) was derived by writing the equation of equilibrium in the longitudinal direction for an element of fluid with a circular cross section. If one derives an equation analogous to Eq. (10.3) for flow in a noncircular conduit in which the shear stress acts on the conduit surface having a perimeter P (such as the perimeter of a rectangular conduit) instead of perimeter $2\pi r$, then τ_0 is given by

$$\tau_0 = \frac{A}{P}\left[-\frac{d}{ds}(p + \gamma z)\right] \tag{10.35}$$

In Eq. (10.3), to which Eq. (10.35) is analogous, the shear stress τ_0 was everywhere constant around the perimeter of the cylindrical element. In Eq. (10.35) for the noncircular conduit, the shear stress is not constant over the perimeter. However, we can still use Eq. (10.21) to relate τ_0 and V, where τ_0 is now the average shear stress on the boundary. Eliminating τ_0 between Eqs. (10.21) and (10.35) and integrating along the pipe yield

$$h_f = \frac{f}{4}\frac{L}{A/P}\frac{V^2}{2g} \tag{10.36}$$

In Eq. (10.36), h_f is the head loss between two points in the conduit, L is the length between the points, and P is the wetted perimeter of the conduit. Thus Eq. (10.36) is the same as the Darcy–Weisbach equation except that D is replaced by $4A/P$. The ratio of the cross-sectional area A to the wetted perimeter P

is defined as the *hydraulic radius* R_h. Experiments have shown that we can solve flow problems involving noncircular conduits, such as rectangular ducts, if we apply the same methods and equations that we did for pipes but use $4R_h$ in place of D. Consequently, the relative roughness is $k_s/4R_h$, and the Reynolds number is defined as $V(4R_h)/\nu$.

EXAMPLE 10.12   Air ($T = 20°C$ and $p = 101$ kPa absolute) flows at a rate of 2.5 m³/s in a commercial steel rectangular duct 30 cm by 60 cm. What is the pressure drop per 50 m of duct?

Solution   First compute Re and $k_s/4R_h$:

$$\text{Re} = \frac{V(4R_h)}{\nu}$$

Here
$$V = \frac{Q}{A} = \frac{2.5 \text{ m}^3/\text{s}}{0.18 \text{ m}^2} = 13.9 \text{ m/s}$$

The hydraulic radius is given by

$$R_h = \frac{A}{P} = \frac{0.18 \text{ m}^2}{1.8 \text{ m}} = 0.10 \text{ m}$$

$$4R_h = 4(0.10 \text{ m}) = 0.40 \text{ m}$$

$$\nu = 1.51 \times 10^{-5} \text{ m}^2/\text{s}$$

Hence
$$\text{Re} = \frac{13.9 \times 0.40}{1.51 \times 10^{-5}} = 3.68 \times 10^5$$

$$\frac{k_s}{4R_h} = \frac{4.6 \times 10^{-5} \text{ m}}{0.40 \text{ m}} = 1.15 \times 10^{-4}$$

Then from Fig. 10.8, $f = 0.015$. Thus

$$h_f = \frac{fL}{4R_h}\frac{V^2}{2g}$$

or
$$\gamma h_f = \Delta p_f = \frac{fL}{4R_h}\rho\frac{V^2}{2}$$

$$\rho = 1.2 \text{ kg/m}^3$$

Finally,

$$\Delta p_f \text{ per 50 m} = \frac{0.015(50 \text{ m})}{0.40 \text{ m}}(1.2 \text{ N} \cdot \text{s}^2/\text{m}^4)\frac{13.9^2}{2} \text{ m}^2/\text{s}^2 = 217 \text{ Pa} \quad \blacktriangleleft$$

Uniform Flow in Open Channels

You may recall from Chap. 4 (p. 103) that for uniform flow the streamlines must be straight and parallel. Therefore, in the case of uniform flow in open channels, the cross-sectional area and shape of the channel will not change from section to section along the channel. Also, the uniform flow condition means that from section to section the velocity and depth will be the same at corresponding points in the sections. Figure 10.21 is an example of uniform flow in a channel having a rectangular cross section. Note here that the velocity varies across the section but does not change from section to section along any given streamline.

The criterion for determining whether the flow in open channels will be laminar or turbulent is similar to that for flow in pipes. Recall that in pipe flow, if the Reynolds number ($VD\rho/\mu = VD/\nu$) is less than 2000, the flow will be laminar, and if it is greater than about 3000, one can expect the flow to be turbulent. The Reynolds number criterion for open-channel flow is the same if we replace D in the Reynolds number by $4R_h$, as we did in the Darcy–Weisbach equation in the preceding subsection. Thus, we can expect laminar flow to occur in open channels if $V(4R_h)/\nu < 2000$. The Reynolds number for open channels is usually defined as VR_h/ν. Therefore, in open channels, if the Reynolds number is less than 500, we will have laminar flow, and if it is greater than about 750, we can expect to have turbulent flow. A brief analysis of this turbulent criterion will show that water flow in channels will usually be turbulent unless the velocity and/or the depth is very small. Example 10.13 illustrates this point.

It should be noted that the wetted perimeter used for calculating the hydraulic radius is the perimeter of the channel that is actually in contact with the flowing liquid. For example, in Fig. 10.21 the hydraulic radius of this channel of rectangular cross section is

$$R_h = \frac{A}{P} = \frac{By}{B + 2y} \tag{10.37}$$

One can see that for very wide, shallow channels the hydraulic radius approaches the depth y.

EXAMPLE 10.13   Water (60°F) flows in a 10-ft-wide rectangular channel at a depth of 6 ft. What is the Reynolds number if the mean velocity is 0.1 ft/s? With this velocity, at what maximum depth can we be assured of having laminar flow?

FIGURE 10.21

Open-channel relations.

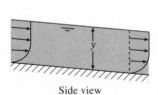

Side view

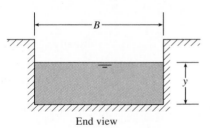

End view

Solution $\text{Re} = VR_h/\nu$

where $V = 0.1$ ft/s

$$R_h = A/P = By/(B + 2y)$$
$$= (10 \times 6)/(10 + 2 \times 6)$$
$$= 2.73 \text{ ft}$$
$$\nu = 1.22 \times 10^{-5} \text{ ft}^2/\text{s (from Table A.5)}$$

then $\text{Re} = (0.1 \text{ ft/s}) (2.73 \text{ ft})/(1.22 \times 10^{-5} \text{ ft}^2/\text{s}) = 22{,}377$ ◄

The maximum Reynolds number at which we can expect to have laminar flow in open channels is 500. Thus, for this limit of Re and for a water velocity of 0.10 ft/s, we can solve for the depth at which this condition will prevail:

$$\text{Re} = VR_h/\nu = (0.10 \text{ ft/s})R_h/(1.22 \times 10^{-5} \text{ ft}^2/\text{s}) = 500$$

Solving for R_h yields $R_h = (500) (1.22 \times 10^{-5} \text{ ft}^2/\text{s})/(0.10 \text{ ft/s}) = 0.061$ ft

For rectangular channels $R_h = (By)/(B + 2y)$

Thus $(By)/(B + 2y) = (10y)/(10 + 2y) = 0.061$ ft

$$y = 0.062 \text{ ft}$$ ◄

Example 10.13 shows that indeed the velocity and/or depth must be very small to yield laminar flow of water in a channel. Note also, the depth and hydraulic radius are virtually the same for this case where the depth is very small relative to the width of the channel.

To determine the head loss for uniform flow in open channels, we utilize Eq. (10.36). That is, the Darcy–Weisbach equation with D replaced by $4R_h$ is used.

EXAMPLE 10.14   Estimate the discharge of water that a concrete channel 10 ft wide can carry if the depth of flow is 6 ft and the slope of the channel is 0.0016.

Solution We use the Darcy–Weisbach equation:

$$h_f = \frac{fL}{4R_h} \frac{V^2}{2g} \quad \text{or} \quad \frac{h_f}{L} = \frac{f}{4R_h} \frac{V^2}{2g}$$

When we have uniform open-channel flow, the slope of the EGL, $S_f = h_f/L$, is the same as the channel slope S_0. Therefore, $h_f/L = S_0$. The foregoing equation then reduces to $V^2/2g = 4R_h S_0/f$, or

$$V = \sqrt{\frac{8g}{f} R_h S_0}$$

Assume $k_s = 0.005$ ft. Then the relative roughness is

$$\frac{k_s}{4R_h} = \frac{0.005 \text{ ft}}{4(60 \text{ ft}^2/22 \text{ ft})} = \frac{0.005 \text{ ft}}{4(2.73 \text{ ft})} = 0.00046$$

Using $k_s/4R_h = 0.00046$ as a guide and referring to Fig. 10.8, we assume that $f = 0.016$. Thus

$$V = \sqrt{\frac{8(32.2 \text{ ft/s}^2)(2.73 \text{ ft})(0.0016)}{0.016}} = \sqrt{70.6 \text{ ft}^2/\text{s}^2} = 8.39 \text{ ft/s}$$

Then $\qquad \text{Re} = V\frac{4R_h}{\nu} = \frac{8.39 \text{ ft/s}(10.9 \text{ ft})}{1.2(10^{-5} \text{ ft}^2/\text{s})} = 7.62 \times 10^6$

Using this new value of Re and with $k_s/4R_h = 0.00046$, we read f as 0.016. Our initial guess was good; and now that the velocity is known, we can compute Q:

$$Q = VA = 8.39 \text{ ft/s}(60 \text{ ft}^2) = 503 \text{ cfs} \qquad \blacktriangleleft$$

For rock-bedded channels such as those in some natural streams or un-lined canals, the larger rocks produce most of the resistance to flow, and essentially none of this resistance is due to viscous effects. Thus, the friction factor is independent of the Reynolds number. For this type of channel, Limerinos (24) has shown that the resistance coefficient f can be given in terms of the size of rock in the stream bed as

$$f = \frac{1}{\left[1.2 + 2.03 \log\left(\dfrac{R_h}{d_{84}}\right)\right]^2} \qquad (10.38)$$

where d_{84} is a measure of the rock size.*

EXAMPLE 10.15   Determine the value of the resistance coefficient, f, for a natural rock-bedded channel that is 100 ft wide and has an average depth of 4.3 ft. The d_{84} size of boulders in the stream bed is 0.72 ft.

Solution   This is a fairly wide channel relative to its depth; therefore, R_h is approximated by its depth of 4.3 ft.

---

*Most river-worn rocks are somewhat elliptical in shape. Limerinos (24) showed that the intermediate dimension correlates best with f. The d_{84} refers to the size of rock (intermediate dimension) for which 84% of the rocks in the random sample are smaller than the d_{84} size. Details for choosing the sample are given by Wolman (42). The basic procedure entails sampling at least 100 rocks on the channel bottom. For example, a grid with 100 points could be laid out on the channel bottom, and the rock under each grid point would be a rock of the sample from which the d_{84} is determined.

$$f = \frac{1}{\left[1.2 + 2.03 \log\left(\dfrac{4.3}{0.72}\right) \right]^2} = 0.130 \qquad \blacktriangleleft$$

The Chezy Equation and Manning Equation

Leaders in open-channel research have recommended the adoption of the methods already presented (involving the Reynolds number and relative roughness) for channel design (9). However, many engineers continue to use the older, more traditional methods for designing open channels. Because of this current use and for historic reasons, the Chezy equation and Manning equation are presented here.

If Eq. (10.36) is rewritten in slightly different form, we have

$$\frac{h_f}{L} = \frac{f}{4R_h} \frac{V^2}{2g} \qquad L \tag{10.39}$$

For uniform flow, the depth—called *normal depth*—is constant. Consequently, h_f/L is the slope S_0 of the channel, and Eq. (10.39) can be written as

$$R_h S_0 = \frac{f}{8g} V^2$$

or

$$V = C\sqrt{R_h S_0} \tag{10.40}$$

where

$$C = \sqrt{8g/f} \tag{10.41}$$

Since $Q = VA$, we can express the discharge in a channel as

$$Q = CA\sqrt{R_h S_0} \tag{10.42}$$

This equation is known as the *Chezy equation* after a French engineer of that name. For practical application the coefficient C must be determined, and the usual design formula for C in the SI system of units is given as

$$C = \frac{R_h^{1/6}}{n} \tag{10.43}$$

When we insert this expression for C into Eq. (10.42), we obtain a common form of the discharge equation for uniform flow in open channels for SI units:

$$Q = \frac{1.0}{n} A R_h^{2/3} S_0^{1/2} \tag{10.44}$$

In Eq. (10.44), n is a resistance coefficient called Manning's n, which has different values for different types of boundary roughness. Table 10.3 gives n for various types of boundary surfaces. The major limitation of this approach is that the viscous or relative roughness effects are not present in the design formula. Hence, application outside the range of normal-sized channels carrying water is not recommended.

TABLE 10.3 TYPICAL VALUES OF THE ROUGHNESS COEFFICIENT n	
Lined Canals	n
Cement plaster	0.011
Untreated gunite	0.016
Wood, planed	0.012
Wood, unplaned	0.013
Concrete, troweled	0.012
Concrete, wood forms, unfinished	0.015
Rubble in cement	0.020
Asphalt, smooth	0.013
Asphalt, rough	0.016
Corrugated metal	0.024
Unlined Canals	
Earth, straight and uniform	0.023
Earth, winding and weedy banks	0.035
Cut in rock, straight and uniform	0.030
Cut in rock, jagged and irregular	0.045
Natural Channels	
Gravel beds, straight	0.025
Gravel beds plus large boulders	0.040
Earth, straight, with some grass	0.026
Earth, winding, no vegetation	0.030
Earth, winding, weedy banks	0.050
Earth, very weedy and overgrown	0.080

[handwritten: 0.015   Channel]

Manning Equation—Traditional System of Units

It can be shown that, in converting from SI to the traditional system of units, one must apply a factor equal to 1.49 if the same value of n is used in the two systems. Thus in the traditional system the discharge equation using Manning's n is given as

$$Q = \frac{1.49}{n} A R_h^{2/3} S_0^{1/2} \qquad (10.45)$$

EXAMPLE 10.16   If the channel of Example 10.15 has a slope of 0.0030, what is the discharge in the channel and what is the numerical value of Manning's n for this channel?

Solution   From Example 10.15, we have an f value of 0.130, and the approximate value of R_h is 4.3 ft. In Example 10.14 it was shown that

$$V = \sqrt{\frac{8g}{f}} \sqrt{R_h S_0}$$

Therefore $V = \left[\sqrt{\dfrac{(8) \, (32.2 \text{ ft/s}^2)}{0.130}} \right] \left[\sqrt{(4.3 \text{ ft}) \, (0.0030)} \right] = 5.06 \text{ ft/s}$

But $Q = VA = (5.06) \, (100 \times 4.3) = 2176 \text{ cfs}$

We solve for n using Eq. (10.41). [*Note:* we are in the traditional system of units; therefore, we use Eq. (10.45) rather than Eq. (10.44)].

$$n = \frac{1.49}{Q} A R_h^{2/3} S_0^{1/2}$$

$$n = \left(\frac{1.49}{2176 \text{ ft}^3/\text{s}} \right) (100 \times 4.3 \text{ ft}^2) \, (4.3 \text{ ft})^{2/3} (0.003)^{1/2}$$

$$n = 0.0426 \qquad \blacktriangleleft$$

EXAMPLE 10.17   Using the Chezy equation with Manning's n, compute the discharge in the channel described in Example 10.14.

Solution
$$Q = \frac{1.49}{n} A R_h^{2/3} S_0^{1/2}$$

From Example 10.14,

$$R_h = \frac{60}{22} = 2.73 \text{ ft} \qquad R_h^{2/3} = 1.95$$

$$S_0 = 0.0016 \qquad S_0^{1/2} = 0.04$$

$$A = 6(10) = 60 \text{ ft}^2$$

$$n = 0.015 \qquad \text{(Table 10.3)}$$

Then $Q = \dfrac{1.49}{0.015} (60) \, (1.96) \, (0.04) = 467 \text{ cfs} \qquad \blacktriangleleft$

10.14 got Q = 503 cfs.

The two results (Examples 10.14 and 10.17) are within expected engineering accuracy for this type of problem. For a more complete discussion of the choice of Manning's n values for use in design or analysis, refer to Chow (8).

Best Hydraulic Section

The quantity $AR^{2/3}$ in Manning's equation [Eqs. (10.44) and (10.45)] is called the section factor in which $R = A/P$; therefore, the section factor relating to uniform flow is given by $A(A/P)^{2/3}$. Thus, for a channel of given resistance and slope, the discharge will increase with increasing cross-sectional area but decrease with increasing wetted perimeter P. For a given area, A, and a given shape of channel, for example, rectangular cross section, there will be a certain ratio of depth to width (y/B) for which the section factor will be maximum. Such a ratio establishes the *best hydraulic section*. That is, the best hydraulic section is the channel proportion that yields a minimum wetted perimeter for a given cross-sectional area.

EXAMPLE 10.18   Determine the best hydraulic section for a rectangular channel.

Solution   For the rectangular channel, $A = By$, and $P = B + 2y$. Let A be constant, then let us minimize P. But

$$P = B + 2y$$

or

$$P = \frac{A}{y} + 2y$$

Thus, we see that the perimeter varies only with y for the given conditions. If we differentiate P with respect to y and set the differential equal to zero, we will have the condition for minimizing P for the given area A:

$$\frac{dP}{dy} = \frac{-A}{y^2} + 2 = 0$$

or

$$\frac{A}{y^2} = 2$$

But

$$A = By, \text{ so}$$

$$\frac{By}{y^2} = 2 \quad \text{or} \quad y = \frac{1}{2}B \qquad \blacktriangleleft$$

Thus, the best hydraulic section for a rectangular channel occurs when the depth is one-half the width of the channel.

It can be shown that the best hydraulic section for a trapezoidal channel is half a hexagon; for the circular section, it is the half circle; and for the triangular section, it is half of a square. Of all the various shapes, the half circle has the best hydraulic section.

The best hydraulic section can be relevant to the cost of the channel. For example, if a trapezoidal channel were to be excavated and if the water surface were to be at adjacent ground level, the minimum amount of excavation (and excavation cost) would result if the channel of best hydraulic section were used.

Flood Flows

When a river is in flood stage, there is flow in the main channel of the river and also in that part of the flood plain that is flooded. That part of the flood plain over which flow is occurring is called the *overbank area* (see Fig. 10.22).

The channel characteristics of the main channel and the overbank area are distinctly different. For example, the depth in the main channel is much greater than in the overbank area. Also, the overbank area is often covered with vegetation, such as crops, brush, and trees, whereas the bed of the main channel may be of rock. Thus, the resistance to flow in the overbank area is usually much greater than in the main channel. See Table 10.4 for typical n values for overbank areas. Because of the different resistance characteristics of the different channel parts, it is customary to consider each part separately when making discharge calculations. The procedure is explained in the next paragraph.

The total discharge for the river in flood stage will be given as

$$Q_T = Q_1 + Q_m \qquad (10.46)$$

where
Q_1 = discharge in overbank area

Q_m = discharge in the main channel

TABLE 10.4 VALUES OF MANNING'S n TO BE USED FOR OVERBANK AREAS ALONG STREAMS OR RIVERS

Channel Description	n
Grass land	
Short grass	0.030
Tall grass	0.035
Cultivated ground	
Bare ground	0.030
Mature row crops	0.035
Mature field crops	0.040
Brushy areas	
Dense weeds and sparse brush	0.050
Brush-covered with some trees (winter)	0.050
Brush-covered with some trees (summer)	0.060
Dense brush (winter)	0.070
Dense brush (summer)	0.100
Forested	
Densely covered with willows (summer)	0.150
Cleared land with stumps; no new growth	0.040
Cleared land with stumps; dense new growth	0.060
Dense stands of large trees; flood stage below branches	0.100
Dense stands of large trees; flood stage reaching branches	0.120

FIGURE 10.22

Cross section of a river in flood stage.

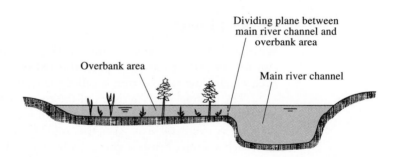

However, Q can be given by the Chezy equation [Eq. (10.42)]; therefore, Eq. (10.46) can be written as

$$Q_T = C_1 A_1 \sqrt{R_{h1} S_{01}} + C_m A_m \sqrt{R_{hm} S_{0m}} \qquad (10.47)$$

The slopes S_{01} and S_{0m} will be equal because a line following the water surface across the channel is assumed to be horizontal and continuous over the main channel as well as the overbank area; therefore, the change in elevation of the water surface between given cross sections will be the same for both the main channel and the overbank area. Because the change in elevation of the water surface for the main and overbank channels will be the same, the slopes of the water surfaces for both will be equal. Then Eq. (10.47) may be given as

$$Q_T = C_1 A_1 \sqrt{R_{h1} S_0} + C_m A_m \sqrt{R_{hm} S_0}$$
$$= (C_1 A_1 \sqrt{R_{h1}} + C_m A_m \sqrt{R_{hm}}) \sqrt{S_0} \qquad (10.48)$$

$$= \sqrt{S_0} \sum C A \sqrt{R_h} \qquad (10.49)$$

If we are using traditional units (ft-lb-s), Eq. (10.49) will be given as

$$Q_T = 1.49 \sqrt{S_0} \sum \frac{1}{n} A R_h^{2/3} \qquad (10.50)$$

The corresponding equation for the SI system is

$$Q_T = \sqrt{S_0} \sum \frac{1}{n} A R_h^{2/3} \qquad (10.51)$$

When using Eq. (10.50) or Eq. (10.51), certain decisions must be made regarding the demarcation between the main channel and the overbank area. It is customary to simply assume a vertical separation plane between them, such as is shown by the vertical dashed line in Fig. 10.22. The placement of this plane will be an arbitrary decision by the engineer making the calculation. Once the separation plane has been chosen, the areas and wetted perimeters associated with each channel can then be calculated. In regard to the wetted perimeters, it is customary to include only the perimeter in contact with the channel bottoms and walls—the perimeter associated with the "imaginary surface" between the two channels is not included.

EXAMPLE 10.19   The cross section of a river in flood stage has the dimensions shown below.

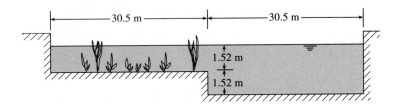

Estimate the total discharge if Manning's n is 0.05 for the overbank area and 0.03 for the main channel, and the channel slope is 0.00167. Also, what is the mean velocity for each section?

Solution   In this simplified problem, it is obvious where to locate the separation plane between the main channel and the overbank area. Therefore,

$$A_m = (30.5 \text{ m}) (3.04 \text{ m}) = 92.72 \text{ m}^2$$

$$A_1 = (30.5 \text{ m}) (1.52 \text{ m}) = 46.36 \text{ m}^2$$

$$P_m = 1.52 \text{ m} + 30.5 \text{ m} + 3.04 \text{ m} = 35.06 \text{ m}$$

$$P_1 = 30.5 \text{ m} + 1.52 \text{ m} = 32.02 \text{ m}$$

Then $$R_{hm} = \frac{A_m}{P_m} = 2.645 \text{ m} \quad \text{and} \quad R_{h1} = 1.448 \text{ m}$$

Solving Eq. (10.51) with the computed and given quantities, we obtain

$$Q_T = 290.0 \text{ m}^3/\text{s} \qquad \blacktriangleleft$$

Also $$Q_m = 241.5 \text{ m}^3/\text{s}$$

$$Q_1 = 48.5 \text{ m}^3/\text{s}$$

Thus $$V_m = \frac{Q_m}{A_m} = 2.605 \text{ m/s} \quad \text{and} \quad V_1 = 1.046 \text{ m/s} \qquad \blacktriangleleft$$

In the development of Eqs. (10.50) and (10.51) and also in Example 10.19, only one overbank area was considered, but it is easy to envision several overbank areas, as shown in Fig. 10.23. In any event, the application of Eqs. (10.50) and (10.51) to a channel with several overbank areas simply involves extra terms in the summation process.

FIGURE 10.23

Multiple overbank areas.

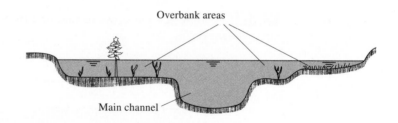

Overbank areas

Main channel

Uniform Flow in Culverts and Sewers

Sewers are conduits that are used to carry domestic, commercial, or industrial waste (called sewage) from households, businesses, and factories to sewage disposal sites. These conduits are often circular in cross section, but elliptical and rectangular conduits are also used. The volume rate of sewage varies throughout the day and season, but of course sewers are designed to carry the maximum design discharge flowing full or nearly full. At discharges less than the maximum, the sewers will operate as open channels.

Sewage usually consists of about 99% water and 1% solid waste. Because most sewage is so dilute, it is assumed that it has the same physical properties as water for purposes of discharge computations. However, if the velocity in the sewer is too small, the solid particles may settle out and cause blockage of the flow. Therefore, sewers are usually designed to have a minimum velocity of about 2 ft/s (0.60 m/s) at maximum flow condition. This condition is met by choosing a slope on the sewer line to achieve the desired velocity.

EXAMPLE 10.20 A sewer line is to be constructed of concrete pipe and to be laid on a slope of 0.006. If $n = 0.013$ and if the design discharge is 110 cfs, what size pipe (commercially available) would you choose for a full flow condition? What will be the mean velocity in the sewer pipe for these conditions? (It should be noted that concrete pipe is readily available in commercial sizes of 8-in., 10-in., and 12-in. diameter and then in 3-in. increments up to 36-in. diameter. From 36-in. diameter up to 144-in. the sizes are available in 6-in. increments.)

Solution

$$Q = \frac{1.49}{n} A R^{2/3} S_0^{1/2}$$

where $Q = 110 \ \text{ft}^3/\text{s}$

$n = 0.013$

$S_0 = 0.006$ (assume atmospheric pressure along the pipe)

Then

$$AR^{2/3} = \frac{(110 \text{ ft}^3/\text{s})\,(0.013)}{(1.49)\,(0.006)^{1/2}} = 12.39 \text{ ft}^{8/3}$$

But

$$R = \frac{A}{P} \qquad R^{2/3} = \left(\frac{A}{P}\right)^{2/3}$$

Then

$$AR^{2/3} = \frac{A^{5/3}}{P^{2/3}} = 12.39 \text{ ft}^{8/3}$$

For a pipe flowing full, $A = \pi D^2/4$ and $P = \pi D$, or

$$\frac{(\pi D^2/4)^{5/3}}{(\pi D)^{2/3}} = 12.39 \text{ ft}^{8/3}$$

Solving for diameter yields $D = 3.98$ ft $= 47.8$ in. Use the next commercial size larger, which is $D = 48$ in. ◄

$$A = \frac{\pi D^2}{4} = 50.3 \text{ ft}^2 \text{ (for pipe flowing full)}$$

Thus

$$V = \frac{Q}{A} = \frac{(110 \text{ ft}^3/\text{s})}{(50.3 \text{ ft}^2)} = 2.19 \text{ ft/s} \qquad ◄$$

A culvert is a conduit placed under a fill such as a highway embankment. It is used to convey streamflow from the uphill side of the fill to the downhill side. Figure 10.24 shows the essential features of a culvert. Culverts are designed to pass the design discharge without adverse effects on the fill. That is, the culvert should be able to convey runoff from a design storm without overtopping the fill and without erosion of the fill at either the upstream or downstream end of the culvert. The design storm, for example, might be the maximum storm that could be expected to occur once in 50 years at the particular site.

The flow in a culvert is a function of many variables, including cross-sectional shape (circular or rectangular), slope, length, roughness, entrance design, and exit design. Flow in a culvert may occur as an open channel throughout its

FIGURE 10.24

Culvert under a highway embankment.

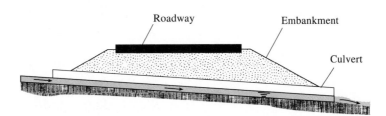

length, it may occur as a completely full pipe, or it may occur as a combination of both. The complete design and analysis of culverts are beyond the scope of this text; therefore, only a simple example is included here. For more extensive treatment of culverts, please refer to Chow (8), Henderson (15), and American Concrete Pipe Assoc. (1).

EXAMPLE 10.21   A 54-in.-diameter culvert laid under a highway embankment has a length of 200 ft and a slope of 0.01. This was designed to pass a 50-year flood flow of 225 cfs under full flow conditions (see the figure below). For these conditions, what head H is required? When the discharge is only 50 cfs, what will be the uniform flow depth in the culvert? Assume $n = 0.012$.

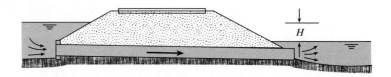

Solution   For the flood flow of 225 cfs, one must consider the entrance and exit head losses as well as the head loss in the pipe itself; therefore, use the energy equation to solve this example.

$$\frac{p_1}{\gamma} + \frac{V_1^2}{2g} + z_1 = \frac{p_2}{\gamma} + \frac{V_2^2}{2g} + z_2 + \sum h_L$$

Let points 1 and 2 be at the upstream and downstream water surfaces, respectively.

Thus, $p_1 = p_2 = 0$ gage   and $V_1 = V_2 = 0$

Also, $z_1 - z_2 = H$

Then we have $H = \sum h_L$

$H = $ pipe head loss + entrance head loss + exit head loss

$$H = \frac{V^2}{2g}(K_e + K_E) + \text{pipe head loss}$$

Assume $K_e = 0.50$ (from Table 10.2)

$K_E = 1.00$ (from Table 10.2)

For the pipe head loss, use Eq. (10.45):

$$Q = \frac{1.49}{n} A R_h^{2/3} S_0^{1/2} \tag{10.45}$$

where
$$Q = 225 \text{ ft}^3/\text{s}$$

$$A = \frac{\pi D^2}{4} = 15.90 \text{ ft}^2$$

$$R_h = \frac{A}{P} = \frac{\pi D^2/4}{\pi D} = \frac{D}{4} = 1.125 \text{ ft}$$

$$R_h^{2/3} = (1.125 \text{ ft})^{2/3} = 1.0817 \text{ ft}^{2/3}$$

$$S_0 = \frac{h_f}{L}$$

Then Eq. (10.45) is written as

$$225 = \frac{1.49}{0.012}(15.90 \text{ ft}^2)(1.0817 \text{ ft}^{2/3})\left(\frac{h_f}{200}\right)^{1/2}$$

$$h_f = 2.22 \text{ ft}$$

$$V = \frac{Q}{A} = \frac{225 \text{ ft}^3/\text{s}}{15.90 \text{ ft}^2/\text{s}} = 14.15 \text{ ft/s}$$

Solving for H:

$$H = \frac{14.15^2}{64.4}(0.50 + 1.0) + 2.22$$

$$H = 4.66 \text{ ft} + 2.22 \text{ ft} = 6.88 \text{ ft} \qquad \blacktriangleleft$$

For $Q = 50$ cfs, we need to use Eq. (10.45):

$$50 = \frac{1.49}{0.012}AR_h^{2/3}(0.01)^{1/2}$$

However, this culvert will flow only partly full with a Q of 50 cfs. Therefore, the physical relationship will be as shown below.

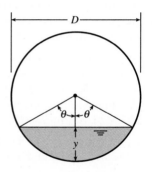

Thus, if the angle θ is given in degrees, the cross-sectional flow area will be given as

$$A = \left[\left(\frac{\pi D^2}{4} \right) \left(\frac{2\theta}{360°} \right) \right] - \left(\frac{D}{2} \right)^2 (\sin \theta \cos \theta)$$

The wetted perimeter will be $P = \pi D (\theta/180°)$, or

$$R_h = \frac{A}{P} = \left(\frac{D}{4} \right) \left[1 - \left(\frac{\sin \theta \cos \theta}{\pi (\theta/180°)} \right) \right]$$

Substituting these relations for A and R_h into the discharge equation and solving for θ yields

$$\theta = 70°$$

Depth of flow:

$$y = \frac{D}{2} - \frac{D}{2} \cos \theta = \left(\frac{54 \text{ in.}}{2} \right) (1 - 0.342) = 17.8 \text{ in.} \quad \blacktriangleleft$$

Problems

10.1 Consider the mean-velocity profiles for flow in the pipes shown. Match the profiles with the following: a) turbulent flow, b) obviously a case of hypothetical flow (zero viscosity), c) laminar case, d) $\alpha = 1.0$, e) $\alpha = 1.05$, f) $\alpha = 2.00$.

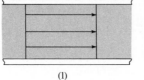 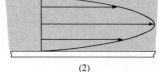 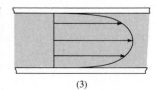

      (1)   (2)   (3)

PROBLEM 10.1

10.2 Liquid in the pipe shown in the figure has a specific weight of 10 kN/m³. The acceleration of the liquid is zero. Is the liquid stationary, moving upward, or moving downward in the pipe? If the pipe diameter is 1 cm and the liquid viscosity is 3.125×10^{-3} N · s/m², what is the magnitude of the mean velocity in the pipe?

10.3 A viscous oil is contained in this cylinder/nozzle system that has a vertical orientation. A valve is instantaneously opened to let the oil drain out of the cylinder. Below are listed words that might characterize the flow at point A. Which ones are valid characterizations at the time when the oil surface reaches the level of section 2? a) steady, b) unsteady, c) rotational, d) irrotational, e) uniform, f) nonuniform.

10.4 Oil ($S = 0.97$, $\mu = 10^{-2}$ lb-s/ft²) is pumped through a 2-in. pipe at the rate of 0.20 cfs. What is the pressure drop per 100 ft of level pipe?

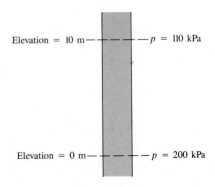

Elevation = 10 m — p = 110 kPa

Elevation = 0 m — p = 200 kPa

PROBLEM 10.2

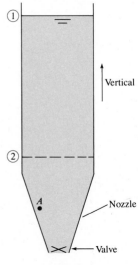

Vertical

Nozzle

A

Valve

PROBLEM 10.3

10.5 Liquid flows downward in a 1-cm, vertical, smooth pipe with a mean velocity of 2.0 m/s. The liquid has a density of 1000 kg/m³ and a viscosity of 0.10 N · s/m³. If the pressure at a given section is 600 kPa, what will be the pressure at a section 10 m below that section?

10.6 A liquid (ρ = 1000 kg/m³, μ = 10^{-2} N · s/m², ν = 10^{-5} m²/s) flows uniformly with a mean velocity of 1 m/s in a pipe with a diameter of 10 mm. For this condition, will the velocity distribution be logarithmic or parabolic? What will be the ratio of the shear stress at 1 mm from the wall to the shear stress of the wall?

10.7 Glycerin at a temperature of 30°C flows at a rate of 8 × 10^{-6} m³/s through a horizontal tube with a 30-mm diameter. What is the pressure drop in pascals per 100 m?

10.8 Kerosene (S = 0.80 and T = 68°F) flows from the tank shown and through the $\frac{1}{4}$-in.-diameter (ID) tube. Determine the mean velocity in the tube and the discharge.

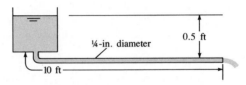

PROBLEM 10.8

10.9   Oil (S = 0.94, μ = 0.048 N · s/m²) is pumped through a horizontal 5-cm pipe at the rate of 2.0 × 10⁻³ m³/s. What is the pressure drop per 100 m of pipe?

10.10   SAE 10W-30 oil is pumped through a 5-m length of 1-cm-diameter drawn tubing at a discharge of 7.85 × 10⁻⁴ m³/s. There is a pump in the line as shown. The pipe is horizontal, and the pressures at points 1 and 2 are equal. Find the power necessary to operate the pump, assuming the pump has an efficiency of 100%. Properties of SAE 10W-30 oil: kinematic viscosity = 7.6 × 10⁻⁵ m²/s, specific weight = 8630 N/m³.

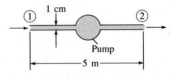

PROBLEM 10.10

10.11   Oil (S = 0.9; μ = 10⁻² lbf-s/ft²; ν = 0.0057 ft²/s) flows downward at a rate of 0.0157 ft³/s in the pipe, which is 0.10 ft in diameter and has a slope of 30° with the horizontal. What is the pressure gradient (dp/ds) along the pipe?

PROBLEM 10.11

10.12   A fluid (μ = 10⁻² N · s/m²; ρ = 800 kg/m³) flows with a mean velocity of 10 cm/s in a 10-cm smooth pipe. Answer the following questions relating to the given flow conditions.

    a. What is the magnitude of the maximum velocity in the pipe?
    b. What is the magnitude of the resistance coefficient f?
    c. What is the shear velocity for these flow conditions?
    d. What is the shear stress at a radial distance of 25 mm from the center of the pipe?

10.13   Kerosene (20°C) flows at a rate of 0.02 m³/s in a 20-cm pipe. Would you expect the flow to be laminar or turbulent?

10.14   In the pipe system for a given discharge, the ratio of the head loss in a given length of the 1-m pipe to the head loss in the same length of the 2-m pipe is a) 2, b) 4, c) 16, d) 32.

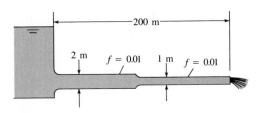

PROBLEM 10.14

10.15 Glycerin ($T = 68°F$) flows in a pipe with a 1-ft diameter at a mean velocity of 2 ft/s. Is the flow laminar or turbulent? Plot the velocity distribution across the flow section.

10.16 Glycerine ($T = 20°C$) flows through a funnel as shown. Calculate the mean velocity of the glycerine exiting the tube.

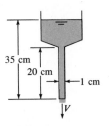

PROBLEM 10.16

10.17 What size of steel pipe should be used to carry 0.1 cfs of castor oil at 90°F a distance of 1 mi with an allowable friction loss of 20 psi? Assume S = 0.85.

10.18 Mercury at 20°C flows downward in a long circular tube that is open to the atmosphere at the top and bottom. The tube is vertically oriented. Find the tube diameter for which the flow would just become turbulent (Re = 2000).

10.19 Glycerin (20°C) flows in a 4-cm steel tube with a mean velocity of 40 cm/s. Is the flow laminar or turbulent? What is the shear stress at the center of the tube and at the wall? If the tube is vertical and the flow is downward, will the pressure increase or decrease in the direction of flow? At what rate?

10.20 A 4-cm pipe 7 m long is supported vertically over a tank and discharges oil (S = 0.8, $\mu = 10^{-1}$ N · s/m^2) at a mean velocity of 3 m/s as shown. If the pipe weighs 200 N, what force is required to support it?

10.21 Velocity measurements are made across a 1-ft pipe. The velocity at the center is found to be 3 fps, and the velocity distribution is seen to be parabolic. If the pressure drop is found to be 16 psf per 100 ft of pipe, what is the kinematic viscosity ν of the fluid? Assume that the fluid's specific gravity is 0.90.

10.22 Velocity measurements are made in a 30-cm pipe. The velocity at the center is found to be 1.5 m/s, and the velocity distribution is observed to be parabolic. If the pressure drop is found to be 1.9 kPa per 100 m of pipe, what is the kinematic viscosity ν of the fluid? Assume that the fluid's specific gravity is 0.90.

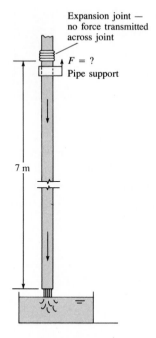

PROBLEM 10.20

10.23 Water is pumped through a heat exchanger consisting of tubes 5 mm in diameter and 5 m long. The velocity in each tube is 10 cm/s. The water temperature increases from 20°C at the entrance to 30°C at the exit. Calculate the pressure difference across the heat exchanger, neglecting entrance losses but accounting for the effect of temperature change.

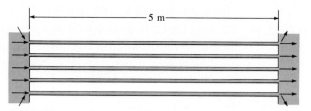

PROBLEM 10.23

10.24 The velocity of oil (S = 0.8) through the 2-in. smooth pipe is 5 ft/s. Here $L = 30$ ft, $z_1 = 2$ ft, $z_2 = 4$ ft, and the manometer deflection is 4 in. Determine the flow direction, the resistance coefficient f, whether the flow is laminar or turbulent, and the viscosity of the oil.

10.25 The velocity of oil (S = 0.8) through the 5-cm smooth pipe is 1.2 m/s. Here $L = 10$ m, $z_1 = 1$ m, $z_2 = 2$ m, and the manometer deflection is 10 cm. Determine the flow direction, the resistance coefficient f, whether the flow is laminar or turbulent, and the viscosity of the oil.

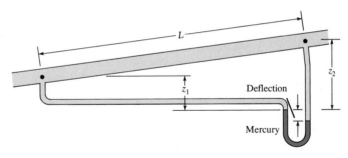

PROBLEMS 10.24, 10.25

10.26   Flow of a liquid in a smooth 3-cm pipe is found to yield a head loss of 2 m per meter of pipe when the mean velocity is 1 m/s. If the rate of flow were doubled, would the head loss also be doubled? Explain.

10.27   In a 12-in. smooth pipe, f is 0.019 when oil having a specific gravity of 0.82 flows with a mean velocity of 6 ft/s. If τ_{app} is found to be 0.10 psf at a distance of 1 in. from the wall of the pipe, what is the viscous shear stress on the wall?

10.28   Consider the flow of oil ($\rho = 900$ kg/m^3; $\mu = 10^{-1}$ N $\cdot$ s/m^2) in a 10-cm smooth pipe and the flow of a gas ($\rho = 1.0$ kg/m^3; $\mu = 10^{-5}$ N $\cdot$ s/m^2) in a 10-cm smooth pipe. Both the oil and the gas flow with a mean velocity of 1 m/s. Will the ratio of the maximum velocity in the oil to the maximum velocity in the gas ($V_{\text{max,oil}}/V_{\text{max,gas}}$) be a) less than 1, b) equal to 1, or c) greater than 1?

10.29   At a Reynolds number of 100,000, what is the nominal thickness of the viscous sublayer in a 12-cm smooth pipe? Also, what is f?

10.30   Consider laminar flow in this pipe with rounded entrance. Then consider the ratio of the shear stress at point A to that at point B, τ_{0A}/τ_{0B}. The value of this ratio will be a) less than 1, b) equal to 1, c) greater than 1.

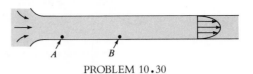

PROBLEM 10.30

10.31   Water (70°F) flows through a 10-in. smooth pipe at the rate of 3 cfs. What is the resistance coefficient f?

10.32   Water (10°C) flows through a 25-cm smooth pipe at a rate of 0.06 m^3/s. What is the resistance coefficient f?

10.33   Air flows in a 3-cm smooth tube at a rate of 0.012 m^3/s. If $T = 20$°C and $\rho = 110$ kPa absolute, what is the pressure drop per meter of length of tube?

10.34   Glycerine at 20°C flows at 0.5 m/s in the 2-cm commercial steel pipe. Two standpipes are used as shown to measure the piezometric head. The distance along the pipe between the

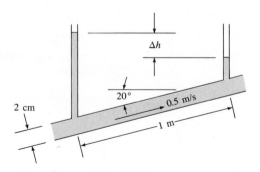

PROBLEM 10.34

standpipes is 1 m. The inclination of the pipe is 20°. What is the height difference Δh between the glycerine in the two standpipes?

10.35   Air flows in a 1-in. smooth tube at a rate of 20 cfm. If $T = 80°F$ and $p = 15$ psia, what is the pressure drop per foot of length of tube?

10.36   A pipe can be used to measure the viscosity of a fluid. A liquid flows in a 1-cm smooth pipe 1 m long with an average velocity of 2 m/s. A head loss of 30 cm is measured. Find the kinematic viscosity.

10.37   Water flows in the pipe shown, and the manometer deflects 80 cm. What is f for the pipe if $V = 4$ m/s?

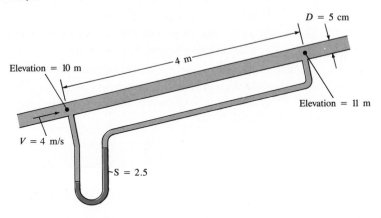

PROBLEM 10.37

10.38   What is f for the flow of water at 10°C through a 30-cm cast-iron pipe with a mean velocity of 3 m/s? Plot the velocity distribution for this flow.

10.39   Consider flow in a long, uniform-diameter pipe, $k_s = 10^{-4}$ ft. If the mean velocity is equal to 1 ft/s, if $d = 0.10$ ft, and if $\nu = 10^{-4}$ ft²/s, which of the following values is the correct resistance coefficient f? a) 0.064, b) 0.044, c) 0.034, d) 0.020.

10.40   Water at 20°C flows through a 6-cm smooth brass tube at a rate of 0.003 m³/s. What is f for this flow?

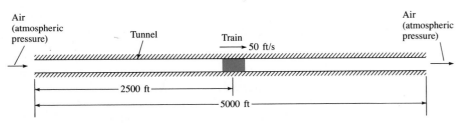

PROBLEM 10.41

10.41   A train travels through a tunnel as shown. The train and tunnel are circular in cross section. Clearance is small, causing all air to be pushed from the front of the train and discharged from the tunnel. The tunnel is 10 ft in diameter and is concrete. The train speed is 50 fps.

   a. Determine the change in pressure between the front and rear of the train that is due to *pipe-friction* effects.
   b. Sketch the energy and hydraulic grade lines for the train position shown.
   c. What power is required to produce the air flow in the tunnel?

10.42   Estimate the elevation required in the upper reservoir to produce a water discharge of 10 cfs in the system. Carefully draw the HGL and the EGL for the system. Where is the point of minimum pressure in the pipe, and what is the magnitude of the pressure at that point? What kind of pipe do you think this is?

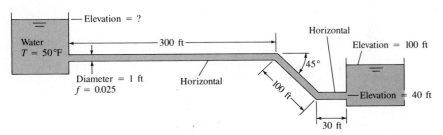

PROBLEM 10.42

10.43   A fluid ($\mu = 10^{-2}$ N · s/m²; $\rho = 800$ kg/m³) flows with a mean velocity of 100 mm/s in a 100-mm smooth pipe. Answer the following questions relating to the given flow conditions.

   a. What is the magnitude of the maximum velocity in the pipe?
   b. What is the magnitude of the resistance coefficient f?
   c. What is the shear velocity?
   d. What is the shear stress at a radial distance of 25 mm from the center of the pipe?
   e. If the discharge is doubled, will the head loss per length of pipe also be doubled?

$$\frac{V^2}{2g} + \frac{p}{\gamma} + z = \frac{V^2}{2g} + \frac{p}{\gamma} + z + h_f$$

10.44   Which of the following statements are true of the energy grade line for steady flow in pipes where no pumps or turbines are present?

a. The EGL always slopes downward in the direction of flow.

b. The EGL is steeper for a rough pipe than for a smooth pipe if the pipe diameter and discharge are equal.

c. In a given smooth pipe ($D = 1$ m and $\nu = 10^{-6}$ m^2/s), the slope of the EGL is doubled when the velocity is changed from 1 m/s to 2 m/s.

d. In a given smooth pipe ($D = 1$ m and $\nu = 10^{-2}$ m^2/s), the slope of the EGL is doubled when the velocity is changed from 1 m/s to 2 m/s.

e. In a given rough pipe ($D = 1$ m, $\nu = 10^{-6}$ m^2/s, $k_s = 1$ cm), the slope of the EGL is doubled when the velocity is changed from 1 m/s to 2 m/s.

10.45   Referring to the figure, what do you think are at A and C? What do you think is at B? Beyond D, complete the physical setup that could yield the EGL and HGL shown. What other information is indirectly revealed by the EGL and HGL?

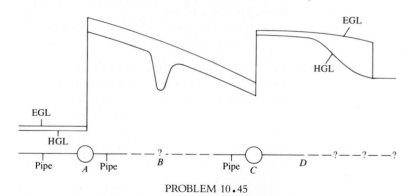

PROBLEM 10.45

10.46   Water (20°C) flows in a 15-cm cast-iron pipe at a rate of 0.04 m^3/s. For these conditions, determine or estimate the following:

a. Shear stress at the wall, τ_0
b. Shear stress 1 cm from the wall
c. Velocity 1 cm from the wall

10.47   Water flows from reservoir A to reservoir B. Reservoir A has a water-surface elevation of 100 m, and reservoir B has a water-surface elevation of 70 m. The two reservoirs are 100 m apart. Devise a conduit system to transport water from A to B in such a way that part of the conduit will have subatmospheric pressure in it. Assume the system contains neither pumps nor valves. Also, draw the EGL and the HGL for the system, and determine the magnitude of the minimum pressure.

10.48   Water is pumped through a vertical 10-cm new steel pipe to an elevated tank on the roof of a building. The pressure on the discharge side of the pump is 1.5 MPa. What pressure can be expected at a point in the pipe 80 m above the pump when the flow is 0.02 m^3/s? Assume $T = 20$°C.

10.49 Suppose that it is possible to prevent turbulence from developing in the flow in a pipe. What would be the ratio of the head loss for the laminar flow to that for the turbulent flow when kerosene (S = 0.82, $\nu = 2 \times 10^{-6}$ m²/s) is pumped through a 3-cm smooth pipe with an average velocity of 4 m/s?

10.50 In a 4-in. uncoated cast-iron pipe, 0.03 cfs of water flows at 60°F. Determine f from Fig. 10.8.

10.51 Determine the head loss in 1500 ft of a concrete pipe with a 6-in. diameter (k_s = 0.0002 ft) carrying 0.75 cfs of fluid. The properties of the fluid are $\nu = 3.33 \times 10^{-2}$ ft²/s and $\rho = 1.5$ slug/ft³.

10.52 Points A and B are 1 km apart along a 15-cm new steel pipe. Point B is 20 m higher than A. With a flow from A to B of 0.03 m³/s of crude oil (S = 0.82) at 10°C ($\mu = 10^{-2}$ N · s/m²), what pressure must be maintained at A if the pressure at B is to be 350 kPa?

10.53 Water flows from a tank through a 2.6-m length of galvanized-iron pipe 26 mm in diameter. At the end of the pipe is an angle valve that is wide open. The height of the water level in the tank above the valve exit is 10 m. Calculate the water velocity in the pipe (in meters per second).

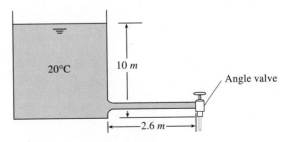

PROBLEM 10.53

10.54 Points A and B are 3 mi apart along a 24-in. new cast-iron pipe carrying water ($T = 50°F$). Point A is 30 ft higher than B. The pressure at B is 20 psi greater than that at S. Determine the direction and amount of flow.

10.55 If a flow of 0.10 m³/s of water is to be maintained in the system shown, what power must be added to the water by the pump? The pipe is made of steel and is 15 cm in diameter. Draw the EGL and the HGL for the system.

10.56 Water at 10°C flows from a reservoir (water surface at 120 m elevation) through 10 km of concrete pipe of 1.0-m diameter and discharges into another reservoir (water surface at 20 m elevation). Neglecting inlet and exit losses, what is the discharge? If riveted steel pipe of the same diameter were used, what would be the discharge? What pump power (in megawatts) would be required to produce a discharge of 2.8 m³/s in the reverse direction (uphill) in the concrete pipe?

10.57 A fluid with $\nu = 10^{-6}$ m²/s and $\rho = 900$ kg/m³ flows through the 8-cm galvanized-iron pipe. Estimate the flow rate for the conditions shown in the figure.

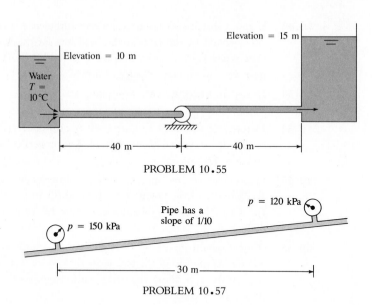

PROBLEM 10.55

PROBLEM 10.57

10.58   What power must the pump supply to the system to pump the oil from the lower reservoir to the upper reservoir at a rate of 0.20 m³/s? Sketch the HGL and the EGL for the system.

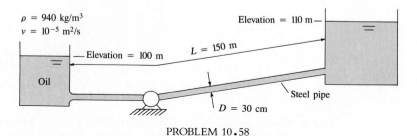

PROBLEM 10.58

10.59   For a 40-cm pipe the resistance coefficient f was found to be 0.06 when the mean velocity was 3 m/s and the kinematic viscosity was 10^{-5} m²/s. If the velocity were doubled, would you expect the head loss per meter of length of pipe to double, triple, or quadruple?

10.60   A cast-iron pipe 1.0 ft in diameter and 200 ft long joins two water reservoirs. The upper reservoir has a water-surface elevation of 80.0 ft, and the lower one has a water-surface elevation of 40.0 ft. The pipe exits from the side of the upper reservoir at an elevation of 70.0 ft and enters the lower reservoir at an elevation of 35.0 ft. There are two wide-open gate valves in the pipe. Draw the EGL and the HGL for the system, and determine the discharge in the pipe.

10.61   What diameter of cast-iron pipe is needed to carry water at a rate of 10 cfs between two reservoirs if the reservoirs are 2 mi apart and the elevation difference between the water surfaces in the reservoirs is 20 ft?

10.62   Determine the diameter of commercial steel pipe required to convey 300 cfs of water at 60°F with a head loss of 1 ft per 1000 ft of pipe. Assume pipes are available in the even sizes when the diameters are expressed in inches (that is, 10 in., 12 in., etc.).

10.63   A pipeline is to be designed to carry crude oil (S = 0.93, $\nu = 10^{-5}$ m²/s) with a discharge of 0.10 m³/s and a head loss per kilometer of 30 m. What diameter of steel pipe is needed? What power output from a pump is required to maintain this flow?

10.64   Design a pipe to carry water at a rate of 15 cfs between two reservoirs if the reservoirs are 3 mi apart and the elevation difference between their water surfaces is to be 30 ft. Assume pipes are available in even sizes (that is, diameters of 16 in., 18 in., etc.).

10.65   Determine the discharge of kerosene (20°C) through the 15-cm smooth brass pipe if $h = 3$ m and $L = 70$ m.

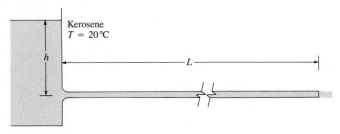

PROBLEM 10.65

10.66   The water-surface elevation in a reservoir is 120 ft. A straight pipe 100 ft long and 6 in. in diameter conveys water from the reservoir to an open drain. The pipe entrance (it is abrupt) is at elevation 100 ft, and the pipe outlet is at elevation 80 ft. At the outlet the water discharges freely into the air. The water temperature is 50°F. If the pipe is asphalted cast iron, what will be the discharge rate in the pipe? Consider all head losses. Also draw the HGL and the EGL for this system.

10.67   The water-surface elevation in a reservoir is 600 ft. A straight pipe 1200 ft long and 2 ft in diameter conveys water from the reservoir to an open drain. The pipe entrance (it is rounded) is at elevation 570 ft, and the pipe outlet is at elevation 200 ft. At the outlet the water discharges freely into the air. The water temperature is 50°F. The discharge of water in the pipe is 157 cfs. Determine the minimum pressure in the pipe and the equivalent sand roughness, k_s, of the pipe.

10.68   A heat exchanger is being designed as a component of a geothermal power system in which heat is transferred from the geothermal brine to a "clean" fluid in a closed-loop power cycle. The heat exchanger, a shell and tube type, consists of 100 galvanized-iron tubes 2 cm in diameter and 5 m long, as shown. The temperature of the fluid is 200°C, the density is 860 kg/m³, and the viscosity is 1.35×10^{-4} N · s/m². The total mass-flow rate through the exchanger is 50 kg/s.

a. Calculate the power required to operate the heat exchanger, neglecting entrance and outlet losses.

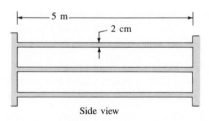

Side view

PROBLEM 10.68

b. After continued use, 2 mm of scale develops on the inside surfaces of the tubes. This scale has an equivalent roughness of 0.5 mm. Calculate the power required under these conditions.

10.69 The heat exchanger shown consists of drawn tubing 2 cm in diameter and 20 m long with 19 return bends. Water is pumped through the system at 3×10^{-4} m^3/s, entering at 20°C and exiting at 80°C. The elevation difference between the entrance and the exit is 0.8 m. Calculate the pump power required to operate the heat exchanger if there is no pressure change between 1 and 2. Use the viscosity corresponding to the average temperature in the heat exchanger.

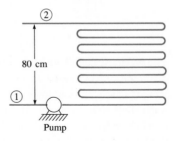

PROBLEM 10.69

10.70 Determine the discharge of water through the system shown. In addition,

    a. Draw the HGL and the EGL for the system.
    b. Locate the point of maximum pressure.
    c. Locate the point of minimum pressure.
    d. Calculate the maximum and minimum pressures in the system.

10.71 Gasoline ($T = 50°F$) is pumped from the gas tank of an automobile to the carburetor through a $\frac{1}{4}$-in. fuel line of drawn tubing 10 ft long. The line has five 90° smooth bends with an r/d of 6. The gasoline discharges through a $\frac{1}{32}$-in. jet in the carburetor to a pressure of 14 psia. The pressure in the tank is 14.7 psia. The pump is 80% efficient. What power must be supplied by the pump if the automobile is accelerating and consuming fuel at the rate of 0.1 gpm? Obtain gasoline properties from Figs. A.2 and A.3 in the Appendix.

10.72 Find the loss coefficient K_v of the partially closed valve that is required to reduce the discharge to 50% of the flow with the valve wide open as shown.

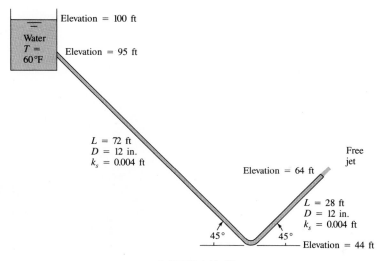

PROBLEM 10.70

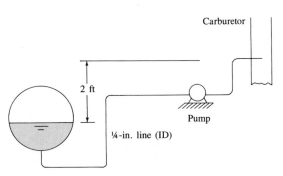

PROBLEM 10.71

10.73 The pressure at a water main is 300 kPa gage. What size of pipe is needed to carry water from the main at a rate of 0.025 m³/s to a factory that is 140 m from the main? Assume that galvanized-steel pipe is to be used and that the pressure required at the factory is 60 kPa gage at a point 10 m above the main connection.

10.74 Two reservoirs with a difference in water-surface elevation of 11 ft are joined by 45 ft of 1-ft steel pipe and 30 ft of 6-in. steel pipe in series. The 1-ft line contains 3 bends ($r/d = 1$), and the 6-in. line contains 2 bends ($r/d = 4$). If the 1-ft and 6-in. lines are joined by an abrupt contraction, determine the discharge. Assume $T = 60°F$.

10.75 The 10-cm galvanized-steel pipe is 1000 m long and discharges water into the atmosphere. The pipeline has an open globe valve and four threaded elbows; $h_1 = 3$ m and $h_2 = 15$ m. What is the discharge, and what is the pressure at A, the midpoint of the line?

10.76 In the laboratory setup shown, water at 50°F is pumped at a rate of 5 cfs from the lower tank to the upper one through an 8-in. pipe. Water returns from the upper tank to the

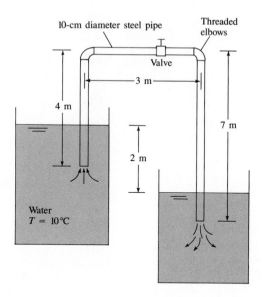

PROBLEM 10.72

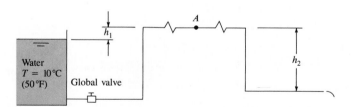

PROBLEM 10.75

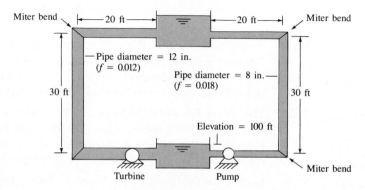

PROBLEM 10.76

lower one through a 12-in. pipe and turbine. Assuming that steady flow prevails (water in both tanks is kept at a constant level), what power is given up to the turbine if the head supplied by the pump is 50 ft?

10.77 The piping system in a typical dwelling is shown in the accompanying figure. The exit diameter of the faucet is $\frac{1}{2}$ in. The resistance coefficient for all of the 1-in. pipe is 0.025 and for the $\frac{1}{2}$-in. pipe is 0.030. The main pressure is 40 psig. Determine the time required to fill a 20-gal bathtub.

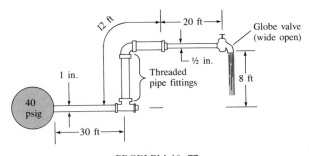

PROBLEM 10.77

10.78 What power must be supplied by the pump to the flow if water ($T = 20°C$) is pumped through the 200-mm steel pipe from the lower tank to the upper one at a rate of 0.314 m³/s?

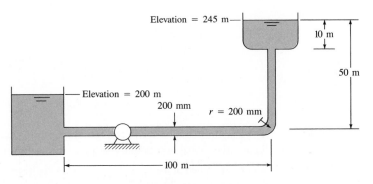

PROBLEM 10.78

10.79 If the pump for Fig. 10.17 is installed in the system of Prob. 10.78, what will be the rate of discharge of water from the lower tank to the upper one?

10.80 The 3-in.-diameter injector pipe is intended to make the system operate like a jet pump. If water velocity in the injector pipe is 60 ft/s, will the system operate as a pump? In other words, will water be drawn from the lower reservoir and discharged into the upper reservoir? Give computations and/or statements to support your conclusion.

10.81 A pump that has the characteristic curve shown in the accompanying graph is to be installed as shown. What will be the discharge of water in the system?

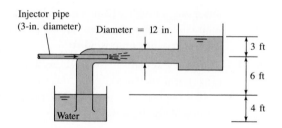

Injector pipe
(3-in. diameter)   Diameter = 12 in.

3 ft

6 ft

4 ft

Water

PROBLEM 10.80

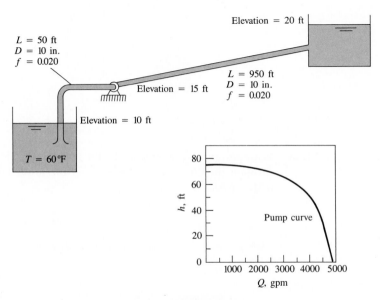

Elevation = 20 ft

L = 50 ft
D = 10 in.
f = 0.020

L = 950 ft
D = 10 in.
f = 0.020

Elevation = 15 ft

Elevation = 10 ft

T = 60°F

Pump curve

h, ft

80
60
40
20
0

1000 2000 3000 4000 5000

Q, gpm

PROBLEMS 10.81, 10.82

10.82 If the liquid of Prob. 10.81 is a superliquid (zero head loss occurs with the flow of this liquid), then what will be the pumping rate, assuming that the pump curve is the same?

10.83 Water is pumped at a rate of 15 m³/s from the reservoir and out through the pipe, which has a diameter of 1.50 m. What power must be supplied to the water to effect this discharge?

10.84 Both pipes shown have an equivalent sand roughness k_s of 0.10 mm and a discharge of 0.1 m³/s. Also, D_1 = 15 cm, L_1 = 50 m, D_2 = 30 cm, and L_2 = 150 m. Determine the difference in the water-surface elevation between the two reservoirs.

10.85 Estimate the discharge of oil through the pipe shown. Also, draw the HGL and the EGL for the system.

10.86 If the discharge through the system shown is 2.0 cfs, what horsepower is the pump supplying to the water? Draw the HGL and the EGL for the system, and determine the water pressure at the midpoint of the long pipe. The four bends have a radius of 12 in., and the 6-in. pipe is smooth.

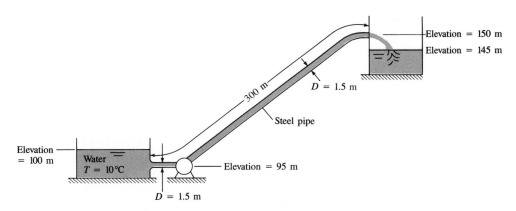

PROBLEM 10.83

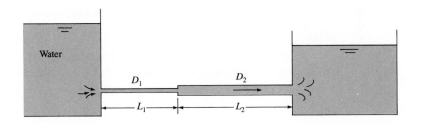

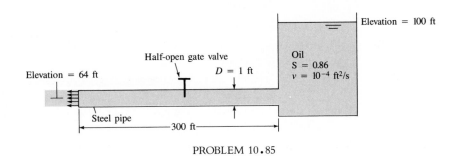

PROBLEM 10.85

PROBLEM 10.86

10.87   If the pump efficiency is 70%, what power must be supplied to the pump in order to pump fuel oil (S = 0.94) at a rate of 1 m³/s up to the high reservoir? Assume that the conduit is a steel pipe and $\nu = 5 \times 10^{-5}$ m²/s.

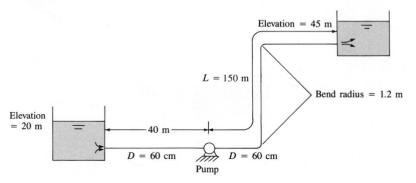

PROBLEM 10.87

10.88   Determine the elevation of the water surface in the upstream reservoir if the discharge in the system is 0.15 m³/s. Carefully sketch the HGL and the EGL, showing relative magnitudes and slopes. Label $V^2/2g$, p/γ, and z at section A–A.

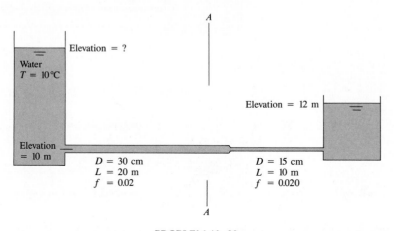

PROBLEM 10.88

10.89   Liquid discharges from a tank through the piping system shown. There is a venturi section at A and a sudden contraction at B. The liquid discharges to the atmosphere. Sketch the energy and hydraulic gradelines. Where might cavitation occur?

10.90   The steel pipe shown carries water from the main pipe A to the reservoir and is 2 in. in diameter and 240 ft long. What must be the pressure in A to provide a flow of 45 gpm?

10.91   If the elevation in reservoir B is 100 m, what must the elevation in reservoir A be if a flow of 0.03 m³/s is to occur in the cast-iron pipe? Draw the HGL and the EGL, including relative slopes and changes in slope.

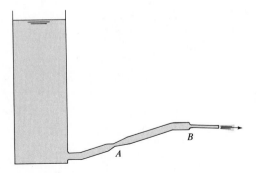

PROBLEM 10.89

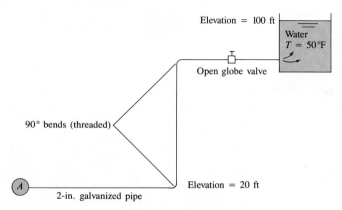

PROBLEM 10.90

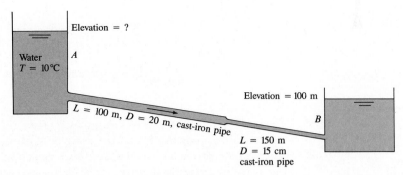

PROBLEM 10.91

10.92 Design a pipe system to supply water flow from the elevated tank to the reservoir at a discharge of 2.5 m³/s. Write a computer program to solve this type of problem, and then run it to get the desired solution.

10.93 Design a system to supply water flow from the reservoir to the elevated tank of Prob. 10.92 at a discharge of 2.0 m³/s.

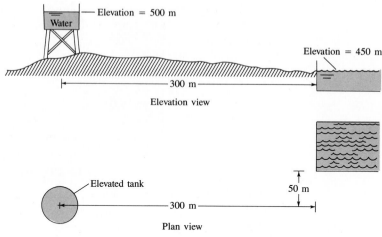

PROBLEM 10.92

10.94   An engineer is designing a water flow system consisting of two supply pipes that are connected by a tee section to a single pipe. All pipes are galvanized iron, 1/2 in. in diameter.
One line supplies water at 150°F and the other water at 50°F. The outlet pressure is atmospheric, and the valves are used for controlling the flow rate. The length of each line up
to the tee is 10 ft, and the length after the tee to the exit is 5 ft. There is an elbow in each
supply line as shown. Originally the globe valves are wide open. First calculate the temperature of the water coming out the exit. Then consider the case where the valve on the
hot water supply is partially closed so that the loss coefficient is increased from 10 to 20.
Find the temperature of the exiting water. Assume all pipes lie in a horizontal plane.

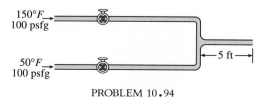

PROBLEM 10.94

10.95   The head loss for a discharge of 1 m³/s of water flow in a 50-cm cast-iron pipe was found
to be excessive, so another 50-cm cast-iron pipe of the same length was connected in parallel with the first. Then for the same discharge of 1 m³/s, the parallel system should
have what fraction of the head loss of the single-pipe system? Assume the same f for both
systems.

10.96   In the parallel system shown, pipe 1 has a length of 1000 m and is 50 cm in diameter.
Pipe 2 is 1500 m long and 40 cm in diameter. The pipe is commercial steel. What is the
division of the flow of water at 10°C if the total discharge is to be 1.0 m³/s?

10.97   A flow is divided into two branches as shown. A gate valve, half open, is installed in one
line, and a globe valve fully open is installed in the other branch. The head loss due to

friction in each branch is negligible compared to the head loss across the valves. Find the ratio of the velocity in the branch with the gate valve to that in the branch with the globe valve (include elbow losses for threaded pipe fittings).

PROBLEM 10.97

10.98 Pipes 1 and 2 (see Fig. for Probs. 10.96 and 10.98) are both the same kind (cast-iron pipe), but pipe 2 is four times as long as pipe 1. Both are the same diameter (1 ft). If the discharge of water in pipe 2 is 1 cfs, then what will be the discharge in pipe 1? Assume the same value of f in both pipes.

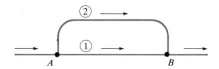

PROBLEMS 10.96, 10.98

10.99 Water flows from left to right in this parallel pipe system. The pipe having the greatest velocity is a) pipe A, b) pipe B, c) pipe C.

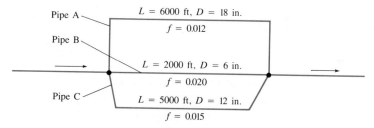

PROBLEM 10.99

10.100 Two pipes are connected in parallel. One pipe is twice the diameter of the other and three times as long. Assume that f in the larger pipe is 0.010 and f in the smaller one is 0.013. Determine the ratio of the discharges in the two pipes.

10.101 With a total flow of 14 cfs, determine the division of flow and the head loss from A to B.

10.102 The pipes shown in the system are all concrete. With a flow of 20 cfs of water, find the head loss and the division of flow in the pipes from A to B. Assume $f = 0.030$ for all pipes.

10.103 A parallel pipe system is set up as shown. Flow occurs from A to B. To augment the flow, a pump having the characteristics shown in Fig. 10.17 is installed at point C. For a total

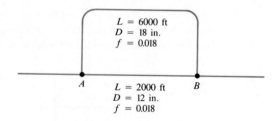

PROBLEM 10.101

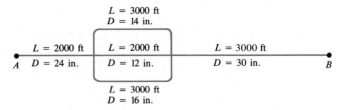

PROBLEM 10.102

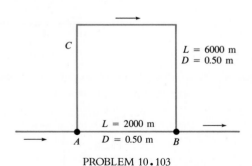

PROBLEM 10.103

discharge of 0.60 m³/s, what will be the division of flow between the pipes and what will be the head loss between A and B? Assume commercial steel pipe.

10.104 Frequently in the design of pump systems, a bypass line will be installed in parallel to the pump so that some of the fluid can recirculate as shown. The bypass valve then controls the flow rate in the system. Assume that the head-versus-discharge curve for the pump is given by

$$h_p = 100 - 100Q \qquad \text{(meters)}$$

where Q is in m³/s. The bypass line is 10 cm in diameter. Assume the only head loss is that due to the valve, which has a head-loss coefficient of 0.2. The discharge leaving the system is 0.2 m³/s. Find the discharge through the pump and bypass line.

10.105 Consider air flow in a square duct (4 ft × 4 ft in cross section) and water flow in an open channel (4 ft wide and with a depth of flow of 2 ft). Let the hydraulic radius for the air flow be R_A, and let the hydraulic radius for the water flow be R_W. For these conditions, indicate which one of the following statements relating to hydraulic radius is true: a) $R_A = R_W$, b) $R_A = 1.5R_W$, c) $R_A = 0.67R_W$, d) $R_A = 0.50R_W$.

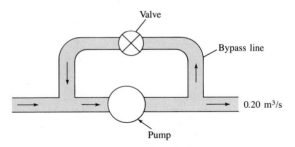

PROBLEM 10.104

10.106 Air at 60°F and atmospheric pressure flows in a horizontal duct with a cross section corresponding to an equilateral triangle (all sides equal). The duct is 10 ft long, and the dimension of a side is 6 in. The duct is constructed of galvanized iron ($k_s = 0.0005$ ft). The mean velocity in the duct is 10 ft/s. What is the pressure drop over the 10-ft length?

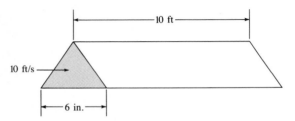

PROBLEM 10.106

10.107 Consider uniform flow of water in these two channels. They both have the same slope, the same wall roughness, and the same cross-sectional area. Then one can conclude that a) $Q_A = Q_B$, b) $Q_A < Q_B$, c) $Q_A > Q_B$.

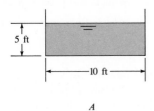
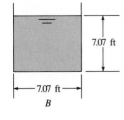

PROBLEM 10.107

10.108 A cold-air duct 100 cm by 15 cm in cross section is 200 m long and made of galvanized iron. This duct is to carry air at a rate of 5 m³/s at a temperature of 15°C and atmospheric pressure. What is the power loss in the duct?

10.109 An air conditioning system is designed to have a duct with a rectangular cross section 1 ft by 2 ft, as shown. During construction, a truck driver backed into the duct and made it a trapezoidal section, as shown. The contractor, behind schedule, installed it anyway. For the same pressure drop along the pipe, what will be the ratio of the velocity in the trapezoidal duct to that in the rectangular duct? Assume the Darcy–Weisbach resistance coefficient is the same for both ducts.

PROBLEM 10.109

10.110 Water ($\nu = 10^{-6}$ m²/s) flows full through a concrete ($k_s = 10^{-3}$ m) duct (no free surface) that has a rectangular cross section (70 cm by 100 cm). If the velocity of flow is 10 m/s, the resistance coefficient f will be about a) 0.01, b) 0.02, c) 0.03, d) 0.04.

10.111 This wood flume has a slope of 0.001. What will be the discharge of water in it for a depth of 1 m?

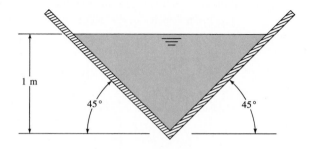

PROBLEM 10.111

10.112 Estimate the discharge in a rock-bedded stream ($d_{84} = 30$ cm) that has an average depth of 2.21 m, a slope of 0.0037, and a width of 46 m. Assume $k_s = d_{84}$.

10.113 Estimate the discharge of water ($T = 10°C$) that flows 1.5 m deep in a long rectangular concrete channel that is 3 m wide and is on a slope of 0.004.

10.114 A rectangular concrete channel is 12 ft wide and has uniform water flow. If the channel drops 10 ft in a length of 8000 ft, what is the discharge? Assume $T = 60°F$. The depth of flow is 4 ft.

10.115 Consider channels of rectangular cross section carrying 100 cfs of water flow. The channels have a slope of 0.001. Determine the cross-sectional areas required for widths of 2 ft, 4 ft, 6 ft, 8 ft, 10 ft, and 15 ft. Plot A versus y/b, and see how the results compare with the accepted result for the best hydraulic section.

10.116 A concrete sewer pipe 4 ft in diameter is laid so it has a drop in elevation of 0.90 ft per 1000 ft of length. If sewage (assume the properties are the same as those of water) flows at a depth of 2 ft in the pipe, what will be the discharge?

10.117 Determine the discharge in a 5-ft-diameter concrete sewer pipe on a slope of 0.01 that is carrying water at a depth of 4 ft.

10.118 Water flows at a depth of 6 ft in the trapezoidal, concrete-lined channel shown. If the channel slope is 1 ft in 2000 ft, what is the average velocity and what is the discharge?

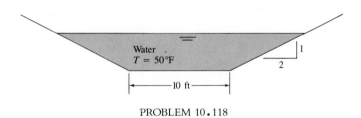

PROBLEM 10.118

10.119 What will be the depth of flow in a trapezoidal concrete-lined channel that has a water discharge of 1000 cfs? The channel has a slope of 1 ft in 500 ft. The bottom width of the channel is 10 ft, and the side slopes are 1 vertical to 1 horizontal.

10.120 What discharge of water will occur in a trapezoidal channel that has a bottom width of 10 ft and side slopes of 1 vertical to 1 horizontal if the slope of the channel is 5 ft/mi and the depth is to be 5 ft? The channel will be lined with concrete.

10.121 A rectangular concrete channel 4 m wide on a slope of 0.004 is designed to carry a water ($T = 10°C$) discharge of 25 m^3/s. Estimate the uniform flow depth for these conditions. The channel has a rectangular cross section.

10.122 A rectangular troweled concrete channel 12 ft wide with a slope of 10 ft in 8000 ft is designed for a discharge of 600 cfs. For a water temperature of 40°F, estimate the depth of flow.

10.123 A concrete-lined trapezoidal channel having a bottom width of 10 ft and side slopes of 1 vertical to 2 horizontal is designed to carry a flow of 3000 cfs. If the slope of the channel is 0.001, what will be the depth of flow in the channel?

10.124 This channel has a slope of 0.001. Assume the overbank areas are covered with long grass and the main channel is a straight gravel-bed river. If the depth of flow in the main channel is 26.0 ft, estimate the flood discharge.

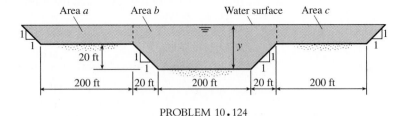

PROBLEM 10.124

10.125 Design a canal having a trapezoidal cross section to carry a design discharge of irrigation water of 900 cfs. The slope of the canal is to be 0.002. The canal is to be lined with concrete, and it is to have the best hydraulic section for the design flow.

10.126 A river with the cross section shown experienced a 50,750-cfs flood and produced the depth shown. Determine Manning's n for the main channel and overbank areas, assuming they are equal. The channel has a uniform slope of 0.45%.

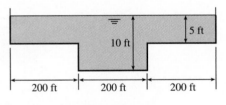

PROBLEM 10.126

10.127   If, in Prob. 10.126, Manning's *n* is known to be 0.020 in the main channel, determine the value of *n* for the overbank areas. What is the flow rate in the main channel and in the overbank areas? What is the flow velocity in the main channel and in the overbank areas?

10.128   The compound cross section shown has Manning's *n* values of 0.025 for the main channel and 0.070 for the overbank areas. The uniform channel slope is 0.004. Determine the depth and flow rate in the main channel, the left overbank area, and the right overbank area. The flood flow is 45,400 cfs.

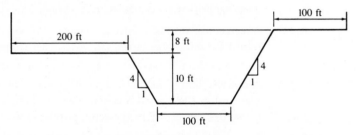

PROBLEM 10.128

10.129   For the cross section shown, determine the depth of flow for a summer flood of 100,000 cfs. Assume a uniform channel slope of 0.09%. Further assume that the overbank area is brush-covered with some trees and the main channel is straight with a gravel bed.

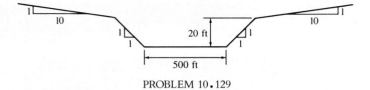

PROBLEM 10.129

10.130   For the given source and loads shown below, how will the flow be distributed in the simple network, and what will be the pressures at the load points if the pressure at the source is 60 psi? Assume horizontal pipes and *f* = 0.012 for all pipes.

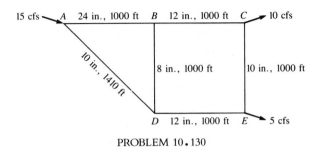

PROBLEM 10.130

References

1. American Concrete Pipe Assoc. *Concrete Pipe Design Manual.* American Concrete Pipe Assoc., Vienna, Va., 1980.

2. ASHRAE. "ASHRAE Handbook, 1977 Fundamentals." Am. Soc. of Heating, Refrigerating and Air Conditioning Engineers, Inc., New York, 1977.

3. Barbin, A. R., and J. B. Jones. "Turbulent Flow in the Inlet Region of a Smooth Pipe." *Trans. ASME, Ser. D: J. Basic Eng.,* 85, no. 1 (March 1963).

4. Beaty, W. R., et al. "New High-Performance Flow Improver Offers Alternatives to Pipelines." *Oil and Gas Journal,* Aug. 9, 1982.

5. Beij, K. H. "Pressure Losses for Fluid Flow in 90° Pipe Bends." *J. Res. Nat. Bur. Std.,* 21 (1938). Information cited in Streeter (36).

6. Berman, N. S. "Drag Reduction by Polymers." *Ann. Rev. of Fluid Mechanics,* vol. 10, pp. 47–64, 1978.

7. Calhoun, Roger G. "A Statistical Roughness Model for Computation of Large Bed-Element Stream Resistance." M.S. thesis, Washington State University, Pullman, Wash., 1975.

8. Chow, Ven Te. *Open Channel Hydraulics.* McGraw-Hill, New York, 1959.

9. Committee on Hydromechanics of the Hydraulics Division of American Society of Civil Engineers. "Friction Factors in Open Channels." *J. Hydraulics Div., Am. Soc. Civil Eng.* (March 1963).

10. Crane Co. "Flow of Fluids Through Valves, Fittings and Pipe," Technical Paper No. 410, Crane Co. (1978). (Available only through distributor.)

11. Cross, Hardy. "Analysis of Flow in Networks of Conduits or Conductors." *Univ. Illinois Bull.,* 286 (November 1936).

12. Daily, James W., and Donald R. F. Harleman. *Fluid Dynamics.* Addison-Wesley, Reading, Mass., 1966.

13. Eldridge, J. R. "An Analytical Method for Predicting Resistance to Flow in Rough Pipes and Open Channels." M.S. thesis, Washington State University, Pullman, Wash., 1983.

14. Hamilton, J. B. "The Suppression of Intake Losses by Various Degrees of Rounding." *Univ. Wash. Expt. Sta. Bull.,* 51 (1929). Information cited in Streeter (36).

15. Henderson, F. M. *Open Channel Flow.* Macmillan, New York, 1966.

16. Hoag, Lyle N., and Gerald Weinberg. "Pipeline Network Analysis by Digital Computers." *J. Am. Water Works Assoc.,* 49 (1957).

17. Hoyt, J. W. "The Effect of Additives on Fluid Friction." *Trans. of ASME, Journal of Basic Engr.,* vol. 94D, p. 258 (1972).

18. Hydraulic Institute. "Pipe Friction Manual." Hydraulic Institute, 122 E. 42nd St., New York, N.Y. 10017.

19. Idel'chik, I. E. *Handbook of Hydraulic Resistance-Coefficients of Local Resistance and of Friction.* Trans. A. Barouch. Israel Program for Scientific Translations, 1966.

20. Jeppson, Roland W. *Analysis of Flow in Pipe Networks.* Ann Arbor Science Publishers, Ann Arbor, Mich., 1976.

21. Kumar, S. "An Analytical Model for Computation of Rough Pipe Resistance." Ph.D. dissertation, Washington State University, Pullman, Wash., 1979.

22. Laufer, John. "The Structure of Turbulence in Fully Developed Pipe Flow." *NACA Rept.,* 1174 (1954).

23. Lester, C. D. Four-part series on drag-reducing agents. *Oil and Gas Journal,* Feb. 4, 18; Mar. 4, 11 (1985).

24. Limerinos, J. T. "Determination of the Manning Coefficient from Measured Bed Roughness in Natural Channels." Water Supply Paper 1898-B, U.S. Geological Survey, Washington, D.C., 1970.

25. Lumley, J. L. *Annual Review of Fluid Mechanics,* vol. 1, p. 367 (W. R. Sears and M. VanDyke, eds., Palo Alto, Annual Review Inc., 1969).

26. Marlow, Thomas A., et al. "Improved Design of Fluid Networks with Computers." *J. Hydraulics Div, Am. Soc. Civil Eng.* (July 1966).

27. McComb, W. D., and K. T. Chan. "Laser-Doppler Anemometer Measurements of Turbulent Structure in Drag-Reducing Fibre Suspensions." *J. Fluid Mech.,* 152 (1985).

28. Moody, Lewis F. "Friction Factors for Pipe Flow." *Trans. ASME,* 671 (November 1944).

29. Nikuradse, J. "Strömungsgesetze in rauhen Rohren." *VDI-Forschungsh.,* no. 361 (1933). Also translated in *NACA Tech. Memo,* 1292.

30. Prandtl, L. "Über die ausgebildete Turbulenz." *Zamm.,* 5 (1925), pp. 136–139. Also summarized in Schlichting (35).

31. Reynolds, O. "An Experimental Investigation of the Circumstances Which Determine Whether the Motion of Water Shall Be Direct or Sinuous and of the Law of Resistance in Parallel Channels." *Phil. Trans. Roy. Soc. London,* 174, part III (1883).

32. Roberson, John A., and C. K. Chen. "Flow in Conduits with Low Roughness Concentration." *J. Hydraulics Div, Am. Soc. Civil Eng.,* 96, no. HY4 (April 1970).

33. Rouse, Hunter, and Simon Ince. *History of Hydraulics.* Iowa Institute of Hydraulic Research, University of Iowa, 1957.

34. Schlichting, Hermann. "Experimentelle Untersuchugen zum Rauhigkietsproblem." *Ingr.-Arch.,* 7 (1936), pp. 1–34.

35. Schlichting, Hermann. *Boundary Layer Theory,* 7th ed. McGraw-Hill, New York, 1979.

36. Streeter, V. L. (ed.) *Handbook of Fluid Dynamics.* McGraw-Hill, New York, 1961.

37. Streeter, V. L., and E. B. Wylie. *Fluid Mechanics,* 7th ed. McGraw-Hill, New York, 1979.

38. Swamee, P. K., and A. K. Jain. "Explicit Equations for Pipe-Flow Problems." *Journal of the Hydraulic Division of the ASCE,* 102, no. HY5 (May 1976).

39. Toms, B. A. *Proc. of 1st Internat. Congress of Rheology* (North Holland, Amsterdam), 2, p. 135 (1948).

40. Tulin, M. P., and J. Wu. "Additive Effects on Free Turbulent Flow." *Physics of Fluids,* 20, no. 10, pt. 11 (October 1977).

41. U.S. Bureau of Reclamation. "Friction Factors for Large Conduits Flowing Full." Engineering Monograph No. 7, U.S. Govt. Printing Office, 1965.

42. Wolman, M. G. "The Natural Channel of Brandywine Creek, Pennsylvania." Prof. Paper 271, U.S. Geological Survey, Wahington D.C., 1954.

43. Wright, S. J. "A Theory for Predicting the Resistance to Flow in Conduits with Nonuniform Roughness." M. S. thesis, Washington State University, Pullman, Wash., 1973.

Drag and Lift

The black-browed albatross has a wingspan of nearly eight feet. The extremely large aspect ratio of its wings yields a very high lift-to-drag ratio (about 40) that allows it to soar effortlessly over the oceans.

A body immersed in a flowing fluid is acted on by both pressure and viscous forces from the flow. The sum of the forces (pressure, viscous, or both) that acts normal to the free-stream direction is the *lift*, and the sum that acts parallel to the free-stream direction is the *drag*. Buoyant or weight forces may also act on the body; however, lift and drag forces are limited by definition to those forces produced by the dynamic action of the flowing fluid. We will first consider the mathematical formulation for lift and drag in terms of the viscous stresses and pressure; then we will consider each more intensively in subsequent sections.

11.1 Basic Considerations

Consider the forces acting on the *airfoil* in Fig. 11.1. The vectors normal to the surface of the airfoil are normal forces per unit area, referred to simply as pressure. As shown here, the pressure is referenced to the free-stream pressure.

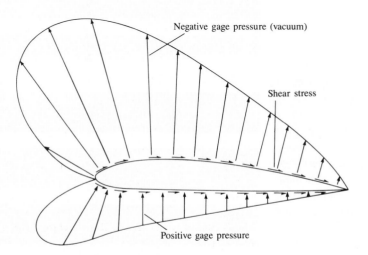

FIGURE 11.1

Pressure and shear stress acting on an airfoil.

Because the velocity of the flow over the top of the airfoil is greater than the free-stream velocity, the pressure over the top is negative, or less than the free-stream pressure. This follows directly from application of Bernoulli's equation. Because the velocity along the underside of the wing is less than the free-stream velocity, the pressure there is positive, or greater than the free-stream pressure. Hence both the negative pressure over the top and the positive pressure along the bottom contribute to the lift.

The vectors in Fig. 11.1 that are parallel to the surface of the airfoil represent the shear forces per unit area, referred to simply as shear stress. Except on the front of the airfoil, shear stress acts essentially parallel to the free-stream direction. Hence the shear stress contributes largely to the drag of the airfoil.

The mathematical formulation for lift and drag in terms of the pressure and shear stress can be derived with the aid of Fig. 11.2. Here the pressure and viscous forces acting on a differential area of the surface of the airfoil are shown. The magnitude of the pressure force is $dF_p = p\,dA$, and the magnitude of the viscous force is $dF_V = \tau\,dA$. However, we are interested in separating the forces into components that are normal and parallel to the free-stream direction to determine lift and drag, respectively. Hence the differential lift force is*

$$dF_L = -p\,dA\,\sin\theta - \tau\,dA\,\cos\theta$$

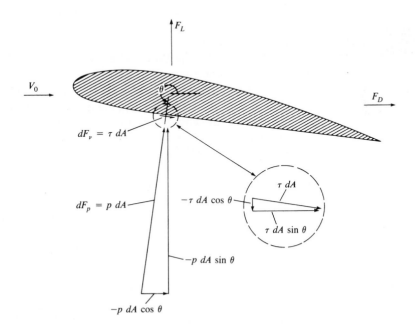

FIGURE 11.2

Pressure and viscous forces acting on a differential element of area.

* *The sign convention on τ is such that a clockwise sense of $\tau\,dA$ on the surface of the foil signifies a positive sign for τ.*

and the differential drag is

$$dF_D = -p\,dA\,\cos\theta + \tau\,dA\,\sin\theta$$

Then the total lift and drag on the airfoil are obtained by integration of the respective differential forces over the entire surface of the airfoil:

$$F_L = \int (-p\,\sin\theta - \tau\,\cos\theta)\,dA \qquad (11.1)$$

$$F_D = \int (-p\,\cos\theta + \tau\,\sin\theta)\,dA \qquad (11.2)$$

Equations (11.1) and (11.2) are for a two-dimensional flow; that is, there is no velocity component in the direction normal to the page, so the shear-stress and pressure-force vectors lie in the plane of the page. The same basic principle (separation of forces into directions parallel and normal to the free-stream direction) can easily be extended to three-dimensional flows.

In this chapter we shall refer to two-dimensional and three-dimensional bodies. By a two-dimensional body, we mean a body over which the flow is two-dimensional. For example, a very long cylinder that has flow approaching it from a normal direction is classified as a two-dimensional body because the flow around the ends does not affect the flow pattern and pressure distribution over the central part of the body. However, a short cylinder is classified as a three-dimensional body because the end effects are significant. For a two-dimensional body, the aerodynamic forces and representative areas are sometimes based on a unit length of the body. The two-dimensional body is identified by cross-hatching on the figures.

Still another classification is the axisymmetric body. Here, if the approach flow is uniform and parallel to the axis of symmetry, then in effect the resulting flow is two-dimensional. That is, for an x-r coordinate system, where x is measured along the axis of symmetry and r is the normal radial distance from the axis, the velocity components exist only in the x and r directions.

Equations (11.1) and (11.2) can be used to evaluate F_L and F_D when the pressure and shear stresses are obtained either analytically or experimentally. However, it is also common to obtain the overall drag and lift by force-dynamometer measurements in a wind tunnel. The next section considers the direct application of the pressure and shear-stress variation over the surface of a plate to determine the drag on the plate.

11.2 Drag of Two-Dimensional Bodies

Drag of a Thin Plate

To illustrate the relative effect of pressure and viscous forces on drag, we shall consider the drag of a plate first oriented parallel to the flow and then oriented normal to the flow. In the parallel position, the only force acting is viscous shear in the direction of flow. Hence, from our considerations of surface resistance in Chapter 9, the drag for both sides of the plate is given as

$$F_D = 2C_f b\ell\rho\frac{V_0^2}{2}$$

When the plate is turned normal to the flow, as in Fig. 11.3, both pressure and viscous forces act on the plate. However, the viscous forces act only in the transverse direction and, in addition, are symmetrical about the midpoint of the plate. Consequently, the viscous forces do not directly contribute to the lift or drag of the plate. Because the pressure on the plate acts to produce a force only in a direction parallel to the flow, the pressure force contributes totally to the drag of the body. Hence Eq. (11.2) applied to the plate reduces to

$$F_D = \int (-p \cos \theta)\, dA$$

The pressures on the front and rear sides of the plate can be obtained experimentally and are usually given in terms of C_p, as shown in Fig. 11.4 for flow with a relatively high value of the Reynolds number ($V_0 b/\nu > 10^4$).

Since the pressure on the rear side is essentially constant,

$$p = p_0 - 1.2\rho\frac{V_0^2}{2}$$

and since $\theta = 0$, the contribution to drag for the rear side is

$$F_{D,\text{rear}} = -\left(p_0 - 1.2\rho\frac{V_0^2}{2}\right)b\ell = -p_0 b\ell + 1.2\rho\frac{V_0^2}{2}b\ell$$

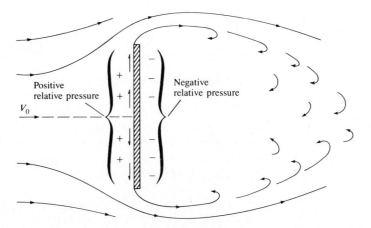

FIGURE 11.3

Flow past a flat plate.

FIGURE 11.4

Pressure distribution on a plate normal to the approach flow for Re > 10⁴.

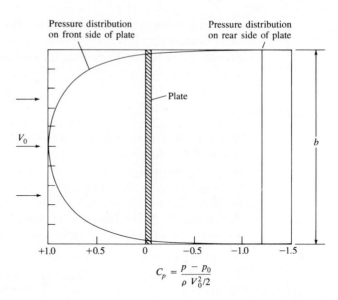

$$C_p = \frac{p - p_0}{\rho V_0^2/2}$$

where ℓ is the length of the plate normal to the plane of the paper, and by definition of a two-dimensional body, $\ell \gg b$. For the front side, $\theta = \pi$. Hence $\cos \theta = -1$, and the contribution to drag due to pressure on the front side is

$$F_{D,\text{front}} = \int_{-b/2}^{b/2} \left(p_0 + C_p \rho \frac{V_0^2}{2} \right) \ell \, dy = p_0 b \ell + \rho \frac{V_0^2}{2} \ell \int_{-b/2}^{b/2} C_p \, dy$$

Then the total drag of the plate is given by

$$F_D = F_{D,\text{front}} + F_{D,\text{rear}}$$

$$= \rho \frac{V_0^2}{2} \ell \left(\int_{-b/2}^{b/2} C_p \, dy + 1.2b \right) \tag{11.3}$$

Evaluation of the first term inside the parentheses on the right-hand side of Eq. (11.3) yields a magnitude of approximately $0.80b$. Thus the drag of this plate is given as

$$F_D = \rho \frac{V_0^2}{2} b \ell (0.80 + 1.2) \tag{11.4}$$

We must take note of the pure numbers inside the parentheses of Eq. (11.4). The number 0.8 in Eq. (11.4) represents the average pressure coefficient C_p over the front side of the plate. In fact, the sum inside the parentheses (0.8 + 1.2) of this equation reflects the manner in which the pressure is distributed over the front and rear of the body. Because the drag varies directly with the magnitude of this quantity, it has been appropriately defined as the *coefficient of drag* C_D. Thus Eq. (11.4) can be written as

$$F_D = C_D A_p \rho \frac{V_0^2}{2} \tag{11.5}$$

where C_D is the coefficient of drag, A_p is the projected area of the body, ρ is the fluid density, and V_0 is the free-stream velocity. The projected area A_p is the silhouetted area that would be seen by a person looking at the body from the direction of flow. For example, the projected area of the above plate normal to the flow is $b\ell$, and the projected area of a cylinder with its axis normal to the flow is $d\ell$. In Chapter 8 we saw that C_p is a function of the Reynolds number Re; and because $C_D = f(C_p)$, C_D is also a function of Re. When the drag of bodies is due solely to the shear stress on the body, C_D is still a function of Re because τ is also a function of Re.

Coefficients of Drag for Various Two-Dimensional Bodies

We have already seen that C_D can be determined if the pressure and shear-stress distribution around a body are known. The coefficient of drag can also be calculated if the total drag is measured, for example, by means of a force dynamometer in a wind tunnel. Then C_D is calculated using Eq. (11.5) written as follows:

$$C_D = \frac{F_D}{A_p \rho V_0^2/2} \tag{11.6}$$

Much of the data (C_D versus Re) found in the literature is obtained in this manner.

The coefficient of drag for the flat plate normal to the free stream and for other two-dimensional bodies, for a wide range of Reynolds numbers, is given in Fig. 11.5. In general, the total drag of a blunt body is partly due to viscous resistance and partly due to pressure variation. The pressure drag is largely a function of the form or shape of the body; hence it is called *form drag*. The viscous drag is often called *skin-friction drag*.

EXAMPLE 11.1 A television-transmitting antenna is on top of a pipe 30 m high (98.4 ft) and 30 cm (11.8 in.) in diameter, which is on top of a tall building. What will be the total drag of the pipe and the bending moment at the base of the pipe in a 35-m/s (115-ft/s) wind at normal atmospheric pressure and a temperature of 20°C (68°F)?

Solution For the conditions given, the viscosity and density of the air are obtained from the Appendix:

$$\mu = 1.81 \times 10^{-5} \text{ N} \cdot \text{s/m}^2 \ (3.78 \times 10^{-7} \text{ lbf-s/ft}^2)$$

$$\rho = 1.20 \text{ kg/m}^3 \ (0.00234 \text{ slugs/ft}^3)$$

Next the Reynolds number is calculated:

$$\text{Re} = \frac{V_0 d\rho}{\mu} = \frac{35 \text{ m/s} \times 0.30 \text{ m} \times 1.20 \text{ kg/m}^3}{(1.81 \times 10^{-5} \text{ N} \cdot \text{s/m}^2)} = 7.0 \times 10^5$$

FIGURE 11.5

Coefficient of drag versus Reynolds number for two-dimensional bodies. [Data sources: Bullivant (5), Defoe (7), Goett and Bullivant (10), Jacobs (13), Jones (15), and Lindsey (19)]

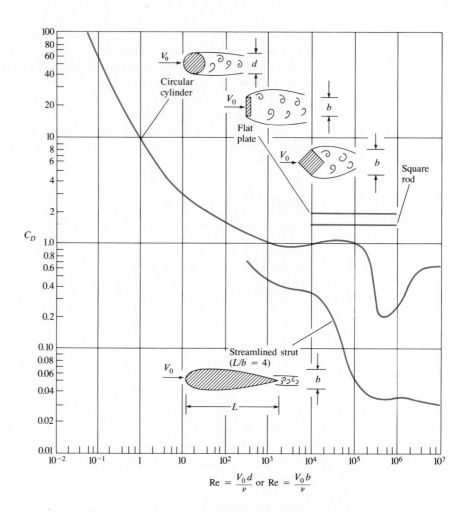

Then, from Fig. 11.5, $C_D = 0.20$. Now we compute the total drag:

$$F_D = \frac{C_D A_p \rho V_0^2}{2}$$

$$= \frac{(0.2)\,(30 \text{ m})\,(0.3 \text{ m})\,(1.20 \text{ kg/m}^3)\,(35^2 \text{ m}^2/\text{s}^2)}{2} = 1323 \text{ N} \qquad \blacktriangleleft$$

Assuming that the resultant drag force acts midway up the pole, the moment is

$$M = F_D\left(\frac{L}{2}\right) = (1323 \text{ N})\left(\frac{30}{2} \text{ m}\right) = 19{,}845 \text{ N} \cdot \text{m} \qquad \blacktriangleleft$$

Traditional units:

$$F_D = 0.2(98.4 \text{ ft})\left(\frac{11.8}{12} \text{ ft}\right)(0.00234 \text{ slugs/ft}^3)\left(\frac{115^2 \text{ ft}^2/\text{s}^2}{2}\right) = 299 \text{ lbf} \blacktriangleleft$$

$$M = (299 \text{ lbf})\left(\frac{98.4}{2} \text{ ft}\right) = 14{,}700 \text{ ft-lbf} \blacktriangleleft$$

Discussion of C_D for Two-Dimensional Bodies

At low Reynolds numbers, C_D changes with the Reynolds number. The change is due to the relative change in viscous resistance, which has already been mentioned in Chapter 8. Above Re $= 10^4$, the flow pattern remains virtually unchanged, thereby producing constant values of C_p over the body. Constancy of C_p at high Reynolds numbers is reflected in the constancy of C_D. This characteristic, the constancy of C_D at high values of Re, is representative of most bodies that have angular form. However, certain bodies with rounded form, such as circular cylinders, show a remarkable decrease in C_D with an increase in Re from about 10^5 to 5×10^5.

This reduction in C_D at a Reynolds number of approximately 10^5 is due to a change in the flow pattern triggered by a change in the character of the boundary layer. For Reynolds numbers less than 10^5, the boundary layer is laminar, and separation occurs about midway between the front and rear of the cylinder (Fig. 11.6). Hence the entire rear half of the cylinder is exposed to a relatively low pressure, which in turn produces a relatively high value for C_D. When the Reynolds number is increased to about 10^5, the boundary layer on the surface of the cylinder becomes turbulent, which causes higher-velocity fluid to be mixed into the region close to the wall of the cylinder. As a consequence of the presence of this high-velocity, high-momentum fluid in the boundary layer, the flow proceeds farther downstream along the surface of the cylinder against the adverse pressure before separation occurs (Fig. 11.7). Hence the flow pattern

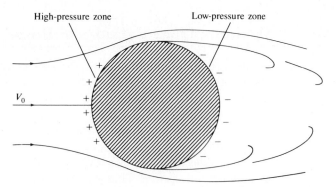

FIGURE 11.6

Flow pattern around a cylinder for
$10^3 < Re < 10^5$.

FIGURE 11.7

Flow pattern around a cylinder for Re > 5 × 10⁵.

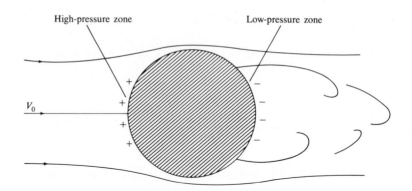

causes C_D to be reduced for the following reason: with the turbulent boundary layer, the streamlines downstream of the cylinder midsection diverge somewhat before separation, and hence a decrease in velocity occurs before separation. According to Bernoulli's equation, the decrease in velocity produces a pressure at the point of separation that is greater than the pressure at the midsection. Thus the pressure at the point of separation, and also in the zone of separation, is significantly greater under these conditions than when separation occurs farther upstream. Therefore, the pressure difference between the front and rear surfaces of the cylinder is less at high values of Re, yielding a lower drag and a lower C_D.

Because the boundary layer is so thin, it is also very sensitive to other conditions. For example, if the surface of the cylinder is slightly roughened upstream of the midsection, the boundary layer will be forced to become turbulent at lower Reynolds numbers than those for a smooth cylinder surface. The same trend can also be produced by creating abnormal turbulence in the approach flow. The effects of roughness are shown in Fig. 11.8 for cylinders that were roughened with sand grains of size k. A small to medium size of roughness ($10^{-3} < k/d < 10^{-2}$) on a cylinder triggers an early onset of reduction of C_D. However, when the relative roughness is quite large ($10^{-2} < k/d$), the characteristic dip in C_D is absent.

11.3 Vortex Shedding from Cylindrical Bodies

Figures 11.6 and 11.7 show the average (temporal mean) flow pattern around a cylinder. The phenomenon becomes more complex, however, when we observe the detailed flow pattern as time passes. Observations show that above Re ≈ 50, vortices are formed and shed periodically downstream of the cylinder. Hence, at a given time, the detailed flow pattern might appear as in Fig. 11.9. In this figure a vortex is in the process of formation near the top of the cylinder. Below and to the right of the first vortex is another vortex, which was formed and shed a short time before. Thus the flow process in the wake of a cylinder involves the formation and shedding of vortices alternately from one side and then the other.

FIGURE 11.8

Effects of roughness on C_D for a cylinder. [After Miller et al. (20)]

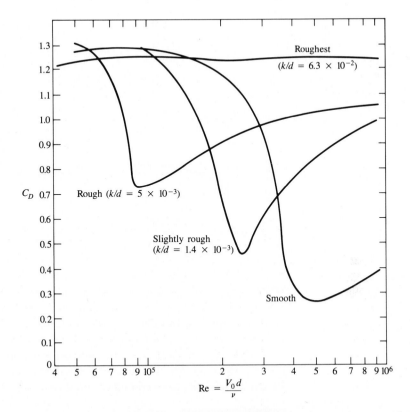

This phenomenon is of major importance in engineering design, because the alternate formation and shedding of vortices also creates a regular change in pressure with consequent periodicity in side thrust on the cylinder. Vortex shedding was the primary cause of failure of the Tacoma Narrows suspension bridge in the state of Washington in 1940. Another, more commonplace, effect of vortex shedding is the "singing" of wires in the wind.

If the frequency of the vortex shedding is in resonance with the natural frequency of the member that produces it, large amplitudes of vibration with resulting large stresses can develop. Experiments show that the frequency of shedding is given in terms of the Strouhal number S, and this in turn is a function of the Reynolds number. Here the Strouhal number is defined as

$$S = \frac{nd}{V_0} \tag{11.7}$$

FIGURE 11.9

Formation of a vortex behind a cylinder.

FIGURE 11.10

Strouhal number versus Reynolds number for flow past a circular cylinder. [After Jones (15) and Roshko (25)]

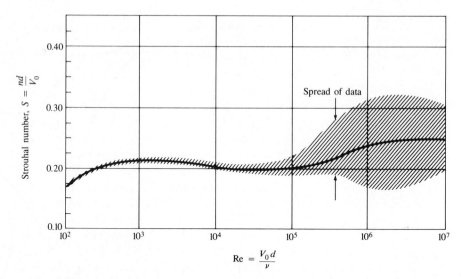

where n is the frequency of shedding of vortices from one side of cylinder, in Hz, d is the diameter of cylinder, and V_0 is the free-stream velocity.

The relationship between the Strouhal number and the Reynolds number for vortex shedding from a circular cylinder is given in Fig. 11.10.

Other cylindrical and two-dimensional bodies also shed vortices. Consequently, the engineer should always be alert to vibration problems when designing structures that are exposed to wind or water flow.

EXAMPLE 11.2 For the cylinder and conditions of Example 11.1, at what frequency will the vortices be shed?

Solution We compute the frequency n from the Strouhal number, which is given in Fig. 11.10 as a function of the Reynolds number. Thus with a Reynolds number of 7.0×10^5 from Example 11.1, we read a Strouhal number of 0.23. But

$$S = \frac{nd}{V_0}$$

so $\qquad n = \frac{SV_0}{d} = \frac{0.23 \times 35 \text{ m/s}}{0.30 \text{ m}} = 27 \text{ Hz} \qquad \blacktriangleleft$

11.4 Effect of Streamlining

For Reynolds numbers greater than 10^3, the drag of a cylinder is predominantly due to the pressure variation around the cylinder. The pressure difference between the front and rear sides of the cylinder is the primary cause of drag, and this pressure difference is due largely to separation. Hence, if the separation can be eliminated, the drag will be reduced. That is exactly what streamlining does. Streamlining reduces the extreme curvature on the downstream side of the body, and this process reduces or eliminates separation. Therefore, the coefficient of drag is greatly reduced, as seen in Fig. 11.5 (C_D for the streamlined shape is only about 10% of C_D for the circular cylinder when Re $\approx 5 \times 10^5$).

When a body is streamlined by elongating it and reducing its curvature, the pressure drag is reduced. However, the viscous drag is increased because there is a greater amount of surface on the streamlined body than on the nonstreamlined body. Consequently, when a body is streamlined to produce minimum drag, there is an optimum condition to be sought. The optimum condition results when the sum of surface drag and pressure drag is minimum.

In this discussion of streamlining, it is interesting to note that streamlining to produce minimum drag at high Reynolds numbers will probably not produce minimum drag at very low Reynolds numbers. For Reynolds numbers less than unity, the majority of the drag of a cylinder is due to the viscous shear stress on the wall of the cylinder. Hence, if the cylinder is streamlined, the viscous shear stress is simply magnified and C_D may actually increase for this range of Re where the viscous resistance is predominant.

At high values of the Reynolds number, another advantage of streamlining is that the periodic formation of vortices is eliminated.

EXAMPLE 11.3   Compare the drag of the cylinder of Example 11.1 with the drag of the streamlined shape shown in Fig. 11.5. Assume that they both have the same projected area and that the streamlined shape is oriented for minimum drag.

Solution   Because $F_D = C_D A_p \rho V_0^2 / 2$, the drag of the streamlined shape will be

$$F_{DS} = F_{DC}\left(\frac{C_{DS}}{C_{DC}}\right)$$

where F_{DS} is the drag of the streamlined shape, F_{DC} is the drag of the cylinder, C_{DS} is the coefficient of drag of the streamlined shape, and C_{DC} is the coefficient of drag of the cylinder. But $C_{DS} = 0.034$, from Fig. 11.5 with Re $= 7.0 \times 10^5$. Then

$$F_{DS} = F_{DC}\left(\frac{0.034}{0.20}\right) = (1323 \text{ N})\left(\frac{0.034}{0.20}\right)$$

$$F_{DS} = 225 \text{ N} \qquad \blacktriangleleft$$

$$\frac{F_{DS}}{F_{DC}} = 0.17 \qquad \blacktriangleleft$$

11.5 Drag of Axisymmetric and Three-Dimensional Bodies

The same principles that apply to the drag of two-dimensional bodies also apply to that of axisymmetric and three-dimensional bodies. That is, at very low values of the Reynolds number, the coefficient of drag is given by exact equations relating C_D and Re. At high values of Re, the coefficient of drag becomes constant for angular bodies, whereas rather abrupt changes in C_D occur for rounded bodies. All of these characteristics can be seen in Fig. 11.11, where C_D is plotted against Re for several axisymmetric bodies.

For Reynolds numbers less than 0.5, the flow around the sphere is laminar and amenable to analytical solutions. An exact solution by Stokes yielded the following equation, which is called *Stokes' law,* for the drag of a sphere:

$$F_D = 3\pi\mu V_0 d \tag{11.8}$$

Note that the drag for this laminar-flow condition varies directly with the first power of V_0. This is characteristic of all laminar-flow processes. For com-

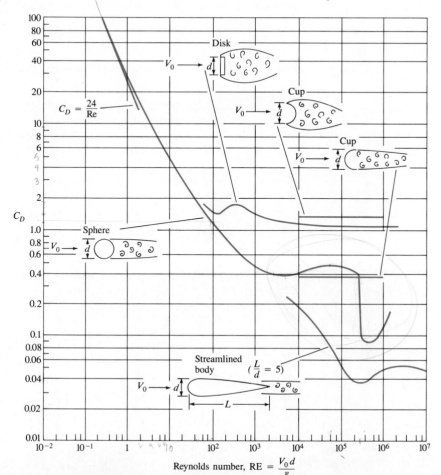

FIGURE 11.11

Coefficient of drag versus Reynolds number for axisymmetric bodies. [Data sources: Abbott (1), Brevoort and Joyner (4), Freeman (9), and Rouse (26)]

pletely turbulent flow, the drag is a function of the velocity to the second power. When Eqs. (11.8) and (11.6) are solved simultaneously, we get the coefficient of drag corresponding to Stokes' law:

$$C_D = \frac{24}{\text{Re}} \tag{11.9}$$

Thus for flow past a sphere, when Re $\leq$ 0.5, one may use the direct relation for C_D given above. Values for C_D for other axisymmetric and three-dimensional bodies at high Reynolds numbers (Re $>$ 10^4) are given in Table 11.1.

EXAMPLE 11.4   What is the drag of a 12-mm sphere that drops at a rate of 8 cm/s in oil ($\mu = 10^{-1}$ N $\cdot$ s/m^2, S = 0.85)?

Solution

$$\rho = 0.85 \times 1000 \text{ kg/m}^3$$

Then $$\text{Re} = \frac{Vd\rho}{\mu} = \frac{(0.08 \text{ m/s}) (0.012 \text{ m}) (850 \text{ kg/m}^3)}{10^{-1} \text{ N} \cdot \text{s/m}^2} = 8.16$$

Next, from Fig. 11.11,

$$C_D = 5.3$$

$$F_D = \frac{C_D A_p \rho V_0^2}{2}$$

Hence $$F_D = \frac{(5.3) (\pi/4) (0.012^2 \text{ m}^2) (850 \text{ kg/m}^3) (0.08^2 \text{ m}^2/\text{s}^2)}{2}$$

$$= 1.63 \times 10^{-3} \text{ N} \qquad \blacktriangleleft$$

EXAMPLE 11.5   A bicyclist travels on a level path. The frontal area of the cyclist and bicycle is 2 ft^2 and the drag coefficient is 0.8. The density of the air is 0.075 lbm/ft^3. The physiological limit of power for this well-trained cyclist is 0.5 hp. Neglecting rolling friction, find the maximum speed of the cyclist.

Solution   The power is the product of the drag force and the velocity.

$$P = F_D V = (1/2) C_D A_p \rho V^3$$

Solving for the velocity, we get

$$V = (2P/\rho C_D A_p)^{1/3}$$

$$= \left[\frac{2 \times 0.5 \text{ hp} \times 550 \text{ ft-lbf/s hp}}{0.075 \text{ lbm/ft}^3 \times (1/32.2) \text{ slugs/lbm} \times 0.8 \times 2 \text{ ft}^2} \right]^{1/3}$$

$$= 52.8 \text{ ft/s} = 36.0 \text{ mph} \qquad \blacktriangleleft$$

TABLE 11.1 APPROXIMATE C_D VALUES FOR VARIOUS BODIES				
	Type of Body	**Length Ratio**	**Re**	C_D
	Rectangular plate	$l/b = 1$	$>10^4$	1.18
		$l/b = 5$	$>10^4$	1.20
		$l/b = 10$	$>10^4$	1.30
		$l/b = 20$	$>10^4$	1.50
		$l/b = \infty$	$>10^4$	1.98
	Circular cylinder— axis // to flow	$l/d = 0$ (disk)	$>10^4$	1.17
		$l/d = 0.5$	$>10^4$	1.15
		$l/d = 1$	$>10^4$	0.90
		$l/d = 2$	$>10^4$	0.85
		$l/d = 4$	$>10^4$	0.87
		$l/d = 8$	$>10^4$	0.99
	Square rod	∞	$>10^4$	2.00
	Square rod	∞	$>10^4$	1.50
	Triangular cylinder	∞	$>10^4$	1.39
	Semicircular shell	∞	$>10^4$	1.20
	Semicircular shell	∞	$>10^4$	2.30
	Hemispherical shell		$>10^4$	0.39
	Hemispherical shell		$>10^4$	1.40
	Cube		$>10^4$	1.10
	Cube		$>10^4$	0.81
	Cone—60° vertex		$>10^4$	0.49
	Parachute		$\approx 3 \times 10^7$	1.20

SOURCES: Brevoort and Joyner (4), Lindsey (19), Morrison (22), Roberson et al. (24), Rouse (26), and Scher and Gale (28).

11.6 Terminal Velocity

The engineer is often required to use drag data in the computation of the *terminal velocity* of a body. When a body is first dropped in the atmosphere or in water, it accelerates under the action of its weight. Then, as the speed of the body increases, the drag increases. Finally, the drag reaches a magnitude such that the sum of all external forces on the body is zero. Hence acceleration ceases, and the body has attained its terminal velocity. Thus the terminal velocity is the maximum velocity attained by a falling body. This assumes that the fluid through which it falls has constant properties over the path of descent. The following two examples illustrate the method of computing the terminal velocity of a body— first in air, then in water.

EXAMPLE 11.6   A 50-mm solid plastic sphere (S = 1.3) is dropped from an airplane. Assuming standard atmospheric conditions ($T = 20°C$, $p_{atm} = 101 \text{ kPa}$ absolute), what will be the terminal velocity of the sphere.

Solution   The free-body diagram of the sphere is shown in the accompanying figure, where F_B is the buoyant force, F_D is the drag, and W is the weight.

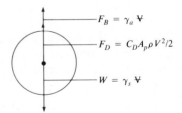

Obviously, the sphere will drop at a high rate of speed. Therefore, we make a first guess at C_D by choosing a value from the high range of Re. Assume $C_D = 0.15$ (based on Fig. 11.11). Then

$$F_D = \frac{C_D A_p \rho V_0^2}{2} \qquad \text{or} \qquad V_0 = \left(\frac{2F_D}{C_D A_p \rho} \right)^{1/2}$$

Here F_D is the net weight of the sphere, or

$$F_D = (\gamma_s - \gamma_a)\left(\frac{4}{3}\right)\pi r^3 = (\gamma_s - \gamma_a)\left(\frac{4}{3}\right)\pi\left(\frac{d^3}{8}\right) = \frac{1}{6}\pi d^3(\gamma_s - \gamma_a)$$

where γ_s is the specific weight of the sphere and γ_a is the specific weight of air.

Also
$$A_p = \frac{\pi}{4}d^2$$

The specific weight of the sphere, γ_s, is the specific gravity of the sphere times γ_{water}, or $\gamma_s = 1.3 \times 9.81$ kN/m³ = 12.7 kN/m³. Also, $\rho_{\text{air}} = 1.20$ kg/m³ and $\gamma_a = 1.20 \times 9.81$ N/m³. Thus $\gamma_a \ll \gamma_s$. Hence

$$V_0 = \left(\frac{2\gamma_s(1/6)\pi d^3}{C_D(\pi/4)d^2\rho_{\text{air}}} \right)^{1/2} = \left(\frac{\gamma_s(4/3)d}{C_D\rho_{\text{air}}} \right)^{1/2}$$

$$= \left(\frac{12.7(10^3 \text{ N/m}^3)(4/3)(0.05 \text{ m})}{C_D(1.20 \text{ kg/m}^3)} \right)^{1/2} = \left(\frac{706 \text{ m}^2/\text{s}^2}{C_D} \right)^{1/2} = \frac{26.6 \text{ m/s}}{(C_D)^{1/2}}$$

With our first guess of $C_D = 0.15$, we get a velocity of

$$V_0 = \frac{26.6 \text{ m/s}}{(0.15)^{1/2}} = 68.6 \text{ m/s}$$

Now we compute Re based on this velocity of 68.6 m/s, so that we can get a better value of C_D.

$$\text{Re} = \frac{Vd\rho}{\mu}$$

$$= \frac{(68.6 \text{ m/s})(0.05 \text{ m})(1.20 \text{ kg/m}^3)}{1.81 \times 10^{-5} \text{ N} \cdot \text{s/m}^2} = 2.3 \times 10^5$$

Then a better estimate of C_D based on Re = 2.3×10^5 is $C_D = 0.40$. Our first guess for C_D was too low. Hence we make another calculation of V_0 based on the more precise value of C_D:

$$V_0 = \frac{26.6 \text{ m/s}}{(0.40)^{1/2}} = 42.1 \text{ m/s}$$

Another calculation based on the new velocity of 42.1 m/s yields a C_D of 0.42:

$$V_0 = \frac{26.6 \text{ m/s}}{(0.42)^{1/2}} = 41.0 \text{ m/s}$$

One more computation for Re followed by determination of C_D yields Re = 1.4×10^5 and $C_D = 0.42$.

Therefore, the terminal velocity is $V_0 = 41.0$ m/s. ◀

EXAMPLE 11.7   A 50-mm plastic sphere (S = 1.3) is dropped in water. Determine its terminal velocity. Assume $T = 20°C$.

Solution   We approach this problem in the same manner as Example 11.6:

$$F_D = (\gamma_s - \gamma_w)(1/6)\pi d^3$$

Then
$$V_0 = \left[\frac{(\gamma_s - \gamma_w)\,(4/3)d}{C_D\,\rho_w}\right]^{1/2}$$

$$= \left[\frac{(12.7 - 9.79)\,(10^3 \text{ N/m}^3)\,(4/3)\,(0.05 \text{ m})}{C_D \times 10^3 \text{ kg/m}^3}\right]^{1/2}$$

$$= \left(\frac{0.194}{C_D}\right)^{1/2} = \frac{0.44}{C_D^{1/2}} \text{ m/s}$$

Now, the velocity of fall will not be as great as in air, so a reasonable first guess at C_D is 0.48. With this value of C_D, the velocity is

$$V_0 = \frac{0.44 \text{ m/s}}{0.48^{1/2}} = 0.63 \text{ m/s}$$

Then Re $= Vd\rho/\mu$, where $\rho = 10^3$ kg/m^3 and $\mu = 1.00 \times 10^{-3}$ N $\cdot$ s/m^2.

$$\text{Re} = \frac{(0.63 \text{ m/s})\,(0.05 \text{ m})\,(10^3 \text{ kg/m}^3)}{1.00(10^{-3} \text{ N} \cdot \text{s/m}^2)} = 3.2 \times 10^4$$

From Fig. 11.11, $C_D = 0.48$. Therefore

$$V_0 = 0.63 \text{ m/s} \qquad \blacktriangleleft$$

11.7 Effect of Compressibility on Drag

The variation of drag coefficient with Mach number for three axisymmetric bodies is shown in Fig. 11.12. In each case, the drag coefficient increases only slightly with the Mach number at low Mach numbers and then increases sharply as transonic flow (M $\approx$ 1) is approached. Note that the rapid increase in drag coefficient occurs at a higher Mach number (closer to unity) if the body is slender with a pointed nose. The drag coefficient reaches a maximum at a Mach number somewhat larger than unity and then decreases as the Mach number is further increased.

The slight increase in drag coefficient with low Mach numbers is attributed to an increase in form drag due to compressibility effects on the pressure distribution. However, as the flow velocity is increased, the maximum velocity on the body finally becomes sonic. The Mach number of the free-stream flow at which sonic flow first appears on the body is called the *critical Mach number*. Further increases in flow velocity result in local regions of supersonic flow (M > 1), which lead to wave drag due to shock-wave formation and an appreciable increase in drag coefficient.

The critical Mach number for a sphere is approximately 0.6. One notes in Fig. 11.12 that the drag coefficient begins to rise sharply at about this Mach

FIGURE 11.12

Drag characteristics of projectile, sphere, and cylinder with compressibility effects. [After Rouse (26)]

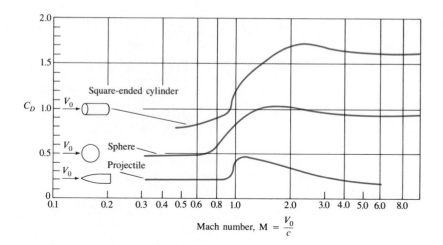

number. The critical Mach number for the pointed body is larger, and, correspondingly, the rise in drag coefficient occurs at a Mach number closer to unity.

The drag-coefficient data for the sphere shown in Fig. 11.12 are for a Reynolds number of the order of 10^4. The data for the sphere shown in Fig. 11.11, on the other hand, are for very low Mach numbers. The question then arises about the general variation of the drag coefficient of a sphere with both Mach number and Reynolds number. Information of this nature is often needed to predict the trajectory of a body through the upper atmosphere, and it is sometimes needed to analyze the flows transporting solid-phase particles.

A contour plot of the drag coefficient of a sphere versus both Reynolds and Mach numbers based on available data (6) is shown in Fig. 11.13. One notices the C_D-versus-Re curve from Fig. 11.11 in the $M = 0$ plane. Correspondingly, one sees the C_D-versus-M curve from Fig. 11.12 in the Re $= 10^4$ plane. We see, then, that at low Reynolds numbers C_D decreases with increasing Mach number, whereas at high Reynolds numbers the opposite trend is observed. Using this figure, the engineer can determine the drag coefficient of a sphere at any combination of Re and M. Of course, corresponding C_D contour plots can be generated for any body, provided the data are available.

11.8 Lift

In Sec. 11.1 it was shown that a differential pressure between the top and bottom of a body causes a lateral force or lift to be imposed on the body. However, no explanation was given for the cause of the differences in velocity that produce such a pressure distribution. In this section we will consider circulation, which is the basic cause of lift. Then we will consider the lift and drag characteristics of typical airfoils.

FIGURE 11.13

*Contour plot of the drag
coefficient of the sphere
versus Reynolds and Mach
numbers. [After Crowe
et al. (6)]*

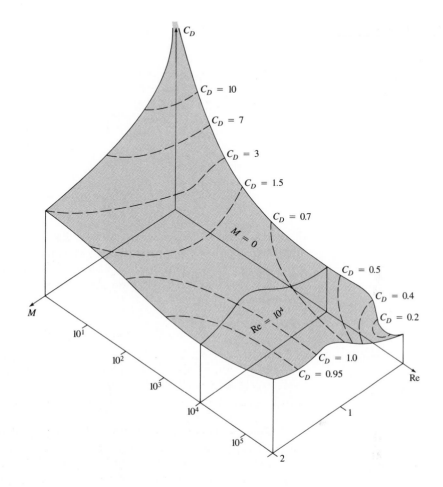

Circulation

Consider flow along a closed path such as is shown in Fig. 11.14. Along any differential segment of the path, the velocity can be resolved into components that are tangent and normal to the path. Let us signify the tangential component of velocity as V_L. If we integrate $V_L\,dL$ around the curve, the resulting quantity is called *circulation,* which is represented by the Greek letter Γ (capital gamma). Hence we have

$$\Gamma = \oint V_L\,dL \tag{11.10}$$

Sign convention dictates that in applying Eq. (11.10), we take tangential velocity vectors that have a counterclockwise sense around the curve as negative and take

FIGURE 11.14

Concept of circulation.

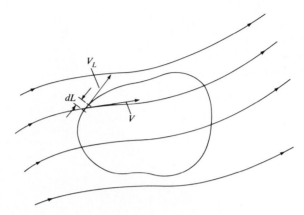

those that have a clockwise direction as having a positive contribution.* The circulation for an irrotational vortex is determined as follows. The tangential velocity at any radius is C/r, where a positive C means a clockwise rotation. Therefore, if we evaluate the circulation about a curve with radius r, the differential circulation is

$$d\Gamma = V_L \, dL = \frac{C}{r_1} r_1 \, d\theta = C \, d\theta \qquad (11.11)$$

Then, when we integrate this around the entire circle, we obtain

$$\Gamma = \int_0^{2\pi} C \, d\theta = 2\pi C \qquad (11.12)$$

One way to induce circulation physically is to rotate a cylinder about its axis. In Fig. 11.15a we see the flow pattern produced by such action. The velocity of the fluid next to the surface of the cylinder is equal to the velocity of the cylinder surface itself because of the nonslip condition that must prevail between the fluid and solid. At some distance from the cylinder, however, the velocity decreases with r, much like it does for the irrotational vortex. In the next section we will see how circulation produces lift.

Combination of Circulation and Uniform Flow Around a Cylinder

If we now superpose the velocity field produced for uniform flow around a cylinder, Fig. 11.15b, onto a velocity field with circulation around a cylinder, Fig. 11.15a, we see that the velocity is reinforced on the top side of the cylinder and reduced on the other side (Fig. 11.15c). We also observe that the stagnation points have both moved toward the low-velocity side of the cylinder. Consistent with the Bernoulli principle (assuming irrotational flow through-

The sign convention is the opposite of that for the mathematical definition of a line integral.

FIGURE 11.15

*Ideal flow around a
cylinder. (a) Circulation.
(b) Uniform flow.
(c) Combination of
circulation and uniform flow.*

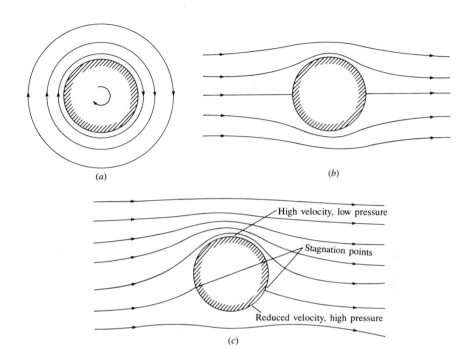

(a)

(b)

High velocity, low pressure

Stagnation points

Reduced velocity, high pressure

(c)

out), we find that the pressure on the high-velocity side is lower than the pressure on the low-velocity side. Hence a pressure differential exists that causes a side thrust, or lift, on the cylinder. According to ideal-flow theory, the lift per unit length of an infinitely long cylinder is given by $F_L/\ell = \rho V_0 \Gamma$, where F_L is the lift on the segment of length ℓ. For this ideal irrotational flow there is no drag on the cylinder. For the real-flow case, separation and viscous stresses do produce drag, and the same viscous effects will reduce the lift somewhat. Even so, the lift is significant when flow occurs past a rotating body or when a body is translating and rotating through a fluid. Hence we have the reason for the "curve" on a pitched baseball or the "drop" on a Ping-Pong ball that is given a fore spin. This phenomenon of lift produced by rotation of a solid body is called the *Magnus effect* after a nineteenth-century German scientist who made early studies of the lift on rotating bodies. A paper by Mehta (21) offers an interesting account of the motion of rotating sports balls.

Coefficients of lift and drag for the rotating cylinder with end plates are shown in Fig. 11.16. In this figure, the parameter $r\omega/V_0$ is the ratio of cylinder-surface speed to the free-stream velocity, where r is the radius of the cylinder and ω is the angular speed in radians per second. The corresponding curves for the rotating sphere are given in Fig. 11.17. Note that the lift is given as $F_L = C_L A_p \rho V_0^2/2$, where A_p is the projected area.

FIGURE 11.16

*Coefficients of lift and drag
as functions of rω/V₀ for a
rotating cylinder. [After
Rouse (26)]*

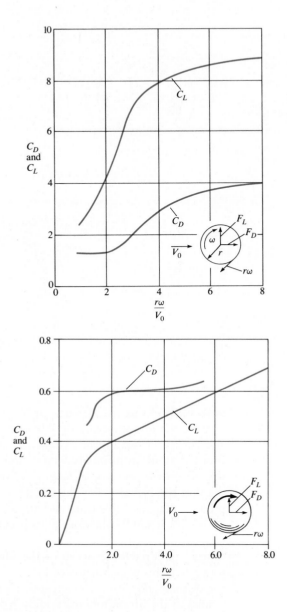

FIGURE 11.17

*Coefficients of lift and drag
for a rotating sphere. [After
Barkla et al. (3)]*

FIGURE 11.16

*Coefficients of lift and drag
as functions of rω/V₀ for a
rotating cylinder. [After
Rouse (26)]*

FIGURE 11.17

*Coefficients of lift and drag
for a rotating sphere. [After
Barkla et al. (3)]*

Lift of an Airfoil

Let us first consider the motion of an airfoil through an ideal (nonviscous) fluid.
With such a fluid, the flow past an airfoil is irrotational, as shown in Fig. 11.18a.
Here, as for irrotational flow past a cylinder, the lift and drag are zero. There is
a stagnation point on the bottom side near the leading edge, and another on the
top side near the trailing edge of the foil. In the real-flow (viscous-fluid) case,

FIGURE 11.18

Patterns of flow around an airfoil. (a) Ideal flow—no circulation. (b) Real flow — circulation.

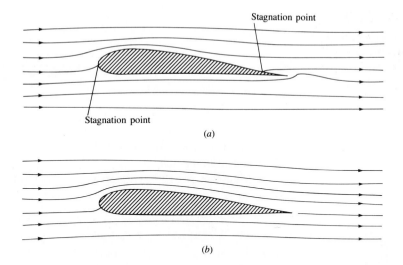

the flow pattern around the front half of the foil is plausible. However, the flow pattern in the region of the trailing edge, as shown in Fig. 11.18*a*, cannot occur. A stagnation point on the upper side of the foil indicates that fluid must flow from the lower side around the trailing edge and then toward the stagnation point. Such a flow pattern implies an infinite acceleration of the fluid particles as they turn the corner around the trailing edge of the wing. This is a physical impossibility, and as we have seen in previous sections of the text, separation occurs at the sharp edge. As a consequence of the separation, the rear stagnation point moves to the trailing edge. Flow from both the top and bottom sides of the airfoil in the vicinity of the trailing edge then leaves the airfoil smoothly and essentially parallel to these surfaces at the trailing edge (Fig. 11.18*b*).

To bring theory into line with the physically observed phenomenon, it was hypothesized that a circulation around the airfoil must be induced in just the right amount so that the downstream stagnation point is moved all the way back to the trailing edge of the airfoil, thus allowing the flow to leave the airfoil smoothly at the trailing edge. This is called the *Kutta condition* (18), named after a pioneer in aerodynamic theory. When analyses are made with this simple assumption concerning the magnitude of the circulation, very good agreement occurs between theory and experiment for the flow pattern and the pressure distribution, as well as for the lift on a two-dimensional airfoil section (no end effects). Ideal-flow theory then shows that the magnitude of the circulation required to maintain the rear stagnation point at the trailing edge (the Kutta condition) of a symmetric airfoil with a small angle of attack is given by

$$\Gamma = \pi c V_0 \alpha \qquad (11.13)$$

FIGURE 11.19

Definition sketch for an airfoil section.

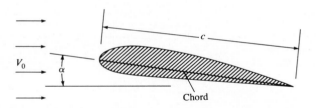

where Γ is the circulation, c is the chord length of the airfoil, and α is the angle of attack of the chord of the airfoil with the free-stream direction (see Fig. 11.19 for a definition sketch).

Like that for the cylinder, the lift per unit length for an infinitely long wing is

$$F_L/\ell = \rho V_0 \Gamma$$

The planform area for the length segment ℓ is ℓc. Hence the lift on segment ℓ is

$$F_L = \rho V_0^2 \pi c \ell \alpha \qquad (11.14)$$

For an airfoil we define the coefficient of lift as

$$C_L = \frac{F_L}{S\rho V_0^2/2} \qquad (11.15)$$

where S is the planform area of the wing—that is, the area seen from the plan view. On combining Eqs. (11.14) and (11.15) and identifying S as the area associated with length segment ℓ, we find that C_L for irrotational flow past a two-dimensional airfoil is given by

$$C_L = 2\pi\alpha \qquad (11.16)$$

Equations (11.14) and (11.16) are the theoretical lift equations for an infinitely long airfoil at a small angle of attack. Flow separation near the leading edge of the airfoil produces deviations (high drag and low lift) from the ideal-flow predictions at high angles of attack. Hence experimental wind-tunnel tests are always made to evaluate the performance of a given type of airfoil section. For example, the experimentally determined values of lift coefficient versus α for two NACA airfoils are shown in Fig. 11.20. Note in this figure that the coefficient of lift increases with the angle of attack, α, to a maximum value and then decreases with further increase in α. This condition, where C_L starts to decrease with a further increase in α, is called *stall*. Stall occurs because of the onset of separation over the top of the airfoil, which changes the pressure distribution in such a way as not only to decrease lift but also to increase drag. Data for many other airfoil sections are given by Abbott and Von Doenhoff (2).

FIGURE 11.20

*Values of C_L for two
NACA airfoil sections.
[After Abbott (1)]*

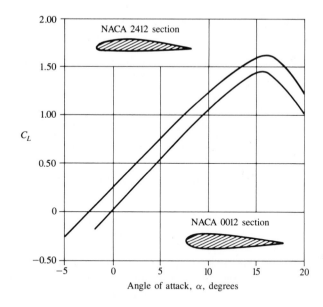

Airfoils of Finite
Length—Effect on Drag and Lift

The drag of a two-dimensional foil at a low angle of attack (no end effects) is
primarily viscous drag. However, wings of finite length also have an added drag
and a reduced lift associated with vortices generated at the wing tips. These vortices occur because the high pressure below the wing and the low pressure on
top cause fluid to circulate around the end of the wing from the high- to the
low-pressure zone, as shown in Fig. 11.21. This induced flow has the effect of
adding a downward component of velocity, w, to the approach velocity V_0.
Hence, the "effective" free-stream velocity is now at an angle ($\phi \approx w/V_0$) to the
direction of the original free-stream velocity, and the resultant force is tilted
back as shown in Fig. 11.22. Thus the effective lift is smaller than the lift for
the infinitely long wing because the effective angle of incidence is smaller. This

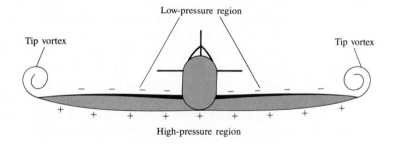

FIGURE 11.21

Formation of tip vortices.

FIGURE 11.22

Definition sketch for induced-drag relations.

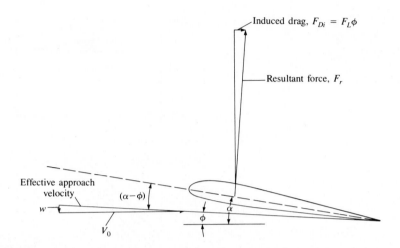

resultant force has a component parallel to V_0 that is called the *induced drag* and is given by $F_L\phi$. Prandtl (23) showed that the induced velocity w for an elliptical spanwise lift distribution is given by the following equation:

$$w = \frac{2F_L}{\pi\rho V_0 b^2} \qquad (11.17)$$

where b is the total length (or span) of the finite wing. Hence

$$F_{Di} = F_L\phi = \frac{2F_L^2}{\pi\rho V_0^2 b^2} = \frac{C_L^2}{\pi}\frac{S^2}{b^2}\frac{\rho V_0^2}{2} \qquad (11.18)$$

From Eq. (11.18) it can be easily shown that the coefficient of induced drag, C_{Di}, is given by

$$C_{Di} = \frac{C_L^2}{\pi(b^2/S)} \qquad (11.19)$$

which happens to represent the minimum induced drag for any wing planform. Here the ratio b^2/S is called the aspect ratio Λ of the wing, and S is the planform area of the wing. Thus, for a given wing section (constant C_L and constant chord c), longer wings (larger aspect ratios) have smaller induced-drag coefficients. The induced drag is a significant portion of the total drag of an airplane at low velocities and must be given careful consideration in airplane design. Aircraft (such as gliders) and even birds (such as the albatross and seagull) that are required to be airborne for long periods of time with minimum energy expenditure are noted for their long, slender wings. Such a wing is more efficient because the induced drag is small. To illustrate the effect of finite span, we turn to Fig. 11.23, which shows C_L and C_D versus α for wings with several aspect ratios.

FIGURE 11.23

*Coefficients of lift and drag
for three wings with aspect
ratios of 3, 5, and 7. [After
Prandtl (23)]*

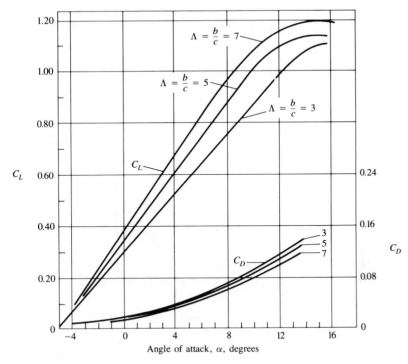

The total drag of a rectangular wing is computed by

$$F_D = (C_{D0} + C_{Di}) \frac{bc\rho V_0^2}{2}$$

where C_{D0} is the coefficient of form drag of the wing section and C_{Di} is the co-efficient of induced drag.

A graph showing C_L and C_D versus α is given in Fig. 11.24. Note in this graph that C_D is separated into the induced-drag coefficient C_{Di} and the form-drag coefficient C_{D0}.

EXAMPLE 11.8 A light plane (weight = 10 kN) has a wingspan of 10 m and a chord length of 1.5 m. If the lift characteristics of the wing are like those given in Fig. 11.23, what must be the angle of attack for a takeoff speed of 140 km/h? What is the stall speed? Assume two passengers at 800 N each and standard atmospheric conditions.

Solution The lift must be 10 kN + 1.6 kN = 11.6 kN. Hence

$$C_L = \frac{F_L}{S\rho V_0^2/2}$$

$$= \frac{11,600 \text{ N}}{(15 \text{ m}^2)(1.2 \text{ kg/m}^3)[(140,000/3600)^2 \text{ m}^2/\text{s}^2]/2} = 0.852$$

FIGURE 11.24

Coefficients of lift and drag for a wing with an aspect ratio of 5. [After Prandtl (23)]

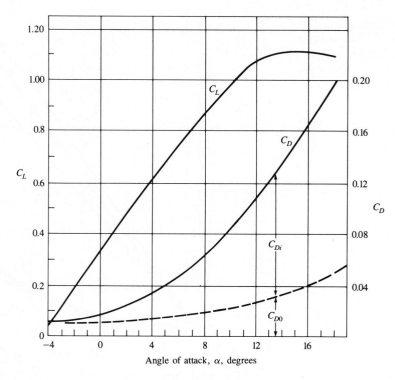

Also

$$\frac{b}{c} = \frac{10}{1.5} = 6.67$$

Hence from the curves of Fig. 11.23, we get

$$\alpha = 7°$$

Stall will occur at

$$C_L = 1.18$$

Then

$$11,600 = 1.18(15)\left(\frac{1.2}{2}\right)(V_{stall})^2$$

$$V_{stall} = 33.0 \text{ m/s} = 119 \text{ km/h}$$

Drag and Lift on Road Vehicles

Early in the development of cars, aerodynamic drag was a minor factor in performance because normal highway speeds were quite low. Thus in the 1920s, coefficients of drag for cars were around 0.80. As highway speeds increased and the science of metal forming became more advanced, cars took on a less angular shape, so that by the 1940s drag coefficients were 0.70 and lower. In the 1970s the average C_D for U.S. cars was approximately 0.55. In the early 1980s the average C_D for American cars dropped to 0.45, and currently auto manufacturers are giving even more attention to reducing drag in designing their cars. All major U.S., Japanese, and European automobile companies now have models with C_D's of about 0.33, and some companies even report C_D's as low as 0.29 on new models. European manufacturers were the leaders in the streamlining of cars because European gasoline prices (including tax) have been for a number of years about twice those in the United States. Table 11.2 shows the C_D for a 1932 Fiat and for other, more contemporary car models.

TABLE 11.2   COEFFICIENTS OF DRAG FOR CARS

Make and Model	Profile	C_D
1932 Fiat Balillo		0.60
Volkswagen "Bug"		0.46
Toyota Previa		0.34
Plymouth Laser		0.33
Ford Taurus		0.32
Volvo 850 GLT		0.32
Oldsmobile Cutlass		0.29
Ford Probe V (concept car)		0.14
GM Sunraycer (experimental solar vehicle)		0.12

Great strides have been made in reducing the drag coefficients for passenger cars. However, significant future progress will be very hard to achieve. One of the most streamlined cars was the "Bluebird," which set a world land-speed record in 1938. Its C_D was 0.16. The minimum C_D of well-streamlined racing cars is about 0.20. Thus, lowering the C_D for passenger cars below 0.30 will require exceptional design and workmanship. For example, the underside of most cars is aerodynamically very rough (axles, wheels, muffler, fuel tank, shock absorbers, and so on). One way to smooth the underside is to add a panel to the bottom of the car. But then clearance may become a problem, and adequate dissipation of heat from the muffler may be hard to achieve. Other basic features of the automobile that contribute to drag but are not very amenable to drag-reduction modifications are interior air-flow systems for engine cooling, wheels, exterior features such as rear-view mirrors and antennas, and other surface protrusions. The reader is directed to two books on road vehicle aerodynamics, (14) and (29), which address all aspects of the drag and lift of road vehicles in considerably more detail than is possible here.

To produce low-drag vehicles, the basic tear-drop shape is an idealized starting point. This shape can be altered to accommodate the necessary functional features of the vehicle. For example, the rear end of the tear-drop shape must be lopped off to yield an overall vehicle length that will be manageable in traffic and will fit in our garages. Also, the shape should be more wide than high. Wind-tunnel tests are always helpful in producing the most efficient design. One such test was done on a $\frac{3}{8}$-scale model of a typical notchback sedan. Wind-tunnel test results for such a sedan are shown in Fig. 11.25. Here the centerline pressure distribution (distribution of C_p) for the conventional sedan is shown by a solid line and that for a sedan with a 68-mm rear-deck lip is shown by a dashed line. Clearly the rear-deck lip causes the pressure on the rear of the car to increase (C_p is less negative), thereby reducing the drag on the car itself. It also decreases the lift, thereby improving traction. Of course, the lip itself produces some drag, and these tests show that the optimum lip height for greatest overall drag reduction is about 20 mm.

FIGURE 11.25

Effect of rear-deck lip on model surface. Pressure coefficients are plotted normal to the surface. [After Schenkel (27)]

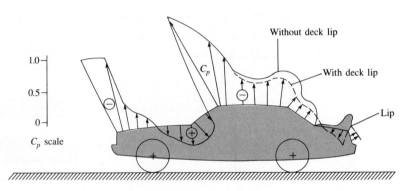

The drag of trucks can be reduced by installing vanes near the corners of the truck body to deflect the flow of air more sharply around the corner, thereby reducing the degree of separation. This in turn creates a higher pressure on the rear surfaces of the truck, which reduces the drag of the truck. Problem 11.31 addresses this issue of drag reduction by the use of vanes.

Another application of aerodynamics to road vehicles is the use of negative-lift vanes on racing cars, as shown in Fig. 11.26. Here the negative lift produced by the vanes improves the stability and traction of the racing car.

FIGURE 11.26

Racing car with negative-lift devices.

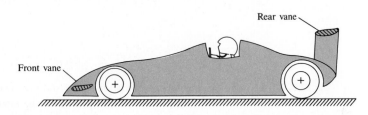

EXAMPLE 11.9   The rear vane installed on the racing car of Fig. 11.26 is at an angle of attack of 8° and has characteristics like those given in Fig. 11.23. Estimate the downward thrust (negative lift) and drag from the vane that is 1.5 m long and has a chord length of 250 mm. Assume the racing car travels at a speed of 270 km/h on a track where normal atmospheric pressure and a temperature of 30°C prevail.

Solution

$$F_L = C_L \ell c \rho V_0^2 / 2$$

where

$$\ell = 1.5 \text{ m} \quad c = 0.25 \text{ m} \quad \ell/c = 6.0$$

$$\rho = 1.17 \text{ kg/m}^3 \quad V_0 = 270 \text{ km/h} = 75 \text{ m/s}$$

$$C_L = 0.93 \quad \text{(from Fig. 11.23)}$$

Then negative lift is

$$F_L = 0.93 \times 1.5 \times 0.25 \times 1.17 \times (75)^2 / 2 = 1148 \text{ N} \quad \blacktriangleleft$$

Also

$$C_D = 0.070 \quad \text{(from Fig. 11.23)}$$

Therefore

$$F_D = (0.070/0.93) \times 1148 = 86.4 \text{ N} \quad \blacktriangleleft$$

Problems

11.1   A hypothetical pressure-coefficient variation over a long (length normal to the page) plate is shown. What is the coefficient of drag for the plate in this orientation and with the given pressure distribution?

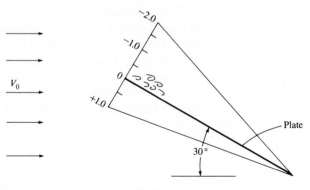

PROBLEM 11.1

11.2   Flow is occurring past the square rod shown. The pressure-coefficient values are as shown. From which direction do you think the flow is coming? a) SW direction, b) SE direction, c) NW direction, d) NE direction.

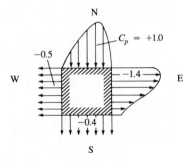

PROBLEM 11.2

11.3   The hypothetical pressure distribution on a rod of triangular (equilateral) cross section is shown, where flow is from left to right. That is, C_p is maximum and equal to $+1.0$ at the leading edge and decreases linearly to zero at the trailing edges. The pressure coefficient on the downstream face is constant with a value of -0.5. Neglecting skin-friction drag, find C_D for the rod.

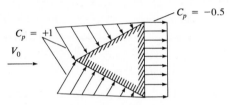

PROBLEM 11.3

11.4   The pressure distribution on a rod having a triangular (equilateral) cross section is shown, where flow is from left to right. What is C_D for the rod?

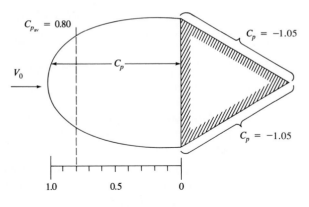

11.5 Based upon the C_p distribution shown, which rod will have the greater coefficient of drag? a) square b) rectangular.

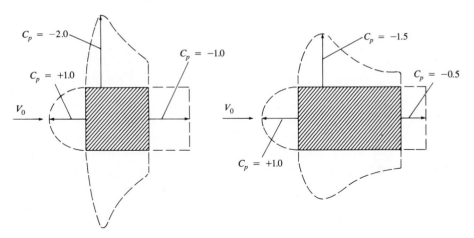

11.6 For the rectangular rod of Prob. 11.5, which one of the following would be a reasonable C_D? a) $C_D = 0.25$, b) $C_D = 1.30$, c) $C_D = 1.50$, d) $C_D = 2.0$.

11.7 Compute the overturning moment exerted by a 30-m/s wind on a smokestack that has a diameter of 3 m and a height of 80 m. Assume that the air temperature is 20°C and that p_a is 99 kPa absolute.

11.8 What is the moment at the bottom of a flag pole 35 m (115 ft) high and 10 cm in diameter in a 25-m/s wind? The atmospheric pressure is 100 kPa (14.5 psia), and the temperature is 20°C (68°F).

11.9 A target meter is a technique sometimes used to measure the velocity of a flow. It consists of a hemispherical cup mounted in the flow as shown; the force on the cup is measured

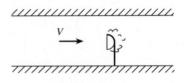

PROBLEM 11.9

by strain gages. The cup is 1 cm in diameter mounted in a duct with airflow at a temperature of 100°C and a pressure of 200 kPa absolute. A force of 0.50 N is measured. What is the velocity (in meters per second) of the air in the duct?

11.10   A cooling tower, used for cooling recirculating water in a modern steam power plant, is 350 ft high and about 250 ft in diameter, as shown. Estimate the drag on the cooling tower in a 200-ft/s wind ($T = 60°F$).

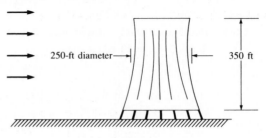

PROBLEM 11.10

11.11   A cylindrical rod of diameter d and length L is rotated about its midpoint in a horizontal plane. Assume the drag force at each section of the rod can be calculated assuming a two-dimensional flow with an oncoming velocity equal to the relative velocity component normal to the rod. Assume C_D is constant along the rod.

   a. Derive an expression for the average power needed to rotate the rod.
   b. Calculate the power for $\omega = 100$ rad/s, $d = 2$ cm, $L = 1$ m, $\rho = 1.2$ kg/m^3, and $C_D = 1.2$.

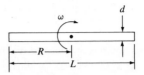

PROBLEM 11.11

11.12   At what frequency will vortices be shed from the smokestack of Prob. 11.7?

11.13   At what frequency will vortices be shed from the flag pole of Prob. 11.8?

11.14   Estimate the wind force on a billboard 10 ft high and 40 ft wide when a 60-mph wind ($T = 60°F$) is blowing normal to it.

11.15   Determine the drag of a 5 ft × 5 ft plate held at right angles to a stream of air (60°F and standard atmospheric pressure) having a velocity of 40 fps.

11.16   Estimate the ratio of the drag of a large square plate (2 m by 2 m) when it is towed through water (10°C) in a normal orientation to that when it is towed edgewise. Assume a towing speed of about 1 m/s.

11.17   What drag is produced when a disk 1 m in diameter is submerged and towed behind a boat at a speed of 3 m/s? Assume orientation of the disk so that maximum drag is produced.

11.18   A circular billboard having a diameter of 7 m is mounted so as to be freely exposed to the wind. Estimate the total force exerted on the structure by a wind that has a direction normal to the structure and a speed of 30 m/s. Assume $T = 10°C$ and $p = 101$ kPa absolute.

11.19   Estimate the moment at ground level on a signpost supporting a sign measuring 2 m by 2 m if the wind is normal to the surface and has a speed of 35 m/s and the center of the sign is 2 m above the ground. Neglect the wind load on the post itself. Assume $T = 10°C$ and $p = 100$ kPa absolute.

11.20   Estimate the additional power required for the truck when it is carrying the rectangular sign at a speed of 20 m/s over that required when it is traveling at the same speed but is not carrying the sign.

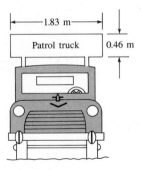

PROBLEM 11.20

11.21   Estimate the added power required for the car when the cartop carrier is used and the car is driven at 80 km/h in a 20-km/h head wind over that required when the carrier is not used in the same conditions.

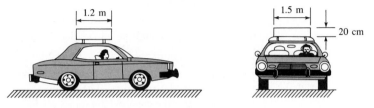

PROBLEM 11.21

11.22   The resistance to motion of an automobile consists of rolling resistance and aerodynamic drag. The rolling resistance is the product of the weight and the coefficient of rolling friction. The weight of an automobile is 3000 lb, and it has a frontal area of 20 ft². The drag coefficient is 0.40, and the coefficient of rolling friction is 0.10. Determine the percentage savings in gas mileage that one achieves when one drives at 55 mph instead of 65 mph on a level road. Assume an air temperature of 60°F.

11.23   An automobile is traveling on a level road into a head wind of 10 m/s. The speed of the automobile with respect to the ground is 30 m/s. The drag coefficient is 0.4, and the projected area is 2 m². The air density is 1.2 kg/m³. The car weighs 10,000 N, and the coefficient of rolling friction is 0.02. What power (in kilowatts) is needed to overcome both aerodynamic drag and rolling resistance?

11.24   Estimate the wind force that would act on *you* if you were standing on top of a tower in a 30-m/s (115-ft/s) wind on a day when the temperature was 20°C (68°F) and the atmospheric pressure was 96 kPa (14 psia).

11.25   Windstorms sometimes blow empty boxcars off their tracks. The dimensions of one type of boxcar are shown. What minimum wind velocity normal to the side of the car would be required to blow the car over?

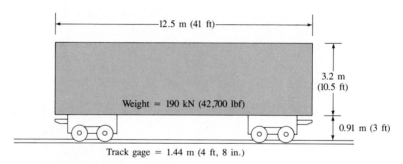

PROBLEM 11.25

11.26   A bicyclist is coasting down a hill with a slope of 10° into a head wind (measured with respect to the ground) of 5 m/s. The mass of the cyclist and bicycle is 60 kg, and the coefficient of rolling friction is 0.02. The drag coefficient is 0.5, and the projected area is 0.5 m². The air density is 1.2 kg/m³. Find the speed of the bicycle in meters per second.

11.27   A bicyclist is capable of delivering 100 W of power to the wheels. The rolling resistance of a bicycle is negligible, so the power equals the aerodynamic drag times the velocity. How fast can the bicyclist travel in a 5-m/s head wind if his or her projected area is 0.5 m², the drag coefficient is 0.4, and the air density is 1.2 kg/m³?

11.28   The drag coefficient of a sports car with a convertible roof is 0.3 when the roof is closed and increases to 0.42 when the roof is open. The resistance to motion consists of rolling resistance and aerodynamic drag. The rolling resistance is the product of the coefficient of rolling friction and the weight. The sports car has a mass of 800 kg, a frontal area of 4 m², and a coefficient of rolling friction of 0.10. The maximum power delivered to the wheels is 80 kW. Assuming a temperature of 20°C and a level road, determine the maximum possible speed with the roof closed and with it open.

11.29   A motorist has an automobile that has a drag coefficient of 0.30 and a frontal area (on which the drag coefficient is based) of 2 m². The total mass of driver and automobile is

500 kg. The coefficient of rolling friction is 0.1. Traveling into a head wind one day, the driver notes that the gasoline consumption is 20% higher than it is on a still day when she travels at the same speed, 90 km/h. The terrain is flat, and the air density is 1.2 kg/m^3. Find the velocity of the head wind.

11.30 Assume that the horsepower of the engine in the original 1932 Fiat Balillo (see Table 11.2, page 525) was 40 bhp (brake horsepower) and that the maximum speed at sea level was 60 mph. Also assume that the projected area of the automobile is 30 ft^2. Assume that the automobile is now fitted with a modern 220-bhp motor with a weight equal to the weight of the original motor; thus the rolling resistance is unchanged. What is the maximum speed of the "souped up" Balillo at sea level?

11.31 One way to reduce the drag of a blunt object is to install vanes to suppress the amount of separation. Such a procedure was used on model trucks in a wind-tunnel study by Kirsch and Bettes (16). For tests on a van-type truck, they noted that without vanes the C_D was 0.78. However, when vanes were installed around the top and side leading edges of the truck body (see the figure), a 25% reduction in C_D was achieved. For a truck with a projected area of 8.36 m^2, what reduction in drag force will be effected by installation of the vanes when the truck travels at 93 km/h? Assume standard atmospheric pressure and a temperature of 20°C.

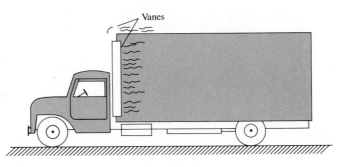

PROBLEM 11.31

11.32 A dirigible flies at 30 ft/s at an altitude where the specific weight of the air is 0.07 lbf/ft^3 and the kinematic viscosity is 1.3×10^{-4} ft^2/s. The dirigible has a length-to-diameter ratio of 5 and has a drag coefficient corresponding to the streamlined body in Fig. 11.11. The diameter of the dirigible is 100 ft. What is the power required to propel the dirigible at this speed?

11.33 For the truck of Prob. 11.31 assume that the total resistance is given by $R = F_D + C$, where F_D is the air drag and C is the resistance due to bearing friction, etc. If C is constant at 450 N for the given truck, what fuel savings (percentage) will be effected by installation of the vanes when the truck travels at 80 km/h and at 100 km/h?

11.34 If the train of Prob. 9.63 has a form-drag coefficient (for the locomotive and irregular undercarriage) of 0.80 and a constant bearing resistance of 3000 N, what will be the total resistance for the train at a speed of 100 km/h and at a speed of 200 km/h? Assume the projected area of 9 m^2 and standard atmospheric conditions prevail. What percentage of the resistance is due to bearing resistance, form drag, and skin-friction drag at each of the two speeds?

11.35 The coefficient of drag of a tear-drop shape is less than the drag of a sphere having the same diameter, at a Reynolds number of about 10^5, because: a) the skin-friction drag is less for the tear-drop shape than for the sphere; b) the degree of separation is less for the tear-drop shape than for the sphere; c) the pressure coefficient on the front of the tear-drop shape is less than that for the sphere.

11.36 A sphere 1 ft in diameter weighs 27 lbf in air. If it is released from rest 15 ft below the surface of a tank 30 ft deep that is filled with very viscous oil ($S = 0.85$, $\mu = 1.0$ lbf $\cdot$ s/ft^2), will its terminal velocity be in the laminar-flow range? What will be the terminal velocity?

11.37 A sphere 2 cm in diameter rises in oil at a velocity of 3 cm/s. What is the specific weight of the sphere if the oil density is 900 kg/m^3 and the dynamic viscosity is 0.096 N $\cdot$ s/m^2?

11.38 Estimate the terminal velocity of a 2.5-mm plastic sphere in oil. The oil has a specific gravity of 0.95 and a kinematic viscosity of 10^{-4} m^2/s. The plastic has a specific gravity of 1.2. The volume of a sphere is given by $\pi D^3/6$.

11.39 Consider a small air bubble (approximately 4 mm diameter) rising in a very tall column of liquid. Will the bubble accelerate or decelerate as it moves upward in the liquid? Will the drag of the bubble be largely skin-friction drag or form drag? Explain.

11.40 A sphere 5 cm in diameter with a specific gravity of 0.20 rises at a terminal velocity of 0.5 cm/s in a fluid with a specific gravity of 0.66. Find the kinematic viscosity of the fluid.

11.41 If Stokes' law is considered valid below a Reynolds number of 0.5, what is the largest raindrop that will fall in accordance with this equation?

11.42 What is the terminal velocity of a 1-cm hailstone in air that has an atmospheric pressure of 96 kPa absolute and a temperature of 0°C? Assume that the hailstone has a specific weight of 6 kN/m^3.

11.43 A spherical rock weighs 45 N in air and 25 N in water. Estimate its terminal velocity as it falls in water (20°C).

11.44 A paratrooper and parachute weigh 800 N. What rate of descent will they have if the parachute is 7 m in diameter and the air has a density of 1.20 kg/m^3?

11.45 A cylinder of wood 80 cm long and 20 cm in diameter is weighted with lead at one end so that its total weight in air is 260 N. If this cylinder is released at a depth of 100 m in a lake 200 m deep, what will be the terminal velocity of the cylinder? Assume $T = 10$°C.

11.46 This cube is weighted so that it will fall with one edge down as shown. The cube weighs 20.8 N in air. What will be its terminal velocity in water?

11.47 If a balloon weighs 0.05 N and is inflated with helium to a diameter of 30 cm, what will be its terminal velocity in air (standard atmospheric conditions)?

11.48 If a balloon weighs 0.01 lb and is inflated with helium to a diameter of 12 in., what will be its terminal velocity in air at 60°F?

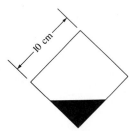

PROBLEM 11.46

11.49 A cylindrical anchor (vertical axis) made of concrete ($\gamma = 15$ kN/m^3) is reeled in at a rate of 1.5 m/s by a man in a boat. If the anchor is 30 cm in diameter and 30 cm long, what tension must be applied to the rope to pull it up at this rate? Neglect the weight of the rope.

11.50 Determine the terminal velocity of a spherical pebble $\frac{1}{4}$ in. in diameter that is falling in 75°F water. Assume that the specific gravity of the pebble equals 2.94 and that the kinematic viscosity of the water equals 1×10^{-5} ft^2/s.

11.51 Determine the terminal velocity in water ($T = 10$°C) of a 15-cm ball that weighs 15 N in air.

11.52 A spherical balloon 2 m in diameter that is used for meteorological observations is filled with helium. The balloon itself weighs 3 N. What velocity of ascent will it attain under standard atmospheric conditions?

11.53 A meteor with a density 2500 kg/m^3 enters the atmosphere with a terminal velocity of Mach 1 in air at 20 kPa and −55°C. The meteor is spherical. What is the diameter of the meteor?

11.54 Determine the characteristics of a sphere that is to have a terminal velocity of 0.5 m/s when falling in water (20°C) and a diameter no smaller than 10 cm and no larger than 20 cm.

11.55 A sphere of diameter 0.2 ft rotating at a rate of 60 rad/s is placed in a stream of water (density = 1.94 slugs/ft^3) with velocity 3 ft/s. Determine the lift force on the sphere.

11.56 The baseball is thrown from west to east with a spin about its vertical axis as shown. Under these conditions it will "break" toward the a) north, b) south, c) neither.

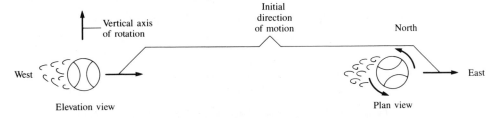

PROBLEM 11.56

11.57 Analyses of pitched baseballs indicate that C_L of a rotating baseball is approximately three times that shown in Fig. 11.17. This greater C_L is due to the added circulation caused by the seams of the ball. What is the lift of a ball pitched at a speed of 85 mph and with a spin rate of 35 rps? Also, how much will the ball be deflected from its original path by the time it gets to the plate as a result of the lift force? *Note:* The mound-to-plate distance is 60 ft, the weight of the baseball is 5 oz, and the circumference is 9 in. Assume standard atmospheric conditions, and assume that the axis of rotation is vertical.

11.58 This circular cylinder is positioned between the vertical walls of a wind tunnel and is rotated as shown. If the wind tunnel has air flow from right to left, choose which force vector (a, b, c, or d) best represents the force that must be applied to the cylinder to hold it in position. Neglect the weight of the cylinder.

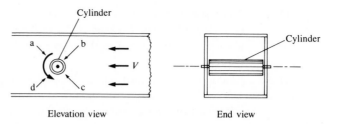

PROBLEM 11.58

11.59 A semiautomatic popcorn popper is shown. After the unpopped corn is placed in screen S, the fan F blows air past the heating coils C and then past the popcorn. When the corn pops, its projected area increases; thus it is blown up and into a container. Unpopped corn has a mass of about 0.15 g per kernel and an average diameter of approximately 6 mm. When the corn pops, its average diameter is about 18 mm. Within what range of airspeeds in the chamber will the device operate properly?

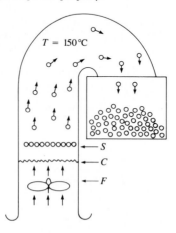

PROBLEM 11.59

11.60 Hoerner (12) presents data that show that fluttering flags of moderate-weight fabric have a drag coefficient (based on the flag area) of about 0.14. Thus the total drag is about 14 times the skin-friction drag alone. Design a flag pole that is 100 ft high and is to fly an American flag 6 ft high. Make your own assumptions regarding other required data.

11.61 For the conditions of Prob. 11.1 what is the C_L?

11.62   An airplane wing having the characteristics shown in Fig. 11.23 (p. 523) is to be designed to lift 2000 lb when the airplane is cruising at 200 ft/s and with an angle of attack of 4°. If the chord length is to be 4 ft, what length of wing is required? Assume $\rho = 0.0024$ slugs/ft^3.

11.63   A boat of the hydrofoil type has a lifting vane with an aspect ratio of 4 that has the characteristics shown in Fig. 11.23. If the angle of attack is 4° and the weight of the boat is 5 tons, what foil dimensions are needed to support the boat at a velocity of 60 fps?

11.64   An airplane that has a weight of 10,000 lbf is flying at 600 ft/s at 36,000 ft, where the pressure is 3.3 psia and the temperature is $-67°$F. The lift coefficient is 0.2. Calculate the area (in square feet) of the wing.

11.65   One wing (wing A) is identical (same cross section) to another wing (wing B) except that wing B is twice as long as wing A. Then for a given wind speed past both wings and with the same angle of attack, one would expect the total lift of wing B to be a) the same as that of wing A, b) less than that of wing A, c) double that of wing A, d) more than double that of wing A.

11.66   The landing speed of an airplane is 5 m/s faster than its stalling speed. The lift coefficient at landing speed is 1.2, and the maximum lift coefficient (stall condition) is 1.4. Calculate both the landing speed and the stalling speed.

11.67   An airplane has a rectangular-planform wing that has an elliptical spanwise lift distribution. The airplane has a mass of 1000 kg, a wing area of 20 m^2, and a wingspan of 14 m, and it is flying at 60 m/s at 3000-m altitude in a standard atmosphere. If the form-drag coefficient is 0.01, calculate the total drag on the wing and the power ($P = F_D V$) necessary to overcome the drag.

11.68   The figure shows the relative pressure distribution for a Göttingen 387-FB lifting vane (17) when the angle of attack is 8°. If such a vane with a 20-cm chord were used as a hydrofoil at a depth of 70 cm, at what speed in 10°C fresh water would cavitation begin? Also, estimate the lift per unit of length of foil at this speed.

11.69   Consider the distribution of C_p as given for the wing section in Prob. 11.68. For this distribution of C_p, the lift coefficient C_L will fall within which range of values? a) $0 < C_L < 1.0$, b) $1.01 < C_L < 2.0$, c) $2.01 < C_L < 3.0$, d) $3.0 < C_L$.

11.70   The total drag coefficient for a wing with an elliptical lift distribution is

$$C_D = C_{D_0} + \frac{C_L^2}{\pi \Lambda}$$

where Λ is the aspect ratio. Derive an expression for C_L that corresponds to minimum C_D/C_L (maximum C_L/C_D) and the corresponding C_L/C_D.

11.71   A glider at 1000-m altitude has a mass of 200 kg and a wing area of 20 m^2. The glide angle is 1.7°, and the air density is 1.2 kg/m^3. If the lift coefficient of the glider is 1.0, how many minutes will it take to reach sea level on a calm day?

11.72   The wing loading on an airplane is defined as the aircraft weight divided by the wing area. An airplane with a wing loading of 2000 N/m^2 has the aerodynamic characteristics given by Fig. 11.24. Under cruise conditions the lift coefficient is 0.4. If the wing area is 10 m^2, find the drag force.

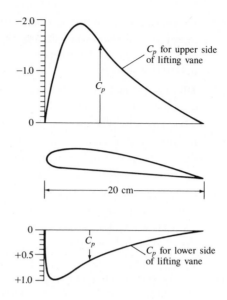

PROBLEM 11.68

11.73 An ultralight airplane has a wing with an aspect ratio of 5 and with lift and drag coefficients corresponding to Fig. 11.23. The planform area of the wing is 200 ft². The weight of the airplane and pilot is 400 lbf. The airplane flies at 50 ft/s in air with a density of 0.002 slugs/ft³. Find the angle of attack and the drag force on the wing.

References

1. Abbott, I. H. "The Drag of Two Streamline Bodies as Affected by Protuberances and Appendages." *NACA Rept.,* 451 (1932).

2. Abbott, I. H., and A. E. Von Doenhoff. *Theory of Wing Sections.* Dover Publications, New York, 1949.

3. Barkla, H. M., et al. "The Magnus or Robins Effect on Rotating Spheres." *J. Fluid Mech.,* 47, part 3 (1971).

4. Brevoort, M. J., and U. T. Joyner. "Experimental Investigation of the Robinson-Type Cup Anemometer." *NACA Rept.,* 513 (1935).

5. Bullivant, W. K. "Tests of the NACA 0025 and 0035 Airfoils in the Full Scale Wind Tunnel." *NACA Rept.,* 708 (1941).

6. Crowe, C. T., et al. "Drag Coefficient for Particles in Rarefied, Low Mach-Number Flows." In *Progress in Heat and Mass Transfer,* vol. 6, pp. 419–431. Pergamon Press, New York, 1972.

7. DeFoe, G. L. "Resistance of Streamline Wires." *NACA Tech. Note,* 279 (March 1928).

8. Fage, A., and J. H. Warsap. "The Effects of Turbulence and Surface Roughness on the Drag of a Circular Cylinder." *Rept. Mem., Brit. Aeronaut. Res. Comm.,* 1283 (October 1929).

9. Freeman, H. B. "Force Measurements on a 1/40-Scale Model of the U.S. Airship 'Akron'." *NACA Rept.,* 432 (1932).

10. Goett, H. J., and W. K. Bullivant. "Tests of NACA 0009, 0012, and 0018 Airfoils in the Full Scale Tunnel." *NACA Rept.,* 647 (1938).

11. Goldstein, S. *Modern Developments in Fluid Dynamics.* Dover Publications, New York, 1965.

12. Hoerner, S. F. *Fluid Dynamic Drag.* Published by the author, 1958.

13. Jacobs, E. N. "The Drag of Streamline Wires." *NACA Tech. Note,* 480 (December 1933).

14. Hucho, Wolf-Heinrich, ed. *Aerodynamics of Road Vehicles.* Butterworth, London, 1987.

15. Jones, G.W., Jr. "Unsteady Lift Forces Generated by Vortex Shedding About a Large, Stationary, and Oscillating Cylinder at High Reynolds Numbers." *Symp. Unsteady Flow, ASME* (1968).

16. Kirsch, J.W., and W. H. Bettes. "Feasibility Study of the S^3 Air Vane and Other Truck Drag Reduction Devices."*Proc. of Conference on Reduction of the Aerodynamic Drag of Trucks, NSF RANN* (October 1974).

17. Knight, Montgomery, and Oscar Loeser, Jr. "Pressure Distribution over a Rectangular Monoplane Wing Model up to 90° Angle of Attack." *NACA Rept.,* 288 (1928).

18. Kuethe, A. M., and J. D. Schetzer. *Foundations of Aerodynamics.* John Wiley, New York, 1967.

19. Lindsey, W. F. "Drag of Cylinders of Simple Shapes." *NACA Rept.,* 619 (1938).

20. Miller, B. L., J. F. Mayberry, and I. J. Salter. "The Drag of Roughened Cylinders at High Reynolds Numbers." *NPL Rept. MAR Sci.,* R132 (April 1975).

21. Mehta, R. D. "Aerodynamics of Sports Balls." *Annual Review of Fluid Mechanics,* 17, p. 151 (March 1985).

22. Morrison, R. B. (ed.). *Design Data for Aeronautics and Astronautics.* John Wiley, New York, 1962.

23. Prandtl, L. "Applications of Modern Hydrodynamics to Aeronautics." *NACA Rept.,* 116 (1921).

24. Roberson, J. A., et al. "Turbulence Effects on Drag of Sharp-Edged Bodies." *J. Hydraulics Div., Am. Soc. Civil Eng.* (July 1972).

25. Roshko, A. "Turbulent Wakes from Vortex Streets." *NACA Rept.,* 1191 (1954).

26. Rouse, H. *Elementary Mechanics of Fluids.* John Wiley, New York, 1946.

27. Schenkel, Franz K. "The Origins of Drag and Lift Reductions on Automobiles with Front and Rear Spoilers." *SAE Paper,* no. 770389 (February 1977).

28. Scher, S. H., and L. J. Gale. "Wind Tunnel Investigation of the Opening Characteristics, Drag, and Stability of Several Hemispherical Parachutes." *NACA Tech. Note,* 1869 (1949).

29. Scibor-Rylski, A. J. *Road Vehicle Aerodynamics.* Pentech Press, London, 1975.

30. White, R. G. S. "A Method of Estimating Automobile Drag Coefficients." *SAE Paper,* no. 690189 (1969).

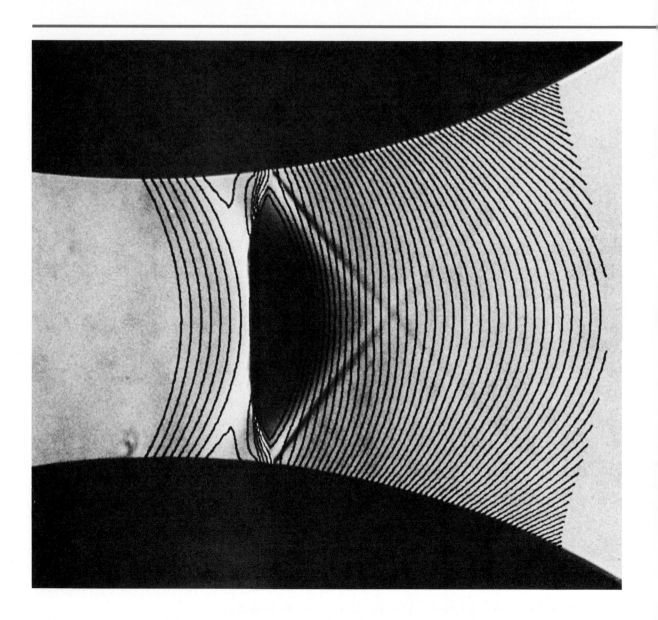

Compressible Flow

Shown here are constant Mach number lines and shock wave structure for two-dimensional transonic flow in a de Laval nozzle with heat release due to vapor condensation. This case is for an initial relative humidity of 64.8%. (Courtesy G. Schnerr, Institut für Strömungslehre and Strömungsmaschinen, Universität (TH) Karlsruhe, Germany)

U p to this point in our study of fluid flow, we have assumed that the density of the fluid is constant. This assumption is well founded for the flow of liquids because very large, atypical pressure changes effect only small density variations. For example, an increase in pressure of 20 MPa changes the density of water by less than 1%. On the other hand, pressure changes normally encountered in gas-flow problems can cause significant density changes. For example, if the same 20 MPa pressure change were applied to air initially at atmospheric pressure, its density would change by 4370%. A change of this magnitude cannot be regarded as negligible!

The variables that describe the state of a flowing liquid are velocity, pressure, and temperature. The equations available to solve for these variables derive from the conservation of mass, momentum, and energy. If, in addition, density is a variable, one more equation is necessary. This is the equation of state, which, for an ideal gas, is

$$p = \rho RT \tag{12.1}$$

and which was introduced in Chapter 2. The developments in this chapter will be based on ideal-gas relationships.

12.1 Wave Propagation in Compressible Fluids

The speed of a flowing liquid is typically much less than the speed at which a pressure disturbance is propagated through the liquid. Gas flows, on the other hand, can achieve speeds that are comparable to and even exceed the speed at which pressure disturbances are propagated. In this situation, with compressible fluids, the propagation speed is an important parameter and must be incorporated into the flow analysis. In this section we will show how the speed of an infinitesimal pressure disturbance can be evaluated and what its significance is to flow of a compressible fluid.

Speed of Sound

Everyone has had the experience during a thunderstorm of seeing lightning flash and hearing the accompanying thunder an instant later. Obviously, the sound was produced by the lightning, so the sound wave must have traveled at a finite speed. If the air were totally incompressible (if that were possible), we would hear the thunder and see the flash simultaneously, because all disturbances propagate at infinite speed through incompressible media.* It is analogous to striking one end of a bar of incompressible material and recording instantaneously the response at the other end. Actually, all materials are compressible to some degree and propagate disturbances at finite speeds.

The *speed of sound* is defined as the rate at which an *infinitesimal* disturbance (pressure pulse) propagates in a medium with respect to the frame of reference of that medium. Actual sound waves, comprised of pressure disturbances of finite amplitude, such that the ear can detect them, travel only slightly faster than the "speed of sound." We found in Chapter 6 that the speed at which a pressure wave travels through a fluid depends on the bulk modulus of the fluid and its density. We performed that analysis by considering the unsteady flow within a control volume as the wave passed through the control volume. Here we will derive an equation for the speed of sound assuming the control volume moves with the wave, thereby analyzing a steady-flow problem.

Let us consider a small section of a pressure wave as it propagates at velocity c through a medium, as depicted in Fig. 12.1. As the wave travels through the gas at pressure p and density ρ, it produces infinitesimal changes of Δp, $\Delta \rho$, and ΔV. We realize that these changes must be related through the laws of conservation of mass and momentum. Let us draw a control surface around the wave and let the control volume travel with the wave. The velocities, pressures, and densities relative to the control volume (which is assumed to be very thin) are shown in Fig. 12.2. Conservation of mass in a steady flow requires that the net mass flux across the control surface be zero. Thus

$$-\rho c A + (\rho + \Delta \rho)(c - \Delta V)A = 0 \tag{12.2}$$

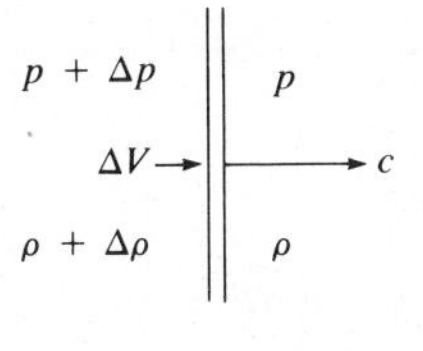

FIGURE 12.1

Section of a sound wave.

$$p + \Delta p \quad \| \quad p$$
$$\Delta V \rightarrow \quad | \quad \longrightarrow c$$
$$\rho + \Delta \rho \quad \| \quad \rho$$

Actually, the thunder would be heard before the lightning was seen, because light also travels at a finite, though very high, speed! However, this would violate one of the basic tenets of relativity theory. No medium can be completely incompressible and propagate disturbances exceeding the speed of light.

FIGURE 12.2

Flow relative to the sound wave.

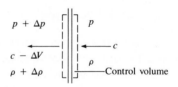

where A is the cross-sectional area of the control volume. Neglecting products of higher-order terms ($\Delta\rho\,\Delta V$) and dividing by the area reduce the conservation-of-mass equation to

$$-\rho\,\Delta V + c\,\Delta\rho = 0 \qquad (12.3)$$

The momentum equation for steady flow,

$$\sum \mathbf{F} = \sum \mathbf{v}\rho\mathbf{V} \cdot \mathbf{A} \qquad (12.4)$$

applied to the control volume containing the pressure wave gives

$$(p + \Delta p)A - pA = (-c)(-\rho Ac) + (-c + \Delta V)\rho Ac \qquad (12.5)$$

where the direction to the right is defined as positive. The momentum equation reduces to

$$\Delta p = \rho c\,\Delta V \qquad (12.6)$$

Substituting the expression for ΔV obtained from Eq. (12.3) into Eq. (12.6) gives

$$c^2 = \frac{\Delta p}{\Delta\rho} \qquad (12.7)$$

which shows how the speed of propagation is related to the pressure and density change across the wave. We immediately see from this equation that if the flow were ideally incompressible, $\Delta\rho = 0$, the propagation speed would be infinite, which confirms the argument presented earlier.

Equation (12.7) gives us an expression for the speed of a general pressure wave. The sound wave is a special type of pressure wave. By definition, a sound wave produces only infinitesimal changes in pressure and density, so it can be regarded as a reversible process. There is also negligibly small heat transfer, so one can assume the process is *adiabatic*. A reversible, adiabatic process is an *isentropic* process; thus the resulting expression for the speed of sound is

$$c^2 = \left.\frac{\partial p}{\partial\rho}\right|_s \qquad (12.8)$$

This equation is valid for the speed of sound in any substance. However, for many substances the relationship between p and ρ at constant entropy is not very well known.

To reiterate, the speed of sound is the speed at which an infinitesimal pressure disturbance travels through a fluid. Waves of finite strength (finite pressure change across the wave) travel faster than sound waves. Sound speed is the *minimum* speed at which a pressure wave can propagate through a fluid.

We shall now determine the speed of sound in an ideal gas. It can be shown from thermodynamics—see Oswatitsch (4), for example—that the following relationship between pressure and density holds for an isentropic process.

$$\frac{p}{\rho^k} = \text{constant} \tag{12.9}$$

Here k is the ratio of specific heats—that is, the ratio of specific heat at constant pressure to that at constant volume.

$$k = \frac{c_p}{c_v} \tag{12.10}$$

The values of k for some commonly used gases are given in Table A.2 in the Appendix. Taking the derivative of Eq. (12.9) to obtain $\partial p/\partial \rho|_s$ results in

$$\left.\frac{\partial p}{\partial \rho}\right|_s = \frac{kp}{\rho} \tag{12.11}$$

However, from the equation of state for an ideal gas,

$$\frac{p}{\rho} = RT$$

so the speed of sound is given by

$$c = \sqrt{kRT} \tag{12.12}$$

Thus we find that the speed of sound in an ideal gas varies with the square root of the temperature. Using this equation to predict sound speeds in real gases at standard conditions gives results very near the measured values. Of course, if the state of the gas is far removed from ideal conditions (high pressures, low temperatures), then using Eq. (12.12) is not advisable.

EXAMPLE 12.1 Calculate the speed of sound in air at 15°C.

Solution From Table A.2 we find that $k = 1.4$ and $R = 287$ J/kg K for air. Using Eq. (12.12), we calculate

$$c = [(1.4)(287 \text{ J/kg K})(288 \text{ K})]^{1/2} = 340 \text{ m/s} \quad \blacktriangleleft$$

Mach Number

We shall now demonstrate in a very simple way the significance of sound in a compressible flow. Consider the airfoil traveling at speed V in Fig. 12.3. As this airfoil travels through the fluid, the pressure disturbance generated by the airfoil's motion propagates as a wave at sonic speed ahead of the airfoil. These pressure disturbances travel a considerable distance ahead of the airfoil before being attenuated by the viscosity of the fluid, and they "warn" the upstream fluid that the airfoil is coming (the Paul Reveres of fluid flow!). In turn, the fluid particles begin to move apart in such a way that there is a smooth flow over the airfoil by the time it arrives. If a pressure disturbance created by the airfoil is essentially attenuated in time Δt, then the fluid at a distance $\Delta t(c - V)$ ahead is alerted to prepare for the airfoil's impending arrival.

What happens as the speed of the airfoil is increased? Obviously, the relative velocity $c - V$ is reduced, and the upstream fluid has less time to prepare for the airfoil's arrival. The flow field is modified by smaller streamline curvatures, and the form drag on the airfoil is increased. If the airfoil speed increases to the speed of sound or greater, the fluid has no warning whatsoever that the airfoil is coming and cannot prepare for its arrival. Nature, at this point, resolves the problem by creating a shock wave that stands off the leading edge, as shown in Fig. 12.4. As the fluid passes through the shock wave near the leading edge, it is decelerated to a speed less then sonic speed and therefore has time to divide and flow around the airfoil. Shock waves will be treated in more detail in Sec. 12.3.

Another approach to appreciating the significance of sound propagation in a compressible fluid is to consider a point source of sound moving in a quiescent fluid, as shown in Fig. 12.5. The sound source is moving at a speed less than the local sound speed in Fig. 12.5a and faster than the local sound speed in Fig. 12.5b. At time $t = 0$ a sound pulse is generated and propagates radially outward at the local speed of sound. At time t_1 the sound source has moved a distance Vt_1, and the circle representing the sound wave emitted at $t = 0$ has a radius of ct_1. The sound source emits a new sound wave at t_1 that propagates ra-

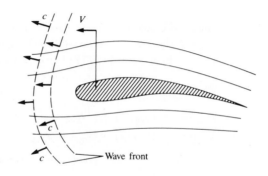

FIGURE 12.3

Propagation of a sound wave by an airfoil.

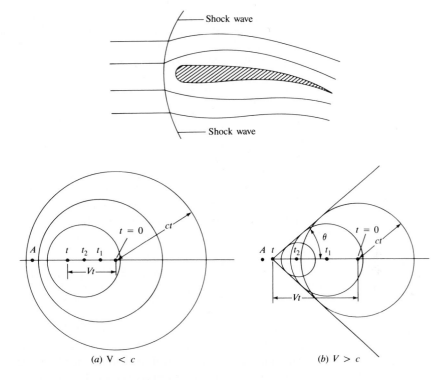

FIGURE 12.4

Standing shock wave in front of an airfoil.

FIGURE 12.5

Sound field generated by a moving point source of sound.

dially outward. At time t_2 the sound source has moved to Vt_2, and the sound waves have moved outward as shown.

When the sound source moves at a speed less than the speed of sound, the sound waves form a family of nonintersecting eccentric circles, as shown in Fig. 12.5a. For an observer stationed at A the frequency of the sound pulses would appear higher than the emitted frequency because the sound source is moving toward the observer. In fact, the observer at A will detect a frequency of

$$f = f_0/(1 - V/c)$$

where f_0 is the emitting frequency of the moving sound source. This change in frequency is known as the *Doppler effect*.

When the sound source moves faster than the local sound speed, the sound waves intersect and form the locus of a cone with a half-angle of

$$\theta = \sin^{-1}(c/V)$$

The observer at A will not detect the sound source until it has passed. In fact, only an observer within the cone is aware of the moving sound source.

In view of the physical arguments given above, it is apparent that an important parameter relating to sound propagation and compressibility effects is the ratio V/c. This parameter, already introduced in Chapter 8, was first proposed by Ernst Mach, an Austrian scientist, and bears his name. The Mach number is defined as

$$\mathrm{M} = \frac{V}{c} \tag{12.13}$$

The conical wave surface depicted in Fig. 12.5b is known as a *Mach wave* and the conical half-angle as the *Mach angle.*

Besides the heuristic argument presented above for Mach number, we also recall from Chapter 8 that the Mach number is the ratio of the inertial to elastic forces acting on the fluid. If the Mach number is small, the inertial forces are ineffective in compressing the fluid and the fluid can be regarded as incompressible.

Compressible flows are characterized by their Mach-number regimes as follows:

$$\mathrm{M} < 1 \qquad \text{Subsonic flow}$$

$$\mathrm{M} \approx 1 \qquad \text{Transonic flow}$$

$$\mathrm{M} > 1 \qquad \text{Supersonic flow}$$

Flows with Mach numbers exceeding 5 are sometimes referred to as *hypersonic.* Airplanes designed to travel near sonic speeds and faster are equipped with Mach meters because of the significance of the Mach number with respect to aircraft performance.

EXAMPLE 12.2   The Concorde is traveling at 1400 km/h at an altitude of 12,000 m, where the temperature is $-56°C$. Determine the Mach number at which the airplane is flying and characterize the flow.

Solution   First we must determine the speed of sound. Using Eq. (12.12), we calculate

$$c = \sqrt{kRT} = [(1.4)\,(287\ \mathrm{J/kg\ K})\,(217\ \mathrm{K})]^{1/2} = 295\ \mathrm{m/s}$$

The speed of the airplane in meters per second is

$$V = (1400\ \mathrm{km/h})\left(\frac{1}{3600}\ \mathrm{h/s}\right)(1000\ \mathrm{m/km}) = 389\ \mathrm{m/s}$$

The Mach number is therefore

$$\mathrm{M} = \frac{389}{295} = 1.32 \qquad \blacktriangleleft$$

and the flow is supersonic.

12.2   Mach-Number Relationships

We are already familiar with Bernoulli's equation and the energy equation and with their utility in determining fluid properties along streamlines in liquid flow. In this section, we shall learn how the Mach number is used to determine fluid properties in compressible flows. Let us consider a control volume bounded by two streamlines in a steady compressible flow, as shown in Fig. 12.6. Applying the energy equation, Eq. (7.14), to this control volume, realizing that the shaft work is zero, gives

$$-\dot{m}_1\left(h_1 + \frac{V_1^2}{2} + gz_1\right) + \dot{m}_2\left(h_2 + \frac{V_2^2}{2} + gz_2\right) = \dot{Q} \qquad (12.14)$$

As pointed out in Chapter 7, the elevation terms (z_1 and z_2) can usually be neglected for gaseous flows. If the flow is adiabatic ($\dot{Q} = 0$), the energy equation reduces to

$$\dot{m}_1\left(h_1 + \frac{V_1^2}{2}\right) = \dot{m}_2\left(h_2 + \frac{V_2^2}{2}\right) \qquad (12.15)$$

From the principle of continuity, the mass-flow rate is constant, $\dot{m}_1 = \dot{m}_2$, so

$$h_1 + \frac{V_1^2}{2} = h_2 + \frac{V_2^2}{2} \qquad (12.16)$$

Since positions 1 and 2 are arbitrary points on the same streamline, we say that

$$h + \frac{V^2}{2} = \text{constant along a streamline in an adiabatic flow} \qquad (12.17)$$

The constant in this expression is called the *total enthalpy*, h_t. It is the enthalpy that would arise if the flow velocity were brought to zero in a reversible, adia-

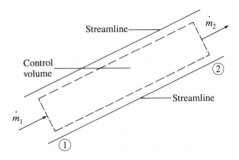

FIGURE 12.6

Control volume enclosed by streamlines.

batic process. Thus the energy equation along a streamline under adiabatic conditions is

$$h + \frac{V^2}{2} = h_t \tag{12.18}$$

If h_t is the same for all streamlines, the flow is *homenergic.*

It is instructive at this point to compare Eq. (12.18) with Bernoulli's equation. Expressing the specific enthalpy as the sum of the specific internal energy and p/ρ, Eq. (12.18) becomes

$$u + \frac{p}{\rho} + \frac{V^2}{2} = \text{constant}$$

If the fluid is incompressible and there is no heat transfer, the specific internal energy is constant and the equation reduces to Bernoulli's equation (excluding the hydrostatic pressure terms).

Temperature

The enthalpy of an ideal gas can be written as

$$h = c_p T \tag{12.19}$$

where c_p is the specific heat at constant pressure. Substituting this relation into Eq. (12.18) and dividing by $c_p T$, we obtain

$$1 + \frac{V^2}{2c_p T} = \frac{T_t}{T} \tag{12.20}$$

where T_t is the total temperature. From thermodynamics (4) we know that for an ideal gas

$$c_p - c_v = R \tag{12.21}$$

or

$$k - 1 = \frac{R}{c_v} = \frac{kR}{c_p}$$

Therefore

$$c_p = \frac{kR}{k - 1} \tag{12.22}$$

Substituting this expression for c_p back into Eq. (12.20) and realizing that kRT is the speed of sound squared result in

$$T_t = T\left(1 + \frac{k - 1}{2}\mathrm{M}^2\right) \tag{12.23}$$

The temperature T is called the *static temperature*—the temperature that would be registered by a thermometer moving with the flowing fluid. Total temperature is analogous to total enthalpy in that it is the temperature that would arise if the velocity were brought to zero isentropically. If the flow is adiabatic, the total temperature is constant along a streamline. If not, the total temperature varies according to the amount of thermal energy transferred.

EXAMPLE 12.3   An aircraft is flying at $M = 1.6$ at an altitude where the atmospheric temperature is $-50°C$. The temperature on the aircraft's surface is approximately the total temperature. Estimate the surface temperature, taking $k = 1.4$.

Solution   This problem can be visualized as the aircraft being stationary and an airstream with a static temperature of $-50°C$ flowing past the aircraft at a Mach number of 1.6. The static temperature in absolute temperature units is

$$T = 223 \text{ K}$$

Using Eq. (12.23) to calculate the total temperature gives

$$T_t = 223[1 + 0.2(1.6)^2] = 337 \text{ K or } 64°C \qquad \blacktriangleleft$$

Pressure

If the flow is isentropic, thermodynamics shows that the following relationship for pressure and temperature of an ideal gas between two points on a streamline is valid (4):

$$\frac{p_1}{p_2} = \left(\frac{T_1}{T_2}\right)^{k/(k-1)} \qquad (12.24)$$

Isentropic flow means that there is no heat transfer, so the total temperature is constant along the streamline. Therefore

$$T_t = T_1\left(1 + \frac{k-1}{2}M_1^2\right) = T_2\left(1 + \frac{k-1}{2}M_2^2\right) \qquad (12.25)$$

Solving for the ratio T_1/T_2 and substituting into Eq. (12.24) shows that the pressure variation with the Mach number is given by

$$\frac{p_1}{p_2} = \left\{\frac{1 + [(k-1)/2]M_2^2}{1 + [(k-1)/2]M_1^2}\right\}^{k/(k-1)} \qquad (12.26)$$

In that the equation of state is used to derive Eq. (12.24), absolute pressures must always be used in calculations with these equations.

The total pressure in a compressible flow is defined as

$$p_t = p\left(1 + \frac{k-1}{2}M^2\right)^{k/(k-1)} \qquad (12.27)$$

which is the pressure that would result if the flow were decelerated to zero speed reversibly and adiabatically. Unlike total temperature, total pressure may not be constant along streamlines in adiabatic flows. For example, we will discover that flow through a shock wave, though adiabatic, is not reversible and, therefore, not isentropic. The total pressure variation along a streamline in an adiabatic flow can be obtained by substituting Eqs. (12.27) and (12.25) into Eq. (12.26) to give

$$\frac{p_{t_1}}{p_{t_2}} = \frac{p_1}{p_2} \left\{ \frac{1 + [(k-1)/2]\mathrm{M}_1^2}{1 + [(k-1)/2]\mathrm{M}_2^2} \right\}^{k/(k-1)} = \frac{p_1}{p_2} \left(\frac{T_2}{T_1} \right)^{k/(k-1)} \qquad (12.28)$$

Unless the flow is also reversible and Eq. (12.24) is applicable, the total pressures at points 1 and 2 will not be equal. However, if the flow is isentropic, total pressure is constant along streamlines.

Density

Analogous to the total pressure, the total density in a compressible flow is given by

$$\rho_t = \rho \left(1 + \frac{k-1}{2} \mathrm{M}^2 \right)^{1/(k-1)} \qquad (12.29)$$

where ρ is the local or static density. If the flow is isentropic, then ρ_t is a constant along streamlines and Eq. (12.29) can be used to determine the variation of gas density with the Mach number.

In literature dealing with compressible flows, one often finds reference to "stagnation" conditions—that is, "stagnation temperature" and "stagnation pressure." By definition, *stagnation* refers to the conditions that exist at a point in the flow where the velocity is zero, regardless of whether or not the zero velocity has been achieved by an adiabatic, or reversible process. For example, if one were to insert a Pitot tube into a compressible flow, strictly speaking one would measure stagnation pressure, not total pressure, since the deceleration of the flow would not be reversible. In most cases, however, the difference between stagnation and total pressure is negligibly small.

Kinetic Pressure

The kinetic pressure, $q = \rho V^2/2$, is often used, as we have seen in Chapter 11, to calculate aerodynamic forces with the use of appropriate coefficients. It can also be related to the Mach number. Using the equation of state for an ideal gas to replace ρ gives

$$q = \frac{1}{2} \frac{p V^2}{RT} \qquad (12.30)$$

Then using the equation for the speed of sound, Eq. (12.12), results in

$$q = \frac{k}{2} p M^2 \tag{12.31}$$

where p must always be an absolute pressure since it derives from the equation of state.

EXAMPLE 12.4   The drag coefficient for a sphere at a Mach number of 0.7 is 0.95. Determine the drag force on a sphere 10 mm in diameter in air if $p = 101$ kPa.

Solution   The drag force on a sphere is

$$F_D = \frac{1}{2} \rho V^2 C_D A_p = q C_D A_p$$

where A_p is the projected area. The kinetic pressure is

$$q = \frac{1.4}{2} (101 \text{ kPa}) (0.7)^2 = 34.6 \text{ kPa}$$

The drag force is calculated to be

$$F_D = 0.95 \left(34.6 \times 10^3 \frac{\text{N}}{\text{m}^2} \right) \left(\frac{\pi}{4} \right) (10^{-2})^2 \text{ m}^2 = 2.6 \text{ N} \qquad \blacktriangleleft$$

There is one very important fact about compressible flow that must be stressed: Bernoulli's equation is not valid for compressible flows! Let us see what would happen if one decided to measure the Mach number of a high-speed air flow with a Pitot-static tube, assuming that Bernoulli's equation was valid. Let us say a total pressure of 180 kPa and a static pressure of 100 kPa were measured. By Bernoulli's equation the kinetic pressure is equal to the difference between the total and static pressures, so

$$\frac{1}{2} \rho V^2 = p_t - p \qquad \text{or} \qquad \frac{k}{2} p M^2 = p_t - p$$

Solving for the Mach number,

$$M = \sqrt{\frac{2}{k} \left(\frac{p_t}{p} - 1 \right)}$$

and substituting in the measured values, one obtains

$$M = 1.07$$

Now, what should have been done? The expression relating the total and static pressures in a compressible flow is Eq. (12.27). Solving that equation for the Mach number gives

$$M = \left\{ \frac{2}{k-1} \left[\left(\frac{p_t}{p} \right)^{(k-1)/k} - 1 \right] \right\}^{1/2} \qquad (12.32)$$

and substituting in the measured values yields

$$M = 0.96$$

Thus applying Bernoulli's equation, would have led one to say that the flow was supersonic, whereas the flow was actually subsonic. In the limit of low velocities $(p_t/p \to 1)$, Eq. (12.32) reduces to the expression derived from Bernoulli's equation, which is indeed valid for very low $(M \ll 1)$ Mach numbers.

It is instructive to see how the pressure coefficient at the stagnation (total pressure) condition varies with Mach number. The pressure coefficient is given by

$$C_p = \frac{p_t - p}{\frac{1}{2}\rho V^2}$$

Using Eq. (12.31) for the kinetic pressure enables us to express C_p as a function of Mach number and the ratio of specific heats.

$$C_p = \frac{2}{kM^2} \left[\left(1 + \frac{k-1}{2}M^2 \right)^{k/(k-1)} - 1 \right]$$

The variation of C_p with Mach number is shown in Fig. 12.7. At a Mach number of zero, the pressure coefficient is unity, which corresponds to incompressible flow. The pressure coefficient begins to depart significantly from unity at a Mach number of about 0.3. From this observation we infer that compressibility effects in the flow field are unimportant for Mach numbers less than 0.3.

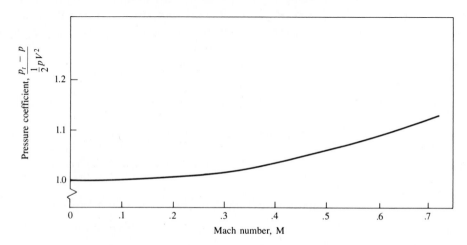

FIGURE 12.7

Variation of the pressure coefficients with Mach number.

12.3   Normal Shock Waves

Normal shock waves are wave fronts normal to the flow across which a supersonic flow is decelerated to a subsonic flow with an attendant increase in static temperature, pressure, and density. The normal shock wave is analogous to the water hammer introduced in Chapter 6 and somewhat analogous to the hydraulic jump, which will be introduced in Chapter 15.

Change in Flow Properties
Across a Normal Shock Wave

The most straightforward way to analyze a normal shock wave is to draw a control surface around the wave, as shown in Fig. 12.8, and write down the continuity, momentum, and energy equations. The net mass flux into the control volume is zero because the flow is steady. Therefore

$$-\rho_1 V_1 A + \rho_2 V_2 A = 0 \tag{12.33}$$

where A is the cross-sectional area of the control volume. Equating the net pressure forces acting on the control surface to the net efflux of momentum from the control volume gives

$$\rho_1 V_1 A(-V_1 + V_2) = (p_1 - p_2)A \tag{12.34}$$

The energy equation can be expressed simply as

$$T_{t_1} = T_{t_2} \tag{12.35}$$

because the temperature gradients on the control surface are assumed negligible and thus heat transfer is neglected (adiabatic).

Using the equation for the speed of sound, Eq. (12.12), and the equation of state for an ideal gas, the continuity equation can be rewritten to include the Mach number as follows:

$$\frac{p_1}{RT_1} M_1 \sqrt{kRT_1} = \frac{p_2}{RT_2} M_2 \sqrt{kRT_2} \tag{12.36}$$

FIGURE 12.8

Control volume enclosing a normal shock wave.

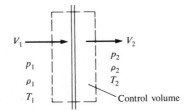

The Mach number can be introduced into the momentum equation in the following way:

$$\rho_2 V_2^2 - \rho_1 V_1^2 = p_1 - p_2$$

$$p_1 + \frac{p_1}{RT_1} V_1^2 = p_2 + \frac{p_2}{RT_2} V_2^2$$

$$p_1(1 + k\mathrm{M}_1^2) = p_2(1 + k\mathrm{M}_2^2) \qquad (12.37)$$

Rearranging Eq. (12.37) for the static-pressure ratio across the shock wave results in

$$\frac{p_2}{p_1} = \frac{(1 + k\mathrm{M}_1^2)}{(1 + k\mathrm{M}_2^2)} \qquad (12.38)$$

As we shall show later, the Mach number of a normal shock wave is always greater than unity upstream and less than unity downstream, so the static pressure always increases across a shock wave.

We can rewrite the energy equation in terms of the temperature and Mach number, as we did in Eq. (12.23), by utilizing the fact that $T_{t_2}/T_{t_1} = 1$:

$$\frac{T_2}{T_1} = \frac{\{1 + [(k - 1)/2]\mathrm{M}_1^2\}}{\{1 + [(k - 1)/2]\mathrm{M}_2^2\}} \qquad (12.39)$$

Substituting Eqs. (12.38) and (12.39) into Eq. (12.36) yields the following relationship for the Mach numbers upstream and downstream of a normal shock wave:

$$\frac{\mathrm{M}_1}{1 + k\mathrm{M}_1^2}\left(1 + \frac{k - 1}{2}\mathrm{M}_1^2\right)^{1/2} = \frac{\mathrm{M}_2}{1 + k\mathrm{M}_2^2}\left(1 + \frac{k - 1}{2}\mathrm{M}_2^2\right)^{1/2} \quad (12.40)$$

Then, solving this equation for M_2 as a function of M_1, we obtain two solutions. One solution is trivial, $\mathrm{M}_1 = \mathrm{M}_2$, which corresponds to no shock wave in the control volume. The other solution is

$$\mathrm{M}_2^2 = \frac{(k - 1)\mathrm{M}_1^2 + 2}{2k\mathrm{M}_1^2 - (k - 1)} \qquad (12.41)$$

Note: Because of the symmetry of Eq. (12.40), we can also use Eq. (12.41) to solve for M_1 given M_2, by simply interchanging the subscripts on the Mach numbers.

Setting $\mathrm{M}_1 = 1$ in Eq. (12.41) results in M_2 also being equal to unity. Equations (12.38) and (12.39) also show that there would be no pressure or temperature increase across such a wave. In fact, the wave corresponding to $\mathrm{M}_1 = 1$ is the sound wave across which, by definition, pressure and temperature changes are infinitesimal. Thus the sound wave represents a degenerate normal shock wave.

EXAMPLE 12.5 A normal shock wave occurs in air flowing at a Mach number of 1.5. The static pressure and temperature of the air upstream of the shock wave are 100 kPa absolute and 15°C. Determine the Mach number, pressure, and temperature downstream of the shock wave.

Solution We use Eq. (12.41) to calculate the Mach number downstream of the shock wave:

$$M_2^2 = \frac{(0.4)(1.5)^2 + 2}{(2.8)(1.5)^2 - 0.4} = 0.49$$

$$M_2 = 0.7 \qquad \blacktriangleleft$$

Equations (12.38) and (12.39) provide the downstream pressure and temperature:

$$p_2 = p_1 \left(\frac{1 + kM_1^2}{1 + kM_2^2} \right)$$

$$= (100 \text{ kPa}) \left[\frac{1 + (1.4)(1.5)^2}{1 + (1.4)(0.7)^2} \right] = 246 \text{ kPa, absolute} \qquad \blacktriangleleft$$

$$T_2 = T_1 \left\{ \frac{1 + [(k-1)/2]M_1^2}{1 + [(k-1)/2]M_2^2} \right\}$$

$$= (288 \text{ K}) \left[\frac{1 + (0.2)(2.25)}{1 + (0.2)(0.49)} \right] = 380 \text{ K or } 107°\text{C} \qquad \blacktriangleleft$$

Note that absolute pressures and temperatures must always be used in carrying out these calculations. The changes in flow properties across a shock wave are presented in Table A.1 in the Appendix for a gas, such as air, for which $k = 1.4$.

A shock wave is an adiabatic process in which no shaft work is done. Thus for ideal gases the total temperature (and total enthalpy) is unchanged across the wave. The total pressure, however, does change across a shock wave. The total pressure upstream of the wave in Example 12.5 is

$$p_{t_1} = p_1 \left(1 + \frac{k-1}{2} M_1^2 \right)^{k/(k-1)}$$

$$= 100 \text{ kPa}[1 + (0.2)(2.25)]^{3.5} = 367 \text{ kPa}$$

The total pressure downstream of the same wave is

$$p_{t_2} = p_2 \left(1 + \frac{k-1}{2} M_2^2 \right)^{k/(k-1)}$$

$$= 246 \text{ kPa}[1 + (0.2)(0.49)]^{3.5} = 341 \text{ kPa}$$

Thus we see that the total pressure decreases through the wave, which, as we will see later, is because the flow through the shock wave is not an isentropic process. Total pressure remains constant along streamlines only in isentropic flow. Values for the ratio of total pressure across a normal shock wave are also provided in Table A.1 in the Appendix.

Existence of Shock Waves Only in Supersonic Flows

Let us look back at Eq. (12.41), which gives the Mach number downstream of a normal shock wave. If one were to substitute a value for M_1 less than unity, it is easy to see that one would obtain a value for M_2 larger than unity. For example, if $M_1 = 0.5$ in air, then

$$M_2^2 = \frac{(0.4)(0.5)^2 + 2}{(2.8)(0.5)^2 - 0.4}$$

$$M_2 = 2.65$$

Is it possible to have a shock wave in a subsonic flow across which the Mach number becomes supersonic? We also find that the total pressure would increase across such a wave; that is,

$$\frac{p_{t_2}}{p_{t_1}} > 1$$

The existence of such a wave would be a significant scientific discovery!

The only way to determine whether such a solution is possible is to invoke the second law of thermodynamics, which states that for any process the entropy of the universe must remain unchanged or increase.

$$\Delta s_{univ} \geq 0 \qquad (12.42)$$

Because the shock wave is an adiabatic process, there is no change in the entropy of the surroundings; thus the entropy of the system must remain unchanged or increase.

$$\Delta s_{sys} \geq 0 \qquad (12.43)$$

The entropy change of an ideal gas between pressures p_1 and p_2 and temperatures T_1 and T_2 is given by Van Wylen and Sonntag (9):

$$\Delta s_{1 \to 2} = c_p \ln \frac{T_2}{T_1} - R \ln \frac{p_2}{p_1} \qquad (12.44)$$

Using the relationship between c_p and R, Eq. (12.22), we can express the entropy change as

$$\Delta s_{1 \to 2} = R \ln \left[\frac{p_1}{p_2} \left(\frac{T_2}{T_1} \right)^{k/(k-1)} \right] \qquad (12.45)$$

Note that the quantity in the square brackets is simply the total pressure ratio as given by Eq. (12.28). Therefore the entropy change across a shock wave can be rewritten as

$$\Delta s = R \ln \frac{p_{t_1}}{p_{t_2}} \qquad (12.46)$$

A shock wave across which the Mach number changes from subsonic to supersonic would give rise to a total pressure ratio less than unity and a corresponding decrease in entropy,

$$\Delta s_{sys} < 0$$

which violates the second law of thermodynamics. Therefore shock waves can exist only in supersonic flow.

The total pressure ratio approaches unity for sound waves, which conforms with the definition that they are isentropic ($\ln 1 = 0$).

EXAMPLE 12.6   Find the entropy increase across the shock wave considered in Example 12.5.

Solution
$$\Delta s = R \ln \frac{p_{t_1}}{p_{t_2}}$$

$$p_{t_1} = 367 \text{ kPa} \qquad \text{and} \qquad p_{t_2} = 341 \text{ kPa}$$

$$\Delta s = (287 \text{ J/kg K}) \left(\ln \frac{367}{341} \right) = 21 \text{ J/kg K} \qquad \blacktriangleleft$$

More examples of shock waves will be given in the next section. We shall conclude this section by qualitatively discussing other features of shock waves.

Besides the normal shock waves studied here, there are oblique shock waves that are inclined with respect to the flow direction. Let us look once again at the shock-wave structure in front of a blunt body, as depicted qualitatively in Fig. 12.9. The portion of the shock wave immediately in front of the body behaves like a normal shock wave. As the shock wave bends in the free-stream direction, oblique shock waves result. The same relationships derived above for the normal shock waves are valid for the velocity components normal to oblique waves. The oblique shock waves continue to bend in the downstream direction until the Mach number of the velocity component normal to the wave is unity. Then the oblique shock has degenerated into a so-called Mach wave across which changes in flow properties are infinitesimal.

FIGURE 12.9

*Shock-wave structure in
front of a blunt body.*

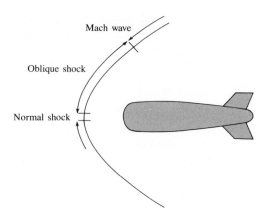

FIGURE 12.9

*Shock-wave structure in
front of a blunt body.*

The familiar sonic booms are the result of weak oblique shock waves that reach ground level. One can appreciate the damage that would ensue from stronger oblique shock waves if aircraft were permitted to travel at supersonic speeds near ground level.

Isentropic Compressible Flow
12.4 Through a Duct with Varying Area

We are already familiar with incompressible flow through ducts of varying cross-sectional area, such as the venturi tube. As the flow approaches the throat (smallest area), the velocity increases and the pressure decreases; then as the area again increases, the velocity decreases. The same velocity–area relationship is not always found for compressible flows.

Dependence of the Mach
Number on Area Variation

Consider the duct of varying area shown in Fig. 12.10. It is assumed that the flow is isentropic and that the flow properties at each section are uniform. This

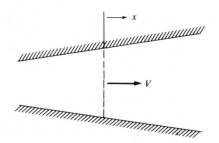

FIGURE 12.10

Duct with variable area.

type of analysis, in which the flow properties are assumed to be uniform at each section yet in which the cross-sectional area is allowed to vary (nonuniform), is still classified as "one-dimensional."

The mass flow through the duct is given by

$$\dot{m} = \rho A V \tag{12.47}$$

where A is the duct's cross-sectional area. Since the mass flow is constant along the duct, we have

$$\frac{d\dot{m}}{dx} = \frac{d(\rho A V)}{dx} = 0 \tag{12.48}$$

which can be written as*

$$\frac{1}{\rho}\frac{d\rho}{dx} + \frac{1}{A}\frac{dA}{dx} + \frac{1}{V}\frac{dV}{dx} = 0 \tag{12.49}$$

The flow is assumed to be inviscid, so Euler's equation for steady flow is applicable:

$$\rho V\frac{dV}{dx} + \frac{dp}{dx} = 0 \tag{12.50}$$

Making use of Eq. (12.8), which relates $dp/d\rho$ to the speed of sound in an isentropic flow, gives

$$\frac{-V}{c^2}\frac{dV}{dx} = \frac{1}{\rho}\frac{d\rho}{dx} \tag{12.51}$$

This equation is now used to eliminate ρ in Eq. (12.49). The result is

$$\frac{1}{V}\frac{dV}{dx} = \frac{(1/A)(dA/dx)}{M^2 - 1} \tag{12.52}$$

which, though simple, leads to the following important, far-reaching conclusions.

Subsonic Flow

For subsonic flow, $M^2 - 1$ is negative, which means that a decreasing area leads to an increasing velocity, and, correspondingly, an increasing area leads to a decreasing velocity. This velocity–area relationship agrees with our experience relating to flow through pipes with section changes.

* This step can easily be seen by first taking the logarithm of Eq. (12.47):

$$\ln(\rho A V) = \ln \rho + \ln A + \ln V$$

and then taking the derivative of each term:

$$\frac{d}{dx}[\ln(\rho A V)] = 0 = \frac{1}{\rho}\frac{d\rho}{dx} + \frac{1}{A}\frac{dA}{dx} + \frac{1}{V}\frac{dV}{dx}$$

Supersonic Flow

For supersonic flow, $M^2 - 1$ is positive, so a decreasing area leads to a decreasing velocity, and an increasing area leads to an increasing velocity. Thus the velocity at the minimum area of a duct with supersonic compressible flow is a minimum. This is the principle underlying the operation of diffusers on jet engines for supersonic aircraft, as shown in Fig. 12.11. The purpose of the diffuser is to decelerate the flow so that there is sufficient time for combustion in the chamber. Then the diverging nozzle accelerates the flow again to achieve a larger kinetic energy of the exhaust gases and an increased engine thrust.

FIGURE 12.11

Engine for supersonic aircraft.

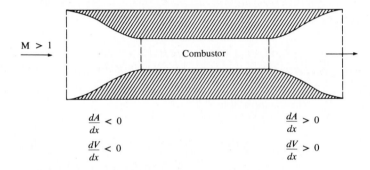

$$\frac{dA}{dx} < 0 \qquad\qquad \frac{dA}{dx} > 0$$

$$\frac{dV}{dx} < 0 \qquad\qquad \frac{dV}{dx} > 0$$

Transonic Flow (M ≈ 1)

Stations along a duct corresponding to $dA/dx = 0$ represent either a local minimum or a local maximum in the duct's cross-sectional area, as illustrated in Fig. 12.12. If at these stations the flow were either subsonic ($M < 1$) or supersonic ($M > 1$), then by Eq. (12.52) $dV/dx = 0$, so the flow velocity would have either a maximum or a minimum value. In particular, if the flow were supersonic through the duct of Fig. 12.12a, then the velocity would be a minimum at the throat; if subsonic, a maximum.

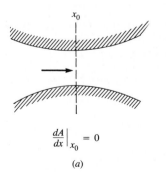

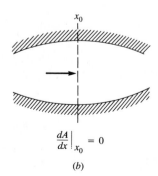

FIGURE 12.12

Duct contours for which dA/dx is zero.

$$\frac{dA}{dx}\Big|_{x_0} = 0 \qquad\qquad\qquad \frac{dA}{dx}\Big|_{x_0} = 0$$

(a)   (b)

Now, what happens if the Mach number is unity? Equation (12.52) tells us that if the Mach number is unity and dA/dx is not equal to zero, the velocity gradient dV/dx is infinite—a physically impossible situation. Therefore, dA/dx must be zero where the Mach number is unity in order for a finite, physically reasonable velocity gradient to exist.*

We can argue one step further here to show that sonic flow can occur only at a minimum area. Consider Fig. 12.12*a*. If the flow is initially subsonic, the converging duct accelerates the flow toward a sonic velocity. If the flow is initially supersonic, the converging duct decelerates the flow toward a sonic velocity. Using this same reasoning, one can prove that sonic flow is impossible in the duct depicted in Fig. 12.12*b*. If the flow is initially supersonic, the diverging duct increases the Mach number even more. If the flow is initially subsonic, the diverging duct decreases the Mach number; thus sonic flow cannot be achieved at a maximum area. Hence the Mach number in a duct of varying cross-sectional area can be unity only at a local area minimum (throat). This does not imply, however, that the Mach number must always be unity at a local area minimum.

Laval Nozzle

The Laval nozzle is a duct of varying area that produces supersonic flow. The nozzle is named after its inventor, de Laval (1845–1913), a Swedish engineer. According to the foregoing discussion, the nozzle must consist of a converging section to accelerate the subsonic flow, a throat section for transonic flow, and a diverging section to further accelerate the supersonic flow. Thus the shape of the Laval nozzle is as shown in Fig. 12.13.

One very important application of the Laval nozzle is the supersonic wind tunnel, which has been an indispensable tool in the development of supersonic aircraft. Basically, the wind tunnel, as illustrated in Fig. 12.14, consists of a high-pressure source of gas, a Laval nozzle to produce supersonic flow, and a test section. The high-pressure source may be from a large pressure tank, which is connected to the Laval nozzle through a regulator valve to maintain a constant

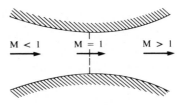

FIGURE 12.13

Laval nozzle.

*Actually, the velocity gradient is indeterminate because the numerator and denominator are both zero. It can be shown by application of L'Hôpital's rule, however, that the velocity gradient is finite.

FIGURE 12.14

Wind tunnel.

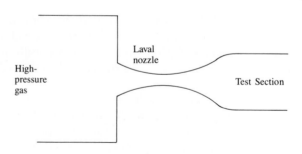

upstream pressure, or from a pumping system, which provides a continuous high pressure supply of gas.

The equations relating to the compressible flow through a Laval nozzle have already been developed. Since the mass-flow rate is the same at every cross section along the nozzle, we have

$$\rho VA = \text{constant}$$

and the constant is usually evaluated corresponding to those conditions that exist when the Mach number is unity. Thus

$$\rho VA = \rho_* A_* V_* \tag{12.53}$$

where the asterisk signifies conditions wherein the Mach number is equal to unity. Rearranging Eq. (12.53) gives

$$\frac{A}{A_*} = \frac{\rho_* V_*}{\rho V}$$

However, the velocity is the product of the Mach number and the local speed of sound. Therefore

$$\frac{A}{A_*} = \frac{\rho_*}{\rho} \frac{M_* \sqrt{kRT_*}}{M\sqrt{kRT}} \tag{12.54}$$

By definition $M_* = 1$, so

$$\frac{A}{A_*} = \frac{\rho_*}{\rho} \left(\frac{T_*}{T} \right)^{1/2} \frac{1}{M} \tag{12.55}$$

Because the flow in a Laval nozzle is assumed to be isentropic, the total temperature and total pressure (and total density) are constant throughout the nozzle. From Eq. (12.29), we have

$$\frac{\rho_*}{\rho} = \left\{ \frac{1 + [(k-1)/2]M^2}{(k+1)/2} \right\}^{1/(k-1)}$$

and from Eq. (12.25), the temperature ratio is given by

$$\frac{T_*}{T} = \frac{1 + [(k-1)/2]M^2}{(k+1)/2}$$

Substituting these expressions into Eq. (12.55) yields the following relationship between area and Mach number in a Laval nozzle:

$$\frac{A}{A_*} = \frac{1}{M}\left\{\frac{1 + [(k-1)/2]M^2}{(k+1)/2}\right\}^{(k+1)/2(k-1)} \qquad (12.56)$$

This equation is valid, of course, for all Mach numbers—subsonic, transonic, and supersonic. The area ratio A/A_* is the ratio of the area at the station where the Mach number is M to the area where M is equal to unity. Many supersonic wind tunnels are designed to maintain the same test-section area and to vary the Mach number by varying the throat area.

EXAMPLE 12.7 Suppose we are designing a supersonic wind tunnel to operate with air at a Mach number of 3. If the throat area is 10 cm², what must the cross-sectional area of the test section be?

Solution Putting $k = 1.4$ for air and $M = 3$ in Eq. (12.56) gives

$$\frac{A}{A_*} = \frac{1}{3}\left[\frac{1 + (0.2)3^2}{1.2}\right]^3 = 4.23$$

Thus the area of the test section must be 42.3 cm². ◀

Example 12.7 demonstrates that it is a straightforward task to calculate the area ratio given the Mach number and ratio of specific heats. However, in practice, one usually knows the area ratio and wishes to determine the Mach number. It is not possible to solve Eq. (12.56) for the Mach number as an explicit function of the area ratio. For this reason, compressible flow tables have been developed that allow one to obtain the Mach number easily given the area ratio.

Let us look again at Table A.1 in the Appendix. This table has been developed for a gas, such as air, for which $k = 1.4$. The symbols that head each column are defined at the beginning of the table. Tables for both subsonic and supersonic flow are provided.

EXAMPLE 12.8 A wind tunnel using air has an area ratio of 10. The absolute total pressure and temperature are 4 MPa and 350 K. Find the Mach number, pressure, temperature, and air velocity in the test section.

Solution From the table for supersonic flow we find that the Mach number must be between 3.5 and 4.0. Interpolating between the two points

M	A/A_*
3.5	6.79
4.0	10.72

gives M = 3.91 at $A/A_* = 10.0$ ◀

The compressible flow tables can also be used to interpolate for T/T_t and p/p_t, or these ratios can be calculated using Eqs. (12.23) and (12.27). The results are

$$\frac{p}{p_t} = 0.00743 \qquad \text{and} \qquad \frac{T}{T_t} = 0.246$$

In the test section,

$$p = 29.7 \text{ kPa} \qquad \text{and} \qquad T = 86 \text{ K} \qquad \blacktriangleleft$$

The velocity in the test section is obtained from

$$V = Mc = M\sqrt{kRT} = 727 \text{ m/s} \qquad \blacktriangleleft$$

Mass-Flow Rate through a Laval Nozzle

An important consideration in the design of a supersonic wind tunnel is size. A large wind tunnel requires a large mass-flow rate, which, in turn, requires a large pumping system for a continuous-flow tunnel or a large tank for sufficient run time in an intermittent tunnel. The easiest station at which to calculate the mass-flow rate is the throat, because at this station the Mach number is unity.

$$\dot{m} = \rho_\ast A_\ast V_\ast = \rho_\ast A_\ast \sqrt{kRT_\ast}$$

It is more convenient, however, to express the mass flow in terms of total conditions. The local density and static temperature at sonic velocity are related to the total density and temperature by

$$\frac{T_\ast}{T_t} = \left(\frac{2}{k+1}\right)$$

$$\frac{\rho_\ast}{\rho_t} = \left(\frac{2}{k+1}\right)^{1/(k-1)}$$

which, when substituted into the foregoing equation, give

$$\dot{m} = \rho_t \sqrt{kRT_t} A_\ast \left(\frac{2}{k+1}\right)^{(k+1)/2(k-1)} \tag{12.57}$$

Usually, the total pressure and temperature are known. Using the equation of state for an ideal gas to eliminate ρ_t, we have

$$\dot{m} = \frac{p_t A_\ast}{\sqrt{RT_t}} k^{1/2} \left(\frac{2}{k+1}\right)^{(k+1)/2(k-1)} \tag{12.58}$$

For gases with a ratio of specific heats of 1.4,

$$\dot{m} = 0.685 \frac{p_t A_*}{\sqrt{RT_t}} \tag{12.59}$$

and for gases with $k = 1.67$,

$$\dot{m} = 0.727 \frac{p_t A_*}{\sqrt{RT_t}} \tag{12.60}$$

EXAMPLE 12.9   A supersonic wind tunnel with a square test section 15 cm by 15 cm is being designed to operate at a Mach number of 3 using air. The static temperature and pressure in the test section are $-20°C$ and 50 kPa. Calculate the mass-flow rate.

Solution   From Example 12.7 the area ratio for a Mach-3 wind tunnel is 4.23. Thus the area of the throat must be

$$A_* = \frac{225}{4.23} = 53.2 \text{ cm}^2 = 0.00532 \text{ m}^2$$

The total pressure is obtained from Eq. (12.27).

$$p_t = p\left(1 + \frac{k-1}{2}M^2\right)^{k/(k-1)} = 50(36.7) = 1836 \text{ kPa} = 1.836 \text{ MPa}$$

The total temperature is

$$T_t = T\left(1 + \frac{k-1}{2}M^2\right) = 253(2.8) = 708 \text{ K}$$

Finally the mass-flow rate from Eq. (12.59) is

$$\dot{m} = \frac{(0.685)[1.836(10^6 \text{ N/m}^2)](0.00532 \text{ m}^2)}{[(287 \text{ J/kg K})(708 \text{ K})]^{1/2}} = 14.8 \text{ kg/s} \quad \blacktriangleleft$$

A pump capable of moving air at this rate against a 1.8-MPa pressure would require over 6000 kW of power input. Such a system would be large and costly to build and to operate.

Classification of Nozzle Flow by Exit Conditions

Let us now take a qualitative look at the pressure distribution in a Laval nozzle. Consider the Laval nozzle depicted in Fig. 12.15 with the corresponding pressure and Mach-number distributions plotted beneath it. The pressure at the nozzle entrance is very near the total pressure, because the Mach number is small. As the area decreases toward the throat, the Mach number increases and

FIGURE 12.15

*Distributions of static
pressure and Mach number
in a Laval nozzle.*

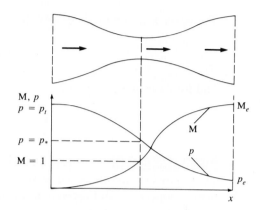

the pressure decreases. The static-to-total-pressure ratio at the throat, where conditions are sonic, is called the *critical pressure ratio.* It has a value of

$$\frac{p_\star}{p_t} = \left(\frac{2}{k+1}\right)^{k/(k-1)}$$

which for air is

$$\frac{p_\star}{p_t} = 0.528$$

It is called a critical pressure ratio because to achieve sonic flow with air in a nozzle, it is necessary that the exit pressure be at least less than 0.528 of the total pressure. The pressure continues to decrease until it reaches the exit pressure that corresponds to the nozzle exit-area ratio. Similarly, the Mach number monotonically increases with distance down the nozzle.

Now what happens if the nozzle exit pressure p_e is different from the back pressure (the pressure to which the nozzle exhausts)? If the exit pressure is higher than the back pressure, an expansion wave exists at the nozzle exit, as shown in Fig. 12.16a. These waves, which will not be studied here, effect a turning and further acceleration of the flow to achieve the back pressure. As one watches the exhaust of a rocket motor as it rises through the ever-decreasing pressure of higher altitudes, one can see the plume fan out as the flow turns more to achieve the lower pressure. A nozzle for which the exit pressure is larger than the back pressure is called an *underexpanded nozzle* because the flow could have expanded further.

If the exit pressure is less than the back pressure, shock waves occur. If the exit pressure is only slightly less than the back pressure, then pressure equalization can be obtained by oblique shock waves at the nozzle exit, as shown in Fig. 12.16b.

If, however, the difference between back pressure and exit pressure is larger than can be accommodated by oblique shock waves, a normal shock wave

FIGURE 12.16

Conditions at a nozzle exit. (a) Expansion waves. (b) Oblique shock waves. (c) Normal shock wave.

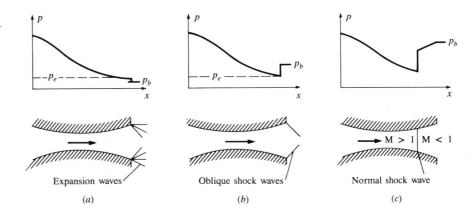

Expansion waves	Oblique shock waves	Normal shock wave
(a)	(b)	(c)

will occur in the nozzle, as shown in Fig. 12.16c. A pressure jump occurs across the normal shock wave. The flow becomes subsonic and decelerates in the remaining portion of the diverging section in such a way that the exit pressure is equal to the back pressure. As the back pressure is further increased, the shock wave moves toward the throat region until, finally, there is no region of supersonic flow. A nozzle in which the exit pressure corresponding to the exit-area ratio of the nozzle is less than the back pressure is called an *overexpanded nozzle*. Any flow that exits from a duct (or pipe) subsonically must always exit at the local back pressure.

A nozzle with supersonic flow in which the exit pressure is equal to the back pressure is *ideally expanded*.

EXAMPLE 12.10 The total pressure in a nozzle with an area ratio (A/A_*) of 4 is 1.3 MPa. Air is flowing through the nozzle. If the back pressure is 100 kPa, is the nozzle overexpanded, ideally expanded, or underexpanded?

Solution Interpolating between two area-ratio values in the supersonic flow part of Table A.1 of the Appendix:

M	A/A_*
2.90	3.850
3.0	4.235

gives M = 2.94 at $A/A_* = 4.0$. The corresponding pressure ratio is

$$\frac{p}{p_t} = 0.0298$$

so
$$p = 38.7 \text{ kPa}$$

Therefore the nozzle is overexpanded. ◄

EXAMPLE 12.11 The Laval nozzle shown in the figure has an expansion ratio of 4 (exit area/throat area). Air flows through the nozzle, and a normal shock wave occurs where the area ratio is 2. The total pressure upstream of the shock is 1 MPa. Determine the static pressure at the exit.

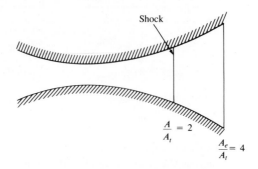

Solution From the supersonic-flow part of Table A.1, we find that the Mach number corresponding to an area ratio of 2 is 2.20. From the same line in that table, we find that the Mach number downstream of the shock is 0.547 and the ratio of total pressures across the shock wave is

$$\frac{p_{t_2}}{p_{t_1}} = 0.6281$$

Thus the total pressure downstream of the wave is

$$p_{t_2} = 0.6281 \times 1 \text{ MPa} = 628 \text{ kPa}$$

From the subsonic part of Table A.1, the value for $A/A_\star$ corresponding to the Mach number behind the shock wave is 1.26. Thus the ratio of nozzle area to throat area that would be needed to develop sonic flow downstream of the shock wave (if that were to be done) is

$$\frac{A_\star}{A_t} = \frac{A/A_t}{A/A_\star} = \frac{2}{1.26} = 1.59$$

Hence the ratio of the exit area to the area needed to develop sonic flow downstream of the shock wave is

$$\frac{A_e}{A_\star} = \frac{A_e/A_t}{A_\star/A_t} = \frac{4}{1.59} = 2.52$$

From Table A.1, the subsonic Mach number corresponding to this area ratio is

$$M_e = 0.24$$

The static pressure corresponding to this Mach number and a total pressure of 628 kPa is obtained from Eq. (12.27):

$$p_e = \frac{628 \text{ kPa}}{[1 + (0.2)(0.24)^2]^{3.5}} = 603 \text{ kPa} \qquad \blacktriangleleft$$

Mass Flow through a Truncated Nozzle

The *truncated nozzle* is a Laval nozzle cut off at the throat, as shown in Fig. 12.17. The nozzle exits to a back pressure p_b. This type of nozzle is important to engineers because of its frequent use as a flow-metering device for compressible flows.

To calculate the mass flow, we must first determine whether the flow at the exit is sonic or subsonic. Of course, the flow at the exit could never be supersonic, since the nozzle area does not diverge. First we calculate the value of the critical pressure ratio

$$\frac{p_\star}{p_t} = \left(\frac{2}{k + 1}\right)^{k/(k-1)}$$

which, for air, is 0.528. We then evaluate the ratio of back pressure to total pressure, p_b/p_t, and compare it with the critical pressure ratio:

1. If $p_b/p_t < p_\star/p_t$, the exit pressure is higher than the back pressure, so the exit flow must be sonic. Pressure equilibration is achieved after exit by a series of expansion waves. The mass flow is calculated using Eq. (12.58), where $A_\star$ is the area at the truncated station.

2. If $p_b/p_t > p_\star/p_t$, the flow exits subsonically. If we were to irrationally assume that the flow exited at the speed of sound, then the exit pressure $p_\star$ would be less than p_b. There can be no shock waves in a sonic flow (only sound waves) to raise the exit pressure to the back pressure. Therefore, the flow adjusts itself to the back pressure by exiting subsonically.

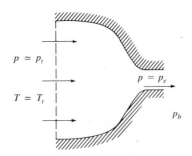

FIGURE 12.17

Truncated nozzle.

In case 2, one must first determine the Mach number at the exit by using Eq. (12.32):

$$M_e = \sqrt{\frac{2}{k-1}\left[\left(\frac{p_t}{p_b}\right)^{(k-1)/k} - 1\right]}$$

Then, using this value for Mach number, one calculates the static temperature and speed of sound at the exit:

$$T_e = \frac{T_t}{\{1 + [(k-1)/2]M_e^2\}}$$

$$c_e = \sqrt{kRT_e}$$

The gas density at the nozzle exit is determined by using the exit temperature and back pressure

$$\rho_e = \frac{p_b}{RT_e}$$

Finally, the mass flow is given by

$$\dot{m} = \rho_e A_e M_e c_e$$

where A_e is the area at the truncated section.

EXAMPLE 12.12   Air exhausts through a truncated nozzle 3 cm in diameter from a reservoir at a pressure of 160 kPa and a temperature of 80°C. Calculate the mass-flow rate if the back pressure is 100 kPa.

Solution   First we must determine the nature of the flow at the nozzle exit by evaluating the pressure ratio $p_b/p_t = 100/160 = 0.625$. Because 0.625 is larger than the critical pressure ratio for air (0.528), the flow at the nozzle exit must be subsonic. The Mach number is

$$M_e^2 = \frac{2}{k-1}\left[\left(\frac{p_t}{p_b}\right)^{(k-1)/k} - 1\right]$$

$$M_e = 0.85$$

The static temperature at the exit is

$$T_e = \frac{T_t}{\{1 + [(k-1)/2]M_e^2\}} = 308 \text{ K}$$

Correspondingly, the density at the exit is

$$\rho_e = \frac{p_b}{RT_e} = \frac{100 \times 10^3 \text{ N/m}^2}{(287 \text{ J/kg K})(309 \text{ K})} = 1.13 \text{ kg/m}^3$$

The speed of sound at the exit is

$$c_e = [(1.4)(287 \text{ J/kg K})(309 \text{ K})]^{1/2} = 352 \text{ m/s}$$

Finally, we calculate the mass-flow rate to be

$$\dot{m} = (1.13 \text{ kg/m}^3)(0.785)(0.03^2 \text{ m}^2)(0.85)(352 \text{ m/s}) = 0.239 \text{ kg/s} \quad \blacktriangleleft$$

Had p_b/p_t been less than 0.528, then we would have used Eq. (12.58) to calculate the mass-flow rate.

12.5   Compressible Flow in a Pipe with Friction

The flow of liquid through a pipe was studied in Chapter 10. The analysis of compressible flow in a pipe is somewhat more difficult because of the dependence of density and pressure on temperature. The problem is also complicated by the fact that wall friction and heat transfer cannot be simply combined into a single head-loss parameter because of their distinct effect on the Mach-number distribution along the pipe. In most engineering problems, however, the effect of wall friction is the most significant parameter. Thus it will be studied here. Two problems will be considered: first, flow in an insulated pipe in which the fluid is treated as an adiabatic system, and second, an isothermal flow that approximates flow in long pipelines.

Adiabatic Flow

The conservation-of-mass equation for uniform flow through a constant-area duct is

$$\rho V = \text{constant}$$

which, expressed in differential form, becomes

$$\frac{dV}{V} + \frac{d\rho}{\rho} = 0 \tag{12.61}$$

The conservation-of-energy equation, Eq. (12.17), can be written as

$$h + \frac{V^2}{2} = \text{constant}$$

since the flow is adiabatic. Using the relationships between enthalpy and temperature for an ideal gas, Eqs. (12.19) and (12.22), we can rewrite the energy equation in differential form as follows:

$$\frac{kR\,dT}{k-1} + V\,dV = 0 \tag{12.62}$$

The conservation-of-momentum equation can be obtained by applying the momentum equation to a control volume of length Δx contained in a pipe, as shown in Fig. 12.18. Equating the forces acting on the system to the net efflux of momentum from the control volume results in

$$A[p - (p + \Delta p)] - \tau_0 C \Delta x = \rho V A (-V + V + \Delta V) \qquad (12.63)$$

where C is the circumference of the pipe and τ_0 is the shear stress at the wall. Introducing the Darcy–Weisbach resistance coefficient for τ_0, Eq. (10.21),

$$\tau_0 = \frac{f \rho V^2}{8}$$

and simplifying, we can rewrite the momentum equation in differential form as follows:

$$\rho V dV + dp + \frac{f \rho V^2 dx}{2D} = 0 \qquad (12.64)$$

where D is the pipe diameter.

FIGURE 12.18

Control volume in a pipe.

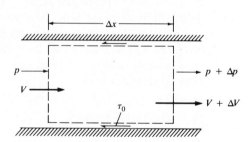

Mach-Number Distribution Along a Pipe

Our goal now is to combine the foregoing conservation equations with the equation of state, Eq. (12.1), to obtain an expression for the Mach-number distribution along a pipe. Dividing each term in Eq. (12.64) by the pressure p, and realizing that

$$\frac{p}{\rho} = RT = \frac{c^2}{k}$$

from Eqs. (12.1) and (12.2), we obtain

$$k M^2 \frac{dV}{V} + \frac{dp}{p} + \frac{k f M^2 dx}{2D} = 0 \qquad (12.65)$$

The equation of state can be written in differential form:

$$\frac{dp}{p} = \frac{d\rho}{\rho} + \frac{dT}{T} \qquad (12.66)$$

Using the continuity equation, Eq. (12.61), to replace ρ by V and the energy equation, Eq. (12.62), to replace T by V yields

$$\frac{dp}{p} = \frac{-dV}{V} - (k - 1)M^2\frac{dV}{V} \qquad (12.67)$$

which when substituted in the momentum equation, Eq. (12.65), results in

$$(M^2 - 1)\frac{dV}{V} + \frac{kfM^2}{2}\frac{dx}{D} = 0 \qquad (12.68)$$

The Mach number is defined as

$$M = \frac{V}{(kRT)^{1/2}}$$

which can be written in differential form as follows:

$$\frac{dM}{M} = \frac{dV}{V} - \frac{1}{2}\frac{dT}{T} \qquad (12.69)$$

Again, using Eq. (12.62) to eliminate T yields

$$\frac{dM}{M} = \frac{dV}{V}\left[1 + \frac{(k - 1)M^2}{2}\right] \qquad (12.70)$$

Using this equation to eliminate V in Eq. (12.68) results in the following differential equation for the Mach number and distance:

$$\frac{(1 - M^2)\,dM}{M^3\{1 + [(k - 1)/2]M^2\}} = \frac{kf\,dx}{2D} \qquad (12.71)$$

This equation tells us that if the flow is subsonic, then $dM/dx > 0$ and the Mach number increases with distance along the pipe. Conversely, if the flow is supersonic, then $dM/dx < 0$ and the Mach number decreases along the pipe. Thus the effect of wall friction is always to cause the Mach number to approach unity. It is impossible for the Mach number of a compressible flow in a pipe to change from subsonic to supersonic. Consequently, the maximum Mach number that an initially subsonic flow can attain is unity, and this can be reached only at the end of the pipe. Shock waves, of course, can occur in the pipe to change an initially supersonic flow to a subsonic flow.

From Chapter 10 we know that the resistance coefficient f is a function of the Reynolds number and the relative roughness of the pipe. The Reynolds number is constant along the length of a pipe transporting a liquid. The continuity equation requires also that ρV be constant along a pipe transporting a compressible fluid. Temperature, however, may vary by as much as 20% for the subsonic flow of air in a pipe, which corresponds to a viscosity variation of approximately 10%. Referring back to Fig. 10.8 on p. 428, one notes that a 10% change in Reynolds number gives rise to a considerably smaller change in f if the flow is turbulent, which is usually the case. Thus it is reasonable to assume when integrating Eq. (12.71) that f is a constant and equal to the average value, $\bar{f}$, in the pipe.

We are now ready to integrate Eq. (12.71) to determine the variation of the Mach number with distance along the pipe. The left-hand side of the equation can be reduced to a sum of partial fractions to facilitate integration:

$$\left(\frac{1}{M^3} - \frac{k+1}{2M} + \frac{(k+1)(k-1)M}{4\{1 + [(k-1)/2]M^2\}}\right)dM = \frac{k\bar{f}\,dx}{2D} \tag{12.72}$$

Integrating each side gives

$$\frac{-1}{2M^2} - \frac{k+1}{2}\ln M + \frac{k+1}{4}\ln\left(1 + \frac{k-1}{2}M^2\right) = \frac{k\bar{f}x}{2D} + C \tag{12.73}$$

where C is the integration constant. It is convenient to evaluate C by defining $x_\bullet$ as the distance corresponding to a Mach number of unity.

$$C = \frac{-k\bar{f}x_\bullet}{2D} - \frac{1}{2} + \frac{k+1}{4}\ln\left(\frac{k+1}{2}\right) \tag{12.74}$$

Substituting this expression for C into Eq. (12.73) results in

$$\frac{1 - M^2}{kM^2} + \frac{k+1}{2k}\ln\left[\frac{(k+1)M^2}{2 + (k-1)M^2}\right] = \frac{\bar{f}(x_\bullet - x_M)}{D} \tag{12.75}$$

where x_M is the distance corresponding to a Mach number M.

EXAMPLE 12.13   The initial Mach number of the flow of air in a pipe is 0.2. The average value of f is 0.015. Calculate the distance along the pipe (in pipe diameters) required to achieve sonic flow and to achieve a Mach number of 0.8.

Solution   Substituting M = 0.2 and k = 1.4 into Eq. (12.75) gives

$$\frac{\bar{f}(x_\bullet - x_{0.2})}{D} = 14.53$$

Thus the number of pipe diameters to reach sonic flow is

$$\frac{x_\bullet - x_{0.2}}{D} = 969 \qquad \blacktriangleleft$$

Substituting M = 0.8 into Eq. (12.75) gives

$$\frac{x_\bullet - x_{0.8}}{D} = 0.07 \qquad \blacktriangleleft$$

The distance required to increase the Mach number from 0.2 to 0.8 can be obtained by subtraction.

$$\frac{\overline{f}(x_{0.8} - x_{0.2})}{D} = \frac{\overline{f}(x_\star - x_{0.2})}{D} - \frac{\overline{f}(x_\star - x_{0.8})}{D} = 14.53 - 0.07 = 14.46$$

$$\frac{x_{0.8} - x_{0.2}}{D} = 964 \qquad \blacktriangleleft$$

Note that the Mach number increases very rapidly near the end of the pipe.

Thus we see that it is relatively easy to solve for distance along a pipe once the Mach number is known. However, it is more difficult to solve for the change in Mach number given a distance along the pipe. For this reason a plot of Mach number versus $\overline{f}(x_\star - x_M)/D$ to use in solving problems of this type is presented in Fig. 12.19.

EXAMPLE 12.14   Air flows at 60 m/s into a commercial steel pipe that has a 5-cm diameter. The pressure and temperature of the air are 1 MPa and 100°C. Determine the Mach number at a distance of 50 m down the pipe.

Solution   First, we must evaluate f. Referring to Table A.3 in the Appendix, we find that the dynamic viscosity at 100°C is 2.17×10^{-5} N · s/m². Using the ideal-gas equation, we calculate

$$\rho = \frac{p}{RT} = \frac{10^6 \text{ N/m}^2}{(287 \text{ J/kg K})(373 \text{ K})} = 9.34 \text{ kg/m}^3$$

The Reynolds number, then, is

$$\text{Re} = \frac{(60)(9.34)(0.05)}{2.17(10^{-5})} = 1.29(10^6)$$

Referring again to Figs. 10.8 and 10.9, we determine that $k_s/D = 0.001$ and that the corresponding value of f is 0.0195.

The speed of sound at the entrance to the pipe is

$$c = (kRT)^{1/2} = [1.4(287 \text{ J/kg K})(373 \text{ K})]^{1/2} = 387 \text{ m/s}$$

Thus the initial Mach number is $M = 60/387 = 0.16$, and reference to Fig. 12.19 shows that

$$\frac{\overline{f}(x_\star - x_{0.16})}{D} = 24$$

FIGURE 12.19

Variation of $\overline{f}(x_\bullet - x_M)/D$ with Mach number.

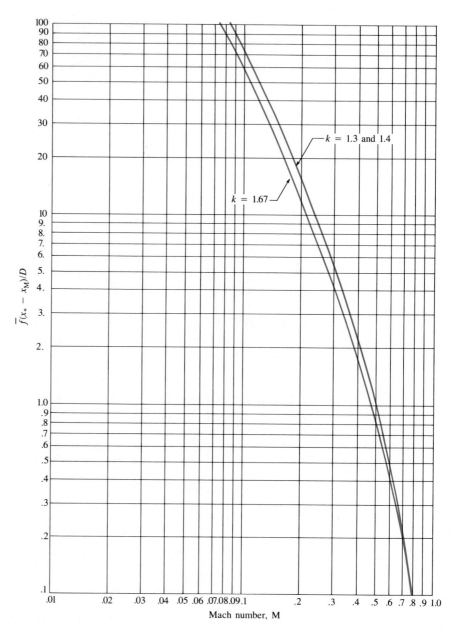

Now we can write

$$\frac{\overline{f}(x_M - x_{0.16})}{D} = \frac{\overline{f}(x_\bullet - x_{0.16})}{D} - \frac{\overline{f}(x_\bullet - x_M)}{D}$$

where $x_M - x_{0.16}$ is the distance along the pipe. Substituting the foregoing values for distance, diameter, and average resistance coefficient into this equation yields

$$\frac{\bar{f}(x_{\bullet} - x_{\text{M}})}{D} = 24 - \frac{(50)\,(0.0195)\ \text{m}}{0.05\ \text{m}} = 4.5$$

which, from Fig. 12.19, corresponds to a Mach number of

$$\text{M} = 0.32 \qquad \blacktriangleleft$$

This calculation was done by taking the initial value of f as the average value. We now ask ourselves how valid this procedure is. The total temperature at the entrance to the tube can be calculated using Eq. (12.23):

$$T_t = (373\ \text{K})\left[1 + \frac{0.4}{2}(0.16^2)\right] = 375\ \text{K}$$

Since it is assumed that the flow is adiabatic, the total temperature does not change along the pipe. Therefore the static temperature 50 m along the pipe is

$$T = \frac{375\ \text{K}}{1 + (0.4/2)\,(0.32^2)} = 367\ \text{K}$$

Thus the temperature changes approximately 8 K, and the viscosity change resulting from this temperature change would be approximately 1%. Therefore, the change in Reynolds number is negligible, and the initial friction factor can be used for the average value.

Variation of Pressure with Distance

The differential equation for pressure as a function of velocity and Mach number is Eq. (12.67),

$$\frac{dp}{p} = \frac{-dV}{V}[1 + (k - 1)\text{M}^2]$$

Using Eq. (12.70) we can obtain a differential expression for pressure as a function solely of Mach number:

$$\frac{dp}{p} = \frac{-d\text{M}}{\text{M}}\left\{\frac{1 + (k - 1)\text{M}^2}{1 + [(k - 1)/2]\text{M}^2}\right\} \qquad (12.76)$$

From this expression we conclude that $dp/d\text{M} < 0$. This means that pressure decreases with increasing Mach number and increases with decreasing Mach number. Thus for subsonic flow in a pipe, the pressure decreases with distance, the negative pressure gradient providing the force to overcome the wall shear force and accelerate the fluid.

Carrying out the division of the factors contained in the brackets in Eq. (12.76) and dividing by M, we arrive at

$$\frac{dp}{p} = \left\{-\frac{1}{\text{M}} - \frac{[(k - 1)/2]\text{M}}{1 + [(k - 1)/2]\text{M}^2}\right\}d\text{M} \qquad (12.77)$$

Integrating each side, we obtain

$$\ln p = -\ln M - \frac{1}{2}\ln\left(1 + \frac{k-1}{2}M^2\right) + C \qquad (12.78)$$

We can evaluate the constant of integration C by setting $p = p_\star$ at $M = 1$, so we have

$$\ln p_\star = -\frac{1}{2}\ln\left(\frac{k+1}{2}\right) + C$$

which, when substituted back into Eq. (12.78), gives

$$\frac{p_M}{p_\star} = \frac{1}{M}\left[\frac{k+1}{2 + (k-1)M^2}\right]^{1/2} \qquad (12.79)$$

where p_M is the pressure at a Mach number M. The variation of M with $p_M/p_\star$ is plotted in Fig. 12.20.

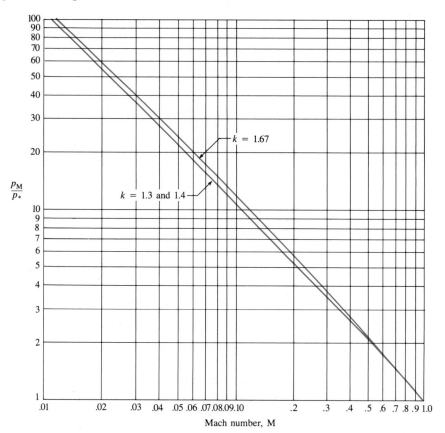

FIGURE 12.20

Variation of $p_M/p_\star$ *with Mach number for adiabatic viscous flow in a constant-area duct.*

EXAMPLE 12.15   Calculate the pressure at 50 m down the pipe of Example 12.14.

Solution   The pressure ratio corresponding to the initial and final Mach numbers can be written as

$$\frac{p_{0.32}}{p_{0.16}} = \frac{p_{0.32}}{p_*}\frac{p_*}{p_{0.16}}$$

Using Fig. 12.20, we find

$$\frac{p_{0.32}}{p_{0.16}} = \frac{3.4}{6.8} = 0.50$$

Therefore the pressure at the 50-m distance is

$$p = 0.50(10^6) \text{ Pa} = 500 \text{ kPa} \qquad \blacktriangleleft$$

Let us now consider how the pressure and Mach number vary along a pipe that discharges to the atmosphere as the upstream static pressure is increased. Consider the qualitative distributions of static pressure and Mach number along the pipe shown in Fig. 12.21. The pipe discharges to a pressure p_0. Cases A through D represent a continuous increase in the upstream static pressure:

Case A   The static pressure is uniform along the pipe and equal to p_0. Thus there is no flow in the pipe, and the Mach number is everywhere zero.

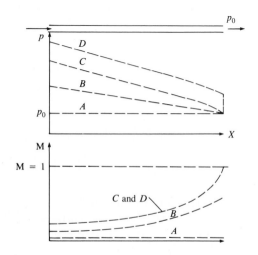

FIGURE 12.21

Distribution of static pressure and Mach number along a pipe.

Case B   The static pressure uniformly decreases to p_0, while the Mach number increases along the pipe.

Case C   The static pressure decreases more rapidly and causes the flow to accelerate to a Mach number of unity at the exit. In this case $p_0 = p_*$.

Case D   The static-pressure distribution is nearly identical to that in case C but shifted to a higher value. Thus the pipe discharges at a pressure higher than p_0, and pressure equilibration is achieved by a series of expansion waves. The Mach-number distribution differs little from that for case C, the flow continuing to discharge at sonic speed.

EXAMPLE 12.16   A brass tube is 3 cm in diameter and 8 m long. The total temperature of the airflow in the tube is 300 K. The pipe discharges to the atmosphere, where the pressure is 100 kPa. Calculate the mass flow in the tube (a) when the inlet static pressure is 120 kPa and (b) when it is 400 kPa.

Solution   The approach used to solve this problem depends on whether the flow exits subsonically (case *B*) or sonically (case *D*). If the flow exits subsonically, then the mass-flow rate must be such that the exit pressure is equal to the atmospheric pressure. On the other hand, if the flow exits sonically, the mass-flow rate is determined by finding the pressure level that gives the correct resistance coefficient f in the pipe.

In order to establish which approach to use, we first determine the upstream pressure corresponding to case *C* for which the flow exits sonically at atmospheric pressure. The static temperature and speed of sound at the exit for case *C* are

$$T_e = (300 \text{ K})\frac{2}{k+1} = 250 \text{ K}$$

$$c_e = [(1.4)(287 \text{ J/kg K})(250 \text{ K})]^{1/2} = 317 \text{ m/s}$$

The gas density at the exit is found to be

$$\rho_e = \frac{100 \times 10^3 \text{ N/m}^2}{(287 \text{ J/kg K})(250 \text{ K})} = 1.39 \text{ kg/m}^3$$

Thus the Reynolds number at the exit for case *C* has a value of

$$\text{Re} = \frac{(317 \text{ m/s})(1.39 \text{ kg/m}^3)(0.03 \text{ m})}{1.5(10^{-5} \text{ N} \cdot \text{s/m}^2)} = 8.85 \times 10^5$$

Referring to Fig. 10.9, we find that the relative roughness for the 3-cm brass tube is 0.00005. The corresponding value of the resistance coefficient is found in Fig. 10.8 to be 0.013. Evaluating the parameter $\bar{f}(x_* - x_M)/D$ and entering

Fig. 12.19, we find that the initial Mach number would be 0.35. Finally, Fig. 12.20 tells us that the pressure ratio corresponding to this Mach number is 3.1. Thus for case C, the static pressure 8 m upstream would be 310 kPa. From this we conclude that part (*a*) corresponds to case B and part (*b*) to case D.

a. This problem must be solved using an iterative approach. We shall assume an initial Mach number, calculate the exit Mach number as done in Example 12.14, and then determine the exit pressure as done in Example 12.15. The initial Mach number is varied until the exit pressure matches the atmospheric pressure. The most straightforward approach is to generate a table such as that shown, where M_e is the exit Mach number.

M	T, K	Re	$\overline{f}$	M_e	p/p_e	p_e, kPa
0.1	299	5.4×10^4	0.020	0.104	1.04	115
0.2	298	1.08×10^5	0.0185	0.24	1.21	99
0.19	298	1.02×10^5	0.0190	0.23	1.19	101

By interpolation we find that $p_e = 100$ kPa when $M = 0.195$. The temperature, density, and speed of sound at this Mach number are

$$T = 298 \text{ K} \qquad \rho = 1.40 \text{ kg/m}^3 \qquad c = 346 \text{ m/s}$$

Thus the mass-flow rate is

$$\dot{m} = (0.195)(346 \text{ m/s})(1.40 \text{ kg/m}^3)(0.785)(0.03^2 \text{ m}^2) = 0.067 \text{ kg/s} \quad \blacktriangleleft$$

b. The way to solve this part is to use the iterative approach again but on $\overline{f}$. We begin by assuming an $\overline{f}$, calculating the upstream Mach number and Reynolds number, and finding a new $\overline{f}$, which is used as the assumed value for the next iteration. The solution is the value at which $\overline{f}$ no longer changes. A check is always made to be sure that $p_\star$ exceeds p_0. The iteration using $\overline{f} = 0.02$ as the initial assumed value is demonstrated in the accompanying table.

f	$\overline{f}(x_\star - x_M)/D$	M	Re	$\overline{f}$	$p/p_\star$	p/p_0
0.02	5.33	0.3	5.4×10^5	0.0138	3.6	1.11
0.0138	3.67	0.34	6.1×10^5	0.0135	3.25	1.23
0.0135	3.59	0.35	6.3×10^5	0.0135	3.15	1.26

Thus the initial Mach number is 0.35, the exit Mach number is unity, and the exit pressure exceeds the atmospheric pressure. The temperature, density, and speed of sound at $M = 0.35$ are

$$T = 293 \text{ K} \qquad \rho = 4.77 \text{ kg/m}^3 \qquad c = 342 \text{ m/s}$$

The mass-flow rate is

$$\dot{m} = (0.35)(342 \text{ m/s})(4.77 \text{ kg/m}^3)(0.785)(0.03^2 \text{ m}^2) = 0.403 \text{ kg/s} \quad \blacktriangleleft$$

Isothermal Flow

The analysis of isothermal flow in a constant-area duct is simplified by the fact that the energy equation is

$$T = \text{constant}$$

This also means that the speed of sound in the duct is constant.

The momentum equation for flow in the duct is given by Eq. (12.65):

$$k\text{M}^2 \frac{dV}{V} + \frac{dp}{p} + \frac{kf\text{M}^2 \, dx}{2D} = 0$$

Because the temperature is constant, the pressure is proportional to the density [see Eq. (12.66)]. Thus

$$\frac{dp}{p} = \frac{d\rho}{\rho} \tag{12.80}$$

Using the continuity equation, Eq. (12.16), to relate ρ and V gives

$$\frac{d\rho}{\rho} = -\frac{dV}{V} \tag{12.81}$$

which, substituted back into Eq. (12.65), yields

$$\frac{d\text{M}}{dx} = \frac{f}{2D} \frac{k\text{M}^3}{1 - k\text{M}^2} \tag{12.82}$$

One notes that if the Mach number is less than $1/\sqrt{k}$, then it increases with distance, whereas the opposite trend is noted for Mach numbers exceeding $1/\sqrt{k}$. Thus the Mach number must always approach $1/\sqrt{k}$ for isothermal flows, compared to unity for adiabatic flows.

Mach-Number Distribution
Along a Constant-Area Duct

The Mach-number distribution along a constant-area duct is determined by integrating Eq. (12.82) for Mach number as a function of distance. Rewriting Eq. (12.82) as

$$d\text{M} \frac{(1 - k\text{M}^2)}{k\text{M}^3} = \frac{f \, dx}{2D} \tag{12.83}$$

or
$$\frac{dM}{kM^3} - \frac{dM}{M} = \frac{f\,dx}{2D} \qquad (12.84)$$

we can integrate each side to obtain

$$\frac{1}{-2kM^2} - \ln M = \frac{fx}{2D} + C \qquad (12.85)$$

When we set x_T as the maximum length at which $M = 1/\sqrt{k}$, the constant of integration becomes

$$-\frac{1}{2} - \ln\left(\frac{1}{\sqrt{k}}\right) - \frac{fx_T}{2D} = C \qquad (12.86)$$

and substituting this constant in Eq. (12.85) gives

$$\frac{f(x_T - x_M)}{D} = \ln(kM^2) + \frac{(1 - kM^2)}{kM^2} \qquad (12.87)$$

The variation of $f(x_T - x_M)/D$ with kM^2 is plotted in Fig. 12.22.

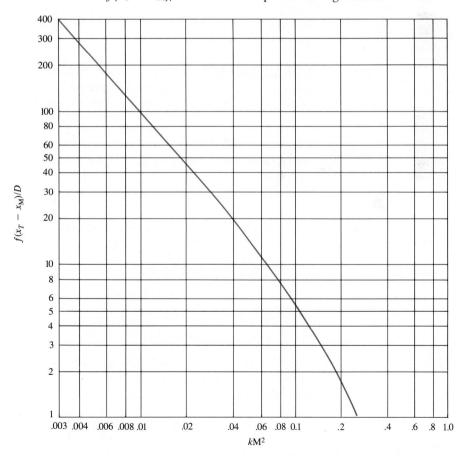

FIGURE 12.22

Variation of
$f(x_T - x_M)/D$ with kM^2
for isothermal viscous flow
in a constant-area duct.

Pressure Variation

In isothermal flow the speed of sound is constant, so velocity is directly proportional to the Mach number. Also, the pressure is directly proportional to the density. Substituting these relationships into Eq. (12.81), we obtain

$$\frac{dp}{p} = -\frac{d\mathrm{M}}{\mathrm{M}} \tag{12.88}$$

Integrating Eq. (12.88) yields

$$\ln p = -\ln \mathrm{M} + C \tag{12.89}$$

Letting p_T be the pressure corresponding to the maximum distance, we find the constant of integration and substitute it back into Eq. (12.89) to give

$$\frac{p_M}{p_T} = \frac{1}{\sqrt{k}\,\mathrm{M}} \tag{12.90}$$

Similar expressions can be found for the variation of density, total pressure, and total temperature along the duct.

EXAMPLE 12.17   Methane is to be transported at 15°C in 50-cm commercial steel pipe that is 1 km long. The pressure at the pipe exit is 100 kPa. Determine the maximum flow rate through the pipe and the pressure at the pipe entrance.

Solution   From Table A.2 in the Appendix, the value of k for methane is 1.31, the gas constant is 518 J/kg K, and the kinematic viscosity at 15°C is 1.59×10^{-5} m²/s.

The Mach number at the pipe exit is

$$\mathrm{M} = \frac{1}{\sqrt{k}} = \frac{1}{\sqrt{1.31}} = 0.874$$

The corresponding velocity is

$$V = \mathrm{M}c = 0.874 \times \sqrt{1.31 \times 518 \times 288} = 386 \text{ m/s}$$

The corresponding mass flow, which is the maximum flow rate, is

$$\dot{m} = \rho AV = \left(\frac{10^5}{518 \times 288}\right)(0.5)^2\left(\frac{\pi}{4}\right)(386) = 50.8 \text{ kg/s} \quad \blacktriangleleft$$

The Reynolds number at the exit is

$$\mathrm{Re} = \frac{VD}{\nu} = \frac{(386)(0.5)}{1.59 \times 10^{-5}} = 1.2 \times 10^7$$

Since ρV is constant along the pipe (continuity equation) and μ is constant (depends primarily on temperature), the Reynolds number is the same everywhere along the pipe. Therefore, the resistance coefficient does not change. Referring to Figs. 10.8 and 10.9, we find that f is 0.012. Thus the left-hand side of Eq. 12.87 has the value

$$\frac{f(x_T - x_M)}{D} = \frac{(0.012)\,(10^3)}{0.5} = 24$$

Using Fig. 12.22, we find

$$kM^2 = 0.035$$

The pressure at the entrance is found using Eq. 12.90:

$$p_M = \frac{100 \text{ kPa}}{\sqrt{0.035}} = 535 \text{ kPa} \qquad \blacktriangleleft$$

In this section we have treated the adiabatic and isothermal flow of a viscous compressible gas in a constant-area duct. The reader is referred to more specialized texts on compressible flow, such as that by Owczarek (5), for studies of other types of flows, such as the effect of heat addition due to chemical reaction.

Problems

12.1 How fast (in meters per second) will a sound wave travel in methane at 20°C?

12.2 Calculate the speed of sound in helium at 50°C.

12.3 Calculate the speed of sound in hydrogen at 75°F.

12.4 How much faster will a sound wave propagate in helium than in nitrogen if the temperature of both gases is 15°C?

12.5 Determine what the equation for the speed of sound in an ideal gas would be if the sound wave were an isothermal process.

12.6 The relationship between pressure and density for the propagation of a sound wave through a fluid is

$$p - p_0 = E_v \ln (\rho/\rho_0)$$

where p_0 and ρ_0 are the reference pressure and density (constants) and E_v is the bulk modulus of elasticity. Determine the equation for the speed of a sound wave in terms of E_v and ρ. Calculate the sound speed for water with $\rho = 1000$ kg/m^3 and $E_v = 2.20$ GN/m^2.

12.7   A supersonic aircraft is flying at Mach 2 through air at $-30°C$. What temperature could be expected on exposed aircraft surfaces? What is the airspeed behind the shock?

12.8   What is the temperature on the nose of a supersonic fighter flying at Mach 2 through air at 273 K?

12.9   A high-performance aircraft is flying at a Mach number of 1.5 at an altitude of 10,000 m, where the temperature is $-44°C$ and the pressure is 30.5 kPa.

     a. How fast is the aircraft traveling in kilometers per hour?

     b. The total temperature is an estimate of surface temperature on the aircraft. What is the total temperature under these conditions?

     c. Calculate the total pressure under these conditions.

     d. If the aircraft slows down, at what speed (kilometers per hour) will the Mach number be unity?

12.10   An airplane travels at 800 km/h at sea level where the temperature is 15°C. How fast would the airplane be flying at the same Mach number at an altitude where the temperature was $-40°C$?

12.11   An airplane flies at a Mach number of 0.9 at a 10,000-m altitude, where the static temperature is $-44°C$ and the pressure is 30 kPa absolute. The lift coefficient of the wing is 0.05. Determine the wing loading (lift force/wing area).

12.12   An object is immersed in an airflow with a static pressure of 200 kPa absolute, a static temperature of 20°C, and a velocity of 250 m/s. What are the pressure and temperature at the stagnation point?

12.13   An airflow at $M = 0.5$ passes through a conduit with a cross-sectional area of 65 cm^2. The total absolute pressure is 340 kPa, and the total temperature is 10°C. Calculate the mass-flow rate through the conduit.

12.14   Oxygen flows from a reservoir in which the temperature is 200°C and the pressure is 300 kPa absolute. Assuming isentropic flow, calculate the velocity, pressure, and temperature when the Mach number is 0.9.

12.15   One problem in creating high-Mach-number flows is condensation of the oxygen component in the air when the temperature reaches 50 K. If the temperature of the reservoir is 300 K and the flow is isentropic, at what Mach number will condensation of oxygen occur?

12.16   Hydrogen flows from a reservoir where the temperature is 20°C and the pressure is 500 kPa absolute to a section 2 cm in diameter where the velocity is 300 m/s. Assuming isentropic flow, calculate the temperature, pressure, Mach number, and mass-flow rate at the 2-cm section.

12.17   The total pressure in a Mach-2 wind tunnel operating with air is 600 kPa absolute. A sphere 1 cm in diameter, positioned in the wind tunnel, has a drag coefficient of 0.95. Calculate the drag of the sphere.

12.18   Using Eq. (12.27), develop an expression for the pressure coefficient at stagnation conditions—that is, $C_p = (p_t - p)/(\frac{1}{2}\rho V^2)$—in terms of Mach number and ratio of specific heats, $C_p = f(k, M)$. Evaluate C_p at $M = 0$, 2, and 4 for $k = 1.4$. What would its value be for incompressible flow?

12.19 For low velocities, the total pressure is only slightly larger than the static pressure. Thus one can write $p_t/p = 1 + \varepsilon$, where ε is a small positive number ($\varepsilon \ll 1$). Using this approximation, show that, as $\varepsilon \to 0$ ($M \to 0$), Eq. (12.32) reduces to

$$M = \left[\frac{2(p_t/p - 1)}{k}\right]^{1/2}$$

12.20 A normal shock wave exists in a 500-m/s stream of nitrogen having a static temperature of $-40°C$ and a static pressure of 70 kPa. Calculate the Mach number, pressure, and temperature downstream of the wave and the entropy increase across the wave.

12.21 A normal shock wave exists in a Mach-3 stream of air having a static temperature and pressure of 45°F and 30 psia. Calculate the Mach number, pressure, and temperature downstream of the shock wave.

12.22 A Pitot-static tube is used to measure the Mach number on a supersonic aircraft. The tube, because of its bluntness, creates a normal shock wave as shown. The absolute total pressure downstream of the shock wave (p_{t_2}) is 150 kPa. The static pressure of the free stream ahead of the shock wave (p_1) is 40 kPa and is sensed by the static pressure tap on the probe. Determine the Mach number (M_1) graphically.

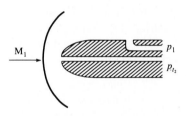

PROBLEM 12.22

12.23 A shock wave occurs in a methane stream in which the Mach number is 2, the static pressure is 100 kPa absolute, and the static temperature is 20°C. Determine the downstream Mach number, static pressure, static temperature, and density.

12.24 The Mach number downstream of a shock wave in helium is 0.8, and the static temperature is 100°C. Calculate the velocity upstream of the wave.

12.25 Show that the lowest Mach number possible downstream of a normal shock wave is

$$M_2 = \sqrt{\frac{k - 1}{2k}}$$

and that the largest density ratio possible is

$$\frac{\rho_2}{\rho_1} = \frac{k + 1}{k - 1}$$

What are the limiting values of M_2 and ρ_2/ρ_1 for air?

12.26   Show that the Mach number downstream of a weak wave ($M \simeq 1$) is approximated by

$$M_2^2 = 2 - M_1^2$$

[*Hint:* Let $M_1^2 = 1 + \varepsilon$ where $\varepsilon \ll 1$ and expand Eq. (12.41) in terms of ε.] Compare values for M_2 obtained using this equation with values for M_2 from Table A.1 for $M_1 = 1$, 1.05, 1.1, and 1.2.

12.27   The truncated nozzle shown in the figure is used to meter the mass flow of air in a pipe. The area of the nozzle is 3 cm². The total pressure and total temperature measured upstream of the nozzle in the pipe are 300 kPa absolute and 20°C. The pressure downstream of the nozzle (back pressure) is 90 kPa absolute. Calculate the mass-flow rate.

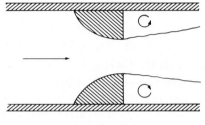

PROBLEM 12.27

12.28   The truncated nozzle shown in Prob. 12.27 is used to monitor the mass-flow rate of methane. The area of the nozzle is 3 cm², and the area of the pipe is 12 cm². The upstream total pressure and total temperature are 150 kPa absolute and 30°C. The back pressure is 100 kPa.

     a. Calculate the mass-flow rate of methane.

     b. Calculate the mass-flow rate assuming Bernoulli's equation is valid, the density being the density of the gas at the nozzle exit.

12.29   A truncated nozzle with an exit area of 5 cm² is used to measure a mass flow of air of 0.25 kg/s. The static temperature of the air at the exit is 10°C, and the back pressure is 100 kPa. Determine the total pressure.

12.30   A truncated nozzle with a 12-cm² exit area is supplied from a helium reservoir in which the absolute pressure is first 130 kPa and then 350 kPa. The temperature in the reservoir is 28°C, and the back pressure is 100 kPa. Calculate the mass-flow rate of helium for the two reservoir pressures.

12.31   A sampling probe is used to draw gas samples from a gas stream for analysis. In sampling, it is important that the velocity entering the probe equal the velocity of the gas stream (isokinetic condition). Consider the sampling probe shown, which has a truncated nozzle inside it to control the mass-flow rate. The probe has an inlet diameter of 4 mm and a truncated nozzle diameter of 2 mm. The probe is in a hot-air stream with a static temperature of 600°C, a static pressure of 100 kPa absolute, and a velocity of 50 m/s. Calculate the pressure required in the probe (back pressure) to maintain the isokinetic sampling condition.

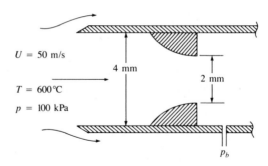

PROBLEM 12.31

12.32 A wind tunnel is designed to have a Mach number of 2.5, a static pressure of 1.5 psia, and a static temperature of −10°F in the test section. Determine the area ratio of the nozzle required and the reservoir conditions that must be maintained if air is to be used.

12.33 A Laval nozzle is to be designed to operate supersonically and expand ideally to an absolute pressure of 30 kPa. If the stagnation pressure in the nozzle is 1 MPa, calculate the nozzle area ratio required. Determine the nozzle throat area for a mass flow of 5 kg/s and a stagnation temperature of 550 K. Assume that the gas is nitrogen.

12.34 A rocket nozzle with an area ratio of 4 is operating at a total absolute pressure of 1.3 MPa and exhausting to an atmosphere with an absolute pressure of 35 kPa. Determine whether the nozzle is overexpanded, underexpanded, or ideally expanded. Assume $k = 1.4$.

12.35 A Laval nozzle with an exit area ratio of 1.688 exhausts air from a large reservoir into ambient conditions at $p = 100$ kPa.

  a. Show that the reservoir pressure must be 782.5 kPa to achieve ideally expanded exit conditions at M = 2.
  b. What are the static temperature and pressure at the throat if the reservoir temperature is 17°C with the pressure as in (a)?
  c. If the reservoir pressure were lowered to 700 kPa, what would be the exit condition (overexpanded, ideally expanded, underexpanded, subsonic flow in entire nozzle)?
  d. What reservoir pressure would cause a normal shock to form at the exit?

12.36 Determine the Mach number and area ratio at which the dynamic pressure is maximized in a Laval nozzle with air. [*Hint:* Express q in terms of p and M, and use Eq. (12.27) for p. Differentiate with respect to M and equate to zero.]

12.37 A rocket motor operates at an altitude where the atmospheric pressure is 20 kPa. The expansion ratio of the nozzle is 4 (exit area/throat area). The chamber pressure of the motor (total pressure) is 1.2 MPa, and the chamber temperature (total temperature) is 3000°C. The ratio of the specific heats of the exhaust gas is 1.2, and the gas constant is 400 J/kg K. The throat area of the rocket nozzle is 100 cm².

  a. Determine the Mach number, density, pressure, and velocity at the nozzle exit.
  b. Determine the mass-flow rate.

c. Calculate the thrust of the rocket using [see Eq. (6.13)]

$$T = \dot{m}V_e + (p_e - p_0)A_e$$

d. What would the chamber pressure of the rocket have to be to have an ideally expanded nozzle? Calculate the rocket thrust under this condition.

12.38   A rocket motor is being designed to operate at sea level, where the pressure is 100 kPa absolute. The chamber pressure (total pressure) is 1.8 MPa, and the chamber temperature (total temperature) is 3300 K. The throat area of the nozzle is 10 cm². The ratio of the specific heats (k) of the exhaust gas is 1.2, and the gas constant is 400 J/kg K.

a. Determine the nozzle expansion ratio that is required to achieve an ideally expanded nozzle, and determine the nozzle thrust under these conditions (see Prob. 12.37 for the thrust equation).

b. Determine the thrust that would be obtained if the expansion ratio were reduced by 10% to achieve an underexpanded nozzle.

12.39   Air flows through a Laval nozzle with an expansion ratio of 4. The total pressure of the air entering the nozzle is 200 kPa, and the back pressure is 100 kPa. Determine the area ratio at which the shock wave occurs in the expansion section of the nozzle. (*Hint:* This problem can be solved graphically by calculating the exit pressure corresponding to different shock wave locations and finding the location where the exit pressure is equal to the back pressure.)

12.40   A rocket nozzle has the configuration shown. The diameter of the throat is 4 cm, and the exit diameter is 8 cm. The half-angle of the expansion cone is 15°. Gases with a specific heat of 1.2 flow into the nozzle with a total pressure of 250 kPa. The back pressure is 100 kPa. First, using an iterative or graphical method, determine the area ratio at which the shock occurs. Then determine the shock wave's distance from the throat in centimeters.

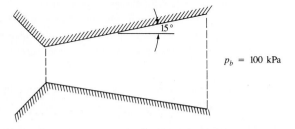

PROBLEM 12.40

12.41   A normal shock wave occurs in a nozzle at an area ratio of 4. Determine the entropy increase if the gas is hydrogen.

12.42   Consider airflow in the variable-area channel shown. Determine the Mach number, static pressure, and stagnation pressure at station 3. Assume isentropic flow except for normal shock waves.

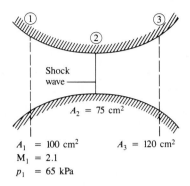

$A_1 = 100 \text{ cm}^2$
$M_1 = 2.1$
$p_1 = 65 \text{ kPa}$

$A_2 = 75 \text{ cm}^2$

$A_3 = 120 \text{ cm}^2$

PROBLEM 12.42

12.43 An engineer is designing a piping system for airflow. The pipe must be 10 m long, must be fabricated from brass, and must carry a mass flow of 0.2 kg/s. The total temperature of the flow in the pipe is 373 K, and the tolerable pressure loss is 140 kPa. The pipe discharges to a pressure of 100 kPa. Determine the diameter of the pipe.

12.44 Determine the atmospheric pressure necessary for the shock wave to position itself as shown. The fluid is air.

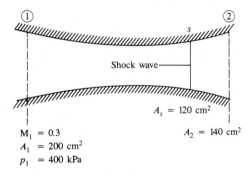

$M_1 = 0.3$
$A_1 = 200 \text{ cm}^2$
$p_1 = 400 \text{ kPa}$

$A_s = 120 \text{ cm}^2$

$A_2 = 140 \text{ cm}^2$

PROBLEM 12.44

12.45 The velocity of air entering a commercial steel pipe with a 1-in. diameter is 150 ft/s. The static temperature and pressure are 67°F and 30 psia. Calculate the length of pipe needed to achieve sonic flow; also calculate the pressure at the end of the pipe. Assume the viscosity of the air is independent of pressure.

12.46 The Mach number of air flowing out the exit of a 3-cm brass tube is 0.8. The total temperature of the airflow is 373 K, and the atmospheric pressure is 100 kPa. How far upstream is the Mach number equal to 0.2?

12.47 The Mach numbers at the inlet and exit of a pipe 0.5 in. in diameter and 20 ft long are 0.2 and 0.6, respectively. Calculate $\bar{f}$ in the pipe for $k = 1.4$.

12.48 Oxygen flows through a wrought-iron pipe 2.5 cm in diameter and 10 m long. It discharges to the atmosphere, where the pressure is 100 kPa. If the absolute static pressure at the beginning of the pipe is 300 kPa and the total temperature is 293 K, calculate the mass flow through the pipe.

12.49 Repeat Prob. 12.48, using an absolute static pressure of 500 kPa at the beginning of the pipe.

12.50 A pressure hose 10 ft long is to be connected to the outlet of the regulator valve on a nitrogen bottle. The hose must deliver 0.06 lbm/s when the regulator pressure is 45 psia. The hose exhausts to a back pressure of 7 psia. The total temperature is 100°F. Assume the hose has a roughness equivalent to that of galvanized iron, and calculate the required hose diameter.

12.51 An air blower and pipe system is to be designed to convey agricultural products through a 20-cm steel pipe. The pipe is to be 150 m long, and the outlet velocity is to be 50 m/s. If the pipe discharges into the atmosphere (100 kPa, 15°C), what will be the pressure, velocity, and density of the air at the inlet end of the pipe? Assume that the ratio of specific heats for the particle-laden flow is 1.4.

12.52 Methane is pumped into a 15-cm steel pipe at a pressure of 1 MPa and a temperature of 320 K ($\mu = 1.5 \times 10^{-5}$ N $\cdot$ s/m^2) and with a velocity of 20 m/s. What is the pressure 3000 m downstream?

12.53 Hydrogen is transported in an underground pipeline. The pipe is 50 m long and 10 cm in diameter; it is maintained at a temperature of 15°C. The initial pressure and velocity are 200 kPa and 200 m/s. Determine the pressure drop in the pipe.

12.54 Helium flows in a 5-cm brass tube 100 m long that is maintained at a temperature of 15°C. The entrance pressure is 120 kPa, and the exit pressure is 100 kPa. Determine the mass-flow rate in the pipe. (This problem requires an iterative solution.)

References

1. Anderson, J. D., Jr. *Modern Compressible Flow with Historical Perspective.* McGraw-Hill, New York, 1982.

2. Chapman, Alan J., and William F. Walker. *Introductory Gas Dynamics.* Holt, Rinehart and Winston, New York, 1971.

3. Crocco, L. "One-Dimensional Treatment of Steady Gas Dynamics." In H. W. Emmons (ed.), *Fundamentals of Gas Dynamics,* vol. 3. Princeton University Press, Princeton, N.J., 1958.

4. Oswatitsch, K. *Gas Dynamics.* Trans. G. Kuerti. Academic Press, New York, 1956.

5. Owczarek, J. A. *Fundamentals of Gas Dynamics.* McGraw-Hill, New York, 1971.

6. Pope, Alan. *Aerodynamics of Supersonic Flight.* Pitman, New York, 1950.

7. Shapiro, A. H. *The Dynamics and Thermodynamics of Compressible Fluid Flow.* Ronald, New York, 1953.

8. Thompson, P. A. *Compressible-Fluid Dynamics.* McGraw-Hill, New York, 1971.

9. Van Wylen, G. J., and R. E. Sonntag. *Fundamentals of Classical Thermodynamics.* John Wiley, New York, 1965.

10. Von Mises, R. *Mathematical Theory of Compressible Fluid Flow.* Academic Press, New York, 1958.

Flow Measurements

This flow pattern occurs downstream of an array of symmetric airfoils placed in a low-speed wind tunnel. The smoke used for visualization of the flow was made by bubbling air through a jar of liquid containing a mixture of titanium tetrachloride and carbon tetrachloride. The resulting chemical reaction between the liquid and moisture in the air produced a dense white smoke. (Courtesy B. Bienkiewicz and J. E. Cermak, Colorado State University)

The most common flow measurements are pressure, rate of flow, and velocity. In Chapter 3 several methods were presented for measuring pressure, and in Chapter 5 the basic theory of stagnation and Pitot tubes for the measurement of velocity was given. In this chapter we will consider the Pitot, static, and stagnation tubes in more detail, and we will describe other ways of measuring velocity. Finally, several methods will be presented for measuring the rate of flow of fluids.

13.1 Instruments for the Measurement of Velocity and Pressure

Stagnation Tube

In Chapter 5, where the stagnation tube was first introduced, it was assumed that viscous effects were negligible. This assumption is valid for tubes of normal size when we are measuring moderately high-velocity air or water. However, if the velocity to be measured is very low or the tube is very small, viscous effects become significant, and a correction must be applied to the basic equation. That is, as the Reynolds number decreases, the viscous effects become significant. This influence is shown in Fig. 13.1, where the pressure coefficient $C_p = (p - p_0)/(\rho V_0^2/2)$ is plotted as a function of the Reynolds number.

In Fig. 13.1 it is seen that when the Reynolds number for the circular stagnation tube is greater than 60, the error in measured velocity is less than 1%. Also in Fig. 13.1 are data for stagnation tubes that are flattened at the end and are often used to measure boundary-layer velocities. By flattening the end of the tube, the velocity measurement can be taken nearer the boundary than if a circular tube were used. For these flattened tubes, note that the pressure coefficient remains near unity for a Reynolds number as low as 30.

FIGURE 13.1

Viscous effects on C_p. [After Macmillan (13)]

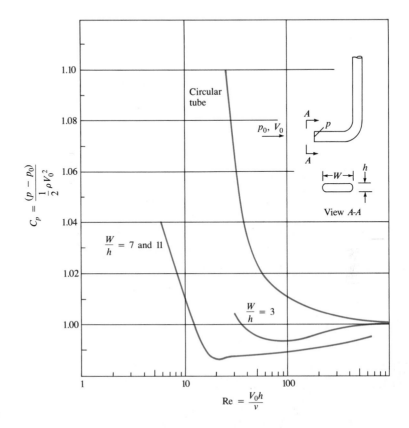

Flow Direction with Stagnation-Type Tubes

When the direction of flow is not known, simple pressure-type *yaw meters* can be used to sense the flow direction. Three types of yaw meters are shown in Fig. 13.2. The first two can be used for two-dimensional flow, where flow direction in only one plane needs to be found, and the third is used for determining flow direction in three dimensions. In all these devices, the tube is turned until the pressure on symmetrically opposite openings is equal. This pressure is sensed by a differential pressure gage or manometer connected to the openings in the yaw meter. The flow direction is sensed when a null reading is indicated on the differential gage.

Static Tube

At times one wants to measure the static pressure in a flowing fluid. This is accomplished by sensing the pressure at a point along the static tube where the pressure is the same as the free-stream pressure. Such a tube is shown in

FIGURE 13.2

Various types of yaw meters.
(a) Cylindrical-tube yaw
meter. (b) Two-tube yaw
meter. (c) Three-dimensional
yaw meter.

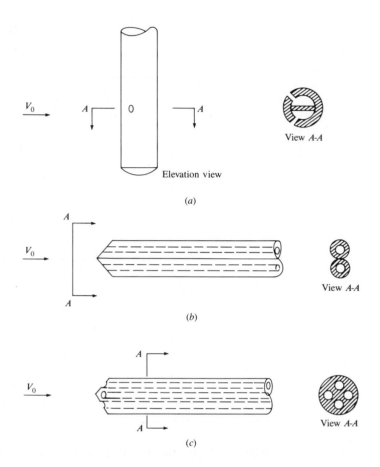

Fig. 13.3. This is like the Pitot tube except that the stagnation pressure tap is omitted. The placement of the holes along the probe is important for sensing the static pressure, because the rounded nose on the tube causes some decrease of pressure along the tube and the downstream stem causes an increase in pressure in front of it. Hence the location for sensing the static pressure must be at the point where these two effects cancel each other. Experiments reveal that the optimum location is at a point approximately six diameters downstream of the front of the tube and eight diameters upstream from the stem.

The Vane or Propeller Anemometer

The basic vane or propeller anemometer has been used in various applications for a number of years to measure the velocity of either gases or liquids. Basically, the type used for airflows, Fig. 13.4, consists of vanes (propeller blades) attached to a rotor that drives a low-friction gear train that, in turn, drives a pointer that in-

FIGURE 13.3

Static tube.

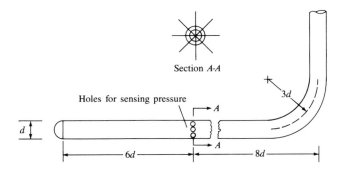

dicates feet on a dial. Thus if the anemometer is held in an airstream for 1 min and the pointer indicates a 300-ft change on the scale, the average airspeed is 300 ft/min.

Another type of vane anemometer is used for measuring the velocity of flowing water. The usual commercial anemometer of this type is connected to an electrical circuit in such a manner that an electrical signal is triggered for a given number of revolutions. Thus the frequency of rotation of the vane is obtained, which is directly converted to velocity by a calibration curve.

Vane anemometers are also used inside pipes, where they are calibrated to indicate the flow rate directly in cubic meters per second, or in any other appropriate units desired.

Another common anemometer used in meteorological measurements is an ordinary propeller attached to the forward part of a wind vane, all of which is mounted on a mast. The propeller drives an electromagnetic generator, the voltage from which is proportional to the wind speed.

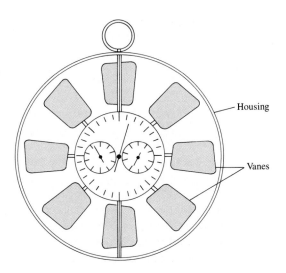

FIGURE 13.4

Vane anemometer for measuring velocity of airflows.

Cup Anemometer

The most common type of cup anemometer is that used by meteorologists to measure wind velocity (Fig. 13.5). However, hydraulic engineers also use a cup anemometer to measure the velocity of flow in streams and rivers. Again, the frequency of rotation is directly related to the velocity of flow by appropriate calibration data.

FIGURE 13.5

Cup anemometer.
(a) Elevation view.
(b) Top view.

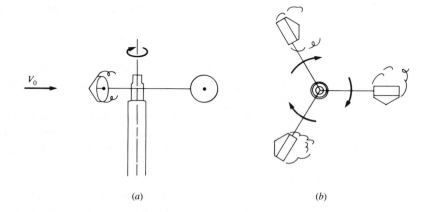

V_0

(a)   (b)

Hot-Wire and Hot-Film Anemometers

The velocity-measuring devices we have already considered are suitable for measuring velocity that either is steady or changes slowly with time. However, if we want to measure the velocity fluctuations due to the eddies in turbulent flow, the response of the aforementioned instruments is too slow to record the rapid changes in velocity. These instruments are also too large for local velocity measurements, which one might wish to make in a thin boundary layer. The hot-wire anemometer is an instrument that is very sensitive to rapid fluctuations in velocity, and its sensing element is small so that its presence does not seriously disturb the nature of the surrounding flow. Another advantage of the hot-wire anemometer is that it is sensitive to low-velocity flows, a characteristic lacking in the Pitot tube. The main disadvantages of the hot-wire anemometer are its delicate nature (the sensor wire is easily broken) and its relatively high cost (approximately $2000 and up).

The basic principle of the hot-wire anemometer is described as follows: A wire of very small diameter—the sensing element of the hot-wire anemometer—is welded to supports as shown in Fig. 13.6. In operation the wire either is heated by a fixed flow of electric current (the constant-current anemometer) or is maintained at a constant temperature by adjusting the current (the constant-temperature anemometer).

FIGURE 13.6

*Probe for hot-wire
anemometer (enlarged).*

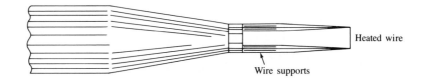

Heated wire

Wire supports

A flow of fluid past the hot wire causes the wire to cool because of convective heat transfer. In the constant-current anemometer, the cooling of the wire causes its resistance to change, and a corresponding voltage change occurs across the wire. Because the rate of cooling is a function of the speed of flow past the heated wire, the voltage across the wire is correlated with the flow velocity. The more popular type of anemometer, the constant-temperature anemometer, operates by varying the current in such a manner as to keep the resistance (and temperature) constant. The flow of current is correlated with the speed of the flow: the higher the speed, the greater the current needed to maintain a constant temperature. Typically, the wires are 1 to 2 mm in length and heated to 150°C. The wires may be 10 μm or less in diameter; the time response improves with the smaller wire. The lag of the wire's response to a change in velocity (thermal inertia) can be compensated for more easily, using modern electronic circuitry, in constant-temperature anemometers than in constant-current anemometers. The signal from the hot wire is processed electronically to give the desired information, such as mean velocity or the root-mean-square of the velocity fluctuation.

To illustrate the versatility of these instruments, note that the hot-wire anemometer can measure accurately gas-flow velocities from 30 cm/s to 150 m/s; it can measure fluctuating velocities with frequencies up to 100,000 Hz; and it has been used satisfactorily for both gases and liquids.

The single hot wire mounted normal to the mean-flow direction measures the fluctuating component of velocity in the mean-flow direction. Other probe configurations and electronic circuitry can be used to measure other components of velocity.

For velocity measurements in liquids or dusty gases, where wire breakage is a problem, the hot-film anemometer is more suitable. This anemometer consists of a thin conducting metal film (less than 0.1 μm thick) mounted on a ceramic support, which may be 50 μm in diameter. The hot film operates in the same fashion as the hot wire. Recently, the split film has been introduced. It consists of two semicylindrical films mounted on the same cylindrical support and electrically insulated from each other. The split film provides both speed and directional information.

For more detailed information on the hot-wire and hot-film anemometers, see Bradshaw (2) and King (8).

Laser-Doppler Anemometer

The laser-Doppler anemometer (LDA) is a relatively new instrument for taking fluid-velocity measurements. The major advantage of the LDA over the Pitot tube and hot-wire anemometer is that the flow field is not disturbed by the presence of a probe or wire support. A further advantage is that the velocity is measured in a very small flow volume, providing excellent spatial resolution.

There are several different configurations for the LDA, depending on the properties and accessibility of the flow. One configuration, known as the dual-beam mode, is shown in Fig. 13.7. The laser beam, which is a highly coherent, monochromatic light source, is first split into two parallel beams and then passed through a converging lens. The point where the two beams cross is the measuring volume, which might best be described as an ellipsoid that is typically 0.3 mm in diameter and 2 mm long, illustrating the excellent spatial resolution achievable. The interference of the two beams generates a series of light and dark fringes in the measuring volume perpendicular to the plane of the two beams. As a particle passes through the fringe pattern, light is scattered and a portion of the scattered light passes through the collecting lens toward the photodetector. A typical signal obtained from the photo-detector is shown in the figure.

FIGURE 13.7

Dual-beam laser-Doppler anemometer.

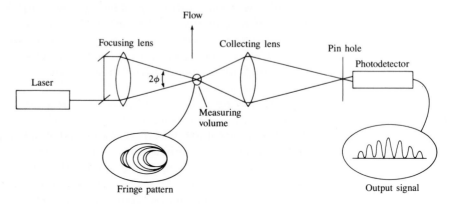

It can be shown from optics theory that the spacing between the fringes is given by

$$\Delta x = \frac{\lambda}{2 \sin \phi}$$

where λ is the wavelength of the laser beam and ϕ is the half-angle between the crossing beams. By suitable electronic circuitry, the frequency of the signal (f) is measured, so the velocity is given by

$$U = \frac{\Delta x}{\Delta t} = \frac{\lambda f}{2 \sin \phi}$$

The operation of the laser-Doppler anemometer depends on the presence of particles in the flow to scatter the light, particles sufficiently small that they always move at the fluid velocity. In liquid flows, the impurities of the fluid typically serve as scattering centers. In gaseous flows, however, it is sometimes necessary to "seed" the flow with small particles. Smoke is often used for this seeding.

The laser-Doppler anemometer has specific application in flow measurements requiring fine spatial resolution and minimum disturbance of the flow. Currently, many industrial applications of the LDA are being found to measure velocities of materials such as melted glass, in which it is not possible to mount a mechanical probe. The LDA is also used in biomedical research dealing with the flow of blood.

Laser-Doppler anemometers that provide two or three velocity components of a particle traveling through the measuring volume are now available. This is accomplished by using laser-beam pairs of different colors (wavelengths). The measuring volumes for each color are positioned at the same physical location but oriented differently to measure a different component. The signal-processing system can discriminate the signals from each color and thereby provide component velocities.

Another recent technological advance in laser-Doppler anemometry is the use of fiber optics. The fiber optics transmit the laser beams from the laser to a probe that contains optical elements to cross the beams and generate a measuring volume. Thus measurements at different locations can be made by moving the probe and without moving the laser.

Marker Methods

The procedure for the marker method for determining velocity stems from the basic concept of the pathline (see Chapter 4). Identifiable particles are placed in the stream. Then, by analyzing the motion of these particles, one can deduce the velocity of the flow itself. Of course, this requires that the markers follow virtually the same path as the surrounding fluid elements. It means, then, that the marker must have nearly the same density as the fluid or that it must be so small that its motion relative to the fluid is negligible. Thus for water flow it is common to use colored droplets from a liquid mixture that has nearly the same density as the water. For example, Macagno (12) used a mixture of n-butyl phthalate and xylene with a bit of white paint to yield a mixture that had the same density as water and could be photographed effectively. Solid particles, such as plastic beads, that have densities near that of the liquid being studied can also be used as markers.

Hydrogen bubbles have also been used for markers in water flow. Here an electrode placed in flowing water causes small bubbles to be formed and swept downstream, thus revealing the motion of the fluid. The wire must be very small so that the resulting bubbles do not have a significant rise velocity with respect to the water. By pulsing the current through the electrode, it is possible to add a time frame to the visualization technique, thus making it a useful tool for velocity measurements. Fig. 13.8 shows patches of tiny hydrogen bubbles that were released with a pulsing action from noninsulated segments of a wire located to the left of the picture. Flow is from left to right, and the necked-down section of the flow passage has higher water velocity. Therefore, the patches are longer in that region. Next to the walls the patches of bubbles are shorter, indicating less distance traveled per unit of time. Other details concerning the marker methods of flow visualization are described by Macagno (12).

Smoke is often used as a marker in gaseous flows. One technique is to suspend a wire vertically across the flow field and allow oil to flow down the wire. The oil tends to accumulate in droplets along the wire. Applying a voltage to the wire vaporizes the oil, creating streaks from the droplets. Figure 13.9 is an example of a flow pattern revealed by such a method. Smoke generators that provide smoke by heating oils are also commercially available. It is also possible to position a thin sheet of laser light through the smoke field to obtain an improved spatial definition of the flow field indicated by the smoke.

FIGURE 13.8 _____

Combined time-streak markers (hydrogen bubbles); flow is from left to right. [After Kline (9)] (Courtesy of Education Development Center, Inc., Newton, MA.)

13.2 Instruments and Procedures for Measurement of Flow Rate

The methods of flow measurement can be classified in a broad sense as either direct or indirect. Direct methods involve the actual measurement of the quantity of flow (volume or weight) for a given time interval. Indirect methods involve the measurement of a pressure change (or some other variable), which in turn is directly related to the rate of flow. *Venturi meters, orifices,* and *flow nozzles* are all devices that apply indirect methods to measure the rate of flow in closed

FIGURE 13.9

Flow pattern in the wake of a flat plate.

conduits. *Weirs* are devices that employ indirect means to obtain flow rates in open channels. Still another indirect device is the *electromagnetic flow meter,* which operates on the principle that a voltage is generated when a conductor moves in a magnetic field. All of these methods and several others, as well as the *velocity–area integration* of flow measurement, will be discussed in this section.

Direct Volume or Weight Measurements

One of the most accurate methods of obtaining flow rates of liquids is to collect a sample of the flowing fluid over a given period of time t. Then the sample is weighed, and the average weight rate of flow is W/t, where W is the weight of the sample. The volume of a sample can also be measured (usually in a calibrated tank), and from this the average volume rate of flow is calculated as Ψ/t, where Ψ is the volume of the sample.

Velocity–Area Integration

If the velocity of flow in a pipe is symmetrical, the distribution of the velocity along a radial line can be used to determine the volume rate of flow (discharge) in the pipe. The discharge is obtained by numerically or graphically integrating $V\,dA$ over the cross-sectional area of the pipe. Thus a Pitot tube or hot-wire-anemometer velocity traverse across the flow section provides the primary data from which the discharge is evaluated. One procedure for evaluating this discharge is given in the next paragraph.

From test data of V versus r, one computes $2\pi Vr$ for various values of r. Then when $2\pi Vr$ versus r is plotted, the area under the resulting curve, Fig. 13.10, is equal to the discharge. This is so because $dQ = V\,dA = V(2\pi r\,dr)$, which is given by an elemental strip of area in Fig. 13.9. Hence the total area is equivalent to the total discharge.

FIGURE 13.10

*Graphical integration of
VdA in a pipe.*

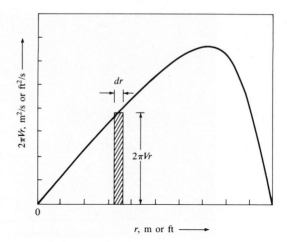

The data given in the accompanying table are for a velocity traverse of airflow in a pipe 100 cm in diameter. What is the volume rate of flow in cubic meters per second?

r, cm	V, m/s
0.00	50.0
5.00	49.5
10.00	49.0
15.00	48.0
20.00	46.5
25.00	45.0
30.00	43.0
35.00	40.5
40.00	37.5
45.00	34.0
47.50	25.0
50.00	0.0

Solution   First make a table of $2\pi Vr$ versus r as shown on the facing page. Now plot $2\pi Vr$ versus r. This plot is shown in the figure below the table. The area under the curve is measured to be 29.3 m³/s. Hence

$$Q = 29.3 \text{ m}^3/\text{s}$$

r, cm	$2\pi Vr$, m^2 s
0.00	0.0
5.00	15.6
10.00	30.8
15.00	45.2
20.00	58.4
25.00	70.7
30.00	81.1
35.00	89.1
40.00	94.2
45.00	96.1
47.50	74.6
50.00	0.0

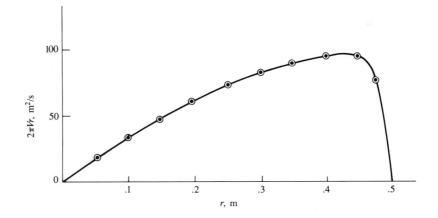

The foregoing procedure involving velocity–area integration is applicable to pipes where the velocity distribution is symmetrical with the axis of the pipe. However, even for flows that are unsymmetrical, it should be obvious that by summing $V\Delta A$ over a flow section one can obtain the total flow rate. Such a procedure is commonly used to obtain the discharge in streams and rivers.

Orifices

A restricted opening through which fluid flows is an *orifice*. If the geometric characteristics of the orifice plus the properties of the fluid are known, then the orifice can be used to measure flow rates. Consider flow through the sharp-edged pipe orifice shown in Fig. 13.11. Note that the streamlines continue to converge a short distance downstream of the plane of the orifice. Hence

FIGURE 13.11

Flow through a sharp-edged pipe orifice.

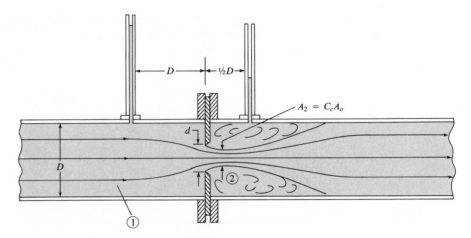

the minimum-flow area is actually smaller than the area of the orifice. To relate the minimum-flow area, often called the contracted area of the jet, or *vena contracta,* to the area of the orifice A_o, we use the contraction coefficient, which is defined as

$$A_j = C_c A_o$$

$$C_c = \frac{A_j}{A_o}$$

Then, for a circular orifice,

$$C_c = \frac{(\pi/4)d_j^2}{(\pi/4)d^2} = \left(\frac{d_j}{d}\right)^2$$

Because d_j and d_2 are identical, we also have $C_c = (d_2/d)^2$. At low values of the Reynolds number, C_c is a function of the Reynolds number. However, at high values of the Reynolds number, C_c is only a function of the geometry of the orifice. For d/D ratios less than 0.3, C_c has a value of approximately 0.62. However, as d/D is increased to 0.8, C_c increases to a value of 0.72.

We begin the derivation of the discharge equation for the orifice by writing Bernoulli's equation between section 1 and section 2 in Fig. 13.11:

$$\frac{p_1}{\gamma} + \frac{V_1^2}{2g} + z_1 = \frac{p_2}{\gamma} + \frac{V_2^2}{2g} + z_2$$

V_1 is eliminated by means of the continuity equation $V_1 A_1 = V_2 A_2$. Then solving for V_2 gives

$$V_2 = \left\{ \frac{2g[(p_1/\gamma + z_1) - (p_2/\gamma + z_2)]}{1 - (A_2/A_1)^2} \right\}^{1/2} \tag{13.1a}$$

However, $A_2 = C_c A_o$ and $h = p/\gamma + z$; so Eq. (13.1a) reduces to

$$V_2 = \sqrt{\frac{2g(h_1 - h_2)}{1 - C_c^2 A_o^2/A_1^2}} \tag{13.1b}$$

Our primary objective is to obtain an expression for discharge in terms of h_1, h_2, and the geometric characteristics of the orifice. The discharge is given by $V_2 A_2$. Hence, when we multiply both sides of Eq. (13.1b) by $A_2 = C_c A_o$, we obtain the desired result:

$$Q = \frac{C_c A_o}{\sqrt{1 - C_c^2 A_o^2 / A_1^2}} \sqrt{2g(h_1 - h_2)} \tag{13.2}$$

Equation (13.2) is the discharge equation for the flow of an incompressible inviscid fluid through an orifice. However, it is valid only at relatively high Reynolds numbers. For low and moderate values of the Reynolds number, viscous effects are significant, and an additional coefficient called the *coefficient of velocity* C_v must be applied to the discharge equation to relate the ideal to the actual flow.* Thus for viscous flow through an orifice, we have the following discharge equation:

$$Q = \frac{C_v C_c A_o}{\sqrt{1 - C_c^2 A_o^2 / A_1^2}} \sqrt{2g(h_1 - h_2)}$$

The product $C_v C_c$ is called the *discharge coefficient* C_d, and the combination $C_v C_c / (1 - C_c^2 A_o^2 / A_1^2)^{1/2}$ is called the *flow coefficient K*. Thus we have $Q = KA_o \sqrt{2g(h_1 - h_2)}$, where

$$K = \frac{C_d}{\sqrt{1 - C_c^2 A_o^2 / A_1^2}}$$

If Δh is defined as $h_1 - h_2$, then the final form of the discharge equation for an orifice reduces to

$$Q = KA_o \sqrt{2g\,\Delta h} \tag{13.3}$$

Note that if a differential pressure transducer is connected across the orifice, it will sense a change in pressure that is equivalent to $\gamma \Delta h$. Therefore, in this application one simply uses $\Delta p / \gamma$ in place of Δh in Eq. (13.3) and in the parameter at the top of Fig. 13.12. Experimentally determined values of K as a function of d/D and Reynolds number based on orifice size are given in Fig. 13.12. If Q is given, Re_d is equal to $4Q/\pi\,dv$. Then K is obtained from Fig. 13.12 (using the vertical lines and the bottom scale), and Δh is computed from Eq. (13.3). However, we are often confronted with the problem of determining the discharge Q when a certain value of Δh is given. When Q is to be determined, we do not have a direct way to obtain K by entering Fig. 13.12 with Re because Re is a function of the flow rate, which is still unknown. Hence another scale, which does not involve Q, is constructed on the graph of Fig. 13.12. The variables for

*At low Reynolds numbers the coefficient of velocity may be quite small; however, at Reynolds numbers above 10^5, C_v typically has a value close to 0.98. See Lienhard (10) for C_v analyses.

FIGURE 13.12

Flow coefficient K *and* Re$_d$/K *versus the Reynolds number for orifices, nozzles, and venturi meters. [After Johansen (5) and ASME (1)]*

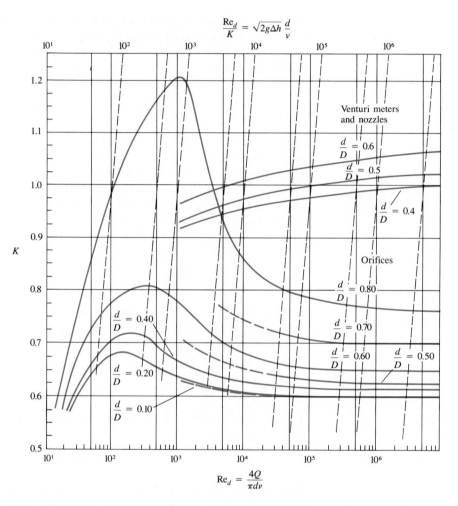

this scale are obtained in the following manner: Because Re$_d$ = 4Q/π dν and Q = Kπd^2/4$\sqrt{2g\,\Delta h}$, we can write Re$_d$ in terms of Δh as

$$\mathrm{Re}_d = K\sqrt{2g\,\Delta h}\,\frac{d}{\nu}$$

or

$$\frac{\mathrm{Re}_d}{K} = \sqrt{2g\,\Delta h}\,\frac{d}{\nu}$$

Thus the slanted dashed lines and the top scale are used in Fig. 13.12 when Δh is known and the flow rate is to be determined.

The literature on orifice flow contains numerous discussions concerning the optimum placement of pressure taps on both the upstream side and the downstream side of an orifice. The data given in Fig. 13.12 are for "corner taps." That is, on the upstream side the pressure readings were taken immediately upstream of the orifice plate (at the corner of the orifice plate and the pipe wall), and the downstream tap was at a similar downstream location. However, pressure data from flange taps (1 in. upstream and 1 in. downstream) and from the taps shown in Fig. 13.11 all yield virtually the same values for K—the differences are no greater than the deviations involved in reading Fig. 13.12. For more precise values of K with specific types of taps, see the ASME report on fluid meters (1).

Head Loss for Orifices

Some head loss occurs between the upstream side of the orifice and the vena contracta. However, this head loss is very small compared to the head loss that occurs downstream of the vena contracta. This downstream portion of the head loss is like that for an abrupt expansion. If we neglect all head loss except that due to the expansion of the flow, we have

$$h_L = \frac{(V_2 - V_1)^2}{2g}$$

where V_2 is the velocity at the vena contracta and V_1 is the velocity in the pipe. It can be shown that the ratio of this expansion loss, h_L, to the change in head across the orifice, Δh, is given as

$$\frac{h_L}{\Delta h} = \frac{\dfrac{V_2}{V_1} - 1}{\dfrac{V_2}{V_1} + 1} \tag{13.4}$$

Table 13.1 shows how the ratio increases with increasing values of V_2/V_1. It is obvious that an orifice is very inefficient from the standpoint of energy conservation.

TABLE 13.1 RELATIVE HEAD LOSS FOR ORIFICES

$V_2/V_1 \rightarrow$	1	2	4	6	8	10
$h_L/\Delta h \rightarrow$	0	0.33	0.60	0.71	0.78	0.82

EXAMPLE 13.2   A 15-cm orifice is located in a horizontal 24-cm water pipe, and a water-mercury manometer is connected to either side of the orifice. When the deflection on the manometer is 25 cm, what is the discharge in the system, and what head loss is produced by the orifice? Assume the water temperature is 20°C.

Solution The discharge is given by Eq. (13.3): $Q = KA_o\sqrt{2g\,\Delta h}$. To either enter Fig. 13.12 or use Eq. (13.3), we need first to evaluate Δh, the change in piezometric head in meters of fluid that is flowing. This is obtained by applying the equation of hydrostatics to the manometer shown in the figure.

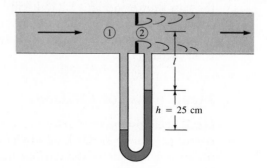

Writing the manometer equation from point 1 to point 2, we have

$$p_1 + \gamma_w l + \gamma_w h - \gamma_{\text{Hg}} h - \gamma_w l = p_2$$

Then $\qquad \dfrac{p_1 - p_2}{\gamma_w} = \Delta h = \dfrac{h(\gamma_{\text{Hg}} - \gamma_w)}{\gamma_w} = h\left(\dfrac{\gamma_{\text{Hg}}}{\gamma_w} - 1\right)$

For this example,

$$\Delta h = (0.25 \text{ m})(13.6 - 1) = 3.15 \text{ m of water}$$

The kinematic viscosity of water at 20°C is 1.0×10^{-6} m²/s. We now can compute $d\sqrt{2g\,\Delta h}/\nu$, the parameter that we need to enter Fig. 13.12:

$$\frac{d\sqrt{2g\,\Delta h}}{\nu} = \frac{0.15 \text{ m}\sqrt{2(9.81 \text{ m/s}^2)(3.15 \text{ m})}}{1.0 \times 10^{-6} \text{ m}^2/\text{s}} = 1.2 \times 10^6$$

From Fig. 13.12 with $d/D = 0.625$, we read K to be 0.66 (interpolated). Hence

$$Q = 0.66 A_o \sqrt{2g\,\Delta h}$$

$$= 0.66\,\frac{\pi}{4}d^2\sqrt{2(9.81 \text{ m/s}^2)(3.15 \text{ m})}$$

$$= 0.66(0.785)(0.15^2 \text{ m}^2)(7.86 \text{ m/s}) = 0.092 \text{ m}^3/\text{s}$$

The head loss is given by $h_L = (V_2 - V_1)^2/2g$. We can solve for V_2 if we know the coefficient of contraction $[V_2 = Q/(C_c A_o)]$. We can solve for the coefficient of contraction C_c from the definition of the flow coefficient K:

$$K = \frac{C_d}{1 - C_c^2 A_o^2/A_1^2}$$

where for this example $K = 0.66$. The ratio $(A_o/A_1)^2 = (0.625)^4 = 0.1526$ and $C_d = C_v C_c$.

Assuming $C_v = 0.98$ (see the footnote on page 611) and solving for C_c, we obtain $C_c = 0.633$. Then

$$V_2 = Q/(C_c A_o) = (0.092 \text{ m}^3/\text{s})/[(0.633)(\pi/4)(0.15^2 \text{ m}^2)] = 8.23 \text{ m/s}$$

$$V_1 = Q/A_{\text{pipe}} = (0.092 \text{ m}^3/\text{s})/[(\pi/4)(0.24^2 \text{ m}^2)] = 2.03 \text{ m/s}$$

and

$$h_L = (V_2 - V_1)^2/2g = (8.23 - 2.03)^2/(2 \times 9.81)$$

$$= 1.96 \text{ m} \qquad \blacktriangleleft$$

EXAMPLE 13.3   An air–water manometer is connected to either side of an 8-in. orifice in a 12-in. water pipe. If the maximum flow rate is 5 cfs, what is the deflection on the manometer? The water temperature is 60°F.

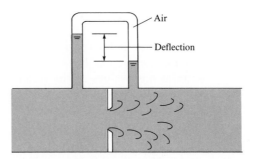

Solution   We first compute the Reynolds number $\text{Re}_d = 4Q/\pi \, d\nu$ so that we can enter Fig. 13.12 to obtain the flow coefficient K. Then K will be used in Eq. (13.3) to compute the deflection Δh.

$$\text{Re}_d = \frac{4Q}{\pi \, d\nu} = \frac{(4)(5)}{3.14(8/12)(1.22)(10^{-5})} = 7.8 \times 10^5$$

From Fig. 13.12, by interpolating between curves of $d/D = 0.6$ and $d/D = 0.7$ for $d/D = 8/12 = 0.667$, we read K to be approximately 0.68. Then from $Q = KA_o\sqrt{2g \, \Delta h}$ we obtain

$$\Delta h = \frac{Q^2}{2gK^2 A_o^2} = \frac{25}{64.4(0.68^2)[(\pi/4)(8/12)^2]^2} = 6.9 \text{ ft} \qquad \blacktriangleleft$$

The foregoing examples involved the determination of either Q or Δh for a given size of orifice. Another type of problem is determination of the diameter of the orifice for a given Q and Δh. For this type of problem a trial-and-error procedure is required. Because one knows the approximate value of K, that is guessed first. Then the diameter is solved for, after which a better value of K can be determined, and so on.

Venturi Meter

The orifice is a simple and accurate device for the measurement of flow; however, the head loss for an orifice is quite large. A device that operates on the same principle as the orifice but with a much smaller head loss is the *venturi meter.* The lower head loss results from streamlining the flow passage, as shown in Fig. 13.13. Such streamlining eliminates any jet contraction beyond the smallest flow section. Consequently, the coefficient of contraction has a value of unity, and the basic discharge equation for the venturi meter is

$$Q = \frac{A_2 C_d}{\sqrt{1 - (A_2/A_1)^2}} \sqrt{2g(h_1 - h_2)} \qquad (13.5)$$

$$= KA_2\sqrt{2g\,\Delta h} \qquad (13.6)$$

Note that the discharge equation for the venturi meter, Eq. (13.6), is the same as that for the orifice, Eq. (13.3). However, K for the venturi meter approaches unity at high values of the Reynolds number and small d/D ratios. This trend can be seen in Fig. 13.12, where values of K for the venturi meter are plotted along with similar data for the orifice.

FIGURE 13.13

Typical venturi meter.

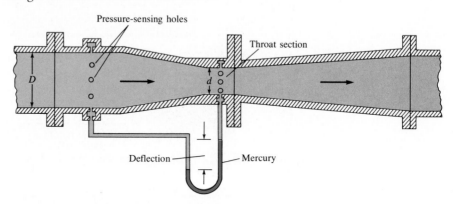

Pressure-sensing holes

Throat section

D

d

Deflection — Mercury

EXAMPLE 13.4 The pressure difference between the taps of a horizontal venturi meter carrying water is 35 kPa. If $d = 20$ cm and $D = 40$ cm, what is the discharge of water at 10°C?

Solution First we compute Δh. No elevation change occurs between the upstream section and the downstream section. Therefore Δh is

$$\Delta h = \frac{\Delta p}{\gamma_{\text{water}}} = \frac{35,000 \text{ N/m}^2}{9810 \text{ N/m}^3} = 3.57 \text{ m of water}$$

From Table A.5 in the Appendix we get $\nu = 1.31 \times 10^{-6}$ m²/s. Then we compute

$$\frac{d\sqrt{2g\,\Delta h}}{\nu} = \frac{0.20\sqrt{2(9.81)\,(3.57)}}{1.31(10^{-6})} = 1.28 \times 10^6$$

Then $$K = 1.02$$

Now we compute the discharge

$$Q = 1.02 A_2 \sqrt{2g\,\Delta h}$$

$$= 1.02(0.785)\,(0.20^2)\sqrt{2(9.81)\,(3.57)} = 0.268 \text{ m}^3/\text{s}$$ ◄

Flow Nozzles

The plate orifice must have a sharp upstream edge to yield the flow coefficients given in Fig. 13.12. If the upstream edge becomes rounded because of chemical or mechanical action, the orifice will not be a reliable flow-measuring device (C_d will increase). The *flow nozzle,* Fig. 13.14, has the advantage that it is less susceptible to wear. The flow nozzle has approximately the same flow coefficients as the venturi meter when the pressure taps are connected as in Fig. 13.14. However, the overall head loss across a flow nozzle is like that for the sharp-edged orifice. That is, the overall head loss is that of an abrupt expansion.

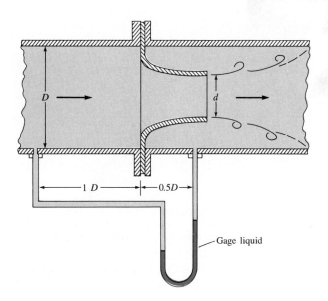

FIGURE 13.14

Typical flow nozzle.

Electromagnetic Flow Meter

All of the flow meters described so far require that some sort of obstruction be placed in the flow. The obstruction may be the rotor of a vane anemometer or the reduced cross section of an orifice or venturi meter. A meter that neither obstructs the flow nor requires pressure taps, which are subject to clogging, is the *electromagnetic flow meter*. Its basic principle is that a conductor that moves in a magnetic field produces an electromotive force. Hence liquids having a degree of conductivity generate a voltage between the electrodes, as in Fig. 13.15, and this voltage is proportional to the velocity of flow in the conduit. It is interesting to note that the basic principle of the electromagnetic flow meter was investigated by Faraday in 1832. However, practical application of the principle was not made until approximately a century later, when it was used to measure blood flow. Recently, with the need for a meter to measure the flow of liquid metal in nuclear reactors and with the advent of sophisticated electronic signal detection, this type of meter has found extensive commercial use.

    The main advantages of the electromagnetic flow meter are that the output signal varies linearly with the flow rate and that the meter causes no resistance to the flow. The major disadvantages are its high cost and its unsuitability for measuring gas flow.

    For a summary of the theory and application of the electromagnetic flow meter, the reader is referred to Shercliff (18). This reference also includes a comprehensive bibliography on the subject.

FIGURE 13.15

Electromagnetic flow meter.

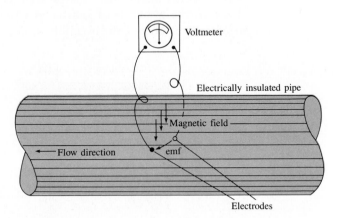

Ultrasonic Flow Meter

Another form of nonintrusive flow meter that is used in diverse applications ranging from blood-flow measurement to open-channel flow is the *ultrasonic flow meter.* Basically, there are two different modes of operation for ultrasonic flow meters. One mode involves measuring the difference in travel time for a sound wave traveling upstream and downstream between two measuring stations. The difference in travel time is proportional to flow velocity. The second mode of operation is based on the Doppler effect. When an ultrasonic beam is projected into an inhomogeneous fluid, some acoustic energy is scattered back to the transmitter at a different frequency (Doppler shift). The measured frequency difference is related directly to the flow velocity.

Turbine Flow Meter

The *turbine flow meter* consists of a wheel with a set of curved vanes (blades) mounted inside a duct. The volume rate of flow through the meter is related to the rotational speed of the wheel. This rotational rate is generally measured by a blade passing an electromagnetic pickup mounted in the casing. The meter must be calibrated for the given flow conditions. The turbine meter is versatile in that it can be used for either liquids or gases. It has an accuracy of better than 1% over a wide range of flow rates, and it operates with small head loss. The turbine flow meter is used extensively in monitoring flow rates in fuel-supply systems.

Vortex Flow Meter

The *vortex flow meter* consists of a cylinder mounted across a duct. This cylinder sheds vortices and gives rise to an oscillatory flow field. Proper design of the cylindrical element yields a constant Strouhal number for vortex shedding for a range of Reynolds numbers from 10^4 to 10^6. Over this flow range the fluid velocity and volume flow rate are directly proportional to the frequency of oscillation, which can be measured by several different methods. An advantage of this meter is that it has no moving parts (reliability), but it does give rise to a head loss comparable to that from other obstruction-type meters.

Rotameter

The *rotameter* consists of a vertical tapered tube through which the fluid flows upward and inside of which is located the rotor or active element of the meter (Fig. 13.16). Vanes cause the rotor to rotate slowly about the axis of the tube, thus keeping it centered within the tube. Because the velocity is lower at the top of the tube (greater flow section there) than at the bottom, the rotor seeks a neutral position where the drag on it just balances its weight. Thus the rotor "rides" higher or lower in the tube, depending on the rate of flow. A calibrated scale on the side of the tube indicates the rate of flow. Although venturi and orifice meters have better accuracy (approximately 1% of full scale) than the rotameter (approximately 5% of full scale), the rotameter offers other advantages, such as simplicity of design.

FIGURE 13.16

Rotameter.

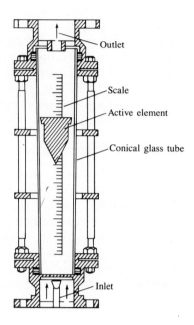

Rectangular Weir

A *weir* is an obstruction in an open channel over which liquid flows. The discharge over the weir is a function of the weir geometry and of the *head* on the weir. Consider flow over the weir in a rectangular channel, shown in Fig. 13.17. The head H on the weir is defined as the vertical distance between the weir crest and the liquid surface taken far enough upstream of the weir to avoid local free-surface curvature (see Fig. 13.17).

The basic discharge equation for the weir is derived by integrating $V dA = VL\, dh$ over the total head on the weir. Here L is the length of the weir and V is the velocity at any given distance h below the free surface. Neglecting streamline curvature and assuming negligible velocity of approach upstream of the weir, we obtain an expression for V by writing Bernoulli's equation between a point upstream of the weir and a point in the plane of the weir (see Fig. 13.18). This equation is

$$\frac{p_1}{\gamma} + H = (H - h) + \frac{V^2}{2g} \qquad (13.7)$$

Here the reference elevation is the elevation of the crest of the weir, and the reference pressure is atmospheric pressure. Therefore $p_1 = 0$, and Eq. (13.7) reduces to

$$V = \sqrt{2gh}$$

FIGURE 13.17
_____

*Definition sketch for a
sharp-crested weir. (a) Plan
view. (b) Elevation view.*

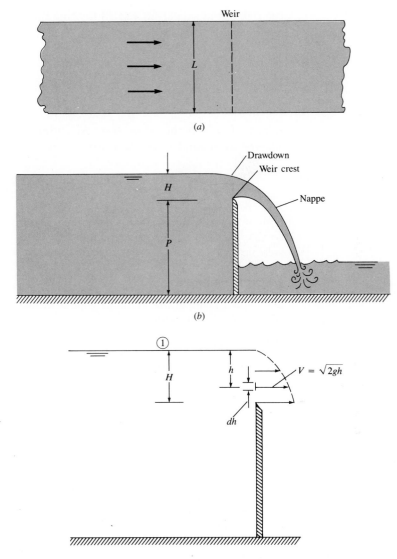

Then $dQ = \sqrt{2gh}\,L\,dH$, and the discharge equation becomes

$$Q = \int_0^H \sqrt{2gh}\,L\,dh$$

$$= \tfrac{2}{3}L\sqrt{2g}\,H^{3/2} \qquad\qquad (13.8)$$

In the case of actual flow over a weir, the streamlines converge downstream of the plane of the weir, and viscous effects are not entirely absent. Consequently, a discharge coefficient C_d must be applied to the basic expression on the right-hand side of Eq. (13.8) to bring the theory in line with the actual flow rate. Thus we have

$$Q = \tfrac{2}{3}C_d\sqrt{2g}\,LH^{3/2}$$

$$= K\sqrt{2g}\,LH^{3/2} \tag{13.9}$$

For low-viscosity liquids, the flow coefficient K is primarily a function of the relative head on the weir, H/P. An empirically determined equation for K adapted from Kindsvater and Carter (6) is

$$K = 0.40 + 0.05\frac{H}{P} \tag{13.10}$$

This is valid up to an H/P value of 10 as long as the weir is well ventilated so that atmospheric pressure prevails on both the top and the bottom of the weir nappe.

EXAMPLE 13.5 The head on a rectangular weir that is 60 cm high in a rectangular channel that is 1.3 m wide is measured to be 21 cm. What is the discharge of water over the weir?

Solution

$$Q = K\sqrt{2g}\,LH^{3/2} \quad \text{where } K = 0.40 + 0.05\frac{H}{P}$$

Hence
$$K = 0.40 + 0.05\left(\frac{21}{60}\right) = 0.417$$

Then
$$Q = 0.417\sqrt{2(9.81)}\,(1.3)\,(0.21^{3/2}) = 0.23 \text{ m}^3/\text{s} \quad \blacktriangleleft$$

When the rectangular weir does not extend the entire distance across the channel, as in Fig. 13.19, additional end contractions occur. Therefore, K will be smaller than for the weir without end contractions. The reader is referred to King and Brater (7) for additional information on flow coefficients for weirs.

Triangular Weir

A definition sketch for the triangular weir is shown in Fig. 13.20. The primary advantage of the triangular weir is that it has a higher degree of accuracy over a much wider range of flow than does the rectangular weir, because the average width of the flow section increases as the head increases.

FIGURE 13.19

Rectangular weir with end contractions. (a) Plan view. (b) Elevation view.

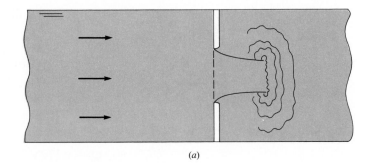

(a)

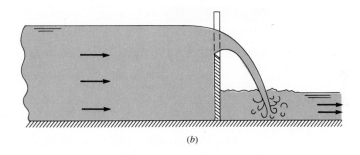

(b)

FIGURE 13.20

Definition sketch for the triangular weir.

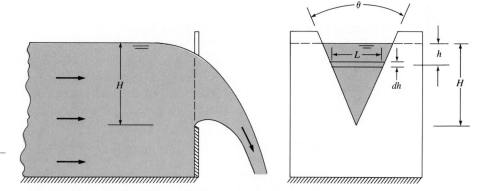

The basic discharge equation for the triangular weir is derived in the same manner as that for the rectangular weir. The differential discharge $dQ = VdA = VL\,dh$ is integrated over the total head on the weir. Thus we have

$$Q = \int_0^H \sqrt{2gh}\,(H - h)2\,\tan\!\left(\frac{\theta}{2}\right)dh$$

which integrates to

$$Q = \frac{8}{15}\sqrt{2g}\,\tan\!\left(\frac{\theta}{2}\right)H^{5/2}$$

However, a coefficient of discharge must still be used with the basic equation. Hence we have

$$Q = \frac{8}{15} C_d \sqrt{2g} \tan\left(\frac{\theta}{2}\right) H^{5/2} \qquad (13.11)$$

Experimental results with water flow over weirs with $\theta = 60°$ and $H > 2$ cm indicate that C_d has a value of 0.58. Hence the discharge equation for the triangular weir with these limitations is

$$Q = 0.179\sqrt{2g}\, H^{5/2} \qquad (13.12)$$

EXAMPLE 13.6 The head on a 60° triangular weir is measured to be 43 cm. What is the flow of water over the weir?

Solution We use the discharge equation $Q = 0.179\sqrt{2g}\, H^{5/2}$. Hence

$$Q = 0.179 \times \sqrt{2 \times 9.81} \times (0.43)^{5/2} = 0.096 \text{ m}^3/\text{s} \qquad \blacktriangleleft$$

More details about flow-measuring devices for incompressible flow can be found in references (14) and (17).

13.3 Measurement in Compressible Flow

Many of the measuring techniques used to determine the velocities and flow rates of liquids can be applied to compressible flows. However, since Bernoulli's equation is invalid for compressible flow, alternative relations must be used to correlate velocity and discharge with pressure difference.

Pressure Measurements

Static-pressure measurements can be made using the conventional static-pressure taps of a probe. However, if the boundary layer is disturbed by the presence of a shock wave in the vicinity of the pressure tap, the reading may not give the correct static pressure. The effect of the shock wave on the boundary layer is smaller if the boundary layer is turbulent. Therefore an effort is sometimes made to trip the boundary layer and ensure a turbulent boundary layer in the region of the pressure tap.

The stagnation pressure can be measured with a stagnation tube aligned with the local velocity vector. If the flow is supersonic, however, a shock wave forms around the tip of the probe, as shown in Fig. 13.21, and the stagnation pressure measured is that downstream of the shock wave and not that of the free

FIGURE 13.21

*Stagnation tube in
supersonic flow.*

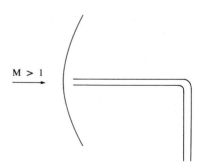

$$M > 1$$

stream. The stagnation pressure in the free stream can be calculated using the normal shock relationships, provided the free-stream Mach number is known.

Mach Number and Velocity Measurements

A Pitot-static tube can be used to measure Mach numbers in compressible flows. Taking the measured stagnation pressure as the total pressure, one can calculate the Mach number in subsonic flows from the total-to-static-pressure ratio according to Eq. (12.32):

$$M = \left\{ \frac{2}{k-1} \left[\left(\frac{p_t}{p} \right)^{(k-1)/k} - 1 \right] \right\}^{1/2}$$

It is interesting to note here that one must measure the stagnation and static pressures separately to determine the pressure ratio, whereas one needs only the pressure difference to calculate the velocity of a liquid flow.

If the flow is supersonic, then the indicated stagnation pressure is the pressure behind the shock wave standing off the tip of the tube. By taking this pressure as the total pressure downstream of a normal shock wave and the measured static pressure as the static pressure upstream of the shock wave, one can determine the Mach number of the free stream (M_1) from the static-to-total-pressure ratio (p_1/p_{t_2}) according to the expression

$$\frac{p_1}{p_{t_2}} = \frac{\{[2k/(k+1)]M_1^2 - [(k-1)/(k+1)]\}^{1/(k-1)}}{\{[(k+1)/2]M_1^2\}^{k/(k-1)}} \qquad (13.13)$$

which is called the *Rayleigh supersonic Pitot formula*. Note, however, that M_1 is an implicit function of the pressure ratio and must be determined graphically or by some numerical procedure. Many normal-shock tables, such as those in Reference (15), have p_1/p_{t_2} tabulated versus M_1, which enables one to find M_1 quite easily by interpolation.

Once the Mach number is determined, more information is needed to evaluate the velocity—namely, the local speed of sound. This can be done by a probe into the flow to measure total temperature and then calculating the static temperature using Eq. (12.23):

$$T = \frac{T_t}{1 + [(k-1)/2]\mathrm{M}_1^2}$$

The local speed of sound is then determined by Eq. (12.12):

$$c = \sqrt{kRT}$$

and the velocity is calculated from

$$V = \mathrm{M}_1 c$$

The hot-wire anemometer can also be used to measure velocity in compressible flows, provided it is calibrated to account for Mach-number effects.

Mass-Flow Measurement

Measuring the flow rate of a compressible fluid using a truncated nozzle was discussed in some detail in Chapter 12. Basically, the flow nozzle is a truncated nozzle located in a pipe, so the equations developed in Chapter 12 can be used to determine the flow rate through the flow nozzle. Strictly speaking, the flow rate so calculated should be multiplied by the discharge coefficient. For the high Reynolds numbers characteristic of compressible flows, however, the discharge coefficient can be taken as unity. If the flow at the throat of the flow nozzle is sonic, it is conceivable that the complex flow field existing downstream of the nozzle will make the reading from the downstream pressure tap difficult to interpret. That is, there can be no assurance that the measured pressure is the true back pressure. In such a case, it is advisable to use a venturi meter because the pressure is measured directly at the throat.

The mass-flow rate of a compressible fluid through a venturi meter can easily be analyzed using the equations developed in Chapter 12. Consider the venturi meter shown in Fig. 13.22. Writing the energy equation, Eq. (12.16), for the flow of an ideal gas between stations 1 and 2 gives

$$\frac{V_1^2}{2} + \frac{kRT_1}{k-1} = \frac{V_2^2}{2} + \frac{kRT_2}{k-1} \qquad (13.14)$$

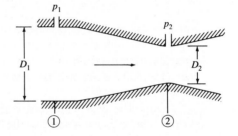

FIGURE 13.22

Venturi meter.

By conservation of mass, the velocity V_1 can be expressed as

$$V_1 = \frac{\rho_2 A_2 V_2}{\rho_1 A_1}$$

Substituting this result into Eq. (13.14), using the ideal-gas law to eliminate temperature, and solving for V_2 give

$$V_2 = \left\{ \frac{[2k/(k-1)][(p_1/\rho_1) - (p_2/\rho_2)]}{1 - (\rho_2 A_2/\rho_1 A_1)^2} \right\}^{1/2} \tag{13.15}$$

Assuming that the flow is isentropic,

$$\frac{p_1}{p_2} = \left(\frac{\rho_1}{\rho_2} \right)^k$$

the equation for the velocity at the throat can be rewritten as

$$V_2 = \left\{ \frac{[2k/(k-1)](p_1/\rho_1)[1 - (p_2/p_1)^{(k-1)/k}]}{1 - (p_2/p_1)^{2/k}(D_2/D_1)^4} \right\}^{1/2} \tag{13.16}$$

The mass flow is obtained by multiplying V_2 by $\rho_2 A_2$. This analysis, however, has been based on a one-dimensional flow, and two-dimensional effects can be accounted for by the discharge coefficient C_d. Thus we finally have

$$\dot{m} = C_d \rho_2 A_2 V_2 = C_d A_2 \left(\frac{p_2}{p_1} \right)^{1/k} \left\{ \frac{[2k/(k-1)]p_1 \rho_1[1 - (p_2/p_1)^{(k-1)/k}]}{1 - (p_2/p_1)^{2/k}(D_2/D_1)^4} \right\}^{1/2} \tag{13.17}$$

This equation is valid for all flow conditions, subsonic or supersonic, provided no shock waves occur between station 1 and station 2. It is good design practice to avoid supersonic flows in the venturi meter in order to prevent the formation of shock waves and the attendant total pressure losses. Also, the discharge coefficient can generally be taken as unity if no shock waves occur between 1 and 2.

EXAMPLE 13.7   Calculate the mass-flow rate of air through a venturi meter with a throat having a 1-cm diameter (D_2) in a pipe having a 3-cm diameter (D_1). The upstream static pressure is 150 kPa, and the throat pressure is 100 kPa. The static temperature of the air in the pipe is 27°C.

Solution   From Table A.2 in the Appendix, we find that $k = 1.4$ and $R = 287$ J/kg K. The gas density in the pipe is

$$\rho_1 = \frac{p_1}{RT_1} = \frac{150 \times 10^3 \text{ N/m}^2}{(287 \text{ J/kg K})(300 \text{ K})} = 1.74 \text{ kg/m}^3$$

Substituting the appropriate values in Eq. (13.17), we find that

$$\dot{m} = 1 \times 0.785 \times 10^{-4} \text{ m}^2 \left(\frac{1}{1.5}\right)^{0.714}$$

$$\times \left\{\frac{7 \times 150 \times 10^3 \text{ N/m}^2 \times 1.74 \text{ kg/m}^3[1 - (1/1.5)^{0.286}]}{[1 - (1/1.5)^{1.43}(1/3)^4]}\right\}^{1/2}$$

$$= 0.0264 \text{ kg/s} \qquad \blacktriangleleft$$

A square-edged orifice can also be used to measure the flow rate of compressible fluids. The discharge equation for liquids is multiplied by an empirical factor that accounts for compressibility effects (2). The resulting equation is

$$\dot{m} = YA_o K\sqrt{2\rho_1(p_1 - p_2)}$$

where K is the flow coefficient [see Eq. (13.3)], A_o is the orifice area, and Y is called the compressibility factor and is given by

$$Y = 1 - \left\{\frac{1}{k}\left(1 - \frac{p_2}{p_1}\right)\left[0.41 + 0.35\left(\frac{A_o}{A_1}\right)^2\right]\right\}$$

One must remember when using the equations above for Mach number and flow rate that absolute pressures, not gage pressures, must always be used.

Shock-Wave Visualization

When studying the qualitative features of a supersonic flow in a wind tunnel, it is important to be able to locate and identify the shock-wave pattern. Unfortunately, shock waves cannot be seen with the naked eye, so the application of some type of optical technique is necessary. There are three techniques by which shock waves can be seen: the shadowgraph, the interferometer, and the schlieren system. Each technique has its special application related to the type of information on density variation that is desired. The schlieren technique, however, finds frequent use in shock-wave visualization.

An illustration of the essential features of the schlieren system is given in Fig. 13.23. Light from the source s is collimated by lens L_1 to produce a parallel-light beam. The light then passes through a second lens L_2 and produces an image of the source at plane f. A third lens L_3 focuses the image on the display screen. A sharp edge, usually called the *knife edge*, is positioned at plane f so as to block out a portion of the light.

If a shock wave occurs in the test section, the light is refracted by the density change across the wave. As illustrated by the dashed line in Fig. 13.23, the refracted ray escapes the blocking effect of the knife edge and the shock wave appears as a lighter region on the screen. Of course, if the beam is refracted

FIGURE 13.23

Schlieren system.

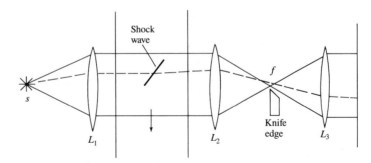

in the other direction, the knife edge blocks out more light, and the shock wave appears as a darker region. The contrast can be increased by intercepting more light with the knife edge.

Interferometry

The *interferometer* allows one to map contours of constant density and to measure the density changes in the flow field. The underlying principle is the phase shift of a light beam on passing through media of different densities. The system now employed almost universally is the Mach–Zender interferometer, shown in Fig. 13.24. Light from a common source is split into two beams as it passes through the first half-silvered mirror. One beam passes through the test section, the other through the reference section. The two beams are then recombined and projected onto a screen or photographic plate. If the density in the test section and that in the reference section are the same, there is no phase shift between the two beams, and the screen is uniformly bright. However, a change of

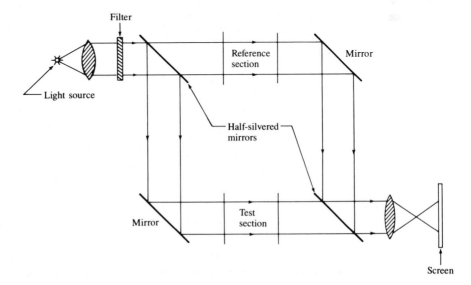

FIGURE 13.24

Schematic diagram of a Mach–Zender interferometer.

density in the test section changes the light speed of the test-section beam, and a phase shift is generated between the two beams. Upon recombination of the beams, this phase shift gives rise to a series of dark and light bands on the screen. Each band represents a uniform density contour, and the change in density across each band can be determined for a given system.

Problems

13.1 Without exceeding an error of 1%, what is the minimum air velocity that can be obtained using a 2-mm circular stagnation tube if the formula

$$V = \sqrt{2\,\Delta p_{\text{stag}}/\rho} = \sqrt{2gh_{\text{stag}}}$$

is used for computing the velocity? Assume standard atmospheric conditions.

13.2 Without exceeding an error of 1%, what is the minimum water velocity that can be obtained using a 2-mm circular stagnation tube if the formula

$$V = \sqrt{2\,\Delta p_{\text{stag}}/\rho} = \sqrt{2gh_{\text{stag}}}$$

is used for computing the velocity? Assume the water temperature is 20°C.

13.3 A stagnation tube 2 mm in diameter is used to measure the velocity in a stream of air as shown. What is the air velocity if the deflection on the air–water manometer is 0.80 mm? Air temperature = 10°C, and $p = 100$ kPa.

13.4 If the velocity in an airstream ($p_a = 98$ kPa, $T = 10$°C) is 13 m/s, what deflection will be produced on an air–water manometer if the stagnation tube is 2 mm in diameter?

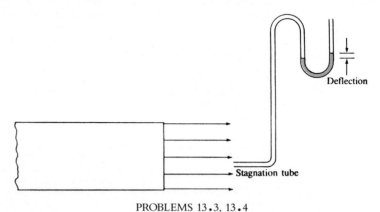

Deflection

Stagnation tube

PROBLEMS 13.3, 13.4

13.5 What would be the error in velocity determination if one used a C_p value of 1.00 for a circular stagnation tube instead of the true value? Assume the measurement is made with a stagnation tube 2 mm in diameter that is measuring air ($T = 25$°C, $p = 100$ kPa absolute) velocity for which the stagnation pressure reading is 5.00 Pa.

13.6 A velocity-measuring probe used frequently for measuring stack-gas velocities is shown. The probe consists of two tubes bent away from and toward the flow direction

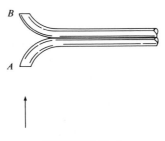

PROBLEM 13.6

and cut off on a plane normal to the flow direction, as shown. Assume the pressure coefficient is 1.0 at A and -0.4 at B. The probe is inserted in a stack where the temperature is 300°C and the pressure is 100 kPa absolute. The gas constant of the stack gases is 410 J/kg K. The probe is connected to a water manometer, and a 1.1-cm deflection is measured. Calculate the stack-gas velocity.

13.7 Water from a pipe is diverted into a tank for 3 min. If the weight of diverted water is measured to be 10,000 N, what is the discharge in cubic meters per second? Assume the water temperature is 20°C.

13.8 Water from a test apparatus is diverted into a calibrated volumetric tank for 6 min, 42 s. If the volume of diverted water is measured to be 78 m³, what is the discharge in cubic meters per second, gallons per minute, and cubic feet per second?

13.9 A velocity traverse in a 24-cm oil pipe yields the data in the table. What are the discharge, mean velocity, and ratio of maximum to mean velocity? Does the flow appear to be laminar or turbulent?

r, cm	0	1	2	3	4	5	6	7	8	9	10	10.5	11.0	11.5
V, m/s	8.7	8.6	8.4	8.2	7.7	7.2	6.5	5.8	4.9	3.8	2.5	1.9	1.4	0.7

13.10 A velocity traverse inside a 16-in. circular air duct yields the data in the table. What is the rate of flow in cubic feet per second and cubic feet per minute? What is the ratio of V_{max} to V_{mean}? Does it appear that the flow is laminar or turbulent? If $p = 14.3$ psia and $T = 70°F$, what is the mass-flow rate?

y^*	0.0	0.1	0.2	0.4	0.6	1.0	1.5	2.0	3.0	4.0	5.0	6.0	7.0	8.0
V, ft/s	0	72	79	88	93	100	106	110	117	122	126	129	132	135

*Distance from pipe wall, in.

13.11 The asymmetry of the flow in stacks means that flow velocity must be measured at several locations on the cross-flow plane. Consider the cross section of the cylindrical stack shown. The two access holes through which probes can be inserted are separated by 90°. Velocities can be measured at the five points shown (five-point method).

a. Determine the ratio r_m/D such that the areas of the five measuring segments are equal.

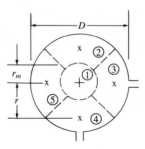

PROBLEM 13.11

b. Determine the ratio r/D (probe location) that corresponds to the centroid of the segment.

c. The data in the table are taken for a stack 2 m in diameter in which the gas temperature is 300°C, the pressure is 110 kPa absolute, and the gas constant is 400 J/kg K. The data represent the deflection on a water manometer connected to a conventional Pitot tube located at the measuring stations. Calculate the mass-flow rate.

Station	Δh, cm
1	1.2
2	1.1
3	1.1
4	0.9
5	1.05

13.12 Repeat Prob. 13.11 for the case in which three access holes are separated by 60° and seven measuring points are used. The diameter of the stack is 1.5 m, the gas temperature is 250°C, the pressure is 115 kPa absolute, and the gas constant is 420 J/kg K. The data in the table represent the deflection of a water manometer connected to a conventional Pitot tube at the measuring stations. Calculate the mass-flow rate.

Station	Δh, mm
1	8.2
2	8.6
3	8.2
4	8.9
5	8.0
6	8.5
7	8.4

13.13 Theory and experimental verification indicate that the mean velocity along a vertical line in a wide stream is closely approximated by the velocity at 0.6 depth. If the indicated velocities at 0.6 depth in a river cross section are measured, what is the discharge in the river?

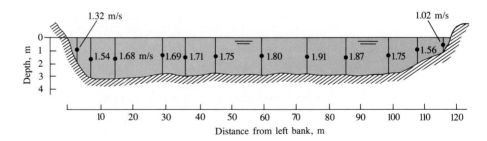

PROBLEM 13.13

13.14   A laser-Doppler anemometer (LDA) system is being used to measure the velocity of air in a tube. The laser is a ruby laser with a wavelength of 6394 angstroms. The angle between the laser beams is 20°. The time interval is determined by measuring the time between five spikes, as shown, on the signal from the photodetector. The time interval between the five spikes is 8 microseconds. Find the velocity.

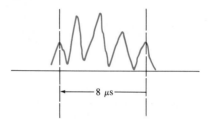

PROBLEM 13.14

13.15   For the jet and orifice shown, determine C_v, C_c, and C_d.

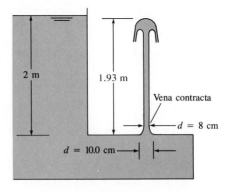

PROBLEM 13.15

13.16   A fluid jet discharging from a 2-cm orifice has a diameter of 1.73 cm at its vena contracta. What is the coefficient of contraction?

13.17 An orifice 6 in. in diameter ($d = 6$ in.) discharges water from the very large tank shown into the atmosphere. The surface of the water is under a pressure p_0, and $h = 6$ ft. If the discharge is 3.9 cfs, what is the pressure p_0?

13.18 An orifice 20 cm in diameter ($d = 20$ cm) discharges water from the large tank shown into the atmosphere. The surface of the water is under a pressure p_0 and is located 5 m above the center of the orifice ($h = 5$ m). If the discharge is 0.36 m³/s, what is the pressure p_0?

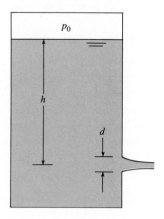

PROBLEMS 13.17, 13.18

13.19 A 6-in. orifice is placed in a 10-in. pipe, and a mercury manometer is connected to either side of the orifice. If the flow rate of water (60°F) through this orifice is 3 cfs, what will be the manometer deflection?

13.20 Determine the discharge of water through this 6-in. orifice that is installed in a 12-in. pipe.

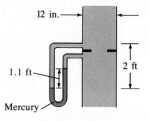

PROBLEM 13.20

13.21 Determine the rate of flow of water ($T = 10°C$) through the orifice shown if $h = 3$ m, $D = 10$ cm, and $d = 5$ cm.

13.22 Determine the discharge of water ($T = 60°F$) through the orifice shown if $h = 6$ ft, $D = 4$ in., and $d = 3$ in.

13.23 A pressure transducer is connected across an orifice to measure the flow rate of kerosene at 20°C. The pipe diameter is 2 cm, and the ratio of orifice diameter to pipe diameter is 0.6. The pressure differential as indicated by the transducer is 10 kPa. What is the mean velocity of the kerosene in the pipe?

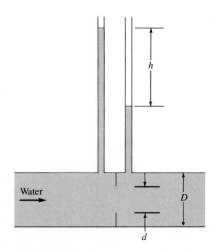

PROBLEMS 13.21, 13.22

13.24 The 10-cm orifice in the horizontal 30-cm pipe shown is the same size as the orifice in the vertical pipe. The manometers are mercury–water manometers, and water ($T = 20°C$) is flowing in the system. The gages are Bourdon-tube gages. The flow, at a rate of 0.1 m³/s, is to the right in the horizontal pipe and therefore downward in the vertical pipe. Is Δp as indicated by gages A and B the same as Δp as indicated by gages D and E? Determine their values. Is the deflection on manometer C the same as the deflection on manometer F? Determine the deflections.

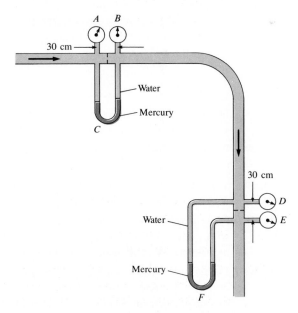

PROBLEM 13.24

13.25   A 15-cm plate orifice at the end of a 30-cm pipe is enlarged to 20 cm. With the same pressure drop across the orifice (approximately 50 kPa), what will be the percentage increase in discharge?

13.26   If water (20°C) is flowing through this 10-cm orifice, estimate the rate of flow.

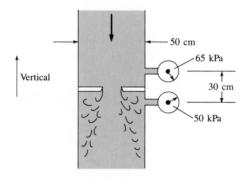

PROBLEM 13.26

13.27   Water ($T = 50°F$) is pumped at a rate of 20 cfs through the system shown in the figure. What differential pressure will occur across the orifice? What power must the pump supply to the flow for the given conditions? Also, draw the HGL and the EGL for the system. Assume $f = 0.015$ for the pipe.

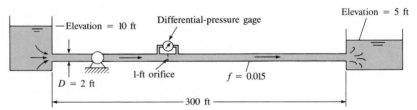

PROBLEM 13.27

13.28   Determine the size of orifice required in a 15-cm pipe to measure 0.03 m³/s of water with a deflection of 1 m on a mercury–water manometer.

13.29   What is the discharge of gasoline ($S = 0.68$) in a 10-cm horizontal pipe if the differential pressure across a 6-cm orifice in the pipe is 35 kPa?

13.30   What size orifice is required to produce a change in head of 8 m for a discharge of 2 m³/s of water in a pipe 1 m in diameter?

13.31   An orifice is to be designed to have a change in pressure of 50 kPa across it (measured with a differential-pressure transducer) for a discharge of 3.0 m³/s of water in a pipe 1.2 m in diameter. What diameter should the orifice have to yield the desired results?

13.32   Hemicircular orifices such as the one shown are sometimes used to measure the flow rate of liquids that also transport sediments. The opening at the bottom of the pipe allows free passage of the sediment. Derive a formula for Q as a function of Δp, D, and other relevant variables associated with the problem. Then, using that formula and guessing any unknown data, estimate the water discharge through such an orifice when Δp is read as 80 kPa and flow is in a 30-cm pipe.

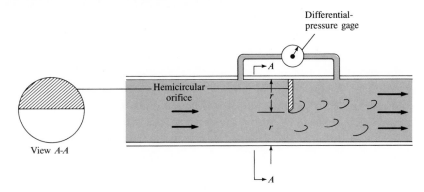

PROBLEM 13.32

13.33 Water flows through a venturi meter that has a 30-cm throat. The venturi meter is in a 60-cm pipe. What deflection will occur on a mercury–water manometer connected between the upstream and throat sections if the discharge is 0.59 m³/s? Assume $T = 20°C$.

13.34 What is the throat diameter required for a venturi meter in a 200-cm horizontal pipe carrying water with a discharge of 10 m³/s if the differential pressure between the throat and the upstream section is to be limited to 200 kPa at this discharge?

13.35 Estimate the rate of flow of water through the venturi meter shown.

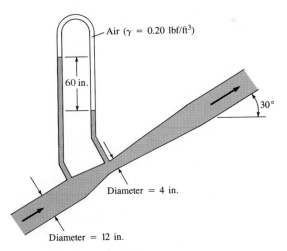

PROBLEM 13.35

13.36 An orifice and a venturi meter are placed in series in a 12-in. pipe as shown. Both are the same size ($d_o = 6$ in. $= d_v$). For a given discharge of water, one would expect a) $h_o = h_v$, b) $h_o > h_v$, c) $h_o < h_v$.

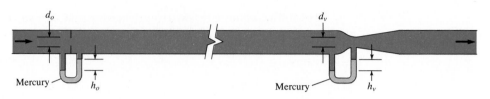

PROBLEM 13.36

13.37 In which direction is flow occurring in Prob. 13.36?

13.38 When no flow occurs through the venturi meter, the indicator on the differential-pressure gage is straight up and indicates a Δp of zero. When 5 cfs of water flow to the right, the differential-pressure gage indicates $\Delta p = +10$ psi. If the flow is now reversed

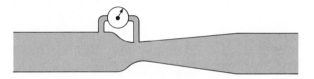

PROBLEM 13.38

and 5 cfs flow to the left through the venturi meter, in which range would Δp fall? a) $\Delta p < -10$ psi, b) -10 psi $< \Delta p < 0$, c) $0 < \Delta p < 10$ psi, d) $\Delta p = 10$ psi.

13.39 The pressure differential across this venturi meter is 110 kPa. What is the discharge of water through it?

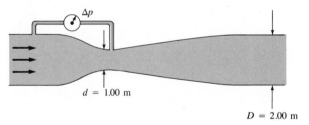

PROBLEM 13.39

13.40 Engineers are calibrating a not very well-designed venturi meter for the flow of an incompressible liquid by relating the pressure difference between taps 1 and 2 to the discharge. By applying Bernoulli's equation and assuming a quasi-one-dimensional flow (velocity uniform across every cross section), the engineers find that

$$Q_0 = A_2[2(p_1 - p_2)/\rho]^{0.5}[1 - (d/D)^4]^{-0.5}$$

where D and d are the duct diameters at stations 1 and 2. However, they realize that the flow is not quasi-one-dimensional and that the pressure at tap 2 is not equal to the average pressure in the throat because of streamline curvature. Thus the engineers introduce a correction factor K into the foregoing equation to yield

$$Q = KQ_0$$

Use your knowledge of pressure variation across curved streamlines to decide whether K is larger or smaller than unity, and support your conclusion by presenting a rational argument.

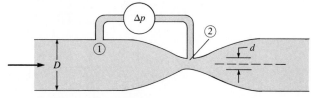

PROBLEM 13.40

13.41 The differential-pressure gage on the venturi meter shown reads 6.0 psi, $h = 30$ in., $d = 6$ in., and $D = 12$ in. What is the discharge of water in the system? Assume $T = 50°F$.

13.42 The differential-pressure gage on the venturi meter reads 50 kPa, $d = 20$ cm, $D = 40$ cm, and $h = 80$ cm. What is the discharge of gasoline ($S = 0.69$, $\mu = 3 \times 10^{-4}$ N · s/m²) in the system?

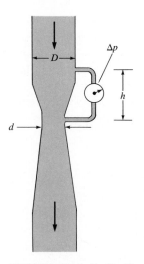

PROBLEMS 13.41, 13.42

13.43 A flow nozzle has a throat diameter of 2 cm and a beta ratio (d/D) of 0.5. Water flows through the nozzle, creating a pressure difference across the nozzle of 10 kPa. The viscosity of the water is 10^{-6} m²/s, and the density is 1000 kg/m³. Find the discharge.

13.44 Water flows through an annular venturi consisting of a body of revolution mounted inside a pipe. The pressure is measured at the minimum area and upstream of the body. The pipe is 5 cm in diameter, and the body of revolution is 2.5 cm in diameter. A head difference of 1 m is measured across the pressure taps. Find the discharge in cubic meters per second.

13.45 What is the head loss in terms of $V_0^2/2g$ for the flow nozzle shown?

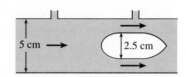

PROBLEM 13.44

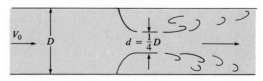

PROBLEM 13.45

13.46 A vortex flow meter is used to measure the discharge in a duct 5 cm in diameter. The diameter of the shedding element is 1 cm. The Strouhal number based on the shedding frequency from one side of the element is 0.2. A signal frequency of 50 Hz is measured by a pressure transducer mounted downstream of the element. What is the discharge in the duct?

13.47 A rotameter operates by aerodynamic suspension of a weight in a tapered tube. The scale on the side of the rotameter is calibrated in scfm of air—that is, cubic feet per minute at standard conditions ($p = 1$ atm and $T = 68°F$). By considering the balance of weight and aerodynamic force on the weight inside the tube, determine how the readings would be corrected for nonstandard conditions. In other words, how would the actual cubic feet per minute be calculated from the reading on the scale, given the pressure, temperature, and gas constant of the gas entering the rotameter?

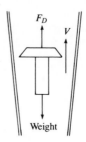

PROBLEM 13.47

13.48 A rotameter is used to measure the flow rate of a gas with a density of 0.8 kg/m³. The scale on the rotameter indicates 3 liters/s. However the rotameter is calibrated for a gas with a density of 1.2 kg/m³. What is the actual flow rate of the gas (in liters per second)?

13.49 One mode of operation of ultrasonic flow meters is to measure the travel times between two stations for a sound wave traveling upstream and then downstream with the flow. The downstream propagation speed with respect to the measuring stations is $c + V$, where c is the sound speed and V is the flow velocity. Correspondingly, the upstream propagation speed is $c - V$.

a. Derive an expression for the flow velocity in terms of the distance between the two stations, L; the difference in travel times, Δt; and the sound speed.

b. The sound speed is typically much larger than V ($c \gg V$). With this approximation, express V in terms of L, c, and Δt.

c. A 10-ms time difference is measured for waves traveling 20 m in a gas where the speed of sound is 300 m/s. Calculate the flow velocity.

13.50 Water flows over a rectangular weir that is 3 m wide and 30 cm high. If the head on the weir is 13 cm, what is the discharge in cubic meters per second?

13.51 The head on a 60° triangular weir is 30 cm. What is the discharge over the weir in cubic meters per second?

13.52 Water flows over two rectangular weirs. Weir A is 5 ft long in a channel 10 ft wide; weir B is 5 ft long in a channel 5 ft wide. Both weirs are 2 ft high. If the head on both weirs is 1.00 ft, then one can conclude that a) $Q_A = Q_B$, b) $Q_A > Q_B$, c) $Q_A < Q_B$.

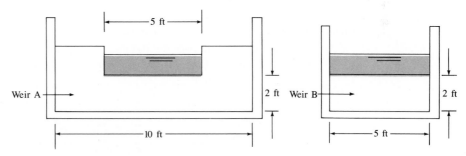

PROBLEM 13.52

13.53 A 1-ft-high rectangular weir (weir 1) is installed in a 2-ft-wide rectangular channel, and the head on the weir is observed for a discharge of 10 cfs. Then the 1-ft weir is replaced by a 2-ft-high rectangular weir (weir 2), and the head on the weir is observed for a discharge of 10 cfs. The ratio H_1/H_2 should be a) equal to 1.00, b) less than 1.00, c) greater than 1.00.

13.54 A 3-m-long rectangular weir is to be constructed in a 3-m-wide rectangular channel, as shown (a). The maximum flow in the channel will be 4 m³/s. What should be the height P of the weir to yield a depth of water of 2 m in the channel upstream of the weir?

13.55 Consider the rectangular weir described in Prob. 13.54. When the head is doubled, the discharge is a) doubled, b) less than doubled, c) more than doubled.

13.56 At one end of a rectangular tank 1 m wide is a sharp-crested rectangular weir 1 m high. In the bottom of the tank is a 10-cm sharp-edged orifice. If 0.10 m³/s of water flows into the tank and leaves the tank both through the orifice and over the weir, what depth will the water in the tank attain?

13.57 What is the discharge over a rectangular weir 1 m high in a channel 2 m wide if the head on the weir is 21 cm?

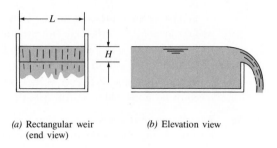

(a) Rectangular weir
(end view)

(b) Elevation view

PROBLEMS 13.54, 13.55

13.58   What is the water discharge over a rectangular weir 2 ft high and 10 ft long in a rectangular channel 10 ft wide if the head on the weir is 1.0 ft?

13.59   A reservoir drains through a rectangular weir that is 2 ft long. The initial head on the weir is 4 in. The surface area of the reservoir is 400 ft^2. The flow coefficient is 0.4. How long will it take the level of the water in the reservoir to lower by 2 in.?

13.60   At a particular instant water flows into the tank shown through pipes *A* and *B*, and it flows out of the tank over the rectangular weir at *C*. The tank width and weir length (dimensions normal to page) are 2 ft. Then, for the given conditions, is the water level in the tank rising or falling?

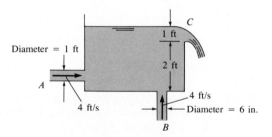

PROBLEM 13.60

13.61   Water flows from the first reservoir to the second over a rectangular weir with a width-to-head ratio of 3. The height *P* of the weir is twice the head. The water from the second reservoir flows over a 60° triangular weir to a third reservoir. The discharge across both weirs is the same. Find the ratio of the head on the rectangular weir to the head on the triangular weir.

13.62   A rectangular irrigation canal 3 m wide carries water with a discharge of 6 m^3/s. What height of rectangular weir installed across the canal will raise the water surface to a level 2 m above the canal floor?

13.63   The head on a 60° triangular weir is 1.6 ft. What is the discharge of water over the weir?

13.64   An engineer is designing a triangular weir for measuring the flow rate of a stream of water that has a discharge of 10 cfm. The weir has an included angle of 45° and a coefficient of discharge of 0.6. Find the head on the weir.

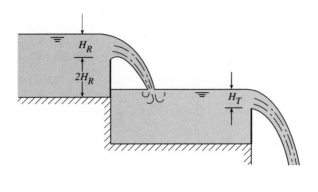

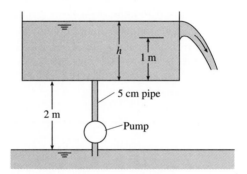

PROBLEM 13.65

13.65 A pump is used to deliver water at 10°C from a well to a tank. The bottom of the tank is 2 m above the water surface in the well. The pipe is commercial steel 2.5 m long with a diameter of 5 cm. The pump develops a head of 20 m. A triangular weir with an included angle of 60° is located in a wall of the tank with the bottom of the weir 1 m above the tank floor. Find the level of the water in the tank above the floor of the tank.

13.66 A Pitot tube is used to measure the Mach number in a compressible subsonic flow of air. The stagnation pressure is 140 kPa, and the static pressure is 100 kPa. The total temperature of the flow is 300 K. Determine the Mach number and the flow velocity.

13.67 Use the normal shock-wave relationships developed in Chapter 12 to derive the Rayleigh supersonic Pitot formula.

13.68 The static and stagnation pressures measured by a Pitot tube in a supersonic airflow are 54 kPa and 200 kPa, respectively. The total temperature is 350 K. Determine the Mach number and the velocity of the free stream.

13.69 A venturi meter is used to measure the flow of helium in a pipe. The pipe is 1 cm in diameter, and the throat diameter is 0.5 cm. The measured upstream and throat pressures are 120 kPa and 80 kPa, respectively. The static temperature of the helium in the pipe is 17°C. Determine the mass-flow rate.

13.70   The mass-flow rate of methane is measured with a square-edged orifice. The pipe diameter is 2 cm, and the orifice diameter is 0.8 cm. The upstream and downstream pressure taps measure 150 kPa and 110 kPa absolute, respectively. The static temperature in the pipe is 300 K. Determine the mass-flow rate.

13.71   The mass-flow rate of air is measured by an orifice with a diameter of 1 cm in a 2-cm pipe. The upstream pressure is 150 kPa, and the downstream pressure is 100 kPa (absolute). The upstream air density is 1.8 kg/m^3, and the kinematic viscosity is $1.8 \cdot 10^{-5}$ m^2/s. The ratio of specific heats, k, is 1.4. Calculate the mass-flow rate.

13.72   Hydrogen at atmospheric pressure and 15°C flows through a sharp-edged orifice with a beta ratio, d/D, of 0.5 in a 2-cm pipe. The pipe is horizontal, and the pressure change across the orifice is 1 kPa. The flow coefficient is 0.62. Find the mass flow (in kilograms per second) through the orifice.

13.73   A hole 0.25 in. in diameter is accidently punctured in a line carrying natural gas (methane). The pressure in the pipe is 50 psig, and the atmospheric pressure is 14 psia. The temperature in the line is 70°F. What is the rate at which the methane leaks through the hole (in lbm/s)? The hole can be treated as a truncated nozzle.

References

1. ASME. *Fluid Meters—Their Theory and Applications.* ASME, 1959.

2. Bradshaw, P. *An Introduction to Turbulence and Its Measurement.* Pergamon Press, New York, 1971.

3. Goldstein, R. J. (ed.). *Fluid Mechanics Measurements.* Hemisphere Publishing, New York, 1983.

4. Holman, J. P. *Experimental Methods for Engineers.* McGraw-Hill, New York, 1971.

5. Johansen, F. C. *Proc. Roy. Soc. London, Ser. A,* 125 (1930).

6. Kindsvater, Carl E., and R. W. Carter. "Discharge Characteristics of Rectangular Thin-Plate Weirs." *Trans. Am. Soc. Civil Eng.,* 124 (1959), 772–822.

7. King, H. W., and E. F. Brater. *Handbook of Hydraulics.* McGraw-Hill, New York, 1963.

8. King, L. V. *Phil. Trans. Roy. Soc. London, Ser. A,* 14 (1914), 214.

9. Kline, J. J. "Flow Visualization." In *Illustrated Experiments in Fluid Mechanics, The NCFMF Book of Film Notes.* Educational Development Center, 1972.

10. Lienhard, J. H., V, and J. H. Lienhard, IV. "Velocity Coefficients for Free Jets from Sharp-Edged Orifices." *Jour. Fluids Engineering,* Transactions of ASME, 106, (March 1984).

11. Liepmann, H. W., and A. Roshko. *Elements of Gasdynamics.* John Wiley, New York, 1957.

12. Macagno, Enzo O. "Flow Visualization in Liquids." *Iowa Inst. Hydraulic Res. Rept.,* 114 (1969).

13. Macmillan, F. A. "Viscous Effects on Flattened Pitot Tubes at Low Speeds." *J. Roy. Aeronaut. Soc.,* technical note (December 1954).

14. Miller, R. W. *Flow Measurement Engineering Handbook.* McGraw-Hill, New York, 1983.

15. NACA. "Equations, Tables, and Charts for Compressible Flow." TR 1135 (1953).

16. Ower, E., and R. C. Pankhurst. *The Measurement of Air Flow.* Pergamon Press, New York, 1966.

17. Scott, R. W. W., ed. *Developments in Flow Measurement—1.* Applied Science Publisher, Englewood, N.J., 1982.

18. Shercliff, J. A. *Electromagnetic Flow-Measurement.* Cambridge University Press, New York, 1962.

19. Wendt, R. E., Jr. (ed.). *Flow Measuring Devices.* "Its Measurement and Control in Sciences and Industry," part 2. Instrument Society of America, 1974.

CHAPTER

14

Turbomachinery

Water with high velocity discharges from the nozzle (far left side of photo). The water jet strikes the moving buckets (deflecting vanes) on the periphery of the impulse turbine wheel. The force acting on the moving buckets produces the desired hydropower. (Courtesy of Sulzer–Escher Wyss Ltd., Switzerland)

A ll turbomachines include either a rotating propeller or rotating vanes that change flow velocity and/or pressure within the fluid. In the process, either the machine does work on the fluid or the fluid does work on the machine. Examples of machines that do work on fluids are *pumps, blowers,* and *compressors,* whereas machines that have work done on them are called *turbines.* In this chapter we shall first consider the characteristics of propellers; then we shall see how propellers are incorporated into certain types of pumps in which the flow is essentially parallel to the axis of the pump shaft. These are called *axial-flow pumps.* Next we will study *radial-flow pumps* in which the fluid flows radially outward from the shaft axis. Finally, we will consider the elementary characteristics and theory of turbines.

14.1 Propeller Theory

The design of a propeller is based on the fundamental principles of airfoil theory. For example, if we observe a section of the propeller in Fig. 14.1, we see the analogy between the lifting vane and the propeller. This propeller is rotating at an angular speed ω, and the speed of advance of the airplane and propeller is V_0. If we focus on an elemental section of the propeller, Fig. 14.1c, we note that the given section has a velocity made of components V_0 and V_t. Here V_t is the tangential velocity, $V_t = r\omega$, resulting from the rotation of the propeller. Reversing and adding the velocity vectors V_0 and V_t yield the velocity of the air relative to the particular propeller section (Fig. 14.1d). Thus we have a flow situation that is directly analogous to that for the airfoil.

The propeller is designed to produce thrust, and since the greatest contribution to thrust comes from the lift force F_L, the goal is to maximize lift and minimize drag, F_D. For a given shape of propeller section, the optimum angle of attack can be determined from data such as are given in Fig. 11.23. Because the angle θ (Fig. 14.1d) decreases with an increase in r, for a well-designed propeller

FIGURE 14.1

Propeller motion.
(a) Airplane motion.
(b) View A–A.
(c) View B–B.
(d) Velocity relative to blade element.

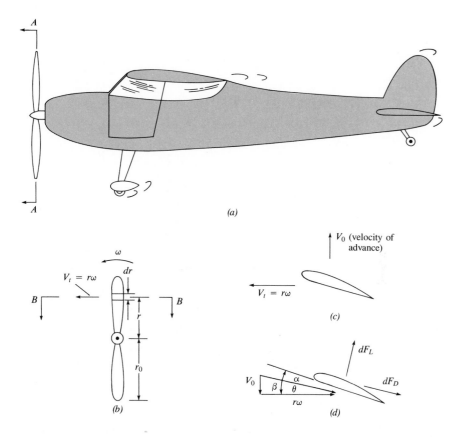

the propeller blade must be warped in order to obtain the optimum angle of attack along the length of the blade.

Blade Analysis

To analyze the forces on a blade element, we will consider an enlarged view of Fig. 14.1*d* as shown in Fig. 14.2. By definition, the lift is normal to the relative air velocity V_R, and the drag is the force acting parallel to the same velocity. Thus one component of the lift force, $dF_L \cos \theta$, produces a positive thrust component, and a component of the drag, $dF_D \sin \theta$, produces a negative thrust component:

$$dF_{\text{thrust}} = dF_L \cos \theta - dF_D \sin \theta \qquad (14.1)$$

In a similar manner, it can be shown that the tangential force for this blade element is given by

$$dF_{\text{tang}} = dF_L \sin \theta + dF_D \cos \theta \qquad (14.2)$$

FIGURE 14.2

*Definition sketch for
propeller-blade element.*

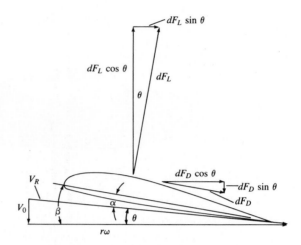

The torque T opposing rotation of the propeller is given by

$$dT = (dF_L \sin \theta + dF_D \cos \theta)r \qquad (14.3)$$

From Chapter 11 the lift per unit length for a two-dimensional airfoil is given as $F_L/\ell = \frac{1}{2}C_L c\rho V_0^2$ and the drag per unit length is $F_D/\ell = \frac{1}{2}C_D c\rho V_0^2$. Hence the incremental lift and drag on a radial element dr are $dF_L = \frac{1}{2}C_L c\rho V_R^2\, dr$ and $dF_D = \frac{1}{2}C_D c\rho V_R^2\, dr$, respectively. The incremental thrust and torque for the blade element are, respectively,

$$dF_{\text{thrust}} = \frac{1}{2}(C_L \cos \theta - C_D \sin \theta)c\rho V_R^2\, dr$$

$$dT = \frac{1}{2}(C_L \sin \theta + C_D \cos \theta)c\rho V_R^2 r\, dr$$

The total thrust is obtained by integrating the incremental thrust over the length of the propeller:

$$F_{\text{thrust}} = \frac{1}{2}N \int_{r_h}^{r_0} (C_L \cos \theta - C_D \sin \theta)c\rho V_R^2\, dr \qquad (14.4)$$

In a similar manner, the total torque producing rotation of the propeller is given as

$$T = \frac{1}{2}N \int_{r_h}^{r_0} (C_L \sin \theta + C_D \cos \theta)c\rho V_R^2 r\, dr \qquad (14.5)$$

Note that in Eqs. (14.4) and (14.5), N is the number of blades and r_h is the radius of the propeller hub.

Thrust and Power Relationships

In Eq. (14.4), ρ is assumed to be constant and $V_R = r\omega/\cos\theta$, so that Eq. (14.4) becomes

$$F_{\text{thrust}} = \tfrac{1}{2}\rho N\omega^2 \int_{r_h}^{r_0} \left(\frac{C_L}{\cos\theta} - C_D\frac{\tan\theta}{\cos\theta}\right) cr^2\,dr \tag{14.6}$$

$$\frac{F_{\text{thrust}}}{\tfrac{1}{2}N\rho r_0^2\omega^2} = \int_{r_h}^{r_0} \left(\frac{C_L}{\cos\theta} - C_D\frac{\tan\theta}{\cos\theta}\right)\left(\frac{r}{r_0}\right)^2 c\,dr \tag{14.7}$$

Since $c\,dr$ is the differential area dA of the propeller, Eq. (14.7) is written as

$$\frac{F_{\text{thrust}}}{\tfrac{1}{2}N\rho r_0^2\omega^2} = \int_A \left(\frac{C_L}{\cos\theta} - C_D\frac{\tan\theta}{\cos\theta}\right)\left(\frac{r}{r_0}\right)^2 dA \tag{14.8}$$

The right-hand side of Eq. (14.8) can be expressed in functional form as

$$\int_A \left(\frac{C_L}{\cos\theta} - C_D\frac{\tan\theta}{\cos\theta}\right)\left(\frac{r}{r_0}\right)^2 dA = C_1 A$$

Here C_1 is a dimensionless coefficient that depends on the shape of the propeller blade, the speed of rotation ω, and the speed of advance. The area A is proportional to D^2, where D is the diameter of the propeller. Then, for a particular propeller, Eq. (14.8) can be expressed as

$$F_{\text{thrust}} = \tfrac{1}{2}\rho r_0^2\omega^2 C_2 D^2$$

or

$$= C_T\rho n^2 D^4 \tag{14.9}$$

where n is the speed of rotation, rps (revolutions per second); D is the diameter; and C_T is the thrust coefficient. The *thrust coefficient* is a function of the geometry of the propeller and of the *advance diameter ratio, V_0/nD*. This ratio in effect establishes the angle of attack for each section of a given propeller. Hence we have defined the thrust coefficient:

$$C_T = \frac{F_{\text{thrust}}}{\rho D^4 n^2} \tag{14.10}$$

where

$$C_T = f\left(\frac{V_0}{nd}\right)$$

The thrust coefficient can be considered a similarity parameter for propellers in the same way in which the lift coefficient C_L was used for wings.*

In a manner similar to the development of the thrust coefficient, it can be shown that the torque is given by

$$T = CN\rho r_0^3\omega^2 D^2$$

---

*C_T is also a function of Re and M. However, these effects are usually small compared to the effect of the advance diameter ratio.

However, our main interest in the torque is as a means for evaluating the power that must be supplied to the propeller. The power input is $T\omega$. Hence we can express power as

$$P = C_P \rho D^5 n^3$$

or

$$C_P = \frac{P}{\rho D^5 n^3} \tag{14.11}$$

Here C_P is the *power coefficient*. Like C_T, it is a function of the propeller geometry and V_0/nD. The functional relationship between C_P and V_0/nD for a particular propeller is shown along with C_T in Fig. 14.3. Once the relationships between C_T and V_0/nD and between C_P and V_0/nD are found (usually experimentally) for a given propeller, they can be used to predict the operation of geometrically similar propellers over a wide range of speed and size. Example 14.1 illustrates such an application.

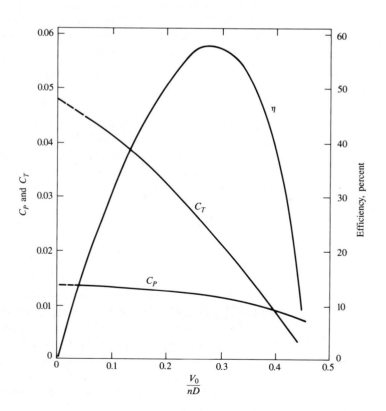

FIGURE 14.3

Dimensionless performance curves for a typical propeller; D = 2.90 m, n = 1400 rpm. [After Weick (16)]

The efficiency of a propeller is defined as the ratio of the power output — that is, thrust times velocity of advance—to the power input. Hence the efficiency η is given as

$$\eta = \frac{F_T V_0}{P} = \frac{C_T \rho D^4 n^2 V_0}{C_P \rho D^5 n^3}$$

which reduces to $\eta = C_T/C_P$ times the advance diameter ratio. Figure 14.3 includes the efficiency for the given propeller.

As a point of interest, the curves of C_T and C_P are obtained from performance characteristics of a given propeller operated with different values of V_0, as shown in Fig. 14.4. Even though the data for the curves are obtained for a given propeller size and a given angular speed, the curves can be applied by similarity principles to the same propeller operating at different angular speeds and to geometrically similar propellers of different sizes and at different angular speeds. There will be some change in characteristics with different conditions, but the deviation is usually small. See Weick (16) for examples of the effect of angular speed on propeller performance.

FIGURE 14.4

Power and thrust of a propeller 2.90 m in diameter at a rotational speed of 1400 rpm. [After Weick (16)]

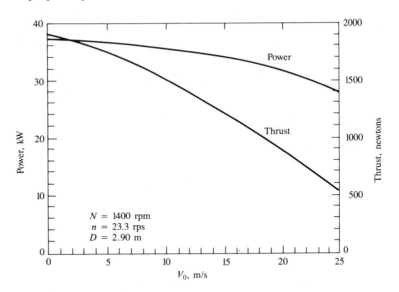

$N = 1400$ rpm
$n = 23.3$ rps
$D = 2.90$ m

V_0, m/s

EXAMPLE 14.1 A propeller having the characteristics shown in Fig. 14.3 is to be used to drive a swamp boat. If the propeller is to have a diameter of 2 m and a rotational speed of $N = 1200$ rpm, what should be the thrust starting from rest? If the boat resistance (air and water) is given by the empirical equation $F_D = 0.003\rho V_0^2/2$, where V_0 is the boat speed in meters per second, F_D is the drag, and ρ is the mass density of the water, what will be the maximum speed of the boat and what power will be required to drive the propeller? Assume $\rho_{\text{air}} = 1.20 \text{ kg/m}^3$, or $1.20 \text{ N} \cdot \text{s}^2/\text{m}^4$.

Solution   First we evaluate the thrust starting from rest:

$$\frac{V_0}{nD} = \frac{(0 \text{ m/s})}{(20 \text{ rps}) (2 \text{ m})} = 0$$

Then, referring to Fig. 14.3, we see that $C_T = 0.048$ for $V_0/nD = 0$. Since $C_T = F_T/\rho_a D^4 n^2$, we can compute F_T:

$$F_T = C_T \rho_a D^4 n^2 = 0.048(1.20 \text{ N} \cdot \text{s}^2/\text{m}^4) (2\text{m})^4 (20 \text{ rps})^2 = 369 \text{ N} \quad \blacktriangleleft$$

To obtain the maximum speed, we should recognize that

$$F_T = F_D$$

or
$$C_T \rho_a D^4 n^2 = 0.003 \rho_w \left(\frac{V_0^2}{2}\right)$$

Here ρ_w is the mass density of water, 1000 kg/m^3, or $1.0 \text{ kN} \cdot \text{s}^2/\text{m}^4$. To determine the equilibrium condition where $F_T = F_D$, we plot F_T and F_D versus V_0. Where the two curves intersect is the point of equilibrium. The data we need to construct such curves are included in the following table. When we plot F_T and F_D against V_0, we obtain the graph shown. The curves intersect at $V_0 = 11 \text{ m/s}$. Hence the boat will attain a maximum speed of 11 m/s. $\quad \blacktriangleleft$

V_0	V_0/nD	C_T	$F_T = C_T \rho_a D^4 n^2$	$F_D = 0.003 \rho_w V_0^2/2$
5 m/s	0.125	0.040	307 N	37.5 N
10 m/s	0.250	0.027	207 N	150 N
15 m/s	0.375	0.012	92 N	337 N

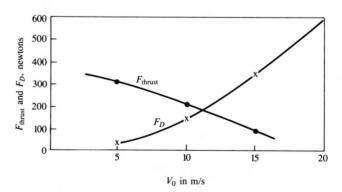

The input power is $P = C_P \rho_a D^5 n^3$; and since the maximum C_P is 0.014 when $V_0/nD = 0$, we compute the power input as

$$P = 0.014(1.20 \text{ N} \cdot \text{s}^2/\text{m}^4) (2 \text{ m})^5 (20 \text{ rps})^3$$

$$= 4300 \text{ m} \cdot \text{N/s} = 4.30 \text{ kW} \quad \blacktriangleleft$$

14.2 Axial-Flow Pumps

Pressure Changes

Axial-flow pumps (or blowers) are designed so that the impeller (much like the propeller discussed in Sec. 14.1) is enclosed within a housing (see Fig. 14.5). The action of the propeller applies a force to the fluid that causes a pressure change between the sections upstream and downstream of the pump. Because engineers working with pumps are usually more interested in the change in head than in the thrust, the thrust coefficient is replaced by the head coefficient, and the ratio V_0/nD is replaced by a more useful parameter involving the discharge.

Head and Discharge Coefficients for Pumps

The thrust coefficient is defined as $F_T/\rho D^4 n^2$, and if the same variables are applied to flow in an axial pump, the thrust can be expressed as $F_T = \Delta pA = \gamma \Delta HA$ or

$$C_T = \frac{\gamma \Delta HA}{\rho D^4 n^2} = \frac{\pi}{4} \frac{\gamma \Delta HD^2}{\rho D^4 n^2} = \frac{\pi}{4} \frac{g \Delta H}{D^2 n^2} \tag{14.12}$$

Now if we define a new parameter, the *head coefficient* C_H, using the variables of Eq. (14.12), we have

$$C_H = \frac{4}{\pi} C_T = \frac{\Delta H}{D^2 n^2/g} \tag{14.13}$$

The independent parameter relating to propeller operation is V_0/nD; however, for pump operation it is convenient to substitute $Q/A = Q/(\pi D^2/4)$

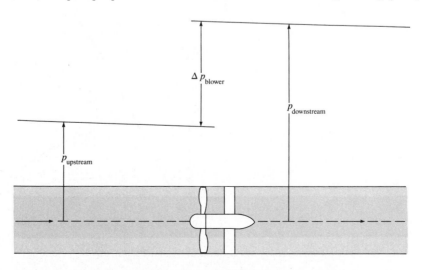

FIGURE 14.5

Axial-flow blower in a duct.

for V_0 and let the numerical factor be absorbed in the functional relationship. Thus the independent parameter for pump similarity studies is Q/nD^3, and this is termed the *discharge coefficient* C_Q. The power coefficient used for pumps is exactly like the power coefficient used for propellers. Summarizing, the dimensionless parameters used in similarity analyses of pumps are as follows:

$$C_H = \frac{\Delta H}{D^2 n^2/g} \qquad (14.14)$$

$$C_P = \frac{P}{\rho D^5 n^3} \qquad (14.15)$$

$$C_Q = \frac{Q}{nD^3} \qquad \bullet \qquad (14.16)$$

where C_H and C_P are functions of C_Q for a given type of pump.

Figure 14.6 is a set of curves of C_H and C_P versus C_Q for a typical axial-flow pump. Also plotted on this graph is the efficiency of the pump as a function of C_Q. The dimensional curves (head and power versus Q for a constant speed of rotation) from which Fig. 14.6 was developed are shown in Fig. 14.7. Because curves like those shown in Fig. 14.6 or Fig. 14.7 characterize pump performance, they are often called *characteristic curves* or *performance curves.* These curves are obtained by experiment.

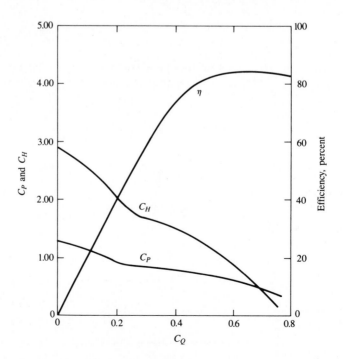

FIGURE 14.6

Dimensionless performance curves for a typical axial-flow pump. [After Stepanoff (14)]

FIGURE 14.7

*Performance curves for a
typical axial-flow pump.
[After Stepanoff (14)]*

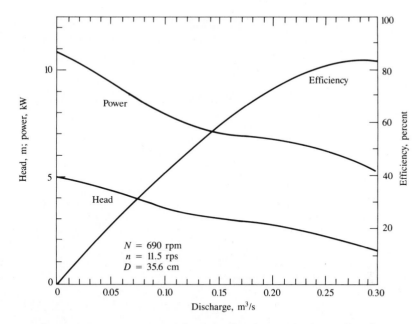

FIGURE 14.7

*Performance curves for a
typical axial-flow pump.
[After Stepanoff (14)]*

Performance curves are used to predict prototype operation from model tests or the effect of change of speed of the pump. The following are examples of these applications.

EXAMPLE 14.2   For the pump represented by Figs. 14.6 and 14.7, what discharge of water in cubic meters per second will occur when the pump is operating against a 2-m head and at a speed of 600 rpm? What power in kilowatts is required for these conditions?

Solution   First compute C_H. Here

$$D = 35.6 \text{ cm} \qquad \text{and} \qquad n = 10 \text{ rps}$$

Then

$$C_H = \frac{2 \text{ m}}{(0.356 \text{ m})^2 (10^2 \text{ s}^{-2})/(9.81 \text{ m/s}^2)} = 1.55$$

Next, using a value of 1.55 for C_H, read a value of 0.40 for C_Q from Fig. 14.6. Hence Q is calculated as follows:

$$C_Q = 0.40 = \frac{Q}{nD^3}$$

or

$$Q = 0.40(10 \text{ s}^{-1})(0.356 \text{ m})^3 = 0.180 \text{ m}^3/\text{s} \qquad \blacktriangleleft$$

From Fig. 14.6 the value of C_P is 0.72 for $C_Q = 0.40$. Then

$$P = 0.72\rho D^5 n^3$$

$$= 0.72(1.0 \text{ kN} \cdot \text{s}^2/\text{m}^4)(0.356 \text{ m})^5 (10 \text{ s}^{-1})^3$$

$$= 4.12 \text{ km} \cdot \text{N/s} = 4.12 \text{ kJ/s} = 4.12 \text{ kW} \qquad \blacktriangleleft$$

EXAMPLE 14.3 If a 30-cm axial-flow pump having the characteristics shown in Fig. 14.6 is operated at a speed of 800 rpm, what head ΔH will be developed when the water-pumping rate is 0.127 m³/s? What power is required for this operation?

Solution First compute $C_Q = Q/nD^3$, where

$$Q = 0.127 \text{ m}^3/\text{s}$$

$$n = \frac{800}{60} = 13.3 \text{ rps}$$

$$D = 30 \text{ cm}$$

Then

$$C_Q = \frac{0.127 \text{ m}^3/\text{s}}{(13.3 \text{ s}^{-1})(0.30 \text{ m})^3} = 0.354$$

Now enter Fig. 14.6 with a value for C_Q of 0.354 and read a value of 1.70 for C_H and a value of 0.80 for C_P. Then

$$\Delta H = \frac{C_H D^2 n^2}{g} = \frac{1.70(0.30 \text{ m})^2(13.3 \text{ s}^{-1})^2}{(9.81 \text{ m/s}^2)} = 2.76 \text{ m} \quad \blacktriangleleft$$

and

$$P = C_P \rho D^5 n^3$$

$$= 0.80(1.0 \text{ kN} \cdot \text{s}^2/\text{m}^4)(0.30 \text{ m})^5(13.3 \text{ s}^{-1})^3 = 4.57 \text{ kW} \quad \blacktriangleleft$$

Range of Application of Axial-Flow Machines

In practical applications, axial-flow machines are best suited for relatively low heads and high rates of flow. Hence pumps used for dewatering lowlands, such as those behind dikes, are almost always of the axial-flow type. Water turbines in low-head dams (less than 30 m) where the flow rate and power production are large are also generally of the axial type. For larger heads, radial- or mixed-flow machines are more efficient. These are discussed in the next section.

14.3 | Radial-Flow Machines

Centrifugal Pumps

In Fig. 14.8 the type of impeller that is used for many radial-flow pumps is shown. Such pumps are often called *centrifugal pumps*. Fluid from the inlet pipe enters the pump through the eye of the impeller and then travels outward between the vanes of the impeller to its edge, where the fluid enters the casing of

FIGURE 14.8

Centrifugal pump.

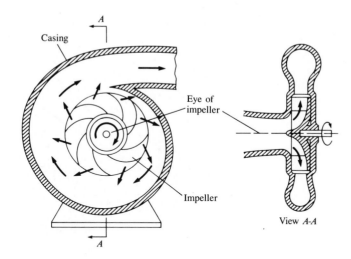

Casing

Eye of
impeller

Impeller

View *A-A*

the pump and is then conducted to the discharge pipe. The principle of the radial-flow pump is different from that of the axial-flow pump in that the change in pressure results in large part by rotary action (pressure increases outward like that in the rotating tank in Sec. 5.2) produced by the rotating impeller. Additional pressure increase is produced in the radial-flow pump when the high-velocity flow leaving the impeller is reduced in the expanding section of the casing.

Although the basic designs are different for radial- and axial-flow pumps, it can be shown that the same similarity parameters (C_Q, C_P, and C_H) apply for both types. Thus the methods that have already been discussed for relating size, speed, and discharge in axial-flow machines also apply to radial-flow machines.

The major practical difference betwen axial- and radial-flow pumps so far as the user is concerned is the difference in the performance characteristics of the two. In Fig. 14.9 the dimensional performance curves for a typical radial-flow pump operating at a constant speed of rotation are shown. In Fig. 14.10 the dimensionless performance curves for the same pump are shown. Note that the power required at shutoff flow is less than that required for flow at maximum efficiency. Normally, the motor to drive the pump is chosen for conditions of maximum pump efficiency. Hence the flow can be throttled between the limits of shutoff condition and normal operating conditions without any chance of overloading the pump motor. Such is not the case for an axial-flow pump, as seen in Fig. 14.6. In that case, when the pump flow is throttled below maximum-efficiency conditions, the required power increases with decreasing flow, thus leading to the possibility of overloading at low-flow conditions. For very large installations, special operating procedures are followed in order to avoid such overloading. For instance, the valve in the bypass from the pump discharge back to the pump inlet can be adjusted to maintain a constant flow

FIGURE 14.9

*Performance curves for a
typical centrifugal pump;
D = 37.1 cm. [After
Daugherty and Franzini
(3)]*

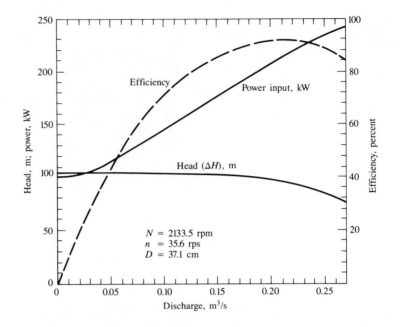

FIGURE 14.10

*Dimensionless performance
curves for a typical
centrifugal pump, from data
given in Fig. 14.9. [After
Daugherty and Franzini
(3)]*

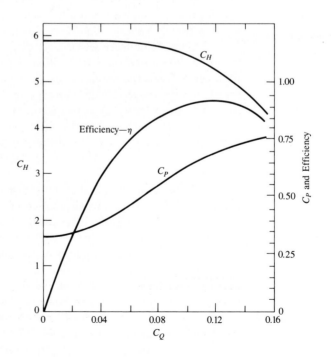

through the pump. However, for small-scale applications, it is often desirable to have complete flexibility in flow control without the complexity of special operating procedures. In this latter case, a radial-flow pump offers a distinct advantage. Radial-flow pumps are manufactured in sizes from 1 hp or less and heads of 50 or 60 ft to thousands of horsepower and heads of several hundred feet. Figure 14.11 shows a cutaway view of a single-suction, single-stage, horizontal-shaft radial pump. Another common design has flow entering the impeller from both sides, as shown in Figs. 14.12 and 14.13. Such a *double-suction impeller* is equivalent to two single-suction impellers placed back to back and made as a single casting. This arrangement gives balanced end thrust on the shaft of the impeller.

FIGURE 14.11

Cutaway view of a single-suction, single-stage, horizontal-shaft radial pump. (Courtesy of Ingersol Rand Co.)

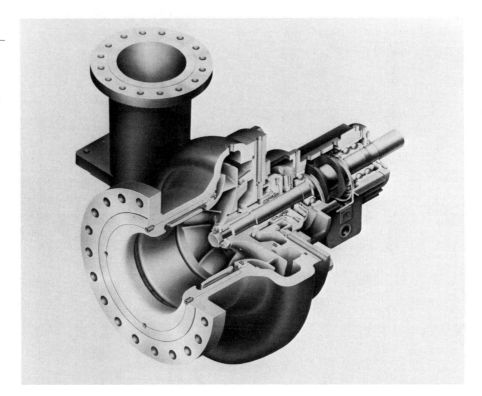

EXAMPLE 14.4   A pump that has the characteristics given in Fig. 14.9 when operated at 2133.5 rpm is to be used to pump water at maximum efficiency under a head of 76 m. At what speed should the pump be operated, and what will the discharge be for these conditions?

FIGURE 14.12

Sectional view of double-suction, single-stage, horizontal-shaft split-casing pump. (Courtesy of Dresser Pump Division)

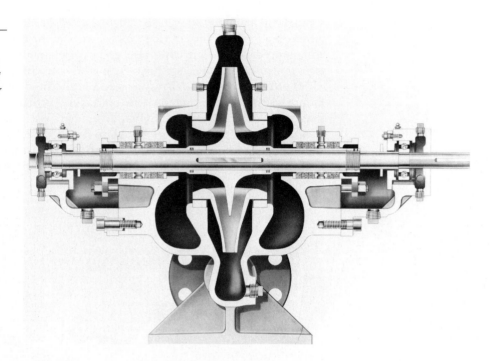

Solution   Since the diameter is fixed, the only change that will occur results from the change in speed (assuming negligible change due to viscous effects). The C_H, C_P, C_Q, and η for this pump operating at maximum efficiency against a head of 76 m are the same as for its operation at maximum efficiency with a speed of rotation of 2133.5 rpm, since both operating conditions correspond to the point of maximum efficiency in Fig. 14.10. Thus we can write

$$(C_H)_N = (C_H)_{2133.5 \text{ rpm}}$$

Here N refers to the speed of rotation with $\Delta H = 76$ m. The graph of Fig. 14.9 indicates that $\Delta H = 90$ m and $Q = 0.225$ m^3/s at maximum efficiency for $N = 2133.5$ rpm. Thus

$$\frac{76 \text{ m}}{N^2} = \frac{90 \text{ m}}{(2133.5)^2}$$

$$N^2 = (2133.5)^2 \left(\frac{76}{90}\right)$$

$$N = 2133.5 \left(\frac{76}{90}\right)^{1/2} = 1960 \text{ rpm} \qquad \blacktriangleleft$$

FIGURE 14.13

Overall outside view of a double-suction, single-stage, horizontal-shaft split-casing pump. (Courtesy of Dresser Pump Division)

Using $(C_Q)_{1960} = (C_Q)_{2133.5 \text{ rpm}}$ and solving for the ratio of discharge, we have

$$\frac{Q_{1960}}{Q_{2133.5}} = \frac{1960}{2133.5} = 0.919$$

$$Q_{1960} = 0.207 \text{ m}^3/\text{s} \qquad \blacktriangleleft$$

EXAMPLE 14.5 The pump having the characteristics shown in Figs. 14.9 and 14.10 is a model of a pump that was actually used in one of the pumping plants of the Colorado River Aqueduct [see Daugherty and Franzini (3)]. For a prototype that is 5.33 times larger than the model and operates at a speed of 400 rpm, what head, discharge, and power are to be expected at maximum efficiency?

Solution   From Fig. 14.10 we find values of 0.12, 5.2, and 0.69 for C_Q, C_H, and C_P, respectively, for the maximum-efficiency condition. Then for $n = (400/60)$ rps and $D = 0.371 \times 5.33 = 1.98$ m, we solve for P, ΔH, and Q:

$$P = C_p \rho D^5 n^3 = 0.69(1.0 \text{ kN} \cdot \text{s}^2/\text{m}^4)(1.98 \text{ m})^5 \left(\frac{400}{60} \text{ s}^{-1} \right)^3 = 6200 \text{ kW} \blacktriangleleft$$

$$\Delta H = \frac{C_H D^2 n^2}{g} = \frac{5.2(1.98 \text{ m})^2 (400/60 \text{ s}^{-1})^2}{(9.81 \text{ m/s}^2)} = 92.4 \text{ m} \blacktriangleleft$$

$$Q = C_Q n D^3 = 0.12 \left(\frac{400}{60} \text{ s}^{-1} \right) (1.98 \text{ m})^3 = 6.21 \text{ m}^3/\text{s} \blacktriangleleft$$

Centrifugal Compressors

Centrifugal compressors are similar in design to centrifugal pumps. Because the density of the air or gases used is much less than the density of a liquid, the compressor must turn at much higher speeds than the pump does to effect a sizable pressure increase. If the compression process were isentropic and the gases ideal, the power necessary to compress the gas from p_1 to p_2 would be

$$P_{\text{theo}} = \frac{k}{k-1} Q_1 p_1 \left[\left(\frac{p_2}{p_1} \right)^{(k-1)/k} - 1 \right] \tag{14.17}$$

where Q_1 is the volume flow rate into the compressor. The power calculated using Eq. (14.17) is referred to as the *theoretical adiabatic power.* The efficiency of a compressor with no water cooling is defined as the ratio of the theoretical adiabatic power to the actual power required at the shaft. Ordinarily, the efficiency improves with higher inlet-volume flow rates, increasing from a typical value of 0.60 at 0.6 m³/s to 0.74 at 40 m³/s. Higher efficiencies are obtainable with more expensive design refinements.

EXAMPLE 14.6   Determine the shaft power required to operate a compressor that compresses air at the rate of 1 m³/s from 100 kPa to 200 kPa. The efficiency of the compressor is 65%.

Solution   We first calculate the theoretical adiabatic power, using Eq. (14.17) with $k = 1.4$.

$$P_{\text{theo}} = \frac{k}{k-1} Q_1 p_1 \left[\left(\frac{p_2}{p_1} \right)^{(k-1)/k} - 1 \right]$$

$$= (3.5)(1 \text{ m}^3/\text{s})(10^5 \text{ N/m}^2)[(2)^{0.286} - 1]$$

$$= 0.767 \times 10^5 \text{ N} \cdot \text{m/s} = 76.7 \text{ kW}$$

The shaft power required is

$$P_{shaft} = \frac{76.7}{0.65} \text{ kW} = 118 \text{ kW} \qquad \blacktriangleleft$$

Cooling is necessary for high-pressure compressors because of the high gas temperatures resulting from the compression process. Cooling can be achieved through the use of water jackets or intercoolers that cool the gases between stages. The efficiency of water-cooled compressors is based on the power required to compress ideal gases isothermally, or

$$P_{theo} = p_1 Q_1 \ln \frac{p_2}{p_1} \qquad (14.18)$$

which is usually called the *theoretical isothermal power*. The efficiencies of water-cooled compressors are generally lower than those of noncooled compressors. If a compressor is cooled by water jackets, its efficiency characteristically ranges between 55 and 60%. The use of intercoolers results in efficiencies from 60 to 65%.

14.4 | Specific Speed

From the discussion in preceding sections we have seen that a pump's performance is given by the values of its power and head coefficients (C_P and C_H) for a range of values of the discharge coefficient C_Q. It has been noted that certain types of machines are best suited for certain head and discharge ranges. For example, an axial-flow machine is best suited for low heads and high discharges, whereas a radial-flow machine is best suited for higher heads and lower discharges. The parameter used to pick the type of pump (or turbine) best suited for a given application is specific speed n_s. Specific speed is obtained by combining both C_H and C_Q in such a manner that the diameter D is eliminated:

$$n_s = \frac{C_Q^{1/2}}{C_H^{3/4}} = \frac{(Q/nD^3)^{1/2}}{[\Delta H/(D^2 n^2/g)]^{3/4}} = \frac{nQ^{1/2}}{g^{3/4} \Delta H^{3/4}}$$

Thus specific speed relates different types of pumps without reference to their sizes.

When the actual efficiencies of different types of pumps are plotted against n_s, it is seen that certain types of pumps have higher efficiencies for certain ranges of n_s. In fact, in the range between the completely axial-flow machine and the completely radial-flow machine, there is a gradual change in impeller shape to accommodate the particular flow conditions with maximum efficiency (see Fig. 14.14).

FIGURE 14.14

Optimum efficiency and impeller design versus specific speed.

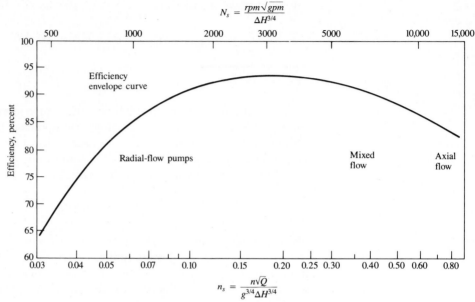

(a) Optimum efficiency and impeller designs versus specific speed n_s

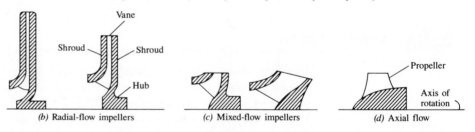

(b) Radial-flow impellers   (c) Mixed-flow impellers   (d) Axial flow

It should be noted that the specific speed traditionally used for pumps in the United States is defined as $N_s = NQ^{1/2}/\Delta H^{3/4}$. Here the speed N is in revolutions per minute, Q is in gallons per minute, and ΔH is in feet. This form is not dimensionless. Therefore its values are much larger than those found for n_s (the conversion factor is 17,200). Most texts and references published before the introduction of the SI system of units use this traditional definition for specific speed.

14.5 | Suction Limitations of Pumps

Because most centrifugal pumps used in a given range of n_s have about the same shape and performance characteristics, it is possible to establish certain general limitations based on the flow conditions on the suction side of the pump. Such limitations are needed to prevent cavitation, which can cause a loss of efficiency or even structural damage. These limitations are published by the Hydraulic

Institute (6) and are given in terms of maximum ΔH versus n_s for different suction lifts or suction heads. Here suction lift or suction head is defined as the negative or positive gage pressure, respectively, on the suction side of the pump, converted to head. A chart for single-suction, mixed-flow, and axial-flow pumps is shown in Fig. 14.15. The use of the chart in Fig. 14.15 is illustrated in Example 14.7.

FIGURE 14.15
_____

Limitations on specific speed
for single-suction,
mixed-flow, and axial-flow
pumps (pumping clear water,
30°C, at sea level).
[Adapted from the
Hydraulic Institute
Standards (6)]

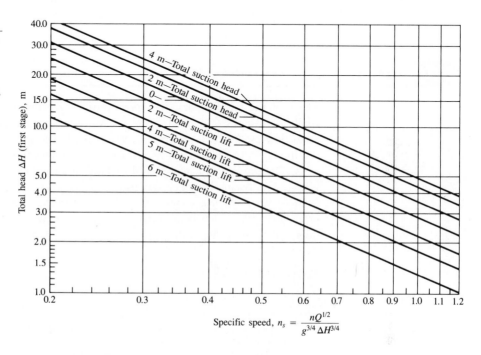

Specific speed, $n_s = \dfrac{nQ^{1/2}}{g^{3/4}\Delta H^{3/4}}$

EXAMPLE 14.7 An axial-flow pump is to be used to lift water from a main irrigation canal to a smaller irrigation canal at a higher level. If the total head (elevation difference plus head losses in the pipe) is to be 11 m and if the total suction head is to be 2 m (the impeller is below the water level in the main canal), what is the safe upper limit of specific speed? Would it be safe to operate a pump at a speed of 1200 rpm and with a discharge of 0.5 m³/s under these conditions?

Solution We enter Fig. 14.15 with a total head of 11 m and a total suction head of 2 m and read a value of 0.51 for n_s. Thus the safe upper limit of specific speed is 0.51.

By definition we have

$$n_s = \frac{nQ^{1/2}}{g^{3/4}\Delta H^{3/4}}$$

Hence for $N = 1200$ rpm or $n = 20$ rps, $Q = 0.50$ m³/s, and $\Delta H = 11$ m, we compute n_s as follows:

$$n_s = \frac{20(0.50)^{1/2}}{(9.81)^{3/4}(11)^{3/4}} = 0.42$$

The n_s computed here is less than the allowable n_s. Consequently, the stated operating conditions are *within the safe range*. ◀

14.6   Turbines

Much of the basic theory and most similarity parameters used for pumps also apply to turbines. However, there are some differences in physical features and terminology. Also, we have not yet considered details of the flow through the impellers of radial-flow machines. These topics will now be discussed.

    The two broad categories of turbines are the *impulse turbine* and the *reaction turbine*. For hydroelectric installations the latter is further subdivided into the Francis type, which is characterized by a radial-flow impeller, and the Kaplan or propeller type, which is an axial-flow machine. We consider first the basic elements of the impulse turbine and then the details of the Francis turbine.

Impulse Turbine

In the impulse turbine a jet of fluid issuing from a nozzle impinges on vanes of the turbine wheel or *runner,* thus producing power as the runner rotates (see Fig. 14.16). Fig. 14.17 shows a runner for the Henry Borden hydroelectric plant

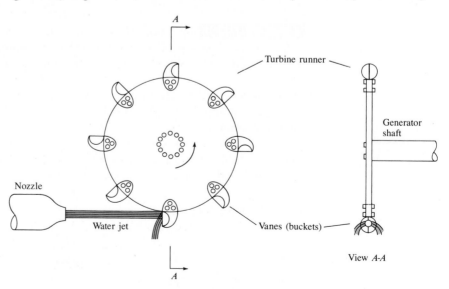

FIGURE 14.16

Impulse turbine.

FIGURE 14.17

FIGURE 14.17

Spare runner for the Henry Borden power plant in Brazil. (Courtesy of Voith Hydro Inc.)

in Brazil. The primary feature of the impulse turbine with respect to fluid mechanics is the power production as the jet is deflected by the moving vanes. When the momentum equation is applied to this deflected jet, it can be shown [see Daugherty and Franzini (3)] for idealized conditions that the maximum power will be developed when the vane speed is one-half of the initial jet speed. With such conditions the exiting jet speed will be zero—all of the kinetic energy of the jet will have been expended in driving the vane. Thus if we apply the energy equation, Eq. (7.23), between the incoming jet and the exiting fluid (assuming negligible head loss and negligible kinetic energy at exit), we find that the head given up to the turbine is $h_t = V_j^2/2g$ and the power thus developed is

$$P = Q\gamma h_t \qquad (14.19)$$

where Q is the discharge of the incoming jet, γ is the specific weight of jet fluid, and $h_t = V_j^2/2g$, or the velocity head of the jet. Thus Eq. (14.19) reduces to

$$P = \rho Q \frac{V_j^2}{2} \qquad (14.20)$$

To obtain the torque on the turbine shaft, we apply the angular-momentum equation, Eq. (6.27a), to a control volume, as shown in Fig. 14.18.

FIGURE 14.18

Control-volume approach for the impulse turbine using the angular-momentum principle.

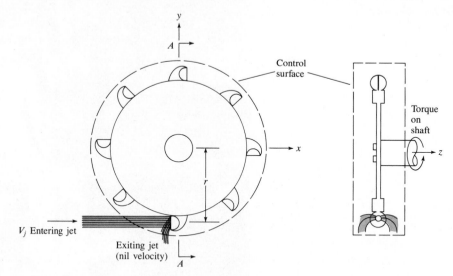

Then for steady flow we have

$$\sum \mathbf{M} = \sum_{\text{cs}} (\mathbf{r} \times \mathbf{v})\rho\mathbf{V} \cdot \mathbf{A}$$

or

$$\mathbf{T}_{\text{shaft}} = \sum_{\text{cs}} (\mathbf{r} \times \mathbf{v})\rho\mathbf{V} \cdot \mathbf{A} \tag{14.21}$$

Now if we consider the angular momentum about the axis of the turbine shaft, we see that $\mathbf{r} \times \mathbf{v}$ for the entering jet will be simply $rV_j\mathbf{k}$ and that $\mathbf{V} \cdot \mathbf{A}$ for the entering jet will be $-V_jA_j$ or $-Q$. In addition, we are assuming that the exiting jet has negligible angular momentum. Hence the torque acting on the system— that is, the torque on the shaft at the section where the control surface passes through the shaft—is given by $\mathbf{T} = -\rho QVr\,\mathbf{k}$. The absolute value of the torque is simply

$$T = \rho QV_j r \tag{14.22}$$

The power developed by the turbine is $T\omega$, or

$$P = \rho QV_j r\omega \tag{14.23}$$

Furthermore, if the velocity of the turbine vanes is $\frac{1}{2}V_j$ for maximum power, as noted earlier, we have $P = \rho QV_j^2/2$, the same as Eq. (14.20).

EXAMPLE 14.8 What power in kilowatts can be developed by the impulse turbine shown if the turbine efficiency is 85%? Assume that the resistance coefficient f of the penstock is 0.015 and the head loss in the nozzle itself is negligible. What will be the angular speed of the wheel, assuming ideal conditions ($V_j = 2V_{bucket}$), and what torque will be exerted on the turbine shaft?

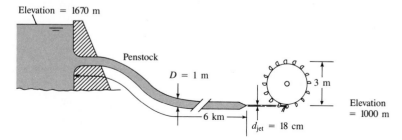

Solution   First determine the jet velocity by applying the energy equation from the reservoir to the free jet before it strikes the turbine buckets.

$$\frac{p_1}{\gamma} + \frac{V_1^2}{2g} + z_1 = \frac{p_j}{\gamma} + \frac{V_j^2}{2g} + z_j + h_L$$

where

$$p_1 = 0 \qquad\qquad p_j = 0$$

$$z_1 = 1670 \text{ m} \qquad z_j = 1000 \text{ m}$$

$$V_1^2/2g = 0 \qquad\qquad \gamma = 9810 \text{ N/m}^3 \text{ at } 10°\text{C (assumed)}$$

The penstock water velocity is

$$V_{penstock} = \frac{V_j A_j}{A_{penstock}} = 0.0324 V_j$$

Then

$$h_L = \frac{fL}{D}\frac{V^2}{2g} = \frac{0.015 \times 6000}{1}(0.0324)^2\frac{V_j^2}{2g} = 0.094\frac{V_j^2}{2g}$$

Now, solving the energy equation for V_j yields

$$V_j = \left(\frac{2g \times 670}{1.094}\right)^{1/2} = 109.6 \text{ m/s}$$

The gross power is

$$P = Q\gamma\frac{V_j^2}{2g} = \frac{\gamma A_j V_j^3}{2g}$$

$$= \frac{9810(\pi/4)(0.18)^2(109.6)^3}{2 \times 9.81} = 16{,}750 \text{ kW}$$

The power output of the turbine is

$$P = 16{,}750 \times \text{efficiency} = 14{,}238 \text{ kW}$$

The tangential bucket speed will be $\frac{1}{2}V_j$. Therefore

$$V_{\text{bucket}} = \tfrac{1}{2}(109.6 \text{ m/s}) = 54.8 \text{ m/s}$$

or

$$r\omega = 54.8 \text{ m/s}$$

Thus

$$\omega = \frac{54.8 \text{ m/s}}{1.5 \text{ m}} = 36.53 \text{ rad/s}$$

The wheel speed is

$$N = (36.53 \text{ rad/s})\frac{1 \text{ rev}}{2\pi \text{ rad}}(60 \text{ s/min}) = 349 \text{ rpm} \qquad \blacktriangleleft$$

Finally, $\qquad\qquad\qquad\qquad\qquad\qquad$ Power $= T\omega$

Thus $\qquad\qquad\qquad T = \dfrac{\text{power}}{\omega} = \dfrac{14{,}238 \text{ kW}}{36.53 \text{ rad/s}} = 390 \text{ kN} \cdot \text{m} \qquad \blacktriangleleft$

Characteristics of the Reaction Turbine

In contrast to the impulse turbine, where a jet under atmospheric pressure impinges on only one or two vanes at a time, flow in a reaction turbine is under pressure. Also, this flow completely fills the chamber in which the impeller is located (see Fig. 14.19). There is a drop in pressure from the outer radius of the impeller, r_1, to the inner radius, r_2. This is another point of difference with the impulse turbine, in which the pressure is the same for the entering and exiting flow. The original form of the reaction turbine, first extensively tested by J. B. Francis, had a completely radial-flow impeller (Fig. 14.20). That is, the flow passing through the impeller had velocity components only in a plane normal to the axis of the runner. However, more recent impeller designs, such as the mixed-flow and axial-flow types, are still called reaction turbines.

Torque and Power
Relations for the Reaction Turbine

As we did for the impulse turbine, we will use the angular-momentum equation to develop formulas for the torque and power for the reaction turbine. The segment of turbine runner shown in Fig. 14.20 depicts the flow conditions that occur for the entire runner. We can see that guide vanes outside the runner itself cause the fluid to have a tangential component of velocity around the entire circumference of the runner. Thus the fluid has an initial amount of angular momentum with respect to the turbine axis when it approaches the turbine runner. As the fluid passes through the passages of the runner, the runner vanes effect a

FIGURE 14.19

Schematic view of a reaction-turbine installation. (a) Elevation view. (b) Plan view, section A–A.

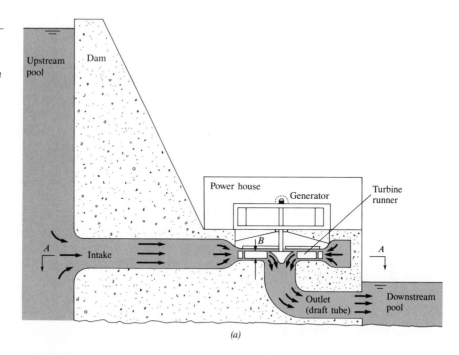

(a)

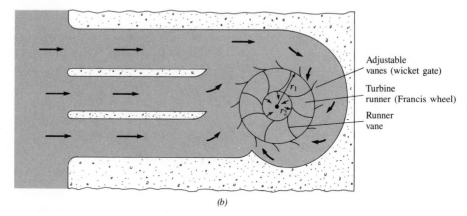

(b)

change in the magnitude and direction of its velocity. Thus the angular momentum of the fluid is changed, which produces a torque on the runner. This torque drives the runner, which, in turn, generates power.

To quantify the above, we let V_1 and α_1 represent the incoming velocity and the angle of the velocity vector with respect to a tangent to the runner, respectively. Similar terms at the inner-runner radius are V_2 and α_2. Applying the

FIGURE 14.20

Velocity diagrams for the impeller for a Francis turbine.

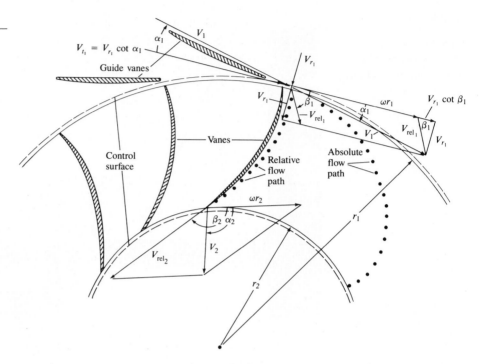

angular-momentum equation for steady flow, Eq. (14.21), to the control volume shown in Fig. 14.20 yields

$$T = (-r_1 V_1 \cos \alpha_1)\rho(-Q) + (-r_2 V_2 \cos \alpha_2)\rho(+Q)$$

$$= \rho Q(r_1 V_1 \cos \alpha_1 - r_2 V_2 \cos \alpha_2) \tag{14.24}$$

The power from this turbine will be $T\omega$, or

$$P = \rho Q \omega (r_1 V_1 \cos \alpha_1 - r_2 V_2 \cos \alpha_2) \tag{14.25}$$

Equation (14.25) shows that the power production is a function of the directions of the flow velocities entering and leaving the impeller—that is, α_1 and α_2.

It is interesting to note that even though the pressure varies within the flow in a reaction turbine, it does not enter into the expressions we have derived using the angular-momentum equation. The reason it does not appear is that the outer and inner control surfaces we chose are concentric with the axis about which we are evaluating the moments and angular momentum. The pressure forces acting on these surfaces all pass through the given axis; therefore they do not produce moments about the given axis.

Vane Angles

It should be apparent that the head loss in a turbine will be less if the flow enters the runner with a direction tangent to the runner vanes than if the flow approaches the vane with an angle of attack. In the latter case, separation will occur with consequent head loss. Thus vanes of an impeller designed for a given speed and discharge and with fixed guide vanes will have a particular optimum blade angle β_1. However, if the discharge is changed from the condition of the original design, the guide vanes and impeller vane angles will not "match" the new flow condition. Most turbines for hydroelectric installations are made with movable guide vanes on the inlet side to effect a better match at all flows. Thus α_1 is increased or decreased automatically through governor action to accommodate fluctuating power demands on the turbine.

To relate the incoming-flow angle α_1 and the vane angle β_1, we first assume that the flow entering the impeller is tangent to the blades at the periphery of the impeller. Likewise, the flow leaving the stationary guide vane is assumed to be tangent to the guide vane. In developing the desired equations, we will consider both the radial and the tangential components of velocity at the outer periphery of the wheel ($r = r_1$). We can easily compute the radial velocity, given Q and the geometry of the wheel, by the continuity equation:

$$V_{r_1} = \frac{Q}{2\pi r_1 B} \tag{14.26}$$

where B is the height of the turbine blades. The tangential (tangent to the outer surface of the runner) velocity of the incoming flow is

$$V_{t_1} = V_{r_1} \cot \alpha_1 \tag{14.27}$$

However, this tangential velocity is equal to the tangential component of the relative velocity in the runner, $V_{r_1} \cot \beta_1$, plus the velocity of the runner itself, ωr_1. Thus the tangential velocity, when viewed with respect to the runner motion, is

$$V_{t_1} = r_1 \omega + V_{r_1} \cot \beta_1 \tag{14.28}$$

Now, by eliminating V_{t_1} between Eqs. (14.27) and (14.28), we have

$$V_{r_1} \cot \alpha_1 = r_1 \omega + V_{r_1} \cot \beta_1 \tag{14.29}$$

Equation (14.29) can be rearranged to yield

$$\alpha_1 = \text{arccot}\left(\frac{r_1 \omega}{V_{r_1}} + \cot \beta_1\right) \tag{14.30}$$

EXAMPLE 14.9 A Francis turbine is to be operated at a speed of 600 rpm and with a discharge of 4.0 m³/s. If $r_1 = 0.60$ m, $\beta_1 = 110°$, and the blade height B is 10 cm, what should be the guide vane angle α_1 for a nonseparating flow condition at the runner entrance?

Solution
$$\alpha_1 = \text{arccot}\left(\frac{r_1 \omega}{V_{r_1}} + \cot \beta_1\right)$$

where $\quad r_1\omega = 0.6 \text{ m} \times 600 \text{ rpm} \times 2\pi \text{ rad/rev} \times 1/60 \text{ min/s} = 37.7 \text{ m/s}$

$$V_{r_1} = \frac{Q}{2\pi r_1 B} = \frac{4.00 \text{ m}^3/\text{s}}{2\pi \times 0.6 \text{ m} \times 0.10 \text{ m}} = 10.61 \text{ m/s cot } \beta_1 = -0.364$$

Then $\qquad \alpha_1 = \text{arccot}(3.55 - 0.364) = 17.4°$ ◀

Specific Speed for Turbines

Because of the attention focused on the production of power by turbines, the specific speed for turbines is defined in terms of power:

$$n_s = \frac{nP^{1/2}}{g^{3/4}\gamma^{1/2}h_t^{5/4}}$$

It should also be noted that large water turbines are innately more efficient than pumps. The reason for this is that as the fluid leaves the impeller of a pump, it decelerates appreciably over a relatively short distance. Also, because guide vanes are generally not used in the flow passages with pumps, large local velocity gradients develop, which in turn cause intense mixing and turbulence, thereby producing large head losses. In most turbine installations, the flow that exits the turbine runner is gradually reduced in velocity through a gradually expanding *draft tube,* thus producing a much smoother flow situation and less head loss than for the pump. For additional details of hydropower turbines, see Daugherty and Franzini (3).

Gas Turbines

The conventional gas turbine consists of a compressor that pressurizes the air entering the turbine and delivers it to a combustion chamber. The high-temperature, high-pressure gases resulting from combustion in the combustion chamber expand through a turbine, which both drives the compressor and delivers power. The theoretical efficiency (power delivered/rate of energy input) of a gas turbine depends on the pressure ratio between the combustion chamber and the intake; the higher the pressure ratio, the higher the efficiency. The reader is directed to Cohen et al. (2) for more detail.

Wind Turbines

Extraction of energy from the wind by a wind turbine is discussed frequently as an alternative energy source. In essence, the wind turbine is just a reverse application of the process of introducing energy into an airstream to derive a propulsive force. The wind turbine extracts energy from the wind to produce power. There is one significant difference, however. The theoretical upper limit of efficiency of a propeller supplying energy to an airstream is 100%; that is, it is theoretically possible, neglecting viscous and other effects, to convert all the energy supplied to a propeller into energy of the airstream. On the other hand, the theoretical upper limit of wind-turbine efficiency as given by Glauert (4) is 16/27, or 59.3%. Thus the theoretical maximum power deliverable by a wind turbine with capture area A is

$$P_{\max} = \frac{16}{27}\left(\frac{1}{2}\rho U^3 A\right)$$

where ρ is the air density and U is the wind speed. The capture area is the area swept by the wind turbine viewed from the wind direction. Other factors, such as swirl of the airstream and viscous effects, further reduce the wind turbine's efficiency.

The conventional wind turbine consists of a propeller mounted on a horizontal axis with a vane, or other device, to align the propeller shaft in the wind direction. In recent years considerable effort has been devoted to assessment of the Savonius rotor and the Darrieus turbine, both of which are vertical-axis turbines, as shown in Fig. 14.21. The Savonius rotor consists of two curved blades forming an S-shaped passage for the airflow. The Darrieus turbine consists of two or three airfoils attached to a vertical shaft; the unit resembles an egg beater. The advantage of vertical-axis turbines is that their operation is independent of wind direction. The Darrieus wind turbine is considered superior in performance but has a disadvantage in that it is not self-starting. Frequently, a Savonius rotor is mounted on the axis of a Darrieus turbine to provide the starting torque. For more information on wind turbines, refer to *Proceedings of the Cambridge Symposium on Wind Energy Systems* (10).

FIGURE 14.21

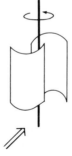
(a) Savonius rotor

(b) Darrieus turbine

14.7   Viscous Effects

In the foregoing sections, we developed similarity parameters to predict prototype results from model tests, neglecting viscous effects. The latter assumption is not necessarily valid, especially if the model is quite small. To minimize the viscous effects in modeling pumps, the *Hydraulic Institute Standards* (6) recommend that the size of the model be such that the model impeller is not less than 30 cm in diameter. These same standards [B-146(e)] state that "the model should have complete geometric similarity with the prototype, not only in the pump proper, but also in the intake and discharge conduits."

Even with complete geometric similarity, one can expect the model to be less efficient than the prototype. An empirical formula proposed by Moody (9) is used for estimating prototype efficiencies of radial- and mixed-flow pumps and turbines from model efficiencies. That formula is

$$\frac{1 - e_1}{1 - e} = \left(\frac{D}{D_1}\right)^{1/5} \qquad (14.31)$$

Here e_1 is the efficiency of the model and e is the efficiency of the prototype.

EXAMPLE 14.10   A model having an impeller diameter of 45 cm is tested and found to have an efficiency of 85%. If a geometrically similar prototype has an impeller diameter of 1.80 m, estimate its efficiency when it is operating under conditions that are dynamically similar to those in the model test $(C_{Q,\,\text{model}} = C_{Q,\,\text{prototype}})$.

Solution   We apply Eq. (14.31) with the conditions that $e_1 = 0.85$ and $D/D_1 = 4$. Then

$$e = 1 - \frac{1 - e_1}{(D/D_1)^{1/5}} = 1 - \frac{0.15}{1.32} = 1 - 0.11 = 0.89$$

The efficiency of the prototype is estimated to be 89%.   ◀

Problems

14.1   What thrust is obtained from a propeller 4.0 m in diameter that has the characteristics given in Fig. 14.3 when the propeller is operated at an angular speed of 1400 rpm and an advance velocity of zero? Assume $\rho = 1.05$ kg/m³.

14.2   What thrust is obtained from a propeller 3 m in diameter that has the characteristics given in Fig. 14.3 when the propeller is operated at an angular speed of 1400 rpm and an advance velocity of 80 km/h? What power is required to operate the propeller under these conditions? Assume $\rho = 1.1$ kg/m³.

14.3   A propeller 8 ft in diameter has the characteristics shown in Fig. 14.3. What thrust is produced by the propeller when it is operating at an angular speed of 1000 rpm and a forward speed of 30 mph? What power input is required under these operating conditions? If the forward speed is reduced to zero, what is the thrust? Assume $\rho = 0.0024$ slugs/ft³.

14.4   A propeller 6 ft in diameter and like the one for which characteristics are given in Fig. 14.3 is to be used on a swamp boat and is to operate at maximum efficiency when cruising. If the cruising speed is to be 30 mph, what should the angular speed of the propeller be?

14.5 For the propeller and conditions described in Prob. 14.4, determine the thrust and the power input.

14.6 A propeller is being selected for an airplane that will cruise at 2000 m altitude where the pressure is 60 kPa absolute and the temperature is 0°C. The mass of the airplane is 1000 kg, and the planform area of the wing is 10 m². The lift-to-drag ratio is 30:1. The lift coefficient is 0.4. The engine speed at cruise conditions is 3000 rpm. The propeller is to operate at maximum efficiency, which corresponds to a thrust coefficient of 0.025. Calculate the diameter of the propeller and the speed of the aircraft.

14.7 If the tip speed of a propeller is to be kept below 0.9c, where c is the speed of sound, what is the maximum allowable angular speed of propellers having diameters of 2 m (6.56 ft), 3 m (9.84 ft), and 4 m (13.12 ft)? Take the speed of sound as 335 m/s (1099 ft/s).

14.8 A propeller 2.0 m in diameter and like the one for which characteristics are given in Fig. 14.3 is to be used on a swamp boat and is to operate at maximum efficiency when cruising. If the cruising speed is to be 50 km/h, what should the angular speed of the propeller be?

14.9 For the propeller and conditions described in Prob. 14.8, determine the thrust and the power input. Assume $\rho = 1.1$ kg/m³.

14.10 A propeller 2.0 m in diameter and like the one for which characteristics are given in Fig. 14.3 is used on a swamp boat. If the angular speed is 1000 rpm and if the boat and passengers have a combined mass of 300 kg, estimate the initial acceleration of the boat when starting from rest. Assume $\rho = 1.1$ kg/m³.

14.11 If a pump having the characteristics shown in Fig. 14.6 has a diameter of 40 cm and is operated at a speed of 1000 rpm, what will be the discharge when the head is 3 m?

14.12 If a pump that is geometrically similar to the one characterized in Fig. 14.7 is operated at the same speed (690 rpm) but is twice as large, $D = 71.2$ cm, what will be the water discharge and the power demand when the head is 10 m?

14.13 The pump used in the system shown has the characteristics given in Fig. 14.7. What discharge will occur under the conditions shown, and what power is required?

14.14 If the conditions are the same as in Prob. 14.13 except that the speed is increased to 900 rpm, what discharge will occur, and what power is required for the operation?

14.15 For a pump having the characteristics given in Fig. 14.6 or 14.7, what water discharge and head will be produced at maximum efficiency if the pump diameter is 24 in. and the angular speed is 1100 rpm? What power is required under these conditions?

14.16 A pump has the characteristics given by Fig. 14.6. What discharge and head will be produced at maximum efficiency if the pump size is 50 cm and the angular speed is 45 rps? What power is required when pumping water under these conditions?

14.17 For a pump having the characteristics of Fig. 14.6, plot the head-discharge curve if the pump is 14 in. in diameter and is operated at a speed of 900 rpm.

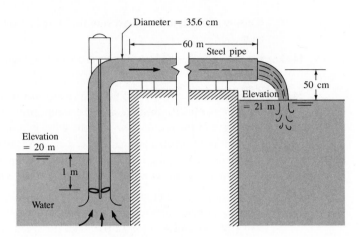

PROBLEMS 14.13, 14.14

14.18 For a pump having the characteristics of Fig. 14.6, plot the head-discharge curve if the pump diameter is 60 cm and the speed is 690 rpm.

14.19 If a pump having the characteristics given in Fig. 14.9 is doubled in size but halved in speed, what will be the head and discharge at maximum efficiency?

14.20 An axial fan 2 m in diameter is used in a wind tunnel (test section 1 m in diameter; test section velocity of 60 m/s). The rotational speed of the fan is 1800 rpm. Assume the density of the air is constant at 1.2 kg/m³. There are negligible losses in the tunnel. The performance curve of the fan is identical to that shown in Fig. 14.6 on page 656. Calculate the power needed to operate the fan.

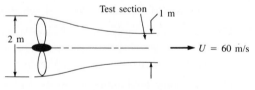

PROBLEM 14.20

14.21 A pump having the characteristics given in Fig. 14.9 pumps water from a reservoir at an elevation of 366 m to a reservoir at an elevation of 450 m through a 36-cm steel pipe. If the pipe is 610 m long, what will be the discharge through the pipe?

14.22 If a pump having the characteristics given in Fig. 14.9 or 14.10 is operated at a speed of 1500 rpm, what will be the discharge when the head is 160 ft?

14.23 If a pump having the performance curve shown is operated at a speed of 1500 rpm, what will be the maximum possible head developed?

14.24 If a pump having the characteristics given in Fig. 14.9 is operated at a speed of 30 rps, what will be the shutoff head?

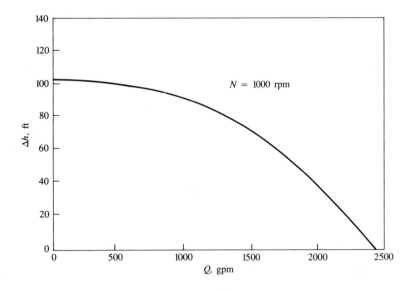

PROBLEM 14.23

14.25   If a pump having the characteristics given in Fig. 14.10 is 40 cm in diameter and is operated at a speed of 25 rps, what will be the discharge when the head is 50 m?

14.26   A centrifugal pump 20 cm in diameter is used to pump kerosene at a speed of 5000 rpm. Assume that the pump has the characteristics shown in Fig. 14.10. Calculate the flow rate, the pressure rise across the pump, and the power required if the pump operates at maximum efficiency.

14.27   For a pump having the characteristics shown in Fig. 14.10, plot the head-discharge curve if the pump diameter is 1.52 m and the speed is 500 rpm.

14.28   What is the specific speed for the pump that is operating under the conditions given in Prob. 14.13? Is this a safe operation with respect to susceptibility to cavitation?

14.29   What type of pump should be used to pump water at a rate of 12 cfs and under a head of 25 ft? Assume $N = 1500$ rpm.

14.30   For most efficient operation, what type of pump should be used to pump water at a rate of 0.30 m^3/s and under a head of 8 m? Assume $n = 25$ rps.

14.31   What type of pump should be used to pump water at a rate of 0.40 m^3/s and under a head of 70 m? Assume $N = 1100$ rpm.

14.32   What type of pump should be used to pump water at a rate of 12 cfs and under a head of 600 ft? Assume $N = 1100$ rpm.

14.33   An axial-flow pump is to be used to lift water against a head (friction and static) of 5.0 m. If the discharge is to be 0.40 m^3/s, what maximum speed in revolutions per minute is allowed if the suction head is 1.5 m?

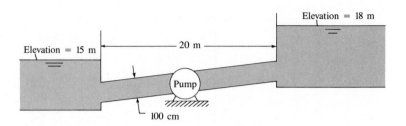

PROBLEM 14.34

14.34 You want to pump water at a rate of $1.0 \text{ m}^3/\text{s}$ from the lower to the upper reservoir shown in the figure. What type of pump would you use for this operation if the impeller speed is to be 600 rpm?

14.35 An axial-flow blower is used for a wind tunnel that has a test section measuring 60 cm by 60 cm and is capable of airspeeds up to 30 m/s. If the blower is to operate at maximum efficiency at the highest speed and if the rotational speed of the blower is 2000 rpm at this condition, what are the diameter of the blower and the power required? Assume that the blower has the characteristics shown in Fig. 14.6.

14.36 An axial-flow blower is used to air-condition an office building that has a volume of 10^5 m^3. It is decided that the air in the building must be completely changed every 15 min. Assume that the blower operates at 600 rpm at maximum efficiency and has the characteristics shown in Fig. 14.6. Calculate the diameter and power requirements for two blowers operating in parallel.

14.37 Methane flowing at the rate of 1 kg/s is to be compressed by a noncooled centrifugal compressor from 100 kPa to 150 kPa. The temperature of the methane entering the compressor is 27°C. The efficiency of the compressor is 65%. Calculate the shaft power necessary to run the compressor.

14.38 A 10-kW (shaft output) motor is available to run a noncooled compressor for carbon dioxide. The pressure is to be increased from 90 kPa to 140 kPa. If the compressor is 60% efficient, calculate the volume flow rate into the compressor.

14.39 A water-cooled centrifugal compressor is used to compress air from 100 kPa to 400 kPa at the rate of 1 kg/s. The temperature of the inlet air is 15°C. The efficiency of the compressor is 50%. Calculate the necessary shaft power.

14.40 A penstock 1 m in diameter and 10 km long carries water from a reservoir to an impulse turbine. If the turbine is 83% efficient, what power can be produced by the system if the upstream reservoir elevation is 650 m above the turbine jet and the jet diameter is 16.0 cm? Assume that $f = 0.016$ and neglect head losses in the nozzle. What should the diameter of the turbine wheel be if it is to have an angular speed of 360 rpm? Assume ideal conditions for the bucket design $(V_{\text{bucket}} = \frac{1}{2}V_j)$.

14.41 Consider an idealized bucket on an impulse turbine that turns the water through 180°. Prove that the bucket speed should be one-half the incoming jet speed for maximum power production. (*Hint:* Set up the momentum equation to solve for the force on the bucket in terms of V_j and V_{bucket}; then the power will be given by this force times V_{bucket}. You can use your mathematical talent to complete the problem.)

14.42 Consider a single jet of water striking the buckets of the impulse wheel as shown. Assume ideal conditions for power generation ($V_{bucket} = \frac{1}{2}V_j$ and the jet is turned through 180° of arc). With the foregoing conditions, solve for the jet force on the bucket and then solve for the power developed. Note that this power is not the same as that given by Eq. (14.20)! Study the figure to resolve this discrepancy.

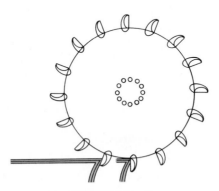

PROBLEM 14.42

14.43 a. For a given Francis turbine, $\beta_1 = 60°$, $\beta_2 = 90°$, $r_1 = 5$ m, $r_2 = 3$ m, and $B = 1$ m. What should α_1 be for a nonseparating flow condition at the entrance to the runner when the discharge rate is 126 m³/s and $N = 60$ rpm?
b. What is the maximum attainable power with the conditions noted?
c. If you were to redesign the turbine blades of the runner, what changes would you suggest to increase the power production if the discharge and overall dimensions are to be kept the same?

14.44 A Francis turbine is to be operated at a speed of 60 rpm and with a discharge of 4.0 m³/s. If $r_1 = 1.5$ m, $r_2 = 1.20$ m, $B = 30$ cm, $\beta_1 = 85°$, and $\beta_2 = 165°$, what should α_1 be for nonseparating flow to occur through the runner? What power and torque should result with this operation?

14.45 A Francis turbine is to be operated at a speed of 120 rpm and with a discharge of 113 m³/s. If $r_1 = 2.5$ m, $B = 0.90$ m, and $\beta_1 = 45°$, what should α_1 be for nonseparating flow at the runner inlet?

14.46 Shown is a preliminary layout for a proposed small hydroelectric project. The initial design calls for a discharge of 8 cfs through the penstock and turbine. Assume 80% turbine efficiency. For this setup, what power output could be expected from the power plant? Draw the HGL and EGL for the system.

14.47 Calculate the maximum power derivable from a conventional, horizontal-axis wind turbine with a propeller 2 m in diameter in a 50-km/h wind whose density is 1.2 kg/m³.

14.48 Calculate the minimum possible capture area necessary for a windmill that is to operate five 100-watt bulbs if the wind velocity is 20 km/h and the density is 1.2 kg/m³.

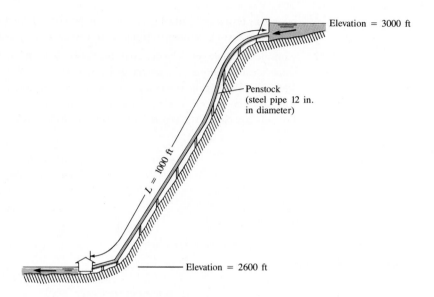

Elevation = 3000 ft

Penstock
(steel pipe 12 in.
in diameter)

L = 1000 ft

Elevation = 2600 ft

PROBLEM 14.46

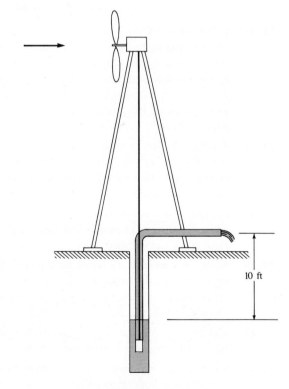

10 ft

PROBLEM 14.49

14.49 A windmill is connected directly to a mechanical pump that is to pump water from a well 10 ft deep. The windmill is a conventional horizontal-axis type with a fan diameter of 10 ft. The efficiency of the mechanical pump is 80%. The density of the air is 0.07 lbm/ft^3. Assume the windmill delivers the maximum power available. What would the discharge of the pump be (in gallons per minute) for a 30-mph wind? (1 cfm = 7.48 gpm)

References

1. Church, A. H. *Centrifugal Pumps and Blowers.* John Wiley, New York, 1944.

2. Cohen, H., G. F. C. Rogers, and H. I. H. Saravanamuttoo. *Gas Turbine Theory.* John Wiley, New York, 1972.

3. Daugherty, Robert L., and Joseph B. Franzini. *Fluid Mechanics with Engineering Applications.* McGraw-Hill, New York, 1957.

4. Glauert, H. "Airplane Propellers." *Aerodynamic Theory,* vol. IV, ed. W. F. Durand. Dover Publications, New York, 1963.

5. Hicks, T. G. *Pump Operation and Maintenance.* McGraw-Hill, New York, 1958.

6. Hydraulic Institute. *Hydraulic Institute Standards,* 12th ed. Hydraulic Institute, New York, 1969.

7. Karassick, I. J., and R. Carter. *Centrifugal Pumps.* F.W. Dodge Company, a division of McGraw-Hill, New York, 1960.

8. Marks, Lionel S. (ed.). *Mechanical Engineers' Handbook.* McGraw-Hill, New York, 1951.

9. Moody, L. F. "Hydraulic Machinery." In *Handbook of Applied Hydraulics,* ed. C.V. Davis. McGraw-Hill, New York, 1942.

10. *Proceedings of the Cambridge Symposium on Wind Energy Systems.* BHRA, Cranfield, England, 1976.

11. Sorensen, H. A. *Gas Turbines.* Ronald, New York, 1951.

12. Spannhake, W. *Centrifugal Pumps, Turbines, and Propellers.* The Technology Press of the Massachusetts Institute of Technology, Cambridge, Mass., 1934.

13. Spannhake, W. "Problems of Modern Pump and Turbine Design." *Trans. ASME,* 56, no. 4 (1934), 225.

14. Stepanoff, A. J. *Centrifugal and Axial Flow Pumps,* 2nd ed. John Wiley, New York, 1957.

15. Weick, F. E. *Aircraft Propeller Design.* McGraw-Hill, New York, 1930.

16. Weick, Fred E. "Full Scale Tests on a Thin Metal Propeller at Various Pit Speeds." *NACA Report,* 302 (January 1929).

Varied Flow in Open Channels

Flood waters flow over the spillway of Wanapum Dam on the Columbia River in central Washington State. (Courtesy Harza Engineering Company, Chicago, Illinois)

The primary difference between flow in closed conduits and flow in open channels is that in open channels there is a free surface (liquid surface is exposed to the atmosphere), and that surface is at atmospheric pressure, whereas in closed-conduit flow there is no free surface. As noted in Chapter 7, the hydraulic grade line (HGL) is a line showing where $p/\gamma = 0$ gage. Thus, for open-channel flow, the HGL is coincident with the free surface (see pages 295–297 for examples of this characteristic).

Almost all cases of open-channel flow are with water or wastewater as the flowing liquid. The Reynolds number for flow in open channels is given as $\mathrm{Re} = VR_h/\nu$. If this Re is greater than 750, one can expect the flow to be turbulent. Because water has a kinematic viscosity of about 10^{-5} ft^2/s (in the traditional system of units), the flow of water in a channel will be turbulent if $VR_h \approx 750 \times 10^{-5}$ ft^2/s $= 0.0075$ ft^2/s. For most engineering problems involving the flow of water in channels, the velocity is rarely less than 1 ft/s and the hydraulic radius is seldom less than 0.1 ft. Thus the lower limit of VR_h for most engineering problems is 0.1 ft^2/s, which is an order of magnitude greater than the limit for onset of turbulence with water flow. Thus most open-channel flow with water as the medium will be turbulent (see Example 10.13 on page 450, which illustrates this point).

In Chapter 10 (pages 450 to 464) we considered the case of steady uniform flow in open channels, that is, where the channel is straight and where the depth is uniform along the length of the channel. The depth for these uniform-flow conditions is called *normal* depth and is designated by y_n. The next more complicated classification of flow is steady nonuniform flow. By definition, this classification includes cases where the velocity is constant with respect to time but changes from section to section along the channel. The velocity change may be due simply to a change in depth along a straight prismatic channel, or it may also be a result of a change in channel configuration, such as a bend and/or change in cross-sectional shape and/or change in channel slope. This chapter focuses on the theory for and examples of steady nonuniform flow in open channels, usually referred to as varied flow.

Sections 15.1, 15.2, and 15.3 address cases of nonuniform flow in which the depth and velocity change markedly over a relatively short distance (this type of flow is often called *rapidly varied flow*). For these cases one can neglect the resistance of the channel walls and bottom. In Section 15.4 we consider cases where significant changes of depth and velocity occur over long reaches of the channel, and for this reason surface resistance *is* a significant variable in the flow process. This type of flow is called *gradually varied flow.*

The most complicated open-channel flow is unsteady nonuniform flow. An example of this is a breaking wave on a sloping beach. Theory and analysis of unsteady nonuniform flow are reserved for more advanced courses.

In Chapter 8 we noted that when gravity influences the flow pattern, the Froude number is a significant correlating parameter. Up to this point in the text, except for problems involving dimensional analysis and flow over weirs, all of the problems considered have been ones for which either the Reynolds number or the Mach number was the significant correlating parameter. In this chapter, we will focus on liquid flow in open channels, for which the force of gravity is a very significant variable. It will be shown that the Froude number is indeed a significant parameter for such flows.

15.1   Energy Relations in Open Channels

The Energy Equation Applied to Open-Channel Flow

The one-dimensional energy equation for open channels (see Fig. 15.1) is

$$\frac{p_1}{\gamma} + \alpha_1 \frac{V_1^2}{2g} + z_1 = \frac{p_2}{\gamma} + \alpha_2 \frac{V_2^2}{2g} + z_2 + h_L \qquad (15.1)$$

We see from Fig. 15.1 that the following equalities hold:

$$\frac{p_1}{\gamma} + z_1 = y_1 + S_0 \Delta x \qquad \text{and} \qquad \frac{p_2}{\gamma} + z_2 = y_2$$

FIGURE 15.1

Definition sketch for flow in open channels.

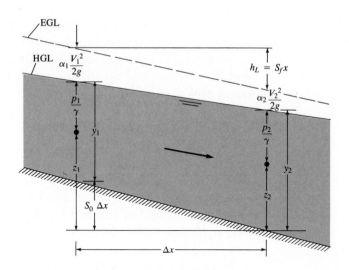

Here S_0 is the slope of the channel bottom, and y is the depth of flow. Then if we assume $\alpha_1 = \alpha_2 = 1.0$, we can write Eq. (15.1) as

$$y_1 + \frac{V_1^2}{2g} + S_0 \Delta x = y_2 + \frac{V_2^2}{2g} + h_L \qquad (15.2)$$

Now, if we consider the special case where the channel bottom is horizontal ($S_0 = 0$) and the head loss is zero ($h_L = 0$), Eq. (15.2) becomes

$$y_1 + \frac{V_1^2}{2g} = y_2 + \frac{V_2^2}{2g} \qquad (15.3)$$

Specific Energy

The sum of the depth of flow and the velocity head is defined as specific energy:

$$E = y + \frac{V^2}{2g} \qquad (15.4)$$

Thus Eq. (15.3) states that the specific energy at section 1 is equal to the specific energy at section 2, or $E_1 = E_2$. The continuity equation between sections 1 and 2 is

$$A_1 V_1 = A_2 V_2 = Q \qquad (15.5)$$

Therefore, Eq. (15.3) can be expressed as

$$y_1 + \frac{Q^2}{2gA_1^2} = y_2 + \frac{Q^2}{2gA_2^2} \qquad (15.6)$$

FIGURE 15.2

Relation between depth and specific energy.

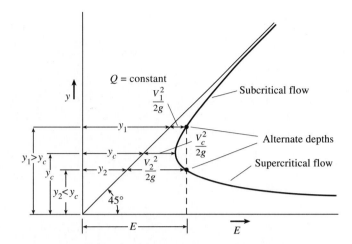

Because A_1 and A_2 are both functions of the depth y, the magnitude of the specific energy at section 1 or 2 is solely a function of the depth at each section. If, for a given channel and given discharge, one plots depth versus specific energy, a relationship such as that shown in Fig. 15.2, is obtained. By studying Fig. 15.2 for a given value of specific energy, we can see that the depth may be either large or small. In a physical sense, this means that for the small depth, the bulk of the energy of flow is in the form of kinetic energy $(Q^2/2gA^2)$; whereas for a larger depth, most of the energy is in the form of potential energy. Flow under a *sluice gate* (Fig. 15.3) is an example of flow in which two depths occur for a given value of specific energy. The large depth and low kinetic energy occurs upstream of the gate; the low depth and large kinetic energy occurs downstream. The depths as used here are called *alternate depths.* That is, for a given value of E, the large depth is alternate to the low depth, or vice versa. Returning to the flow under the sluice gate, we find that if we maintain the same rate of flow but set the gate with a larger opening, as in Fig. 15.3*b*, the upstream depth will drop, and the downstream depth will rise. Thus we have different alternate depths and a smaller value of specific energy than before. This is consistent with the diagram in Fig. 15.2.

Finally, it can be seen in Fig. 15.2 that a point will be reached where the specific energy is minimum and only a single depth occurs. At this point, the flow is termed *critical.* Thus one definition of critical flow is the flow that occurs when the specific energy is minimum for a given discharge. The flow for which the depth is less than critical (velocity is greater than critical) is termed *supercritical flow,* and the flow for which the depth is greater than critical (velocity is less than critical) is termed *subcritical flow.* Using this terminology, we can see that subcritical flow occurs upstream and supercritical flow occurs downstream of the sluice gate in Figure 15.3. It should be noted that some engineers refer to subcritical and supercritical flow as *tranquil* and *rapid* flow, respectively. We will consider other aspects of critical flow in the next section.

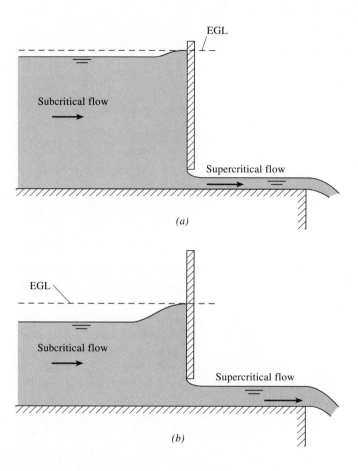

Characteristics of Critical Flow

We have already seen that critical flow occurs when the specific energy is minimum for a given discharge. The depth for this condition may be determined if we solve for dE/dy from $E = y + Q^2/2gA^2$ and set dE/dy equal to zero:

$$\frac{dE}{dy} = 1 - \frac{Q^2}{gA^3} \cdot \frac{dA}{dy} \tag{15.7}$$

However, $dA = T\,dy$, where T is the width of the channel at the water surface, as shown in Fig. 15.4. Then Eq. (15.7), with $dE/dy = 0$, will reduce to

$$\frac{Q^2 T_c}{gA_c^3} = 1 \tag{15.8}$$

or

$$\frac{A_c}{T_c} = \frac{Q^2}{gA_c^2} \tag{15.9}$$

FIGURE 15.4

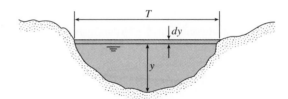

If we define the hydraulic depth D as A/T, then Eq. (15.9) will become

$$D_c = \frac{Q^2}{gA_c^2} \tag{15.10}$$

$$D_c = \frac{V^2}{g} \tag{15.11}$$

Upon dividing Eq. (15.11) by D_c and taking the square root of it, we get

$$1 = \frac{V}{\sqrt{gD_c}} \tag{15.12}$$

Note: $V/\sqrt{gD_c}$ is the Froude number. Therefore, it has been shown that the Froude number is equal to unity when critical flow prevails.

EXAMPLE 15.1   Determine the critical depth in this trapezoidal channel for a discharge of 500 cfs. The width of the channel bottom is B = 20 ft, and the sides slope upward at an angle of 45°.

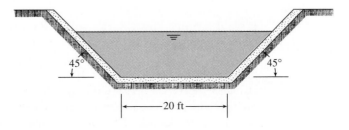

Solution   Starting with Eq. (15.8),

$$\frac{Q^2 T_c}{gA_c^3} = 1$$

or

$$\frac{A_c^3}{T_c} = \frac{Q^2}{g}$$

Then, for $Q = 500$ cfs,

$$\frac{A_c^3}{T_c} = \frac{500^2}{32.2} = 7764 \text{ ft}^2$$

For this channel, $A = y(B + y)$ and $T = B + 2y$. Then by iteration (choose y and compute A^3/T), we can find y that will yield an A^3/T equal to 7764 ft^2. Such a solution yields $y_c = 2.57$ ft.   ◀

If a channel is of rectangular cross section, then A/T is the actual depth, and $Q^2/A^2 = q^2/y^2$, so the formula for critical depth [Eq. (15.9)] becomes

$$y_c = \left(\frac{q^2}{g}\right)^{1/3} \tag{15.13}$$

where q is the discharge per unit width of channel.

Critical flow may also be examined in terms of how the discharge in a channel varies with depth for a given specific energy. For example, consider flow in a rectangular channel where

$$E = y + \frac{Q^2}{2gA^2}$$

or

$$E = y + \frac{Q^2}{2gy^2B^2}$$

If we consider a unit width of the channel and let $q = Q/B$, then the above equation becomes

$$E = y + \frac{q^2}{2gy^2}$$

If we determine how q varies with y for a constant value of specific energy, we see that critical flow occurs when the discharge is maximum (Figure 15.5).

Originally, the term *critical flow* probably related to the unstable character of the flow for this condition. If we refer to Fig. 15.2, we see that only a slight change in specific energy will cause the depth to increase or decrease a significant amount; this is a very unstable condition. In fact, observations of critical flow in open channels show that the water surface consists of a series of standing waves. Because of the unstable nature of the depth in critical flow, designing canals so that normal depth is either well above or well below critical depth is usually best. The flow in canals and rivers is usually subcritical; however, the flow in steep chutes or over spillways is supercritical.

In this section, various characteristics of critical flow have been explored. The main ones can be summarized as follows:

1. Critical flow occurs when

$$\frac{A^3}{T} = \frac{Q^2}{g}$$

FIGURE 15.5

Variation of q *and* y *with constant specific energy.*

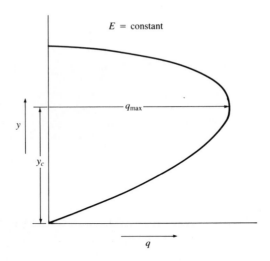

2. Critical flow occurs when Fr = 1.

3. Critical flow occurs when the specific energy is minimum for a given discharge.

4. Critical flow occurs when the discharge is maximum for a given specific energy.

5. For rectangular channels, critical depth is given as $y_c = (q^2/g)^{1/3}$.

Occurrence of Critical Depth

Critical flow occurs when a liquid passes over a broad-crested weir (Fig. 15.6a). The principle of the broad-crested weir is illustrated by first considering a closed sluice gate that prevents water from being discharged from the reservoir (Fig. 15.6b). If the gate is opened a small amount (gate position a'–a'), the flow upstream of the gate will be subcritical and the flow downstream will be supercritical (like the condition first introduced in Fig. 15.3). As the gate is opened further, a point is finally reached where the depths immediately upstream and downstream of the gate are the same. This is the critical condition. At this gate opening and beyond, the gate has no influence on the flow; this is the condition shown in Fig. 15.6a, the broad-crested weir. If the depth of flow over the weir is measured, the rate of flow can easily be computed from Eq. (15.13):

$$q = \sqrt{gy_c^3}$$

or
$$Q = L\sqrt{gy_c^3} \qquad (15.14)$$

where L is the length of the weir crest normal to the flow direction.

FIGURE 15.6

Flow over a broad-crested weir.

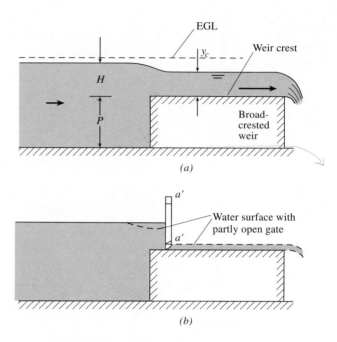

Because $y_c/2 = V_c^2/2g$ [from Eq. (15.11)], it is easily shown that $y_c = \frac{2}{3}E$, where E is the total head above the crest $(H + V_{approach}^2/2g)$; hence Eq. (15.14) can be rewritten as

$$Q = L\sqrt{g}\left(\frac{2}{3}\right)^{3/2} E^{3/2}$$

or $$Q = 0.385L\sqrt{2g}\,E_c^{3/2} \qquad (15.15)$$

For high weirs, the upstream velocity of approach is almost zero. Hence Eq. (15.15) can be expressed as

$$Q_{theor} = 0.385L\sqrt{2g}\,H^{3/2} \qquad (15.16)$$

If the height P of the broad-crested weir is relatively small, then the velocity of approach may be significant, and the discharge produced will be greater than that given by Eq. (15.16). Also, head loss will have some effect. To account for these effects, a discharge coefficient C is defined as

$$C = Q/Q_{theor} \qquad (15.17)$$

Then $$Q = 0.385CL\sqrt{2g}\,H^{3/2} \qquad (15.18)$$

where Q is the actual discharge over the weir. The value of C has been determined by experiments, and these results are shown in Fig. 15.7. The curve in Fig. 15.7 is for a weir with a vertical upstream face and a sharp corner at the intersection of the upstream face and the weir crest. If the upstream face is sloping at a 45° angle, the discharge coefficient should be increased 10% over that given

FIGURE 15.7

Discharge coefficient for a
broad-crested weir for
0.1 < H/L < 0.8 (8).

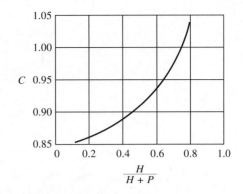

in Fig. 15.7. Rounding of the upstream corner will also produce a coefficient of discharge as much as 3% greater.

Equation (15.18) reveals a definite relationship for Q as a function of the head, H. This type of discharge-measuring device is in the broad class of discharge meters called *critical-flow flumes.* Another very common critical-flow flume is the *Venturi flume,* which was developed and calibrated by Parshall (5). Figure 15.8 shows the essential features of the Venturi flume. The discharge equation for the Venturi flume is in the same form as Eq. (15.18), the only difference being that the experimentally determined coefficient C will have a different value from the C for the broad-crested weir. For more details on the Venturi flume, you may refer to Roberson et al. (9), Parshall (5), and Chow (2). The Venturi flume is especially useful for discharge measurement in irrigation systems because little head loss is required for its use and sediment is easily flushed through if the water happens to be silty.

The depth also passes through a critical stage in channel flow where the slope changes from a mild one to a steep one. A *mild slope* is defined as a slope for which the normal depth y_n is greater than y_c. Likewise, a *steep slope* is one for

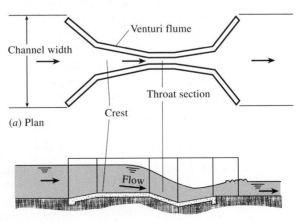

FIGURE 15.8

Flow through a Venturi
flume.

FIGURE 15.9

*Critical depth at a break
in grade.*

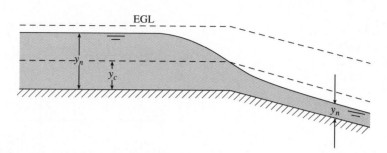

which $y_n < y_c$. This condition is shown in Fig. 15.9. Note that y_c is the same for both slopes in the figure because y_c is a function of the discharge only. However, normal depth (uniform flow depth) for the mild upstream channel is greater than critical, whereas the normal depth for the steep downstream channel is less than critical; hence it is obvious that the depth must pass through a critical stage. Experiments show that critical depth occurs a very short distance upstream of the intersection of the two channels.

Another place where critical depth occurs is upstream of a free overfall at the end of a channel with a mild slope (Fig. 15.10). Critical depth will occur at a distance of 3 to $4y_c$ upstream of the brink. Such occurrences of critical depth (at a break in grade or at a brink) are useful in computing surface profiles because they provide a point for starting surface-profile calculations.*

Channel Transitions

Whenever a channel's cross-sectional configuration (shape or dimension) changes along its length, the change is termed a *transition.* We will use the basic concepts already presented in the first part of this chapter to show how the flow depth changes when the floor of a rectangular channel is increased in elevation or when the width of the channel is decreased. In these developments we assume negligible energy losses. We will look first at the case where the floor of the channel is raised (an upstep). Later in this section we will look at configurations of transitions used for subcritical flow from a rectangular to a trapezoidal channel.

Consider the rectangular channel shown in Fig. 15.11, where the floor rises an amount Δz. To help us in evaluating depth changes, we use a diagram of specific energy versus depth, which is similar to Fig. 15.2. This diagram is placed both at the section upstream of the transition and at the section just downstream of the transition. Because the discharge, Q, is the same at both sections, the given diagram is valid at both sections. As noted in Fig. 15.11, the

*The procedure for making these computations starts on page 710.

FIGURE 15.10

*Critical depth at a free
overfall.*

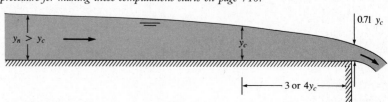

FIGURE 15.11

*Change in depth with
change in bottom elevation of
a rectangular channel.*

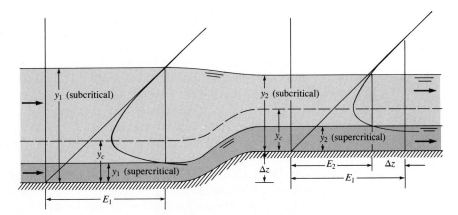

depth of flow at section 1 can be either large (subcritical) or small (supercritical) if the specific energy E_1 is greater than that required for critical flow. It can also be seen in Fig. 15.11 that when the upstream flow is subcritical, a decrease in depth occurs in the region of the elevated channel bottom. This occurs because the specific energy at this section, E_2, is less than that at section 1 by the amount Δz. Therefore, the specific-energy diagram indicates that y_2 will be less than y_1. In a similar manner it can be seen that when the upstream flow is supercritical, the depth as well as the actual water-surface elevation increases from section 1 to section 2. A further note should be made about the effect on flow depth of a change in bottom-surface elevation. If the channel bottom at section 2 is at an elevation greater than that just sufficient to establish critical flow at section 2, then there is not enough head at section 1 to cause flow to occur over the rise under steady-flow conditions. Instead, the water level upstream will rise until it is just sufficient to reestablish steady flow.

When the channel bottom is kept at the same elevation but the channel is decreased in width, then the discharge per unit of width between sections 1 and

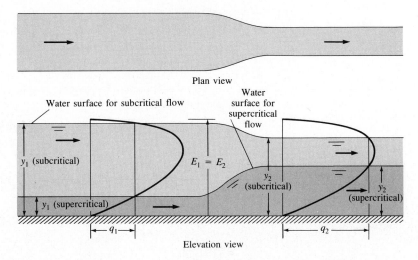

FIGURE 15.12

*Change in depth with
change in channel width.*

2 increases, but the specific energy E remains constant. Thus when we utilize the diagram of q versus depth for the given specific energy E, we note that the depth in the restricted section increases if the upstream flow is supercritical and decreases if it is subcritical (see Fig. 15.12).

The foregoing paragraphs describe gross effects for the simplest transitions. In practice, it is more common to find transitions between a channel of one shape (rectangular cross section, for example) and a channel having a different cross section (trapezoidal, for example). A very simple transition between two such channels consists of two straight vertical walls joining the two channels, as shown by half section in Fig. 15.13.

This type of transition can work, but it will produce excessive head loss because of the abrupt change in cross section and the ensuing separation that will occur. To reduce the head losses, a more gradual type of transition is used. Figure 15.14 is a half section of a transition similar to that of Fig. 15.13, but with the angle θ much greater than 90°. This is called a *wedge transition*.

The *warped-wall* transition shown in Fig. 15.15 will yield even smoother flow than either of the other two, and it will thus have less head loss. In the practical design and analysis of transitions, engineers usually use the complete energy equation, including the kinetic energy factors α_1 and α_2 as well as a head loss term h_L, to define velocity and water-surface elevation through the transition. Analyses of transitions utilizing the one-dimensional form of the energy equation are applicable only if the flow is subcritical. If the flow is supercritical, then a much more involved analysis is required. For more details on the design and analysis of transitions, you are referred to Hinds (4), Chow (2), U.S. Bureau of Reclamation (11), and Rouse (10).

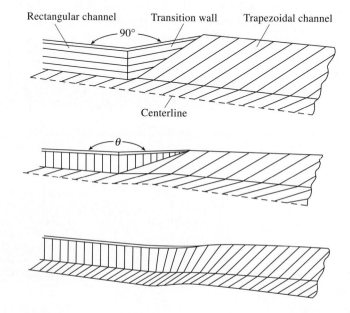

FIGURE 15.13

Simplest type of transition between a rectangular channel and a trapezoidal channel.

FIGURE 15.14

Half section of a wedge transition.

FIGURE 15.15

Half section of a warped-wall transition.

Wave Celerity

Wave celerity is the velocity at which an infinitesimally small wave travels relative to the velocity of the fluid in which it is traveling. The following is a derivation of the wave celerity c.

Consider a small solitary wave moving with a velocity c in an otherwise calm body of liquid of small depth (Fig. 15.16a). Because the velocity in the liquid changes with time, this is a condition of unsteady flow. However, if we refer all velocities to a reference frame moving with the wave, the shape of the wave would be fixed and the flow would be steady. Then the flow is amenable to analysis with Bernoulli's equation. The steady-flow condition is shown in Fig. 15.16b. When Bernoulli's equation is written between a point on the surface of the undisturbed fluid and a point at the wave crest, we have

$$\frac{c^2}{2g} + y = \frac{V^2}{2g} + y + \Delta y \tag{15.19}$$

In Eq. (15.19), V is the velocity of the liquid in the section where the crest of the wave is located. From the continuity equation we have $cy = V(y + \Delta y)$. Hence

$$V = \frac{cy}{y + \Delta y}$$

and

$$V^2 = \frac{c^2 y^2}{(y + \Delta y)^2} \tag{15.20}$$

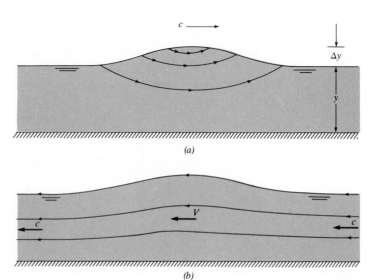

FIGURE 15.16

Solitary wave (exaggerated vertical scale). (a) Unsteady flow. (b) Steady flow.

When Eq. (15.20) is substituted into Eq. (15.19), we obtain

$$\frac{c^2}{2g} + y = \frac{c^2 y^2}{2g[y^2 + 2y\Delta y + (\Delta y)^2]} + y + \Delta y \qquad (15.21)$$

Solving Eq. (15.21) for c after discarding terms with $(\Delta y)^2$, since we are assuming an infinitesimally small wave, we obtain

$$c = \sqrt{gy} \qquad (15.22)$$

It has thus been shown that the speed of a small solitary wave is equal to the square root of the product of the depth and g.

15.2 | The Hydraulic Jump

Occurrence of the Hydraulic Jump

When the flow is supercritical in an upstream section of a channel and is then forced to become subcritical in a downstream section (the change in depth can be forced by a sill in the downstream part of the channel or just by the prevailing depth in the stream further downstream), a rather abrupt change in depth usually occurs, and considerable energy loss accompanies the process. This flow phenomenon, called the *hydraulic jump* (Fig. 15.17), is often considered in the design of open channels and spillways of dams. For example, many spillways are designed so that a jump will occur on an *apron* of the spillway, thereby reducing the downstream velocity so that objectionable erosion of the river channel is prevented. If a channel is designed to carry water at supercritical velocities, the designer must be certain that the flow will not become subcritical prematurely. If it did, overtopping of the channel walls would undoubtedly occur, with consequent failure of the structure. Because the energy loss in the hydraulic jump is initially not known, the energy equation is not a suitable tool for analysis of the velocity–depth relationships. Therefore, the momentum equation is applied to the problem.

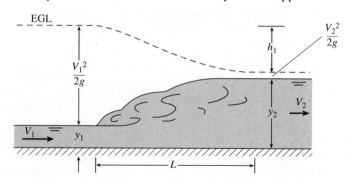

FIGURE 15.17

Definition sketch for the hydraulic jump.

Derivation of Depth Relationships

Consider flow as shown in Fig. 15.17. Here it is assumed that uniform flow occurs both upstream and downstream of the jump and that the resistance of the channel bottom is negligible. The derivation is for a horizontal channel, but experiments show that the results of the derivation will apply to all channels of moderate slope ($S_0 < 0.02$). We start the derivation by applying the momentum equation in the x direction to the control volume shown in Fig. 15.18:

$$\sum F_x = \sum_{cs} V_x \rho \mathbf{V} \cdot \mathbf{A}$$

The forces are the hydrostatic forces on each end of the system; thus the following is obtained:

$$\bar{p}_1 A_1 - \bar{p}_2 A_2 = \rho V_1(-V_1 A_1) + \rho V_2(V_2 A_2)$$

or

$$\bar{p}_1 A_1 + \rho Q V_1 = \bar{p}_2 A_2 + \rho Q V_2 \qquad (15.23)$$

In Eq. (15.23), $\bar{p}_1$ and $\bar{p}_2$ are the pressures at the centroids of the respective areas A_1 and A_2.

A representative problem might be to determine the downstream depth y_2 given the discharge and upstream depth. The left-hand side of Eq. (15.23) would be known because V, A, and p are all functions of y and Q, and the right-hand side is a function of y_2; therefore, y_2 can be solved for.

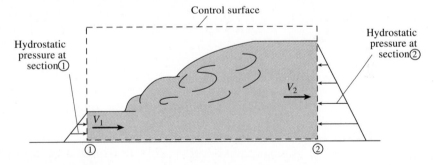

FIGURE 15.18

Control-volume analysis for the hydraulic jump.

EXAMPLE 15.2 Water flows in a trapezoidal channel at a rate of 300 cfs. The channel has a bottom width of 10 ft and side slopes of 1 vertical to 1 horizontal. If a hydraulic jump is forced to occur where the upstream depth is 1.00 ft, what will be the downstream depth and velocity? What are the values of Fr_1 and Fr_2?

Solution For the upstream section, the cross-sectional flow area is 11 ft². Therefore, the mean velocity $V_1 = Q/A_1 = 27.3$ ft/s. The hydraulic depth $D_1 = A_1/T_1 = 11$ ft²/12 ft = 0.9167 ft.

Then

$$\text{Fr}_1 = \frac{V_1}{\sqrt{gD_1}} = \frac{27.3 \text{ ft/s}}{\sqrt{32.2 \text{ ft/s}^2 \times 0.9167 \text{ ft}}} = 5.02 \qquad \blacktriangleleft$$

The location of the centroid of the area A_1 can be obtained by taking moments of the subareas about the water surface:

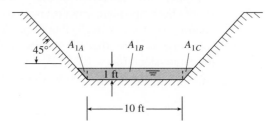

$$A_1 \bar{y}_1 = A_{1A} \times 0.333 \text{ ft} + A_{1B} \times 0.500 \text{ ft} + A_{1C} \times 0.333 \text{ ft}$$

$$(11 \text{ ft}^2)\bar{y}_1 = (0.333 \text{ ft}) (0.500 \text{ ft}^2 \times 2) + (0.50 \text{ ft}) (10.00 \text{ ft}^2)$$

$$\bar{y}_1 = 0.485 \text{ ft}$$

Thus, the pressure at the centroid will be 62.4 lb/ft^3 × 0.485 ft = 30.264 lb/ft^2. Substituting the appropriate quantities into Eq. (15.23) yields

$$30.26 \times 11 + 1.94 \times 300 \times 27.3 = \bar{p}_2 A_2 + \rho Q V_2$$

or
$$\bar{p}_2 A_2 + \rho Q V_2 = 16{,}221 \text{ lbf}$$

$$\gamma \bar{y}_2 A_2 + \frac{\rho Q^2}{A_2} = 16{,}221$$

where
$$\bar{y}_2 = \frac{\sum A_i \gamma_i}{A_2} = \frac{B y_2^2 / 2 + y_2^3 / 3}{A_2}$$

Then
$$\gamma \left(\frac{B y_2^2}{2} + \frac{y_2^3}{3} \right) + \frac{\rho Q^2}{(By + y^2)} = 16{,}221 \text{ lb}$$

Using $\gamma = 62.4$ lb/ft^3, $B = 10$ ft, $\rho = 1.94$ lb · s^2/ft^4, and $Q = 300$ ft^2/s and solving the above equation for y_2 yields

$$y_2 = 5.75 \text{ ft} \qquad \blacktriangleleft$$

$$V_2 = \frac{Q}{A_2} = \frac{300}{57.5 + 33.06} = 3.31 \text{ ft/s}$$

Also
$$D_2 = \frac{A}{T} = \frac{90.56}{21.5} = 4.21 \text{ ft}$$

$$\text{Fr}_2 = \frac{V}{\sqrt{gD}} = \frac{3.31}{\sqrt{32.2 \times 4.21}} = 0.284 \qquad \blacktriangleleft$$

Hydraulic Jump in Rectangular Channels

If we write Eq. (15.23) for a unit width of a rectangular channel where $\bar{p}_1 = \gamma y_1/2$, $\bar{p}_2 = \gamma y_2/2$, $Q = q$, $A_1 = y_1$, and $A_2 = y_2$, we will have

$$\gamma \frac{y_1^2}{2} + \rho q V_1 = \gamma \frac{y_2^2}{2} + \rho q V_2 \tag{15.24}$$

but $q = Vy$, so Eq. (15.24) can be rewritten as

$$\frac{\gamma}{2}(y_1^2 - y_2^2) = \frac{\gamma}{g}(V_2^2 y_2 - V_1^2 y_1) \tag{15.24}$$

The above equation can be further manipulated to yield

$$\frac{2V_1^2}{g y_1} = \left(\frac{y_2}{y_1}\right)^2 + \frac{y_2}{y_1} \tag{15.25}$$

The term on the left-hand side of Eq. (15.25) will be recognized as twice Fr_1^2. Hence Eq. (15.25) is written as

$$\left(\frac{y_2}{y_1}\right)^2 + \frac{y_2}{y_1} - 2\mathrm{Fr}_1^2 = 0 \tag{15.26}$$

By use of the quadratic formula, it is easy to solve for y_2/y_1 in terms of the upstream Froude number. Thus we obtain

$$\frac{y_2}{y_1} = \frac{1}{2}(\sqrt{1 + 8\mathrm{Fr}_1^2} - 1) \tag{15.27}$$

or

$$y_2 = \frac{y_1}{2}(\sqrt{1 + 8\mathrm{Fr}_1^2} - 1) \tag{15.28}$$

The other solution of Eq. (15.26) gives a negative downstream depth, which is not relevant here. Hence we have expressed the downstream depth in terms of the upstream depth and the upstream Froude number. In Eqs. (15.27) and (15.28), the depths y_1 and y_2 are said to be *conjugate* or *sequent* (both terms are in common use) to each other, in contrast to the alternate depths obtained from the energy equation. Numerous experiments show that the relation represented by Eqs. (15.27) and (15.28) is valid over a wide range of Froude numbers. Although no theory has been developed to predict the length of a hydraulic jump, experiments [see Chow (2)] show that the relative length of the jump, L/y_2, is approximately 6 for a range of Fr_1 from 4 to 18.

Head Loss in a Hydraulic Jump

In addition to determining the geometric characteristics of the hydraulic jump, it is often desirable to determine the head loss produced by it. This is obtained by comparing the specific energy before the jump to that after the jump, the head loss being the difference between the two specific energies. It can be shown that this head loss for a jump in a rectangular channel is

$$h_L = \frac{(y_2 - y_1)^3}{4y_1 y_2} \tag{15.29}$$

For more information on the hydraulic jump, see Chow (2).

EXAMPLE 15.3   Water flows in a rectangular channel at a depth of 30 cm and with a velocity of 16 m/s, as shown in the figure. If a downstream sill (not shown in the figure) forces a hydraulic jump, what will be the depth and velocity downstream of the jump? What head loss is produced by the jump?

Solution   To solve the problem we must know Fr_1, which is computed first:

$$Fr_1 = \frac{V}{\sqrt{gy_1}} = \frac{16}{\sqrt{9.81(0.30)}} = 9.33$$

Next we compute y_2, using Eq. (15.28):

$$y_2 = \frac{0.30}{2}[\sqrt{1 + 8(9.33)^2} - 1] = 3.81 \text{ m}$$

Then $$V_2 = \frac{q}{y_2} = \frac{(16 \text{ m/s})(0.30 \text{ m})}{3.81 \text{ m}} = 1.26 \text{ m/s}$$

The head loss is found by use of Eq. (15.29):

$$h_L = \frac{(3.81 - 0.30)^3}{4(0.30)(3.81)} = 9.46 \text{ m}$$

We can check the validity of Eq. (15.29) since the head loss is equal to $E_1 - E_2$, or

$$h_L = \left(0.30 + \frac{16^2}{2 \times 9.81}\right) - \left(3.81 + \frac{1.26^2}{2 \times 9.81}\right) = 9.46 \text{ m}$$

The answers check.

Use of the Hydraulic Jump on Downstream End of Dam Spillway

We have already shown that the transition from supercritical to subcritical flow produces a hydraulic jump, and that the relative height of the jump (y_2/y_1) is a function of Fr_1. Because flow over the spillway of a dam invariably results in supercritical flow at the lower end of the spillway, and because flow in the chan-

FIGURE 15.19

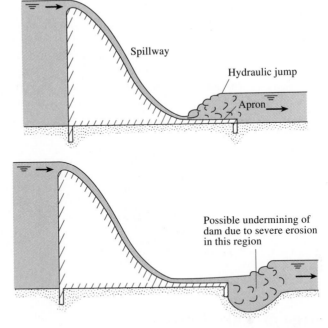

FIGURE 15.20

*Hydraulic jump occurring
downstream of spillway
apron.*

nel downstream of a spillway is usually subcritical, it is obvious that a hydraulic
jump must form near the base of the spillway (see Fig. 15.19). The downstream
portion of the spillway, called the spillway *apron,* must be designed so that the
hydraulic jump always forms on the concrete structure itself. If the hydraulic
jump were allowed to form beyond the concrete structure, as in Fig. 15.20,
severe erosion of the foundation material as a result of the high-velocity supercritical flow could undermine the dam and cause complete failure of it. One way
to solve this problem might be to incorporate a long, sloping apron into the
design of the spillway, as shown in Fig. 15.21. A design like this would work
very satisfactorily from the hydraulics point of view. For all combinations of Fr_1
and water-surface elevation in the downstream channel, the jump would always

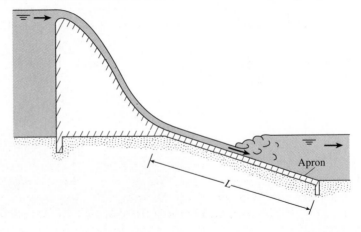

FIGURE 15.21

Long sloping apron.

form on the sloping apron. However, its main drawback is cost of construction. Construction costs will be reduced as the length, L, of the stilling basin is reduced. Much research has been devoted to the design of stilling basins that will operate properly for all upstream and downstream conditions and yet be relatively short to reduce construction cost. Research by the U.S. Bureau of Reclamation (12) has resulted in sets of standard designs that can be used. These designs include sills, baffle piers, and chute blocks, as shown in Fig. 15.22.

FIGURE 15.22

Spillway with stilling basin Type III as recommended by the U.S.B.R. (12).

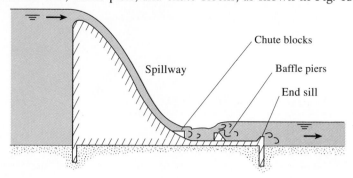

15.3   Surge or Tidal Bore

Tides are generally low enough so that the waves they produce are smooth and nondestructive. However, in some parts of the world the tides are so high that their entry into shallow bays or mouths of rivers causes surges to be formed that may be very hazardous to small boats. A *surge* is actually a moving hydraulic jump. Hence the same analytical methods that we used for the jump can be used to solve for the speed of the surge. In Fig. 15.23 a surge is shown coming into an otherwise still body of water. As indicated in Fig. 15.23, the flow is unsteady because throughout the surge itself the velocities are changing with time. However, if all velocities are measured in a coordinate system moving with the surge front, then a steady-flow pattern is obtained. Figure 15.24 shows the steady-flow pattern and control volume. Now the problem is directly analogous to the hydraulic-jump problem. We simply replace V_1 in Eq. (15.25) by V_s to yield

$$\frac{V_s}{\sqrt{gy_1}} = \left[\frac{y_2}{2y_1} \left(\frac{y_2}{y_1} + 1 \right) \right]^{1/2} \qquad (15.30)$$

FIGURE 15.23

Surge moving into still water.

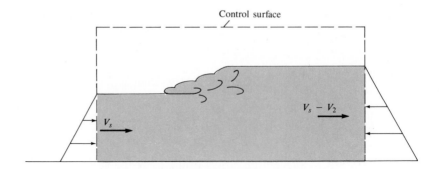

FIGURE 15.24

*Control-volume analysis
applied to the surge.*

Control surface

$V_s - V_2$

V_s

15.4   **Gradually Varied Flow in Open Channels**

In Secs. 15.1, 15.2, and 15.3 we considered cases of rapidly varied flow, and in those examples the channel resistance was assumed to be negligible. For gradually varied flow, however, channel resistance is usually a significant factor in the flow process. We utilize the energy equation for the solution of this type of problem.

Basic Differential Equation
for Gradually Varied Flow

There are a number of cases of open-channel flow in which the change in water-surface profile is so gradual that it is possible to integrate the relevant differential equation from one section to another to obtain the desired change in depth. This may be either an analytical integration or, more commonly, a numerical integration. In Sec. 15.1, the energy equation was written between two sections of a channel Δx distance apart. Because the only head loss here is the channel resistance, we can write h_L as Δh_f, and Eq. (15.2) becomes

$$y_1 + \frac{V_1^2}{2g} + S_0\,\Delta x = y_2 + \frac{V_2^2}{2g} + \Delta h_f \tag{15.31}$$

The friction slope S_f is defined as the slope of the EGL, or $\Delta h_f/\Delta x$. Then $\Delta h_f = S_f \Delta x$, and if we let $\Delta y = y_2 - y_1$ and

$$\frac{V_2^2}{2g} - \frac{V_1^2}{2g} = \frac{d}{dx}\left(\frac{V^2}{2g}\right)\Delta x \tag{15.32}$$

Eq. (15.31) becomes

$$\Delta y = S_0\,\Delta x - S_f\Delta x - \frac{d}{dx}\left(\frac{V^2}{2g}\right)\Delta x$$

Dividing through by Δx and taking the limit as Δx approaches zero gives us

$$\frac{dy}{dx} + \frac{d}{dx}\left(\frac{V^2}{2g}\right) = S_0 - S_f \tag{15.33}$$

The second term is rewritten as $[d(V^2/2g)/dy] \, dy/dx$, so that Eq. (15.33) simplifies to

$$\frac{dy}{dx} = \frac{S_0 - S_f}{1 + d(V^2/2g)/dy} \tag{15.34}$$

To put Eq. (15.34) in a more usable form, we express the denominator in terms of the Froude number. This is accomplished by observing that

$$\frac{d}{dy}\left(\frac{V^2}{2g}\right) = \frac{d}{dy}\left(\frac{Q^2}{2gA^2}\right) \tag{15.35}$$

After differentiating the right side of Eq. (15.35), the equation becomes

$$\frac{d}{dy}\left(\frac{V^2}{2g}\right) = \frac{-2Q^2}{2gA^3} \cdot \frac{dA}{dy}$$

But $dA/dy = T$ (top width), and $A/T = D$ (hydraulic depth); therefore,

$$\frac{d}{dy}\left(\frac{V^2}{2g}\right) = \frac{-Q^2}{gA^2D}$$

or

$$\frac{d}{dy}\left(\frac{V^2}{2g}\right) = -\text{Fr}^2$$

Hence, when the expression for $d(V^2/2g)/dy$ is substituted into Eq. (15.34), we obtain

$$\frac{dy}{dx} = \frac{S_0 - S_f}{1 - \text{Fr}^2} \tag{15.36}$$

This is the general differential equation for gradually varied flow. It is used to describe the various types of water-surface profiles that occur in open channels. Note that, in the derivation of the equation, S_0 and S_f were taken as positive when the channel and energy grade lines, respectively, were sloping downward in the direction of flow. Also note that y is measured from the bottom of the channel. Therefore, $dy/dx = 0$ if the slope of the water surface is equal to the slope of the channel bottom, and dy/dx is positive if the slope of the water surface is less than the channel slope.

Introduction to Water-Surface Profiles

In the design of projects involving the flow in channels (rivers or irrigation canals, for example), the engineer must often estimate the *water-surface profile* (elevation of the water surface along the channel) for a given discharge. For example, when a dam is being designed for a river project, the water-surface profile in the river upstream must be defined so that the project planners will know how much land to acquire to accommodate the upstream pool. The first step in defining a water-surface profile is to locate a point or points along the channel where the depth can be computed for a given discharge. For example, at a change in slope from mild to steep, we know that critical depth will occur just upstream of the break in grade (see Sec. 15.1, page 689). At that point we can solve for y_c with Eq. (15.8) or Eq. (15.13). Also, for flow over the spillway of a dam, there will be a discharge equation for the spillway from which we can calculate the water-surface elevation in the reservoir at the face of the dam. Such points where there is a unique relationship between discharge and water-surface elevation are called *controls*. Once the water-surface elevations at these controls are determined, then the water-surface profile can be extended upstream or downstream from the control points to define the water-surface profile for the entire channel. The completion of the profile is done by numerical integration. However, before this integration is performed, it is usually helpful to the engineer if he or she will sketch in the profiles. To assist in the process of sketching the possible profiles, the engineer can refer to different categories of profiles (water-surface profiles have unique characteristics depending upon the relationship between normal depth, critical depth, and the actual depth of flow in the channel). This initial sketching of the profiles helps the engineer to scope the problem and to obtain a solution or solutions in a minimum amount of time. The next section describes the various types of water-surface profiles.

Types of Water-Surface Profiles

There are 12 different types of water-surface profiles for gradually varied flow in channels, and these are shown schematically in Fig. 15.25 (page 712). Each profile is identified by a letter and number designator. For example, the first water-surface profile in column 1 of Fig. 15.25 is identified as an M1 profile. The letter designator is an indicator of the type of slope of the channel—that is, whether the slope is mild (M designator), critical (C designator), steep (S designator), horizontal (H), or adverse (A). The slope is defined as mild if the uniform flow depth, y_n, is greater than the critical flow depth, y_c. Conversely, if y_n is less than y_c, the channel would be termed steep. Or if $y_n = y_c$, this would be a channel with critical slope. The designation M, S, or C is determined by computing y_n and y_c for the given channel for a given discharge. Equation (10.40) or (10.41) is used to compute y_n, and Eq. (15.8) or (15.13) is used to compute y_c. Figure 15.26 shows the relationship between y_n and y_c for the M, S, and C designations. As the name implies, a horizontal slope is one where the channel actually has a zero slope, and an adverse slope is one where the slope of the channel is upward in the direction of flow. Normal depth does not exist for these two cases (for example, water cannot flow at uniform depth in either a horizontal channel or one with adverse slope); therefore, they are given the special designations H and A, respectively.

FIGURE 15.25

Classification of water-surface profiles of gradually varied flow. [Adapted from Open Channel Hydraulics *by Chow (2). Copyright © 1959, McGraw-Hill Book Company, New York; used with permission of McGraw-Hill Book Company.]*

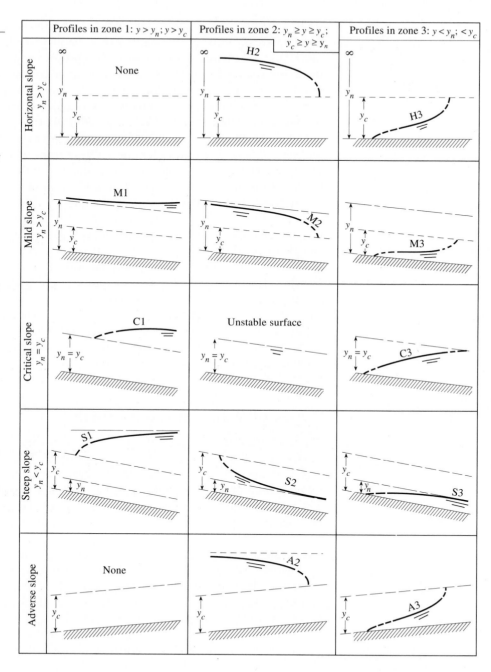

FIGURE 15.26

Letter designators as a function of the relationship between y_n *and* y_c.

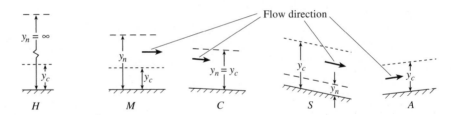

The number designator for the type of profile relates to the position of the *actual* water surface in relation to the position of the water surface for uniform and critical flow in the channel. If the actual water surface is above that for uniform and critical flow ($y > y_n; y > y_c$), then that condition is given a 1 designation. If the actual water surface is between those for uniform and critical flow, then it is given a 2 designation; and if the actual water surface lies below those for uniform and critical flow, then it is given a 3 designation. Figure 15.27 depicts these conditions for mild and steep slopes.

Figure 15.28 shows how different water-surface profiles can develop in certain field situations. More specifically, if we consider in detail the flow downstream of the sluice gate (see Fig. 15.29), we see that the discharge and slope are such that the normal depth is greater than the critical depth; therefore the slope is termed mild. The actual depth of flow shown in Fig. 15.29 is less than either y_c or y_n. Hence a type 3 water-surface profile exists. The complete classification of the profile in Fig. 15.29, therefore, is a mild type 3 profile, or simply an M3 profile. Using these designations, we would categorize the profile upstream of the sluice gate as type M1.

EXAMPLE 15.4 Classify the water-surface profile for the flow downstream of the sluice gate in Fig. 15.3 if the slope is horizontal, and that for the flow immediately downstream of the break in grade in Fig. 15.9.

Solution In Fig. 15.3 the actual depth is less than critical; thus the profile is type 3. The channel is horizontal; hence the profile is designated type H3. ◄

FIGURE 15.27

Number designator as a function of the location of the actual water surface in relation to y_n *and* y_c.

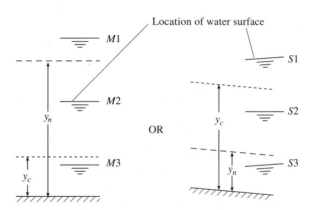

FIGURE 15.28

Water-surface profiles associated with flow behind a dam, flow under a sluice gate, and flow in a channel with a change in grade.

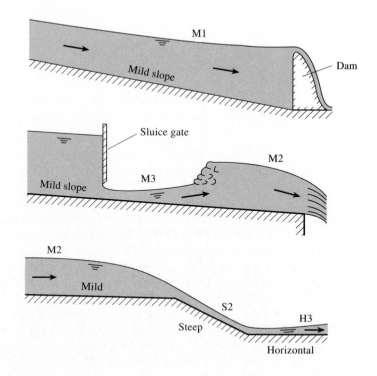

In Fig. 15.9 the actual depth is greater than normal but less than critical, so the profile is type 2. The uniform-flow depth (normal depth y_n) is less than the critical depth; hence the slope is steep. Therefore the water-surface profile is designated type S2. ◄

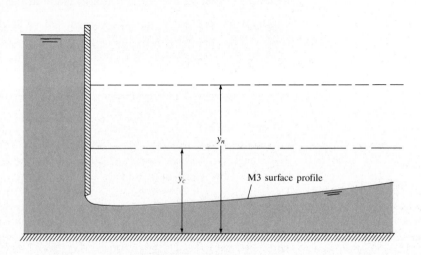

FIGURE 15.29

Water-surface profile, M3 type.

With the foregoing introduction to the classification of water-surface profiles, we can now refer to Eq. (15.36) to describe the shapes of the profiles. Again, for example, if we consider the M3 profile, we know that Fr > 1 because the flow is supercritical ($y < y_c$), and that $S_f > S_0$ because the velocity is greater than normal velocity. Hence a head loss greater than that for normal flow must exist. Inserting these relative values into Eq. (15.36), we see that both the numerator and the denominator are negative. Thus dy/dx must be positive (the depth increases in the direction of flow), and as critical depth is approached, the Froude number approaches unity. Hence the denominator of Eq. (15.36) approaches zero. Therefore, as the depth approaches critical depth, $dy/dx \rightarrow \infty$. What actually occurs in cases where the critical depth is approached in supercritical flow is that a hydraulic jump forms and a discontinuity in profile is thereby produced.

Certain general features of profiles, as shown in Fig. 15.25, are evident. First, as the depth becomes very great, the velocity of flow approaches zero. Hence Fr $\rightarrow 0$ and $S_f \rightarrow 0$, and dy/dx approaches S_0 because $dy/dx = (S_0 - S_f)/(1 - \text{Fr}^2)$. In other words, the depth increases at the same rate at which the channel bottom drops away from the horizontal. Thus the water surface approaches the horizontal. The profiles that show this tendency are types M1, S1, and C1. A physical example of the M1 type is the water-surface profile upstream of a dam, as shown in Fig. 15.28. The second general feature of several of the profiles is that those that approach normal depth do so asymptotically. This is shown in the S2, S3, M1, and M2 profiles. Also note in Fig. 15.25 that profiles that approach critical depth are shown by dashed lines. This is done because, near critical depth, either discontinuities develop (hydraulic jump) or the streamlines are very curved (such as near a brink). These profiles cannot be accurately predicted by Eq. (15.36) because this equation is based on one-dimensional flow, which, in these regions, is invalid.

Quantitative Evaluation of the Water-Surface Profile

In practice, most water-surface profiles are generated by numerical integration, that is, by dividing the channel into short reaches and carrying the computation for water-surface elevation from one end of the reach to the other. For one method, called the *direct step method,* the depth and velocity are known at a given section of the channel (one end of the reach), and one arbitrarily chooses the depth at the other end of the reach. Then the length of the reach is solved for. The applicable equation for quantitative evaluation of the water-surface profile is the energy equation written for a finite reach of channel, Δx:

$$y_1 + \frac{V_1^2}{2g} + S_0 \Delta x = y_2 + \frac{V_2^2}{2g} + S_f \Delta x$$

From this equation we obtain

$$\Delta x(S_f - S_0) = \left(y_1 + \frac{V_1^2}{2g}\right) - \left(y_2 + \frac{V_2^2}{2g}\right)$$

or

$$\Delta x = \frac{(y_1 + V_1^2/2g) - (y_2 + V_2^2/2g)}{S_f - S_0} = \frac{(y_1 - y_2) + (V_1^2 - V_2^2)/2g}{S_f - S_0}$$

$$(15.37)$$

The procedure for evaluation of a profile is first to ascertain which type applies to the given reach of channel (we use the methods of the preceding subsection). Then, starting from a known depth, we compute a finite value of Δx for an arbitrarily chosen change in depth. The process of computing Δx, step by step, up (negative Δx) or down (positive Δx) the channel is repeated until the full reach of channel has been covered. Usually small changes of y are taken, so that the friction slope is approximated by the following equation:

$$S_f = \frac{h_f}{\Delta x} = \frac{fV^2}{8gR}$$

$$(15.38)$$

Here V is the mean velocity in the reach and R is the mean hydraulic radius in the reach. That is, $V = (V_1 + V_2)/2$ and $R_h = (R_{h1} + R_{h2})/2$. It is obvious that a numerical approach of this type is ideally suited for solution by computer.

EXAMPLE 15.5   Water discharges from under a sluice gate into a horizontal rectangular channel at a rate of 1 m³/s per meter of width as shown. What is the classification of the water-surface profile? Quantitatively evaluate the profile downstream of the gate and determine whether it will extend all the way to the abrupt drop 80 m downstream. Make the simplifying assumption that the resistance factor f is equal to 0.02 and that the hydraulic radius R_h is equal to the depth y.

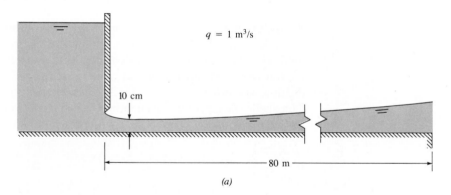

$q = 1$ m³/s

10 cm

80 m

(a)

SOLUTION TO EXAMPLE 15.5

Section Number Downstream of Gate	Depth y, m	Velocity at Section V, m/s	Mean Velocity in Reach $(V_1 + V_2)/2$	V^2	Mean Hydraulic Radius $R_m = (y_1 + y_2)/2$	$S_f = \dfrac{fV^2_{mean}}{8gR_m}$	$\Delta x = \dfrac{(y_1 - y_2) + \dfrac{(V_1^2 - V_2^2)}{2g}}{(S_f - S_0)}$	Distance from Gate x, m
1 (at gate)	0.1	10	...	100	...	...	...	0
	...	...	8.57	73.4	0.12	0.156	15.7	15.7
2	0.14	7.14	...	51.0	...	...	...	15.7
	...	...	6.35	40.3	0.16	0.064	15.3	
3	0.18	5.56	...	30.9	...	...	...	31.0
	...	...	5.05	25.5	0.20	0.032	15.1	
4	0.22	4.54	...	20.6	...	...	...	46.1
	...	...	4.19	17.6	0.24	0.019	13.4	
5	0.26	3.85	...	14.8	...	...	...	59.5
	...	...	3.59	12.9	0.28	0.012	12.4	
6	0.30	3.33	...	11.1	...	...	...	71.9
	...	...	3.13	9.8	0.32	0.008	10.9	
7	0.34	2.94	...	8.6	...	...	...	82.8

Solution   First determine the critical depth y_c:

$$y_c = (q^2/g)^{1/3} = [(1^2 \text{ m}^4/\text{s}^2)/(9.81 \text{ m/s}^2)]^{1/3} = 0.467 \text{ m}$$

The depth of flow from the sluice gate is less than the critical depth. Hence the water-surface profile is classified as type H3.   ◄

To solve for the depth versus distance along the channel, we apply Eqs. (15.37) and (15.38), using a numerical approach. The results of the computations are given in the table on page 717. From the numerical results we plot the profile shown in the accompanying figure, and we see that the *profile extends to the abrupt drop.*   ◄

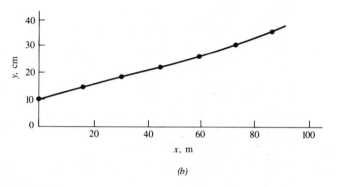

(b)

Problems

15.1   Water flows at a depth of 4 in. with a velocity of 18 ft/s in a rectangular channel. Is the flow subcritical or supercritical? What is the alternate depth?

15.2   The water discharge in a rectangular channel 16 ft wide is 900 cfs. If the depth of water is 3 ft, is the flow subcritical or supercritical?

15.3   The discharge in a rectangular channel 15 ft wide is 350 cfs. If the water velocity is 3 ft/s, is the flow subcritical or supercritical?

15.4   Water flows at a rate of 12 m³/s in a rectangular channel 3 m wide. Determine the Froude number and the type of flow (subcritical, critical, or supercritical) for depths of 30 cm, 1.0 m, and 2.0 m. What is the critical depth?

15.5   For the discharge and channel of Prob. 15.4, what is the alternate depth to the 30-cm depth? What is the specific energy for these conditions?

15.6   Water flows at the critical depth with a velocity of 6 m/s. What is the depth of flow?

15.7   Water flows uniformly at a rate of 9.0 m³/s in a rectangular channel that is 4.0 m wide and has a bottom slope of 0.005. If n is 0.014, is the flow subcritical or supercritical?

15.8 The discharge in a trapezoidal channel is 10.0 m³/s. The bottom width of the channel is 3.0 m, and the side slopes are 1 vertical to 1 horizontal. If the depth of flow is 1.0 m, is the flow supercritical or subcritical?

15.9 For the channel of Prob. 15.8, determine the critical depth for a discharge of 20 m³/s.

15.10 A rectangular channel is 6 m wide, and the discharge of water in it is 18 m³/s. Plot depth versus specific energy for these conditions. Let specific energy range from E_{min} to $E = 7$ m. What are the alternate and sequent depths to the 30-cm depth?

15.11 A long rectangular channel that is 3 m wide and has a mild slope ends in a free outfall. If the water depth at the brink is 0.350 m, what is the discharge in the channel?

15.12 A rectangular channel that is 15 ft wide and has a mild slope ends in a free outfall. If the water depth at the brink is 1.20 ft, what is the discharge in the channel?

15.13 A horizontal rectangular channel 15 ft wide carries a discharge of water of 500 cfs. If the channel ends with a free outfall, what is the depth at the brink?

15.14 What discharge of water will occur over a 2-ft-high, broad-crested weir that is 10 ft long if the head on the weir is 1.5 ft?

15.15 What discharge of water will occur over a 2-m-high, broad-crested weir that is 6 m long if the head on the weir is 60 cm?

15.16 The crest of a high, broad-crested weir has an elevation of 100.00 m. If the weir is 10 m long and the discharge of water over the weir is 25 m³/s, what is the water-surface elevation in the reservoir upstream?

15.17 The crest of a high, broad-crested weir has an elevation of 300.00 ft. If the weir is 50 ft long and the discharge of water over the weir is 1500 cfs, what is the water-surface elevation in the reservoir upstream?

15.18 Water flows with a velocity of 3 m/s and at a depth of 3 m in a rectangular channel. What is the change in depth and in water-surface elevation produced by a gradual upward change in bottom elevation (upstep) of 30 cm? What would be the depth and elevation changes if there were a gradual downstep of 30 cm? What is the maximum size of upstep that could exist before upstream depth changes would result?

15.19 Water flows with a velocity of 2 m/s and at a depth of 3 m in a rectangular channel. What is the change in depth and in water-surface elevation produced by a gradual upward change in bottom elevation (upstep) of 60 cm? What would be the depth and elevation changes if there were a gradual downstep of 15 cm? What is the maximum size of upstep that could exist before upstream depth changes would result?

15.20 Assuming no energy loss, what is the maximum value of Δz that will permit the unit flow rate of 6 m²/s to pass over the hump without increasing the upstream depth. Sketch carefully the water-surface shape from section 1 to section 2. On the sketch give values for Δz, the depth, and the amount of rise or fall in the water surface from section 1 to section 2.

15.21 Water flows with a velocity of 3 m/s in a rectangular channel 3 m wide at a depth of 3 m. What is the change in depth and in water-surface elevation produced when a gradual contraction in the channel to a width of 2.6 m takes place? Determine the greatest contraction allowable without altering the specified upstream conditions.

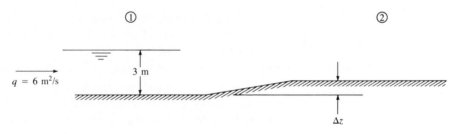

PROBLEM 15·20

15.22   Because of the increased size of ships, the phenomenon called "ship squat" has produced serious problems in harbors where the draft of vessels approaches the depth of the ship channel. When a ship steams up a channel, the resulting flow situation is analogous to open-channel flow in which a constricting flow section exists (the ship reduces the cross-sectional area of the channel). The problem may be analyzed by referencing the water velocity to the ship and applying the energy equation. Thus, at the section of the channel where the ship is located, the relative water velocity in the channel will be greatest and the water level in the channel will be reduced as dictated by the energy equation. Consequently, the ship itself will be at a lower elevation than if it were stationary; this lowering is referred to as "ship squat." Estimate the squat of the fully loaded supertanker Bellamya when it is steaming at 5 kts (1 kt = 0.515 m/s) in a channel that is 35 m deep and 200 m wide. The draft of the Bellamya when fully loaded is 29 m. Its width and length are 63 m and 414 m, respectively.

15.23   A rectangular channel that is 10 ft wide is very smooth except for a small reach that is roughened with angle irons attached to the bottom. Water flows in the channel at a rate of 200 cfs and at a depth of 1.00 ft upstream of the rough section. Assume frictionless flow except over the roughened part, where the total drag of all roughness (all of the angle irons) is assumed to be 2000 lb. Determine the depth downstream of the roughness for the assumed conditions.

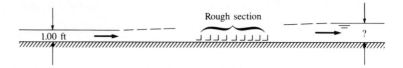

PROBLEM 15·23

15.24   Water flows from a reservoir into a steep rectangular channel that is 4 m wide. The reservoir water surface is 3 m above the channel bottom at the channel entrance. What discharge will occur in the channel?

15.25   A small wave is produced in a pond that is 1 ft deep. What is the speed of the wave in the pond?

15.26   A small wave in a pool of water having constant depth travels at a speed of 2.0 m/s. How deep is the water?

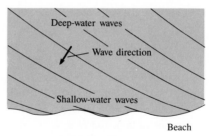

Aerial view of waves

PROBLEM 15•27

15.27 As waves in the ocean approach a sloping beach, they curve so that they are nearly parallel to the beach when they finally break (see the accompanying figure). Explain why the waves curve like this.

15.28 The baffled ramp shown is used as an energy dissipator in a two-dimensional open channel. For a discharge of 18 cfs per foot of width, calculate the head lost, the power dissipated, and the horizontal component of force exerted by the ramp on the water.

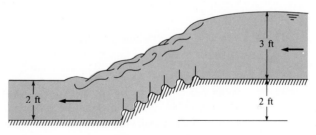

PROBLEM 15•28

15.29 The spillway shown has a discharge of 2.0 m³/s per meter of width occurring over it. What depth y_2 will exist downstream of the hydraulic jump? Assume negligible energy loss over the spillway.

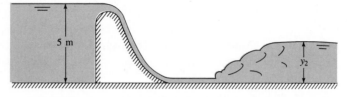

PROBLEM 15•29

15.30 The flow of water downstream from a sluice gate in a horizontal channel has a depth of 30 cm and a flow rate of 1.8 m³/s per meter of width. Could a hydraulic jump be caused to form downstream of this section? If so, what would be the depth downstream of the jump?

$$\frac{y_2}{y_1} = \frac{1}{2}\left[-1 + \sqrt{1 + \frac{Q\,v_1^2}{g\,y'}}\right]$$

15.31 Consider the dam and spillway shown in Fig. 15.19, page 707. Water is discharging over the spillway of the dam as shown. The elevation difference between the upstream pool level and the floor of the apron of the dam is 100 ft. If the head on the spillway is 5 ft and if a hydraulic jump forms on the horizontal apron, what is the depth of flow on the apron just downstream of the jump? Assume that the velocity just upstream of the jump is 95% of the maximum theoretical velocity. *Note:* The discharge over the spillway is given as $Q = KL\sqrt{2g}\,H^{3/2}$, where L is the length of the spillway, K is a coefficient (assume it has a value of 0.5), and H is the head on the spillway.

15.32 It is known that the discharge per unit width is 65 cfs/ft and that the *height* (H) of the hydraulic jump is 14 ft. What is the depth y_1?

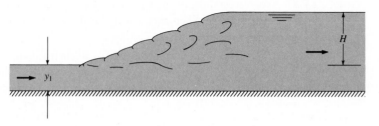

PROBLEM 15.32

15.33 Water flows in a channel at a depth of 40 cm and with a velocity of 5 m/s. An obstruction causes a hydraulic jump to be formed. What is the depth of flow downstream of the jump?

15.34 Water flows in a trapezoidal channel at a depth of 40 cm and with a velocity of 10 m/s. An obstruction causes a hydraulic jump to be formed. What is the depth of flow downstream of the jump? The bottom width of the channel is 5 m, and the side slopes are 1 vertical to 1 horizontal.

15.35 A hydraulic jump occurs in a wide rectangular channel. If the depths upstream and downstream are 0.50 ft and 13 ft, respectively, what is the discharge per foot of width of channel?

15.36 The 20-ft-wide rectangular channel shown has three different reaches. $S_{0_1} = 0.01$; $S_{0_2} = 0.0004$; $S_{0_3} = 0.00317$; $Q = 500$ cfs; $n_1 = 0.015$; normal depth for reach 2 is 5.4 ft and that for reach 3 is 2.7 ft. Determine critical depth and normal depth for reach 1 (use Manning's equation). Then classify the flow in each reach (supercritical, subcritical, critical), and determine whether a hydraulic jump could occur. In which reach(es) might it occur if it does occur?

15.37 Water flows from under the sluice gate as shown and continues on to a free overfall (also shown). Upstream from the overfall the flow soon reaches a normal depth of 1.1 m. The profile immediately downstream of the sluice gate is as it would be if there were no influence from the part nearer the overfall. Will a hydraulic jump form for these conditions? If so, locate its position. If not, sketch the full profile and label each part. Draw the energy grade line for the system.

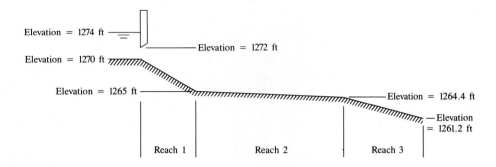

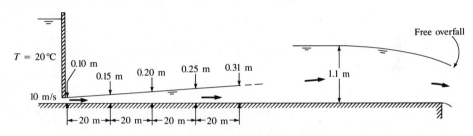

PROBLEM 15.37

15.38 For the conditions of Prob. 15.37, estimate the shear stress on the smooth bottom 0.5 m downstream of the sluice gate. That shear stress τ falls within the range a) $0 < \tau_0 < 1$ N/m², b) $1 < \tau_0 < 10$ N/m², c) $10 < \tau_0 < 40$ N/m², d) 40 N/m² $< \tau_0$.

15.39 Water is flowing as shown under the sluice gate in a horizontal rectangular channel that is 5 ft wide. The depths y_0 and y_1 are 65 ft and 1 ft, respectively. What will be the horse-power lost in the hydraulic jump?

15.40 Water is flowing as shown under the sluice gate in a horizontal rectangular channel that is 2 m wide. The depths y_0 and y_1 are 20 m and 30 cm, respectively. What will be the power lost in the hydraulic jump?

15.41 Water flows uniformly at a depth $y_1 = 40$ cm in the concrete channel, which is 10 m wide. Estimate the height of the hydraulic jump that will form when a sill is installed to force it to form.

15.42 For the derivation of Eq. (15.28) it is assumed that the bottom shearing force is negli-gible. For the conditions of Prob. 15.41, estimate the magnitude of the shearing force F_s associated with the hydraulic jump and then determine F_s/F_H, where F_H is the net hydro-static force on the hydraulic jump.

15.43 The normal depth in the channel downstream of the sluice gate shown is 1 m. What type of water-surface profile occurs downstream of the sluice gate? Also, estimate the shear stress on the smooth bottom at a distance 0.5 m downstream of the sluice gate.

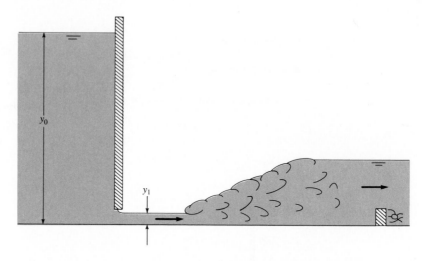

PROBLEMS 15.39, 15.40

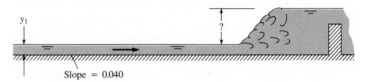

PROBLEMS 15.41, 15.42

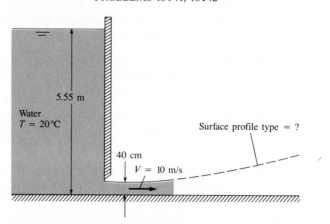

PROBLEM 15.43

15.44 Water flows at a rate of 100 ft^3/s in a rectangular channel 10 ft wide. The normal depth in that channel is 2 ft. The actual depth of flow in the channel is 4 ft. The water-surface profile in the channel for these conditions would be classified as a) S1, b) S2, c) M1, d) M2.

15.45 The water-surface profile labeled with a question mark is a) M2, b) S2, c) H2, d) A2.

PROBLEM 15.45

15.46 The partial water-surface profile shown is for a rectangular channel that is 3 m wide and has water flowing in it at a rate of 5 m³/s. Sketch in the missing part of the water-surface profile and identify the type(s).

PROBLEM 15.46

15.47 A very long 10-ft-wide concrete rectangular channel with a slope of 0.0001 ends with a free overfall. The discharge in the channel is 120 cfs. One mile upstream the flow is uniform. What kind (classification) of water surface occurs upstream of the brink?

15.48 The horizontal rectangular channel downstream of the sluice gate is 10 ft wide, and the water discharge therein is 108 cfs. The water-surface profile was computed by the direct step method. If a 2-ft-high sharp-crested weir is installed at the end of the channel, do you think a hydraulic jump would develop in the channel? If so, approximately where would it be located? Justify your answers by appropriate calculations. Label any water-surface profiles that can be classified.

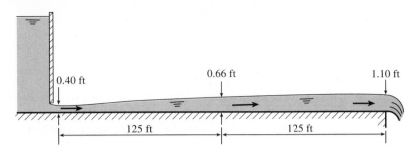

PROBLEM 15.48

15.49 The discharge per foot of width in this rectangular channel is 20 cfs. The normal depths for parts 1 and 3 are 0.5 ft and 1.00 ft, respectively. The slope for part 2 is 0.001 (sloping upward in the direction of flow). Sketch all possible water-surface profiles for flow in this channel, and label each part with its classification.

15.50 Water flows from under a sluice gate into a horizontal rectangular channel at a rate of 3 m³/s per meter of width. The channel is concrete, and the initial depth is 20 cm.

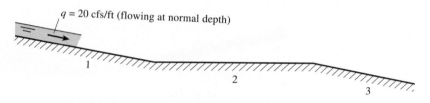

$q = 20$ cfs/ft (flowing at normal depth)

PROBLEM 15•49

Apply Eq. (15.37) to construct the water-surface profile up to a depth of 60 cm. In your solution, compute reaches for adjacent pairs of depths given in the following sequence: $d = 20$ cm, 30 cm, 40 cm, 50 cm, and 60 cm. Assume that f is constant with a value of 0.02. Plot your results.

15.51   A horizontal rectangular concrete channel terminates in a free outfall. The channel is 4 m wide and carries a discharge of water of 12 m³/s. What is the water depth 300 m upstream from the outfall?

15.52   Consider the hydraulic jump shown for the long horizontal rectangular channel. What kind of water-surface profile (classification) is located upstream of the jump? What kind of water-surface profile is located downstream of the jump? If baffle blocks are put on the bottom of the channel in the vicinity of A to increase the bottom resistance, what changes are likely to occur given the same gate opening? Explain and/or sketch the changes.

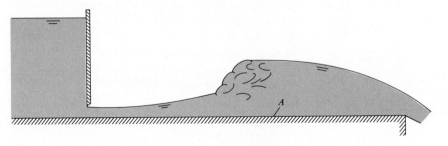

PROBLEM 15•52

15.53   The steep rectangular concrete spillway shown is 4 m wide and 500 m long. It conveys water from a reservoir and delivers it to a free outfall. The channel entrance is rounded and smooth (negligible head loss at the entrance). If the water-surface elevation in the reservoir is 2 m above the channel bottom, what will the discharge in the channel be?

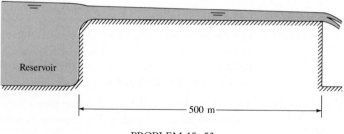

Reservoir

|← 500 m →|

PROBLEM 15•53

15.54 The concrete rectangular channel shown is 3.5 m wide and has a bottom slope of 0.001. The channel entrance is rounded and smooth (negligible head loss at the entrance), and the reservoir water surface is 2.5 m above the bed of the channel at the entrance.

   a. Estimate the discharge in it if the channel is 3000 m long.

   b. Tell how you would solve for the discharge in it if the channel were only 100 m long.

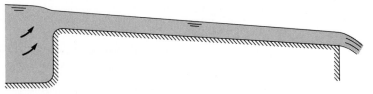

PROBLEM 15.54

15.55 A dam 50 m high backs up water in a river valley as shown. During flood flow, the discharge per meter of width, q, is equal to 10 m³/s. Making the simplifying assumptions that $R = y$ and $f = 0.030$, determine the water-surface profile upstream from the dam to a depth of 6 m. In your numerical calculation, let the first increment of depth change be y_c; use increments of depth change of 10 m until a depth of 10 m is reached; and then use 2-m increments until the desired limit is reached.

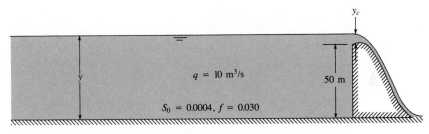

PROBLEM 15.55

15.56 Water flows at a steady rate of 12 cfs per foot of width ($q = 12$ cfs) in the wide rectangular concrete channel shown. Determine the water-surface profile from section 1 to section 2.

PROBLEM 15.56

References

1. Bakhmeteff, B. A. *Hydraulics of Open Channels.* McGraw-Hill, New York, 1932.

2. Chow, Ven Te. *Open Channel Hydraulics.* McGraw-Hill, New York, 1959.

3. Henderson, F. M. *Open Channel Flow.* Macmillan, New York, 1966.

4. Hinds, J. "The Hydraulic Design of Flume and Siphon Transitions." *Trans. ASCE,* 92 (1928), pp. 1423–59.

5. Parshall, R. L. "The Improved Venturi Flume." *Trans. ASCE,* 89 (1926), pp. 841–51.

6. Posey, C. J. "Gradually Varied Open Channel Flow." Chap. IX in H. Rouse (ed.), *Engineering Hydraulics.* John Wiley, New York, 1950.

7. Rajaratnam, N. "Hydraulic Jumps." In V. T. Chow (ed.), *Advances in Hydroscience,* vol. 4. Academic Press, New York, 1967.

8. Raju, K. G. R. *Flow Through Open Channels.* Tata McGraw-Hill, New Delhi, 1981.

9. Roberson, J. A., J. J. Cassidy, and M. H. Chaudhry. *Hydraulic Engineering.* Houghton Mifflin, Boston, 1988.

10. Rouse, H. (ed.). *Engineering Hydraulics.* John Wiley, New York, 1950.

11. U.S. Bureau of Reclamation. *Design of Small Canal Structures.* U.S. Dept. of Interior, U.S. Govt. Printing Office, 1978.

12. U.S. Bureau of Reclamation. *Hydraulic Design of Stilling Basin and Bucket Energy Dissipators.* Engr. Monograph, no. 25, U.S. Supt. of Doc., 1958.

13. U.S. Bureau of Reclamation. "Dams and Control Works." U.S. Govt. Printing Office, 1938.

14. Woodward, S. M., and C. J. Posey. *Steady Flow in Open Channels.* John Wiley, New York, 1941.

Introduction to Computational Fluid Mechanics

The generation of vortices is a very complex flow phenomenon. Numerical models now enable researchers to explore in a fundamental way the effects of the primary variables. This figure shows vorticity contours for a finite difference simulation of the generation of a two-dimensional vortex using the full viscous Navier–Stokes equations. (Courtesy G. Tryggvason, W. J. A. Dahm, and K. Sbeih)

The engineer involved in design and development is often asked to estimate the effects of design changes on the fluid-dynamic features of the design. For example, an engineer working on the design of a new car may be asked to determine the effect of a change in body contour on the drag coefficient of the car. If the automotive engineer has access to an experimental facility, the effects of contour changes on aero-dynamic performance can be measured experimentally. However, if the test facility is not a full-scale one, corrections for Reynolds-number effects have to be estimated. If the facility is a full-scale one, it may be extremely costly to operate, and that cost may prohibit its extensive use. If the flow around the car can be predicted by a numerical model, then design changes can be executed quickly at a reasonable cost, leading to the optimum design. The need for computer models can also be found in the fields of aircraft and building design, open-channel flow, marine hydrodynamics, and many other applications of fluid mechanics.

The fundamental equations describing the motion of a viscous fluid are the Navier–Stokes equations, introduced in Chapter 6. For many years mathematicians and engineers have sought analytic solutions to these equations to meet their specific needs. The equations, however, are nonlinear and not readily amenable to closed-form analytic solution. The finite-difference formulations of the equations have also been known for many years, but the time required to carry out step-by-step solutions of the equations was prohibitive until about 1950, when the electronic computer made high-speed calculations possible. Los Alamos Scientific Laboratory, through the efforts of J. von Neumann, gave considerable impetus to computational fluid mechanics. In 1965, Harlow and Fromm of Los Alamos published an article in *Scientific American* that created widespread awareness of and interest in the utility of computational fluid mechanics. The literature continues to grow, describing improved differencing schemes, iterative methods, and ways to handle discontinuities such as shock waves. The growth of the field is evidenced by the fact that there are now several journals that publish papers solely on the subject of computational fluid mechanics.

However, computational fluid mechanics is not a panacea for all the costs and inconveniences of experimentation. The numerical model is only as good as its ability to simulate the essential physical characteristics of the real flow. A general description of turbulence and Reynolds stresses has not been formulated. Only approximate descriptions that are valid over limited flow conditions are available. Thus one must be wary of predictions of turbulent flow fields. A coordinated program that combines computation and experimentation is essential to modeling turbulent flows.

The subject of computational fluid mechanics is so broad, and is growing so quickly, that one cannot hope to cover it in one chapter. The goal of this chapter is to introduce some of the basic concepts of computational fluid mechanics and thereby provide a background with which the student can pursue the subject at greater depth. A bibliography of reference material on computational fluid mechanics and numerical analysis is included at the end of the chapter.

The numerical approach discussed in this chapter is called the finite-difference approach. The many other numerical approaches for modeling fluid flows include the finite-element and the boundary-element methods.

16.1   Finite-Difference Representation of Differential Equations

The essence of computational fluid mechanics is the representation of the governing equations in an algebraic form suitable for solution by available mathematical techniques. Because the governing equations are generally expressed in the form of differential equations, the first task is to represent the differentials as finite differences and then convert the differential equations to algebraic equations. This is called the *finite-difference approach.*

First-Order Ordinary Differential Equations

The basis for approximating differentials by finite differences is the Taylor series. Consider the variation of the function $y(x)$ with x, as shown in Fig. 16.1a. By the Taylor series, the value of the function as $x_0 + \Delta x$ in terms of the value at x_0 and the derivatives evaluated at x_0 is

$$y(x_0 + \Delta x) = y(x_0) + y'(x_0)\,\Delta x + y''(x_0)\frac{\Delta x^2}{2!} + y'''(x_0)\frac{\Delta x^3}{3!} + \cdots$$

$$(16.1)$$

The inclusion of more terms in the Taylor series continually improves the accuracy of the value for $y(x_0 + \Delta x)$. However, from a pragmatic viewpoint, it is more important to have an estimate of the error incurred by truncating the series after a finite number of terms. Truncating the series after the nth term yields an error in $y(x_0 + \Delta x)$ that is less than

$$\frac{\left| y^{n+1} \right|_{max} \Delta x^{n+1}}{(n + 1)!}$$

where $\left| y^{n+1} \right|_{max}$ is the largest absolute value of the $n + 1$ derivative in the interval Δx. Unfortunately, we do not know the value of $\left| y^{n+1} \right|_{max}$, so we can say only that the magnitude of the error varies as Δx^{n+1}. Thus we say that the error is of the order of Δx^{n+1} and designate it as $\mathcal{O}(\Delta x^{n+1})$.

For purposes of identification, let us say that the point x_0 corresponds to i and the point at $x_0 + \Delta x$ corresponds to $i + 1$, as shown in Fig. 16.1b. Then we can rewrite the Taylor series as

$$y_{i+1} = y_i + y_i' \Delta x + \mathcal{O}(\Delta x^2) \tag{16.2}$$

We can now solve for the derivative of y at i, namely y_i', and obtain

$$y_i' = \frac{y_{i+1} - y_i}{\Delta x} + \mathcal{O}(\Delta x) \tag{16.3}$$

This formulation for the first derivative is described as being *first-order accurate,* which means that terms of the order of Δx and higher are disregarded. Obviously, the formulation for the derivative as expressed by Eq. (16.3) is the result of approximating the function between y_i and y_{i+1} as a straight line.

It is equally valid to apply the Taylor series in the negative x direction and express the derivative y_i' as

$$y_i' = \frac{y_i - y_{i-1}}{\Delta x} + \mathcal{O}(\Delta x) \tag{16.4}$$

FIGURE 16.1

Variation of function y(x)
with x.

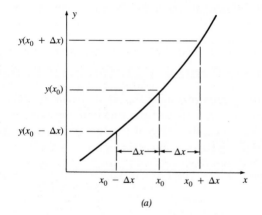

(a)

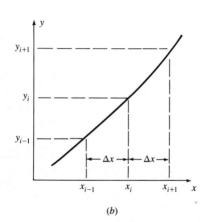

(b)

This representation is also first-order accurate. The use of the forward point $(i + 1)$ to evaluate the derivative is known as the *forward difference*. Correspondingly, Eq. (16.4) represents the *backward difference*.

The forward and backward formulations for the derivatives lead to the definition of implicit and explicit formulations for representation of differential equations by finite differences. Assume that we have an ordinary first-order differential equation represented by

$$\frac{dy}{dx} = f(x, y) \tag{16.5}$$

with the initial condition $y = y_0$ at $x = x_0$. By using the forward-difference representation for the derivative, we can write the differential equation in finite-difference form as

$$\frac{y_{i+1} - y_i}{\Delta x} = f(x_i, y_i) \tag{16.6}$$

or

$$y_{i+1} = y_i + \Delta x [f(x_i, y_i)] \tag{16.7}$$

This is the explicit formulation; that is, the value for y_{i+1} can be determined directly from the values for x and y at location i (x_i and y_i). We would proceed with the numerical solution by starting with the initial condition and using Eq. (16.7), step by step, until the desired range of x was covered. This approach is known as *Euler's method*.

On the other hand, if the backward formulation of the derivative is used, the finite-difference equation becomes

$$\frac{y_i - y_{i-1}}{\Delta x} = f(x_i, y_i) \tag{16.8}$$

or

$$y_i - f(x_i, y_i) \Delta x = y_{i-1} \tag{16.9}$$

In this formulation, y_{i-1} is known (from the previous step or from the initial condition), and y_i must be solved for. This is the implicit formulation of the finite-difference equation.

Both the forward and backward formulations have the same accuracy in representation of the derivatives (first-order). One may favor the explicit over the implicit scheme because of its simplicity. However, if one chooses distance increments (Δx's) that are too large, then severe oscillations (instability) as well as inaccuracies can result. Obviously, the reduction of Δx leads to more accurate representations of the derivatives, but it requires longer computation time.

EXAMPLE 16.1   The equation of motion for a spherical projectile moving horizontally through a quiescent fluid is

$$\frac{dV}{dt} = -\frac{3}{4} \frac{C_D \rho_f |V| V}{D \rho_p}$$

where C_D is the drag coefficient, D is the diameter of the sphere, and ρ_f/ρ_p is the ratio of the density of the fluid to the density of the projectile material. Assuming that C_D is constant, one can write the equation as

$$\frac{dV}{dt} = -k|V|V$$

where k is a constant. Write out the explicit and implicit formulations of the finite-difference equation.

Solution Let the velocity at time t be V_n and the velocity at time $t + \Delta t$ be V_{n+1}. Using the forward-difference formulation for the derivative gives

$$\frac{V_{n+1} - V_n}{\Delta t} = -kV_n|V_n| \quad \text{or} \quad V_{n+1} = (1 - k|V_n|\Delta t)V_n \qquad \blacktriangleleft$$

This is the explicit formulation. That is, V_{n+1} is an explicit function of the known velocity V_n. It is readily apparent that use of a large Δt could lead to a negative value for V_{n+1}, which is physically impossible.

Using the backward-difference formulation for the derivative gives

$$\frac{V_{n+1} - V_n}{\Delta t} = -k|V_{n+1}|V_{n+1} \quad \text{or} \quad V_{n+1}(1 + k|V_{n+1}|\Delta t) = V_n \qquad \blacktriangleleft$$

This is the implicit formulation. That is, this equation has to be solved for V_{n+1}. The use of a large Δt always leads to a positive V_{n+1}, but the accuracy is poor.

_____

The accuracy of the finite-difference formulation can be improved by using central differencing in lieu of forward or backward differencing. Applying the Taylor series for the expansion of the function y from x_0 to $x_0 + \Delta x$ and from x_0 to $x_0 - \Delta x$ gives

$$y_{i+1} = y_i + y_i'\Delta x + y_i''\frac{\Delta x^2}{2!} + y_i'''\frac{\Delta x^3}{3!} + y_i^{iv}\frac{\Delta x^4}{4!} \cdots \qquad (16.10a)$$

and

$$y_{i-1} = y_i - y_i'\Delta x + y_i''\frac{\Delta x^2}{2!} - y_i'''\frac{\Delta x^3}{3!} + y_i^{iv}\frac{\Delta x^4}{4!} \cdots \qquad (16.10b)$$

Subtracting Eq. (16.10b) from Eq. (16.10a) gives

$$y_{i+1} - y_{i-1} = 2y_i'\Delta x + y_i'''\frac{\Delta x^3}{3} \cdots \qquad (16.11)$$

Solving for y_i' yields

$$y_i' = \frac{y_{i+1} - y_{i-1}}{2\Delta x} - y_i'''\frac{\Delta x^2}{3} \cdots \qquad (16.12)$$

or
$$y_i' = \frac{y_{i+1} - y_{i-1}}{2\,\Delta x} + \mathcal{O}(\Delta x^2) \tag{16.13}$$

This finite-difference representation is *second-order accurate*. It is equivalent to fitting a parabola through the three points y_{i-1}, y_i, and y_{i+1} and taking the slope of the parabola at location i. If the grid spacing is not uniform ($\Delta x \neq$ constant), the central-difference formulation yields only first-order accuracy.

Second-Order Ordinary Differential Equations

The Taylor series can also be used to obtain higher-order derivatives. At least three points are needed to represent a second derivative by finite differences. Adding Eqs. (16.10*a*) and (16.10*b*) gives

$$y_{i+1} + y_{i-1} - 2y_i = y_i''\Delta x^2 + y_i^{iv}\frac{\Delta x^4}{12} \cdots \tag{16.14}$$

and solving for y_i'' results in

$$y_i'' = \frac{y_{i+1} + y_{i-1} - 2y_i}{\Delta x^2} + \mathcal{O}(\Delta x^2) \tag{16.15}$$

As with the first derivative, central differencing yields second-order accuracy, provided the grid spacing is uniform.

An alternative approach to obtaining the second derivative is to recognize it as the derivative of the first derivative. That is,

$$\frac{d}{dx}\left(\frac{dy}{dx}\right) \simeq \frac{\Delta(dy/dx)}{\Delta x} \tag{16.16}$$

Consider the grid system shown in Fig. 16.2, where the points $i + \frac{1}{2}$ and $i - \frac{1}{2}$ represent the midway points between i and $i + 1$ and between i and $i - 1$, respectively. The second derivative at i is represented as

$$\frac{d}{dx}\left(\frac{dy}{dx}\right) \simeq \frac{\dfrac{dy}{dx}\bigg|_{i+(1/2)} - \dfrac{dy}{dx}\bigg|_{i-(1/2)}}{\Delta x} \tag{16.17}$$

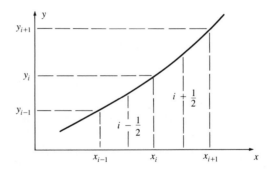

FIGURE 16.2 _____

Grid system with mid-interval points.

Using the central-difference formulation for the first derivatives gives

$$\frac{d}{dx}\left(\frac{dy}{dx}\right) \simeq \frac{\left(\dfrac{y_{i+1} - y_i}{\Delta x}\right) - \left(\dfrac{y_i - y_{i-1}}{\Delta x}\right)}{\Delta x} \tag{16.18}$$

or

$$\frac{d}{dx}\left(\frac{dy}{dx}\right) \simeq \frac{y_{i+1} - 2y_i + y_{i-1}}{\Delta x^2} \tag{16.19}$$

which is the same as Eq. (16.15).

One can apply the same procedure, using the Taylor series for representation of derivatives by finite differences, to nonuniform grid spacing. Consider the function $y = y(x)$ shown in Fig. 16.3. Here the distance from the point i to $i + 1$ is $a\,\Delta x$, where a is a factor representing the degree of nonuniformity. Applying the Taylor series to represent y_{i+1} yields

$$y_{i+1} = y_i + y_i'a\,\Delta x + y_i''\frac{a^2\,\Delta x^2}{2!} + y_i'''\frac{a^3\,\Delta x^3}{3!} \cdots \tag{16.20}$$

Multiplying Eq. (16.10b) by the factor a and adding the result to Eq. (16.20) give

$$y_{i+1} + \alpha y_{i-1} - (a + 1)y_i = y_i''\frac{\Delta x^2}{2}a(a + 1) + y_i'''\frac{\Delta x^3}{3}a(a^2 - 1) \cdots \tag{16.21}$$

Solving for y_i'' gives

$$y_i'' = \frac{2}{a(a + 1)}\left[\frac{y_{i+1} + ay_{i-1} - (a + 1)y_i}{\Delta x^2}\right] + \mathcal{O}(\Delta x) \tag{16.22}$$

Note that the finite-difference representation is now only first-order accurate. Equation (16.22) reduces to Eq. (16.15) when a is equal to unity.

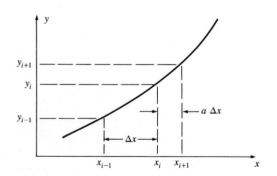

FIGURE 16.3

Nonuniform grid spacing.

Partial Differential Equations

The majority of equations encountered in computational fluid mechanics are partial differential equations. One can use the same approach for partial derivatives as was used to represent ordinary derivatives as finite differences. Consider the function $z(x,y)$ represented as a surface in Fig. 16.4. Consider the point (x_0, y_0) and the neighboring point $(x_0 + \Delta x, y_0)$. The partial derivative of $z(x,y)$ with respect to x is the slope of the curve along the surface in the $y = y_0$ plane. In fact, one can express the variation of z with x along the $y = y_0$ plane, using the Taylor series, as follows:

$$z(x_0 + \Delta x, y_0) = z(x_0, y_0) + \frac{\partial z}{\partial x}\bigg|_0 \Delta x + \frac{\partial^2 z}{\partial x^2}\bigg|_0 \frac{\Delta x^2}{2!} \cdots \qquad (16.23)$$

Designating the point (x_0, y_0) as (i, j), the point $(x_0 + \Delta x, y_0)$ as $(i + 1, j)$, and the point $(x_0 - \Delta x, y_0)$ as $(i - 1, j)$, one can express the partial derivative of $z(x,y)$ with respect to x in finite-difference form by analogy with ordinary derivatives as

$$\frac{\partial z}{\partial x} = \frac{z_{i+1,j} - z_{i,j}}{\Delta x} + \mathcal{O}(\Delta x) \qquad \text{(forward difference)} \qquad (16.24a)$$

$$\frac{\partial z}{\partial x} = \frac{z_{i,j} - z_{i-1,j}}{\Delta x} + \mathcal{O}(\Delta x) \qquad \text{(backward difference)} \qquad (16.24b)$$

$$\frac{\partial z}{\partial x} = \frac{z_{i+1,j} - z_{i-1,j}}{2\,\Delta x} + \mathcal{O}(\Delta x^2) \qquad \text{(central difference)} \qquad (16.24c)$$

One can express the second partial derivative of $z(x,y)$ with respect to x in finite-difference form as

$$\frac{\partial^2 z}{\partial x^2} = \frac{z_{i+1,j} - 2z_{i,j} + z_{i-1,j}}{\Delta x^2} + \mathcal{O}(\Delta x^2) \qquad (16.25)$$

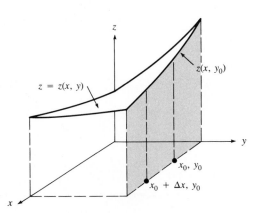

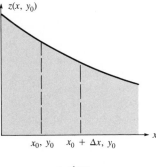

FIGURE 16.4

Surface represented by z(x,y) *with cross-sectional plane at* y_0.

For nonuniform grid spacing with point $(i + 1, j)$ located a distance $a \, \Delta x$ in the positive direction from point (i, j), and point $(i - 1, j)$ located a distance Δx in the negative direction, one can express the partial derivative as

$$\frac{\partial^2 z}{\partial x^2} = \frac{2}{a(a + 1)} \left[\frac{z_{i+1,j} - (1 + a)z_{i,j} + az_{i-1,j}}{\Delta x^2} \right] + \mathcal{O}(\Delta x) \qquad (16.26)$$

Applying the same approach to represent the partial derivatives of z with respect to y in finite-difference form by using the $x = x_0$ plane, one has

$$\frac{\partial^2 z}{\partial y^2} = \frac{z_{i,j+1} - 2z_{i,j} + z_{i,j-1}}{\Delta y^2} + \mathcal{O}(\Delta y^2) \qquad (16.27)$$

for uniform grid spacing. The equation analogous to Eq. (16.26) for nonuniform grid spacing in the y direction is obvious.

We shall now apply the finite-difference formulation to analyze a problem involving Couette flow.

16.2 Couette Flow with a Variable Distribution of Viscosity

The purpose of this section is to show the numerical solution of the finite-difference equations describing a Couette flow. As discussed in Sec. 9.2, Couette flow is the fluid motion produced between two flat plates moving with a relative velocity parallel to each other.

Differential Equation for Couette Flow

Consider the element of fluid shown in Fig. 16.5. By definition, the element is not accelerating, so the sum of the forces acting on the element is zero. Also by definition, the pressure is the same on each end of the element, so the force balance is achieved through shear stress alone. That is,

$$\left(\tau + \frac{d\tau}{dy} \Delta y \right) \Delta x - \tau \Delta x = 0 \qquad (16.28)$$

or
$$\frac{d\tau}{dy} = 0 \qquad (16.29)$$

FIGURE 16.5

Fluid element in Couette flow.

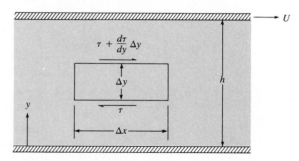

Assuming that the fluid is Newtonian, the shear stress is given by

$$\tau = \mu \frac{du}{dy} \tag{16.30}$$

Thus the governing differential equation is

$$\frac{d}{dy}\left(\mu \frac{du}{dy}\right) = 0 \tag{16.31}$$

If the viscosity is uniform, Eq. (16.31) simplifies to

$$\frac{d^2u}{dy^2} = 0 \tag{16.32}$$

for which the general solution is

$$u = A + By \tag{16.33}$$

Substituting the boundary conditions $u(0) = 0$ and $u(h) = U$ into Eq. (16.33) gives

$$u = \frac{Uy}{h} \tag{16.34}$$

which is the classic solution for velocity distribution in a Couette flow.

For many applications, the assumption of uniform viscosity may not be valid. For example, when a temperature gradient is imposed across the fluid, the viscosity of the fluid, which is a function of temperature, varies with y. When the variation of μ with y can be expressed analytically, it may be possible to integrate Eq. (16.31) directly. In that case there will be no need to resort to numerical methods to obtain the velocity distribution. However, if the variation of viscosity is provided in the form of a table, or if the analytic form of $\mu(y)$ precludes direct integration, numerical methods must be used.

Equation for Variation of Viscosity

As an example of the application of numerical methods to obtain the velocity distribution for a Couette flow with a nonuniform distribution of viscosity, consider the motion of an oil between two plates having unequal temperatures. The temperature of the bottom plate is 38°C, and the temperature of the top plate is 100°C. The variation of viscosity with temperature in this range is represented reasonably well by the equation

$$\mu = 4.2 \times 10^{-3} e^{0.031(100-T)} \qquad (N \cdot s/m^2) \tag{16.35}$$

where T is in degrees Celsius. It is assumed that the temperature of the oil varies linearly with distance between the plates. This can be expressed as

$$T = 38 + \frac{62y}{h} \tag{16.36}$$

Substituting this equation into the equation for viscosity, Eq. (16.35), we have

$$\mu = 4.2 \times 10^{-3}e^{1.92(1-y/h)} \quad (\text{N} \cdot \text{s/m}^2) \tag{16.37}$$

For computational ease, one can nondimensionalize the viscosity and velocity with respect to their values at the top plate. The equation for the viscosity then simplifies to

$$\mu = e^{1.92(1-y/h)} \tag{16.38}$$

and the velocity of the top plate is equal to unity.

Finite-Difference Equations

Let us begin by setting up the grid system shown in Fig. 16.6. There are I equally spaced grid lines, with the letter i identifying each grid line. The stationary bottom wall is grid line $i = 1$ and the top wall is $i = I$. The distance between grid lines, Δy, is equal to $h/(I - 1)$. The finite-difference equation representing the differential equation around grid line i is

$$
\begin{aligned}
\frac{d}{dy}\left(\mu \frac{du}{dy}\right) &= \frac{\mu_{i+(1/2)}\dfrac{du}{dy}\bigg|_{i+(1/2)} - \mu_{i-(1/2)}\dfrac{du}{dy}\bigg|_{i-(1/2)}}{\Delta y} \\
&= \frac{\mu_{i+(1/2)}(u_{i+1} - u_i) - \mu_{i-(1/2)}(u_i - u_{i-1})}{\Delta y^2} = 0
\end{aligned} \tag{16.39}
$$

The resulting set of finite-difference equations is

$$\mu_{i+(1/2)}u_{i+1} - (\mu_{i+(1/2)} + \mu_{i-(1/2)})u_i + \mu_{i-(1/2)}u_{i-1} = 0 \tag{16.40}$$

with the boundary conditions

$$u_1 = 0 \quad \text{and} \quad u_I = 1 \tag{16.41}$$

The problem is reduced to solving a set of $I - 2$ algebraic equations for the velocity u_i. For example, if the value chosen for I is 6 (six grid lines), there will be

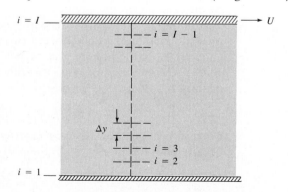

FIGURE 16.6

Grid system for analysis of Couette flow.

four ($i = 2$ through 5) algebraic equations for u_i, which can be expressed in matrix form as follows:

$$
\begin{bmatrix}
-(\mu_{1.5} + \mu_{2.5}) & \mu_{2.5} & 0 & 0 \\
\mu_{2.5} & -(\mu_{2.5} + \mu_{3.5}) & \mu_{3.5} & 0 \\
0 & \mu_{3.5} & -(\mu_{3.5} + \mu_{4.5}) & \mu_{4.5} \\
0 & 0 & \mu_{4.5} & -(\mu_{4.5} + \mu_{5.5})
\end{bmatrix}
\begin{Bmatrix} u_2 \\ u_3 \\ u_4 \\ u_5 \end{Bmatrix}
=
\begin{Bmatrix} 0 \\ 0 \\ 0 \\ -\mu_{5.5} \end{Bmatrix}
$$

$$(16.42)$$

Substituting the values for viscosity provided by Eq. (16.38) results in the following equation set:

$$-9.46u_2 + 3.83u_3 = 0 \tag{16.43a}$$

$$3.83u_2 - 6.44u_3 + 2.61u_4 = 0 \tag{16.43b}$$

$$2.61u_3 - 4.39u_4 + 1.78u_5 = 0 \tag{16.43c}$$

$$1.78u_4 - 2.99u_5 = -1.21 \tag{16.43d}$$

Solving these equations for u_i yields the finite-difference solution for the velocity distribution between the plates. We shall now discuss methods of obtaining the solution to this equation set.

Matrix Inversion

The most obvious and direct way to obtain a solution to Eqs. (16.43a) through (16.43d) is to use matrix methods. One can express the set of equations in matrix form as

$$[A]\{u\} = \{c\} \tag{16.44}$$

By finding the inverse of matrix $[A]$, one can obtain the solution for u from

$$\{u\} = [A]^{-1}\{c\} \tag{16.45}$$

There are several methods of obtaining $[A]^{-1}$ that can be found in general texts on linear algebra. One of the most commonly used methods is Gauss-Jordan elimination. Most computer systems have "canned" programs to perform matrix inversion or simply to yield the solution to the set of simultaneous algebraic equations.

The inverse matrix $[A]^{-1}$ for the set of equations is

$$
[A]^{-1} =
\begin{bmatrix}
-0.163 & -0.142 & -0.112 & -0.066 \\
-0.142 & -0.352 & -0.276 & -0.164 \\
-0.112 & -0.276 & -0.516 & -0.307 \\
-0.066 & -0.164 & -0.307 & -0.517
\end{bmatrix}
\tag{16.46}
$$

The corresponding solution vector for u_i is

$$\{u\} = \begin{Bmatrix} 0.080 \\ 0.199 \\ 0.372 \\ 0.626 \end{Bmatrix} \tag{16.47}$$

Figure 16.7 shows the predicted velocity profile. We notice that the profile is steepest (large du/dy) in the region where μ is smallest, adjacent to the top plate.

The application of general-purpose codes for inversion of large matrices is generally not recommended because of the excessive computer time required and the associated cost.

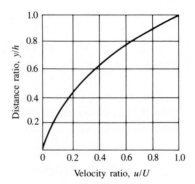

FIGURE 16.7

Velocity profile predicted for Couette flow with nonuniform viscosity.

Tridiagonal Matrix Algorithm

Very efficient schemes are available for solving a banded matrix, which is the type of matrix encountered in this problem. Matrix $[A]$ in this problem is identified as a tridiagonal matrix that consists of the main diagonal and the two adjacent diagonals. The general form of the difference equation is

$$a_i u_{i-1} + b_i u_i + c_i u_{i+1} = d_i \tag{16.48}$$

where a_i, b_i, c_i, and d_i represent the coefficients in the finite-difference equations.

Starting with a recursion relationship for u_i of the form

$$u_{i-1} = e_i u_i + f_i \tag{16.49}$$

and substituting it into the first term of Eq. (16.48), we get

$$(a_i e_i + b_i) u_i + c_i u_{i+1} = d_i - a_i f_i \tag{16.50}$$

or

$$u_i = \frac{-c_i u_{i+1} + d_i - a_i f_i}{a_i e_i + b_i} \tag{16.51}$$

Going back to Eq. (16.49) and replacing i with $i + 1$, we have

$$u_i = e_{i+1}u_{i+1} + f_{i+1} \tag{16.52}$$

Equating the coefficients of u_{i+1} and the constant term in Eqs. (16.51) and (16.52) yields

$$e_{i+1} = \frac{-c_i}{a_i e_i + b_i} \tag{16.53a}$$

$$f_{i+1} = \frac{d_i - a_i f_i}{a_i e_i + b_i} \tag{16.53b}$$

These are recursion relations for e_i and f_i. That is, if you know e_i and f_i, you can calculate e_{i+1} and f_{i+1}. Obviously, if you know the initial values of e_i and f_i, you can obtain all of the others by successive application of Eqs. (16.53). Finally, if you know all values of e_i and f_i, you can find the values of u_i from Eq. (16.49).

The method is most easily explained by applying it to the example presented above. The first equation of the set, Eq. (16.43a), can be written as

$$u_2 = 0.405u_3 \tag{16.54}$$

Thus, by reference to Eq. (16.49), we find that the values for e_3 and f_3 are

$$e_3 = 0.405 \tag{16.55a}$$

$$f_3 = 0 \tag{16.55b}$$

Using the recursion relation, Eqs. (16.53), for e_4 and f_4, or

$$e_4 = \frac{-c_3}{a_3 e_3 + b_3} \tag{16.56a}$$

$$f_4 = \frac{d_3 - a_3 f_3}{a_3 e_3 + b_3} \tag{16.56b}$$

and substituting the values for a_3, b_3, c_3, and d_3, we obtain

$$e_4 = 0.534 \tag{16.57a}$$

$$f_4 = 0 \tag{16.57b}$$

This procedure is continued for the remaining e_i and f_i. Table 16.1 shows the values. The last equation, Eq. (16.43d), has no term with u_6, so the coefficient c_5 is zero. This leads to e_6 being zero, which, from Eq. (16.49), results in u_5 being equal to f_6. Starting with this value for u_5, one can calculate the remaining values of u_i using the coefficients e_i and f_i in Eq. (16.49).

This method is known as the *tridiagonal matrix algorithm* (TDMA), or the *Thomas algorithm*. It is the most efficient scheme by which to solve tridiagonal matrices. One can obtain the same recursion relationships by using Gauss elimination to get rid of one of the "off" diagonals.

i	e_i	f_i	u_i
1	—	—	0
2	—	—	0.080
3	0.405	0	0.199
4	0.534	0	0.372
5	0.594	0	0.626
6	0	0.626	1

TABLE 16.1   VALUES FOR TRIDIAGONAL MATRIX CALCULATION

Gauss–Seidel Iteration

One method that is simple but iterative (the solution is obtained after several trials), is the *Gauss–Seidel iteration* or *direct-substitution method*. One begins this procedure by guessing an initial value for u_i and then using the recursion equations to continually improve the estimate for u_i. The set of equations is first written in the form

$$u_2 = 0.405u_3 \tag{16.58a}$$

$$u_3 = 0.595u_2 + 0.405u_4 \tag{16.58b}$$

$$u_4 = 0.595u_3 + 0.405u_5 \tag{16.58c}$$

$$u_5 = 0.595u_4 + 0.405 \tag{16.58d}$$

Table 16.2 shows how the method proceeds. The first column is the initial guess for u_i, which, for this example, is taken as a linear velocity distribution. One begins the first iteration by using Eq. (16.58a) and the initial guess for u_3 to calculate u_2. Then one calculates u_3 with Eq. (16.58b), using the initial guess for u_4 and the previously calculated value for u_2. One continues the calculation by using the remaining equations to complete the first iteration. The procedure of using the updated value for u_{i-1} in each calculation of u_i is known as *successive*

TABLE 16.2   SOLUTION OF EQUATIONS FOR VELOCITY RATIO USING GAUSS-SEIDEL ITERATION

i	Number of iterations									
	0	1	2	3	4	5	6	7	8	9
1	0	0	0	0	0	0	0	0	0	0
2	0.2	0.162	0.137	0.086	0.089	0.087	0.084	0.083	0.082	0.081
3	0.4	0.339	0.213	0.221	0.214	0.208	0.204	0.202	0.201	0.200
4	0.6	0.526	0.418	0.396	0.386	0.381	0.378	0.375	0.374	0.373
5	0.8	0.718	0.653	0.640	0.635	0.632	0.630	0.628	0.627	0.627
6	1.0	1.0	1.0	1.0	1.0	1.0	1.0	1.0	1.0	1.0

substitution. The iterations are continued until the value for u_i fails to change within a prespecified amount with continued iteration.

We see from Table 16.2 that convergence to the solution is quite slow. The convergence rate can be accelerated by using the method of relaxation. We obtain the new value of u_i from

$$u_i = \lambda u_{i,\text{calc}} + (1 - \lambda)u_{i,\text{old}} \qquad (16.59)$$

where $u_{i,\text{calc}}$ is the value calculated by applying Eq. (16.58), $u_{i,\text{old}}$ is the previous value for u_i, and λ is the relaxation factor. If $0 < \lambda < 1$, we have underrelaxation, which slows the change of u_i with each iteration. If $1 < \lambda < 2$, we have overrelaxation, which accelerates the change of u_i. Obviously if λ equals unity, there is no relaxation. This corresponds to the calculations in Table 16.2. The primary difficulty in using the method of relaxation is that the value for λ is usually not known *a priori* and is usually determined by trial and error.

Accuracy of the Finite-Difference Method

The question arises as to the accuracy of the solution. We know that by choosing more grid lines, we can reduce the distance between grid lines (Δx) and make the finite-difference representation of the differential equation more accurate. Thus when we increase the number of grid lines, the solution should become more accurate. We generally do not know, to start with, how many grid lines we need to establish a given level of accuracy. The commonly accepted practice is to increase the number of grid lines until the result shows no further change.

The Appendix contains a FORTRAN program (Program I) for obtaining the velocity distribution for the foregoing problem. This program uses the tridiagonal matrix algorithm to solve the system of equations for u_i. Table 16.3 shows the output of the program using 6 and then 11 grid lines. One notices that reducing Δx from 0.2 to 0.1 brings about no change in the fifth decimal place. Thus the result obtained using six grid lines is taken as the solution to the differential equation.

TABLE 16.3   DEPENDENCE OF NUMERICAL SOLUTION ON GRID-LINE SPACING

y/h	$N = 6$ ($\Delta x = 0.2$)	$N = 11$ ($\Delta x = 0.1$)
0	0	0
0.2	0.08042	0.08042
0.4	0.19850	0.19850
0.6	0.37185	0.37185
0.8	0.62635	0.62635
1.0	1.0	1.0

In this example, we were fortunate that the solution became independent of the number of grid lines with so few grid lines. This results from the rather "smooth" variation of μ with y in this problem, which gives rise to small values for the higher-order derivatives in the Taylor series and to small error terms. Six grid lines may not suffice in cases for which there are large changes in viscosity. Also, it may be necessary to use finer grid spacing in the areas of steep gradients. The choice of the number and distribution of grid lines depends on the problem.

Comparison of Finite-Difference and Analytic Solutions

This particular problem has an analytic solution that provides a comparison with the finite-difference approach. The analytic solution is

$$\frac{u}{U} = \frac{\exp(1.92y/h) - 1}{\exp(1.92) - 1} \tag{16.60}$$

Table 16.4 compares the values obtained by the numerical and analytic solutions. The agreement is excellent. It encourages confidence in the reliability of the numerical approach.

TABLE 16.4 COMPARISON OF NUMERICAL AND ANALYTIC SOLUTIONS		
y/h	$(u/U)_{\text{num}}$	$(u/U)_{\text{exact}}$
0	0	0
0.2	0.08042	0.08042
0.4	0.19850	0.19850
0.6	0.37185	0.37185
0.8	0.62635	0.62635
1.0	1.0	1.0

16.3 Two-Dimensional Inviscid, Incompressible Flow

Let us now apply the concepts discussed in Sec. 16.2 to calculation of the flow field produced by the steady flow of an incompressible, inviscid fluid. The most convenient parameter by which to represent such a flow is the stream function.

The Stream Function

Consider the two streamlines shown in Fig. 16.8 for a planar flow. These two streamlines are represented by the functional relationship

$$\psi(x, y) = C_1 \tag{16.61a}$$

$$\psi(x, y) = C_2 \tag{16.61b}$$

FIGURE 16.8

Streamlines corresponding to the stream function $\psi(x,y)$.

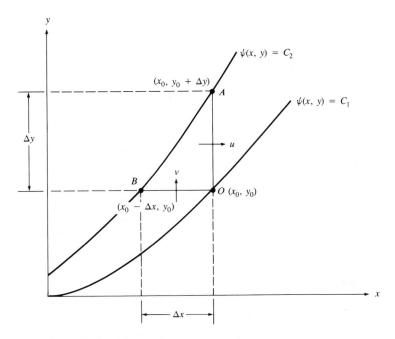

where $\psi(x,y)$ is defined as the *stream function*. For example, the stream function representing a flow with concentric streamlines is

$$\psi(x,y) = x^2 + y^2 \tag{16.62}$$

Setting $\psi(x,y) = R^2$ represents circles with centers at the origin and radii equal to R.

By definition, there is no flow across streamlines, so the discharge per unit depth between the streamlines must be constant. In fact, the difference between the constants C_1 and C_2 is equal to the discharge between the streamlines. Consider the discharge across the line OA, which connects point (x_0, y_0) to $(x_0, y_0 + \Delta y)$. By definition,

$$q \simeq u\,\Delta y \tag{16.63}$$

Also, $$q = C_2 - C_1 = \psi(x_0, y_0 + \Delta y) - \psi(x_0, y_0) \tag{16.64}$$

so $$u \simeq \frac{\psi(x_0, y_0 + \Delta y) - \psi(x_0, y_0)}{\Delta y} \tag{16.65}$$

Taking the limit as $\Delta y \to 0$—that is, calculating the velocity by using streamlines that are progressively closer to the one through point (x_0, y_0)—is represented by the partial derivative of the stream function with respect to y:

$$u = \lim_{\Delta y \to 0} \frac{\psi(x_0, y_0 + \Delta y) - \psi(x_0, y_0)}{\Delta y} = \frac{\partial \psi}{\partial y} \tag{16.66}$$

Similarly, equating the discharge per unit width to the product of v and the distance between streamlines in the x direction (line OB) gives

$$q \simeq v\Delta x \qquad (16.67)$$

As above, q is equal to the change in the stream function. Expressing v as the change in the stream function with respect to x as Δx approaches zero, we have

$$v = \lim_{\Delta x \to 0} \frac{\psi(x_0 - \Delta x, y_0) - \psi(x_0, y_0)}{\Delta x} = -\frac{\partial \psi}{\partial x} \qquad (16.68)$$

Obviously, the relationships between u, v, and the derivatives of the stream function are based on the principle of conservation of mass—uniform discharge between streamlines—so they must satisfy the differential form of the continuity equation. The differential form of the continuity equation was presented in Chapter 4, Eq. (4.31a). For two-dimensional flow, it reduces to

$$\frac{\partial u}{\partial x} + \frac{\partial v}{\partial y} = 0 \qquad (16.69)$$

Substituting the above equations for u and v as derivatives of the stream function into Eq. (16.69) gives

$$\frac{\partial}{\partial x}\left(\frac{\partial \psi}{\partial y}\right) - \frac{\partial}{\partial y}\left(\frac{\partial \psi}{\partial x}\right) = 0$$

$$\frac{\partial^2 \psi}{\partial x\, \partial y} - \frac{\partial^2 x}{\partial x\, \partial y} \equiv 0 \qquad (16.70)$$

Thus the existence of a stream function ensures that the continuity equation is satisfied.

EXAMPLE 16.2 Find the velocity components u and v for the following stream function representing concentric streamlines:

$$\psi(x, y) = x^2 + y^2$$

Solution The u component is given by

$$u = \frac{\partial \psi}{\partial y} = 2y$$

and the v component by

$$v = -\frac{\partial \psi}{\partial x} = -2x$$

This particular stream function represents the case of the forced vortex that was introduced in Chapter 4.

The momentum equations for the steady flow of an inviscid, incompressible fluid are Eqs. (5.11) and (5.12):

$$u\frac{\partial u}{\partial x} + v\frac{\partial u}{\partial y} = -g\frac{\partial h}{\partial x}$$

$$u\frac{\partial v}{\partial x} + v\frac{\partial v}{\partial y} = -g\frac{\partial h}{\partial y}$$

Taking the partial derivative of the first equation above with respect to y and subtracting it from the partial derivative of the second equation with respect to x eliminates the term for piezometric head and yields

$$u\frac{\partial}{\partial x}\left(\frac{\partial v}{\partial x} - \frac{\partial u}{\partial y}\right) + v\frac{\partial}{\partial y}\left(\frac{\partial v}{\partial x} - \frac{\partial u}{\partial y}\right) + \left(\frac{\partial u}{\partial x} + \frac{\partial v}{\partial y}\right)\left(\frac{\partial v}{\partial x} - \frac{\partial u}{\partial y}\right) = 0$$

(16.71)

Recognizing that $(\partial v/\partial x - \partial u/\partial y)$ is the vorticity component Ω_z defined by Eq. (4.35) and that the third term is zero because of the conservation of mass, Eq. (16.69), one can reduce this equation to

$$\frac{d\Omega_z}{dt} = 0$$

(16.72)

This equation tells us that the vorticity is constant along a streamline. If the planar flow of an incompressible, inviscid fluid is initially irrotational ($\Omega_z = 0$), it will remain irrotational and the vorticity will be zero everywhere.

Substituting the relationships between velocities and stream function, Eqs. (16.66) and (16.68), into the irrotationality condition yields

$$\frac{\partial v}{\partial x} - \frac{\partial u}{\partial y} = -\frac{\partial^2 \psi}{\partial x^2} - \frac{\partial^2 \psi}{\partial y^2} = 0$$

or

$$\frac{\partial^2 \psi}{\partial x^2} + \frac{\partial^2 \psi}{\partial y^2} = 0$$

(16.73)

This partial differential equation—known as *Laplace's equation*—must be solved to obtain the stream function. After one obtains the stream function, one can determine the velocity field by differentiating the stream function. The pressure field is obtained by applying Bernoulli's equation.

The Potential Function

Another function encountered in the flow of ideal fluids is the *potential function* ϕ. The velocity components are the negative gradient of the potential function,

$$u = -\frac{\partial \phi}{\partial x}$$

(16.74a)

and
$$v = -\frac{\partial \phi}{\partial y} \qquad (16.74b)$$

The existence of a potential function means that the flow is irrotational. One can illustrate this by substituting Eqs. (16.74a) and (16.74b) into the equation for vorticity, which gives

$$\Omega_z = \frac{\partial v}{\partial x} - \frac{\partial u}{\partial y} = -\frac{\partial^2 \phi}{\partial x \, \partial y} + \frac{\partial^2 \phi}{\partial x \, \partial y} \equiv 0 \qquad (16.75)$$

Thus the potential function bears the same relationship to the irrotationality condition as the stream function does to the conservation-of-mass equation.

Substituting the velocity components as derivatives of the potential function into the conservation-of-mass equation gives

$$\frac{\partial u}{\partial x} + \frac{\partial v}{\partial y} = -\frac{\partial^2 \phi}{\partial x^2} - \frac{\partial^2 \phi}{\partial y^2} = 0$$

or
$$\frac{\partial^2 \phi}{\partial x^2} + \frac{\partial^2 \phi}{\partial y^2} = 0 \qquad (16.76)$$

This, once more, is Laplace's equation. Solving it for either the stream function or the potential function yields the irrotational flow field for an ideal fluid. The streamlines and lines of constant potential are everywhere orthogonal. This system of orthogonal streamlines and potential lines is sometimes referred to as the *flow net*.

Numerical Solution of Laplace's Equation

The numerical solution of the stream function (or the potential function) requires a finite-difference formulation for Laplace's equation. Consider nodal point (i, j) and the four neighboring nodal points, as shown in Fig. 16.9. The distances between the grid lines are Δx and Δy, respectively. One obtains the finite-difference formulation for the second partial derivative of ψ with respect to x by using Eq. (16.25):

$$\frac{\partial^2 \psi}{\partial x^2} = \frac{\psi_{i+1,j} + \psi_{i-1,j} - 2\psi_{i,j}}{\Delta x^2} + \mathcal{O}(\Delta x^2) \qquad (16.77a)$$

The second partial derivative of ψ with respect to y is given by

$$\frac{\partial^2 \psi}{\partial y^2} = \frac{\psi_{i,j+1} + \psi_{i,j-1} - 2\psi_{i,j}}{\Delta y^2} + \mathcal{O}(\Delta y^2) \qquad (16.77b)$$

Adding Eqs. (16.77a) and (16.77b) and setting the sum equal to zero, we obtain the finite-difference formulation for Laplace's equation:

$$(\psi_{i+1,j} + \psi_{i-1,j})\Delta y^2 + (\psi_{i,j+1} + \psi_{i,j-1})\Delta x^2 - 2(\Delta x^2 + \Delta y^2)\psi_{i,j} = 0$$
$$(16.78)$$

FIGURE 16.9

*Nodal and neighboring
points for the finite-difference
formulation of Laplace's
equation.*

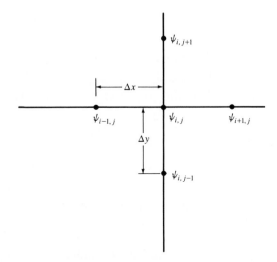

If Δx and Δy are equal (uniform grid spacing), the finite-difference equation simplifies to

$$\psi_{i+1,j} + \psi_{i-1,j} + \psi_{i,j+1} + \psi_{i,j-1} = 4\psi_{i,j} \qquad (16.79)$$

We notice that the value of ψ at a given point depends on the values of ψ at all neighboring points. We also notice that the sum of the coefficients of the stream function at the neighboring points (all are unity in the foregoing equation) is equal to the coefficient of the stream function at the nodal point. This condition holds for nonuniform grid spacing as well.

As an example of the numerical solution of Laplace's equation, we shall obtain the stream function (and velocity and pressure fields) for the converging flow passage illustrated in Fig. 5.7. The passage is 1 meter high upstream and converges to $\frac{1}{2}$ meter high downstream. For convenience, a uniform grid pattern is chosen, as shown in Fig. 16.10. There are 50 grid lines in the direction of flow and 21 lines across the passage. Nodal points lie on the boundaries except at the rounded corners on the outer wall. The flow is assumed to enter the passage with a uniform velocity of 1 m/s. The boundary conditions for the stream function are set as follows.

FIGURE 16.10

*Uniform grid pattern for a
converging-flow passage.*

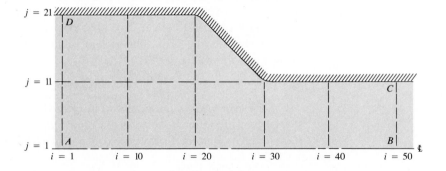

Line AB (centerline)   The stream function is set equal to zero on the center-line. That is,

$$\psi_{i,1} = 0.0 \tag{16.80}$$

Line AD (inlet plane)   Along the inlet plane the flow is uniform and par-allel to the centerline. This means that

$$\frac{\partial \psi}{\partial y} = u = 1 \tag{16.81}$$

and

$$\frac{\partial \psi}{\partial x} = -v = 0 \tag{16.82}$$

Thus along the line $i = 1$, we have

$$\psi = y + \text{constant} \tag{16.83}$$

Using the boundary condition $\psi = 0$ on the centerline gives us

$$\psi_{1,j} = y \tag{16.84}$$

along the inlet plane.

Line BC (exit)   We will assume that the velocity is parallel to the centerline along the exit plane. Thus along line *BC*, we have

$$\frac{\partial \psi}{\partial x} = 0 \tag{16.85}$$

This boundary condition is represented by

$$\left. \frac{\partial \psi}{\partial x} \right|_{i=49} = \frac{\psi_{50,j} - \psi_{49,j}}{2\,\Delta x} = 0 \tag{16.86}$$

or

$$\psi_{50,j} = \psi_{49,j} \tag{16.87}$$

Thus the finite-difference equation along the line $i = 49$ is

$$\psi_{48,j} + \psi_{49,j+1} + \psi_{49,j-1} = 3\psi_{49,j} \tag{16.88}$$

This equation is equivalent to setting the coefficient of $\psi_{50,j}$ equal to zero in the general finite-difference equation and taking the coefficient of $\psi_{49,j}$ as equal to the sum of coefficients of ψ at the neighboring points.

Line CD (wall)   From Eq. (16.84), we find that $\psi = 1$ when $y = 1$, which represents the wall. The stream function remains constant and equal to unity along the wall:

$$\psi_{i,\text{wall}} = 1 \tag{16.89}$$

The two grid points in the vicinity of the corners of the top wall require special consideration. Consider the point $(20, 20)$ in Fig. 16.11. By fitting a third-order polynomial between grid points $(19, 21)$ and $(21, 20)$ to provide a

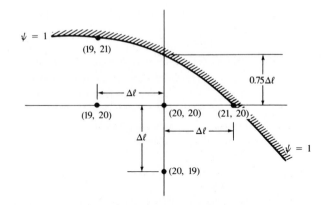

FIGURE 16.11

Locations of grid points surrounding the nodal point (20,20).

smooth transition between the straight wall sections, one finds that the wall is located $0.75\,\Delta\ell$ above the nodal point $(20, 20)$. By using Eq. (16.26), one can write the second partial derivative of ψ with respect to y as

$$\frac{\partial^2 \psi}{\partial y^2} = \frac{1.143\psi_{20,19} - 2.666\psi_{20,20} + 1.523\psi_{\text{wall}}}{\Delta\ell^2} \qquad (16.90)$$

The second partial derivative with respect to x is

$$\frac{\partial^2 \psi}{\partial x^2} = \frac{\psi_{19,20} - 2\psi_{20,20} + \psi_{\text{wall}}}{\Delta\ell^2} \qquad (16.91)$$

Adding Eqs. (16.90) and (16.91) and setting the sum equal to zero give

$$\psi_{19,20} + (1 + 1.523)\psi_{\text{wall}} + 1.143\psi_{20,19} = (2 + 2.666)\psi_{20,20} \qquad (16.92)$$

which represents the finite-difference equation to be used at point $(20, 20)$.

The grid point near the other corner is $(30, 10)$. Using the same approach, we find that the finite-difference equation for ψ at $(30, 10)$ is

$$\psi_{29,11} + (1 + 6.4)\psi_{\text{wall}} + 1.6\psi_{30,10} = (2 + 8.0)\psi_{30,11} \qquad (16.93)$$

The finite-difference equations for ψ at all interior nodes represent 667 simultaneous algebraic equations.

Numerical Solution of Finite-Difference Equations

We can solve the 667 algebraic equations in several ways analogous to the methods used for solving the system of equations generated for the problem involving Couette flow. The most direct approach is to use a software package that solves systems of equations. A general software package may be costly to use. However, the matrix that represents this system of equations is sparse (has many zeros) and the engineer may have access to an efficient package that would yield a fast and accurate solution. (The discussion of such software is beyond the scope of this text.)

The simplest approach is to use Gauss–Seidel iteration. With this approach, one assumes an initial value for ψ at each nodal point. Then one applies the finite-difference equations at each node, node by node, with each calculation using the updated value from the previous node. The calculation may also incorporate the method of relaxation. The calculations are continued until repeated iterations fail to change ψ within a specified amount. Generally, a criterion such as

$$\sum_{\text{All nodes}} |\psi_{\text{new}} - \psi_{\text{old}}| < \varepsilon \tag{16.94}$$

is used for convergence, where ε is prespecified and represents the desired degree of accuracy. The Gauss–Seidel iteration is simple to implement but time-consuming.

An alternative approach is to use the tridiagonal matrix algorithm along each line and to proceed, line by line, along the duct. One can apply the TDMA to an i line by rewriting the finite-difference equation for $\psi_{i,j}$, Eq. (16.79), as

$$\psi_{i,j-1} - 4\psi_{i,j} + \psi_{i,j+1} = -\psi_{i-1,j} - \psi_{i+1,j} \tag{16.95}$$

The right-hand side of the equation represents the constant d in Eq. (16.48); this constant is not known in the beginning. The procedure is initiated by assuming values for ψ at all interior nodes. Starting with the line $i = 2$, one recognizes that the constant d_j in the tridiagonal equations is composed of the known values for $\psi_{1,j}$ and the assumed values for $\psi_{3,j}$. Applying the TDMA yields new values for $\psi_{2,j}$. One then repeats the procedure for the line $i = 3$, using the updated values for $\psi_{2,j}$ and the assumed (or previous) values for $\psi_{4,j}$. The procedure is repeated line by line to the end of the duct. The duct is swept line by line again until convergence, such as that specified by Eq. (16.94), is achieved. This scheme is considerably faster than Gauss–Seidel iteration.

Numerical Results for the Flow Field in a Converging Passage

Program II in the Appendix is a FORTRAN program representing the line-by-line solution for the stream function in the converging-flow passage. The coefficients of the stream function at neighboring nodal points in the finite-difference equation are designated CN, CS, CE, and CW, which stand for the north point, the south point, and so forth. The convergence criterion represented by Eq. (16.94) was implemented with $\varepsilon = 0.01$. It required 109 iterations to achieve convergence.

The output of the program is also provided in the Appendix. Figure 16.12 is a plot of the streamlines. The distance between the streamlines decreases as

FIGURE 16.12

Predicted streamline pattern.

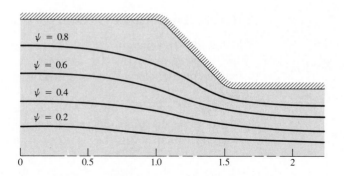

the flow proceeds downstream, effecting an increase in the velocity of flow. The large distance between the $\psi = 0.8$ streamline and the upstream corner indicates a low-velocity, high-pressure region, whereas the nearness of the streamline to the downstream corner leads to a high-velocity, low-pressure region.

Figure 16.13 shows the predicted pressure coefficient along the top wall and centerline. One notes the same trends that appeared in Fig. 5.8. The pressure coefficient monotonically decreases along the centerline. The pressure coefficient along the top wall first increases to a maximum and then decreases to a minimum at the downstream corner.

FIGURE 16.13

Predicted pressure coefficient along centerline and wall.

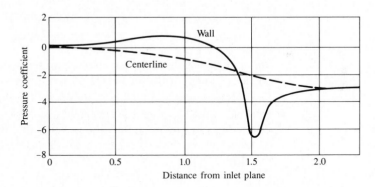

16.4 Flow of Viscous Incompressible Fluids

The fact that we can formulate the equation for the flow of ideal fluid in terms of one dependent variable, the stream function, simplifies the numerical computation. As we found, we can eliminate the pressure in the formulation by combining the component momentum equations to yield the vorticity. For irrotational flows the vorticity is zero, and using the stream function means that we need solve only one partial differential equation—Laplace's equation.

Differential Equations

The continuity equation for two-dimensional flow of a viscous incompressible fluid is the same as that for an inviscid fluid, Eq. (16.69). The existence of a stream function for a flow of viscous fluid automatically ensures conservation of mass.

The momentum equations for an incompressible Newtonian fluid are the Navier–Stokes equations presented in Chapter 5. These equations reduce to the following two equations for planar two-dimensional flows:

$$u\frac{\partial u}{\partial x} + v\frac{\partial u}{\partial y} = -\frac{1}{\rho}\frac{\partial p}{\partial x} + v\left(\frac{\partial^2 u}{\partial x^2} + \frac{\partial^2 u}{\partial y^2}\right) \qquad (16.96a)$$

$$u\frac{\partial v}{\partial x} + v\frac{\partial v}{\partial y} = -\frac{1}{\rho}\frac{\partial p}{\partial y} + v\left(\frac{\partial^2 v}{\partial x^2} + \frac{\partial^2 v}{\partial y^2}\right) \qquad (16.96b)$$

Taking the partial derivative of Eq. (16.96a) with respect to y and the partial derivative of Eq. (16.96b) with respect to x, and then subtracting to eliminate pressure, yield an equation for the vorticity:

$$u\frac{\partial \Omega_z}{\partial x} + v\frac{\partial \Omega_z}{\partial y} = v\left(\frac{\partial^2 \Omega_z}{\partial x^2} + \frac{\partial^2 \Omega_z}{\partial y^2}\right) \qquad (16.97)$$

or

$$\frac{d\Omega_z}{dt} = v\left(\frac{\partial^2 \Omega_z}{\partial x^2} + \frac{\partial^2 \Omega_z}{\partial y^2}\right)$$

We notice from this equation that the presence of viscosity can effect a change in vorticity along a path line, so the flow can no longer be regarded as irrotational.

Replacing u and v in Eq. (16.97) by the derivatives of the stream function yields an equation with two dependent variables, vorticity and the stream function. One obtains an independent relationship between these two variables by substituting the stream-function derivatives for velocity into the equation for Ω_z:

$$\frac{\partial v}{\partial x} - \frac{\partial u}{\partial y} = \Omega_z \qquad (16.98)$$

to yield

$$\frac{\partial^2 \psi}{\partial x^2} + \frac{\partial^2 \psi}{\partial y^2} = -\Omega_z \qquad (16.99)$$

Equations (16.97) and (16.99) represent the two independent equations for the stream function and vorticity. These two equations represent the *vorticity–stream-function formulation* for the flow field.

For convenience in the following discussion, we shall drop the subscript z on Ω_z.

Finite-Difference Equations

Setting up a finite-difference formulation for the vorticity equation yields an important limitation on grid size. To simplify the algebra, let us consider only the terms related to the vorticity change in the x direction. That is,

$$u \frac{\partial \Omega}{\partial x} = v \frac{\partial^2 \Omega}{\partial x^2} \tag{16.100}$$

The term on the left represents the transfer of vorticity by convection, the term on the right that by diffusion. Applying central differencing to represent the derivatives, we have

$$u \left(\frac{\Omega_{i+1} - \Omega_{i-1}}{2 \, \Delta x} \right) = v \left(\frac{\Omega_{i+1} + \Omega_{i-1} - 2\Omega_i}{\Delta x^2} \right) \tag{16.101}$$

Solving for Ω_i, we find

$$\Omega_i = \left(1 - \frac{1}{2} \cdot \frac{u \, \Delta x}{v} \right) \Omega_{i+1} + \left(1 + \frac{1}{2} \cdot \frac{u \, \Delta x}{v} \right) \Omega_{i-1} \tag{16.102}$$

The quantity $u \, \Delta x / v$ looks like a Reynolds number and is called the "cell" Reynolds number, Re_c. It is a Reynolds number based on grid spacing and on local velocity and viscosity. It can take on either a negative or a positive value, depending on the sign (direction) of the velocity u. If the cell Reynolds number is greater than 2, the coefficient of Ω_{i+1} is negative. This means that if Ω_{i+1} were to become a large positive value, then Ω_i would become an increasingly large negative value. It would be unreasonable for the vorticity at adjacent points (Ω_i and Ω_{i+1}) to diverge as the value at one point was increased. This would lead to an instability in the numerical solution. This same observation holds for $u \, \Delta x / v < -2$, in which case the coefficient of Ω_{i-1} is negative.

This instability can be avoided by restricting the cell Reynolds number to values less than 2:

$$|\mathrm{Re}_c| \le 2 \tag{16.103}$$

This requirement imposes stringent limits on the range of flow conditions amenable to flow modeling. For example, assume that one were interested in the flow of air at 10 m/s through a duct 1 m long. The kinematic viscosity of the air is 1.5×10^{-5} m^2/s. The limiting nodal spacing for $|\mathrm{Re}_c| \le 2$ would be

$$\Delta x \le \frac{2v}{u} = \frac{2 \times 1.5 \times 10^{-5}}{10} = 0.3 \times 10^{-5} \text{ m}$$

Thus, to cover 1 meter, more than 30,000 nodal points in the x direction would be needed. Taking the same number of nodal points in the y direction suggests the need for a total of 10^9 grid points—a very large storage requirement for the computation. This limitation in cell size inhibited advancements in computational fluid mechanics for several years.

Upwind Differencing

One can circumvent the problem of the limitation in cell Reynolds number by using "upwind" or "donor-cell" differencing. For example, if the local velocity is positive, then the finite-difference representation of $u\,\partial\Omega/\partial x$ is given by

$$u\frac{\partial\Omega}{\partial x} = u\left(\frac{\Omega_i - \Omega_{i-1}}{\Delta x}\right) + \mathcal{O}(\Delta x) \tag{16.104}$$

or, if u is negative, by

$$u\frac{\partial\Omega}{\partial x} = u\left(\frac{\Omega_{i+1} - \Omega_i}{\Delta x}\right) + \mathcal{O}(\Delta x) \tag{16.105}$$

Assuming that u is positive results in the following finite-difference representation of the convection-diffusion equation, Eq. (16.100),

$$u\left(\frac{\Omega_i - \Omega_{i-1}}{\Delta x}\right) = v\left(\frac{\Omega_{i-1} + \Omega_{i+1} - 2\Omega_i}{\Delta x^2}\right) \tag{16.106}$$

or
$$\left(1 + \frac{u\,\Delta x}{2v}\right)\Omega_i = \frac{\Omega_{i+1}}{2} + \left(1 + \frac{u\,\Delta x}{2v}\right)\Omega_{i-1} \tag{16.107}$$

The coefficients are always positive, they are independent of the magnitude of the cell Reynolds number, and no instability occurs in the solution. The same result is obtained for the case in which u is negative:

$$\left(1 - \frac{u\,\Delta x}{2v}\right)\Omega_i = \frac{\Omega_{i-1}}{2} + \left(1 - \frac{u\,\Delta x}{2v}\right)\Omega_{i+1} \tag{16.108}$$

Since u is negative, all coefficients are positive.

This "fix" of the stability problem is not without its drawback, however. The upwind formulation is only first-order accurate, which introduces "false viscosity" into the numerical result. Using the Taylor series, we can write Ω_{i-1} as

$$\Omega_{i-1} = \Omega_i - \frac{\partial\Omega}{\partial x}\Delta x + \frac{\partial^2\Omega}{\partial x^2}\frac{\Delta x^2}{2} + \mathcal{O}(\Delta x^3) \tag{16.109}$$

Rearranging for $(\Omega_i - \Omega_{i-1})/\Delta x$ gives

$$\frac{\Omega_i - \Omega_{i-1}}{\Delta x} = \frac{\partial\Omega}{\partial x} - \frac{\partial^2\Omega}{\partial x^2}\frac{\Delta x}{2} + \mathcal{O}(\Delta x^2) \tag{16.110}$$

or, expressed another way,

$$\left.\frac{\partial\Omega}{\partial x}\right|_{upwind} = \frac{\partial\Omega}{\partial x} - \frac{\Delta x}{2}\frac{\partial^2\Omega}{\partial x^2} + \mathcal{O}(\Delta x^2) \tag{16.111}$$

When we use upwind differencing—that is, when we represent Eq. (16.100) as

$$u \frac{\partial \Omega}{\partial x}\bigg|_{\text{upwind}} = \nu \frac{\partial^2 \Omega}{\partial x^2} \qquad (16.112)$$

we have, from Eq. (16.111),

$$u \frac{\partial \Omega}{\partial x} = \left(\nu + \frac{u \Delta x}{2} \right) \frac{\partial^2 \Omega}{\partial x^2} \qquad (16.113)$$

When we compare this equation with Eq. (16.100), it appears that the kinematic viscosity is augmented by $u \Delta x/2$. This is known as a *false viscosity*—a viscosity accruing from the use of the finite-difference representation. The numerical solution represents a more viscous fluid than actually exists. This is a fundamental problem in computational fluid mechanics and has been the subject of many studies in the field.

Procedure for Solving Vorticity—Stream-Function Equations

One can formulate equations for the vorticity and the stream function in a finite-difference form in the same way the equations for the stream function for irrotational flow were formulated. The procedure for solving them is the same as well. To begin with, one assumes initial values for stream function and vorticity. One solves for the stream function first and uses the results to obtain the velocity field. One then uses the velocity components u and v to set up the finite-difference equations for the vorticity. These equations are solved to give the vorticity at each nodal point. One then solves for the stream function again, using the updated vorticity. The scheme is repeated until convergence is achieved.

Pressure-Velocity Formulation

The vorticity–stream-function formulation has certain disadvantages. First, the engineer is not accustomed to thinking in terms of stream function and vorticity. He or she is more comfortable interpreting results for the "primitive" variables, velocity and pressure. Also, if one needs the value for pressure—for example, for calculating density using the ideal-gas law—one can obtain it only by integrating the velocity field using the momentum equation. Such a procedure is cumbersome and introduces numerical inaccuracies. For this reason, numerical codes that use the primitive variables have more appeal.

When one uses velocity and pressure as dependent variables in the numerical solution of the fluid-flow equations, one needs a staggered grid—that is, a relative displacement of the pressure and velocity nodes. Let us consider Euler's equation for the one-dimensional steady flow of an inviscid fluid,

$$\rho u \frac{du}{dx} = -\frac{dp_z}{dx} \tag{16.114}$$

where p_z is the piezometric pressure. Assume that we set up a finite-difference formulation using the grid system shown in Fig. 16.14. If we use central differencing, the finite-difference formulation for Eq. (16.114) is

$$\rho u_i \left(\frac{u_{i+1} - u_{i-1}}{2\,\Delta x} \right) = \frac{p_{z,i-1} - p_{z,i+1}}{2\,\Delta x} \tag{16.115}$$

Obviously, this equation would be satisfied with the values shown in the figure, alternate pressures being equal and all velocities constant. This solution is not realistic, however, because the piezometric pressure gradients between neighboring points, i and $i + 1$, would certainly give rise to a change in velocity. One can overcome this problem by locating the velocity nodes midway between the pressure nodes. In this fashion, a pressure gradient always leads to a velocity change at the node.

FIGURE 16.14

Distribution of piezometric pressure and velocity that satisfies finite-difference formulation of Euler's equation.

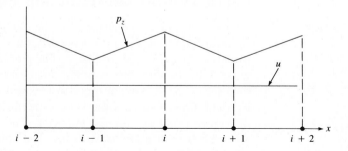

Quasi-One-Dimensional Flow in a Duct

A numerical model for steady, quasi-one-dimensional flow in a duct aptly illustrates the use of a staggered grid, and it requires solution procedures typical of numerical modeling employing the primitive variables. Let us assume that we want to determine the distribution of velocity and pressure along the horizontal duct shown in Fig. 16.15. We know the area distribution through the duct and the pressure at each end of the duct. The fluid is incompressible.

We begin by defining a grid system with a series of nodes along the duct. For this example, we assume that the nodes are equally spaced. The nodes for the pressure are designated as solid lines. The staggered nodes for the velocities are designated as dashed lines and lie to the left of the nodal lines for pressure.

The two equations that we have at our disposal are the equations for continuity and momentum. Let us consider the control volume defined by the duct

FIGURE 16.15

*Grid system and control
volumes for
quasi-one-dimensional flow
in a duct.*

Control volume for
mass conservation

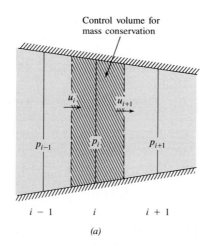

(a)

Control volume for
momentum equation

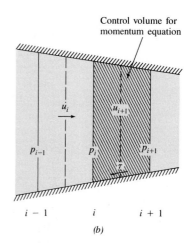

(b)

wall and the velocity nodes i and $i + 1$ as shown in Fig. 16.15a. The continuity equation applied to this control volume is

$$\rho u_i A_i - \rho u_{i+1} A_{i+1} = 0 \qquad (16.116)$$

where A_i is the cross-sectional area of the duct at the velocity node i.

We write the momentum equation for the control volume enclosed by the pressure nodes i and $i + 1$ (Fig. 16.15b). Let us assume, also, that there is a friction force on the fluid, quantified by the local Darcy–Weisbach resistance coefficient f_i. Applying the momentum equation to the control volume gives

$$\dot{m} u_{i+1} - \dot{m} u_i = A_{i+1}(p_i - p_{i+1}) - \frac{f_{i+1}\rho u_{i+1}^2 \Delta x P_{i+1}}{8} \qquad (16.117)$$

where $\dot{m}$ is the mass-flow rate and P_{i+1} is the duct perimeter corresponding to A_{i+1}. Note that the velocities used for the specific momentum (momentum per unit mass) are not located on the control surfaces (pressure nodes) but at the upstream velocity nodes. This corresponds to upwind differencing.

The wall-friction term can also be expressed as

$$\frac{f_{i+1}\rho u_{i+1}^2 P_{i+1} \Delta x}{8} = \frac{f_{i+1}\dot{m} u_{i+1} \Delta x}{2R_{i+1}} \qquad (16.118)$$

where R_{i+1} is the hydraulic radius at velocity node $i + 1$. Substituting this equation into Eq. (16.117) and solving for u_{i+1}, we have

$$u_{i+1} = \frac{\left[u_i + \dfrac{A_{i+1}}{\dot{m}}(p_i - p_{i+1}) \right]}{F_{i+1}} \qquad (16.119)$$

where $F_i = 1 + f_i \Delta x / 2R_i$. This equation represents a recursion relation for u_i, provided $\dot{m}$ and the pressure distribution are known.

One begins the procedure for solving the problem by first assuming a mass-flow rate and a pressure distribution. The assumed mass-flow rate leads to a value for u_1 using $u_1 = \dot{m}/\rho A_1$. One then uses Eq. (16.119) to evaluate the remaining values for u_i. Obviously, this will not yield the correct result because the initial values were only estimates.

One then uses the continuity equation to develop relationships for changes in pressure and velocity that proceed toward satisfying the continuity equation. The development begins by applying a Taylor-series expansion to estimate the effect of pressure changes on the velocity field. Since u_{i+1} is a function of p_i and p_{i+1}, one can relate the new velocity u_{i+1} to the changes in pressure by

$$u_{i+1} = u^o_{i+1} + \frac{\partial u_{i+1}}{\partial p_i}(p_i - p^o_i) + \frac{\partial u_{i+1}}{\partial p_{i+1}}(p_{i+1} - p^o_{i+1}) \qquad (16.120)$$

where u^o_{i+1}, p^o_i, and p^o_{i+1} are the old velocity and pressures corresponding to the previous iteration. From Eq. (16.119), we have

$$\frac{\partial u_{i+1}}{\partial p_i} = \frac{A_{i+1}}{\dot{m}F_{i+1}} \qquad (16.121a)$$

and

$$\frac{\partial u_{i+1}}{\partial p_{i+1}} = -\frac{A_{i+1}}{\dot{m}F_{i+1}} \qquad (16.121b)$$

So we can write the new velocity u_{i+1} as

$$u_{i+1} = u^o_{i+1} + \frac{A_{i+1}}{\dot{m}F_{i+1}}(\Delta p_i - \Delta p_{i+1}) \qquad (16.122)$$

where $\Delta p_i = p_i - p^o_i$. Correspondingly, the equation for u_i is

$$u_i = u^o_i + \frac{A_i}{\dot{m}F_i}(\Delta p_{i-1} - \Delta p_i) \qquad (16.123)$$

Substituting Eqs. (16.122) and (16.123) into the continuity equation, Eq. (16.116), gives

$$\rho u^o_i A_i - \rho u^o_{i+1} A_{i+1} + \frac{\rho A^2_i}{\dot{m}F_i}(\Delta p_{i-1} - \Delta p_i) - \frac{\rho A^2_{i+1}}{\dot{m}F_{i+1}}(\Delta p_i - \Delta p_{i+1}) = 0$$

$$(16.124)$$

or

$$\frac{\rho A^2_i}{\dot{m}F_i}\Delta p_{i-1} - \left(\frac{\rho A^2_i}{\dot{m}F_i} + \frac{\rho A^2_{i+1}}{\dot{m}F_{i+1}}\right)\Delta p_i + \frac{\rho A^2_{i+1}}{\dot{m}F_{i+1}}\Delta p_{i+1} = \rho u^o_{i+1} A_{i+1} - \rho u^o_i A_i$$

$$(16.125)$$

This equation represents a recursion relationship for Δp_i that we can solve by using the tridiagonal matrix algorithm. Since the pressure is fixed at $i = 1$ and $i = N$, the boundary conditions are Δp_1 and $\Delta p_N = 0$. Note that the right-hand side of the equation is the net mass efflux from the control volume calculated by using velocities from the previous iteration. The solution of this equation for Δp_i yields changes in pressure (and velocity) that proceed toward satisfying the continuity equation everywhere in the duct. If the continuity equation is satisfied for all control volumes, then $\Delta p_i = 0$ and the distribution of pressure and velocity represent the final solution.

The numerical solution is carried out in the following manner. First we estimate $\dot{m}$ and p_i, which, in turn, give estimates for u_i°. We apply the momentum equation to obtain u_i. This calculation generally has to be relaxed with a relaxation factor λ as follows:

$$u_{i+1} = (1 - \lambda)u_{i+1}^\circ + \frac{\lambda}{F_{i-1}}\left[u_i^\circ + \frac{A_{i+1}}{\dot{m}}(p_i^\circ - p_{i+1}^\circ)\right] \qquad (16.126)$$

This new velocity field becomes the new estimate u_{i+1}°. We substitute it into Eq. (16.125), which we then solve for Δp_i, using the TDMA. We correct the velocities and pressures by using Eq. (16.123) and by adding Δp_i to p_i°. This new pressure and velocity field represents the new estimates, with which we can repeat the iteration by once again starting with the momentum equation, Eq. (16.119), for u_i. The iteration cycle continues until the sum of the mass-flow residuals is less than a predetermined value.

$$\sum_{i=1}^{N-1} \left| \rho A_{i+1} u_{i+1}^\circ - \rho A_i u_i^\circ \right| < \varepsilon \qquad (16.127)$$

The application of upwind differencing and staggered grids for pressure and velocity has formed the basis for several numerical computations for two- and three-dimensional flows.

Application of the Numerical Model to an Extended-Length Venturi Meter

One can apply the numerical model presented above for quasi-one-dimensional flows to the entrance region of an extended-length venturi meter as shown in Fig. 16.16. The diameter of the duct decreases from 2 cm to 1 cm in a length of 10 cm. The duct is made of cast iron, which has an equivalent sand roughness of 0.26 mm (see Fig. 10.8).

The ultimate objective of the analysis is to predict the flow coefficient K as a function of Reynolds number. Equation (13.3) defines the flow coefficient. Since the venturi meter is oriented horizontally, one can rewrite the equation for the flow coefficient as

$$K = \frac{Q}{A_0(2\,\Delta p/\rho)^{1/2}} \qquad (16.128)$$

FIGURE 16.16

*Geometry of an
extended-length venturi
meter.*

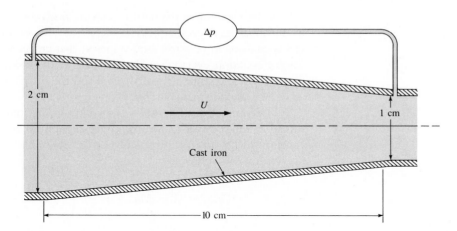

where A_0 is the cross-sectional area at the throat. For the flow of an incompressible inviscid fluid, the flow coefficient is given by

$$K = \frac{1}{(1 - \beta^4)^{1/2}} \tag{16.129}$$

where β is the ratio of throat diameter to pipe diameter. The β-ratio for the venturi meter in Fig. 16.16 is 0.5, so the K value for an ideal fluid is 1.033.

    Program III in the Appendix is the program used for the analysis of this flow field. The value for the Darcy–Weisbach resistance coefficient was calculated using Swamee and Jain's formula, expressed by Eq. (10.26). The initial estimate for pressure in the duct was a linear variation between the upstream and downstream pressures. The Appendix also shows a typical output from the program.

    To determine the effect of grid spacing on the accuracy of the solution, the program was run with $f_i = 0$ (inviscid fluid) and with 11, 51, and 101 grid lines. Table 16.5 shows the results. It compares the flow coefficients predicted by the numerical model with the exact values obtained by applying Bernoulli's equation. We note that the accuracy continually improves when one uses more grid lines. Using 101 grid lines yields an error of about 0.2%. The remaining calculations were performed using 101 grid lines (100 grid spacings).

TABLE 16.5 EFFECT OF NUMBER OF GRID LINES ON ACCURACY OF CALCULATION OF FLOW COEFFICIENT FOR INVISCID FLUID	
Number of grid lines	K/K_{exact}
11	0.954
51	0.993
101	0.998

FIGURE 16.17

Predicted distribution of pressure and velocity along the venturi meter.

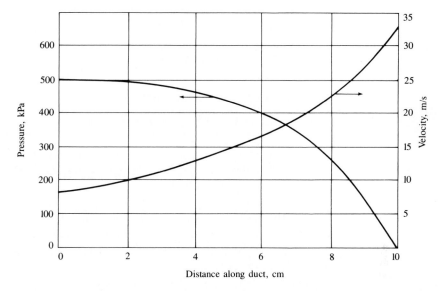

It is also interesting to note that the flow coefficient predicted by the numerical model is uniformly less than the exact value. This trend arises from upwind differencing and the attendant false viscosity, discussed above.

Figure 16.17 shows the predicted distribution of pressure and velocity for a pressure difference of 500 kPa. We note that the changes in velocity and pressure are more rapid near the end of the duct, where the relative change in duct area with distance is the largest.

Figure 16.18 shows the predicted dependence of flow coefficient on Reynolds number. We note that the flow coefficient increases with increasing Reynolds number which corresponds with the curves shown in Fig. 13.11 for the variation of the venturi flow coefficient with Reynolds number.

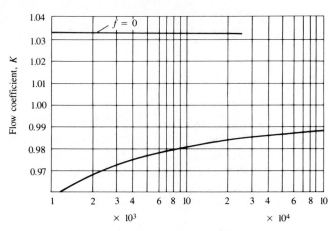

FIGURE 16.18

Predicted flow coefficient as a function of Reynolds number.

The numerical model presented here can be extended to many other applications, such as the flow of non-Newtonian fluids, the flow of compressible fluids, and multiphase flows.

16.5   Numerical Model for Unsteady Motion of a Liquid with a Free Surface

The previous numerical models were developed for steady flow. In this section, we shall illustrate a numerical scheme for the prediction of unsteady motion of water in a rectangular basin.

Figure 16.19 shows the problem we will address. The surface of the water in the basin is initially at rest, as shown. The right end of the channel opens into a reservoir in which the water level is changing with time. The numerical model developed here predicts the water-surface profile and the velocity of the water as a function of time.

FIGURE 16.19

Rectangular basin opening on reservoir.

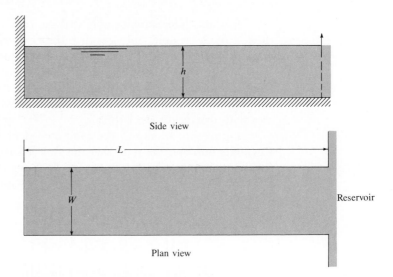

Governing Differential Equations

First let us derive the governing differential equations. Assume that the motion of the water is quasi-one-dimensional. That is, the velocity is uniform across each station of the basin. Consider the control volume shown in Fig. 16.20, with the free surface of the water at height h and the velocities defined at x and $x + \Delta x$. The continuity equation for this control volume is

$$\frac{d}{dt} \int_x^{x+\Delta x} \rho h W \, dx + \rho V W h \bigg|_{x+\Delta x} - \rho V W h \bigg|_x = 0 \qquad (16.130)$$

FIGURE 16.20

Control volume for developing equations for flow in the basin.

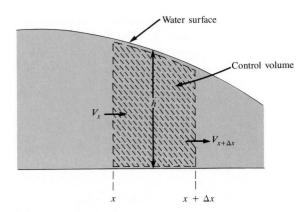

where W is the width of the channel. Since x and $x + \Delta x$ do not vary with time, the derivative of the integral becomes

$$\frac{d}{dt}\int_{x}^{x+\Delta x} \rho h W \, dx = \int_{x}^{x+\Delta x} \frac{\partial}{\partial t}(\rho h W)\, dx \qquad (16.131)$$

Dividing Eq. (16.131) by $\rho W \Delta x$ gives

$$\frac{1}{\Delta x}\int_{x}^{x+\Delta x}\frac{\partial h}{\partial t}\,dx + \frac{Vh\,|_{x+\Delta x} - Vh\,|_{x}}{\Delta x} = 0 \qquad (16.132)$$

and taking the limit as $\Delta x \to 0$ yields

$$\frac{\partial h}{\partial t} + \frac{\partial}{\partial x}(Vh) = 0 \qquad (16.133)$$

This expression is the differential form of the continuity equation.

Applying the momentum equation to the control volume yields

$$\frac{d}{dt}\int_{x}^{x+\Delta x} \rho h V W\, dx + \rho V^2 W h \bigg|_{x+\Delta x} - \rho V^2 W h \bigg|_{x} = \frac{\gamma h^2 W}{2}\bigg|_{x} - \frac{\gamma h^2 W}{2}\bigg|_{x+\Delta x} - F_f \qquad (16.134)$$

where $\gamma h/2$ represents the pressure at the centroid of the control surface, so the pressure force is given by $\gamma h^2 W/2$. The friction force can be related to the shear-stress coefficient by

$$F_f = c_f \frac{\Delta x \rho V |V| P}{2} \qquad (16.135)$$

where P is the wetted perimeter of the duct. Substituting this expression for F_f into Eq. (16.134), dividing by $\rho W \Delta x$, and taking the limit as Δx approaches zero, we obtain

$$\frac{\partial}{\partial t}(hV) + \frac{\partial}{\partial x}\left(hV^2 + \frac{gh^2}{2}\right) = -\frac{c_f V|V|}{2}\left(\frac{P}{W}\right) \qquad (16.136)$$

which is the final form of the momentum equation.

The equations for continuity and momentum represent the two equations to be solved to yield the water-surface profile and the velocity distribution. For convenience, we will introduce new variables for the parametric groupings appearing in the equations:

$$hV = A \qquad (16.137a)$$

$$hV^2 + \frac{gh^2}{2} = B \qquad (16.137b)$$

and

$$\frac{c_f V|V|}{2}\left(\frac{P}{W}\right) = C \qquad (16.137c)$$

In terms of the new variables, the equations simplify to

$$\frac{\partial h}{\partial t} + \frac{\partial A}{\partial x} = 0 \qquad \text{(continuity)} \qquad (16.138)$$

and

$$\frac{\partial A}{\partial t} + \frac{\partial B}{\partial x} = -C \qquad \text{(momentum)} \qquad (16.139)$$

We shall now formulate a method of numerical solution for these equations.

Grid System

Refer to the grid system shown in Fig. 16.21. The spatial lines are designated by i with grid spacing Δx. Lines representing time are designated by n with intervals Δt. The initial conditions are applied at $n = 1$ and for this problem are a level water-surface profile and zero velocity. The procedure is to integrate in time from the initial condition to obtain solutions for increasing time.

There is a definite limit on the magnitude of the time step Δt. Any small disturbance that occurs in the basin, such as local change in elevation of the water surface, travels in both directions as a wave with velocity $\sqrt{gh}$ with respect

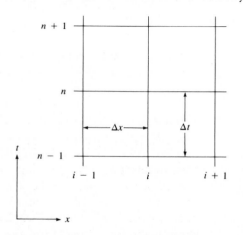

FIGURE 16.21

Grid system for finite-difference equations.

to the water. If, in addition, the water in the basin is moving with velocity V, the wave travels at velocity $(V \pm \sqrt{gh})$ with respect to the basin. Consider the motion of the waves shown in Fig. 16.22, waves that were initiated at time t_0 at location x_0. As the waves propagate from the initial disturbance, their zone of influence continually increases. This zone is represented by the triangular region in the figure. For example, at time level t_1 the water at location A has already experienced the motion induced by the wave, while the water at location B still awaits the wave's arrival. In order to represent the physical problem correctly, the time step corresponding to the distance increment Δx must lie within the zone of influence. This requires that the time step Δt be less than $\Delta x/(V + \sqrt{gh})$, which can be expressed in a more general way as

$$\left(\frac{|V| + \sqrt{gh}}{\Delta x} \right) \Delta t \leq 1 \qquad (16.140)$$

or

$$\Delta t \leq \frac{\Delta x}{|V| + \sqrt{gh}}$$

This restriction in the time step is commonly known as the *Courant condition.*

FIGURE 16.22

Propagation of waves initiated at (x_0, t_0) in the x − t plane.

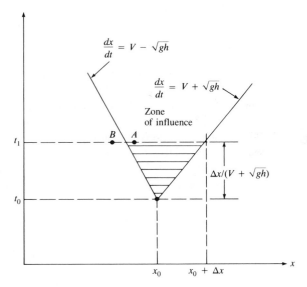

Procedure for Numerical Integration of Equations

Integration of Eqs. (16.138) and (16.139) has long occupied the interest of numerical analysts. Many of the schemes that have been proposed can be found in the literature. We will use the two-step *Lax–Wendroff method,* which has been popular for several years. It introduces an intermediate time step, which we shall

FIGURE 16.23

*Two-step grid system for
Lax–Wendroff integration.*

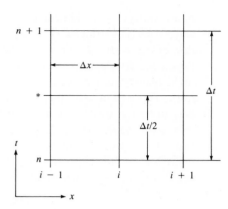

designate by an asterisk, as shown in Fig. 16.23. The intermediate step is located halfway between n and $n + 1$. We can represent the first step from n to $*$ for the continuity equation, Eq. (16.138), in finite-difference form as

$$\frac{h_i^* - h_{i,n}}{\Delta t/2} + \frac{A_{i+1,n} - A_{i-1,n}}{2\,\Delta x} = 0 \qquad (16.141)$$

In the Lax–Wendroff method, $h_{i,n}$ is replaced by

$$h_{i,n} = \tfrac{1}{2}(h_{i+1,n} + h_{i-1,n}) \qquad (16.142)$$

so the value for h_i^* is given by

$$h_i^* = \tfrac{1}{2}(h_{i+1,n} + h_{i-1,n}) - \frac{\Delta t}{4\,\Delta x}(A_{i+1,n} - A_{i-1,n}) \qquad (16.143)$$

Correspondingly, the intermediate step for A_i^* from the momentum equation, Eq. (16.139), is

$$A_i^* = \tfrac{1}{2}(A_{i+1,n} + A_{i-1,n}) - \frac{\Delta t}{4\,\Delta x}(B_{i+1,n} - B_{i-1,n}) - \frac{\Delta t}{2}C_{i,n} \qquad (16.144)$$

At the completion of this time step, we calculate V_i^* from A_i^* and h_i^*, using

$$V_i^* = \frac{A_i^*}{h_i^*} \qquad (16.145)$$

The values for h_i^* and V_i^* at the intermediate time step lead to corresponding values for B_i^* and C_i^*, which we calculate by using Eqs. (16.137b) and (16.137c).

In the second step, we proceed from n to $n + 1$, using the variables at the intermediate time step to evaluate the spatial derivatives. This is known as a "leap-frog" step and suggests a central-difference formulation for the time derivatives. The second step for the continuity equation is represented by

$$\frac{h_{i,n+1} - h_{i,n}}{\Delta t} + \frac{A_{i+1}^* - A_{i-1}^*}{2\,\Delta x} = 0 \qquad (16.146)$$

Rewriting this equation for $h_{i,n+1}$ gives

$$h_{i,n+1} = h_{i,n} - \frac{\Delta t}{2\,\Delta x}(A_{i+1}^{*} - A_{i-1}^{*}) \qquad (16.147)$$

Similarly, we obtain the value for $A_{i,n+1}$ from the finite-difference form for the momentum equation. The result is

$$A_{i,n+1} = A_{i,n} - \frac{\Delta t}{2\,\Delta x}(B_{i+1}^{*} - B_{i-1}^{*}) - C_{i}^{*}\Delta t \qquad (16.148)$$

Solving for $V_{i,n+1}$ using $A_{i,n+1}$ and $h_{i,n+1}$ yields the velocity at the new time level. The integration continues in the same fashion from $n + 1$ to $n + 2$, and so on. The procedure is terminated when the desired time interval has been covered.

Special consideration is necessary at the two ends, where central differencing on A and B is not possible. At the wall, $i = 1$, forward differencing is used to obtain h_{i}^{*} at the intermediate step.

$$h_{1}^{*} = h_{1,n} - \frac{\Delta t}{2\,\Delta x}(A_{2,n} - A_{1,n}) \qquad (16.149)$$

For the second step to $n + 1$, the value for h at the wall is

$$h_{1,n+1} = h_{1,n} - \frac{\Delta t}{\Delta x}(A_{2}^{*} - A_{1}^{*}) \qquad (16.150)$$

The value for h at the reservoir is established by the reservoir level, which is input information for the problem.

Application of the Numerical Model to an Open-Ended Channel

The numerical model was applied to the basin shown in Fig. 16.19. The basin was 50 m long, 10 m wide and, initially, 5 m deep. The surface of the water in the reservoir was assumed to vary sinusoidally with time. The local shear-stress coefficient was assumed to be 0.005.

Program IV in the Appendix is the program listing. The distance L was divided into 100 uniform grid spaces between 101 grid lines.

Figure 16.24 shows the predicted variation in the elevation of the water surface with time at the closed end of the basin. The period of oscillation of the water in the reservoir is 63 s, and the amplitude is 0.5 m. The surface of the water at the closed end of the basin does not respond to the change in elevation of the water in the reservoir until about 7 s have elapsed. The nominal speed of propagation of waves (wave celerity) is 7 m/s, which means that about 7 s are needed for a wave to travel the 50-m length of the basin. The elevation of the water at the end of the basin exceeds that of the reservoir and then decreases rapidly as the water in the basin drains back into the reservoir. The oscillations at

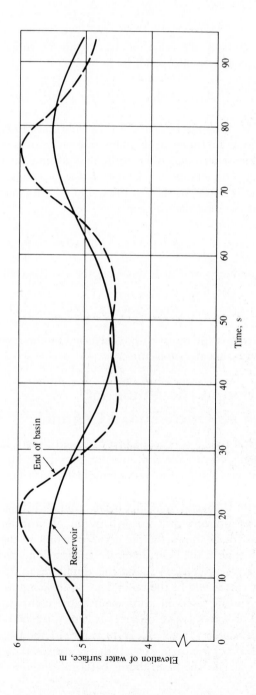

FIGURE 16.24

*Predicted variation with time
of the water surface at the
closed end of the basin for a
reservoir oscillation period of
63 s.*

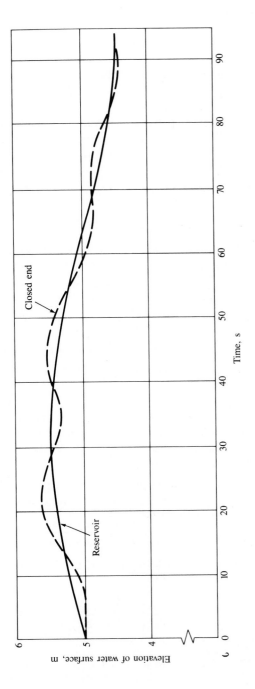

FIGURE 16.25

Predicted variation with time of the water surface at the closed end of the basin for a reservoir oscillation period of 126 s.

the closed end appear to have a natural period of approximately 30 s, which corresponds to the time it takes a wave to traverse the basin back and forth twice.

Figure 16.25 on page 775 shows the response of the elevation of the water surface at the closed end to oscillations in the reservoir with twice the period of those in Fig. 16.24. As before, there is an initial delay before the water surface rises, which is due to the time required for the wave to travel the length of the channel. Then the elevation of the water at the closed end follows the reservoir oscillation quite closely, with a secondary wave pattern corresponding to the natural frequency of the wave motion in the channel.

The analysis demonstrated here can be extended to channels with variable areas and multidimensional flows. Interested students should refer to more advanced texts and pertinent technical journals for information on numerical schemes for analyzing flows of liquids with a free surface.

Problems

16.1   Using the Taylor-series expansion for y_{i+1} and y_{i+2} around y_i with uniform grid spacing, find an expression that is second-order accurate for $dy/dx|_i$ in terms of y_i, y_{i+1}, y_{i+2}, and Δx.

16.2   Write out a central-difference formulation for the equation

$$\frac{d}{dr}\left(r^2 \frac{dv}{dr}\right) - r\frac{dv}{dr} - v = 0$$

in terms of v_{i-1}, v_i, and v_{i+1}, as well as Δr (uniform grid spacing).

16.3   The viscosity of SAE 250W gear oil varies with temperature according to the equation

$$\mu = 6.24 \times 10^{-9} \exp(5900/T) \qquad (N \cdot s/m^2)$$

where T is in degrees Kelvin. Modifying Program I in the Appendix, determine the distribution of velocity in a Couette flow of the gear oil when the temperature of the bottom plate is 40°C and that of the top plate is 100°C. The velocity of the top plate is 1 m/s, and the bottom plate is stationary.

16.4   The values for ψ on the grid system shown in the figure are ψ_{i-1}, ψ_i, and ψ_{i+1}. Using the Taylor series, show that the derivative $d\psi/dx$ at i is given by

$$\left.\frac{d\psi}{dx}\right|_i = \frac{\psi_{i+1} - \psi_{i-1}}{(a+1)\,\Delta x}$$

What is the order of magnitude of the error? Using the approach that

$$\left.\frac{d^2\psi}{dx^2}\right|_i \simeq \frac{\left(\left.\dfrac{d\psi}{dx}\right|_{i+a/2} - \left.\dfrac{d\psi}{dx}\right|_{i-1/2}\right)}{\dfrac{\Delta x}{2}(1+a)}$$

show that one obtains the same result as Eq. (16.22) in the text.

PROBLEM 16.4

16.5 The equation for motion of a viscous fluid between two concentric cylinders is

$$\frac{d}{dr}\left[\mu r^3 \frac{d}{dr}\left(\frac{v}{r}\right)\right] = 0$$

where v is the tangential velocity. This equation can also be written as

$$\frac{d}{dr}\left(\mu r^3 \frac{d\omega}{dr}\right) = 0$$

where ω is the rotational rate of the fluid ($\omega = v/r$). Modify Program I in the Appendix to solve this equation for the following conditions.

a. The viscosity is constant. The inner cylinder has a radius of 0.5 cm and rotates at 10 rad/s. The outer cylinder has a radius of 1 cm and is stationary. Find the variation of v with the radius. Compare the result with the analytic solution.

b. Repeat part (a) with the radius of the inner cylinder equal to 0.95 cm. Compare the result with the analytic solution and with a linear velocity profile for v between the two cylinders (parallel-plate assumption).

c. The radius of the inner cylinder is 0.5 cm; it rotates at 10 rad/s and has a temperature of 100°C. The outer cylinder is stationary, has a radius of 1 cm, and is maintained at a temperature of 40°C. The temperature varies between the two cylinders as

$$T = T_o + (T_i - T_o)\frac{\ln(r/r_o)}{\ln(r_1/r_o)}$$

where T_o is the temperature of the outer cylinder, T_i is the temperature of the inner cylinder, r_o is the radius of the outer cylinder, and r_i is the radius of the inner cylinder. The fluid is oil, with a variation in viscosity with temperature as given in Prob. 16.3. Determine the distribution of velocity and torque on the inner cylinder per centimeter of length. The shear stress is given by

$$\tau = \mu r \frac{d}{dr}\left(\frac{v}{r}\right)$$

16.6 The equation that describes the motion of a fluid between parallel plates is

$$\frac{d}{dy}\left(\mu \frac{du}{dy}\right) = -\frac{dp_z}{ds}$$

where y is measured normal to the plates and dp_z/ds is the gradient in piezometric pressure along the plate. The boundary conditions are $u(0) = 0$ and $u(h) = 0$, where h is the distance between the plates. Modify Program I to carry out the following:

a. Assuming μ is constant, find the velocity distribution in terms of the nondimensional variables

$$y/h \qquad \text{and} \qquad \frac{u\mu}{h^2(-dp_z/ds)}$$

Compare the result with the analytic solution.

b. The temperature of the fluid varies linearly between the bottom plate at 30°C and the top plate at 90°C. The fluid is a silicon-based oil, which has a dependence of viscosity on temperature given by

$$\mu = 5.23 \exp(1770/T) \qquad (\text{N} \cdot \text{s/m}^2)$$

where T is in degrees Kelvin. Find the velocity distribution for u in terms of y/h and $(-dp_z/ds)h^2/\mu_0$, where μ_0 is the viscosity of the oil at the bottom plate. Also find the discharge and the mean velocity.

16.7 An incompressible fluid flows with discharge $Q(t)$ into a conical tank as shown in the figure. The included angle of the tank is 30°, and the height h is measured from the vertex of the cone.

a. Derive an equation for the rate of change of the height h as a function of $Q(t)$ and h.

b. For $Q(t) = Q_0[1 + \cos(\omega t)]$, write out the implicit and explicit forms of the finite-difference formulation of the differential equation.

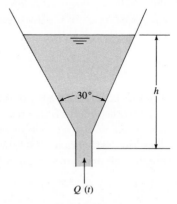

30°

h

$Q\,(t)$

PROBLEM 16.7

16.8 Write the finite-difference form of Laplace's equation for a point in the nonuniform grid system shown in the figure. The east and north points lie at $a\,\Delta x$ and $b\,\Delta y$, respectively, from the nodal point P. Write out the finite-difference formulation for ψ_P in terms of the neighboring values of ψ and a and b.

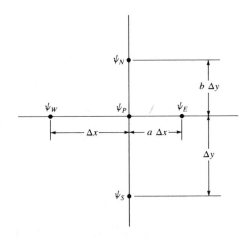

PROBLEM 16.8

16.9   The equation relating stream function and vorticity is

$$\frac{\partial^2 \psi}{\partial x^2} + \frac{\partial^2 \psi}{\partial y^2} = -\Omega_z$$

Assume that the inlet velocity distribution for the converging flow passage shown in Fig. 16.10 is given by

$$u = 0.5 + 0.5y \qquad \text{and} \qquad v = 0$$

This inlet velocity distribution corresponds to a constant vorticity of $\Omega_z = -0.5$. The value for vorticity remains unchanged throughout the flow field. Modify Program II in the Appendix to yield the stream function at each nodal point for this rotational flow field.

16.10   The stream function $\psi = 1/(x^2 + y^2)$ represents the flow of a fluid with concentric streamlines. Find the components of velocity, and determine if the flow is irrotational.

16.11   Plot the streamlines corresponding to $\psi = xy$ and find the equations for the components of velocity.

16.12   Uniform flow of a liquid parallel to the x axis is represented by $u = U$, $v = 0$. Find the stream function for this flow field.

16.13   Modify Program II in the Appendix for flow in the converging passage to extend the rounded corner from node (28, 13) to node (32, 11). The rounded corner is to be a segment of a third-order polynomial. Calculate the pressure coefficient in the region of this corner, and compare the results with those from the standard program.

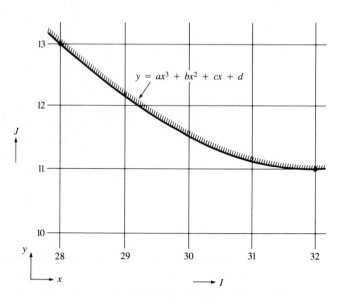

PROBLEM 16.13

16.14   Uniform flow of a liquid at an angle α with respect to the x axis has velocity components

$$u = U \cos \alpha \qquad \text{and} \qquad v = U \sin \alpha$$

Find the stream function for this flow field.

16.15   The stream function for flow of a fluid in the positive x direction past a cylinder of radius a is given by

$$\psi = Uy \left(1 - \frac{a^2}{x^2 + y^2} \right)$$

where U is the free-stream velocity. Find the velocity at the point $(0, a)$ and the corresponding pressure coefficient.

16.16   The gradient of the stream function represents a direction normal to the streamline. The gradient is expressed as

$$\nabla \psi = \mathbf{i} \frac{\partial \psi}{\partial x} + \mathbf{j} \frac{\partial \psi}{\partial y}$$

where $\mathbf{i}$ and $\mathbf{j}$ are unit normal vectors in the x and y directions. Prove that the streamlines and potential lines are mutually orthogonal by showing that

$$\nabla \psi \bullet \nabla \phi = 0$$

where ϕ is the potential function.

16.17   Extend Program II in the Appendix to calculate the components of velocity at every nodal point and the pressure coefficient along the wall and centerline. If facilities are available, also plot the streamlines corresponding to $\psi = 0.2, 0.4, 0.6,$ and 0.8, as well as the profiles representing the centerline and the wall.

16.18 Using the cylindrical element shown in the figure as the control volume, derive the continuity equation for the axisymmetric flow of an incompressible fluid in terms of coordinates r and z and velocity components u_r and u_z.

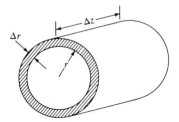

PROBLEM 16.18

16.19 Using the same approach as used in the text to relate velocity components and the derivatives of the stream function, show that the radial and axial components of velocity in a polar coordinate system are given by

$$u_z = \frac{1}{r}\frac{\partial \psi}{\partial r} \quad \text{and} \quad u_r = -\frac{1}{r}\frac{\partial \psi}{\partial z}$$

In developing the relationship between stream function and velocity, you will encounter a factor of 2π in the equations for area. This factor is absorbed in the value for the stream function. By substituting these velocity components into the continuity equation derived in Prob. 16.18, show that continuity is satisfied.

16.20 The irrotationality condition for axisymmetric flow in polar coordinates is expressed by

$$\frac{\partial^2 \psi}{\partial z^2} + r\frac{\partial}{\partial r}\left(\frac{1}{r}\frac{\partial \psi}{\partial r}\right) = 0$$

where r is the radial coordinate and z is the axial coordinate. Modify Program II in the Appendix for axisymmetric flow in the converging flow passage. Careful consideration of the boundary conditions in the vicinity of the axis of symmetry reveals that the lower limit of the computational field is the line $j = 3$ with the boundary condition $\psi_{i,3} = 4\psi_{i,2}$. The velocity on the axis of symmetry is given by

$$u_z = \frac{2\psi_{i,2}}{\Delta y^2}$$

Plot some streamlines and evaluate the pressure coefficient along the axis and the wall.

16.21 For a duct with converging walls, in which the pressure on the sloping wall equals the average of the pressures on each face, show that the net force of the pressure acting on the fluid between stations 1 and 2 is

$$F = (p_1 - p_2)\left(\frac{A_1 + A_2}{2}\right)$$

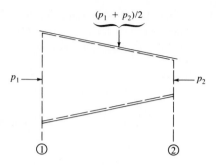

PROBLEM 16.21

16.22 For the isentropic flow of a compressible ideal gas, density and pressure are related by the expression p/ρ^k = constant, where k is the ratio of specific heats. Modify Program III in the Appendix for compressible flow of air (k = 1.4) in the duct, with upstream pressure equal to 100 kPa and exit pressure equal to 60 kPa. The upstream temperature is 20°C. Assume there is no friction. You may find it necessary to include a density correction of the form

$$\rho_i = \rho_i^o + \frac{\partial \rho}{\partial p} \Delta p_i$$

in the pressure formulation of the continuity equation. You may also have to under-relax the pressure change, $\Delta p_i = \lambda (\Delta p_i)_{calc}$. Compare your results for mass flow with Eq. (13.16) in the text.

16.23 Upwind differencing of the convective-acceleration term on a uniform grid is represented by

$$u\frac{\partial u}{\partial x} = u_i \left(\frac{u_i - u_{i-1}}{\Delta x} \right) \quad \text{if } u_i > 0 \quad \text{and} \quad u\frac{\partial u}{\partial x} = u_i \left(\frac{u_{i+1} - u_i}{\Delta x} \right) \quad \text{if } u_i < 0$$

Show that

$$u\frac{\partial u}{\partial x} = \left(\frac{u_i - u_{i-1}}{\Delta x} \right) \left(\frac{|u_i| + u_i}{2} \right) - \left(\frac{u_{i+1} - u_i}{\Delta x} \right) \left(\frac{|u_i| - u_i}{2} \right)$$

always yields the upwind-difference formulation.

16.24 A *Bingham plastic* is a substance that has a threshold shear stress that must be exceeded before deformation is possible. Above the threshold stress, the stress is proportional to the rate of strain, as in the case of a Newtonian fluid. The shear stress on the wall for flow of a Bingham plastic in a duct is

$$\tau_w = \frac{8\eta U}{D} + \frac{4}{3}\tau_0$$

where τ_0 is the threshold stress, U is the velocity of the fluid, D is the diameter of the duct, and η is the slope of the line for stress versus rate of strain. A nuclear-fuel slurry of UO_3–H_2O has a threshold stress of 1270 Pa and a value for η of 60 N · s/m². The density of the fluid is 1500 kg/m³. Modify Program III in the Appendix for flow of this slurry through the duct, and generate a curve for discharge versus pressure drop. You may find it convenient first to calculate the threshold pressure distribution and subsequently to reference all pressures to this distribution. The high value for shear stress suggests that the initial flow rate be estimated by equating the pressure and shear forces.

16.25 The two-dimensional form of the vorticity equation is

$$u \frac{\partial \Omega}{\partial x} + v \frac{\partial \Omega}{\partial y} = \nu \left(\frac{\partial^2 \Omega}{\partial x^2} + \frac{\partial^2 \Omega}{\partial y^2} \right)$$

Write out the finite-difference equation for $\Omega_{i,j}$ as a function of u, v, and Ω at the neighboring nodes, using central differencing on a uniform grid. Show that the coefficients of $\Omega_{i,j}$ and Ω at the neighboring nodes have the same sign only if $u\,\Delta\ell/\nu$ and $v\Delta\ell/\nu$ are less than 2, where $\Delta\ell$ is the grid spacing.

16.26 A screen is placed in the duct modeled by Program III in the Appendix, at the station corresponding to a duct diameter of 1.5 cm. The loss in pressure across the screen is represented by a loss coefficient, k_L, of 0.5. By modifying Program III, find the mass-flow rate and the pressure profile in the duct for a pressure difference of 10 kPa.

16.27 Consider the nodal point in the vicinity of the wall shown in the figure. Show that the boundary condition relating vorticity and the stream function at the wall is

$$\Omega_{\text{wall}} = \frac{\psi_{i,\text{wall}} - \psi_{i,\text{wall}+2}}{2\,\Delta y^2}$$

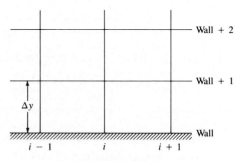

PROBLEM 16.27

16.28 Show that the equation for free-surface flow in a basin yield the wave equation in the form

$$\frac{\partial^2 h}{\partial t^2} = gh \left(\frac{\partial^2 h}{\partial x^2} \right)$$

if C is zero and both $\partial^2(AV)/\partial x^2$ and $g(\partial h/\partial x)^2$ are negligible compared to $gh(\partial^2 h/\partial x^2)$. These assumptions correspond to small-amplitude waves and permit "linearization" to the wave equation.

16.29 Rerun Program IV in the Appendix for a reservoir oscillation near the natural frequency for wave motion in the channel. Describe and interpret your results.

16.30 Subtract the continuity equation for flow in a basin, Eq. (16.133), from the momentum equation, Eq. (16.136). Identify the terms representing the local and convective acceleration.

16.31 By applying the continuity and momentum principles to the control volume on the sloping surface shown in the figure, show that the continuity and momentum equations are

$$\frac{\partial h}{\partial t} + \frac{\partial}{\partial x}(hV) = 0$$

$$\frac{\partial}{\partial t}(hV) + \frac{\partial}{\partial x}\left(hV^2 + \frac{gh^2}{2}\right) = S_0 gh - c_f\left(\frac{|V|VP}{2}\right)$$

where S_0 is the slope of the surface and its magnitude is small compared to unity.

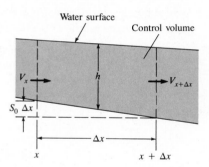

PROBLEM 16.31

16.32 Modify Program IV in the Appendix for a basin with a slope S_0 of 0.05. The initial water-surface profile is level, with a depth of 2.5 m at the closed end and a depth of 5 m at the reservoir. Compare your results with those of the standard program.

16.33 Modify Program IV in the Appendix by using the Chezy equation (see Section 10.7) to evaluate c_f (or f) for a basin having a gravel bed with large boulders ($n = 0.04$). Determine the variation of the water-surface profile with time, and compare the results with those corresponding to $c_f = 0.005$.

References

1. Anderson, Dale A., John C. Tannehill, and Richard H. Plechter. *Computational Fluid Mechanics and Heat Transfer.* McGraw-Hill Book Company, New York, 1984.

2. Carnahan, B., H. A. Luther, and J. O. Wilkes. *Applied Numerical Methods.* John Wiley, New York, 1969.

3. Fox, L. Numerical Solution of Ordinary and Partial Differential Equations. Pergamon Press, London, 1962.

4. Gosman, A. D., et al. *Heat and Mass Transfer in Recirculating Flows.* Academic Press, London, 1969.

5. Holt, M. *Numerical Methods in Fluid Dynamics.* Springer-Verlag, Berlin, 1977.

6. Hornbeck, R. W. *Numerical Methods.* Quantum Publishers, New York, 1975.

7. Pantankar, S. V. *Numerical Heat Transfer and Fluid Flow.* Hemisphere (McGraw-Hill), New York, 1980.

8. Peyret, R., and T. D. Taylor. *Computational Methods for Fluid Flow.* Springer-Verlag, Berlin, 1983.

9. Roache, P. J. *Computational Fluid Mechanics.* Hermosa Publishers, Albuquerque, N.M., 1972.

10. Shoup, T. E. *A Practical Guide to Computer Methods for Engineers.* Prentice-Hall, Englewood Cliffs, N.J., 1979.

Appendix

NOMENCLATURE AND DIMENSIONS		
Symbol	**Dimensions**	**Description**
A	L^2	Area
A_j	L^2	Jet area
A_o	L^2	Orifice area
A_*	L^2	Nozzle area at M = 1
a	L/T^2	Acceleration
B	L	Linear measure
B	$\ldots$	Extensive property
Br		Brinell number
b	L	Linear measure
°C	θ	Temperature, Celsius
C_c	$\ldots$	Coefficient of contraction
C_D	$\ldots$	Coefficient of drag
C_d	$\ldots$	Coefficient of discharge
C_f	$\ldots$	Shear-force coefficient
C_H	$\ldots$	Head coefficient
C_L	$\ldots$	Coefficient of lift
C_P	$\ldots$	Power coefficient
C_p	$\ldots$	Pressure coefficient
C_Q	$\ldots$	Discharge coefficient
C_T	$\ldots$	Thrust coefficient
C_v	$\ldots$	Coefficient of velocity
c	$\ldots$	Centi, multiple = 10^{-2}
c	L/T	Speed of sound
c_f	$\ldots$	Local shear-stress coefficient
c_p	$L^2/T^2\theta$	Specific heat at constant pressure
c_v	$L^2/T^2\theta$	Specific heat at constant volume
D	L	Diameter
D	L	Hydraulic depth
d	L	Diameter
d	L	Depth
E	LF	Energy
E	L	Specific energy
E_v	F/L^2	Elasticity, bulk
e	L^2/T^2	Energy per unit mass
Fr	$\ldots$	Froude number
F	F	Force
°F	θ	Temperature, Fahrenheit
F_D	F	Drag force
F_L	F	Lift force
f	$\ldots$	Resistance coefficient
G	$\ldots$	Giga, multiple = 10^9
g	L/T^2	Acceleration due to gravity

NOMENCLATURE AND DIMENSIONS (CONTINUED)		
Symbol	**Dimensions**	**Description**
g_c	...	Proportionality factor
H	L	Head
h	L	Head
h	L	Piezometric head
h	L^2/T^2	Enthalpy
h_f	L	Friction head loss in pipe
h_L	L	Head loss
h_p	L	Head supplied by pump
h_t	L	Head given up to turbine
I	L^4	Area moment of inertia, centroidal
i	...	Unit vector in x direction
J	FL	Joule, unit of work
j	...	Unit vector in y direction
K	...	Flow coefficient
K	θ	Temperature, Kelvin
k	...	Ratio of specific heats
k	...	Kilo, multiple $= 10^3$
k	...	Unit vector in z direction
k_s	L	Equivalent sand-roughness size
L	L	Linear measure
l	L	Linear measure
ℓ	L	Linear measure
M	...	Mach number
M	FL	Moment
M	FT^2/L	Mass
M	...	Mega, multiple $= 10^6$
m	L	Meter
m	...	Milli, multiple $= 10^{-3}$
$\dot{m}$	FT/L	Mass rate of flow
N	F	Newton, unit of force
N	T^{-1}	Rotational speed
N_s	$L^{3/4}/T^{3/2}$	Specific speed
n	T^{-1}	Frequency in Hertz
n	...	Manning's roughness coefficient
n	T^{-1}	Rotational speed
n_s	...	Specific speed
Pa	F/L^2	Pascal, unit of pressure
p_*	F/L^2	Pressure at M $= 1$
p	F/L^2	Pressure
p_t	F/L^2	Total pressure
p_v	F/L^2	Vapor pressure
Q	L^3/T	Discharge
Q	LF	Heat transferred
q	L^2/T	Discharge per unit width

NOMENCLATURE AND DIMENSIONS (CONTINUED)		
Symbol	**Dimensions**	**Description**
q	F/L^2	Kinetic pressure
R	L	Hydraulic radius
R_h	F	Reaction or resultant force
R	L^2/KT^2	Gas constant
$°R$	θ	Temperature, Rankine
Re	...	Reynolds number
r	L	Linear measure in radial direction
S	L^2	Planform area
S	...	Strouhal number
S_0	...	Channel slope
s	$L^3/FT^4\theta$	Specific entropy
S	...	Specific gravity
s	T	Time, second
s	L	Linear measure
T	LF	Torque
T	θ	Temperature
T_t	θ	Total temperature
T_*	θ	Temperature at M = 1
t	T	Time
U_0	L/T	Free-stream velocity
u	L/T	Velocity component, x direction
u	L^2/T^2	Internal energy per unit of mass
u_*	L/T	Shear velocity
u'	L/T	Velocity fluctuation in x direction
V	L/T	Velocity
V_0	L/T	Free-stream velocity
Ψ	L^3	Volume
v	L/T	Velocity component, y direction
v'	L/T	Velocity fluctuation in y direction
W	LF	Work
W	F	Weight
W	...	Weber number
W	FL/T	Watt, unit of power
w	L/T	Velocity component, z direction
x	L	Linear measure
y	L	Linear measure
y_c	L	Critical depth
y_n	L	Normal depth
z	L	Linear measure

NOMENCLATURE AND DIMENSIONS (CONTINUED)		
Symbol	**Dimensions**	**Description**
GREEK LETTERS		
α	...	Angular measure
α	...	Lapse rate
α	...	Kinetic energy coefficient
α	...	Angle of attack
β	...	Angular measure
β	...	Momentum coefficient
β	...	Intensive property
Γ	L^2/T	Circulation
γ	F/L^3	Specific weight
Δ	...	Increment
δ	L	Boundary-layer thickness
δ'	L	Laminar sublayer thickness
δ'_N	L	Nominal laminar sublayer thickness
η	...	Efficiency
θ	...	Angular measure
κ	...	Turbulence constant
μ	FT/L^2	Viscosity, dynamic
μ	...	Micro, multiple $= 10^{-6}$
τ	F/L^2	Shear stress
ν	L^2/T	Kinematic viscosity
π	...	3.14
ρ	FT^2/L^4	Mass density
ρ_*	FT^2/L^4	Density at M $= 1$
ρ_t	FT^2/L^4	Total density
Ω	T^{-1}	Vorticity
ω	T^{-1}	Angular speed
σ	F/L	Surface tension

PROGRAM I

```
C*********************************************************************
C                          PROGRAM I                                *
C    *PROGRAM FOR FINDING THE VELOCITY DISTRIBUTION IN A COUETTE FLOW *
C     WITH A VARIABLE VISCOSITY DISTRIBUTION                         *
C*********************************************************************
C
      DIMENSION U(101)
      COMMON/TDMA/A(101),B(101),C(101),D(101)
C-----IMAX IS NUMBER OF GRID LINES
      IMAX=6
C-----ISKIP IS THE LINES SKIPPED BETWEEN PRINTOUT
      ISKIP=1
      IM1=IMAX-1
C-----BOUNDARY CONDITIONS
      U(1)=0.0
      U(IMAX)=1.0
C-----EVALUATION OF TDMA COEFFICIENTS
      DY=1./FLOAT(IM1)
      DO 10 I=2,IM1
      AI=FLOAT(I)
      Y=(AI-1.5)*DY
      A(I)=EXP(1.92*(1.-Y))
      Y=(AI-.5)*DY
      C(I)=EXP(1.92*(1.-Y))
      B(I)=-A(I)-C(I)
   10 D(I)=0.0
C-----CALL THE TRIDIAGONAL ALGORITHM TO OBTAIN VELOCITIES
      CALL TRIDIA(1,IMAX,U)
C-----PRINTOUT OF VALUES
      WRITE(1,1000)
      WRITE(1,1001) (I,U(I),I=1,IMAX,ISKIP)
      STOP
 1000 FORMAT(' VELOCITY DISTRIBUTION BETWEEN PLATES *****'
     1/' I        U(I)')
 1001 FORMAT(I4,5X,F10.5)
      END
C
C*****TRIDIAGONAL SUBROUTINE WHERE 'IT' IS THE LAST LINE, 'IB' IS
C     THE FIRST LINE AND 'PHI' IS THE DEPENDENT VARIABLE. THE
C     NOMENCLATURE CORRESPONDS TO EQNS 16-48 THRU 16-53 IN THE
C     TEXT.
      SUBROUTINE TRIDIA(IB,IT,PHI)
      DIMENSION E(101),F(101),PHI(IT)
      COMMON/TDMA/A(101),B(101),C(101),D(101)
      IM1=IT-1
      IP1=IB+1
      IP2=IB+2
C-----SETTING UP BOUNDARY VALUES
      D(IP1)=D(IP1)-A(IP1)*PHI(IB)
      A(IP1)=0.0
      D(IM1)=D(IM1)-C(IM1)*PHI(IT)
      C(IM1)=0.0
      E(IP1+1)=-C(IP1)/B(IP1)
      F(IP1+1)=D(IP1)/B(IP1)
C-----EVALUATION OF E'S AND F'S
      DO 20 K=IP2,IM1
      FAC=1./(A(K)*E(K)+B(K))
      E(K+1)=-C(K)*FAC
   20 F(K+1)=(D(K)-A(K)*F(K))*FAC
```

```
C-----USING E'S AND F'S TO CALCULATE PHI
      PHI(IM1)=F(IT)
      DO 30 L=IP2,IM1
      LL=IT+IP1-L
   30 PHI(LL-1)=E(LL)*PHI(LL)+F(LL)
      RETURN
      END

VELOCITY DISTRIBUTION BETWEEN PLATES *****
      I       U(I)
      1       0.00000
      2       0.08042
      3       0.19850
      4       0.37185
      5       0.62635
      6       1.00000
```

PROGRAM II

```
C***********************************************************************
C                          PROGRAM II                                 *
C        *PROGRAM FOR FINDING VALUES FOR STREAM FUNCTION IN CONVERGING *
C         DUCT BY SOLVING, NUMERICALLY, LAPLACES EQUATION USING THE    *
C         LINE-BY-LINE PROCEDURE.                                      *
C***********************************************************************
C
      DIMENSION PSI(50,21),PSII(21),JTOP(50)
      DIMENSION CN(50,21),CS(50,21),CE(50,21),CW(50,21)
      COMMON/TDMA/A(101),B(101),C(101),D(101)
C-----SET UP J LIMIT ON TOP WALL
C          -APPROACH SECTION
      DO 10 I=1,20
   10 JTOP(I)=20
C          -SLOPING SECTION
      DO 15 I=21,30
   15 JTOP(I)=20-(I-20)
C          -EXIT SECTION
      DO 20 I=31,50
   20 JTOP(I)=10
C          -SPECIAL POINT AT DOWNSTREAM CORNER
      JTOP(30)=11
C-----CONVERGENCE CRITERION
      ERMAX=.01
C-----ZERO FOR ITERATION COUNTER
      NIT=0
C-----INITIALIZE ALL VARIABLES. THE VARIABLES CN,CE,CS AND CW ARE
C     COEFFICIENTS FOR PSI AT NEIGHBOURING POINTS IN FINITE
C     DIFFERENCE FORMULATION OF LAPLACES EQUATION.
      DO 25 I=1,50
      DO 30 J=1,21
      CN(I,J)=1.0
      CS(I,J)=1.0
      CW(I,J)=1.0
      CE(I,J)=1.0
   30 PSI(I,J)=0.0
   25 CONTINUE
C-----SETTING COEFFICIENTS AT (20,20) AND (30,11)
      CN(20,20)=1.523
```

```
              CS(20,20)=1.143
              CN(30,11)=6.4
              CS(30,11)=1.6
C-----SETTING UP BOUNDARY CONDITIONS ON PSI
C          -TOP WALL AND CENTERLINE
              DO 35 I=1,50
              JMAX=JTOP(I)+1
              PSI(I,1)=0.0
           35 PSI(I,JMAX)=1.0
C          -ACROSS INLET
              AJ=1./FLOAT(JTOP(1))
              JT=JTOP(1)
              DO 40 J=1,JT
           40 PSI(1,J)=(J-1)*AJ
C          -EXIT PLANE
              JMAX=JTOP(50)
              DO 45 J=2,JMAX
           45 CE(49,J)=0.0
C-----ESTIMATING INITIAL VALUES FOR PSI
              DO 50 I=2,49
              JT=JTOP(I)
              AJ=1./FLOAT(JT)
              DO 60 J=1,JT
              PSI(I,J)=(J-1)*AJ
           60 CONTINUE
           50 CONTINUE
C-----LINE BY LINE SOLUTION USING TDMA
          199 NIT=NIT+1
              ERROR=0.0
              DO 100 I=2,49
              JMAX=JTOP(I)
              JMP1=JMAX+1
C-----SETTING UP TDMA COEFFICIENTS
              DO 110 J=2,JMAX
              A(J)=CS(I,J)
              C(J)=CN(I,J)
              B(J)=-CN(I,J)-CS(I,J)-CE(I,J)-CW(I,J)
          110 D(J)=-CE(I,J)*PSI(I+1,J)-CW(I,J)*PSI(I-1,J)
              PSII(1)=PSI(I,1)
              PSII(JMP1)=PSI(I,JMP1)
              CALL TRIDIA(1,JMP1,PSII)
C-----EVALUATING THE ERROR AND UPDATING PSI
              DO 105 J=2,JMAX
              ERROR=ERROR+ABS(PSII(J)-PSI(I,J))
          105 PSI(I,J)=PSII(J)
          100 CONTINUE
C-----IF ERROR STILL TOO LARGE, ITERATE AGAIN
              IF(ERROR.GT.ERMAX) GO TO 199
C-----SET FINAL PSI AT EXIT PLANE EQUAL TO NEXT-TO-LAST PLANE
              JMAX=JTOP(49)
              DO 115 J=2,JMAX
          115 PSI(50,J)=PSI(49,J)
C-----PRINT OUT OF PSI VALUES
              WRITE(1,1000)
              WRITE(1,1001) NIT,ERROR
              WRITE(1,1002) (J,J=1,21,2)
              WRITE(1,1003) (I,(PSI(I,J),J=1,21,2),I=1,50)
              STOP
         1000 FORMAT(' *STREAM FUNCTION DISTRIBUTION IN DUCT WITH A CONTRACTION
             1 RATIO OF TWO')
         1001 FORMAT(/'   NIT=',I3,'  ERROR=',E10.2)
         1002 FORMAT(/'J=',1X,11(4X,I3))
```

```
 1003 FORMAT(I2,1X,11F7.3)
      END
C
C*****TRIDIAGONAL SUBROUTINE WHERE 'IT' IS THE LAST LINE, 'IB' IS
C     THE FIRST LINE AND 'PHI' IS THE DEPENDENT VARIABLE. THE
C     NOMENCLATURE CORRESPONDS TO EQNS 16-48 THRU 16-53 IN THE
C     TEXT.
      SUBROUTINE TRIDIA(IB,IT,PHI)
      DIMENSION E(101),F(101),PHI(IT)
      COMMON/TDMA/A(101),B(101),C(101),D(101)
      IM1=IT-1
      IP1=IB+1
      IP2=IB+2
C-----SETTING UP BOUNDARY VALUES
      D(IP1)=D(IP1)-A(IP1)*PHI(IB)
      A(IP1)=0.0
      D(IM1)=D(IM1)-C(IM1)*PHI(IT)
      C(IM1)=0.0
      E(IP1+1)=-C(IP1)/B(IP1)
      F(IP1+1)=D(IP1)/B(IP1)
C-----EVALUATION OF E'S AND F'S
      DO 20 K=IP2,IM1
      FAC=1./(A(K)*E(K)+B(K))
      E(K+1)=-C(K)*FAC
   20 F(K+1)=(D(K)-A(K)*F(K))*FAC
C-----USING E'S AND F'S TO CALCULATE PHI
      PHI(IM1)=F(IT)
      DO 30 L=IP2,IM1
      LL=IT+IP1-L
   30 PHI(LL-1)=E(LL)*PHI(LL)+F(LL)
      RETURN
      END
```

*STREAM FUNCTION DISTRIBUTION IN DUCT WITH A CONTRACTION RATIO OF TWO

NIT=109 ERROR= 0.98E-02

J=	1	3	5	7	9	11	13	15	17	19	21
1	0.000	0.100	0.200	0.300	0.400	0.500	0.600	0.700	0.800	0.900	1.000
2	0.000	0.101	0.201	0.301	0.402	0.502	0.602	0.701	0.801	0.901	1.000
3	0.000	0.101	0.202	0.303	0.403	0.504	0.603	0.703	0.802	0.901	1.000
4	0.000	0.102	0.203	0.304	0.405	0.506	0.605	0.705	0.803	0.902	1.000
5	0.000	0.102	0.204	0.306	0.407	0.508	0.607	0.706	0.805	0.902	1.000
6	0.000	0.103	0.206	0.308	0.409	0.510	0.610	0.708	0.806	0.903	1.000
7	0.000	0.104	0.207	0.310	0.412	0.513	0.612	0.710	0.808	0.904	1.000
8	0.000	0.105	0.209	0.312	0.414	0.515	0.615	0.713	0.810	0.905	1.000
9	0.000	0.105	0.211	0.315	0.418	0.519	0.618	0.716	0.812	0.906	1.000
10	0.000	0.107	0.213	0.318	0.421	0.523	0.622	0.719	0.814	0.908	1.000
11	0.000	0.108	0.215	0.321	0.425	0.527	0.626	0.723	0.817	0.909	1.000
12	0.000	0.109	0.218	0.325	0.430	0.532	0.632	0.728	0.821	0.911	1.000
13	0.000	0.111	0.221	0.329	0.435	0.538	0.638	0.733	0.825	0.913	1.000
14	0.000	0.113	0.224	0.334	0.441	0.545	0.645	0.739	0.830	0.916	1.000
15	0.000	0.115	0.228	0.340	0.448	0.553	0.653	0.747	0.836	0.919	1.000
16	0.000	0.117	0.233	0.346	0.457	0.562	0.662	0.756	0.843	0.924	1.000
17	0.000	0.120	0.238	0.354	0.466	0.573	0.673	0.767	0.852	0.929	1.000
18	0.000	0.122	0.244	0.362	0.477	0.585	0.687	0.780	0.863	0.936	1.000
19	0.000	0.126	0.250	0.372	0.489	0.599	0.702	0.795	0.876	0.945	1.000
20	0.000	0.129	0.257	0.382	0.502	0.616	0.720	0.813	0.893	0.959	1.000
21	0.000	0.133	0.265	0.394	0.518	0.634	0.740	0.834	0.913	0.976	0.000
22	0.000	0.138	0.274	0.407	0.535	0.655	0.764	0.859	0.938	1.000	0.000
23	0.000	0.142	0.283	0.422	0.555	0.679	0.791	0.887	0.967	0.000	0.000
24	0.000	0.147	0.294	0.437	0.576	0.706	0.822	0.920	1.000	0.000	0.000

```
25    0.000   0.153   0.304   0.454   0.600   0.737   0.857   0.958   0.000   0.000   0.000
26    0.000   0.158   0.316   0.472   0.626   0.772   0.898   1.000   0.000   0.000   0.000
27    0.000   0.163   0.327   0.491   0.654   0.811   0.945   0.000   0.000   0.000   0.000
28    0.000   0.169   0.338   0.510   0.684   0.856   1.000   0.000   0.000   0.000   0.000
29    0.000   0.174   0.349   0.528   0.714   0.908   0.000   0.000   0.000   0.000   0.000
30    0.000   0.178   0.359   0.545   0.741   0.967   0.000   0.000   0.000   0.000   0.000
31    0.000   0.183   0.368   0.559   0.762   1.000   0.000   0.000   0.000   0.000   0.000
32    0.000   0.186   0.375   0.570   0.775   1.000   0.000   0.000   0.000   0.000   0.000
33    0.000   0.189   0.381   0.578   0.783   1.000   0.000   0.000   0.000   0.000   0.000
34    0.000   0.192   0.386   0.584   0.789   1.000   0.000   0.000   0.000   0.000   0.000
35    0.000   0.194   0.389   0.588   0.792   1.000   0.000   0.000   0.000   0.000   0.000
36    0.000   0.195   0.392   0.592   0.794   1.000   0.000   0.000   0.000   0.000   0.000
37    0.000   0.197   0.394   0.594   0.796   1.000   0.000   0.000   0.000   0.000   0.000
38    0.000   0.197   0.396   0.595   0.797   1.000   0.000   0.000   0.000   0.000   0.000
39    0.000   0.198   0.397   0.597   0.798   1.000   0.000   0.000   0.000   0.000   0.000
40    0.000   0.199   0.398   0.598   0.798   1.000   0.000   0.000   0.000   0.000   0.000
41    0.000   0.199   0.398   0.598   0.799   1.000   0.000   0.000   0.000   0.000   0.000
42    0.000   0.199   0.399   0.599   0.799   1.000   0.000   0.000   0.000   0.000   0.000
43    0.000   0.199   0.399   0.599   0.799   1.000   0.000   0.000   0.000   0.000   0.000
44    0.000   0.200   0.399   0.599   0.800   1.000   0.000   0.000   0.000   0.000   0.000
45    0.000   0.200   0.399   0.599   0.800   1.000   0.000   0.000   0.000   0.000   0.000
46    0.000   0.200   0.400   0.600   0.800   1.000   0.000   0.000   0.000   0.000   0.000
47    0.000   0.200   0.400   0.600   0.800   1.000   0.000   0.000   0.000   0.000   0.000
48    0.000   0.200   0.400   0.600   0.800   1.000   0.000   0.000   0.000   0.000   0.000
49    0.000   0.200   0.400   0.600   0.800   1.000   0.000   0.000   0.000   0.000   0.000
50    0.000   0.200   0.400   0.600   0.800   1.000   0.000   0.000   0.000   0.000   0.000
```

PROGRAM III

```
C************************************************************************
C                          PROGRAM III                                 *
C     *PROGRAM FOR THE ANALYSIS OF FLUID FLOWING IN AN EXTENDED LENGTH  *
C      VENTURI APPROACH SECTION. PROGRAM PREDICTS FLOW RATE FOR A GIVEN *
C      PRESSURE DIFFERENCE. ALL PARAMETERS ARE IN SI UINTS              *
C************************************************************************
      DIMENSION U(101),P(101),RF(101),DELP(101),AR(101),DIA(101)
     1,FFAC(101),POUT(101)
      COMMON/TDMA/A(101),B(101),C(101),D(101)
      LOGICAL IEND
C-----INLET DIAMETER
      DIA1=.02
C-----EXIT DIAMETER
      DIA2=.01
C-----TOTAL LENGTH
      AL=.1
C-----NUMBER OF GRID LINES
      IMAX=101
C-----LINES SKIPPED IN PRINTOUT
      ISKIP=10
C-----GRID SPACING
      DX=AL/FLOAT(IMAX-1)
C-----ZEROING ITERATION COUNTER
      NITER=0
C-----FLUID DENSITY
      RHO=1000.
C-----KINEMATIC VISCOSITY
      VISK=1.E-06
      RK=.26E-03
```

```
C-----UPSTREAM PRESSURE
      P1=1.E04
C-----DOWNSTREAM PRESSURE
      P2=0.0
      PDIF=(P1-P2)/1000
C-----CONVERGENCE CRITERION
      RESMAX=1.E-04
C-----RELAXATION FACTOR
      RLX=.5
      IEND=.FALSE.
      IM1=IMAX-1
      IM2=IMAX-2
C-----CALCULATE DUCT DIAMETERS
      DIA(1)=DIA1
      DIA(2)=DIA(1)
      SLP=(DIA1-DIA2)/FLOAT(IM2-2)
      DO 10 I=3,IM2
      AI=FLOAT(I)
   10 DIA(I)=DIA1-(AI-2.5)*SLP
      DIA(IM1)=DIA2
      DIA(IMAX)=DIA2
C-----CALCULATE PIPE AREAS AND PARAMETERS USED IN FRICTION FACTOR
      DO 20 I=1,IMAX
      RF(I)=RK/(3.7*DIA(I))
   20 AR(I)=.785*DIA(I)*DIA(I)
C-----INITIAL GUESS ON PRESSURE AND VELOCITY DISTRIBUTION USING
C     BERNOULLI'S EQUATION.
      FLO=AR(1)*SQRT(2*RHO*(P1-P2)/((AR(1)/AR(IMAX))**2-1))
      RESMAX=RESMAX*FLO
      PSLP=(P1-P2)/FLOAT(IM1)
      DO 30 I=1,IMAX
      P(I)=P1-PSLP*(I-1)
      DELP(I)=0.0
   30 U(I)=FLO/(RHO*AR(I))
  100 CONTINUE
      NITER=NITER+1
C-----USE MOMENTUM EQUATION TO OBTAIN U'S
      DO 40 I=2,IMAX
      RE=U(I)*DIA(I)/VISK
C-----FRICTION FACTOR FROM SWAMEE AND JAIN'S EQUATION
      F=1.325/(ALOG(RF(I)+5.74/RE**.9))**2
      FFAC(I)=1.+F*DX/(2*DIA(I))
   40 U(I)=U(I)*(1-RLX)+RLX*(U(I-1)+(P(I-1)-P(I))*AR(I)/FLO)/FFAC(I)
C-----SOLVE FOR PRESSURE CORRECTION USING TDMA
      RES=0.0
      DO 50 I=2,IM1
      A(I)=AR(I)/(U(I)*FFAC(I))
      C(I)=AR(I+1)/(U(I+1)*FFAC(I+1))
      B(I)=-A(I)-C(I)
      D(I)=RHO*U(I+1)*AR(I+1)-RHO*U(I)*AR(I)
   50 RES=RES+ABS(D(I))
C-----CHECK FOR CONVERGENCE
      IF(RES.LT.RESMAX) IEND=.TRUE.
      CALL TRIDIA(1,IMAX,DELP)
C-----CORRECT VELOCITIES AND PRESSURES
      DO 60 I=2,IMAX
      U(I)=U(I)+AR(I)*(DELP(I-1)-DELP(I))/FLO
      P(I)=P(I)+DELP(I)
C-----PRINTOUT IN KPA
   60 POUT(I)=P(I)/1000
      POUT(1)=P(1)/1000
C-----UPDATE FLOW RATE AND VELOCITY AT STATION ONE
```

```
      FLO=U(IMAX)*AR(IMAX)*RHO
      U(1)=FLO/(AR(1)*RHO)
C-----ITERATE AGAIN IF CONVERGENCE NOT ACHIEVED
      IF(.NOT.IEND) GO TO 100
C-----OUTPUT
      WRITE(1,1000) PDIF
      WRITE(1,1001) NITER,FLO,RE,RES
      WRITE(1,1002)
      WRITE(1,1003) (I,DIA(I),U(I),POUT(I),I=1,IMAX,ISKIP)
      STOP
 1000 FORMAT(' ***FLOW RATE, VELOCITY AND PRESSURE DISTRIBUTION FOR'/
     1'     FLOW THROUGH AN EXTENDED LENGTH VENTURI APPROACH SECTION'/
     2'     WITH A PRESSURE DIFFERENCE OF',1PE10.2,' KPA')
 1001 FORMAT(/'  *ITERATION NO=',I3,'    FLOW RATE=',1PE10.3,' KG/S'/
     1'   REYNOLDS NUMBER=',1PE10.2,'   RESIDUAL=',1PE10.2)
 1002 FORMAT(/'   I    DUCT DIA    VELOCITY    PRESSURE'/
     1'         M          M/S         KPA'/'')
 1003 FORMAT(I4,3X,F8.4,3X,F8.2,3X,F8.1)
      END
C
C*****TRIDIAGONAL SUBROUTINE WHERE 'IT' IS THE LAST LINE, 'IB' IS
C     THE FIRST LINE AND 'PHI' IS THE DEPENDENT VARIABLE. THE
C     NOMENCLATURE CORRESPONDS TO EQNS 16-48 THRU 16-53 IN THE
C     TEXT.
      SUBROUTINE TRIDIA(IB,IT,PHI)
      DIMENSION E(101),F(101),PHI(IT)
      COMMON/TDMA/A(101),B(101),C(101),D(101)
      IM1=IT-1
      IP1=IB+1
      IP2=IB+2
C-----SETTING UP BOUNDARY VALUES
      D(IP1)=D(IP1)-A(IP1)*PHI(IB)
      A(IP1)=0.0
      D(IM1)=D(IM1)-C(IM1)*PHI(IT)
      C(IM1)=0.0
      E(IP1+1)=-C(IP1)/B(IP1)
      F(IP1+1)=D(IP1)/B(IP1)
C-----EVALUATION OF E'S AND F'S
      DO 20 K=IP2,IM1
      FAC=1./(A(K)*E(K)+B(K))
      E(K+1)=-C(K)*FAC
   20 F(K+1)=(D(K)-A(K)*F(K))*FAC
C-----USING E'S AND F'S TO CALCULATE PHI
      PHI(IM1)=F(IT)
      DO 30 L=IP2,IM1
      LL=IT+IP1-L
   30 PHI(LL-1)=E(LL)*PHI(LL)+F(LL)
      RETURN
      END


  ***FLOW RATE, VELOCITY AND PRESSURE DISTRIBUTION FOR
    FLOW THROUGH AN EXTENDED LENGTH VENTURI APPROACH SECTION
    WITH A PRESSURE DIFFERENCE OF  1.00E 01 KPA

  *ITERATION NO= 19    FLOW RATE= 3.572E-01 KG/S
   REYNOLDS NUMBER=  4.55E 04    RESIDUAL=  3.50E-05

    I    DUCT DIA   VELOCITY    PRESSURE
         M          M/S         ·KPA
    1    0.0200     1.10        10.0
   11    0.0191     1.21         9.9
   21    0.0181     1.35         9.7
```

31	0.0171	1.52	9.4
41	0.0160	1.72	9.1
51	0.0150	1.96	8.6
61	0.0140	2.26	7.9
71	0.0129	2.64	6.9
81	0.0119	3.11	5.4
91	0.0109	3.73	3.1
101	0.0100	4.41	0.0

PROGRAM IV

```
C***********************************************************************
C                         PROGRAM IV                                  *
C     *PROGRAM FOR THE TIME HISTORY OF THE WATER SURFACE PROFILE IN A  *
C      BASIN OPEN TO A RESERVOIR WITH A SINUSOIDALLY VARYING WATER     *
C      SURFACE LEVEL. LAX-WENDORFF INTEGRATION SCHEME IS USED. INTER-  *
C      MEDIATE TIME LEVEL ASSIGNED TO SECOND ELEMENT OF DOUBLE SCRIPT- *
C      ED ARRAYS.                                                      *
C***********************************************************************
C
      DIMENSION A(101,2),B(101,2),C(101,2),H(101,2),V(101,2)
C-----ACCELERATION DUE TO GRAVITY
      GR=9.81
C-----NUMBER OF SPATIAL GRID LINES
      IMAX=101
      IM1=IMAX-1
C-----SKIP BETWEEN SPATIAL PRINTOUTS
      ISKIP=20
C-----INITIALIZE TIME TO ZERO
      TIME=0.0
C-----TIME INTERVAL BETWEEN PRINTOUT
      PINT=10.0
C-----TIME LIMIT
      TEND=100.
C-----LENGTH OF BASIN
      ALEN=50.
C-----WIDTH OF BASIN
      W=10.0
C-----INITIAL DEPTH OF WATER IN BASIN
      HI=5.0
C-----INITIAL VELOCITY IN BASIN
      VI=0.0
C-----LOCAL SHEAR STRESS COEFFICIENT
      CF=.005
C-----FREQUENCY OF RESERVOIR OSCILLATION (RAD/S)
      FREQ=.1
C-----AMPLITUDE OF RESERVOIR OSCILLATION
      AMP=.5
C-----GRID SPACING
      DX=ALEN/FLOAT(IM1)
      TPRT=TIME+PINT
C-----WRITE TITLE FOR PRINTOUT
      WRITE(1,1000) (I,I=1,IMAX,ISKIP)
C-----SET UP INITIAL VALUES FOR H,V,A,B AND C
      DO 10 I=1,IMAX
      H(I,1)=HI
      V(I,1)=VI
      A(I,1)=H(I,1)*V(I,1)
      B(I,1)=A(I,1)*V(I,1)+GR*H(I,1)*H(I,1)*.5
```

```
      10 C(I,1)=CF*ABS(V(I,1))*V(I,1)*.5*(1.+2*H(I,1)/W)
      20 CONTINUE
C-----ESTABLISH TIME STEP TO SATISFY COURANT CONDITION
         TCAL=0.0
         DO 30 I=1,IMAX
         VPGH=ABS(V(I,1))+SQRT(GR*H(I,1))
      30 TCAL=AMAX1(TCAL,VPGH)
         DT=.6*DX/TCAL
         DTDX=DT/DX
         DTDXI=.5*DTDX
C-----FIRST STEP TO INTERMEDIATE TIME LEVEL
         TIME=TIME+DT*.5
         DO 40 I=2,IM1
         H(I,2)=(H(I-1,1)+H(I+1,1)-DTDXI*(A(I+1,1)-A(I-1,1)))*.5
      40 A(I,2)=(A(I-1,1)+A(I+1,1)-DTDXI*(B(I+1,1)-B(I-1,1))-DT*C(I,1))*.5
C-----BOUNDARY VALUES AT INTERMEDIATE STEP
         H(1,2)=H(1,1)-DTDXI*(A(2,1)-A(1,1))
         H(IMAX,2)=HI+AMP*SIN(FREQ*TIME)
         A(1,2)=0.0
         A(IMAX,2)=A(IMAX,1)-DTDXI*(B(IMAX,1)-B(IM1,1))-C(IMAX,1)*DT*.5
C-----SET UP NEW VALUES FOR B AND C FOR LEAP FROG STEP
         DO 50 I=1,IMAX
         V(I,2)=A(I,2)/H(I,2)
         B(I,2)=A(I,2)*V(I,2)+GR*H(I,2)*H(I,2)*.5
      50 C(I,2)=CF*V(I,2)*ABS(V(I,2))*.5*(1.+H(I,2)/W)
C-----LEAP FROG STEP TO NEW TIME LEVEL
         TIME=TIME+DT*.5
         DO 60 I=2,IM1
         H(I,1)=H(I,1)-DTDX*(A(I+1,2)-A(I-1,2))*.5
      60 A(I,1)=A(I,1)-DTDX*(B(I+1,2)-B(I-1,2))*.5-DT*C(I,2)
C-----BOUNDARY VALUES AT NEW TIME LEVEL
         H(1,1)=H(1,1)-DTDX*(A(2,2)-A(1,2))
         H(IMAX,1)=HI+AMP*SIN(FREQ*TIME)
         A(1,1)=0.0
         A(IMAX,1)=A(IMAX,1)-DTDX*(B(IMAX,2)-B(IM1,2))-DT*C(IMAX,2)
C-----SET UP NEW VALUES FOR NEXT TIME STEP
         DO 70 I=1,IMAX
         V(I,1)=A(I,1)/H(I,1)
         B(I,1)=A(I,1)*V(I,1)+GR*H(I,1)*H(I,1)*.5
      70 C(I,1)=CF*ABS(V(I,1))*V(I,1)*.5*(1.+H(I,1)/W)
C-----CHECK FOR PRINTOUT
         IF(TIME-TPRT) 90,80,80
      80 WRITE(1,1001) TIME
         WRITE(1,1002) (H(I,1),I=1,IMAX,ISKIP)
         WRITE(1,1003) (V(I,1),I=1,IMAX,ISKIP)
         TPRT=TIME+PINT
      90 CONTINUE
C-----CHECK FOR TIME LIMIT
         IF(TIME.LT.TEND) GO TO 20
         STOP
    1000 FORMAT(' ***WATER SURFACE PROFILE OF BASIN OPENING ONTO'/
        1' RESERVOIR WITH SINUSOIDALLY VARYING WATER SURFACE LEVEL'
        2/' I=   ',12(3X,I4))
    1001 FORMAT(/' TIME=',F6.1)
    1002 FORMAT(' H(I) ',11F7.2)
    1003 FORMAT(' V(I) ',11F7.2)
         END

      ***WATER SURFACE PROFILE OF BASIN OPENING ONTO
      RESERVOIR WITH SINUSOIDALLY VARYING WATER SURFACE LEVEL
      I=        1     21     41     61     81    101
```

```
TIME=   10.0
H(I)    5.32    5.31    5.29    5.34    5.38    5.42
V(I)    0.00   -0.19   -0.39   -0.47   -0.53   -0.58

TIME=   20.0
H(I)    6.00    5.97    5.86    5.74    5.60    5.45
V(I)    0.00   -0.01    0.07    0.13    0.18    0.23

TIME=   30.1
H(I)    4.95    4.95    4.95    4.94    4.99    5.07
V(I)    0.00    0.22    0.44    0.64    0.78    0.86

TIME=   40.1
H(I)    4.48    4.48    4.49    4.48    4.54    4.62
V(I)    0.00   -0.01   -0.02   -0.05    0.00    0.07

TIME=   50.1
H(I)    4.63    4.59    4.55    4.52    4.52    4.52
V(I)    0.00    0.00    0.02    0.01   -0.01   -0.04

TIME=   60.1
H(I)    4.63    4.64    4.65    4.68    4.76    4.87
V(I)    0.00   -0.09   -0.17   -0.24   -0.34   -0.46

TIME=   70.2
H(I)    5.58    5.56    5.52    5.43    5.37    5.33
V(I)    0.00   -0.19   -0.37   -0.59   -0.74   -0.82

TIME=   80.2
H(I)    5.81    5.79    5.74    5.66    5.58    5.49
V(I)    0.00    0.20    0.30    0.38    0.44    0.47

TIME=   90.2
H(I)    4.98    4.98    5.01    5.08    5.15    5.20
V(I)    0.00    0.11    0.20    0.22    0.26    0.34
```

FIGURE A.1

Centroids and moments of inertia of plane areas

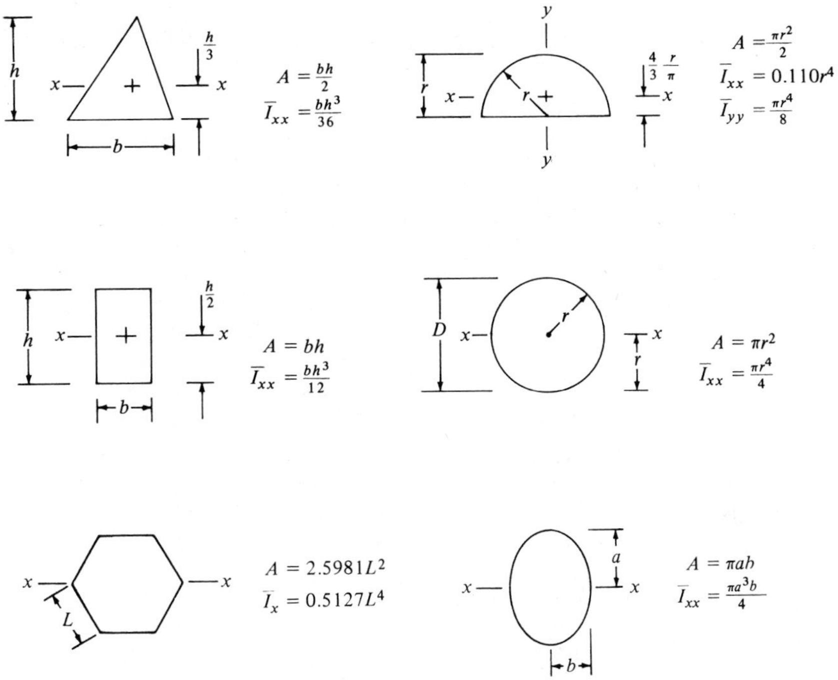

Triangle:

$$A = \frac{bh}{2}$$

$$\bar{I}_{xx} = \frac{bh^3}{36}$$

Semicircle:

$$A = \frac{\pi r^2}{2}$$

$$\bar{I}_{xx} = 0.110 r^4$$

$$\bar{I}_{yy} = \frac{\pi r^4}{8}$$

Rectangle:

$$A = bh$$

$$\bar{I}_{xx} = \frac{bh^3}{12}$$

Circle:

$$A = \pi r^2$$

$$\bar{I}_{xx} = \frac{\pi r^4}{4}$$

Hexagon:

$$A = 2.5981 L^2$$

$$\bar{I}_x = 0.5127 L^4$$

Ellipse:

$$A = \pi ab$$

$$\bar{I}_{xx} = \frac{\pi a^3 b}{4}$$

Volume and Area Formulas:

$$A_{\text{circle}} = \pi r^2 = \pi D^2 / 4$$

$$A_{\text{sphere surface}} = \pi D^2$$

$$V_{\text{sphere}} = \frac{1}{6} \pi D^3$$

TABLE A.1   COMPRESSIBLE FLOW TABLES FOR AN IDEAL GAS WITH $k = 1.4$

M or M_1 = local number or Mach number upstream of a normal shock wave; p/p_t = ratio of static pressure to total pressure; ρ/ρ_t = ratio of static density to total density; T/T_t = ratio of static temperature to total temperature; A/A_* = ratio of local cross-sectional area of an isentropic stream tube to cross-sectional area at the point where M = 1; M_2 = Mach number downstream of a normal shock wave; p_2/p_1 = static-pressure ratio across a normal shock; T_2/T_1 = static-temperature ratio across a normal shock wave; p_{t_2}/p_{t_1} = total pressure ratio across normal shock wave.

		Subsonic Flow		
M	p/p_t	ρ/ρ_t	T/T_t	A/A_*
0.00	1.0000	1.0000	1.0000	∞
0.05	0.9983	0.9988	0.9995	11.5914
0.10	0.9930	0.9950	0.9980	5.8218
0.15	0.9844	0.9888	0.9955	3.9103
0.20	0.9725	0.9803	0.9921	2.9630
0.25	0.9575	0.9694	0.9877	2.4027
0.30	0.9395	0.9564	0.9823	2.0351
0.35	0.9188	0.9413	0.9761	1.7780
0.40	0.8956	0.9243	0.9690	1.5901
0.45	0.8703	0.9055	0.9611	1.4487
0.50	0.8430	0.8852	0.9524	1.3398
0.52	0.8317	0.8766	0.9487	1.3034
0.54	0.8201	0.8679	0.9449	1.2703
0.56	0.8082	0.8589	0.9410	1.2403
0.58	0.7962	0.8498	0.9370	1.2130
0.60	0.7840	0.8405	0.9328	1.1882
0.62	0.7716	0.8310	0.9286	1.1657
0.64	0.7591	0.8213	0.9243	1.1452
0.66	0.7465	0.8115	0.9199	1.1265
0.68	0.7338	0.8016	0.9153	1.1097
0.70	0.7209	0.7916	0.9107	1.0944
0.72	0.7080	0.7814	0.9061	1.0806
0.74	0.6951	0.7712	0.9013	1.0681
0.76	0.6821	0.7609	0.8964	1.0570
0.78	0.6691	0.7505	0.8915	1.0471
0.80	0.6560	0.7400	0.8865	1.0382
0.82	0.6430	0.7295	0.8815	1.0305
0.84	0.6300	0.7189	0.8763	1.0237
0.86	0.6170	0.7083	0.8711	1.0179
0.88	0.6041	0.6977	0.8659	1.0129
0.90	0.5913	0.6870	0.8606	1.0089
0.92	0.5785	0.6764	0.8552	1.0056
0.94	0.5658	0.6658	0.8498	1.0031
0.96	0.5532	0.6551	0.8444	1.0014
0.98	0.5407	0.6445	0.8389	1.0003
1.00	0.5283	0.6339	0.8333	1.0000

TABLE A.1 COMPRESSIBLE FLOW TABLES FOR AN IDEAL GAS WITH $k = 1.4$ (CONTINUED)								
Supersonic Flow					**Normal Shock Wave**			
M_1	p/p_t	ρ/ρ_t	T/T_t	A/A_*	M_2	p_2/p_1	T_2/T_1	p_{t_2}/p_{t_1}
1.00	0.5283	0.6339	0.8333	1.000	1.000	1.000	1.000	1.0000
1.01	0.5221	0.6287	0.8306	1.000	0.9901	1.023	1.007	0.9999
1.02	0.5160	0.6234	0.8278	1.000	0.9805	1.047	1.013	0.9999
1.03	0.5099	0.6181	0.8250	1.001	0.9712	1.071	1.020	0.9999
1.04	0.5039	0.6129	0.8222	1.001	0.9620	1.095	1.026	0.9999
1.05	0.4979	0.6077	0.8193	1.002	0.9531	1.120	1.033	0.9998
1.06	0.4919	0.6024	0.8165	1.003	0.9444	1.144	1.039	0.9997
1.07	0.4860	0.5972	0.8137	1.004	0.9360	1.169	1.046	0.9996
1.08	0.4800	0.5920	0.8108	1.005	0.9277	1.194	1.052	0.9994
1.09	0.4742	0.5869	0.8080	1.006	0.9196	1.219	1.059	0.9992
1.10	0.4684	0.5817	0.8052	1.008	0.9118	1.245	1.065	0.9989
1.11	0.4626	0.5766	0.8023	1.010	0.9041	1.271	1.071	0.9986
1.12	0.4568	0.5714	0.7994	1.011	0.8966	1.297	1.078	0.9982
1.13	0.4511	0.5663	0.7966	1.013	0.8892	1.323	1.084	0.9978
1.14	0.4455	0.5612	0.7937	1.015	0.8820	1.350	1.090	0.9973
1.15	0.4398	0.5562	0.7908	1.017	0.8750	1.376	1.097	0.9967
1.16	0.4343	0.5511	0.7879	1.020	0.8682	1.403	1.103	0.9961
1.17	0.4287	0.5461	0.7851	1.022	0.8615	1.430	1.109	0.9953
1.18	0.4232	0.5411	0.7822	1.025	0.8549	1.458	1.115	0.9946
1.19	0.4178	0.5361	0.7793	1.026	0.8485	1.485	1.122	0.9937
1.20	0.4124	0.5311	0.7764	1.030	0.8422	1.513	1.128	0.9928
1.21	0.4070	0.5262	0.7735	1.033	0.8360	1.541	1.134	0.9918
1.22	0.4017	0.5213	0.7706	1.037	0.8300	1.570	1.141	0.9907
1.23	0.3964	0.5164	0.7677	1.040	0.8241	1.598	1.147	0.9896
1.24	0.3912	0.5115	0.7648	1.043	0.8183	1.627	1.153	0.9884
1.25	0.3861	0.5067	0.7619	1.047	0.8126	1.656	1.159	0.9871
1.30	0.3609	0.4829	0.7474	1.066	0.7860	1.805	1.191	0.9794
1.35	0.3370	0.4598	0.7329	1.089	0.7618	1.960	1.223	0.9697
1.40	0.3142	0.4374	0.7184	1.115	0.7397	2.120	1.255	0.9582
1.45	0.2927	0.4158	0.7040	1.144	0.7196	2.286	1.287	0.9448
1.50	0.2724	0.3950	0.6897	1.176	0.7011	2.458	1.320	0.9278
1.55	0.2533	0.3750	0.6754	1.212	0.6841	2.636	1.354	0.9132
1.60	0.2353	0.3557	0.6614	1.250	0.6684	2.820	1.388	0.8952
1.65	0.2184	0.3373	0.6475	1.292	0.6540	3.010	1.423	0.8760
1.70	0.2026	0.3197	0.6337	1.338	0.6405	3.205	1.458	0.8557
1.75	0.1878	0.3029	0.6202	1.386	0.6281	3.406	1.495	0.8346
1.80	0.1740	0.2868	0.6068	1.439	0.6165	3.613	1.532	0.8127
1.85	0.1612	0.2715	0.5936	1.495	0.6057	3.826	1.569	0.7902
1.90	0.1492	0.2570	0.5807	1.555	0.5956	4.045	1.608	0.7674
1.95	0.1381	0.2432	0.5680	1.619	0.5862	4.270	1.647	0.7442
2.00	0.1278	0.2300	0.5556	1.688	0.5774	4.500	1.688	0.7209
2.10	0.1094	0.2058	0.5313	1.837	0.5613	4.978	1.770	0.6742

| TABLE A.1 COMPRESSIBLE FLOW TABLES FOR AN IDEAL GAS WITH $k = 1.4$ (CONTINUED) |||||||||
| Supersonic Flow | | | | | Normal Shock Wave | | | |
M_1	p/p_t	ρ/ρ_t	T/T_t	A/A_*	M_2	p_2/p_1	T_2/T_1	p_{t_2}/p_{t_1}
2.20	0.9352^{-1}†	0.1841	0.5081	2.005	0.5471	5.480	1.857	0.6281
2.30	0.7997^{-1}	0.1646	0.4859	2.193	0.5344	6.005	1.947	0.5833
2.50	0.5853^{-1}	0.1317	0.4444	2.637	0.5130	7.125	2.138	0.4990
2.60	0.5012^{-1}	0.1179	0.4252	2.896	0.5039	7.720	2.238	0.4601
2.70	0.4295^{-1}	0.1056	0.4068	3.183	0.4956	8.338	2.343	0.4236
2.80	0.3685^{-1}	0.9463^{-1}	0.3894	3.500	0.4882	8.980	2.451	0.3895
2.90	0.3165^{-1}	0.8489^{-1}	0.3729	3.850	0.4814	9.645	2.563	0.3577
3.00	0.2722^{-1}	0.7623^{-1}	0.3571	4.235	0.4752	10.33	2.679	0.3283
3.50	0.1311^{-1}	0.4523^{-1}	0.2899	6.790	0.4512	14.13	3.315	0.2129
4.00	0.6586^{-2}	0.2766^{-1}	0.2381	10.72	0.4350	18.50	4.047	0.1388
4.50	0.3455^{-2}	0.1745^{-1}	0.1980	16.56	0.4236	23.46	4.875	0.9170^{-1}
5.00	0.1890^{-2}	0.1134^{-1}	0.1667	25.00	0.4152	29.00	5.800	0.6172^{-1}
5.50	0.1075^{-2}	0.7578^{-2}	0.1418	36.87	0.4090	35.13	6.822	0.4236^{-1}
6.00	0.6334^{-2}	0.5194^{-2}	0.1220	53.18	0.4042	41.83	7.941	0.2965^{-1}
6.50	0.3855^{-2}	0.3643^{-2}	0.1058	75.13	0.4004	49.13	9.156	0.2115^{-1}
7.00	0.2416^{-3}	0.2609^{-2}	0.9259^{-1}	104.1	0.3974	57.00	10.47	0.1535^{-1}
7.50	0.1554^{-3}	0.1904^{-2}	0.8163^{-1}	141.8	0.3949	65.46	11.88	0.1133^{-1}
8.00	0.1024^{-3}	0.1414^{-2}	0.7246^{-1}	190.1	0.3929	74.50	13.39	0.8488^{-2}
8.50	0.6898^{-4}	0.1066^{-2}	0.6472^{-1}	251.1	0.3912	84.13	14.99	0.6449^{-2}
9.00	0.4739^{-4}	0.8150^{-3}	0.5814^{-1}	327.2	0.3898	94.33	16.69	0.4964^{-2}
9.50	0.3314^{-4}	0.6313^{-3}	0.5249^{-1}	421.1	0.3886	105.1	18.49	0.3866^{-2}
10.00	0.2356^{-4}	0.4948^{-3}	0.4762^{-1}	535.9	0.3876	116.5	20.39	0.3045^{-2}

†x^{-n} means $x \cdot 10^{-n}$

SOURCE: Abridged with permission from R. E. Bolz and G. L. Tuve, *The Handbook of Tables for Applied Engineering Sciences,* CRC Press, Inc., Cleveland, 1973. Copyright 1973 by The Chemical Rubber Co., CRC Press, Inc.

FIGURE A.2

Absolute viscosities of certain gases and liquids [Adapted from Fluid Mechanics, *5th ed., by V. L. Streeter. Copyright © 1971, McGraw-Hill Book Company, New York. Used with permission of the McGraw-Hill Book Company.]*

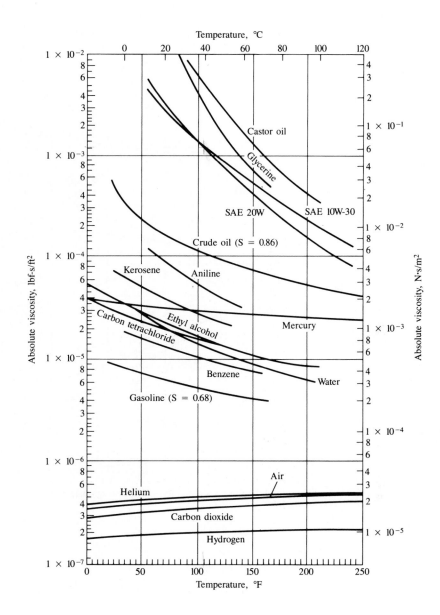

FIGURE A.3

*Kinematic viscosities of
certain gases and liquids.
The gases are at standard
pressure. [Adapted from
Fluid Mechanics, 5th ed.,
by V. L. Streeter.
Copyright © 1971,
McGraw-Hill Book
Company, New York. Used
with permission of the
McGraw-Hill Book
Company.]*

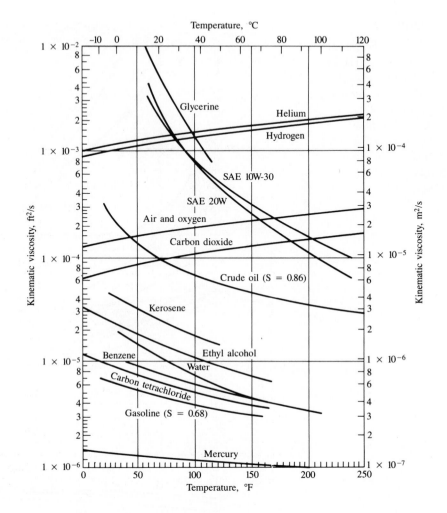

Gas	Density, kg/m³ (slugs/ft³)	Kinematic viscosity, m²/s (ft²/s)	R Gas constant, J/kg K (ft-lbf/slug-°R)	c_p $\dfrac{\text{J}}{\text{kg K}}$ $\left(\dfrac{\text{Btu}}{\text{lbm-°R}}\right)$	$k = \dfrac{c_p}{c_v}$
Air	1.22 (0.00237)	1.46×10^{-5} (1.58×10^{-4})	287 (1716)	1004 (0.240)	1.40
Carbon dioxide	1.85 (0.0036)	7.84×10^{-6} (8.48×10^{-5})	189 (1130)	841 (0.201)	1.30
Helium	0.169 (0.00033)	1.14×10^{-4} (1.22×10^{-3})	2077 (12,419)	5187 (1.24)	1.66
Hydrogen	0.0851 (0.00017)	1.01×10^{-4} (1.09×10^{-3})	4127 (24,677)	14,223 (3.40)	1.41
Methane (natural gas)	0.678 (0.0013)	1.59×10^{-5} (1.72×10^{-4})	518 (3098)	2208 (0.528)	1.31
Nitrogen	1.18 (0.0023)	1.45×10^{-5} (1.56×10^{-4})	297 (1776)	1041 (0.249)	1.40
Oxygen	1.35 (0.0026)	1.50×10^{-5} (1.61×10^{-4})	260 (1555)	916 (0.219)	1.40

TABLE A.2 PHYSICAL PROPERTIES OF GASES AT STANDARD ATMOSPHERIC PRESSURE AND 15°C (59°F)

SOURCES: V. L. Streeter (ed.), *Handbook of Fluid Dynamics,* McGraw-Hill Book Company, New York, 1961; also R. E. Bolz and G. L. Tuve, *Handbook of Tables for Applied Engineering Science,* CRC Press, Inc., Cleveland, 1973; and *Handbook of Chemistry and Physics,* Chemical Rubber Company, 1951.

TABLE A.3 MECHANICAL PROPERTIES OF AIR AT STANDARD ATMOSPHERIC PRESSURE				
Temperature	ρ Density	γ Specific weight	μ Dynamic viscosity	ν Kinematic viscosity
	kg/m³	N/m³	N · s/m²	m²/s
−20°C	1.40	13.7	1.61×10^{-5}	1.16×10^{-5}
−10°C	1.34	13.2	1.67×10^{-5}	1.24×10^{-5}
0°C	1.29	12.7	1.72×10^{-5}	1.33×10^{-5}
10°C	1.25	12.2	1.76×10^{-5}	1.41×10^{-5}
20°C	1.20	11.8	1.81×10^{-5}	1.51×10^{-5}
30°C	1.17	11.4	1.86×10^{-5}	1.60×10^{-5}
40°C	1.13	11.1	1.91×10^{-5}	1.69×10^{-5}
50°C	1.09	10.7	1.95×10^{-5}	1.79×10^{-5}
60°C	1.06	10.4	2.00×10^{-5}	1.89×10^{-5}
70°C	1.03	10.1	2.04×10^{-5}	1.99×10^{-5}
80°C	1.00	9.81	2.09×10^{-5}	2.09×10^{-5}
90°C	0.97	9.54	2.13×10^{-5}	2.19×10^{-5}
100°C	0.95	9.28	2.17×10^{-5}	2.29×10^{-5}
120°C	0.90	8.82	2.26×10^{-5}	2.51×10^{-5}
140°C	0.85	8.38	2.34×10^{-5}	2.74×10^{-5}
160°C	0.81	7.99	2.42×10^{-5}	2.97×10^{-5}
180°C	0.78	7.65	2.50×10^{-5}	3.20×10^{-5}
200°C	0.75	7.32	2.57×10^{-5}	3.44×10^{-5}
	slugs/ft³	lbf/ft³	lbf-s/ft²	ft²/s
0°F	0.00269	0.0866	3.39×10^{-7}	1.26×10^{-4}
20°F	0.00257	0.0828	3.51×10^{-7}	1.37×10^{-4}
40°F	0.00247	0.0794	3.63×10^{-7}	1.47×10^{-4}
60°F	0.00237	0.0764	3.74×10^{-7}	1.58×10^{-4}
80°F	0.00228	0.0735	3.85×10^{-7}	1.69×10^{-4}
100°F	0.00220	0.0709	3.96×10^{-7}	1.80×10^{-4}
120°F	0.00213	0.0685	4.07×10^{-7}	1.91×10^{-4}
150°F	0.00202	0.0651	4.23×10^{-7}	2.09×10^{-4}
200°F	0.00187	0.0601	4.48×10^{-7}	2.40×10^{-4}
300°F	0.00162	0.0522	4.96×10^{-7}	3.05×10^{-4}
400°F	0.00143	0.0462	5.40×10^{-7}	3.77×10^{-4}

Liquid and temperature	Density kg/m^3 (slugs/ft³)	Specific gravity (S) water at 4°C is ref.	Specific weight, N/m^3 (lbf/ft³)	Dynamic viscosity, $N \cdot s/m^2$ (lbf-s/ft²)	Kinematic viscosity, m^2/s (ft²/s)	Surface tension, $N/m*$ (lbf/ft)
Ethyl alcohol[3][1] 20°C (68°F)	799 (1.55)	0.79	7,850 (50.0)	1.2×10^{-3} (2.5×10^{-5})	1.5×10^{-6} (1.6×10^{-5})	2.2×10^{-2} (1.5×10^{-3})
Carbon tetrachloride[3] 20°C (68°F)	1,590 (3.09)	1.59	15,600 (99.5)	9.6×10^{-4} (2.0×10^{-5})	6.0×10^{-7} (6.5×10^{-6})	2.6×10^{-2} (1.8×10^{-3})
Glycerine[3] 20°C (68°F)	1,260 (2.45)	1.26	12,300 (78.5)	6.2×10^{-1} (1.3×10^{-2})	5.1×10^{-4} (5.3×10^{-3})	6.3×10^{-2} (4.3×10^{-3})
Kerosene[2][1] 20°C (68°F)	814 (1.58)	0.81	8,010 (51)	1.9×10^{-3} (4×10^{-5})	2.37×10^{-6} (2.55×10^{-5})	2.9×10^{-2} (2.0×10^{-3})
Mercury[3][1] 20°C (68°F)	13,550 (26.3)	13.55	133,000 (847)	1.5×10^{-3} (3.2×10^{-5})	1.2×10^{-7} (1.3×10^{-6})	4.8×10^{-1} (3.3×10^{-2})
Sea water 10°C at 3.3% salinity	1,026 (1.99)	1.03	10,070 (64.1)	1.4×10^{-3} (3×10^{-5})	1.4×10^{-6} (1.5×10^{-5})	
Oils — 38°C (100°F) SAE 10W[4]	870 (1.69)	0.87	8,530 (54.4)	3.6×10^{-2} (7.4×10^{-4})	4.1×10^{-5} (4.4×10^{-4})	
SAE 10W-30[4]	880 (1.71)	0.88	8,630 (55.1)	6.7×10^{-2} (1.4×10^{-3})	7.6×10^{-5} (8.2×10^{-4})	
SAE 30[4]	880 (1.71)	0.88	8,630 (55.1)	1.0×10^{-1} (2.0×10^{-3})	1.1×10^{-4} (1.2×10^{-3})	

* Liquid–air surface tension values.

SOURCES: (1) V. L. Streeter, *Handbook of Fluid Dynamics*, McGraw-Hill Book Company, New York, 1961; (2) V. L. Streeter, *Fluid Mechanics*, 4th ed., McGraw-Hill Book Company, New York, 1966; (3) J. Vennard, *Elementary Fluid Mechanics*, 4th ed., John Wiley & Sons, Inc., New York, 1961; (4) R. E. Bolz and G. L. Tuve, *Handbook of Tables for Applied Engineering Sciences*, CRC Press, Inc., Cleveland, 1973.

TABLE A.5 APPROXIMATE PHYSICAL PROPERTIES OF WATER* AT ATMOSPHERIC PRESSURE					
Temperature	Density	Specific weight	Dynamic viscosity	Kinematic viscosity	Vapor pressure
	kg/m^3	N/m^3	$N \cdot s/m^2$	m^2/s	N/m^2 abs.
0°C	1000	9810	1.79×10^{-3}	1.79×10^{-6}	611
5°C	1000	9810	1.51×10^{-3}	1.51×10^{-6}	872
10°C	1000	9810	1.31×10^{-3}	1.31×10^{-6}	1230
15°C	999	9800	1.14×10^{-3}	1.14×10^{-6}	1700
20°C	998	9790	1.00×10^{-3}	1.00×10^{-6}	2340
25°C	997	9781	8.91×10^{-4}	8.94×10^{-7}	3170
30°C	996	9771	7.97×10^{-4}	8.00×10^{-7}	4250
35°C	994	9751	7.20×10^{-4}	7.24×10^{-7}	5630
40°C	992	9732	6.53×10^{-4}	6.58×10^{-7}	7380
50°C	988	9693	5.47×10^{-4}	5.53×10^{-7}	12,300
60°C	983	9643	4.66×10^{-4}	4.74×10^{-7}	20,000
70°C	978	9594	4.04×10^{-4}	4.13×10^{-7}	31,200
80°C	972	9535	3.54×10^{-4}	3.64×10^{-7}	47,400
90°C	965	9467	3.15×10^{-4}	3.26×10^{-7}	70,100
100°C	958	9398	2.82×10^{-4}	2.94×10^{-7}	101,300
	$slugs/ft^3$	lbf/ft^3	$lbf\text{-}s/ft^2$	ft^2/s	psia
40°F	1.94	62.43	3.23×10^{-5}	1.66×10^{-5}	0.122
50°F	1.94	62.40	2.73×10^{-5}	1.41×10^{-5}	0.178
60°F	1.94	62.37	2.36×10^{-5}	1.22×10^{-5}	0.256
70°F	1.94	62.30	2.05×10^{-5}	1.06×10^{-5}	0.363
80°F	1.93	62.22	1.80×10^{-5}	0.930×10^{-5}	0.506
100°F	1.93	62.00	1.42×10^{-5}	0.739×10^{-5}	0.949
120°F	1.92	61.72	1.17×10^{-5}	0.609×10^{-5}	1.69
140°F	1.91	61.38	0.981×10^{-5}	0.514×10^{-5}	2.89
160°F	1.90	61.00	0.838×10^{-5}	0.442×10^{-5}	4.74
180°F	1.88	60.58	0.726×10^{-5}	0.385×10^{-5}	7.51
200°F	1.87	60.12	0.637×10^{-5}	0.341×10^{-5}	11.53
212°F	1.86	59.83	0.593×10^{-5}	0.319×10^{-5}	14.70

*Notes: (1) Bulk modulus E_v of water is approximately 2.2 G P_a (3.2×10^5 psi); (2) Water–air surface tension is approximately 7.3×10^{-2} N/m (5×10^{-3} lbf/ft) from 10°C to 50°C.

SOURCE: Reprinted with permission from R. E. Bolz and G. L. Tuve, *Handbook of Tables for Applied Engineering Science,* CRC Press, Inc., Cleveland, 1973: Copyright 1973 by The Chemical Rubber Co., CRC Press, Inc.

Answers to Selected Problems

Chapter 2

2.2 $\rho_{air} = 1.61$ kg/m^3; $\rho_{He} = 0.22$ kg/m^3; $\rho_{CO_2} = 2.44$ kg/m^3

2.4 $\gamma_{He} = 4.26$ N/m^3 2.6 $\rho_W/\rho_a = 872$

2.8 $\gamma_{air} = 37.92$ N/m^3; $\rho_{air} = 3.87$ kg/m^3

2.10 $\nu_a = 1.39 \times 10^{-5}$ m^2/s; $\nu_w = 1.31 \times 10^{-6}$ m^2/s

2.14 $\Delta\nu_a = 3.8 \times 10^{-6}$ m^2/s   2.16 $\nu_a/\nu_w = 15.1$

2.18 $\tau = 0.296$ N/m^2 2.20 $\tau_{max} = 10$ N/m^2

2.24 $u_t = \frac{1}{2}\mu H^2\, dp/ds$ 2.26 $V_{fall} = 0.23$ m/s

2.28 (a) $\tau_2/\tau_3 = 0.667$; (b) $V = 0.06$ m/s; (c) $\tau = 0.30$ N/m^2

2.30 $T = (1/16)\pi\mu\omega D^4/S$ 2.32 $\Delta p = 22$ MN/m^2

2.34 $\Delta p = 97.3$ N/m^2 2.36 $\Delta h_{1/32} = 1.48$ in.

2.38 $p = 292$ N/m^2 2.40 $T = 89.7°C$

Chapter 3

3.2 $F = 530$ N   3.4 $n = 5$ 3.6 "a" is the correct choice

3.8 $p_{abs}/p_{atm} = 4.87$ 3.10 $p = 16.82$ kPa   3.12 $t = 26,615$ ft

3.14 $p = 103.0$ kPa   3.16 $p_{max} = 127.5$ kPa (gage)

3.18 $\Psi = 33.7$ in.^{3} 3.20 $\rho_{20}/\rho_{10} = 1.23$

3.22 $p_{atm} = 104.5$ kPa abs.   3.24 $p_A = -4.81$ psi

3.26 $p_A = 885$ Pa   3.28 $p = 743$ Pa   3.30 $\alpha = 17.3°$

3.32 $p_A = 5.72$ psi $= 39.5$ kPa   3.34 $p_{max} = 172$ psf

3.38 $p_A - p_B = -27.4$ kPa   3.40 $z = 0.594$ ft

3.42 $p_A = 90.45$ kPa   3.44 $p_A - p_B = 4.17$ kPa; $h_A - h_B = -0.50$ m

3.46 $p_A - p_B = 119$ psf; $h_A - h_B = 3.74$ ft   3.48 $d = 4.60$ m

3.52 $T_{3000} = 90°C$ 3.56 28 breaths/min   3.58 $z = 10,430$ ft

3.60 $F = 11,520$ lbf   3.62 $F = 104.4$ kN   3.64 $F = 2384$ kN

3.66 $F_A = 22,338$ lbf   3.68 $T = 545$ kN $\cdot$ m   3.70 $F = 370$ kN

3.72 $h = 1.167l$ 3.74 $T = 26,520$ ft $\cdot$ lbf   3.76 Gate will fall

3.78 $P = 324$ kN   3.80 $R_A = 0.51\ \gamma Wl^2$ 3.82 $P = 95.31$ N

3.84 $M = 18.0$ kN $\cdot$ m/m   3.86 $W = 115.1$ kN   3.88 $h = 6.93$ ft

3.90 $\Psi = 4.74$ ft^3 3.94 $S = 7.82$ 3.96 $\Psi = 0.0309$ m^3

3.98 Δ pond level $= -0.144$ in.   3.100 $h/L = 0.195$

3.102 $L = 2.24$ m   3.104 $d = 3.70$ m   3.106 $\Delta d = 0.0255$ ft

3.108 It will fall   3.110 3011 N

3.114 $\mathbf{F} = (3.12 \times 10^6\ \mathbf{i} + 5.65 \times 10^5\ \mathbf{j})$ lbf

3.116 $\mathbf{F} = (375\ \mathbf{i} - 503\ \mathbf{j})$ lbf   3.118 $F = 27,877$ lbf downward

3.120 $F = 22,518$ lbf in tension   3.122 $\mathbf{F} = (-61.64\ \mathbf{i} - 20.55\ \mathbf{j})$ kN

3.124 $W = 0.0368$ lbf   3.126 $W = 1.499 \times 10^{-2}$ N

3.128 $S = 1.149$ 3.130 Upright   3.132 No

3.134 Not stable about longitudinal axis

Chapter 4

4.14 $Q = 9.425$ cfs $= 4232$ gpm   4.16 $\dot{m} = 0.231$ kg/s

4.18 $D = 1.25$ m   4.20 $\overline{V}/V_0 = 1/3$ 4.22 $Q = 163$ cfs

4.24 $\dot{m} = 6.5$ kg/s   4.26 $Q = 1.86$ m³/s

4.28 $Q = 2.62 \times 10^{-3}$ m³/s   4.30 $V = 2.57$ m/s

4.34 $V = 2.45$ ft/s   4.36 $Q = 63.5$ gpm   4.38 $V_C = 20$ ft/s

4.44 $a = 14.14$ m/s²   4.46 $a_c = 1770$ ft/s²; $a_l = 0$

4.48 $a_l = (8/3)(q_0/t_0)/B$ 4.50 $a_l = 3.56$ ft/s²; $a_c = 37.9$ ft/s²

4.52 $a_l = 7.96$ m/s²; $a_{c,2s} = -1258$ m/s²; $a_{c,3s} = -2842$ m/s²

4.60 $V = 1.29$ ft/s (upward)   4.62 $V = 4.88$ ft/s (upward)

4.64 $Q_{20} = 0.192$ m³/s; $Q_{15} = 0.108$ m³/s

4.66 $V_6 = 5.09$ ft/s; $V_8 = 2.86$ ft/s   4.68 $V_2 = 86.8$ m/s

4.70 $V_{\text{fall}} = 0.0247$ ft/s   4.72 $h_{\text{eq}} = 11.3$ ft; $Q_{\text{dry}} = 67.5$ cfs

4.74 $h = 10.07$ ft   4.76 $V = 5.46$ m/s   4.78 $t = 5$ hr and 50 min.

4.80 $t = 18.4$ s   4.82 Continuity satisfied; irrotational

4.84 Continuity is not satisfied   4.86 $\mathbf{a} = (4\mathbf{i} + 4\mathbf{j})$ m/s²

4.88 $v = \frac{1}{2}A(x^2 - y^2)$ 4.90 $\rho_e = 0.073$ kg/m³

4.92 $d\rho/dt = 249.7$ kg/m³/s

Chapter 5

5.2 $p_2 = 187$ psf   5.4 $\partial p/\partial s = -3000$ N/m³

5.6 $\partial p/\partial z = -59.9$ lbf/ft³   5.8 $a = 116$ ft/s²

5.10 $dp/dx = -4365$ psf/ft   5.12 $\partial p/\partial x = -1260$ Pa/m

5.14 $dp/dx = -4.94\rho$ Pa/m   5.18 $a = 7.36$ m/s²

5.20 $p_C - p_A = 44.15$ kPa; $p_B - p_A = 14.71$ kPa

5.22 $d = (5/6)$ full; $V = 12.8$ m/s   5.24 $p_{\text{max}} = 34.8$ kPa gage

5.26 $p_B = 3.62$ psi   5.28 $S = 0.920$ 5.30 Elev. $= 12.2$ cm

5.32 $p_A = 149.3$ Pa   5.34 $\omega = 9.83$ rad/s   5.36 $\omega = \sqrt{2g/3l}$

5.38 $p_A = -19.94$ kPa gage, $p_B = 6.278$ kPa gage; $p_A = 0$ abs,
    $p_B = 13.18$ kPa gage

5.40 $z = 5.59$ m   5.44 $p_A = 5.36$ psi   5.46 $p_A - p_B = 246.7$ psf

5.48 $p = 65.4$ psi   5.50 $q = 1.63$ m²/s   5.52 $z_2 - z_1 = 0.038$ m

5.54 $p_E = -5.28$ psf   5.56 $\Delta h = 34.0$ cm   5.58 $V = 245$ ft/s

5.60 $V = 6.95$ ft/s   5.64 $V_2 = 80.8$ m/s   5.66 $\theta = 70.5°$

5.68 $Q = 0.157$ m³/s   5.70 $V = 24.3$ ft/s   5.72 $V = 91.9$ ft/s

5.76 $p_A - p_B = 6.195$ kPa   5.78 $V_0 = 13.92$ m/s; $Q = 0.1093$ m³/s

5.80 $V = 8.83$ m/s   5.82 $V = 69.3$ m/s   5.84 $p_{\text{min}} = 6.56$ kPa

5.86 $p_2 - p_1 = 1.76$ kPa   5.88 $V_0 = 11.4$ m/s   5.90 $V_0 = 42.2$ ft/s

5.92 $V_0 = 11.7$ m/s

Chapter 6

6.2 $V = 82.5$ ft/s   6.4 $d = 15.45$ cm   6.6 $T = 788$ lbf

6.10 $F_x = -286$ lbf; $F_y = -74.2$ lbf   6.14 $\mathbf{F} = (1645\,\mathbf{i} + 373\,\mathbf{j})$ N

6.16 $F_x = 5.68$ kN   6.20 Reverse thrust $= 155$ kN

6.22 $d_2 = (b_1/2)(1 - \cos\theta)$; $d_3 = (b_1/2)(1 + \cos\theta)$

6.24 $a = 88.9$ m/s^2 6.28 $F = 70.7$ N   6.32 $F = -1843$ lbf

6.36 $\mathbf{F} = (-27.2\,\mathbf{i} + 1.48\,\mathbf{k})$ kN

6.38 $F_x = -284$ kN   6.40 $F_x = -12{,}820$ lbf

6.42 $F_x = -3179$ lbf   6.44 $F_z = -7167$ lbf

6.46 $\mathbf{F} = (-18.24\,\mathbf{i} - 7.56\,\mathbf{j})$ kN   6.48 $F = -1138$ N

6.50 $\mathbf{F} = (86.4\,\mathbf{i} - 29.2\,\mathbf{j} + 26.0\,\mathbf{k})$ kN   6.52 $p_1 = 78.8$ kPa

6.54 $F_x = -8.03$ lbf   6.56 $\mathbf{F} = (-3.32\,\mathbf{i} - 3.41\,\mathbf{j})$ kN

6.58 $F_x = -434$ lbf   6.60 $F_x = -3622$ lbf   6.62 $F_{\text{bolt}} = 1979$ N

6.64 $F = 159$ N   6.66 $F = 1.38$ kN   6.68 $F = 8167$ N

6.70 $F = 1148$ lbf   6.72 $\mathbf{F} = (-523\,\mathbf{i} - 58.9\,\mathbf{j})$ lbf

6.74 $\mathbf{F} = (-36.8\,\mathbf{i} - 19.0\,\mathbf{j})$ N   6.76 $F_y = 247$ lbf

6.80 $T = 471$ N   6.84 $D = 119$ lbf   6.88 $M_0 = 551$ kg

6.90 $T = 4.3$ MN   6.94 $\Delta p = 2.97$ MPa   6.96 $\Delta p_{\max} = 394$ psi

6.100 $L = 2165$ m; $Q = 1.22$ m^3/s   6.102 $M_z = 1260$ ft-lbf

6.104 $F = 910$ lbf; $\mathbf{M} = (-1819\,\mathbf{i} - 301\,\mathbf{k})$ ft-lbf   6.106 $P = 1.44$ kW

Chapter 7

7.2 $\dot{W}_s = 384$ hp   7.4 $P = 255$ kW   7.6 $\alpha = 1.02$; $V = 0.80 V_{\max}$

7.10 $\alpha = 27/20$ 7.12 $\alpha = 1.08$ 7.14 $K_L = 3.89$

7.16 $p_A = -312$ psf gage   7.18 $p_A - p_B = 42.3$ kPa

7.22 $Q = 0.311$ m^3/s; $p_B = 16.6$ kPa gage

7.24 $Q = 0.383$ m^3/s; $p_B = 45.3$ kPa gage   7.28 $p_2/\gamma = 12.1$ m

7.30 $h = 33.0$ m   7.32 $P = 1.21$ MW   7.34 $P = 26.3$ hp

7.36 $P = 85.6$ kW   7.38 $P = 17.85$ kW   7.40 $F = 10.99$ kN

7.42 $h_L = 3.62$ ft   7.44 $F = 222.5$ lbf   7.46 $h_L = 3.63$ ft

7.48 $Q = 0.303$ m^3/s   7.60 $p = 29.7$ hp   7.62 $z = 120$ ft

7.64 $Q = 0.320$ m^3/s; $p = -78.7$ kPa gage   7.68 $P = 3.40$ kW

7.70 $F = 0.372$ N

Chapter 8

8.2 (a) $[T] = FL = ML^2/T^2$ (b) $[\rho V^2/2] = F/L^2 = M/LT^2$ (c) $[\sqrt{\tau/\rho}] = L/T$
(d) $[Q/ND^3] =$ dimensionless

8.4 $V^4\gamma/(g^2\sigma) = f(h^2\gamma/\sigma)$ 8.6 $F_D/(\mu Vd) = C$

8.8 $\Delta p = f(\mu V \Delta l/D^2)$ 8.10 $V = C\sqrt{\sigma/\rho l}$

8.12 $T/(\omega\mu D^3) = f(S/D)$

8.14 $eV/E = f(\sigma/E, Ed^2/(V\dot{M}_p), ED^2/(\dot{M}_p V), \text{Br})$ 8.18 $V = 5$ m/s

8.20 $nd/V = f(Vd\rho/\mu)$ 8.22 $F_p = 4.10$ lbf   8.24 $\rho_m = 6.23$ kg/m^3
8.26 $V_m = 12.8$ ft/s   8.28 $V_a = 64.6$ m/s   8.30 $V_m = 2.00$ m/s
8.32 $\Delta p_w = 7.33$ kPa   8.34 $\rho_m = 0.024$ slugs/ft^3
8.36 $M_p = 211$ N $\cdot$ m; $V_p = 0.179$ m/s
8.38 $p_m = 1.01$ MPa abs. $= 10$ atm.   8.40 No
8.42 $D = 1.40 \times 10^{-4}$ m   8.46 $V_p = 8$ m/s; $Q_p = 102$ m^3/s
8.48 $Q_m = 1.28$ m^3/s   8.50 $V_p = 39.3$ ft/s; $Q_p = 11{,}030$ cfs
8.52 $Q = 312.5$ m^3/s; $t = 10$ min   8.54 $F_p = 3.83$ MN
8.58 $V_p = 25$ ft/s; $F_p = 40.96$ kN

Chapter 9

9.4 $\mu = 2.85 \times 10^{-3}$ lbf-s/ft^2 9.10 $T = 123$ ft-lbf
9.14 $T = 6.28 \times 10^{-4}$ N $\cdot$ m   9.16 $P = 4.63$ W
9.18 $u_{max} = 0.30$ m/s; $V = 0.20$ m/s $= (2/3)u_{max}$ 9.20 $d = 0.012$ in.
9.22 $u_{max} = 9.0$ mm/s; $F_s = 5.4$ N   9.26 $v_{max} = -0.512$ ft/s (downward)
9.28 $v_{max} = 0.251$ ft/s (flow is upward)   9.30 $dp/ds = -353$ kPa/m
9.32 $q = 8.0 \times 10^{-4}$ m^3/s/m $= 2.88$ m^3/h/m
9.34 $\delta = 0.086$ in.; $\tau_0 = 0.0327$ lbf/ft^2 9.40 $F_s = 7.29$ N
9.42 $u = U_0 = 1$ m/s
9.44 (a) $F_s = 211$ N (b) $P = 11.7$ kW (c) $x = 14$ cm (d) 8.6% increase
9.46 $F_s = 0.219$ N; $\delta = 13.7$ mm   9.50 $F_s = 0.96$ N; $\tau_0 = 0.27$ N/m^2
9.54 Ratio $= 1.82$ 9.56 $T = 72.1$ N   9.58 $U_0 = 0.76$ m/s
9.60 $P = 66.8$ hp   9.62 $F_s = 499$ N
9.64 (a) $\delta = 1.529$ ft (b) $u_{\delta/2} = 27.4$ ft/s (c) $\tau_0 = 1.46$ lbf/ft^2
9.66 $F_s = 1740$ lbf   9.68 $F_p = 3.32 \times 10^4$ lbf
9.70 $\tau_{0,min} = 188$ N/m^2 9.72 $F_D = 131$ kN; $\delta = 1.07$ m

Chapter 10

10.2 $V = 1$ m/s (downward)   10.4 $\Delta p = 73.3$ psi/100 ft
10.6 $\tau/\tau_0 = 0.80$ 10.8 $V = 0.812$ ft/s; $Q = 2.77 \times 10^{-4}$ cfs
10.10 $P = 840$ W   10.16 $V = 0.108$ m/s
10.18 $D = 4.42 \times 10^{-4}$ m   10.20 $F = 253$ N
10.22 $\nu = 7.92 \times 10^{-5}$ m^2/s
10.24 Laminar (downward); $\mu = 1.53 \times 10^{-3}$ lbf $\cdot$ s/ft^2
10.32 $f = 0.015$ 10.34 $\Delta h = 2.02$ m   10.36 $\nu = 7.4 \times 10^{-8}$ m^2/s
10.38 $f = 0.020$ 10.40 $f = 0.019$
10.42 $z = 133$ ft; $p_{min} = -0.6$ psig
10.46 (a) $\tau_0 = 15.3$ N/m^2 (b) $\tau_1 = 13.25$ N/m^2 (c) $u_1 = 2.14$ m/s
10.48 $p = 669$ kPa   10.50 $f = 0.034$ 10.52 $p_A = 723$ kPa
10.54 $Q = 8.60$ cfs (from B to A)
10.56 $Q_c = 2.79$ m^3/s; $Q_s = 2.47$ m^3/s; $P = 5.52$ MW
10.58 $P = 26.7$ kW   10.60 $Q = 16.55$ cfs   10.62 $D = 90$ in.

10.66 $Q = 4.33$ cfs   10.68 (a) $P = 728$ W (b) $P = 3030$ W

10.70 $Q = 17.8$ cfs; $p_{max} = 12.1$ psig; $p_{min} = -3.03$ psig

10.72 $K_v = 18.5$ 10.74 $Q = 3.1$ cfs   10.76 $P = 16.6$ hp

10.78 $P = 332$ kW   10.84 $z_1 - z_2 = 12.74$ m

10.86 $P = 24.4$ hp; $p_m = 22.1$ psig   10.88 $z = 22.3$ m

10.90 $p_A = 42.5$ psig   10.94 $T_e = 100°F$ (case 1); $T_e = 95.25°F$ (case 2)

10.96 $Q_1 = 0.68$ m^3/s; $Q_2 = 0.32$ m^3/s

10.98 $Q_1 = 2$ cfs   10.100 $Q_{large}/Q_{small} = 3.7$

10.102 $Q_{12} = 5.2$ cfs; $Q_{14} = 6.2$ cfs; $Q_{16} = 8.6$ cfs; $h_L = 68.3$ ft

10.104 $Q_{pump} = 0.66$ m^3/s   10.106 $\Delta p = 0.123$ psf

10.108 $P = 47.0$ kW   10.112 $Q = 332$ m^3/s   10.114 $Q = 344$ cfs

10.116 $Q = 24.7$ cfs   10.118 $Q = 684$ cfs   10.120 $Q = 610$ cfs

10.122 $d = 5.6$ ft   10.124 $Q = 100,000$ cfs   10.126 $n = 0.029$

10.128 $y_{main} = 17.0$ ft

Chapter 11

11.4 $C_D = 1.85$ 11.8 $M = 21.8$ kN • m

11.10 $F_D = 2.90 \times 10^6$ lbf   11.12 $n = 2.5$ Hz

11.14 $F_D = 4368$ lbf   11.16 $F_{normal}/F_{edge} = 184$

11.18 $F_D = 25.33$ kN   11.20 $P = 4.85$ kW   11.22 Savings = 6.3%

11.26 $V = 19.6$ m/s   11.28 $V_{open} = 37$ m/s; $V_{closed} = 40.6$ m/s

11.30 $V_{max} = 117$ mph   11.32 $P = 20.9$ hp

11.34 $F_{100} = 7774$ N; $F_{200} = 21,810$ N   11.36 $V = 0.082$ ft/s (upward)

11.38 $V_0 = 8.9$ mm/s   11.40 $\nu = 0.189$ m^2/s   11.42 $V = 12.7$ m/s

11.44 $V = 5.37$ m/s   11.46 $V = 1.39$ m/s

11.48 $V = 7.25$ ft/s (upward)   11.50 $V = 2.08$ ft/s

11.52 $V = 1.33$ m/s (upward)   11.62 $l = 17.7$ ft

11.64 $S = 394$ ft^2 11.66 $V_L = 67.4$ m/s; $V_s = 62.4$ m/s

11.68 $V_0 = 10.5$ m/s; $F_{L/M} \approx 16,000$ N   11.72 $F_D = 1000$ N

Chapter 12

12.2 $c = 1055$ m/s   12.4 $\Delta c = 650.5$ m/s   12.6 $c = 1483$ m/s

12.8 $T = 491$ K $= 218°C$ 12.10 $V = 719$ km/h

12.12 $p_t = 284.6$ kPa   12.14 $T = 407$ K; $p = 177.4$ kPa; $V = 346.4$ m/s

12.16 $T = 289.8$ K; $p = 481.6$ kPa; $\dot{m} = 0.038$ kg/s

12.18 $C_p(0) = 1$; $C_p(2) = 2.43$; $C_p(4) = 13.47$; $C_p(inc.) = 1$

12.20 $M_2 = 0.665$; $p_2 = 200$ kPa; $T_2 = 325$ K; $\Delta s = 34.1$ J/kg K

12.22 $M = 1.59$ 12.24 $V_1 = 1286$ m/s

12.28 (a) $\dot{m} = 0.0733$ kg/s (b) $\dot{m} = 0.0794$ kg/s

12.30 $\dot{m}$ (130 kPa) $= 0.120$ kg/s; $\dot{m}$ (350 kPa) $= 0.386$ kg/s

12.36 $A/A_\star = 1.123$; M $= \sqrt{2}$

12.38 (a) $A/A_\star = 3.38$, $T = 2522$ N (b) $T = 2516$ N

12.40 $A_s/A_\star = 3.25$; $x_s = 5.99$ cm
12.42 $p_t = 499$ kPa; $M_3 = 0.336$; $p_3 = 461$ kPa   12.44 $p_2 = 321$ kPa
12.46 $L = 29.8$ m   12.48 $\dot{m} = 0.149$ kg/s   12.50 $D = 0.445$ in.
12.52 $p = 460$ kPa   12.54 $\dot{m} = 0.0239$ kg/s

Chapter 13

13.2 $V = 0.03$ m/s   13.4 $\Delta h = 10.4$ mm   13.6 $V_0 = 19.0$ m/s
13.8 $Q = 3076$ gpm $= 6.85$ cfs $= 0.194$ m^3/s
13.10 $V_{max}/V_{mean} = 1.24$ (turbulent)
13.12 $r_m/D = 0.189$; $r_c/D = 0.351$; $\dot{m} = 16.4$ kg/s   13.14 $V = 0.92$ m/s
13.16 $C_c = 0.748$ 13.18 $p_0 = 133$ kPa   13.20 $Q = 3.66$ cfs
13.22 $Q = 0.70$ cfs
13.24 $p_A - p_B = 224.9$ kPa; $p_D - p_E = 221.9$ kPa; deflec. $= 1.82$ m (same)
13.26 $Q = 0.0282$ m^3/s   13.28 $d = 6.26$ cm   13.30 $d = 56.9$ cm
13.34 $d = 0.798$ m   13.42 $Q = 0.386$ m^3/s
13.44 $Q = 9.64 \times 10^{-3}$ m^3/s   13.46 $Q = 0.0049$ m^3/s
13.48 $Q = 3.67$ liters/s   13.50 $Q = 0.263$ m^3/s
13.54 $P = 1.215$ m   13.56 $d = 1.124$ m   13.58 $Q = 34.1$ cfs
13.62 $P = 1.00$ m   13.64 $H = 0.476$ ft
13.66 $M = 0.710$; $V = 235$ m/s   13.68 $M = 1.57$; $V = 482$ m/s
13.70 $\dot{m} = 0.00792$ kg/s   13.72 $\dot{m} = 6.35 \times 10^{-4}$ kg/s

Chapter 14

14.2 $F_T = 970$ N; $P = 37.4$ kW   14.4 $N = 1544$ rpm
14.6 $V_0 = 80$ m/s   14.8 $N = 1462$ rpm   14.10 $a = 0.782$ m/s^2
14.12 $Q = 1.66$ m^3/s; $P = 211$ kW   14.14 $Q = 0.32$ m^3/s; $P = 13.5$ kW
14.16 $Q = 3.60$ m^3/s; $\Delta h = 38.7$ m; $P = 1.71$ MW   14.20 $P = 933$ kW
14.22 $Q = 4.73$ cfs   14.24 $H_{30} = 73.8$ m
14.26 $Q = 0.0833$ m^3/s; $\Delta h = 146$ m; $P = 104$ kW
14.28 $n_s = 0.489$ (safe)   14.30 Mixed-flow   14.32 Radial-flow
14.34 Axial-flow   14.36 $D = 2.07$ m; $P = 27.6$ kW
14.38 $Q = 0.143$ m^3/s   14.40 $P = 10.3$ MW; $D = 2.85$ m
14.44 $T = 52{,}780$ N • m; $P = 332$ kW; $\alpha_1 = 8°25'$ 14.46 $P = 271$ hp
14.48 $A_{min} = 8.20$ m^2

Chapter 15

15.2 Supercritical
15.4 $Fr_{0.3} = 7.77$ (super); $Fr_2 = 0.45$ (sub); $y_c = 1.18$ m
15.6 $y_c = 1.12$ ft   15.8 Subcritical
15.10 $y_{alt} = 5.38$ m; $y_{seq} = 2.33$ m   15.12 $Q = 187$ cfs
15.14 $Q = 50.5$ cfs   15.16 Elev. $= 101.44$ m

15.18 $\Delta y = -0.51$ m; $\Delta y = 0.40$ m; max. upstep $= 0.43$ m

15.20 $\Delta z = 0.89$ m   15.22 Ship squat $= 0.30$ m

15.24 $Q = 35.5$ m³/s   15.26 $y = 0.408$ m

15.28 $h_L = 2.30$ ft; $P = 4.70$ hp; $F_x = 51.2$ lbf   15.30 $y_2 = 1.34$ m

15.32 $y_1 = 1.08$ ft   15.34 $y = 2.45$ m   15.40 $P = 1760$ kW

15.54 $Q = 19.5$ m³/s

Chapter 16

16.2

$$v_{i+1}\left(\frac{r_{i+(1/2)}^2}{\Delta r^2} - \frac{r_i}{2\,\Delta r}\right) - v_i\left(\frac{r_{i+(1/2)}^2}{\Delta r^2} + \frac{r_{i-(1/2)}^2}{\Delta r^2} + 1\right) + v_{i-1}\left(\frac{r_{i-(1/2)}^2}{\Delta r^2} + \frac{r_i}{2\,\Delta r}\right) = 0$$

16.4 $\dfrac{d^2\psi}{dx^2} = \dfrac{2}{a(a+1)} \cdot \dfrac{\psi_{i+1} - (1+a)\psi_i + a\psi_{i-1}}{\Delta x^2}$

16.8 $\psi_p = \dfrac{ab\,\Delta x^2\,\Delta y^2}{b\,\Delta y^2 + a\,\Delta x^2}\left(\dfrac{\psi_E + a\psi_W}{a(1+a)\,\Delta x^2} + \dfrac{\psi_N + b\psi_s}{b(1+b)\,\Delta y^2}\right)$

16.10 The flow is rotational.   16.12 $\psi = Uy + $ constant

16.14 $\psi = U(y\cos\alpha - x\sin\alpha) + $ constant

16.18 $\dfrac{1}{r}\dfrac{\partial}{\partial r}(rU_r) + \dfrac{\partial}{\partial z}(U_z) = 0$

16.22 Analytic solution; $\dot{m} = 0.0186$ kg/s

16.24 $\dot{m}$ (kg/s) p (kPa)

$\dot{m}$ (kg/s)	p (kPa)
2×10^{-6}	47.2
2×10^{-4}	48
0.0202	114
0.201	1040

16.28 $\dfrac{\partial^2 h}{\partial t^2} = gh\dfrac{\partial^2 h}{\partial x^2}$

16.30 $\dfrac{\partial V}{\partial t} + V\dfrac{\partial V}{\partial x} = -g\dfrac{\partial h}{\partial x} - c_f\dfrac{V|V|}{2}\left(\dfrac{P}{Wh}\right)$

Index

Torricelli p172

Volume

Unit	Equivalent					
	cubic inches	liters	U.S. gallons	cubic feet	cubic yards	cubic meters
cubic inch	1	0.016 39	0.004 329	578.7×10^{-6}	21.43×10^{-6}	16.39×10^{-6}
liter	61.02	1	0.264 2	0.035 31	0.001 308	0.001
U.S. gallon	231.0	3.785	1	0.133 7	0.004 951	0.003 785
cubic foot	1728	28.32	7.481	1	0.037 04	0.028 32
cubic yard	46,660	764.6	202.0	27	1	0.764 6
meter^3	61,020	1000	264.2	35.31	1.308	1

Discharge (Flow Rate, Volume/Time)

Unit	gallons/minute	liters/second	feet^3/second	meters^3/second
gallons/minute	1	0.063 09	0.002 228	63.09×10^{-6}
liters/second	15.85	1	0.035 31	0.001
acre-feet/day	226.3	14.28	0.504 2	0.014 28
feet^3/second	448.8	28.32	1	0.028 32

Velocity

Unit	feet/day	kilometers/hour	feet/second	miles/hour	meters/second
feet/day	1	12.70×10^{-6}	11.57×10^{-6}	7.891×10^{-6}	3.528×10^{-6}
kilometers/hour	78,740	1	0.911 3	0.621 4	0.277 8
feet/second	86,400	1.097	1	0.681 8	0.304 8
miles/hour	126,700	1.609	1.467	1	0.447 0
meters/second	283,500	3.600	3.281	2.237	1

SOURCE: SI System of Units: Pamphlet prepared for the Universities Council on Water Resources by Peter C. Klingeman, 1976.

$$Re = \frac{VD\rho}{\mu} = \frac{VD}{\nu}$$

kin $\quad \nu = \frac{\mu}{\rho} \quad \frac{m^2}{s}$

dyn $\quad \mu = \frac{FT}{L^2} \cdot \frac{N \cdot s}{m^2}$